# Preface

Published online: 22 November 2006


From our studies of some of the deepest parts of the oceans, the highest mountains, the coldest polar regions and the hottest and most arid deserts, we now know that life exists in numerous parts of the planet that, only a few years ago, were considered incapable of supporting life. In addition, data obtained from space research have provided tantalizing evidence of environments for possibly supporting life elsewhere in our Solar System.

Life in extreme environments is an appealing and exciting subject as it sits at a convergence point for two important questions for mankind: "*How did life appear on our planet?*" and "*Is there life beyond our planet?*" The harshness of Earth-based extreme environments offer the closest approximation to the conditions that probably existed when life first appeared on our planet but also offer potential analogues for conditions on other planetary bodies. Addressing the topic of life in extreme environments is also very relevant for one of today's most crucial issues: the impact of human activity on ecosystems.

B. Andersson (✉)
European Science Foundation (ESF), Strasbourg cedex, France
e-mail: ceo@esf.org

The investigation of life processes in extreme environments has a broad spectrum of relevance, including both societal and economical considerations. These exciting areas of research (whether considering microbes, plants or animals, including humans) are at an early stage and focus on environments that have in the past been difficult to investigate. However they are set to benefit tremendously from the new technological developments (e.g. robotics, information technologies, simulation techniques) as well as from the use of the rapidly developing tools of molecular biology and bioinformatics and are ideal targets for the consideration of species within their whole ecosystems. Actions should be taken to move research in this direction. At the European level the scientific community currently studying extreme environments is significant and well regarded but its structure is relatively fragmented. The benefits resulting from improved coordination and information exchange within this community are clear and implementing greater networking represents a significant challenge to Europe for the future.

In 2004, the European Science Foundation (ESF) launched a call for Expressions of Interest for research topics. The substantial response and high quality of ideas received as well as the broad diversity of domains covered encouraged ESF to set up an informal group of experts in November 2004 and to fund a large interdisciplinary workshop in November 2005: the first integrated interdisciplinary initiative at the European level. It was aimed at identifying the points of view, priorities and recommendations of the European scientific community on these matters and the resulting report is to be published in the first quarter 2006. Alongside this initiative and complementing the report, the European Science Foundation is happy to have provided support to this special issue of Reviews in Environmental Science and Biotechnology on the topic of Life in Extreme Environments.

Professor Bertil Andersson
Chief Executive
European Science Foundation (ESF)

# LIFE IN EXTREME ENVIRONMENTS

# Life in Extreme Environments

*by*

RICARDO AMILS
*Universidad Autónoma de Madrid,*
*Centro de Biología Molecular del CSIC,*
*Madrid, Spain*

CYNAN ELLIS-EVANS
*British Antarctic Survey,*
*Cambridge, United Kingdom,*

and

HELMUT HINGHOFER-SZALKAY
*Medical University,*
*Institute of Physiology,*
*Graz, Austria*

Springer

A C.I.P. Catalogue record for this book is available from the Library of Congress.

ISBN 978-1-4020-6284-1
ISBN 978-1-4020-6285-8

Published by Springer,
P.O. Box 17, 3300 AA Dordrecht, The Netherlands.

*www.springer.com*

*Printed on acid-free paper*

# TABLE OF CONTENTS

**Life strategies of plants in extreme environments**



**Life strategies of animals in extreme environments**



**Human adaptation in extreme environments**

Rev Environ Sci Biotechnol (2007) 6:3–16
DOI 10.1007/s11157-006-0010-3

REVIEW PAPER

# The challenge of the food sufficiency through salt tolerant crops

**Adriana Galvani**

Received: 28 February 2006 / Accepted: 29 May 2006 / Published online: 14 November 2006


**Abstract** This work is focused on deserts, as extreme environments, because the year 2006 has been declared Year of Deserts and Desertification by the United Nations (IYDD Program 2006). The loss of vital resources such as fresh water and soil, and the depletion of biodiversity are emerging hazards, able to transform beneficial situations into extreme environments. Desertification is generated by land degradation: the loss of biological productivity is caused by nature or by human-induced factors and climate change. Nearby the desertification process there is the increasing process of salinisation of soil and water, induced by irrigation itself, or by salt water ingress derived by tsunamis or hurricanes. Increased research on the development of salt-tolerant cultivars could, with appropriate management, result in the broader use of saline soils. Although careful application is necessary, the combination of sand, seawater, sun and salt-tolerant plants presents a valuable opportunity for many developing countries. Cooperation among plant ecologists, plant physiologists, plant breeders, soil scientists, and agricultural engineers could accelerate the development of economic salt tolerant crops. If saline water is available, the introduction of salt tolerant plants in poor regions can improve food or fuel supplies, increase employment, help stem desertification, and contribute to soil reclamation.

**Keywords** 2006 Year of Deserts · Agronomic resources · Demography · Genetic manipulation · Halophytes · Saline lands · Saline soils · Salt tolerant crops · Water management

**Abbreviations**

EC Electric Conductivity
S Siemens
PGI UE Designation of Origin Label

> "As citizens of planet Earth, it is not surprising that we turn to Mother Earth – to life itself – to help our economies to develop in a way which should not just enhance our quality of life, but also maintain it for future generations" Janez Potočnik (EU Science & Research Commissioner)

## Introduction

The agricultural issue is a never-ending story.

Even though agriculture is the first human activity, after years of evolution it still is not able to feed the global population.

Contribution to the monograph "Life in Extreme Environments"

A. Galvani (✉)
Geography Researcher, Department of Economics, University of Bologna, Via Guerrazzi 20, 40125 Bologna, Italy
e-mail: adriana.galvani@unibo.it

Agricultural techniques are increasing very fast, but the global population is increasing even faster and the gap is not filled until now. To solve this gap the demographic path should be reduced and the food production be augmented. Statistics are revealing that setbacks on hunger nearly outweigh progress. Currently there are 852 million food deprived people in the world. In the worst affected regions such as sub-Saharan Africa and Southern Asia the number of hungry people has increased by tens of millions. More than a quarter of the children in the developing world are malnourished. In sub-Saharan Africa the number of underweight children increased from 29 to 37 million between 1990 and 2003 and people with insufficient food was 34% more. Progress was made in Eastern Asia where the number of malnourished children declined from 24 to 10 million and the number of people with insufficient food decreased by 47%. Even in developing and industrialised countries the correlation coefficient, calculated between population growth and per person aliment growth, from 1995–1997 to 2000–2002, is only 0,8 (FAO Statistics 2006).

## The challenge of the sufficient agriculture productivity

To obtain food sufficiency it is necessary to invest more in technology in order not to destroy the lands and natural resources. Effectively, we face an increasingly diminished, if not lost, number of natural habitats and ecosystems. Globally, unparalleled conflicts and disasters exacerbate poverty and hunger.

We witness loss of fertile soil and habitats through conversion by agriculture, infrastructure or urbanisation, overexploitation of resources, introduction of non-native species, pollution and climate change. Climate change could induce significant pest damages that are difficult and costly to address. As Zilberman et al. (2004, p. 375) states

> "Differences in soil conditions and solar radiation across latitudes may also change yield patterns during and after adjustments to climate change" .... "Another prominent secondary affect of climate change concerns water supply and demand. Irrigation systems may undergo extensive chemical evolution-with rising deposition/salinisation in some areas and soil leaching/depletion in others".

Zilberman has documented that changing weather patterns, ecological conditions and pest problems will require significant modifications in crop systems at critical locations and that the challenge entails not only development of knowledge, but also significant commitments to education and technology transfer to promote the right patterns of adaptation/adoption (idem, p. 376).

Solutions require cooperation for the maintenance of life-support systems and people to ensure the access to clean air, clean water, fertile soils, and biodiversity.

According to the economic theory of globalised markets, the variety of agricultural production has decreased until the large utilisation of few species, especially the most profitable. The biodiversity is being rapidly homogenised. The climate change constitutes the major threat to biodiversity and to our ability to extract resources sustainably from many ecosystems. An open mind is necessary to enlarge the range of products that are utilised worldwide for human beings.

The decrease in the number of species used in agriculture and in animal husbandry has also narrowed the genetic base, greatly reducing the options for adapting to change.

> "Of the thousands of plants species deemed edible for humans, some 20 produce the vast majority of the world's food. Staple crops such as wheat, maize, rice, and potatoes are used to feed more people than the next 26 crops combined. Likewise, sheep, goat, cattle and pigs supply nearly all land based protein for human consumption....humans are increasingly reliant on a narrow range of species and then on specific varieties of these species. This process is accelerated by economic policies that encourage production for export markets" (Serageldin 1997, p. 415–16).

The loss of resources and the depletion of biodiversity is an emerging hazard, with the ability to transform fertile conditions into extreme environments, and thus requires suitable solutions. Solutions can be found in technical revolutions or in implementation of a few minor adjustments to the new conditions. Major adjustments are required for marginal areas or extreme environments. Regarding extreme environments, I will focus on deserts since this year, 2006, has been declared Year of Deserts and Desertification by United Nations IYDD Program 2006.

## Desertification and agriculture

Extreme environments are the areas where it is difficult to survive and where it is difficult to obtain essential food. Many documents about this topic are based on conventions or workshops realised by the UN in previous years which have been declared: 2005 the Year of food and 2004 Year of rice. The year 2003 has been celebrated as the UN Year of Water, one of the most important resources for life whose reserves are at stake. In fact, the loss of productivity can derive from an unsustainable use of principal resources such as fresh water and lands. The depletion of water resources is, among others, at the origin of the increasing process of desertification.

A decreasing availability of fresh water, or limited access to water, is a major constraint to increasing crop production.

In this respect, a solution could be to preserve water resources and to exploit drought resistance of crop plants, through dry farming, including cultivars with minimum water requirements, or to improve water management practices; we refer for example to a number of solutions invented for water saving, such as dry farming, techniques, irrigation management strategies and irrigation system control, flood irrigation, spray, drip irrigation..... These methods are very simple and they don't require expensive experiments or materials, but they are based on instinctive solutions.

For example the process of desertification requires:

- Increasing the variety of species;
- Making plant species more tolerant to sand and salt water.

Nevertheless, salt-tolerant species require special approaches that involve breeding and selection of crops.

In saline agriculture, an alternative is to allow the environment to select the crops, to match salt-tolerant plants with desirable characteristics to the available saline resources.

Another major constraint is the reduced amount of arable land, set aside for other functions: lands suffer the concurrence of industrial, residential, tourist, and infra-structural requirements. This progressive loss of farmable land is on a collision course with the expanding global population, which over the next 30 years is expected to require an increase in food production of 20% in developed countries and 60% in developing nations. The result is that the progressively growing number of inhabitants have to face declining resources.

Desertification is generated by land degradation: the loss of the land's biological productivity is caused by human-induced factors or disasters.

> "Disasters will continue to occur, and their social, economic, environmental impact will continue to increase" (Pearce 2005, p. 436).

Desertification affects one third of the earth's surface and over a billion people. Moreover, it has potentially devastating consequences in terms of social and economic costs (UN, YDD).

> "Desertification and drought cause an estimated loss of $42 billion a year from agricultural production, contribute to food insecurity, famine and poverty and can give rise to social, economic, and political tensions that can cause conflicts, further impoverishment and land degradation, according to the Convention's Secretariat.
>
> It is widely recognised that environmental degradation has a role to play in considerations of national security, as well as international stability. Therefore, desertification has been seen as a threat to human security" (UNCCD Executive Secretary Hama Arba Diallo, IYDD Program, 2006).

Throughout the developing world, there are extensive coastal deserts where only seawater or almost brackish water and sand are available.

The disadvantages of sand and saline water for conventional crops become advantages when and where salt-tolerant plants are used: in fact among the 13 mineral nutrients needed by plants, 11 are present in seawater in adequate concentrations for growing crops. The high aeration quality of sand is also valuable. In addition, the infiltration of water through sand reduces salt build up in the root zone when seawater is used for irrigation.

These lands are often located in areas of high nutritional and economic needs as well, so adaptation requires a vulnerability assessment of the population affected and a resilience assessment of the society involved.

Increased research on the development of salt-tolerant cultivars of crop species could, with appropriate management, result in the broader use of saline soils. Conventional food crops can be bred or selected to tolerate mildly saline water.

Although careful application is necessary, the combination of sand, seawater, sun and salt-tolerant plants presents a valuable opportunity for many developing countries.

Anything we can do to help crop plants to cope with environmental stresses will also raise the quality and quantity of food for those who need it most. They will also serve to enlarge the variety of food.

The direct use of seawater for agriculture is probably the most challenging potential application. This would be based on genetic experiments, it may be the only viable solution for nutrition needs worldwide.

Since environmental stress due to salinity is one of the most serious factors limiting the productivity of crops, the innovation of the salt-tolerance gene introduced by Eduardo Blumwald, who led the research team at the University of Toronto, will have significant implications for agriculture worldwide. The gene that controls increased production of the transport protein was taken from *Arabidopsis*, a relative of the cabbage that is commonly used in plant research.

## Salinisation

In addition to the desertification advance there is the increasing process of salinisation of soil and water, induced by human activity, most of all by irrigation itself and the large use of fertilisers; or induced by natural phenomena i.e. salt water ingress caused by inundation, erosion, sea-level rise, subsidence, or derived from extreme events of tectonic origin, such as tsunamis, or from climate change, such as typhoons and hurricanes.

Worldwide, an estimated 24.7 million acres (10 million hectares), of once agriculturally productive land, are being lost annually because of irrigation-induced salinity, according to the US Department of Agriculture. Crop production is limited by salinity on 40% of the world's irrigated land and on 25% of irrigated land in the United States. Crop irrigation is an age-old practice that allows farmers to be less dependent on seasonal rainfall and the uncertainties of the weather. However, irrigation also increases the salinity of soils and water by depositing in the fields soluble salts such as sodium, calcium, magnesium, potassium, sulfate and chloride that the water has picked up from the soils and rocks it has passed through. Eventually, these salts accumulate in the irrigated soils at levels which decrease the vigour and productivity of the crops grown there.

Salty irrigation water wreaks havoc on most plants by upsetting their ability to take in water through their root cells. In fact, if salt concentrations in the soil are very high, the flow of water into the plant is actually reversed and the plant dehydrates, and dies as water is drawn out of its cells.

Salinity stress is one of the most serious factors limiting the productivity of agricultural crops, but previous studies have shown that exogenous fatty acids or nutrient additives enhanced plant performance in saline environment, even if the mechanism remains unclear. Other studies have shown that the application of *Bacillus subtilis* ranked first in alleviating the adverse effects of salinity, if followed by supplemental calcium into a saline nutrient solution (Saleh et al. 2005, p. 30).

To meet future agriculture needs, a solution is an environmentally friendly technique which controls salinisation, uses salt removing crops, chooses halophytes crops for direct salt water irrigation, or selects species that have high salt-removing capacity and commercial value. Some plants can be integrated into rotation programmes or planted as intercrops for perennial plants to control salinisation.

## Solutions

Sometimes solutions are very simple, as demonstrated during the famous green revolution which succeeded in cultivating wheat on highlands and rice in greenhouses, which brought on an excellent production worldwide (García Olmedo 2000). Furthermore I spoke of the "blue revolution" in a paper of mine (Galvani 1994). The blue revolution is considered as the breeding of fish instead of fishing. This particularly occurs in the Po river mouth in Italy where, years ago, a consortium of fishermen imported mussels from the Philippines. Mussels found a better environment than their original and thus originated an unknown activity and a related food industry in one of the poorest areas of Italy. However the excess of proliferation generated anoxia in lagoons low water and market price slow down, and caused a large indebtedness of the new economic enterprises and a related social crisis.

Experiments try to mix different systems deriving from different areas or different fields of production as experimented in Italian grape production. A very fragile cultivar such as the grape has been adapted to face conditions alien to its usual methods of growing. This has been experimented in Italy, a country of great wine production, in order to augment the cultivated areas. Usually, grapes need water in drained soils and so the best area is on the northern hills, however experiments on arid areas in Sicily have also been successful.

As a result, in the framework of the primary sector all experiments are possible: nature has infinite possibilities which humans have to play with. As in children's games, the simplest toys are the best ones, even if the advancements of technology present us the prospects of complicated ways.

Effectively, revolutions in agriculture are based on empirical approaches.

Looking towards the future, sometimes means taking a step to the past, in the sense of becoming more naive in order to discover thousands of possibilities hidden by nature, guided not by ignorance, but by an open mind, enlarged by cultural evolution.

The best practice would be the exchange of good ideas worldwide, as demonstrated by the celebration in the year 2005 of the Microcredit. Simple solutions can be effective in poor countries where technology is not affordable.

> "Progress has always been heralded by paradigm shifts that seemed somehow difficult and dangerous, but moved the world forward into new realm of freedom and prosperity" (Serageldin 1997cit, p. 418).

## Natural disasters

Adverse conditions are caused by natural disasters, such as earthquakes or the big tsunami in 2004 for example. Disasters can also be of anthropogenic origin, directly related to human activity or indirectly, as those climate related; they are accelerated at an increasing pace of destruction caused by environmental pollution and climate change (Serageldin 1997cit, p. 417), for climate change may influence the increase in the magnitude and frequency of disasters (Loster 2003).

Global warming and deforestation are at the origin of the increasing frequency and danger of exceptional events, which are able to change delightful places into deserted areas. These natural events (Fig. 1) include floods, cyclones, droughts, hurricanes, among them Katrina which flattered 150 miles of coastline and Rita which caused $10 billions in damage in the US Gulf Coast in 2005, and the big tsunami in Asia in 2004.

As a consequence, the population faces unforeseen situations and can no longer rely on its traditions, since inhabitants should take steps to minimise damage from impact. They are forced to suddenly abandon previous activities and turn to new ones.

People in areas which are inundated by sea water and remain water logged, may consider to change the land use rather than waiting to reclaim it for agricultural use immediately.

There is little knowledge available on the farmers' recouping capacity following a major disaster,

**Fig. 1** Hurricane's image elaborated by the author

> "nonetheless, if capacity is enhanced for individuals and communities to cope with current natural hazards, they will be better prepared for more frequent occurrence in the future" (Newton et al. 2005, p. 545).

Documentation on the current coping mechanisms of farmers and researchers about the future needs to strengthen the coping strategy of the farmers in changing scenario of hazard intensification and vulnerability context (Ahmad 1987).

In non-irrigated areas, the salt tolerant non-rice crops, like sorghum, chilli, groundnuts etc... can survive; they can be grown in ridges and crops can be planted in the middle of the ridge slopes preferably. Small scale farming can practice this manually.

Traditional farming efforts usually focus on modifying the environment to suit the crop; in saline agriculture, an alternative is to allow the environment to select the crops, to match salt tolerant plants with desirable characteristics to the available saline resources. If saline water is available, the introduction of salt tolerant plants in poor regions can improve food or fuel supplies, increase employment, help stem desertification, and contribute to soil reclamation.

The threat of rapid global climatic change makes the safety net of diversity for species even more important; a species that has many variety is more likely to include individuals that are genetically suited to new conditions than a species that has only one or a few varieties (Hughes et al. 1997, p.73).

It is also important to mention the necessity of the evolution of policies and traditions. Usually, traditions are so strong in agricultural communities that they become impediments to progress and innovations. Food selection, cooking method, flavour, consistency, serving time and place are often established by long tradition, and practitioners are resistant to change.

> "New foods that require significant changes in any of these practices are unlikely to be readily accepted" (Elder 1997, p. 457).

Farming communities should learn how to cope with the conditions of momentary disasters because the same situations can become permanent through:

"Agricultural land affected by sea water ingress (or influx) and crop affected to a large extent, and soil salinity/alkalinity.

- Contamination of domestic water supplying dug wells/tube wells by salt water.
- Failure of standing crop salt intrusion.
- Uprooted trees along coastal lines.
- Loss of productive lands due to salt water logging in embankment areas &/or low lying areas; even the possibility of getting permanently water logged due to large scale embankment erosion.
- Loss of fertile coastal lands due to sediment deposits.
- Loss of farmlands due to sea erosion/scouring.
- Loss of irrigation channels, small embankments, sluices and other infrastructure.
- Loss of human capital due to unforeseen food & nutritional instability.
- Contamination of community water supply.
- Damage of coastal mangroves and shelter belts.

In this work I voluntarily choose of not using the "syndrome approach", expression utilised by the scientists of International Geosphere-Biosphere Program (IGBP) in explaining the earth resources depletion, because it is a medical-neurological term. I prefer to use "integrated approach" because our earth isn't ill, it's only a living system, continually changing -as humans- it only needs interdisciplinary- multi-dimensional studies and cares.

I accept instead the methodology used by the ESSP (Earth System Science Partnership 2006) scientists who pose these questions, in order to grant a food-secure future for those most vulnerable to environmental stress :

- How will global environmental change (GEC) affect the vulnerability of food systems in different regions?
- How can we adapt food systems to cope with GEC and improve food security?
- How will the various adaptation options feed back on environmental and socioeconomic conditions? (2006).

Scientists said they hope to develop crops to feed the world's growing population with new solutions found for extreme environments caused by natural disasters" (IARI; website, 2005).

**Halophytes**

O'Leari stated in 1987:

> "Although economic consideration of halophytes and other salt-tolerant plants is just beginning, they are now receiving increased attention in arid regions where intensive irrigation has led to salt soils or where water shortages are forcing use of marginal resources such as brackish underground water".

The issue should be updated in the new millennium due to the increasing actions causing desertification. Furthermore, production of salt tolerant plants is one of the ways to utilise the waste saline lands around the world (Gallagher 1985).

> "Salt is vital for human body. Iodine daily requirement derives from vegetable, rice or fish. For some people living in the most impoverished areas in high altitude or far from the sea this in not possible. Iodine once present in the soil has long since leached away, causing catastrophic effects for public health. Iodine deficiency can lead to neurological disorders, deafness, psychomotor retardation, brain damage, mental deficiency, and retarded growth. In Tibet raw salt is freely available in salt lakes, but it is the main cause of the nation's high level of retarded growth, goitre, and neurological impairments, because it is non-iodised salt. The Australia's Overseas Aid Program has provided over $2 million to support the elimination of iodine deficiency in Tibet, distributing iodised oil capsules to vulnerable groups" (Focus-AusAID 2006, p. 11).

Halophytes (plants that grow in soils or waters containing significant amounts of inorganic salts) can harness saline resources that are generally neglected and are usually considered impediments rather than opportunities for development (Aronson 1989).

> "In searching for crops for saline agriculture, those that currently comprise the bulk of human food should be considered as models –maize, wheat, rice, potatoes, and barley. If these major crops can be grown using saline resources, or if new tolerant crops that are acceptable substitute can be developed, the world's food supply will have a more diverse and vastly expanded base"(Pasternak 1987, p.275).

Halophytes have been evaluated as potential crops for direct seawater or brackish water irrigation. They can be developed in three areas: in coastal deserts using seawater for irrigation, in inland salt deserts using saline underground or surface water, in arid-zone deserts using brackish drainage water for irrigation. Halophytes grown in desert environments showed levels of biomass and seed production comparable to conventional crops.

Scientists exploring seashores, estuaries, and saline seeps have found thousands of halophytes with potential use as food, fuel, fodder, fibber, and other products. Although the direct consumption of halophytes by humans and animals can be limited, the seeds of many of them are being considered as new sources of grains or vegetable oils (Hinman 1984).

Some conventional crops, such as sugar, fodder, date palm and culinary beets have halophytic ancestors, so they can be irrigated with brackish water.

Barley is the most salt tolerant cereal grain. Rice cells subjected to salt stress and then grown to maturity have progeny with improved salt tolerance, up to 1% salt.

There are also a number of plants that, although not halophytes, have sufficient salt tolerance for use in some saline environments.

The use of water with salt levels equal to, and even exceeding that of seawater for irrigation of various food, fuel, and fodder crops has been reported by many researchers including Boyko (1996), Somers (1975), Iyengar (1982), Epstein (1985), Gallagher (1985), Glenn and O'Leary (1985), Pasternak (1987), Yensen (1988) Aronson (1989). These scientists have produced grains oilseed; grass, tree, and shrub fodder; tree and shrub fuel-wood; and a variety of fibre, pharmaceutical, and other products using highly saline water.

Plants can use salt water at their disposal or they can even be irrigated with it. In India twenty species of trees were planted in a trial using saline water (EC = 4.0–6.1 dS/M) for irrigation and nine of them were growing well after 18 months (Yensen 1988).

It is very difficult to exploit all the possibilities related to salt tolerant species, since it is unlikely that any species will meet all the requirements: although selection is usually based on performance in a similar environment, some species "travel" poorly (Glenn 1985, p.51). Some show extreme variation in regard to source and some perform remarkably well far outside their native climate; some require different levels of salinity during the span of their life: responses are variable at seedling, vegetative or reproductive stages. Adult plants can tolerate salt water better than younger plants, others tolerate medium root salinity and not salt spray, other tolerate salt in leaves and bubble aeration.

Many halophytes have a special and distinguishing feature – their growth is improved by low levels of salt. Other salt-tolerant plants grow well at high salt levels but beyond a certain level, growth is reduced. With salt-sensitive plants, each increment of salt decreases their yield.

Some halophytes require fresh water for germination and early growth but can tolerate higher salt levels during later vegetative and reproductive stages. Some can germinate at high salinities but require lower salinity for maximum growth. Some grow well on permanently wet areas, other's best growth occurs where the soil dries out in the summer.

Some halophytes contain too much salt for consumption, but a solution is to extract leaf protein from the salt-containing foliage (Table 1). Leaf protein can be used to enhance the protein content of many food products, i.e. in Mexico it is used to make fortified spaghetti (Carlsson 1983).

## Salt-tolerant plants

Salt-tolerant plants (Table 2) can also be used to produce economically important materials such as essentials oils, flavours, fragrances, gums, resins, oils, pharmaceuticals, and fibres. They may also be marketed for use in landscape gardening and for their foliage or flowers. In India, peppermint oil and menthol have been produced in saline environments. The salt-tolerant *kewda*, a common species of screw pine, is used to produce perfume and flavouring ingredients (Singh 2005).

*Grindelia camporum*, a salt tolerant resinous shrub which produces aromatic resin, is used commercially in adhesives, varnish, printing inks, soaps, and industrial applications. In Egypt salt tolerant rushes have a potential use in paper making. *Quinoa* is one of the few plants which

**Table 1** Leaf protein composition

| Component | Per 100 g dry matter |
|---|---|
| True protein | 50–60 g |
| Lipids | 10–25 g |
| Beta carotene | 45–150 |
| Starch | 2–5 g |
| Monosaccharides | 1–2 g |
| B-vitamins | 16–22 mg |
| Vitamin E | 15 mg |
| Choline | 220–260 mg |
| Iron | 40–70 mg |
| Calcium | 400–800 mg |
| Phosphorus | 240–570 mg |
| Ash | 5–10 g |

Source: Carlsson, 1988

**Table 2** Salt tolerant crops. Expressed as conductivity (Siemens): mS/cm

| Moderately sensible 0–4 mS/cm | Sensible 4–6 mS/cm | Tolerant 6–8 mS/cm | Highly tolerant 8–12 mS/cm |
|---|---|---|---|
| Carrot | Apricot tree | Oats | Asparagus |
| Cucumber | Onion | Cabbage | Beetroot |
| Water melon | Lettuce | Wheat | Barley |
| Bean | Melon | Sunflower | Rye |
| Strawberry | Potato | Olive tree | |
| Apple tree | Peach | Tomato | |
| Pear tree | Rice | Rice | |
| Plum tree | Soy-bean | Spinach | |
| Radish | Sorghum | | |
| Celery | Grape | | |
| | Pumkin | | |

Source: Saint George Reclamation Consortium, Ferrara (Italy), 2005

can grow in the salt flats of southern Bolivia and northern Chile. Seashore mallow can tolerate 2,5% salinity during growth and has seeds yield ranging 0.8–1.5 tons per hectare, containing 32% protein and 22% oil. Oilseeds of *Salicornia* arrive at a yield of two tons per hectare and its plants arrive at 20 tons when irrigated with seawater. A succulent annual herb, *Mesembryanthemum crystallinum*, native of South Africa, grows on sea coasts and salty deserts (Greenwood 1986).

*Catharanthus* roots, a plant of the coastal sands of India (Fig. 2), contain alkaloids used for leukaemia, and its leaves are reported to lower blood pressure and cholesterol.

Fuel-wood and building materials can be produced from salt-tolerant trees and shrubs using land and water unsuitable for conventional crops. Fuel crop plantations established on saline soils or irrigated with saline water can survive high temperatures of up to 45–50°C which few food crops can withstand, allowing better land and fresh water for food and forage production.

Halophytic grasses, shrubs and trees are all potential sources of fodder for buffalo and goats, these are now being examined in many countries.

## Salt tolerant crops: Tomatoes

Genetically engineered tomato plants also grow in salty water. As the first truly salt-tolerant vegetable, these tomatoes offer hope that other

**Fig. 2** Flowers of *Catharanthus Roseus*

crops can also be genetically modified for planting in many areas of the world that have salty irrigation water and salt-damaged soils.

The tomato is the most widespread crop plant in the whole world. It represents the main ingredient in Mediterranean recipes, thus becoming the major source of vitamin C for the Italian population and contains other phytochemicals with a beneficial effect on our health.

Among EU countries, Spain is the principal producer, Italy being the second, thanks to Sicily which also obtained a PGI designation (Product with Geographical Indication). Italy presents the largest variety of production and this favours a great deal of research among various agronomic centres. The most effective research in specific environmental situations has obtained products of very high quality. The best quality has been obtained in air-ponic cultures or in rock-wool, cocoa fibres, peat, compost substrates, utilising saline water irrigation. The result is a bicolour tomato, more resistant and with a longer life cycle. Biological control measures are efficient and used in glasshouse tomato cultivations but not in the open field, because of technical and economic reasons.

Experiments gave also origin to the grape tomato, such as cherry and cocktail type tomato (Fig. 3), which have been greatly successful. One of the new cultivars is *Naomi* deriving from an Israelite project which utilises an Italian cultivar (Trentini and Piazza 2005).

The moderate resistance of tomatoes allows their utilisation in South Italy where water is more salty than in northern Italy. In Sardinia

**Fig. 3** Cherry tomato

**Fig. 4** The Italian island Sardinia

(Fig. 4), some experiments in *perlite* medium have been very successful, according to the intentional use of salty water in order to obtain a more tasty tomato. The Regional Centre for Agricultural Experimentation (CRAS) obtained the famous tomato *Camone*, an engineered tomato plant, which although with a bigger market price - at even six times that of normal tomatoes - is particularly appreciated by consumers. The high price is due to the costs of experimentation, the nursing in greenhouses, air-ponic techniques over polyethylene substrate, to which marine refined salt has voluntarily been added, in order to examine its effects. The result is a smaller tomato, whose size is due to water loss, but more tasty. Effectively, high salinity decreases the total and marketable yields of fruits, by reducing the size of fruits.

Experiments show that the growth of species on artificial substances with saline soil solutions was inhibited, to a significant extent, and related to an excess of *K* and *Cl* and a lack of other essential nutrients, but stimulated growth and biomass accumulation of some species as *Boletus edulis*.

Effectively, a study on market tomatoes showed that fruits produced under saline conditions were smaller than the controls, but developed a better colour and had a much better taste. For example, seeds from a wild tomato, found on the seashore of the Galapagos Islands, produced tomatoes that were smaller and bitter. When this species was crossed with a commercial tomato cultivar, flavourful fruits, the size and colour of cherry tomatoes, were obtained in 70% seawater.

Dixon (2005) quotes experiments on antioxidant phytonutrients produced by the terpenoid and phenylpropanoid which have been increased simultaneously for the first time, creating an extra-nutritious tomato.

Tomatoes are a principal dietary source of carotenoids and flavonoids, both of which are highly beneficial for human health. Results demonstrate that manipulation of a plant regulatory gene can simultaneously influence the production of several phytonutrients generated from independent biosynthetic pathways and provide a novel example of the use of organ-specific gene silencing to improve the nutritional value of plant-derived products (Davuluri 2005).

Tomato products are the principal dietary sources of lycopene and major source of $\beta$- Carotene, both of which have been shown to benefit human health. To enhance the carotenoid content and profile of tomato fruit, transgenic lines have been produced containing a bacterial carotenoid gene (*crtI*) encoding the enzyme phytoene desaturase, which converts phytoene into lycopene (Römer 2000).

These genetically engineered salt-tolerant plants actually remove salt from the soil, and because their salt-storing activity occurs only in the plants' leaves, the quality of the tomato fruit is maintained (Sirigu et al. 1999, p.55).

Blumwald and colleagues have demonstrated that genetically engineered tomato plants grow and produce fruit even in irrigation water that is about 50 times saltier than normal. The plants were irrigated with water having a salt concentration of 200 mM sodium chloride; more than one third as salty as seawater, which is about 530 mM sodium chloride (Caruso 1993, p. 32).

Salt tolerant crops: Asparagus

Asparagus grows wild on sand and is cultivated experimentally worldwide. In the last decade, the world productivity has doubled, from 620 thousand tons to 1.2 million. The principal producers are: China, Perù, USA, Mexico, Spain, Germany, Italy, Greece, Japan, France. Australia is becoming one of the most important exporting countries (Trentini 2004). Asparagus is also grown in Tunisia where there are yields of up to 8 tons per hectare, when irrigated with salt water; roughly equal to that in areas irrigated with fresh water. In Italy it has a productivity of 5–6 tons per hectare and up to 8 tons in greenhouses.

Asparagus is an excellent crop for developing countries because it is relatively labour intensive.

It is suited to being grown in south Italy where more economic development is needed and intensive labour is possible, due to the higher level of unemployment than in other parts of the country. In northern Italy the greatest area of production is at the Po river delta, where it is demonstrated that the productivity is higher than in non sandy and boggy soils.

Italian green asparagus got the important PGI accreditation by the EU.

Asparagus is a very healthy vegetable, with positive aspects, according to the ayurvedic medicine.

It was appreciated by Egyptians and Romans for its anti-oxidising activity. From a nutritional point of view it has low cholesterol, low calorie content and fats, low sodium content, and a high percentage of alimentary fibres. It is rich in potassium, vitamin B1, B6, folic acid, calcium, magnesium, phosphorus, iron, copper, zinc, manganese and antioxidants, however other characteristics should be further investigated (Bordoni 2004, p.54). Research has found varieties which are adaptable to extreme climates, ranging from the cold to the equatorial, demonstrating the high adaptability of this species to different climates, to the salinity of soil and water; there are in effect 300 hundred varieties. When irrigated, it can produce 20–25% more than in arid climate (Falavigna 2004).

Salt tolerant crops: Rice

In Asia, where a large part of the world's rice is produced, most of the land has been covered with salt water by tsunami. The rice yield under these sub-optimal conditions can be increased by selecting the appropriate varieties, maybe by adding a gene for trehalose – a sugar that helps plants withstand stress – to the rice (Netherlands Organisation for Scientific Research 2005).

Professor Swaminathan of Japan (IARI, website) discovered in the wake of the disastrous tsunami, which ravaged many parts of Asia, that there are many landraces conserved by local communities in rice and other crops which can withstand the fury of the tidal waves and salt water inclusion in the paddy fields.

Some of the rice landraces that survived in Tsunami are:

- Kuzhivedichan
- Kallurundai
- Kundali

Some suggestions in improved management practices may help local communities to take better actions.

It could be possible to interchange the necessary growing stages of plants: i.e. raising the local salt tolerant rice varieties in good nurseries or using some crops during monsoon rainy season.

This could be useful for natural saline soil, but even more so for areas flooded by seawater, as a consequence of natural disasters such as typhoons, hurricanes or tsunamis. Rice is also used for land reclamation on sea marshes

## Research centres

Using salt tolerant crops is not sufficient in itself because a large amount of expertise, research, and laboratory experiments are required, in order to adapt the different stages of a plant's life cycle to its needs or to find the most adapted species to the local environment.

Research can then begin on ways to improve the agronomic qualities of these plants and to utilise their genetic traits. Genotypes with increased tolerance to water and salinity stress have been identified and followed in genetic crosses with conventional genotypes using new techniques in gene mapping and cell physiology. The transfer of these genes from sources in salt-tolerant species to more productive crops will require modifications in cultural practices as well as treatment of the plant products. Interdisciplinary communication is particularly important in research on salt-tolerant plants. Cooperation among plant ecologists, plant physiologists, plant breeders, soil scientists and agricultural engineers could accelerate development of economic crops. Furthermore, universities could introduce special programs to allow extensive study into the special characteristics of saline agriculture, to fulfil growing needs in this field. (Central Soil Salinity Research Institute 2003).

The role of certain specialised research centres around the world such as the Iowa State University, the Agricultural Centre LA Coast in Louisiana, the Netherlands Organisation for Scientific Research, the Forestry Division of the Agricultural Research Organization – Ilanot in Israel, is important. One of the most important is the Indian Agricultural Research Institute (IARI) which is involved in the assessment of the economic losses of degraded land in India. Through multi-disciplinary research programmes, the Institute aims to tap the uncommon and unprecedented opportunities to gain a holistic understanding of crop biology.

It is, most of all, an educational centre which tries to (1) Emphasise utilisation of global plant genetic resources (2) Concentrate on new and emerging cutting edge technologies such as molecular biology and biotechnology and develop inter-disciplinary centres of excellence.

Another internationally recognised Indian centre of excellence in salinity research is the Central Soil Salinity Research Institute (CSSRI) of the Indian Council of Agricultural Research (ICAR) in Karnal, which released a new salt tolerant rice variety (CSR-30) and a new salt tolerant mustard variety (CS-54).

The CSSRI mission is

> "to generate new knowledge and understanding of the process of reclamation and develop technologies for improving and sustaining the productivity of salty landsand waters" (website 2006).

It is specialised in the determination of the sodicity tolerance of crops to develop cropping sequences at various stages of reclamation and in the development of models to study hydrologic behaviour of alkali watersheds.

In Italy, some institutes, such as the Mario Neri ICCE, the CRPV and the VPRC are very renowned in Emilia Romagna region, which is very focused on agro-food industry and where there are also dedicated organisms involved in the reclamation of the delta area in Northern Italy. With a special background in salt tolerant species, there are centres in Sicily and in Sardinia. Many others are in Israel where the climate is very arid and research centres very advanced.

Italian University centres are today focusing especially on olive oil, a basic component of the Mediterranean diet, due to the recent assessments of its essential qualities of anti-oxidation and its ability, it seems, to even reduce body weight.

The olive tree is a century-old tree, resistant to arid climates and poor soils. The extension of vegetable oil production in African countries, for example, can be remunerative through the export to the northern nations with higher GDP, where the diet is abundant in animal fats and butter.

Due to the fact that centres are generally sustained by international funds, the transfer of knowledge towards the poorest countries and the countries where the school system is less advanced is unavoidable. India has a good education platform but many Asian and especially African nations do not have. Exchanging research results would lead to the creation of a global partnership for development.

## Conclusions

Science and technology can provide a sufficient standard of living for the human race, but science and technology alone are not enough; global policies are needed, and policies related to land tenure, forestry and agriculture are critical in this respect. New partnerships and an increasing accessibility to information is challenging the ownership of decision-making and the exchange of innovations.

However scientists have been trying to develop salt-tolerant crop varieties using selective breeding techniques throughout the last century, efforts may be successful only nowadays, in a systemic synergy, which joins the goals of different branches of studies, because it is not an easy task.

Also researchers have to overcome the resistance of farmers' traditions and accepted rules. Several organisations recognise that advancing the biodiversity's sciences and improving public understanding of it requires effective communication between the public and the scientific community (Raven 1997, p. 2). Further experiments on vegetables are required after the several hazards and illnesses tied to breeding. The economic cost of vegetable production is less than that of animal breeding. Furthermore, and more notably, the augmentation of meat intake is progressively dangerous for health, whereas increasing vegetable products in the diet is beneficial. This does not mean avoiding animal derivates, but that only a balanced diet throughout the world permits an inter-and an intra-generational equity, as requested by the concept and practices of sustainability.

Researchers need to substantially adhere to the genes manipulation. If until now GMOs aren't legal or generally accepted in Europe, the former Health Minister of Italy affirms that they can't be dangerous, as every one of us and every species is subject to genes modification for natural causes and by evolution itself. Nonetheless the solution isn't only on gene experiments, able to transplant salt resistant genes, but in interdisciplinary approaches involving different scientists.

To invest in further research on agriculture and fishery is the only solution to join the main scopes of the UN Millennium Goals, so as to eradicate extreme poverty and hunger, and at the same time, to ensure environmental sustainability.

Experiments, deriving from specialised researches, offer us the possibility of production in marginalised soils such as saline soils.

## References


Abrol IP, Yadan JSP, Massoud FI (1988) Salt affected soils and their management. Soils Bull 39:167–177

Adams P (1991) Effects of increasing the salinity of the nutrient solution with major nutrients or sodium chloride on the yield, quality and composition of tomatoes grown in rockwool. J Hortic Sci 64(6):201–207

Adams P (1992) Crop nutrition in hydroponics. Acta Hoeticulturae 323:289–297

Ahmad R (1987) Saline agriculture at coastal sandy belt. University of Karachi–IWMI, Colombo

Aronson JA (1989) Halophites. Salt- tolerant plants of the world. University of Arizona, Tucson, Arizona

AusAID (2006) Focus, 21, N. 1, Jan–Apr:10–11

Bordoni A, Danesi F (2004) Una fonte di benessere, in Il Divulgatore 5:50–57

Bottini L et al (1996) Stress salino: risposta di specie orticole e suggerimenti di tecnica colturale. L'Informatore Agrario 22:15–17

Boyko (1996) Salinity and aridity. New approaches to old problems. Junk Publ., The Hague, Netherlands

Caliandro A, Molinari GP, Fornari L, Mecella G (2001) Qualità delle acque per uso irriguo. Ministero delle Politiche Agricole e Forestali. Metodi di analisi delle acque per uso agricolo e zootec. 8:1–30 Franco Angeli, Milano

Carlsson R (1983) Tropical plants for leaf protein concentrates. In: Telek I, Graham HD (eds) Leaf Protein Concentrate. AVI Publications, Westport, US, pp 193–210

Caruso G (1993) Risposta Fisiologica del Pomodoro (Lycopersicon Esculentum Mill.) alla Concentrazione Salina dell'Acqua d'Irrigazione. Italus Hortus 1:16–19

Central Soil Salinity Research Institute (1998–2000) AICRP on management of salt affected soils and use of saline water in agriculture. CSSRI, Karnal, India

Central Soil Salinity Research Institute (1998) 25 years of research on management of salt-affected soil and use lime water in agriculture. CSSRI, Karnal, India

Central Soil Salinity Research Institute (2000) 50 years of natural resource management research. CSSRI, Karnal, India

Cucci G Rubino P, Caliandro AA (2003) Effects of irrigation water with different salt concentrations and SAR values on soil salinisation and sodification. Ital J of Agr 7(1):41–48

Davuluri GR et al (2005) Fruit-specific RNAi-mediated suppression of *DET1* enhances carotenoid and flavonoid content in tomatoes. Nat Biotechnol 23:890–895

Dixon RA (2005) A two-for-one in tomato nutritional enhancement. Nat Biotechnol 23:825–26

Downton WJS (1984) Salt tolerance of food crops: perspectives of improvement. CRC Critical Rev in Plant Sci 1(3):183–201

Elder J, Ruscello J (1997) The challenge of communicating biodiversity In: Raven PH (ed) Nature and Human Society- The Quest for a Sustainable World. In: Proceeding of the 1997 Forum on Biodiversity. National Academy Press, Washington D.C., pp 455–461

ESSP – (2006) Earth System Science Partnership (of IGBP) (PPT Presentation)
Falavigna A (2004) Poche le varietà interessanti. Il Divulgatore 5:18–27
Gallagher JL (1985) Halophytic crops for cultivation at seawater salinity. Plant Soil 89:323–336
Galvani A (1994) L'acquicoltura – una rivoluzione azzurra. La Pianura 1:20–25
García Olmedo F (2000) La terza rivoluzione verde: piante transgeniche biotecnologie e agricoltura moderna. Il sole 24-ore Pubbl., Milano
Glenn EP, O'Leary W (1985) Productivity and irrigation requirements of halophytes grown with seawater in the sonora desert. J Arid Environ 9(1):81–91
Greenwood EAN (1986) Water use by trees and shrubs for lowering saline groundwater. Reclam Reveget Res 5:423–434
Gupta GN et al (1986) Salt tolerance of some tree species at seedling stage. Indian Forester 112(2):101–113
Hinman CW (1984) New crops for arid lands. Science 225:1445–1448
Hughes JB et al (1997) The loss of population diversity and why it matters. In: Raven PH (ed) Nature and Human Society-Proceedings of the 1997 Forum on Biodiversity. National Academy Press, Washington, pp 71–83
IGBP (2005) International Geosphere Biosphere Programs Science Plan and Implementation Strategy. IGBP Report No. 53/IHDP Report No. 19. IGBP Secretariat, Stockholm, 64 pp
Iyengar EER (1982) Reasearch on Seawater Irriculture in India. In: Sanpietro A (ed) Biosaline Reasearch. Plenum Press, New York, pp 165–175
Logan CEI (1980) Use of Saline Water in Energy Development - Final Report. Utah Water Research Laboratory, College of Engineering, Utah State University
Loster T (2003) Facts, data, trends, topics. Munich 2:20–24
Newton J et al (2005) Climate change and natural hazards in northern canada: integrating indigenous perspectives with government policy. Mitig Adapt Strat Global Change 10:541–571
O'Leary W (1987) Halophytic Food Crops for Arid Lands: In Strategies for Classification and Management of Native Vegetation for Food Production in Arid Areas, Report RM-150, Forest Service, USDA, Colorado
Panel of the Board on Science and Technology for International Development Office of International Affairs - National Research Council - Saline Agriculture (1989). Salt Tolerant Plants for Developing Countries. National Academy Press, Washington, DC
Pasternak D (1987) Salt tolerance and crop production – a comprehensive approach. Annu Rev Phytopathol 25:271–291
Pearce L (2005) The value of public participation during hazard, impact, risk and vulnerability (HIRV) analysis. Mitig Adapt Strat Global Change 10:411–441
Petschel-Held G et al (1999) Syndromes of global change: a qualitative modelling approach to assist global environmental management. Environ Modell Assess 4:295–314
Pomareda C, Hartwich F (2006) Toward Sustainability – A Plan for Collaborative Research on Agriculture and Natural Resource Management. Nat Res Council, Washington, DC
Römer S (2000) Elevation of the pro-vitamin A content of transgenic tomato plants. Nat Biotechnol 18:666–669
Saleh SA, Heuberger H, Schnitzer WH (2005) Alleviation of salinity effect on Artichoke productivity by *Bacillus subtilis* FZB24. J of Appl Botany Food Qual 79(1):24–32
Serageldin IA (1997) World bank perspective. In: Raven PH (ed) Nature and Human Society-Proceedings of the 1997 Forum on Biodiversity. National Academy Press, Washington, DC, pp 413–419
Singh NT (2005) Irrigation and Soil Salinity in the Indian Subcontinent: Past and Present. Lehigh University Press, Bethlehem
Sirigu A, Leoni S, Pisanu B, Gradina R (1999) Concentrazione salina e caratteristiche qualitative del pomodoro camone. L'Informatore Agrario 6:55–58
Stevens MA (1979) Tomato quality: potential for developing cultivars with improved flavour. Acta Hortic 93:317–328
Trentini L (2004) Una nuova geografia del prodotto. Il Divulgatore 27(5):6–17
Trentini L, Piazza R (2005) Varie le proposte, sempre alta la qualità. Il Divulgatore 28(4):15–17
Tyagi NK, Sakthivadivel R, Sharma DK, Ambast SK, Agrawal A (2003) Farmers Decision Making in Irrigation Commands: The Need and Scope for Improvement. CSSRI Synthesis Paper 1, Karnal, India
World Bank (2001) World Water Forum Electronic Resource: from The Hague to Japan, Washington, US
Yensen NP (1988) Plants for Salty Soil, Arid Lands News 27: 3–10, University of Arizona, Tucson, US
Zilberman D et al. (2004) The economics of climate change in agriculture. Mitig Adapt Strat Global Change 9:365–382
http://www.iydd.org
http://www.fao.org
http://www.iari.res.in
http://www.cssri.org
http://www.cras.sardegna.it

Rev Environ Sci Biotechnol (2007) 6:17–31
DOI 10.1007/s11157-006-0011-2

REVIEW PAPER

# Viruses in extreme environments

**Marc Le Romancer · Mélusine Gaillard · Claire Geslin · Daniel Prieur**

Received: 3 March 2006 / Accepted: 30 May 2006 / Published online: 14 September 2006


**Abstract** The tolerance limits of extremophiles in term of temperature, pH, salinity, desiccation, hydrostatic pressure, radiation, anaerobiosis far exceed what can support non-extremophilic organisms. Like all other organisms, extremophiles serve as hosts for viral replication. Many lines of evidence suggest that viruses could no more be regarded as simple infectious "fragments of life" but on the contrary as one of the major components of the biosphere. The exploration of niches with seemingly harsh life conditions as hypersaline and soda lakes, Sahara desert, polar environments or hot acid springs and deep sea hydrothermal vents, permitted to track successfully the presence of viruses. Substantial populations of double-stranded DNA virus that can reach $10^9$ particles per milliliter were recorded. All these viral communities, with genome size ranging from 14 kb to 80 kb, seem to be genetically distinct, suggesting specific niche adaptation. Nevertheless, at this stage of the knowledge, very little is known of their origin, activity, or importance to the in situ microbial dynamics. The continuous attempts to isolate and to study viruses that thrive in extreme environments will be needed to address such questions. However, this topic appears to open a new window on an unexplored part of the viral world.

**Keywords** Bacteriophages · Viral diversity · Viral abundance · Extreme environments · Deep sea subsurface environment · Deserts · Hot springs · Hydrothermal vents · Hypersaline habitats · Polar ecosystems

Marc Le Romancer and Mélusine Gaillard contributed equally to this work.

M. Le Romancer (✉) · M. Gaillard · C. Geslin · D. Prieur
Laboratoire de Microbiologie des Environnements Extrêmes, IUEM, Technopôle Brest Iroise, 29280 Plouzané, France
e-mail: marc.leromancer@univ-brest.fr

## Introduction

Extremophiles include organisms from the three domains of life, Archaea, Bacteria and Eukarya, which thrive in extreme environments that are characterized by physico-chemical conditions close to the limit values in which an organism can live. As bacteria and archaea are almost omnipresent on the planet and have evolved for over 3.5 billion years, "extremophile" conjures up images of prokaryotes, especially from the domain Archaea. Although archaea are present in many moderate environments, they are still primarily considered extremists, flourishing in habitats that brave the physical limits for life, such as sulfur-rich hot acid springs and geysers, deep-sea environment and deep-sea hydrothermal vents, hypersaline and soda lakes or strictly anoxic ecosystems (see review in Rothschild and Mancinelli 2001).

The specific feature of Extremophiles is their remarkable capabilities to adapt to extreme conditions in term of pH, salinity, desiccation, hydrostatic pressure, radiation, anaerobiosis that would be inevitably lethal for non-extremophilic organisms. Extremophiles thrive at temperatures exceeding 80°C and even more than 100°C at hyperbaric pressure (extreme thermophiles in hydrothermal vents) while others live at subzero temperatures (psychrophiles in sea ice). Extreme piezophiles, which can withstand the enormous hydrostatic pressure associated with great depths, grow well in the deep-sea and even in deep subsurface sediments as deep as 1,000 m below the seafloor (mbsf) under anaerobic conditions. Extremophiles are also able to cope with environments of very low water activity and develop well in desert or saturated brines (extreme halophiles) while acidophiles and alkalophiles live in extremely acid and alkaline waters at pH values below 2 and exceeding 10, respectively. In proportion to the multiplicity of physical and geochemical constraints in an extreme ecosystem, organisms that thrive under the seemingly harsh conditions are most of the times polyextremophiles.

Like all other organisms, extremophiles serve as hosts for viral replication. Viruses and virus-like elements (i.e., satellite virus, satellite RNA and viroids) are the smallest infectious biological entities (see http://www.ncbi.nlm.nih.gov/ICTVdb/origin2.htm). Since they are not autonomous—they depend on a cellular host for replication—viruses have been considered as not really alive for a long time. Many lines of evidence have definitely suggested that they could no more be regarded as simple infectious "fragments of life" but on the contrary as one of the major components of the biosphere, who have probably played a key role in the early cellular evolution and that have a profound influence on cellular life (e.g., genome plasticity, biochemical adaptations required to life in extreme environments).

Viruses exist wherever cellular life is found and span the three domains of life. But the extent of viral ubiquity and diversity still remains largely unknown. The recently accumulated knowledge on the number of viruses, from marine environments at least, shows that they probably encompassed all other forms of life in abundance on the earth and represent a vast reservoir of biodiversity (Fuhrman 1999; Wommack and Colwell 2000; Weinbauer 2004; Rohwer 2003; Suttle 2005; Edwards and Rohwer 2005; Breibart and Rohwer 2005). The universal tree of life can thus be considered as immersed into a virtual viral ocean (Bamford 2003). As new niches are explored for life, especially in extreme environments, presence of viruses is readily detected and an amazing number of (new) viruses is discovered.

The aim of this present mini-review, that was inspired following discussions at the workshop entitled "Investigating Life in Extreme Environments," organized by the European Science Foundation (Sant Feliu de Guixols, Spain, 5–8 November 2005), is to give a brief overview of the recent findings about viruses thriving in extreme conditions.

## Extreme halophilic viruses

Liquid water is an absolute requirement for metabolic activity and growth. The high concentration of ions in hypersaline environments is one of the major factors affecting microbial activity because the dissolved substances make the water partly unavailable to microorganisms. Hypersaline habitats, which can vary considerably in ionic composition, are rather common in hot, dry areas throughout the world. Despite seemingly harsh conditions, these environments can be productive ecosystems where halophiles that include a range of organisms (archaea, green algae, cyanobacteria, bacteria) easily cope with osmotic stress and even can withstand in saturated NaCl (Madigan et al. 2003). If the first extremophilic and halophilic virus was discovered fortuitously in 1974, consistent reports on the occurrence of viruses in such extreme habitats raised in the early 1980s from halobacteria (Dyall-Smith et al. 2003).

## Viruses in hypersaline environments

Assessing the viral abundance in the hypersaline Dead Sea where magnesium concentration exceeds 50%, quantities of virus-like particles that reach easily $10^7$ particles $ml^{-1}$ were reported (Oren et al. 1997). In their study of solar salterns, Guixa-Boixareu and co-workers (Guixa-Boixareu et al. 1996) showed that both virus-like particles abundance and diversity increased with salinity and reached about $10^9$ virus particles $ml^{-1}$ at salinities higher than 25%. Hypersaline environments are also important reservoirs of viruses that exhibit a large genomic diversity with genome sizes varying from 10 kb to 533 kb (Sandaa et al. 2003). Pulsed-field electrophoresis analysis showed that the viral population structure vary along a salinity gradient from near seawater (40‰) to saturated sodium chloride brine (370‰). Populations of virus-like genome ranging in size from 32 kb to 340 kb were preponderant within 40‰ to 220‰ salinity gradients, whereas ponds with salinity higher than 220‰ contained virus-like genomes with size ranging from 10 kb to 189 kb. As changes in the total prokaryotic community structure depending on salinity were also recorded, this suggests that viral populations have a dynamic, which probably depends on their hosts' ecology (Sandaa et al. 2003).

Considering the morphological diversity of viruses in hypersaline environments, direct observations with electronic microscope revealed a majority of lemon-shaped particles resembling the archaeal Fuselloviruses, while only some virus-like particles were of head–tail morphology (Oren et al. 1997). In other hand all halophilic viruses isolated from this type of habitat until now infect archaea, most of them (12/15) have a head and tail morphology (Fig. 1, Table 1) reminiscent of bacteriophages belonging to the three main families *Myoviridae*, *Siphoviridae* and *Podoviridae*, highlighting the remarkable morphological similarity between archaeal and bacterial tailed phages. Only three viruses exhibiting different morphotypes more closely related to those of hyperthermophilic archaeoviruses were also characterized. These haloviruses were the spindle-shaped His1 (Fig. 2), His2 which is pleomorphic and the spherical SH1 (Fig. 3). Such differences between direct observations and laboratory specimens suggested that characterized viruses probably did not reflect the real in situ morphological diversity. The bias resides perhaps in the fact that hosts cells easily isolated and cultivated in laboratory are not the dominant species of the natural haloarchaeal flora in hypersaline environments (Dyall-Smith et al. 2003).

All halophilic viruses described until now have genomes which consist of linear double-stranded DNA. Looking at the genome sequences, only little sequence similarity (less than 10%) with bacteria, bacteriophages and eukaryotic viruses were observed. This phenomenon could be partly due to isolation caused by such particular ecosystem (Dyall-Smith et al. 2003). However, there are also strong genetical relationships between different haloviruses, as shown by the haloviruses $\phi$Ch1 and $\phi$H which share up to 97% nucleotide identity, while their hosts, isolated from distinct and geographically distant sites, are phylogenetically different (Klein et al. 2002; Tang et al. 2002). The haloviruses HF1 and HF2 also have genomes that share up to 99% nucleotide identity in the first 60% of their sequence. However, the remainder part shows a significant divergence (87% identity) due to numerous base changes and insertion/deletion events. This significant shift in sequence similarity suggests a recent recombination event between either the two halovirus or with another HF-like halovirus. This recombination occurrence seems to be rather common among viruses from hypersaline waters (Tang et al. 2004; Bath et al. 2006).

## Viruses in alkaline lakes

Even if the water chemistry of soda lakes is similar to hypersaline lakes, solar salt evaporation ponds and deep-sea hypersaline basins, alkaline lakes differ by the high levels of carbonate minerals in the surroundings rocks that maintain pH ranging between 10 and 12. In addition, $Ca^{2+}$ and $Mg^{2+}$ are virtually absent because they precipitate out at high pH and carbonate concentrations (Madigan et al. 2003).

Bacterial abundances and seasonal changes in community composition were recorded in the past decades, but no previous reports on the

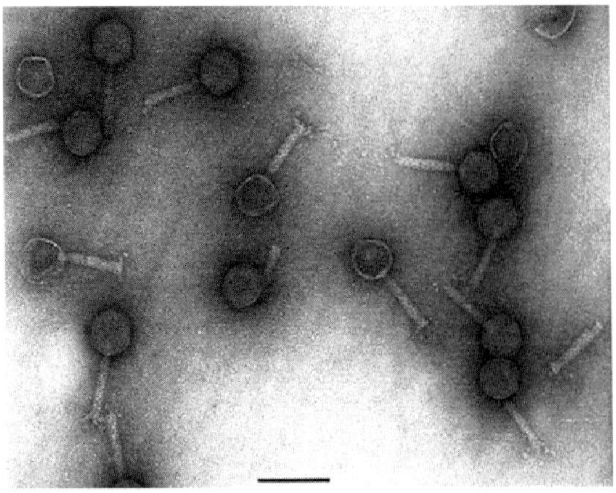

**Fig. 1** Negative stain electron microscopy of head and tail halovirus HF2. Scale bar represents 100 nm. Reprinted from Research in Microbiology, vol. 154, Dyall-Smith M, Tang SL, Bath C, "Haloarchaeal viruses: how diverse are they?", 309–313, Copyright (2003), with permission from Elsevier

occurrence of viruses merged before 2004 from such extreme habitat until Jiang and co-workers tackled the virus populations in Mono Lake, which is a large alkaline (pH ~10), moderate hypersaline lake lying at the western edge of Great Basin in California (Jiang et al. 2004).

In this peculiar environment, viral abundance (from $1.10^8$ to $1.10^9$ $ml^{-1}$) is among the highest observed in any natural aquatic system examined so far. Pulse-field gel electrophoresis revealed length of dsDNA viral genomes ranges from 14 kbp upto 400 kbp, with a majority between 30 kbp and 60 kbp and the analysis of band patterns highlighted at least three dominant clusters of populations defined on the similarities in the viral genome size distribution. Thus, deep-water viral community represents a distinct group from surface and mid-water viral communities suggesting a strong stratification of viral distribution between oxic and anoxic waters.

To date, only one lytic phage, named $\phi$Mono1, has been isolated and partly characterized from this viral population (Table 1). Surprisingly, this virus strain infects a bacterial host, which is closely related to *Idiomarina baltica* previously isolated from surface water of the central Baltic Sea. Using $\phi$Mono1 dsDNA genome as probe in hybridization experiments also revealed seasonal fluctuations in viral communities.

## Viruses in deserts

In deserts, that are extremely dry and exposed to extremes of UV light irradiation and temperature variation, water is always a very limiting factor for life. Nonetheless, eukaryotic and prokaryotic microorganisms have adapted to these extreme conditions and have been found in hot desert such as the Atacama Desert of Chile (Evans and Johansen 1999).

A recent study, carried out on surface sands collected from 13 different locations in the Sahara Desert in Morocco and Tunisia, reported for the first time the presence of virus-like particles. These particles exhibit a great diversity of morphotypes representative of the three major bacteriophage families: *Myoviridae, Siphoviridae* and *Podoviridae* (Table 1). In addition, pulsed-field gel electrophoresis of double-stranded DNA, extracted from the enriched bacteriophages preparations, suggests also a genetic diversity with the presence of at least four potential intact viral genomes ranging in size from 45 kbp to 270 kbp (Prigent et al. 2005).

## Viruses in polar environments

Extreme cold environments such as high-altitude glaciers, polar permafrost, the Dry Valleys of Antarctica, which are the coldest and driest desert on the earth, as well as sea ice, also provide habitats from microbial life (Staley and Gosink 1999). Annual sea ice in the Arctic develops important and dynamic microbial communities (Grossi et al. 1984; Kottmeier et al. 1987; Smith et al. 1989). In Antarctica, microorganisms, including prokaryotes and microeukaryotes thrive in sea ice and cold water (Thomas and Dieckmann 2002). Several well-documented studies reported the presence of viruses and the relationship between viral and bacterial production in Arctic and Antarctic sea ice and in perennially ice-covered lakes located in Taylor Valley, Antarctica.

In Artic sea ice, viral abundance was recorded to be very high as showed by direct counts ($9.10^6$–$3.10^8$ $ml^{-1}$). This value, which was 10- to 100-fold greater than the concentration of viruses in the

**Table 1** Main features of the extremophilic viruses so far characterized in extreme environments

| Environment, source (viral abundance), family, species | Virion morphology, size (nm) (head:tail) or (length × diameter) | | Host (kingdom, genus, species) | Temperate/ lytic /carrier state | dsDNA form, genome size (kb)[a] | | Homologous genes with | Genes with unassigned function (%) | Sequence Acc. No. | References |
|---|---|---|---|---|---|---|---|---|---|---|
| *Hypersaline environments* | | | | | | | | | | |
| Dead Sea ($10^7$ ml$^{-1}$), Solar salterns ($10^9$ ml$^{-1}$) | | | | | | 10–533 | | | | Oren et al. (1997), Guixa-Boixareu et al. (1996) |
| *Myoviridae* | | | | | | | | | | |
| $\phi$H | Isometric head and contractile tail | (64:170) | Archaea: *Halobacterium salinarum* | Temperate | Linear | 59 | $\phi$Ch1 | nd | Genome fragments[b] | Gropp et al. (1989), Schnabel et al. (1982, a,b), Stolt and Zillig (1992, 1993, 1994) |
| $\phi$Ch1 | Isometric head and contractile tail | (70:130) | Archaea: *Natrialba magadii* | Temperate | Linear | 58.5 | $\phi$H, $\psi$M2 | 78.5 | AF440695 | Klein et al. (2002),Witte et al. (1997) |
| HF1 | Isometric head and contractile tail | (58:94) | Archaea: *Haloferax volcanii, Haloarcula Halobacterium salinarium* | Lytic | Linear | 77.7 | HF2, phage RB29, phage T4, phage $\phi$ CTX | 88 | AY190604 | Nuttall and Dyall-Smith (1993), Tang et al. (2004) |
| HF2 | Isometric head and contractile tail | (58:94) | Archaea: *Halobacterium saccharovorum* | Lytic | Linear | 75.9 | HF1, phage RB29, mycophage L5, mycobacteriophage D29 phage T4 | 87 | AF222060 | Nuttall and Dyall-Smith (1993, 1995), Tang et al. (2002) |
| Salterprovirus[d] | | | | | | | | | | |
| His1 | Spindle-shaped | (77 × 44) | Archaea: *Haloarcula, Haloferax, Halobacterium* | Lytic | Linear | 14.4 | His2, SH1, phage PR772 | 94 | AF191796 | Bath and Dyall-Smith (1998) |
| His2 | Spindle-shaped | (67 × 44) | Archaea: *Haloarcula hispanica* | Lytic | Linear | 16.1 | His1, phage GIL16c | 91 | AF191797 | Bath et al. (2006) |
| Unclassified | | | | | | | | | | |
| SH1 | Isometric | (55) | Archaea: *Haloarcula hispanica, Halorubrum sodomense* | Lytic | Linear | 30.9 | $\phi$Ch1, His1 | 78.5 | NC007217 | Bamford et al. (2005), Porter et al. (2005) |
| Alkaline lakes ($10^8$–$10^9$ ml$^{-1}$) | | | | | | 14–400 | | | | Jiang et al. (2004) |
| Unclassified | | | | | | | | | | |
| $\phi$Mono1 | nd | nd | Bacteria: *Idiomarina baltic*[c] | Lytic | nd | nd | nd | nd | nd | Maranger et al. (1994), Borriss et al. (2003) |

**Table 1** continued

| Environment, source (viral abundance), family, species | Virion morphology, size (nm) (head:tail) or (length × diameter) | | Host (kingdom, genus, species) | Temperate/ lytic /carrier state | dsDNA form, genome size (kb)[a] | | Homologous genes with | Genes with unassigned function (%) | Sequence Acc. No. | References |
|---|---|---|---|---|---|---|---|---|---|---|
| *Polar environments* | | | | | | | | | | |
| Arctic sea ice ($9 \times 10^6$–$3 \times 10^8$ ml$^{-1}$) | | | | | | | | | | Borriss et al. (2003) |
| *Myoviridae* | | | | | | | | | | |
| Isolate 1a | Isometric head and contractile tail | (65–73: 93–103) | Bacteria: *Shewanella*[c] | Lytic | nd | 70 | nd | nd | nd | Borriss et al. (2003) |
| *Siphoviridae* | | | | | | | | | | |
| Isolate 11b | Isometric head and non-contractile tail | (37–43: 75–91) | Bacteria: *Flavobacterium*[c] | Lytic | nd | 30 | nd | nd | nd | Borriss et al. (2003) |
| Isolate 21C | Isometric head and non-contractile tail | (45–48: 150–188) | Bacteria: *Colwellia*[c] | Lytic | nd | 40–50 | nd | nd | nd | Borriss et al. (2003) |
| Antarctic sea ice ($5.2 \times 10^6$–$3.5 \times 10^8$ ml$^{-1}$) | | | | | | | | | | Gowing (2003) |
| Unclassified | Icosahedral, spherical and lumpy forms | | Microeucaryotes? | nd | nd | 30–70 | nd | nd | nd | Gowing (2003) |
| Antarctic lakes ($3.4 \times 10^7$ ml$^{-1}$) | | | | | | | | | | Kepner et al. (1998) |
| Unclassified | Large isometric form | | Flagellates? | nd | nd | nd | nd | nd | nd | Kepner et al. (1998) |
| *Desert environments* | | | | | | | | | | |
| Sahara Desert (nd) | | | | | | | | | | |
| *Myoviridae* | Isometric head and contractile tail | (82–185: 129–385) | nd | Temperate? | nd | 270 | nd | nd | nd | Prigent et al. (2005) |
| *Siphoviridae* | Hexagonal head and non-contractile tail | (80–110 × 40: 0–15) | nd | Temperate? | nd | 45–80 | nd | nd | nd | Prigent et al. (2005) |
| *Podoviridae* | Isometric head and non-contractile tail | (50–90: 155–460) | nd | Temperate? | nd | 45–80 | nd | nd | nd | Prigent et al. (2005) |
| *Deep subsurface biosphere* | | | | | | | | | | |
| Deep subsurface sediments ($10^9$ g$^{-1}$) | | nd | nd | nd | nd | nd | nd | nd | nd | Bird et al. (2001) |
| *Extreme thermal environments* | | | | | | | | | | |
| Terrestrial hot springs ($10^6$ ml$^{-1}$) | | | | | | | | | | Breitbart et al. (2004) |
| *Fuselloviridae* | | | | | | | | | | |
| SSV1 | Spindle-shaped | (100 × 60) | Archaea: *Sulfolobus* | Temperate | ccc[f] | 15.5 | SSV2, SS-K1, SSVRH, SIRV1, SIRV2, ARV1 | 88 | XO7234 | Martin et al. (1984), Palm et al. (1991), Schleper et al. (1992) |
| SSV2 | Spindle-shaped | (80 × 55) | Archaea: *Sulfolobus* | Temperate | ccc | 14.8 | SSV2, SS-K1, SSVRH | 88 | AY370762 | Stedman et al. (2003) |

**Table 1** continued

| Environment, source (viral abundance), family, species | Virion morphology, size (nm) (head:tail) or (length × diameter) | | Host (kingdom, genus, species) | Temperate/ lytic /carrier state | dsDNA form, genome size (kb)[a] | | Homologous genes with | Genes with unassigned function (%) | Sequence Acc. No. | References |
|---|---|---|---|---|---|---|---|---|---|---|
| SS-K1 | Spindle-shaped | (90 × 60) | Archaea: *Sulfolobus* | Temperate | ccc | 17.4 | SSV1, SSV2, SSVRH | 92 | AY423772 | Wiedenheft et al. (2004) |
| SSVRH | Spindle-shaped | (90 × 60) | Archaea: *Sulfolobus* | Temperate | ccc | 16.5 | SSV1, SSV2, SS-K1 | 90 | AY388628 | Wiedenheft et al. (2004) |
| *Lipothrixviridae* | | | | | | | | | | |
| TTV1 | Non-flexible rod | (410 × 38) | Archaea : *Thermoproteus tenax* | Temperate | Linear | 15.9 | None | 84 | X14855 | Janekovic et al. (1983) |
| TTV2 | Flexible rod | (1200 × 20) | Archaea : *T. tenax* | Temperate | nd | 16 | nd | nd | nd | Janekovic et al. (1983) |
| TTV3 | Flexible rod | (2500 × 30) | Archaea : *T. tenax* | nd | nd | 27 | nd | nd | nd | Janekovic et al. (1983) |
| SIFV | Flexible rod | (2000 × 24) | Archaea: *Sulfolobus islandicus* | Carrier state | Linear | 40.9 | DAFV, SSV1, SIRV1, SIRV2, ARV1, AFV1, AFV2 | 95 | AF440571 | Arnold et al. (2000b) |
| AFV1 | Flexible rod | (900 × 24) | Archaea: *Acidianus* | Carrier state | Linear | 21.1 | SIFV, SIRV1, SIRV2, SSV1, AFV2 | 88 | AJ567472 | Bettstetter et al. (2003) |
| AFV2 | Flexible rod | (1100 × 24) | Archaea: *Acidianus* | Carrier state | Linear | 31.8 | SIFV, AFV1, SIRV1, SIRV2 | 94 | AJ854042 | Häring et al. (2005b) |
| *Rudiviridae* | | | | | | | | | | |
| SIRV1 | Stiff rod | (930 × 22) | Archaea: *Sulfolobus islandicus* | Temperate | Linear | 32.3 | SIFV, SIRV2, ARV1, AFV1, AFV2, SSV1, poxviruses, ASFV, Chlorella viruses | 89 | AJ414696 | Prangishvili et al. (1999) |
| SIRV2 | Stiff rod | (900 × 23) | Archaea: *Sulfolobus islandicus* | Carrier state | Linear | 35.5 | SIRV1, SIFV, SSV1, AFV1, AFV2, poxviruses, ASFV, Chlorella viruses | 91 | AJ344259 | Prangishvili et al. (1999) |
| ARV1 | Stiff rod | (610 × 22) | Archaea: *Acidianus* | Carrier state | Linear | 24.7 | SIFV, SIRV1, SIRV2, AFV1, SSV1 | 83 | AJ875026 | Vestergaard et al. (2005) |
| *Guttaviridae* | | | | | | | | | | |
| SNDV | Droplet-shaped | (100–185 × 70–95) | Archaea: *Sulfolobus neozelandicus* | Carrier state | nd | | nd | nd | nd | Arnold et al. (2000a) |

**Table 1** continued

| Environment, source (viral abundance), family, species | Virion morphology, size (nm) (head:tail) or (length × diameter) | | Host (kingdom, genus, species) | Temperate/ lytic /carrier state | dsDNA form, genome size (kb)[a] | | Homologous genes with | Genes with unassigned function (%) | Sequence Acc. No. | References |
|---|---|---|---|---|---|---|---|---|---|---|
| '*Globuloviridae*'[e] | | | | | | | | | | |
| PSV | Spherical | (100) | Archaea: *Pyrobaculum, Thermoproteus* | Carrier state | Linear | 28.3 | None | 100 | AJ635162 | Häring et al. (2004) |
| TTSV1 | Spherical | nd | Archaea: *T. tenax* | nd | Linear | 20.9 | PSV VP2 | 100 | AY722806 | Ahn et al. (2004) |
| '*Ampullaviridae*'[e] | | | | | | | | | | |
| ABV | Bottle-shaped | (230 × 75-4) | Archaea: *Acidianus convivator* | Carrier state | Linear | 23.9 | nd | nd | nd | Häring et al. (2005a) |
| '*Bicaudaviridae*'[e] | | | | | | | | | | |
| ATV | Spindle-shaped (110–180 × 70–100) with two tails (total length ~1000) | | Archaea: *Acidianus convivator* | Temperate | ccc | 62.7 | SSV1, ARV1, SIFV, STSV1 | 90 | AJ888457 | Häring et al. (2005c) |
| Unclassified | | | | | | | | | | |
| STIV | Isometric | (60) | Archaea: *Sulfolobus* | nd | ccc | 17.7 | None | 97 | AY569307 | Rice et al. (2004) |
| STSV1 | Spindle-shaped | (230 × 107) | Archaea: *Sulfolobus* | Carrier state | ccc | 75.3 | ATV, SSV1, SSV2, SS-K1, SSVRH | 81 | AJ783769 | Xiang et al. (2005) |
| Deep sea hydrothermal vents ($1.45 \times 10^5$–$9.9 \times 10^7$ ml$^{-1}$) | | | | | | | | | | Ortmann and Suttle (2005) |
| Unclassified | | | | | | | | | | |
| PAV1 | Spindle-shaped | (120 × 80) | Archaea: *Pyrococcus abyssi* | Carrier state | ccc | 18.1 | None | 96 | nd | Geslin et al. (2003a, b) |

nd: not determined

[a] Approximate values

[b] X80163, X80162, X80161, X00805, X52504, AH004327, S63994, 405325, S63993, 405323, S63992

[c] Isolate related to

[d] Floating genus

[e] Taxonomic proposals

[f] Covalently closed, cicular

underlying sea water ($1.1.10^6$ $ml^{-1}$), corresponded with the bacterial abundance in sea ice compared to the water column. Viral proliferation appeared to be enhanced in sea ice relative to open water. Moreover, the virus-to-bacteria ratios were among the highest reported in natural samples, providing the first account of viruses as a dynamic component of sea ice microbial communities (Maranger et al. 1994).

Three distinct phage–host systems (Table 1), which are highly dependent of low temperature conditions, were also isolated and characterized from samples of Artic sea ice collected in north-west of Svalbard. The hosts are psychrophilic bacteria whose closest relatives are *Shewanella frigidimarina*, *Flavobacterium hibernum* and *Colwellia psycherythrae*, respectively. The three phages, which are lytic and host-specific, showed an even more pronounced adaptation to cold temperatures than their hosts did. In fact, phage development was clearly restricted to a lower temperature maximum in comparison to the maximal growth temperature of the host bacterium. Transmission electron microscopy (TEM) observations revealed that these polar phages having a dsDNA genome are morphologically similar to the double-stranded DNA phage families *Siphoviridae* and *Myoviridae* and (Borriss et al. 2003).

Interestingly, samples of Ross Sea pack ice in Antarctic revealed that the range of total viral abundance was similar with the concentration found in Artic sea ice (between $5.2 \times 10^6$ $ml^{-1}$ and $3.5 \times 10^8$ $ml^{-1}$). TEM observations showed that the viruses, which compose the population, are large, with 40% icosahedral, 37% spherical and 23% lumpy forms, and all of them likely infect microeukaryotes (Gowing 2003).

In Antarctic perennially ice-covered lakes, which are microbially dominated ecosystems, virus densities seemed to be less important than in sea ice with a maximum value that reached $3.4 \times 10^7$ $ml^{-1}$. Nevertheless, this virus abundance was higher than in other freshwater or marine systems and the viral population appeared to be highly active in the water column. Many of viruses were found to be large icosahedral specimens, morphologically similar to double-stranded DNA viruses isolated from temperate environments that infect photosynthetic and non-photosynthetic flagellates (Kepner et al. 1998).

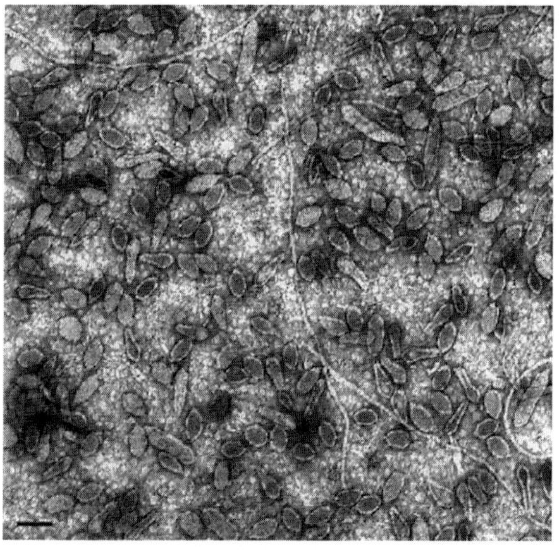

**Fig. 2** Negative stain electron microscopy of spindle-shaped halovirus His1. Scale bar represents 100 nm. Reprinted from Research in Microbiology, vol. 154, Dyall-Smith M, Tang SL, Bath C, "Haloarchaeal viruses: how diverse are they?", 309–313, Copyright (2003), with permission from Elsevier

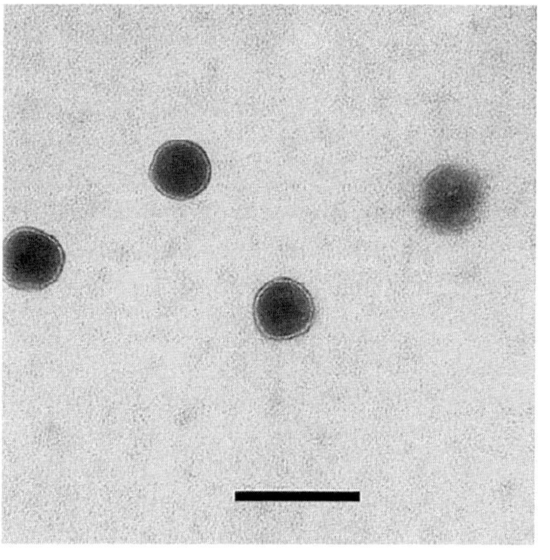

**Fig. 3** Negative stain electron microscopy of spherical halovirus SH1. Scale bar represents 100 nm. Reprinted from Research in Microbiology, vol. 154, Dyall-Smith M, Tang SL, Bath C, "Haloarchaeal viruses: how diverse are they?", 309–313, Copyright (2003), with permission from Elsevier

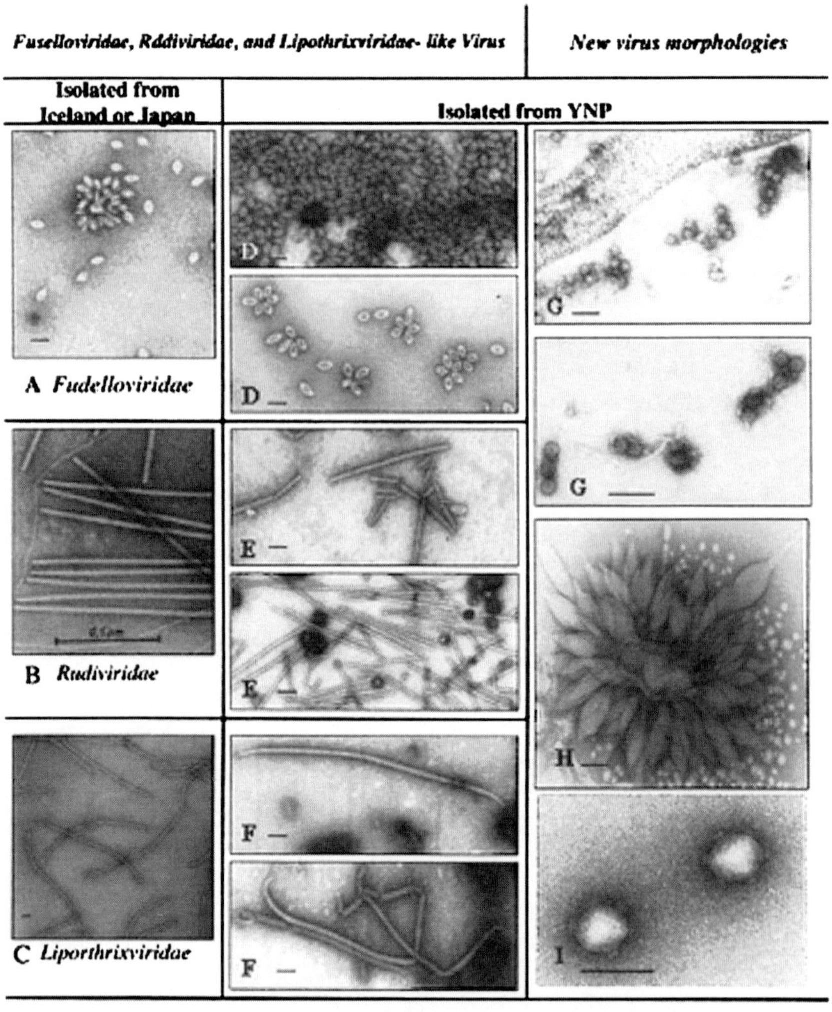


**Fig. 4** Transmission electron microscopy of virus and virus-like particles isolated from Yellowstone National Park. (**A**) SSV1 *Fusellovirus*, (**B**) SIRV *Rudivirus* and (**C**) SIFV *Lipothrixvirus* previously isolated from thermal area of Japan or Iceland. (**D**) SSV-like, (**E**) SIRV-like and (**F**) SIFV-like particle morphologies isolated from Yellowstone National Park thermal features. (*G–I*) Virus-like particles isolated from Yellowstone National Park thermal features. Bars indicate 100 nm. Reprinted from PNAS, vol. 98, Rice G, Stedman K, Snyder J, Wiedenheft B, Willits D, Brumfield S, McDermott T, Young MJ, "Viruses from extreme thermal environments.", 13341–13345, Copyright (2001), with permission from National Academy of Sciences, USA

## Viruses in deep subsurface sediments

The deep subsurface biosphere is one of the least-understood habitats on Earth, even though the huge microbial biomass therein likely plays an important role on global biogeochemical cycles. Recently, the Ocean Drilling Program (ODP) revealed that chemolithotroph microbes thrive in anoxic reducing environments under oceans and continents to depths of >1,000 m despite harsh conditions (i.e., high hydrostatic pressure, anaerobiosis and low concentration in organic nutrients). Prokaryotic biomass in deep marine sediments exceeds $10^5$ microbial cells $cm^{-3}$ even at depths close to 1,000 mbsf (Parkes et al. 1994, 2000).

Presence of viruses in buried marine sediments was investigated recently after drilling a hole at 228.7 m below sea seafloor to a depth at 105.1 mbsf and 118.2 mbsf, near the west Canadian coast (Bird et al. 2001). Analyses revealed the existence of large amounts of viruses. Viral abundances appeared to follow bacterial numbers very closely with an average up to $10^9$ $g^{-1}$ of dry sediment at 105.1 mbsf. Even if microbial communities seemed to be stratified in subseafloor sediments, nothing is known about the viral diversity and the interactions between viral and prokaryotic communities. Nonetheless, given the scarcity of eukaryotic bacterivores in deep marine sediments, the only source of mortality by external agents for the bacterial community lies in phage attack. Thus, considering bacterial and viral abundances being highly correlated, viruses appear to be potential actors of subsurface sediments biogeochemistry.

## Viruses in extreme thermal environments

Live has also adapted to hot temperatures. Given that early branching organisms could have been hyperthermophiles among *Archaea* and thermophiles among *Bacteria*, looking for viruses in such extreme environment may provide interesting information about virus evolution in the early cellular life. The recent observations indicated that terrestrial and oceanic hydrothermal environments represent a bottomless reservoir of a truly remarkable morphological and genomic viral diversity.

### Viruses from terrestrial hot springs

Early studies on viruses of hyperthermophiles were pioneered in the laboratory of Wolfram Zillig in the 1980s. A systematic screening of surface hot springs located in Japan, Iceland, New Zealand, Italy, Russia and the United States led to the isolation of an unprecedented diversity of new viruses (Fig. 4) (Rice et al. 2001; Rachel et al. 2002; Prangishvili and Garrett 2005).

The vast majority of the hyperthermophilic viruses isolated from acidic or neutral hot springs (>80°C) were found to infect a broad spectrum of members of the extremely thermophilic *Crenarchaeaota*, including representatives of the genera *Sulfolobus, Thermoproteus, Acidianus, Pyrobaculum* (Table 1). Based on their exceptional morphology and genomic properties the crenarchaeal viruses were classified in 7 new families which include: lemon-shaped *Fuselloviridae*, filamentous *Lipothrixvidae*, stiff rod-shaped *Rudiviridae*, droplet-shaped *Guttaviridae*, spherical *Globuloviridae*, two tailed spindle-shaped *Bicaudaviridae* and bottle-shaped *Ampullaviridae*. The International Committee of Taxonomy of Viruses has already approved the first four families. The crenarchaeal viruses showed no clear similarities in their morphologies or at the genomic level to either bacterial or eukaryal viruses, except perhaps members of three viral families. The rod-shaped virions of the *Rudiviridae* and *Liphotrixviridae* resemble tobamoviruses and closteroviruses of vascular plants, respectively, while those of the *Globuloviridae* resemble that of viruses of the *Paramyxoviridae*, which infect vertebrates. The 25 hyperthermophilic viruses isolated so far exhibited double-stranded DNA genomes, linear or circular of 15–75 kb, most of them being sequenced and revealing an amazing diversity at the genomic level (Prangishvili et al. 2006). Few significant sequence matches were obtained with either bacterial or eukaryal genes and very few genes have been assigned functions. However, there is some evidence that a 37-kDa coat protein of the *Sulfolobus* turreted icosahedral virus (STIV) can generate a tertiary and quaternary structure similar to that of capsid proteins of bacterial and animal viruses, despite the lack of significance gene similarity. This suggests that some viruses may have a common ancestor that precedes the division into three domains of life (Rice et al. 2004; Khayat et al. 2005). The fact that for most of these viruses, analysis of their genomes showed little or no similarity to genes in the public databases suggests that all these newly discovered viruses employ novel biochemical mechanisms for viral functions.

All viruses of acidophilic hyperthermophiles (except TTV1 and ATV) are non-lytic and persist in host cells in a stable state (pseudolysogeny or "carrier state"). It was hypothesized that such a survival strategy was beneficial for viruses, helping them to avoid direct exposure to the harsh conditions of the host habitat (Prangishvili and Garrett 2004, 2005).

However, hyperthermophilic viral populations, which can reach concentrations of a million viruses per milliliter, were also reported to be resistant to shifts to lower temperature in their natural ecosystem (Breitbart et al. 2004). Breitbart and co-workers showed that more than 75% of phage particles collected from Californian hot springs remained physically intact when incubated on ice. Moreover, they are dynamic and actively produced in situ with a population turnover time of 1 or 2 days. As viruses are the only known microbial predators in this extreme environment, they exert likely an important influence on the microbial community via a high virus-mediated microbial mortality.

## Viruses from deep-sea hydrothermal vents

Deep-sea-vent areas are one of the most extreme habitats on Earth. They are characterized by high hydrostatic pressures, hot (400°C) to warm (10–30°C) temperatures and the hydrothermal fluids are acidic, reduced and enriched with chemicals including heavy metals, methane and hydrogen sulphide (Prieur 1997).

Recently, systematic searches carried out on samples collected in various geographically distant hydrothermal sites revealed high and unexpected abundance and diversity of viruses in deep-sea hydrothermal vents. Viral abundance was recorded to be high as showed by direct counts ($1.45 \times 10^5$–$9.9 \times 10^7$ ml$^{-1}$). High viral abundance at active vents, relative to those in surrounding waters, indicated viral production and hence, virus-mediated microbial mortality (Ortmann and Suttle 2005).

Considering the morphological diversity, direct observations with electronic microscope revealed a great morphological diversity. With the exception of the filamentous and rod-shaped morphotypes which are also known for the *Bacteria*, the morphologies seemed to be characteristic of archaeal viruses. Indeed, the lemon-shaped type prevailed and novel pleomorphic morphologies such as "spoon-shaped" and spindle particles with bipolar expansions were also discovered. The exotic morphological similarities exhibited by viruses from both deep-sea and terrestrial hot environments are very astonishing. For example, the presence of lemon-shaped viruses in diverse extreme environments (salterns, subsurface anaerobic sediments, acidic thermophilic continental solfatara and deep-sea vents) in addition to the fact that this morphotype has never been found among the *Bacteria* or *Eucarya* strengthens the idea of their specificity to the archaeal domain and probably reflects a deep evolutionary history within this domain (Geslin et al. 2003a).

One of these deep-sea hyperthermophilic viruses was successfully purified and was further characterized (Table 1). This virus, named PAV1, is lemon-shaped (120 nm × 80 nm) with a short tail terminated by fibers and infects the hyperthermophilic euryarchaeota *Pyrococcus abyssi*. PAV1 persists in the host strain in a stable carrier state. PAV1 genome consists of a double-stranded circular DNA of 18 kb, which is also present in high copy number in a free form in the host cytoplasm. Viral genome comparisons with all other archaeal, bacterial or eukaryal viruses do not reveal any significant similarity (Geslin et al. 2003b).

## Concluding remarks

Despite the ubiquity of viruses, until recently relatively little was known about viruses in extreme environments because in many instances the extreme growth conditions required by extremophiles have precluded a search for viruses. However, over the past few years our knowledge of viruses in extreme environments considerably increased. Tracking viruses in ecological niches with seemingly harsh conditions has been successful and the presence of virus populations has been consistently detected in all the explored environments. All viral communities appeared to be substantially abundant to the populations rate that are often greater than in standard environments (e.g., $10^9$ ml$^{-1}$ in solar salterns, $3.5 \times 10^8$ ml$^{-1}$ in Antarctic sea ice). All viruses isolated so far from extreme environments are double-stranded DNA viruses with moderate genomic complexity (the genome size range from 14 kb to 80 kb). It is conceivable that this very stable form of genome may be necessary to face harsh constraints of extreme habitats. It could also explain why no RNA virus has been isolated yet, especially from hot environments. However, PFGE analysis used to depict the viral community structure (e.g., in desert and hypersaline habitats environments) produces evidence of a more complex diversity with the recovering of uncharacterized large dsDNA viruses.

The viral communities seem also to be genetically distinct, suggesting specific niche adaptation and great diversity. Nevertheless, at this stage of the knowledge, little is known of their origin, activity, or importance to the in situ microbial dynamics and continuous attempts to isolate and to study viruses that thrive in extreme environments will be needed to address such questions. Moreover, several terrestrial extreme environments are still unexplored, e.g., evaporites, subglacial Antarctic lakes like Lake Vostok, where the DNA signature of a thermophilic bacteria (*Hydrogenophilus* sp.) has been detected (Bulat et al. 2004) or the stratosphere and its airborne biota.

Exploring the virus diversity in extreme environments, the description of an amazing number of new and extraordinary archaeal viruses isolated from terrestrial hot springs especially appears as a benchmark discovery that open a new window on an unexplored and very intriguing part of the viral world (Prangishvili et al. 2006).

More than 85% of the viral genomic sequences lack similarity to previously reported sequences. Thus, the genome of hyperthermophilic viruses and that of any other virus that thrives with extreme conditions probably contains an astronomical number of still unknown proteins. Although some of these proteins could be functional analogues of already known proteins, it would be not surprising to discover proteins encoding novel functions. This exceeds previous results from viral metagenomic analyses (68%) and reinforces the view that viruses represent by far the largest unexplored reservoir of genomic diversity on Earth (Edwards and Rohwer 2005). This constitutes an important issue for further research aimed at understanding the origin of viruses and early life evolution but also for practical purposes such as identification of new enzymatic tools useful for the manipulation of DNA *à façon*.

Extremophiles are probably among the earliest forms of cellular life on Earth that still thrive in a wide range of extreme environments. Therefore, understanding their biology would allow developing hypotheses regarding the conditions required for the origination and early diversification of cellular life on Earth. Even if our perception of the existing viral diversity in extreme ecosystems is still scarce, the recent findings contribute to raise challenging questions about the role of viruses in the early cellular life.

Considering the last updated Forterre's scenario (Forterre 2006) which hypothesized that viruses have played a key role in both RNA-to-DNA transition and in emergence of the three cellular domains presently known, the research on viruses is entering a new exciting stage. The study of the biology and ecology of new viruses isolated from extremophile environments may shed light on the early biological processes as well as on viral evolution.

**Acknowledgements** Many thanks to the Editorial Board of Reviews in Environment Science and Bio/Technology for the invitation to contribute this review. MLR and DP thank the European Science Foundation for the invitation to the ESF workshop on "Investigating life in extreme environments" in Sant Feliu de Guixols, Spain, November 2005. Two anonymous reviewers provided very constructive suggestions that improve the paper. MG is funded through a PhD grant from the Ministère National de l'Enseignement et de la Recherche.

## References

Ahn DG, Kim SI, Rhee JK, Kim KP, Oh JW (2004) TTSV1, a novel globuloviridae family virus isolated from the hyperthermophilic crenarchaeote *Thermoproteus tenax*. GenBank Accession: NC_00655

Arnold HP, Zillig W, Ziese U, Holz I, Crosby M, Utterback T, Weidmann JF, Kristjansson JK, Klenk HP, Nelson KE, Fraser CM (2000a) A novel lipothrixvirus, SIFV, of the extremely thermophilic crenarchaeon Sulfolobus. Virology 267:252–266

Arnold HP, Ziese U, Zillig W (2000b) SNDV, a novel virus of the extremely thermohpilic and acidophilic archaeon *Sulfolobus*. Virology 272:409–416

Bamford DH (2003) Do viruses form lineage across different domains of life? Res Microbiol 154:231–236

Bamford DH, Ravantti JJ, Rönnholm G, Laurinavicius S, Kukkaro P, Dyall-Smith M, Somerharju P, Kalkkinen N, Bamford JKH (2005) Constituents of SH1, a novel lipid-containing virus infecting the halophilic auryarchaeon *Haloarcula hispanica*. J Virol 79(14):9097–9107

Bath C, Dyall-Smith ML (1998) His1, an archaeal virus of the *Fuselloviridae* family that infects *Haloarcula hispanica*. J Virol 72:9392–9395

Bath C, Cukalac T, Porter K, Dyall-Smith ML (2006) His1 and His2 are distantly related, spindle-shaped haloviruses belonging to the novel virus group, Salterprovirus. Virology 350:228–239

Bettstetter M, Peng X, Garrett RA, Prangishvili D (2003) AFV1, a novel virus infecting hyperthermophilic archaea of the genus *Acidianus*. Virology 315:68–79

Bird DF, Juniper SK, Ricciardi-Rigault M, Martineu P, Prairie YT, Calvert SE (2001) Subsurface viruses and bacteria in Holocene/Late Pleistocene sediments of Saanich Inlet, BC: ODP Holes 1033B and 1034B, Leg 169S. Mar Geol 174:227–239

Borriss M, Helmke E, Hanschke R, Schweder T (2003) Isolation and characterization of marine psychrophilic phage–host systems from Arctic sea ice. Extremophiles 7:377–384

Breibart M, Rohwer F (2005) Here a virus, there a virus, everywhere the same virus? Trends Microbiol 13:278–284

Breitbart M, Wegley L, Leeds S, Schoenfeld T, Rohwer F (2004) Phage community dynamics in hot springs. Appl Environ Microbiol 70:1633–1640

Bulat S, Alekhina IA, Blot M, Petit JR, de Angelis M, Wagenbach D, Lipenkov VY, Vasilyeva L, Wloch D, Raynaud D, Lukin VV (2004) DNA signature of thermophilic bacteria from the aged accretion ice of Lake Vostok: implications for searching life in extreme icy environments. Int J Astrobiol 1:1–12

Dyall-Smith M, Tang SL, Bath C (2003) Haloarchaeal viruses: how diverse are they? Res Microbiol 154:309–313

Edwards RA, Rohwer F (2005) Viral metagenomics. Nat Rev Microbiol 3:504–510

Evans RD, Johansen JR (1999) Microbiotic crusts and ecosystem processes. Cri Rev Pl Sci 18:182–225

Forterre P (2006) Three RNA cells for ribosomal lineages and three DNa viruses to replicate their genomes: a hypothesis for the origin of cellular domain. Proc Natl Acad Sci USA 103:3669–3674

Fuhrman JA (1999) Marine viruses and their biogeochemical and ecological effects. Nature 399:541–548

Geslin C, Le Romancer M, Gaillard M, Erauso G, Prieur D (2003a) Observation of virus-like particles in high temperature enrichment cultures from deep-sea hydrothermal vents. Res Microbiol 154:303–307

Geslin C, Le Romancer M, Erauso G, Gaillard M, Perrot G, Prieur D (2003b) PAV1, the first virus-like particle isolated from a hyperthermophilic euryarchaeote, "*Pyrococcus abyssi*". J Bacteriol 185:3888–3894

Gowing MM (2003) Large viruses and infected microeukaryotes in Ross Sea summer pack ice habitats. Mar Biol 142:1029–1040

Gropp F, Palm P, Zillig W (1989) Expression and regulation of *Halobacterium halobium* phage $\phi$H genes. Can J Microbiol 35:182–188

Grossi SM, Kottmeier S, Sullivan CW (1984) Sea ice microbial communities. III. Seasonal abundance of microalgae and associated bacteria, McMurdo Sound, Antarctica. Microb Ecol 10:231–242

Guixa-Boixareu N, Calderon-Paz JI, Heldal M, Bratbak G, Pedros-Alio C (1996) Viral lysis and bacterivory as prokaryotic loss factors along a salinity gradient. Aquat Microb Ecol 11:215–227

Häring M, Peng X, Brügger K, Rachel R, Stetter KO, Garrett RA, Prangishvili D (2004) Morphology and genome organization of the virus PSV of the hyperthermophilic archaeal genera *Pyrobaculum* and *Thermoproteus*: a novel virus family, the *Globuloviridae*. Virology 323:233–242

Häring M, Vestergaard G, Rachel R, Chen L, Garrett RA, Prangishvili D (2005a) Independent virus development outside a host. Nature 436:1101–1102

Häring M, Rachel R, Peng X, Garrett RA, Prangishvili D (2005b) Viral diversity in hot springs of Puzzoli, Italy and characterization of a unique archaeal virus, Acidianus bottle-shaped virus, from a new family, the Ampullaviridae. J Virol 79(15):9904–9911

Häring M, Vestergaard G, Brügger K, Rachel R, Garrett RA, Prangishvili D (2005c) Structure and genome organization of AFV2, a novel archaeal Lipothrixvirus with unusual terminal and core structures. J Bact 187(11):3855–3858

Janekovic D, Wunderl S, Holz I, Zillig W, Gierl A, Neumann H (1983) TTV1, TTV2 and TTV3, a family of viruses of the extremely thermophilic, anaerobic sulfur reducing archaebacterium *Thermoproteus tenax*. Mol Gen Genet 192:39–45

Jiang S, Steward G, Jellison R, Chu W, Choi S (2004) Abundance, distribution and diversity of viruses in alkaline hypersaline Mono Lake, California. Microb Ecol 47:9–17

Khayat R, Tang L, Larson ET, Lawrence CM, Young M, Johnson JE (2005) Structure of an archaeal virus capsid protein reveals a common ancestry to eukaryotic and bacterial viruses. Proc Natl Acad Sci USA 102:1894–1899

Kepner RL, Wharton RA Jr, Suttle CA (1998) Viruses in Antarctic lakes. Limnol Oceanogr 43:1754–1761

Klein R, Baranyi U, Rössler N, Greineder B, Scholz H, Witte A (2002) *Natrialba magadii* virus $\phi$Ch1: first complete nucleotide sequence and functional organization of a virus infecting a haloalkalophilic archaeon. Mol Microbiol 45:851–863

Kottmeier ST, Grossi SM, Sullivan CW (1987) Sea ice microbial communities. VII. Bacterial production in annual sea ice of Mc Murdo Sound, Antartica. Mar Ecol Prog Ser 35:175–186

Madigan MT, Martinko JM, Parker J (2003) Extremely halophilic Archaea. In: Carlson G, Snavely SL, Wechsler DA, Schiaparelli K (eds) Brock biology of microorganisms, 10th edn. Prentice Hall, USA, pp 448–452

Maranger R, Bird DF, Juniper SK (1994) Viral and bacterial dynamics in Arctic sea ice during the spring algal bloom near Resolute, N.W.T., Canada. Mar Ecol Prog Ser 111:121–127

Martin A, Yeats S, Janekovic D, Reiter W-D, Aicher W, Zillig W (1984) SAV1, a temperate u.-v. inducible DNA virus-like particle from the archaebacterium *Sulfolobus acidocaldarius* isolate B12. EMBO J 3(9):2165–2168

Nuttall SD, Dyall-Smith ML (1993) HF1 and HF2: novel bacteriophages of halophilic archaea. Virology 197(2):678–684

Oren A, Bratbak G, Heldal M (1997) Occurrence of virus-like particles in the Dead Sea. Extremophiles 1:143–149

Ortmann AC, Suttle CA (2005) High abundances of viruses in deep-sea hydrothermal vent system indicate viral mediated microbial mortality. Deep-sea Res I 52:1515–1527

Palm P, Schleper C, Grampp B, Yeats S, McWilliam P, Reiter W-D, Zillig W (1991) Complete nucleotide sequence of the virus SSV1 of the archaebacterium *Sulfolobus shibatae*. Virology 185:242–250

Parkes RJ, Cragg BA, Bale SJ, Getliff JM, Goodman K, Rochelle PA, Fry JC, Weightman AJ, Harvey SM (1994) Deep bacterial biosphere in Pacific ocean sediments. Nature 371:410–413

Parkes RJ, Cragg BA, Wellsbury P (2000) Recent studies on bacterial populations and processes in subseafloor sediments: a review. Hydrogeol J 8:11–28

Porter K, Kukkaro P, Bamford JK, Bath C, Kivela HM, Dyall-Smith ML, Bamford DH (2005) SH1: a novel, spherical halovirus isolated from an Australian lake. Virology 335:22–33

Prangishvili D, Arnold HP, Gotz D, Ziese U, Holz I, Kristjansson JK, Zillig W (1999) A novel virus family, the Rudiviridae: structure, virus–host interactions and genome variability of the *Sulfolobus* viruses SIRV1 and SIRV2. Genetics 152(4):1387–1396

Prangishvili D, Garrett RA (2004) Exceptionally diverse morphotypes and genomes of crenarchaeal hyperthermophilic viruses. Bioch Soc Trans 32:204–208

Prangishvili D, Garrett RA (2005) Viruses of hyperthermophilic Crenarchaea. Trends Microbiol 13:535–542

Prangishvili D, Garrett RA, Koonin EV (2006) Evolutionary genomics of archaeal viruses: unique viral genomes in the third domain of life. Virus Res 117: 52–67

Prieur D (1997) Microbiology of deep-sea hydrothermal vents. Trends Biotech 15:242–244

Prigent M, Leroy M, Confalonieri F, Dutertre M, DuBow MS (2005) A diversity of bacteriophage forms and genomes can be isolated from the surface sands of the Sahara Desert. Extremophiles 9:289–296

Rachel R, Bettstetter M, Hedlund BP, Häring M, Kessler A, Stetter KO, Prangishvili P (2002) Remarkable morphological diversity of viruses and virus-like particles in hot terrestrial environments. Arch Virol 147(12):2419–2429

Rice G, Stedman K, Snyder J, Wiedenheft B, Willits D, Brumfield S, McDermott T, Young M (2001) Viruses from extreme thermal environments. Proc Natl Acad Sci USA 98:13341–13345

Rice G, Tang I, Stedman K, Roberto F, Spuhler J, Gillitzer E, Johnson JE, Douglas T, Young M (2004) The structure of a thermophilic archaeal virus shows a double-stranded viral capsid type that spans all domains of life. Proc Natl Acad Sci USA 101:7716–7720

Rohwer F (2003) Global phage diversity. Cell 113:141

Rothschild LJ, Mancinelli RL (2001) Life in extreme environments. Nature 409:1092–1101

Sandaa RA, Skjoldal EF, Bratbak G (2003) Virioplankton community structure along a salinity gradient in a solar saltern. Extremophiles 7:347–351

Schleper C, Kubo K, Zillid W (1992) The particle SSV1 from the extremely thermophilic archaeon *Sulfolobus* is a virus: demonstration of infectivity and of transfection with viral DNA. Proc Natl Acad Sci USA 89:7645–7649

Schnabel H, Schramm E, Schnabel R, Zillig W (1982a) Structural variability in the genome of phage $\phi$H of *Halobacterium halobium*. Mol Gen Genet 188:370–377

Schnabel H, Zillig W, Pfäffle M, Schnabel R, Michel H, Delius H (1982b) *Halobacterium halobium* phage $\phi$H. EMBO J 1:87–92

Smith REH, Clement P, Cota GF (1989) Population dynamics of bacteria in Arctic sea ice. Microb Ecol 17:63–76

Staley JT, Gosink JJ (1999) Poles apart: biodiversity and biogeography of sea ice bacteria. Ann Rev Microbiol 53:189–215

Stedman KM, She Q, Phan H, Arnold HP, Holz I, Garrett RA, Zillig W (2003) Relationships between fuselloviruses infecting the extremely thermophilic archaeon Sulfolobus: SSV1 and SSV2. Res Microbiol 154:295–302

Stolt P, Zillig W (1992) In vivo studies of the effects of immunity genes on early lytic transcription in the *Halobacterium salinarium* phage $\phi$H. Mol Gen Genet 235:197–204

Stolt P, Zillig W (1993) In vivo and in vitro analysis of traznscription of the L region from *Halobacterium salinarium* phage $\phi$H: definition of a repressor-enhancing gene. Virology 195:649–658

Stolt P, Zillig W (1994) Gene regulation in halophage phi-H – more than promoters. Syst Apll Microbiol 16: 591–596

Suttle CA (2005) Viruses in the sea. Nature 437:356–361

Tang SL, Nuttall S, Ngui K, Fisher C, Lopez P, Dyall-Smith M (2002) HF2: a double-stranded DNA tailed haloarcheal virus with a mosaic genome. Mol Microbiol 44:283–296

Tang S-L, Nuttall S, Dyall-Smith M (2004) Haloviruses HF1 and HF2: evidence for a recent and large recombination event. J Bacteriol 186:2810–2817

Thomas DN, Dieckmann GS (2002) Antarctic sea ice – a habitat for extremophiles. Science 295:641–644

Vestergaard G, Häring M, Peng X, Rachel R, Garrett RA, Prangishvili D (2005) A novel rudivirus, ARV1, of the hyperthermophilic archaeal genus *Acidianus*. Virology 336:83–92

Weinbauer MG (2004) Ecology of prokaryotic viruses. FEMS Microbiol Rev 28:127–181

Wiedenheft B, Stedman K, Roberto F, Willits D, Gleske A-K, Zoeller L, Snyder J, Douglas T, Young M (2004) Comparative genomic analysis of the hyperthermophilic archaeal Fuselloviridae viruses. J Virol 78(4):1954–1961

Witte A, Baranyi U, Klein R, Sulzner M, Luo C, Wanner G, Krüger DH, Lubitz W (1997) Characterization of *Natronobacterium magadii* phage $\phi$Ch1, a unique archaeal phage containing DNA and RNA. Mol Microbiol 23(3):603–616

Wommack KE, Colwell RR (2000) Virioplankton: viruses in aquatic ecosystems. Micr Mol Biol Rev 64:69–114

Xiang X, Chen L, Huang X, Luo Y, She Q, Huang L (2005) *Sulfolobus tengchongensis* spindle-shaped virus STSV1: virus–host interactions and genomic features. J Virol 79:8677–8686

Rev Environ Sci Biotechnol (2007) 6:161–179
DOI 10.1007/s11157-006-9109-9

REVIEW PAPER

# Exploration of Ellsworth Subglacial Lake: a concept paper on the development, organisation and execution of an experiment to explore, measure and sample the environment of a West Antarctic subglacial lake

## The Lake Ellsworth Consortium

**M.J. Siegert · A. Behar · M. Bentley · D. Blake · S. Bowden · P. Christoffersen · C. Cockell · H. Corr · D.C. Cullen · H. Edwards · A. Ellery · C. Ellis-Evans · G. Griffiths · R. Hindmarsh · D.A. Hodgson · E. King · H. Lamb · L. Lane · K. Makinson · M. Mowlem · J. Parnell · D.A. Pearce · J. Priscu · A. Rivera · M.A. Sephton · M.R. Sims · A.M. Smith · M. Tranter · J.L. Wadham · G. Wilson · J. Woodward**

Received: 27 February 2006 / Accepted: 1 September 2006 / Published online: 1 November 2006


**Abstract** Antarctic subglacial lakes have, over the past few years, been hypothesised to house unique forms of life and hold detailed sedimentary records of past climate change. Testing this hypothesis requires in situ examinations. The direct measurement of subglacial lakes has been considered ever since the largest and best-known lake, named Lake Vostok, was identified as having a deep water-column. The Subglacial Antarctic Lake Environments (SALE)

M. J. Siegert (✉)
School of GeoSciences, University of Edinburgh,,
West Mains Road, Edinburgh EH9 3JW, UK
e-mail: m.j.siegert@ed.ac.uk

A. Behar · L. Lane
Jet Propulsion Laboratory, NASA, Pasadena, USA
e-mail: alberto.behar@jpl.nasa.gov

L. Lane
e-mail: arthur.l.lane@jpl.nasa.gov

M. Bentley
Department of Geography, Durham University,
Durham, UK
e-mail: m.j.bentley@durham.ac.uk

D. Blake · H. Corr · C. Ellis-Evans · R. Hindmarsh ·
D. A. Hodgson · E. King · K. Makinson ·
D. A.Pearce · A. M. Smith
British Antarctic Survey, Cambridge, UK
e-mail: DMBL@bas.ac.uk

H. Corr
e-mail: HFJC@bas.ac.uk

C. Ellis-Evans
e-mail: JCEL@bas.ac.uk

R. Hindmarsh
e-mail: RCAH@bas.ac.uk

D. A. Hodgson
e-mail: DAHO@bas.ac.uk

E. King
e-mail: ECKI@bas.ac.uk

K. Makinson
e-mail: k.makinson@bas.ac.uk

D. A. Pearce
e-mail: dpearce@bas.ac.uk

A. M. Smith
e-mail: AMSM@bas.ac.uk

programme, set up by the Scientific Committee on Antarctic Research (SCAR) to oversee subglacial lakes research, state that prior exploration of smaller lakes would be a "prudent way forward". Over 145 subglacial lakes are known to exist in Antarctica, but one lake in West Antarctica, officially named Ellsworth Subglacial Lake (referred to hereafter as Lake Ellsworth), stands out as a candidate for early exploration. A consortium of over 20 scientists from seven countries and 14 institutions has been assembled to plan the exploration of Lake Ellsworth. An eight-year programme is envisaged: 3 years for a geophysical survey, 2 years for equipment development and testing, 1 year for field planning and operation, and 2 years for sample analysis and data interpretation. The science experiment is simple in concept but complex in execution. Lake Ellsworth will be accessed using hot water drilling. Once lake access is achieved, a probe will be lowered down the borehole and into the lake. The probe will contain a series of instruments to measure biological, chemical and physical characteristics of the lake water and sediments, and will utilise a tether to the ice surface through which power, communication and data will be transmitted. The probe will pass through the water column to the lake floor. The probe will then be pulled up and out of the lake, measuring its environment continually as this is done. Once at the ice surface, any water samples collected will be taken from the probe for laboratory analysis (to take place over subsequent years). The duration of the science mission, from deployment of the probe to its retrieval, is likely to take between 24 and 36 h. Measurements to be taken by the probe will provide data about the following: depth, pressure, conductivity and temperature; pH levels; biomolecules (using life marker chips); anions (using a chemical analyzer); visualisation of the environment (using cameras and light sources); dissolved gases (using chromatography); and morphology of the lake floor and sediment structures (using sonar). After the probe has been retrieved, a sediment corer may be dropped into the lake to recover material from the lake floor. Finally, if time permits, a thermistor string may be left in the lake water to take time-dependent measurements of the lake's water column over subsequent years. Given that the comprehensive geophysical survey of the lake will take place in two seasons during 2007–2009, a

S. Bowden · J. Parnell
Department of Geology and Petroleum Geology, University of Aberdeen, Aberdeen, UK
e-mail: s.a.bowden@abdn.ac.uk

J. Parnell
e-mail: j.parnell@abdn.ac.uk

P. Christoffersen
Scott Polar Research Institute, University of Cambridge, Cambridge, UK

C. Cockell
CEPSAR, Open University, Milton Keynes, UK
e-mail: c.s.cockell@open.ac.uk

D. C. Cullen
Cranfield Biotechnology Centre, Cranfield University, Cranfield, UK
e-mail: d.cullen@cranfield.ac.uk

H. Edwards
Division of Chemical and Forensic Sciences, Bradford University, Bradford, UK
e-mail: h.g.m.edwards@bradford.ac.uk

A. Ellery
Surrey Space Centre, University of Surrey, Guildford, UK
e-mail: a.ellery@surrey.ac.uk

G. Griffiths · M. Mowlem
National Oceanography Centre, Southampton, University of Southampton, Southampton,, UK
e-mail: gxg@noc.soton.ac.uk

M. Mowlem
e-mail: m.c.mowlem@soton.ac.uk

H. Lamb
Institute of Geography and Earth Sciences, University of Wales, Aberystwyth, Aberystwyth, UK
e-mail: hfl@aber.ac.uk

J. Priscu
Department of Land Resources and Environmental Sciences, Montana State University, Bozeman, USA
e-mail: dpriscu@montana.edu

two-year instrument and logistic development phase from 2008 (after the lake's bathymetry has been assessed) makes it possible that the exploration of Lake Ellsworth could take place at the beginning of the next decade.

**Keywords** Subglacial lakes · Extreme environments · Exploration · Antarctica

## 1 Introduction

Following the discovery that Subglacial Lake Vostok in East Antarctica has a water column over 500 m deep (Kapitsa et al. 1996), there has been widespread scientific and media interest in exploring Antarctic subglacial lake environments (Siegert et al. 2001). Such exploration is driven by the hypotheses that Antarctic subglacial lakes host unique forms of life and hold detailed sedimentary records of past climate change. Testing these hypotheses requires in situ measurement and sampling. Large lakes in central East Antarctica, such as Lake Vostok, are unlikely to be measured and sampled directly in the foreseeable future due to their remote location and lack of detailed comprehension (Siegert 2002) (although Russian Scientists plan to penetrate the existing Vostok borehole into Lake Vostok during 2008/2009 and extract refrozen lake water). Over 145 subglacial lakes known in Antarctica (Fig. 1), but one lake in West Antarctica, officially named Ellsworth Subglacial Lake (Siegert et al. 2004), stands out as a candidate for early exploration for the following reasons:

- Lake Ellsworth (79°S, 90.5°W), being only ~10 km long, can be characterised meaningfully in a short period using seismic and radar surveying (see Fig. 2).
- The lake is logistically accessible through both UK and US scientific field operations and through the commercial operator Antarctic Logistics and Expeditions.
- Lake Ellsworth is representative of other subglacial Antarctic lakes, in terms of pressure and temperature conditions.
- The sediments accumulated across the floor of the lake may yield a record of West Antarctic ice sheet history.
- Lake Ellsworth is located ~20 km from an ice divide, which means that drilling from the ice surface into the lake would not be complicated by ice flow.
- The ice sheet surface over the lake is ~2,000 m above sea level, which is more than a kilometre lower than the ice surface over the vast majority of East Antarctic subglacial lakes (altitude related problems often encountered by scientists at the centre of the East Antarctic Ice Sheet will not, therefore, be as much of an issue during the study of this lake).
- Subglacial access and sampling has precedent in West Antarctica, but not in East Antarctica.

This paper describes the concept behind the exploration of Lake Ellsworth. The two scientific

A. Rivera
Centro de Estudios Cientificos, Valdivia, Chile
e-mail: arivera@cecs.cl

M. A. Sephton
Department of Earth Science and Engineering, Imperial College, London, London, UK
e-mail: m.a.sephton@imperial.ac.uk

M. R. Sims
Space Research Centre, University of Leicester, Leicester, UK
e-mail: mrs@star.le.ac.uk

M. Tranter · J. L. Wadham
Bristol Glaciology Centre, University of Bristol, Bristol, UK
e-mail: m.tranter@bristol.ac.uk

J. L. Wadham
e-mail: j.l.wadham@bristol.ac.uk

G. Wilson
Department of Geology, University of Otago, Dunedin, New Zealand
e-mail: gary.wilson@stonebow.otago.ac.nz

J. Woodward
Division of Geography, Northumbria University, Newcastle, UK
e-mail: john.woodward@unn.ac.uk

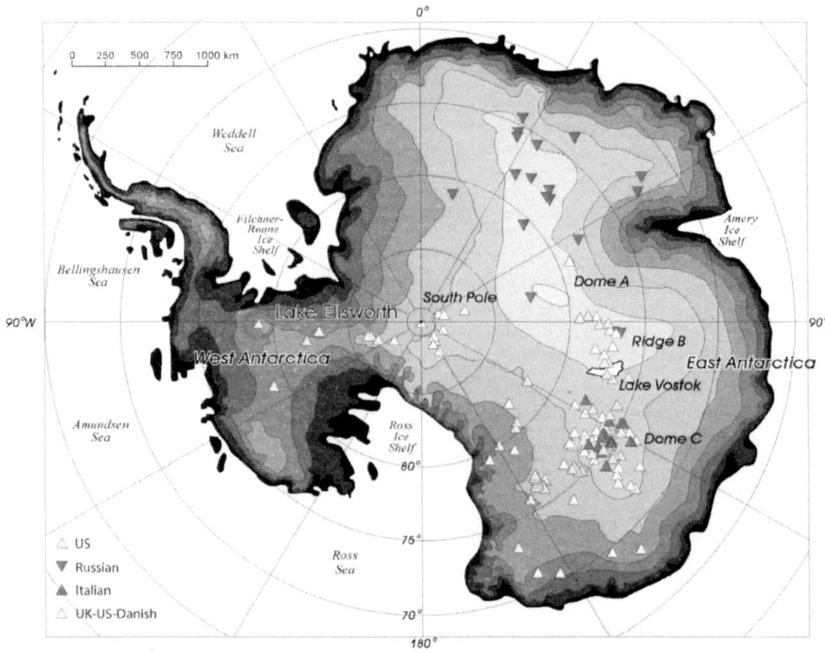


**Fig. 1** The location of 145 Antarctic subglacial lakes (Siegert et al. 2005). Lake Ellsworth is circled

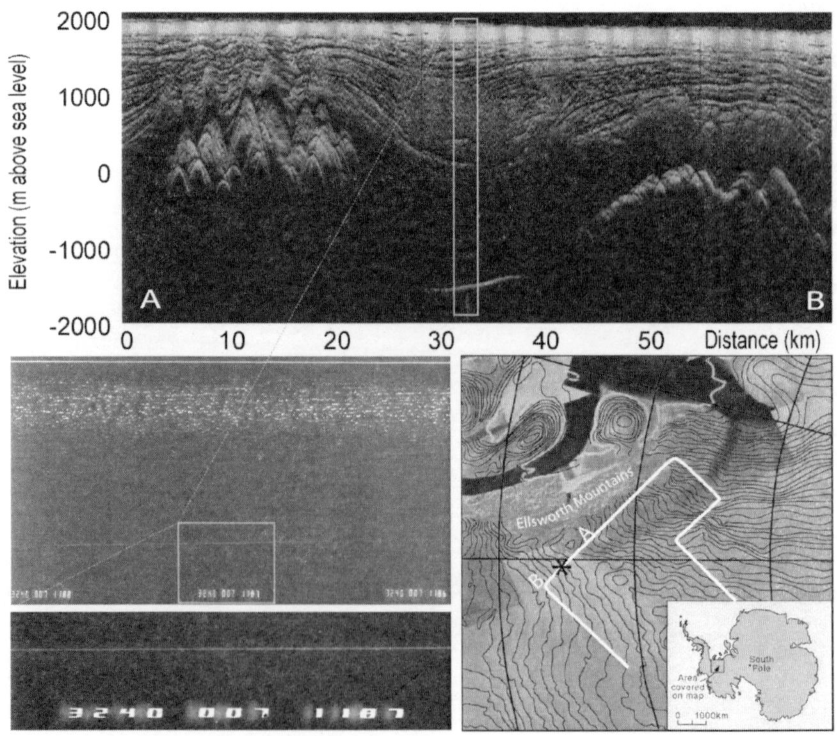


**Fig. 2** Radio Echo Sounding (RES) data from Lake Ellsworth. The lake is located by the bright flat reflector between 28 and 38 km in AB. The insets show magnified sections of the RES data (demonstrating the mirror-like reflection indicative of a subglacial lake) and the location of the lake. Adapted from Siegert et al. (2004)

goals of the project (which are also the goals of subglacial lake exploration in general, see Priscu et al. 2003) are (1) to measure and comprehend life in this extreme environment, and (2) to collect and assess any climate records that exist in the lake floor sediments.

### 1.1 Background to Lake Ellsworth

Subglacial Lake Ellsworth is a 10 km long, 2–4 km wide, lake underneath between 3,240 and 3,320 m of ice near the Ellsworth Mountains in West Antarctica (Fig. 2; Siegert et al. 2004). The ice sheet over the lake is ~20 km from the ice divide and its elevation is ~2 km above sea level, which makes the lake surface over 1.25 km beneath sea level. Although, prior to 2004, only one radio-echo sounding (RES) transect had crossed the lake (Fig. 2), several have been collected by the British Antarctic Survey (BAS) in 2004–2005 and Chilean glaciologists during January 2006. These RES data show the lake to occupy a deep, distinct topographic hollow, which appears similar to an over-deepened section of an ancient fjord.

### 1.2 Planning and meetings to date

An initial group of eight scientists met at BAS on 26th April 2004 to assess the feasibility of a subglacial lake exploration programme. Following the identification of key exploration roles, an expanded group of 16 members met subsequently at the University of Bristol on 1st September 2004 to develop the project further. A third meeting took place at BAS on 8th March 2005 to organise a concept document that detailed how the exploration of Lake Ellsworth can be realised. This paper derives from that concept document. Agenda and minutes of these meetings are available from the project's website (www.geos.ed.ac.uk/ellsworth). As a consequence of the meetings held to date, a consortium of over 30 scientists from seven nations and 14 institutions has been assembled to undertake the proposed programme of activity detailed herein.

### 1.3 Purpose of this paper

This paper provides brief details of the likely requirements of a subglacial lake exploration programme (with specific attention paid to the requirements of Lake Ellsworth). It is important to note that the underlying theme of the proposed project is to undertake 'first access' of a subglacial lake environment and to explore this environment in a simple, efficient and cost-effective manner. Despite this low cost approach no compromise is expected when it comes to contamination of the subglacial environment. More expensive and complex experiments may follow if this initial project is successful, but such experiments are not discussed in this document.

## 2 International collaboration

The Scientific Committee on Antarctic Research (SCAR) commissioned a group of specialists (now a Scientific Research Programme, SRP) named Subglacial Antarctic Lake Environments (SALE) to "consider and recommend mechanisms for the international coordination of a subglacial lake exploration program" (Priscu et al. 2003). Details of the SALE project can be found on its website: http://salepo.tamu.edu/scar_sale. It should be noted that SCAR has no money to fund this level of research. Consequently, its scientific research programmes will not undertake research themselves. Instead, they have been configured to facilitate research and encourage international cooperation through workshops and symposia. Funding for projects within a SCAR SRP must be sought by national funding agencies.

A proposal to undertake a comprehensive geophysical survey of Lake Ellsworth (including RES, seismic surveying and a variety of surface measurements) has been funded through the UK Natural Environment Research Council's Antarctic Funding Initiative (NERC-AFI). The survey is planned in two seasons during the years of the International Polar Year (IPY) 2007–2009 (www.ipy.org). The Lake Ellsworth geophysical campaign is a recognised project of the larger SALE programme, which is officially endorsed by the IPY. Data collected during the IPY will supplement those acquired in 1978, in 2004–2005, and in January 2006. As a consequence of these geophysical results, Lake

Ellsworth and its subglacial catchment will be characterised at a sub-km resolution, which will make planning purposeful direct measurement and sampling possible. Further integration of various international research proposals related to subglacial lakes research into the Lake Ellsworth programme is possible via both the IPY and SALE, and within the project's own management.

SALE will assist the Lake Ellsworth project in terms of environmental planning, data management and dissemination of results.

## 3 Website and media interest

Even though the project is in its planning stages, there has been considerable public and media interest in plans to explore Lake Ellsworth. Details of this interest, and the project in general, can be found from the project website (www.geos.ed.ac.uk/ellsworth), constructed to help schools and colleges follow the progress of work, and to assist the media in understanding the proposed research.

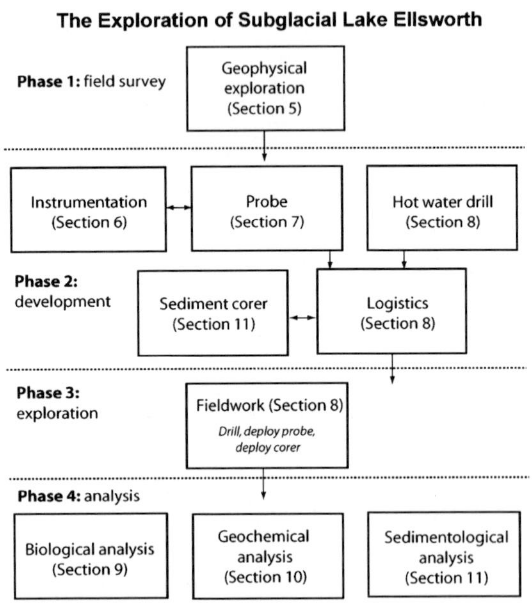


**Fig. 3** Structure of the project

## 4 Structure of the project and objectives

The exploration of Lake Ellsworth requires a multidisciplinary approach. A considerable level of management is required to integrate the various project elements. Currently the project is being managed by a steering committee, and through an ongoing series of project meetings. The project is arranged in four phases each with distinct objectives, which are outlined below (Fig. 3).

It should be noted that a considerable group (in terms of size and breadth of expertise) has been put together to undertake the exploration of Lake Ellsworth. We are collectively known as the 'Lake Ellsworth Consortium'. Given the long-term nature of the project, requiring instrument development, fieldwork and laboratory analysis, publications arising from the direct measurement and sampling of the lake will be authored by the 'Lake Ellsworth Consortium', rather than by individuals.

### 4.1 Phase 1: Geophysical exploration

The first phase of the project involves the measurement of the size and shape of Lake Ellsworth, the flow of the ice sheet over the lake, and the subglacial topography surrounding the lake. This part of the project, which has been funded by the UK NERC-AFI programme and must take place before direct exploration (as it will be essential for fieldwork planning and, later, interpretation of data), will take place during the forthcoming IPY. The objectives of this first phase of the project are as follows (see also Section 5):

- To undertake a comprehensive geophysical survey of Lake Ellsworth and its locale.
- To reveal the water depth of the lake (i.e., lake bathymetry).
- To measure the sediment thickness and structure across the lake's floor.
- To comprehend the wider topography around the lake and quantify the dimensions and morphology of the lake's drainage basin.

### 4.2 Phase 2: Instrument and logistic development

The second phase of the project involves assembling equipment and logistics necessary to undertake the physical exploration of Lake Ellsworth. In this phase the following objectives are required:

- To build, assemble and test instruments to detect life in the lake, to measure the physical and chemical properties of the lake's water and to sample the lake water and sediment (Section 6).
- To construct and test a probe to house the instruments and to allow communication between the probe and the ice surface by which data may be sent back to the ice surface (Section 7).
- To build and test a hot-water drill and organise field logistics (Section 8).

Phase 2 of the project will also require the following objective, if climate records are to be recovered:

- To acquire and test a sediment corer, capable of extracting a 2 m core (or longer) from the floor of Lake Ellsworth (Section 11).

Contamination of subglacial lake environments as a consequence of the lake access experiment must not occur. Hot water drilling has been used on numerous occasions to reach the base of the West Antarctic ice sheet (e.g., Englehardt et al. 1990). The use of hot-water drilling in this project means that all the contamination controls used previously to study the ice sheet base, and surface lakes in Antarctica, will be employed for experiments in Lake Ellsworth. Further assistance on this important issue will be sought from the SCAR-SALE scientific research project prior to fieldwork.

### 4.3 Phase 3: Fieldwork

Once the necessary equipment has been built and tested, we will be in a position to undertake exploration fieldwork. To allow direct measurement and sampling of the lake, the following objective is required:

- To use a hot water drill to gain access to the lake from the ice-sheet surface.

We anticipate a 30 cm wide borehole is required (to pass equipment through), which can be held open for 24–36 h (giving a short window of little more than a day for the actual exploration). Prior to lake entry, 400 m of water within the borehole will be taken out to ensure borehole water does not enter the lake (as the base of the borehole will be at greater pressure than the water within the lake).

Once lake access is achieved, the following objective is required:

- To deploy a probe into the lake capable of measuring the lake's biological, chemical and physical environment.

The probe will be lowered through the water column to the lake floor. Recordings made on this journey could be used to plan sampling on its return up the water column (recordings will also be made on its return journey). The probe will then be retrieved.

Subsequent to the deployment and retrieval of the lake-water probe, and if lake access is ensured for an appropriate length of time, the following objective is required:

- To deploy a sediment corer into the lake to retrieve a 2 m (or longer) sediment core.

In addition, a probable third experiment involves the installation of a permanent station into the lake. No details of such an experiment are provided in this document, but such work is undertaken commonly within Antarctic surface lakes covered by thin ice.

### 4.4 Phase 4: Data analysis and interpretation

Data acquired by the probe will comprise direct measurements and, possibly, samples. The following four objectives are required to assess these data in order to meet the project's goals. The first involves direct measurements, the second and third involve samples of water and sediment and the fourth involves the analysis of sedimentary records.

- To assess data sent by the probe in real-time to the ice surface, to comprehend the physical and chemical structure of the lake, and to ascertain the abundance, distribution and diversity of microbial life in the water column and water-sediment interface.
- To conduct microbiological examination of lake water and sediment in a laboratory (Section 9).

- To undertake geochemical analysis of water and sediment (Section 10).
- To conduct sedimentological laboratory analysis of the lake floor sediment core (Section 11).

If a permanent station is deployed, a means by which data can be stored, retrieved and disseminated will be needed (and will probably require returning to the field site in years subsequent to the exploratory mission).

### 4.5 Project components

The following seven sections provide a 'concept' of the requirements of each phase of the project. The first is the geophysical survey (Section 5). Second is the instrument development for the probe (Section 6). Third is the probe development (Section 7). Fourth is the hot-water drilling and fieldwork (Section 8). Fifth is the biological analysis of water samples (Section 9). Sixth is the geochemical analysis of water samples (Section 10). Seventh is the sedimentological analysis of lake-floor deposits (Section 11).

## 5 Geophysical survey

This component of the project involves the geophysical exploration of Lake Ellsworth. Fieldwork will take place during the IPY 2007–2009 and involves RES of the ice base, seismic surveying of the lake floor, GPS monitoring of ice flow and a series of other ice surface glaciological measurements. Following the fieldwork, numerical modelling will be used to quantify water flow within the lake, ice flow over the lake, and the distribution and rates of melting and freezing at the lake-ice interface. When the fieldwork is completed, the following 11 questions will be answered: (i). What is the water depth of the lake? (ii). What is the topographic setting of the lake? (iii). What is the sediment thickness across the lake floor? (iv). Are the sediments layered? (v). How does the ice sheet flow over the lake? (vi). What is the roughness of the bed around the lake? (vii). What are the rates of subglacial melting and freezing over the lake? (viii). Does the lake have detectable tides? (ix). How is the lake water circulation configured? (x). Where does the melt-water originate? (xi). What is the geothermal setting of the lake?

The outcome of this element of the project will be a detailed characterisation of the subglacial lake and its physiography and the identification of the most appropriate drilling site for direct exploratory research.

## 6 Instrument development

### 6.1 Background

Developing the instrumentation for the probe (Section 7) is focused on measuring the physical and chemical properties of the lake water and identifying life within the lake. Consequently, the following key considerations guide the choice of instrumentation.

- The experiment should be thought of, in the first instance, as a 'demonstration project', and should not aim to accomplish everything that might be possible in subglacial lakes exploration. Some selectivity is therefore required within the instrument specification and design. Once subglacial lake access is achieved and this in situ experiment completed, it may well be appropriate to develop more sophisticated experiments and instrumentation afterwards.
- All instrument items need to be depth (pressure)-rated to 4–5 km of water (corresponding to the likely pressure of Lake Ellsworth).
- The number of items requiring significant development needs to be kept to a minimum, to avoid the project becoming unreasonably expensive, overly ambitious, and difficult to accomplish within its timeframe.
- Instruments need to be miniaturised to fit a casing of < 20 cm diameter (the drill hole will be ~30 cm wide). Pressure-protected instruments may need confining to a 15 cm diameter. It should be noted that some instruments (e.g., Raman spectroscopy) could be deployed from the surface using fibre optic cable to connect to the lake water.

- The first measure of success will be to demonstrate measurements in the lake water. Sample return is very attractive, but will be a bonus, and so should have a lower priority than direct measurements.

### 6.2 Basic instrumentation

The following comprise essential characteristics of the probe's array of instruments:

- A basic off-the-shelf (OTS) camera/light system, and a custom-built CCD (charge-coupled device) system with photon-level resolution and UV light source to detect fluorescence/bioluminescence.
- A water filter system, as life detection will be more efficient if the material within the water is concentrated. Multiple filter devices are required to deploy at different water depths.
- Measurement of ions within the lake water will be made but only using equipment that is OTS.
- Two sonar devices (top and bottom of probe) are needed to allow depth-to-probe and water-depth measurements.
- If a sampling capability is adopted the simplest option is a series of sampling chambers/cavities which may be switched to recover water from the target environment. It should be noted that sample will be limited in volume to a few hundred ml, and that degassing of samples on return to the surface is inevitable.

### 6.3 Detection of life

Defining the equipment needed to detect life within the lake water is critical to the success of the experiment. The following make up the possibilities defined at present:

- An obvious generic marker of active terrestrial life is the molecule adenosine triphosphate (ATP) that is present in all cells. Enzyme-based bioluminescence reagents and technology to detect ATP levels is well established in other applications. This will require re-packaging into appropriate reaction chambers pre-dosed with reagents and filled in situ with filtered and cell-lysed water samples. The reaction chambers will be imaged by the photon-counting camera to detect the resultant emitted ATP dependent bioluminescence. At present four measurements (three at various water depths and one in sediment) are envisaged.
- Endogenous fluorescence will be measured using a UV light source, optical filters and optical detector. If necessary we may measure fluorescence after addition of fluorescent dyes specific to certain cell components.
- Raman spectroscopy could be deployed (from the ice surface) on (i) filtered particulate organic material, and (ii) a surface-enhanced resonance Raman spectroscopy (SERRS) surface, exposed to extract organics from solution. Deployment into sediment may also be possible using a fibre optic probe.
- Life Marker Chip (an appropriately packaged antibody micro-array device currently under development for life and organics detection for planetary exploration—specifically for Martian exploration), using readily available antibodies, is high risk (in terms of the timescale for development) and expensive in terms of development resources. If it can be developed, such equipment would be central to the project's life-detection goal. However, the first three life/organics detection devices, if used in tandem, would result in a multi-pronged approach to the problem. The fourth would be a bonus if it can be developed in time. We note that deployment of this technology will require in situ sample preparation, which may be non trivial.
- Analysis of filtered material from lake water and sediments for recognisable biological particles using traditional microscopy (LM, SEM etc.).

## 7 Probe development

The probe development will require two elements: (a) The construction of electronics to ensure data from the probe's instruments are transferred to the ice surface, and (b) the building of the probe case and tether (capable of withstanding appropriate pressure of the borehole and the lake's water column).

### 7.1 Probe electronics

It is anticipated that the probe will draw power and data through its tether, which will be necessarily highly robust. This will be a potential single point failure mode as the Dante II robot (designed to explore active volcano craters) failed when its winch-deployed rappelling tether snagged and fractured on its maiden descent in 1994. The power distribution bus should be isolated from the data distribution bus in separate wiring harnesses.

The probe will have a small cross-section due to the narrowness of the borehole of <30 cm diameter—similar to sondes used in petroleum wireline logging. The penetrometer shell should be of semi-monocoque construction with thermally isolated internal equipment spaces. Although aluminium alloy may be suitable for the probe structure, wireline logging sondes are generally constructed from titanium alloy. Indeed, the probe may potentially be constructed from compartmentalised segments connected together through standard interfaces at the deployment site as in wireline logging making transport and packaging of the probe easier.

The probe must house the service support subsystems and the relevant scientific instruments. The probe electronics should be housed within the probe connected through each segment by electrical interfaces and a suitable 'bus' architecture (e.g., CANbus). The internal circuitry may be optimised to minimise the length of the wiring harness. Circuit paths should be short with electronics packaged into circuit board stacks linked through an integrating backplane. The electronics comprises:

1. Power microelectronics based on CMOS ASICs (application-specific integrated circuits) for switching voltage regulation and overcurrent protection of onboard devices (an alternative to ASICs is a field programmable gate array (FPGA)).
2. Control and data handling microelectronics (a microcontroller including internal multiple ADC channels) for the storage of telemetry and scientific data and mission sequencing. A transputer (or other co-processors) may be employed to perform image processing (which have been deployed on Surrey Satellite Technology Ltd. micro-sats). Lossless data compression methods may be employed to reduce data rates.

Onboard microprocessors will enable limited data manipulation on the vehicle; will manage data storage and transmission to the surface command station; will control onboard systems; and will calculate attitude from inertial and magnetometer sensors. Attitude data can be combined topside with the sonar images to provide accurate positioning information vital in the measurement of the small currents expected in the subglacial lake.

Although it is anticipated that onboard power will be supplied through a tether, backup primary Li-thionyl chloride batteries designed for a significant period of operation at low temperatures should be included for redundancy. This depends on the duration within the target waters (unknown) and the time for descent and ascent (~20 h). Conductive fins will be required to radiate battery power dissipation. The tether link should provide sufficient capacity to support the telemetered data rate; e.g., 500 kilobytes per second (kbps) depending on the scientific instrument polling rate and data density (e.g., imaging).

Onboard instrumentation should employ dedicated on-chip signal conditioning electronics to allow efficient interfacing to the databus. However, the requirement for minimal volume overhead suggests that tight scientific payload integration into a single system is required to eliminate modular boxes with electrical connectors. This architectural trade-off will require critical examination.

Some bulky scientific instruments may be deployed from the ice surface and exploit the tether using fibre optic cabling (IR Raman spectroscopy is currently being considered for such deployment, and such an approach may also be suitable for optrodes).

### 7.2 Probe casing

The probe concept consists of a negatively buoyant cylindrical vehicle with no propulsive capability. It is essentially a hollow cavity which allows the accommodation of scientific instruments and supporting subsystems. The centre of mass of the probe should be close to the nose, forward of the centre of pressure, to ensure stability during the descent. Thermal control may be achieved through an inside layer of carbon foam or silicon aerogel with a thermal capacity of 300 W/cm$^2$ (multi-layer insulation is likely to be too thick within the limited diameter). Thermal isolation straps may be constructed from low thermally conductive glass fibre reinforced epoxy.

The probe will be lowered under its own weight into the borehole and subsequently to the subglacial lake. The predicted borehole diameter at retrieval limits the vehicle width to less than 200 mm. Length is relatively unconstrained and will be varied to accommodate the required instrumentation and systems. The diameter limit has the most profound effect on the area available for tip mounted instruments i.e., imaging sonar, conductivity-temperature-depth (CTD) instrument, water sampler tube, sediment corer/sampler, video camera and light, as shown in Fig. 4). Water sampling and pumping for analysis can be positioned elsewhere on the vehicle.

Movable surfaces/covers are a possible consideration but are currently avoided by the use of a fixed cage at the rear of the vehicle which guides the cable away from the sonar and encourages the vehicle to centre itself into the borehole for retrieval. A fixed cowl (not shown) protects the tip mounted instrumentation. The generation and use of hot water for both lake entrance and exit is possible but requires further investigation.

At a maximum depth of 4 km below an estimated 3.4 km thick ice sheet, the probe will experience an ambient pressure of ~37 MPa. Systems that are not resistant to this loading will be placed in a pressure resistant cylinder. Pressure tolerant devices that need to be isolated from water will be housed in an oil filled pressure-balanced cylinder with diaphragms to enable differential expansion. This cylinder will have a greater useable diameter due to its thinner sidewalls. All other devices can be exposed to the environment directly or in free flooded areas of the vehicle.

Standard oceanographic CTD cable with optical fibre and dual conductor link (e.g., A303950, Rochester Cables) can be used to lower and retrieve the probe. Four km of 11 mm diameter cable weighs ~2,900 kg (~1,300 kg in water). Using this cable the maximum probe weight is ~4,500 kg. This cable can be used to transmit ~ 100 W (@300 V) to the probe; a similar power is dissipated in transit.

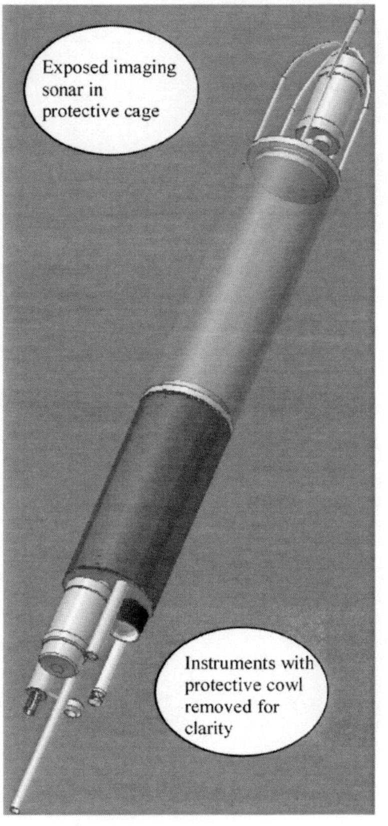


**Fig. 4** Example of probe casing

Trade studies are required to compare power strategies (including raising the voltage, using different conductors or supplying onboard batteries). Primary communication will be via optical fibre, with a modem backup using the cable conductors. The backup system will not be able to transmit high rate video.

Substantial equipment will be required at the surface including a suitable winch (~3 tonnes), power generator, optical and conductor communications link processing, data storage and visualisation with integrated command and control. Testing and verification of components, systems and the integrated vehicle can be undertaken both in the pressure test facility at the National Oceanography Centre Southampton (NOCS), and on field trials in perennially ice covered lakes such as Citadel Bastion Lake (Southern Alexander Island), where solutes and life are in very low concentrations and are, thus, likely to provide for the development and testing of the sensory capability of the probe.

## 8 Drilling, logistics and fieldwork

### 8.1 Background

Exploration of Lake Ellsworth will be undertaken during the 2012/2013 or 2013/2014 season. Hot water drilling equipment will weigh approximately 36 tonnes. A total of 200 drums of fuel will be needed to make two entries into the subglacial lake. The lake is at a depth of 3,400 m. A minimum of 18 personnel are needed on site during the continuous drilling phase (six per shift) to maintain and operate the drill, and for the deployment of probes into the lake. Staff will need to be on site for a period of 10 weeks; 5 weeks to set-up the equipment, 3 weeks to drill two holes and 2 weeks to decommission.

### 8.2 Hot water drilling equipment

The hot water drill will require a hose with a minimum bore of 1.5" to carry 4.5 litres per second of hot water. The temperature of the water will be up to 90°C at the surface, reducing to 55°C at a depth of 3,400 m, thus providing the 1.7 MW needed to melt and maintain the hole. The water will be heated by commercially available heat exchanger units, and pumped using diesel powered units at up to 2,200 psi.

Initially the drill will melt a 36 cm diameter hole but once drilling has ceased, refreezing will reduce the diameter to about 20 cm after 36 h. Drilling such a hole through 3,400 m will require approximately 50 h, and will use around 12,000 l (60 drums) of fuel, assuming no problems. Experience suggests that further reaming of the hole, if needed, will consume approximately 6,000 l of fuel per day.

The pressure loss within the hose is expected to be less than 2,000 psi from the ice surface to the lake ceiling. As the top 300–400 m of the hole is air filled, the hose must be supported using additional strength members, or integral strength member bonded to the hose. A drilling winch capable of handling the hose and additional load of the probe is required.

Biological filtering, using a modular cartridge filter system to remove all bacteria and many viruses (down to 0.045 micron) from the water used during drilling, is required to minimise contamination of the lake water. The filtration plant will be designed to provide the required flow rate and water quality. Prefilters will remove the larger particulates. This system is expected to fit easily in to an airfreight container. Using data from Antarctic ice cores, we can obtain a full analysis of the anticipated dust loading and ionic concentrations for the entire ice column. First estimates suggest that one filter change every 2–3 days (assuming entirely dusty glacial ice) will easily maintain the filtration standards. A more comprehensive analysis of the ice core data would confirm the frequency of the filter changes. If all viruses must be removed then reverse osmosis is required, removing salts and ions as well as viruses. Either system would ideally be housed in small shipping container together with the prefiltering system.

### 8.3 Access and fieldwork

To overcome as many of the logistical problems as possible delivery of equipment, materials and personnel to the drill site, will use chartered aircraft (using Antarctic Logistics and Expeditions). There will be a heavy lift from Chile to the Patriot Hills ice runway, with onward carriage by tractor train (in collaboration with Chilean scientists). All field activities and arrangements will be organised with the assistance of BAS project members, including transportation, tents, clothing, field equipment, communication systems and food.

### 8.4 Fieldwork analysis

The probe will deliver information concerning physical, chemical and biological properties of the lake's water column. Appropriate scientific expertise will be present in the field to (1) manage the probe's sampling strategy, and (2) interpret the probe's results to comprehend the environment of Lake Ellsworth. Data collected by the probe will be recorded on site and made available to project members in the first instance, and the international scientific community via SALE. If samples are recovered, it may be possible to undertake first analysis on site. The bulk of material recovered will be packaged into sterile containers and transferred to laboratories where detailed analysis can take place.

## 9 Biological analysis

### 9.1 Background

Microorganisms have been found in some of the most extreme environments on Earth; some organisms thrive in ice, others flourish in boiling water, acids, water-cores of nuclear reactors, salt crystals and toxic waste (Rothschild and Mancinelli 2001). Lake Ellsworth has the potential to be one of the most extreme environments on Earth with combined stresses of high pressure (>400 atmospheres), low temperature (near the freezing point), permanent darkness and probably oligotrophic (low nutrient) conditions. The identification of significant subglacial bacterial activity in some subglacial environments (e.g., Miteva et al. 2004; Foght et al. 2004), as well as work on permafrost communities, shows that life can survive and grow in low temperature glacial systems. Above Lake Vostok all of the accreted ice samples between 3,541 and 3,611 m contained both prokaryotic and eukaryotic microorganisms (Priscu et al. 1999). In addition, viable microorganisms have been recovered from 1 million year old Antarctic permafrost (Kochkina et al. 2001), which makes it likely that prolonged preservation of viable microorganisms is possible in Antarctic ice-bound habitats. Thus, existing data suggest that the Antarctic ice sheet may at least provide a source of microorganisms for the lake and possibly an environment for their growth.

### 9.2 Does Lake Ellsworth contain life?

Members of the microbial world encompass the three domains of life, the Bacteria, the Archaea and the lower Eukarya. About 4,200 prokaryotic species have been described out of an estimated $10^5$ to $10^6$ prokaryotic species on Earth. The overlying ice core at Lake Vostok is known to contain the full diversity of microbial life: algae, diatoms, bacteria, fungi, yeasts and actinomycetes (Ellis-Evans and Wynn-Williams 1996). A combination of microscopy, biochemical and molecular biological techniques will be studied in 'clean' laboratory facilities to determine the abundance, distribution, and diversity of microorganisms in the lake. We will use standard microbial quantification techniques such as nucleic acid staining (SYTO 9, acridine orange) to obtain microbial numbers. Techniques such as FISH (Fluorescent In situ Hybridisation) will be used to determine the phylogenetic groups of the organisms.

### 9.3 Where is life located within the subglacial lake?

Such a unique environment is expected to harbour significant chemical gradients, including oxygen from gas hydrates released from ice. Microbial life in Lake Ellsworth may, therefore, be pelagic, associated with sediment particles, distributed along gradients, in accretion ice or in overlying meteoric ice. The identification of viable organisms along with a determination of the elemental composition of the surrounding media will provide clues to potential biogeochemical activity and the sources of energy and carbon that would be necessary to sustain metabolically active populations.

### 9.4 What are the characteristics of Lake Ellsworth communities?

While the mere presence of life in itself would be a major scientific discovery, we might expect such organisms to possess special or unique adaptations to this hostile environment. Analysis of the physiological potential using a metagenomic approach and energetics through biochemical pathways of returned samples, will help to gain a fuller understanding of the role of Lake Ellsworth microbes in biogeochemical cycling and the functioning and control of the ecosystem. FISH, CARD-FISH (Catalyzed Reporter Deposition FISH) and similar methods (Table 1) will be used to determine the phylogenetic characteristics of the communities.

### 9.5 Techniques

The following four laboratory techniques are available to investigate microorganisms within samples retrieved. (1) Microscopy; fluorescent and electron microscopy (used with specific gene probes). (2) Biochemistry (biogeochemical cycling). In the absence of light, the microorganisms within Ellsworth must be using either organics or inorganic redox couples to gather energy. We will use gene probes available for different biogeochemical activities to assay the water/samples for the presence of biogeochemical activity. This will include probes for iron cycling (reduction and oxidation), nitrate cycling, manganese reduction and other pathways of dissimilatory metal reduction and oxidation. (3) Molecular biology; genomic DNA (using gene probes coupled with FISH—Fluorescent in situ hybridisation) will be extracted from material obtained and used to construct a metagenomic library to screen for novel physiologies. (4) Infrared Raman (used to detect biomolecules); could reside at a surface station with the sensor head residing in the probe.

## 10 Geochemical experiments

The aims of this Section are to first detail the expected geochemistry of the lake water, and then assess the various measurements and analyses required to quantify the lake's physical and chemical environments (including measurement/sampling strategies and prioritisation of laboratory experiments).

### 10.1 Background: Expected bulk geochemistry of Lake Ellsworth

Let us assume that the dimensions of Lake Ellsworth are 10 km × 10 km × 250 m; then the volume of water is 25 $km^3$. Let us then assume that the melting into the lake equals freezing out of the lake, so that the lake volume is in steady state, and that melting occurs over 50% of the lake at a rate of 10 cm/year. The annual input of melt to the lake is therefore 10 km × 10 km × 0.5 × 10 cm/year or $5 \times 10^{-3}$ $km^3$/year.

The residence time of water is $25/5 \times 10^{-3}$ years or 5 kyr. Assuming the age of Lake Ellsworth to be 400 kyr, there will have been 80 renewals of water.

**Table 1** Some of the techniques that can be applied to detect life in sediments returned from Lake Ellsworth

| Method | Phylogenetic analysis? | Comments |
|---|---|---|
| Microscopy (LM, SEM) | | Preliminary examination of filtered material from the water column and of the sediment for recognisable life forms |
| Nucleic acid staining | No | Determination of total cell numbers in lake sediments |
| FISH, CARD-FISH | YES | Fluorescent In situ Hybridisation and catalyzed reporter deposition-FISH can be used to determine phylogenetic groups of organisms an their metabolic activity. |
| Lipid analysis | Yes | Can be used both for total cell number determinations and phylogenetic identification of groups of organism in sediments. Nb. the technique does not work for Archaea and gram+ bacteria. |
| PCR of functional genes | Gives information on functional groups, which may relate to phylogeny. | Determination of major biogeochemical pathways operating in lake or potentially operating. |
| Genomic analysis | Yes | |

Assuming also that all the solute from the melting ice stays in the lake, since ~0.1% is incorporated into accretion ice. The chemical composition of the lake is therefore 80 times that of the average incoming ice melt chemistry, if no other sources or sinks of ions occurs.

Let us assume finally that the chemistry of ice melt is equivalent to the average chemistry recorded in the Byrd Ice Core, which gives the expected chemistry of lake water in Table 2.

The inferred concentrations of $Ca^{2+}$, $Mg^{2+}$, $Na^{+}$, $K^{+}$ $SO_4^{2-}$ and $HCO_3^{-}$ are probably on the low side, since these ions are generated from interactions between glacial flour and ice melt. We guess that concentrations may be up to a factor of 3–5 higher, putting these concentrations similar to those estimated for Lake Vostok.

The inferred concentrations of $H^{+}$, $NH_4^{+}$ and $NO_3^{-}$ are probably too high, since glacial flour uses up $H^{+}$ in chemical weathering actions, and microbial activity will remove $NH_4^{+}$ and $NO_3^{-}$ (microbial activity may also change levels of $SO_4^{2-}$ and $HCO_3^{-}$). We guess that the pH of the water will be about 6 and that $NO_3^{-}$ and $NH_4^{+}$ concentrations will be around 1 $\mu$eq/l, similar to those in Lake Vostok (Siegert et al. 2003).

## 10.2 Experiment sampling strategy

Given the short but intensive duration of the field experiment (24–36 h), it is essential to formulate a measurement and sampling strategy to acquire appropriate real-time data and quantities of the lake water to ensure its geochemistry can be characterised appropriately. Our three-stage plan for exploration is as follows. (1) Once in the lake, lower the probe continuously to the lake floor taking measurements of lake's physical environment. (2) At the lake floor, collect sediments, deploy life marker sensors into sediment, scrutinise physical-lake data. (3) On the basis of an analysis of data to this point, a sampling strategy should be developed for the return journey up the water column. In this way we can be confident of acquiring data from the most interesting material in the water column. It should be noted that we also intend to sample and analyse ice above the lake (extracted by hot water coring), as this comprises the likely 'input' to the lake system.

## 10.3 Physical environment

Comprehension of the physical environment (flow, temperature, conductivity, density, pH, oxygen saturation etc.) is likely to be one of the first results of the programme. The analysis involves data emplacement within a 3D context (the broad 3D physiography having been established from earlier geophysics (Section 5)).

Questions that can be answered include: Is the lake thermally and/or chemically stratified? What is the rate of water flow? Is the water saline (and does a thermohaline circulation occur)? Does water flow through hydrothermal action? A short period of post-fieldwork processing will be required, in conjunction with 3D visualisation of the lake system.

Sonar may provide fascinating imagery of the lake floor (if side-scan is used), and would complement seismic results of water depths acquired in 2006–2007.

## 10.4 Geochemical environment from laboratory analyses

Assuming that sufficient water is sampled, a strategy to utilise the samples most effectively is required. In the following tables, measurements of various components are prioritised and the minimum amount needed for the measurement is indicated. The first procedure for all samples is to separate water from particulates. Analysis of

**Table 2** Expected bulk chemical composition of water in Lake Ellsworth

| Units: $\mu$eq/l | $H^{+}$ | pH | $Ca^{2+}$ | $Mg^{2+}$ | $Na^{+}$ | $K^{+}$ | $NH_4^{+}$ | $Cl^{-}$ | $SO_4^{2-}$ | $NO_3^{-}$ | $HCO_3^{-}$ |
|---|---|---|---|---|---|---|---|---|---|---|---|
| Average Byrd Ice Core | 1.8 | 5.7 | ~1.0 | 0.4 | 1.5 | 0.05 | 0.13 | 2.0 | 1.0 | 0.7 | ~1.2 |
| Inferred Lake Ellsworth | <140 | >3.9 | >80 | 30 | 120 | 1 | <10 | 160 | 80 | <50 | >100 |
| Lake Ellsworth (including contribution from glacial flour) | ~10 | ~6 | ~420 | ~90 | 120 | 3 | 1 | 160 | 240 | 1 | 240 |

**Table 3** Some of the geochemical analyses that can be applied to the water collected from Lake Ellsworth

| Species | Min volume for analysis (ml) | Cumulative min volume (ml) |
|---|---|---|
| Major cations and anions $Ca^{2+}$, $Mg^{2+}$, $Na^{+}$, $K^{+}$, $SO_4^{2-}$ and $Cl^-$ | 2 | 2 |
| DIC/DOC | 5 | 7 |
| Nutrients ($NO_3^-$, $NH_4^+$ and $PO_4^{3-}$, DON and DOP) | 10 | 17 |
| $\delta^{13}C$-DIC, $\delta^{18}O$-$H_2O$, | 8 | 25 |
| If anoxic, Fe, Mn, $HS^-$ | 5 | 30 |
| CTD and DOX | Water column profile | |
| Should attempt to measure $O_2$, $CO_2$ , $CH_4$ and $N_2$ in the gases in the head space above the water samples | | |

filtered particulates will reveal the following results: particulate concentration, size frequency, lithology, mineralogy and provenance, and the particulate's contribution to solute. Having established the particulate content of lake water, experiments of the water itself can be conducted, with a priority given in Table 3.

### 10.5 Lake-floor sediments

A strategy for analysis of small volumes of lake-floor sediments, acquired by the probe, is required in a similar way to the lake water. A prioritisation of experiments, with amount of material required for each, is given in Table 4.

**Table 4** Some of the sedimentological analyses that can be applied to small sediment samples collected from Lake Ellsworth by the probe

| Species | Min weight of sediment (g) | Cumulative min weight (g) |
|---|---|---|
| Grain size distribution | 0.1 | 0.1 |
| CHNOS | 0.1 | 0.2 |
| SEM/EDAX | 0.1 | 0.3 |
| Mineralogy/ provenance | 1 | 1.3 |
| Sequential extraction for nutrients and Fe | 10 | 11.3 |
| Sulphides | 5 | 16.3 |
| X-ray lamination[a] | Whole core scanner | |
| Magneto-stratigraphy[b] | U-channel or whole core scanner | |

[a] X-ray analysis required use of all material in the sediment core, but will destroy none

[b] Magneto-stratigraphy, see Section 11.2

## 11 Geological experiments

### 11.1 Background

Sediment in Lake Ellsworth will come from two primary sources, namely mineral dust (plus other particulate matter such as propogules, micro-meteorites) from the overlying ice, and sub-glacially-eroded sediment. The dust component is likely to be volumetrically very small, whilst the sub-glacially eroded component is likely to be substantially greater but more spatially restricted to the edges of the lake.

These sediments are likely to be easily sampled using a gravity corer. Subsequent to successful probe deployment, it is feasible that a sediment corer could be dropped to the lake floor to sample and retrieve the sediments.

Retrieval of a shallow (~3 m) sediment core has several potential benefits to understanding the lake's environment, its biota and its long-term history. In ultra-oligotrophic (extreme nutrient-poor) lakes the sediment-water interface is one of the most likely niches for life to exist and so a sample of this environment is necessary to fully test hypotheses of life in the lake. For example in most Antarctic surface lakes the vast majority of biodiversity resides in the sediment. Further, sediments within a short core are likely to contain a potentially long (e.g., tens of thousands of years) history of environmental conditions within the lake, including any records of life such as preserved organic remains (biomolecules etc). If no life is found in the water column then the sediments may show if life ever existed in the lake.

The sediments provide the potential for answering the question of how life got to, and evolved within, the lake. In particular if a sediment core can sample the transition between pre-glacial sediments and those deposited in the lake then the development of the lake and its life may be tracked through time.

Depending on its length, a sediment core from Lake Ellsworth could address three issues concerning the glacial history of West Antarctica: (1) the record of West Antarctic Ice Sheet (WAIS) evolution, particularly the transition to full ice sheet conditions; (2) the timing and duration of periods of WAIS collapse, and (3) the variations in lake 'climate' introduced by glacial-interglacial variations in the ice sheet above.

### 11.2 Sediment core

A cable-operated gravity corer suitable for deployment through a hot water drilled hole has already been developed. A version of the corer could be built for retrieving sediment from Lake Ellsworth and could also be fitted with a percussion head to enable deeper penetration of the sediments. Of particular importance to the construction of the corer will be knowledge of the sediment thickness and water depth: both of which are scientific objectives of the proposed geophysical exploration of the lake (Section 5). Table 5 lists the various sedimentological measurements that could be made on a sediment core from the floor of Lake Ellsworth, and the scientific results that may follow. In particular, magneto-stratigraphy of the core may be essential to understanding the age of sedimentary material recovered from the floor of Lake Ellsworth.

A 'U-channel' cut from the split core half (2.5 × 2.5 cm × length of core), is required for magneto-stratigraphy analysis, but this will be non-destructive and will not modify the sample. All the material will be available for further study post magnetic measurements. It is probably best if the U-channel is cut before the core has the chance to come into any unknown magnetic fields. The coring barrel will be magnetic but this can be compensated for. Exposure to stray magnetic fields from engines, ships, steel objects, etc can modify or overprint the magnetic signal and this can prevent the success of magnetic studies. Shields for U-channels are available to prevent this during transport. If the core is quite condensed (i.e., the sediments are old), the first experiment will be to test whether the Brunhes/Matuyama Boundary (780,000 years ago) can be detected. Assuming the record is complete, it may also be possible to pick up younger field excursions but it would be difficult to work with these unless there is some independent chronology to integrate with. However, the bonus of being proximal to the South Pole is that we can easily detect polarity from inclination records alone.

It should also be possible to recover a secular variation and field intensity record. It is not known whether the secular variation will help with dating as the location is so close to the pole and azimuthal orientation is likely to be absent. However, such a record has never been taken so close to the South Pole, which makes its acquisition important scientifically. Several other geochronological techniques available in sedimentology may be used to establish a time control on the lake-floor core.

## 12 Summary and timetable of activities

The exploration of Lake Ellsworth requires a multidisciplinary effort of considerable size. The scientific justification is based on two important goals: (1) to measure and comprehend life in this extreme environment, and (2) to collect and assess the historical record contained in the lake floor sediments. A multi-staged approach is required to achieve these goals. First, a detailed characterisation of the lake's physiography is required from geophysical measurements. A comprehensive survey of the lake will take place between 2007 and 2009. Subsequently, the specific details of the lake exploration mission will be determined. The project also requires the assembly of scientific equipment to detect life in the lake's water and sediment, and to measure the chemical and physical environment of the lake. The necessary instruments will be held within a probe, fed into the lake via a borehole melted by a hot-water drill. The borehole, ~ 30 cm wide and 3,400 m deep, will be kept open for around 24 h,

**Table 5** Some of the analyses that can be applied to a 2 m (or longer) sediment core from Lake Ellsworth

| Measurement technique | Result | Comments |
|---|---|---|
| Magneto-stratigraphy (see section 11.2) | Age of sediment, e.g., by using magnetic polarity timescale | May be crucial as the most likely way to date the sediment |
| $^{10}Be$, $^{3}He$ etc | Could estimate timing of isolation of lake by ice using cosmogenic isotopes | Dependent on sample size available |
| Other radioactive isotopes | Presence of isotopes with variety of half-lives helps constrain flux of ice/sediment from surface and/or timing of isolation of site (e.g., presence of $^{14}C$ ($t_{1/2}$ = 5,600 years) would be v. difficult to explain without fairly recent exposure or contamination. | Could test ice flow models. |
| U-Th dating | Age of sediment | Requires presence of carbonate precipitate in closed system (may not be. present because of high pressure and low T (high $pCO_2$ therefore high acidity). Upper age limit of *c.* 0.5 Ma. |
| Grain size | Sediment flux/source origin Possibly info on age if grain size follows glacial cycles | |
| SEM | Source of sediment | Distinguish aeolian (dust) vs. subglacial transport of grains |
| XRD, XRF | Mineralogy of sediment | Itrax high-resolution XRF and X-radiography core scanning (sampled at 200 micron intervals) |
| Nd–Sr | Provenance of sediment | |
| Extraterrestrial material (imaging?) | Long-term averages of meteorites etc, testing of hypotheses of links between extraterrestrial flux and climate change | |
| Carbonate isotopes | Environmental conditions in lake | Requires presence of carbonate precipitate in closed system (difficult), but might find ikaite (cf calcite or aragonite) |
| Oxygen isotope analysis of penecontemporaneous minerals | Comparison to marine oxygen isotope stratigraphy—ice flow model could give the age offset | |
| Organic geochemistry | Biomolecule recognition | Molecular constituents or fossils of organisms reveal their presence and mode of living |
| Sedimentology (pre-glacial marine sediment vs. subglacial lake sediment) | Determination of whether the West Antarctic Ice Sheet has ever collapsed | If multiple West Antarctic ice sheet collapses are found in sediments, this raises interesting question of how life in basin adjusts from marine environment to re-glaciation on each occasion |
| | | Question of sediment scouring becomes important if Lake Ellsworth is shallow |
| | Assessment of how (and when) the West Antarctic Ice Sheet developed | |
| Analysis of gas hydrates | Gas composition | If gas hydrates are present a gas-tight system will be deployed in order to facilitate gas capture and later analysis |
| Bedrock sample | Local geology | Highly dependent on sediment thickness (possible, but unlikely that a bedrock sample could be retrieved with a gravity corer) |

in which time the probe must be deployed and retrieved. After the probe is taken out of the lake (possibly with samples) a gravity core will be sent down to recover a short (~3 m) sedimentary core. Finally, if time permits, a permanent thermistor string may be put into the lake, and the borehole sealed. Analysis of data and samples will take place in laboratories during the months following

**Table 6** Timetable for the exploration of Subglacial Lake Ellsworth. Grey boxes indicate planning phases, red boxes are when fieldwork takes place, blue boxes denote computer and laboratory analyses

| Project | 2007/8 | 2008/9 | 2009/10 | 2010/11 | 2011/12 | 2012/13 | 2013/14 |
|---|---|---|---|---|---|---|---|
| **Phase 1:** Geophysical survey | Fieldwork I | Fieldwork II | Modelling | | | | |
| **Phase 2:** Instrument development | | | | | | | |
| **Phase 2:** Probe development | | | | | | | |
| **Phase 2 and 3:** Hot-water drill and fieldwork | | | | | | Fieldwork | |
| **Phase 4:** Biological analysis | | | | | | Fieldwork | Laboratory |
| **Phase 4:** Geochemical analysis | | | | | | Fieldwork | Laboratory |
| **Phase 4:** Sediment analysis | | | | | | Fieldwork | Laboratory |

the fieldwork. It is possible to plan the physical exploration of Lake Ellsworth to take place during the Austral summer of 2012/13, providing: (1) geophysical exploration of Lake Ellsworth is successfully completed in 2007/2008 and 2008/2009, and results are interpreted immediately after fieldwork; (2) funds to develop the instruments, probe and drill are available following the first geophysical results of water depth, at the beginning of 2008, giving four years to build and test equipment; and (3) logistic support for deep hot water drilling at Lake Ellsworth is in place four years after the initial IPY period. In this case, the 8-year timetable of activity for exploration is outlined in Table 6.

## References


Ellis-Evans JC, Wynn-Williams D (1996) Antarctica: a great lake under the ice. Nature 381:644–646

Engelhardt H, Humphrey N, Kamb B, Fahnestock M (1990) Physical conditions at the base of a fast moving Antarctic ice stream. Science 248:57–59

Foght J, Aislabie J, Turner S, Brown CE, Ryburn J, Saul DJ, Lawson W (2004) Culturable bacteria in subglacial sediments and ice from two Southern Hemisphere glaciers. Microb Ecol 47:329–340

Kapitsa A, Ridley JK, de Robin GQ, Siegert MJ, Zotikov I (1996) Large deep freshwater lake beneath the ice of central East Antarctica. Nature 381:684–686

Kochkina GA, Ivanushkina NE, Karasev SG, Gavrish EY, Gurina LV, Evtushenko LI, Spirina EV, Vorob'eva EA, Gilichinskii DA, Ozerskaya SM (2001) Survival of micromycetes and actinobacteria under conditions of long-term natural cryopreservation. Microbiology 70:356–364

Miteva VI, Sheridan PP, Brenchley JE (2004) Phylogenetic and physiological diversity of microorganisms isolated from a deep Greenland glacier ice core. Appl Environ Microbiol 70:202–213

Priscu JC, Adams EE, Lyons WB, Voytek MA, Mogk DW, Brown RL, McKay CP, Takacs CD, Welch KA, Wolf CF, Kirshtein JD and Avci R. (1999) Geomicrobiology of subglacial ice above Lake Vostok. Antarct. Sci. 286:2141–2144

Priscu JC, Bell RE, Bulat SA, Ellis-Evans JC, Kennicutt MC, II, Lukin VV, Petit JR, Powell RD, Siegert MJ, Tabacco IE (2003) An international plan for Antarctic subglacial lake exploration. Polar Geogr 27(1):69–83

Rothschild LJ, Mancinelli RL (2001) Life in extreme environments. Nature 409:1092–1110

Siegert MJ, Ellis-Evans JC, Tranter M, Mayer C, Petit J-R, Salamatin A, Priscu JC (2001) Physical, chemical and biological processes in Lake Vostok and other Antarctic subglacial lakes. Nature 414:603–609

Siegert MJ, Tranter M, Ellis-Evans CJ, Priscu JC, Lyons WB (2003). The hydrochemistry of Lake Vostok and the potential for life in Antarctic subglacial lakes. Hydrol Process 17:795–814

Siegert MJ, Hindmarsh R, Corr H, Smith A, Woodward J, King E, Payne AJ, Joughin I (2004) Subglacial Lake Ellsworth: a candidate for in situ exploration in West Antarctica. Geophys Res Lett 31(23):L23403, 10.1029/2004GL021477

Siegert MJ (2002) Which are the most suitable Antarctic subglacial lakes for exploration?. Polar Geogr 26:134–146

Siegert MJ, Carter S, Tabacco I, Popov S, Blankenship D (2005) A revised inventory of Antarctic subglacial lakes. Antarct Sci 17(3):453–460

Rev Environ Sci Biotechnol (2006) 5:233–242
DOI 10.1007/s11157-006-0009-9

REVIEW PAPER

# Thermostable proteins as probe for the design of advanced fluorescence biosensors

**Marcella de Champdoré · Maria Staiano · Vincenzo Aurilia · Olesia V. Stepanenko · Antonietta Parracino · Mosè Rossi · Sabato D'Auria**

Received: 24 January 2006 / Accepted: 27 May 2006 / Published online: 15 July 2006


**Abstract** In this review we explore the advantages deriving from the use of either enzymes or sugar binding proteins isolated from thermophilic organisms to develop stable fluorescence biosensors. We report on a novel approach to address the consumption of the analyte by enzyme-based biosensors, namely the utilization of apo-enzymes as non-active forms of proteins which are still able to bind the ligand but cannot transform it into product. We also report recent studies in which the fluorescence labeling of a naturally thermostable binding protein allows a quantitative determination of glucose.

**Keywords** Glucose · Proteins · Thermophilic Organisms · Fluorescence · Biosensors

This work is dedicated to Prof. Koki Horikoshi for his outstanding contribution to the knowledge of the world of extremophiles.

M. de Champdoré · M. Staiano · V. Aurilia · O. V. Stepanenko · A. Parracino · M. Rossi · S. D'Auria (✉)
Institute of Protein Biochemistry, Italian National Research Council, Via Pietro Castellino 111, 80131 Naples, Italy
e-mail: s.dauria@ibp.cnr.it

O. V. Stepanenko
Institute of Cytology, Russian Academy of Science, St. Petersburg 194064, Russia

**Abbreviations**

| | |
|---|---|
| IAANS | 2-(4′-(iodoacetamido)anilino) naphthalene- 6-sulfonic acid (IAANS) |
| ANS | 1-(anilino)-naphtalene-8-sulfonate |
| FRET | fluorescence resonance energy transfer |
| LED | light emitting diode |
| GD | glucose dehydrogenase |
| BSGK | glucokinase from *Bacillus stearothermophilus* |
| Ph-SBP | sugar-binding protein from *Pyrococcus horikoshii* |

## Introduction

In recent years, a growing number of articles has been published indicating the use of stable biomolecules as probe for the development of advanced fluorescence biosensors. Fluorescence detection is the dominant analytical approach in medical testing, biotechnology and drug discovery. Starting in the 1980s the first chemically synthesized fluorescence probes for specific analytes became available (Wolfbeis 2000; Lakowicz 1995; Spichiger-Keller 1998). These early probes were designed so as to include in the same molecule both the specific affinity for the ligand and the capacity to change some intrinsic fluorescent property upon binding of the ligand. However, often these chemical probes resulted not soluble in aqueous solution making difficult the realization of the sensing devices.

Modern biotechnology has approached this problem by using proteins and enzymes as analyte recognition element for sensors for biochemical targets (Gilardi et al. 1997; Romoser et al. 1997; Miyawaki et al. 1997). The idea is to exploit the extremely wide range of selective affinities sculpted into the various proteins by biological evolution. The number of potential ligands specifically recognized by different proteins is very large and ranges from small molecules to macromolecules (including protein themselves). The advantages of using proteins as components of biosensors are many and include relatively low costs in design and synthesis, the fact that proteins are, at least in general, soluble in water, and finally, with the progresses of molecular genetics, the possibility of improving/changing some of the properties of the proteins by genetic manipulation. Many of the ligands that are important in clinical medicine and in the food control industry are relatively small. In these cases the enzymes appear to be the class of proteins endowed with the highest specificity and affinity. Other classes of proteins, such as receptors, transporters, antibodies, etc., often present lower specificity although they offer other advantage such as the fact that they can specifically recognize a wide range of much larger ligands. However, a broad use of proteins as probes for the development of sensors requires that some problems be addressed. A crucial concern on the utilization of biomolecules for the design of biosensors is related to the stability of the biomolecules.

Protein stability has been improved using a variety of protocols in the preparation of the sensor (Colacino and Crichton 1997) (immobilization and/or cross linking of the proteins, addition of stabilizing agents) or by utilizing biomolecules isolated from thermophilic organisms (D'Auria et al. 1997, 1998, 1999a, b, 2000; Sun et al. 1999).

In this article we review the recent advanced applications of fluorescence protein-based biosensors showing a few examples of the use of biomolecules isolated from thermophilic organisms as specific probes for the design of advanced sensing systems.

## Thermostable glucose dehydrogenase from *Thermoplasma acidophilum*

The widespread use of proteins as probes for biosensors depends on protocols to enhance the protein stability, such as introduction of changes in the amino acid composition leading to enhanced protein structural stability (Colacino et al. 1997). An alternative method is to use naturally thermostable proteins isolated from thermophilic microorganisms: these macromolecules have intrinsically stable structural features (D'Auria et al. 1999a, b, 2000; Sun et al. 1999), and they can be considered as ideal probes for the realization of stable fluorescence biosensors (D'Auria et al. 2001).

Glucose dehydrogenase (GD) from the thermoacidophilic archaeon *Thermoplasma acidophilum* is a tetramer of about 160 kDa composed of four similar subunits of about 40 kDa each. The enzyme shows a Km value of 10 mM for glucose, and it is resistant to high temperatures and organic solvents. At 55°C, full activity is retained after 9 h, and at 75°C the half-life is approximately 3 h (Smith et al. 1989). Moreover, incubation of the enzyme for up to 6 h at room temperature with 50% (v/v) methanol, acetone or ethanol resulted in no appreciable loss of activity (Smith et al. 1989). We examined the potential of this thermostable GD as a glucose sensor. The enzyme catalyzes the following reaction:

$$\text{Glucose} + \text{NAD(P)}^+ \leftrightarrow 6\ \text{gluconate} + \text{NAD(P)}^+ + \text{H}^+$$

To prevent the glucose oxidation we used the apo-form of GD that is the enzyme without the cofactor which is required for the reaction. We found that the apo-GD still binds glucose with an affinity comparable to the holo-enzyme. GD from *Thermoplasma acidophilum* is a thermophilic protein and can be expected to be rigid under mesophilic conditions. We knew that thermophilic proteins often display increased activity at higher temperatures or in the presence of organic solvents (e.g. methanol, ethanol, 1-propanol) or detergents (e.g. sodium dodecyl sulfate) (D'Auria et al. 1997, 1999a, b), which are conditions expected to increase the protein dynamics. Addition of acetone to the solution containing ANS and GD resulted in a dramatic increase in the ANS intensity as well as in a blue-shift of the emission maximum (Fig. 1). Addition of similar amounts of acetone to ANS in the absence of the protein produced modest fluorescence increase. Hence the increase in the ANS intensity reflects a change in the local protein environment which is due to acetone. To be useful as a glucose sensor the ANS labeled GD must display usefully large spectral changes in the presence of glucose. Addition of glucose to ANS-GD in the presence of 3% acetone resulted in about 25% decrease in intensity.

This seemed to be the optimal acetone concentration because smaller spectral changes were seen at lower and higher acetone concentrations. Apparently at higher acetone concentrations the ANS is already in an environment which results in highest possible increase in quantum yield. At lower acetone concentrations the environment surrounding the ANS changes in response to glucose in a manner which increases the ANS intensity.

In previous reports we described the use of polarization sensing for systems which display changes in intensity, but not lifetime, in response to analytes (Lakowicz and Szmacinski 1993; Szmacinski and Lakowicz 1994; Lakowicz et al. 1999; Gryczynski et al. 1999). Because the intensity changes of ANS-GD in response to glucose are modest, it is important to carefully select the best conditions. Figure 2a shows the emission polarized spectra of ANS-GD at various concentrations of glucose. The polarization decreases at higher glucose concentrations because the emission from this solution is observed through the horizontal polarizer. Moreover, the change in polarization is larger at shorter wavelengths, and this is due to the differences in the emission spectra of reference (ANS in buffer) and sample (ANS+GD) The wavelength dependent changes in polarization were used to create a calibration curve for glucose (Fig. 2b). This curve shows that

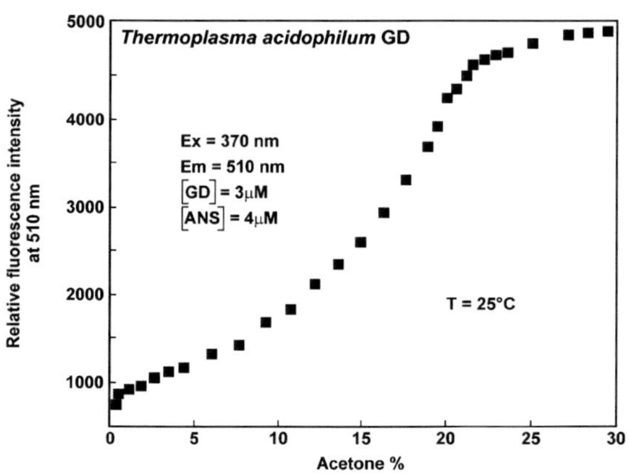


**Fig. 1** ANS-labeled GD fluorescence intensity in the presence of different concentrations of acetone. [GD] = 3 μM. [ANS] = 4 μM. The excitation was at 370 nm, and the emission was monitored at 510 nm

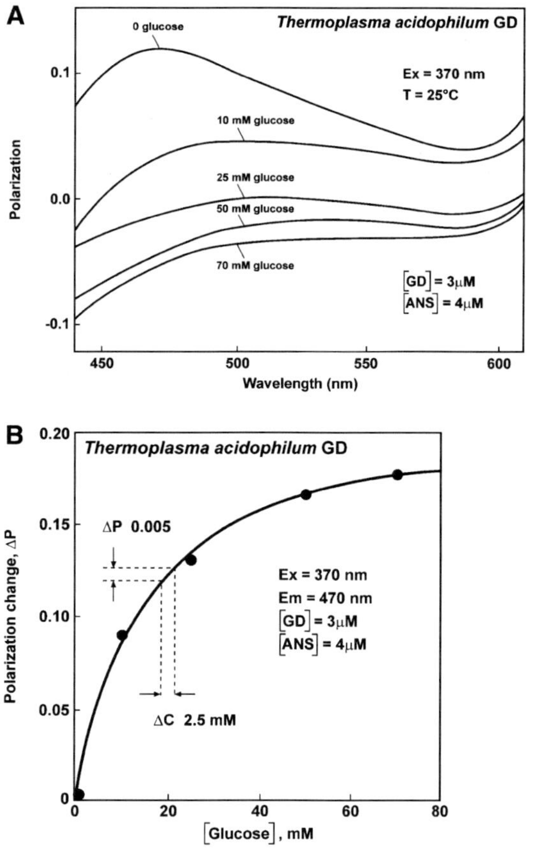


**Fig. 2** (**a**) Fluorescence polarization spectra of ANS-labeled GD in the presence of 3% acetone, and at different concentrations of glucose. Excitation was at 370 nm. [GD] = 3 μM. [ANS] = 4 μM. (**b**) Effect of glucose on the fluorescence polarization of GD in the presence of 3% acetone. The excitation was at 370 nm, and the emission was recorded at 470 nm. [GD] = 3 μM. [ANS] = 4 μM

the present ANS-GD system can yield glucose concentrations accurate to about 2.5 mM, at a glucose concentration near 20 mM.

## A fluorescence competitive assay by using a stable glucokinase

The structure of the hexokinase A from yeast has been determined by X-ray diffraction (Ureta et al. 1987; Bennet and Steitz 1980; Aleshin et al. 1998; Honzatko et al. 1998) (Fig. 3). The polypeptide chain of 485 amino acid residues in the yeast protein is folded into two distinct domains, a smaller N-terminal domain and a larger C-terminal domain. From the high-resolution crystal structures of the enzyme it is evident that in the absence of ligand, the two domains are separated by a deep cleft. This cleft represents the enzyme active site. The binding of glucose causes the small lobe of the molecule to rotate by 12° relative to the large lobe, moving the polypeptide backbone as much as 8°, closing the gap between the two domains. The domain rotation has two effects: the glucose molecule is buried into the interior of the protein and the side chains in the active site are rearranged.

Fluorescence spectral data in the literature suggest that hexokinase can be used as an optical glucose sensor. For instance, glucose binding to the native monomer hexokinase from*Saccharomyces cerevisiae* (this form of protein is found in the protein concentration range 1–10 mg/ml) and dimer hexokinase from*Saccharomyces cerevisiae* (this form of protein is found at protein concentration higher than range 10 mg/ml) was monitored by following the concomitant quenching of the protein fluorescence (Woolfitt et al. 1998; Feldman 1984; McDonald et al. 1979). This enzyme possesses four tryptophan residues: two trp residues are localized on the protein surface, one trp residue is in the cleft (the glucose-quenchable

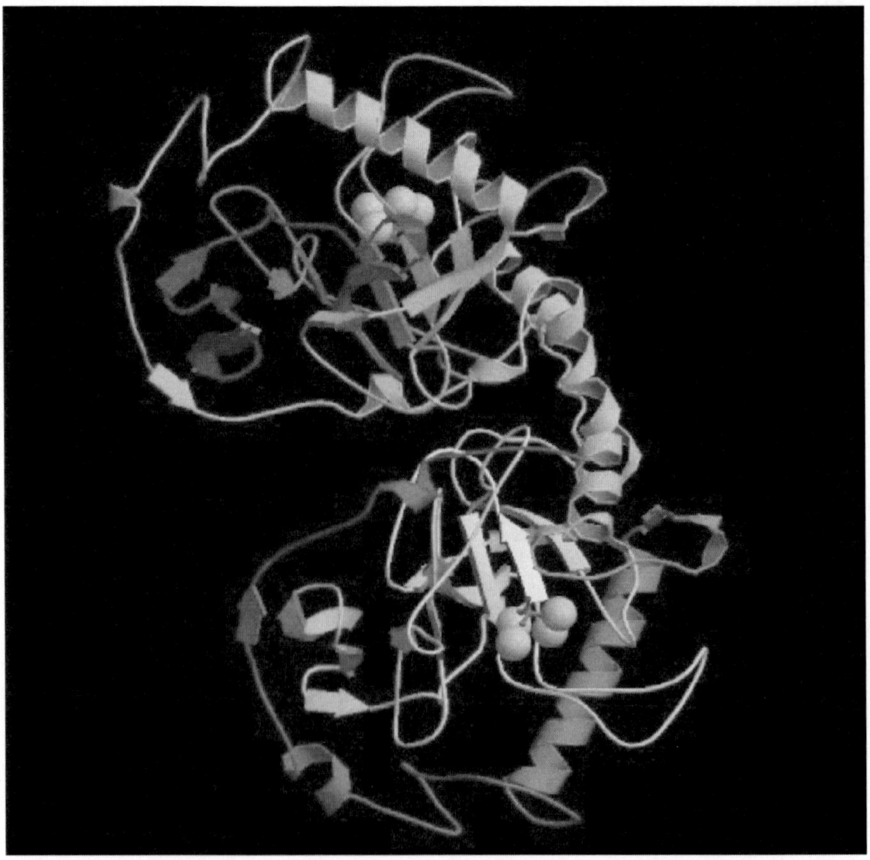

**Fig. 3** Structure of the hexokinase from *Saccharomyces cerevisiae*

residue) and one trp residue is buried in the protein matrix. The maximal quenching induced by glucose was about 25% and the concentration of glucose at half-maximal quenching was 0.4 mM for the monomeric form and 3.4 mM for the dimeric one (Feldman and Norton 1980; Kramp and Feldman 1978).

For use as a sensor a protein should have long term stability. Unfortunately, yeast hexokinase and human hexokinase are unstable and quickly loose activity at room temperature. Thermophilic organisms produce enzymes with unique characteristics such as high temperature-, chemical-, and pH-stability. These enzymes are already in use as biocatalysts in industrial processes (Jaenicke et al. 1996; Sthal 1993). A glucokinase from the thermophilic organism *Bacillus stearothermophilus* has been characterized and is known to display long term stability (Goward et al. 1987; Ishikawa et al. 1987; Tomita et al. 1990). Hence we evaluated the use of this glucokinase (BSGK) in the absence of ATP as a non-consuming glucose sensor.

For use as a sensor the protein must display good long term stability. In order to check the stability properties of BSGK, we incubated a solution of the enzyme (enzyme concentration 1.0 mg/ml) at room temperature. Enzyme aliquots were withdrawn and the enzyme activities as well as the fluorescence intensities were monitored. Figure 4 shows the enzyme activity and

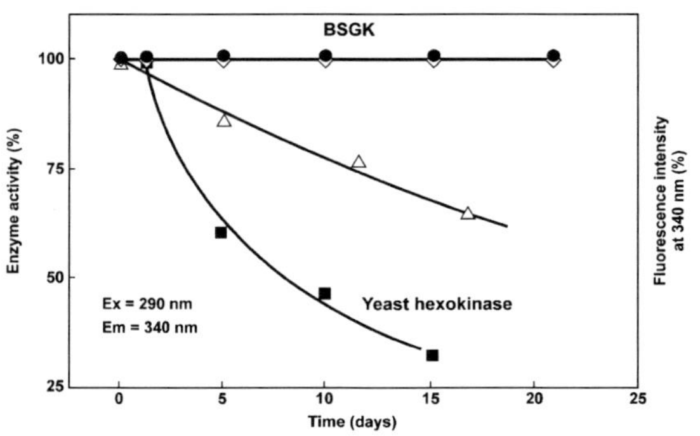


**Fig. 4** Stability of BSGK and YHX at room temperature. Fluorescence measurements (close circle for BSGK and open triangle for yeast hexokinase) and activity measurements (reverse open triangle for BSGK and close square for yeast hexokinase) were performed at room temperature. For fluorescence experiments: Ex = 290 nm; Em = 340 nm. A NATA solution was used as reference

intrinsic fluorescence of BSGK and yeast hexokinase over a period of incubation time at room temperature. Yeast hexokinase loses activity over several days and the fluorescence intensity simultaneously decreases. In contrast BSGK loses no activity over two weeks at room temperature and the fluorescence intensity remains constant. Hence BSGK is a good candidate for a glucose sensing probe.

BSGK has a single cysteine residue located near the active site (Gryczynski et al. 2000). We labeled the residue with a sulphidryl-reactive fluorophore IAANS. The emission of the labeled protein was near 460 nm. The intensity of the IAANS-labeled protein decreased upon addition of glucose. The decreased intensity is consistent with displacement of the water-sensitive IAANS into the aqueous phase upon binding glucose. The change in intensity occurs near 3 mM, which is comparable to the concentration of glucose in blood. The important conclusion from these observations is that BSGK binds glucose in the absence of ATP and can thus serve as a probe for non-consuming glucose sensor.

For highly accurate glucose measurements we were not satisfied with the magnitude of the intensity change. We examined the fluorescence lifetimes to determine if a change occurred upon glucose binding. Unfortunately, IAANS-labeled BSGK displayed no change in lifetime upon glucose binding. Hence we considered alternative methods to use BSGK as probe for a glucose sensor.

Fluorescence resonance energy transfer (FRET) reliably occurs whenever fluorescent donors and acceptors are in close proximity. We developed a FRET-based method to design a new competitive glucose assay. To demonstrate the feasibility of a competitive glucose assay we used the unmodified protein and its intrinsic tryptophan emission as the donor. As the acceptor we used glucose containing the absorbing nitrophenyl group, ONPG (Fig. 5). Addition of ONPG resulted in an approximate 80% decrease in the tryptophan intensity. Addition of glucose resulted in recovery of the fluorescence intensity. At about 6 mM glucose concentration fluorescence intensity returns to its initial value before addition of ONPG (Fig. 6). Further addition of glucose does

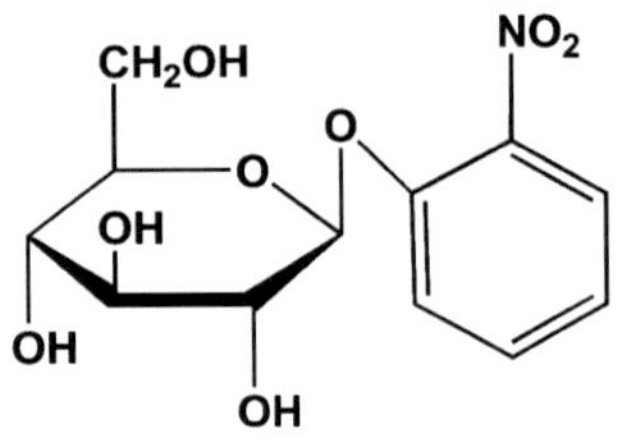


**Fig. 5** Structure of *o*-nitrophenyl-*β*-D-glucopyranoside (ONPG)

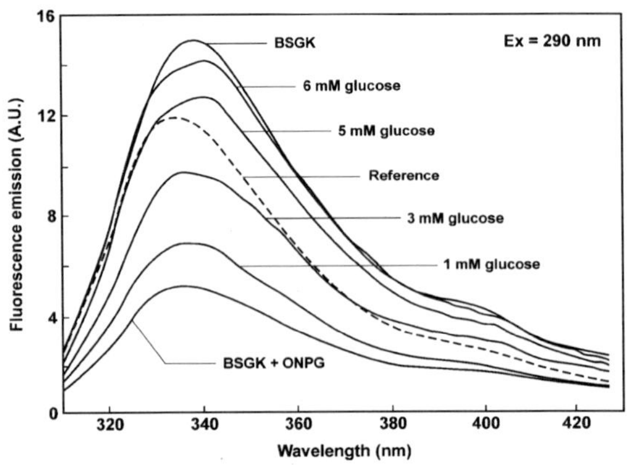


**Fig. 6** Effect of glucose on the intensity emission of BSGK in the presence of 30 μM ONPG. The excitation was at 290 nm, and the emission was recorded at 340 nm. [BSGK] = 3 μM. The reference spectrum was obtained by using a solution [BSGK] = 3 μM, which can be expected to display similar temperature, time or illumination dependent changes as the sample

not change the fluorescence signal. The fact that the intensity was sensitive to glucose demonstrates that the intensity changes are due to a binding event and not due to trivial inner filter effects from ONPG.

In recent publications it was showed how to obtain reliable intensity measurements for sensing purposes which could be used in the absence of useful changes in lifetimes. In particular, it was showed that polarization sensing can be accomplished by constructing a sensor such that a stable intensity reference is observed through one polarizer and the sample is observed through a second orthogonal polarizer (Gryczynski et al. 1999; Lakowicz et al. 1999). In this case as a reference we used a BSGK solution, which can be expected to display similar temperature, time or illumination dependent changes as the sample. To optimize the sensor response the reference intensity was about 65% of the sample response, as calculated for expected a two- to three-fold intensity change. This reference is observed through a horizontally oriented polarizer. The sample contains BSGK, ONPG and various concentrations of glucose, and is observed through a vertically oriented polarizer. The emission from both sides of the sensor is then observed through a vertically and horizontally oriented polarizer in order to measure polarization of the system. Figure 7 shows the observed polarization of the system for BSGK+ONPG and different glucose concentrations. An advantage of polarization measurements for sensing is that they are self-normalized and thus independent of the overall intensity of the sensor. The results shown above demonstrate that a thermostable glucokinase can serve as a glucose sensor. Additional studies are needed to obtain a BSGK-based sensor which displays larger spectral changes. For example, we are hopeful that BSGK labeled with fluorophores other than IAANS will display larger intensity changes, spectral shifts or changes in lifetime. The results in the competitive FRET assay are especially interesting because FRET is a through-space interaction which occurs whenever the donor and acceptor are within the Forster distance ($R_0$), and does not require a conformational change and/or a change in the probe environment. For these reasons we are confident that BSGK can be used with longer wavelength donors and acceptors to develop practical glucose sensors for use in diabetes health care. Since the

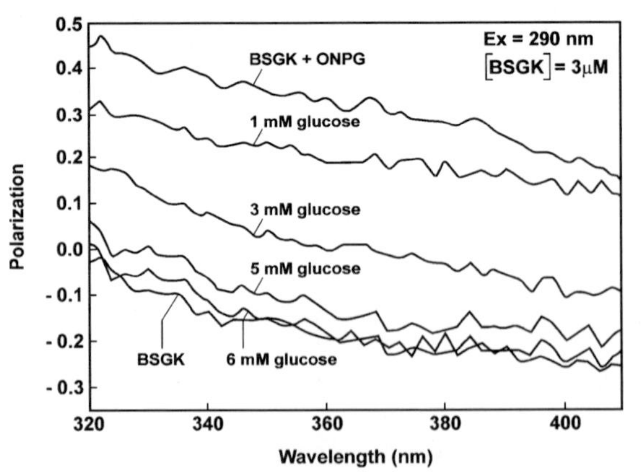


**Fig. 7** Effect of glucose on the polarization spectra of BSGK in the presence of 30 μM ONPG. Excitation was at 290 nm. [BSGK] = 3 μM

measurements through the skin can be easily performed by using a red laser diode or a light emitting diode (LED) as an excitation source, one may envision a polarization based device with an external calibrated standard that will allow non-invasive glucose determinations. The main advantage of using this method is the obtainment of ratiometric polarization measurements that are not influenced by light instability and sample perturbation.

## Thermostable glucose-binding protein as probe for long term stability non-consuming glucose biosensor

The periplasm of Gram-negative bacteria contains a large family of specific binding proteins that are essential primary receptors in transport and, in a few cases, chemotaxis (Boos and Lucht 1995). These proteins usually have a monomeric structure that folds in two main domains linked by three strands commonly referred to as the hinge region. Conformational changes involving the hinge are thought to be necessary for sugars to get in and out of the protein binding site (Quiocho and Ledvina 1996). Differences in the structures of the ligand-bound and ligand-free proteins are essential for their proper recognition by the membrane components (Flocco and Mowbray 1994). This property of binding proteins makes them good candidates as biological recognition elements in the development of biosensors (Tolosa et al. 1999). In fact, in the presence of a specific ligand, these proteins undergo a large conformational change in their global structure to accommodate the ligand inside the binding site (Luck and Falke 1991). Based on this conformational change, sensing systems for maltose and glucose were developed using their respective binding protein (Gilardi et al. 1994; Marvin and Hellinga 1998).

Recently a sugar-binding protein isolated from the archaeon *Pyrococcus horikoshii* (Ph-SBP) has been cloned, expressed and purified (Staiano et al. 2004). The protein is a monomer of 55 kDa that binds glucose molecules (Fig. 8). In this protein, only two Cys residues are present. One of them is localized in the segment defined as "signal sequence" and it is not present in the mature protein (Marabotti et al. 2004). The other Cys residue is located in the C-terminal domain, near (about 10 Å) the sugar binding groove. This is a strategic position, since every variation in the binding site can be detected by a fluorescent probe linked to this Cys residue. Therefore, this amino acid was considered as a good candidate for covalent modifications with fluorescent probes able to detect the binding of the sugar for the development of a stable fluorescence biosensor. The protein was covalently labeled by IAANS to the Cys residue located in the C-terminal domain of Ph-SBP and it was observed that the intensity of the IAANS-protein emission was sensitive to the additions of glucose.

From a practical point of view, to detect biomolecular interactions, one of the partners, the sensor molecule, should be immobilized on a sensor surface. The counterpart molecule, the analyte, is usually dissolved in the liquid phase and interacts with the immobilized sensing molecule. The IAANS–Ph-SBP complex was immobilized on a reactive aldehyde silylated slide. The aldehyde on the silylated slides reacted readily with primary amines on the protein forming a Schiff's base linkage. The immobilized IAANS–Ph-SBP was tested for its capacity to bind glucose. The addition of glucose to the immobilized IAANS–Ph-SBP resulted in the quenching of the emission intensity by about 20% and in a small blue-shift of the emission maximum (Fig. 9a, b) suggesting that the thermostable Ph-SBP can serve as a probe for the development of a stable and non-consuming glucose biosensor.

## Conclusions

There is a wealth of knowledge on enzymes which transform numerous substances of biochemical interest. Hence the possibility that, using non-active apo-enzymes as probes for reversible biosensors will greatly expand the range of biochemically relevant analytes which can be measured using proteins as sensors.

The results described in this article represent a first attempt to use not-active forms of en-

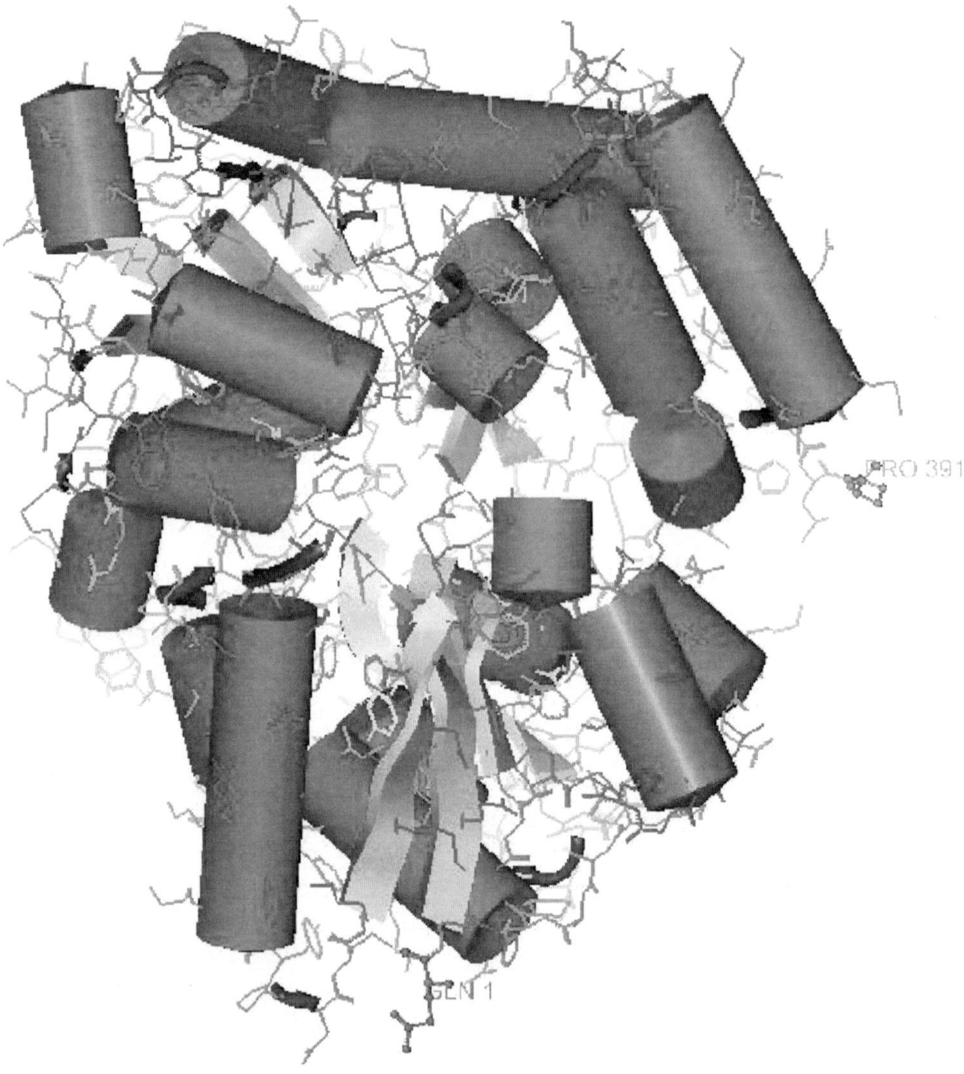


**Fig. 8** Structure of Ph-SBP

zymes as probes for glucose sensor. Larger glucose-dependent spectral change would increase the accuracy of the glucose measurements. In spite of these difficulties we feel that our system demonstrates a useful approach to sensing. The results suggest that the enzymes which use glucose as their substrate can be used as probes for reversible and non-consuming glucose sensors in the absence of required cofactors. The possibility of using not-active apoenzymes for a reversible sensor greatly expands the range of proteins which can be used as probes for sensors, not only for glucose, but for a wide variety of biochemically relevant analytes. Hence one is no longer limited to using signaling proteins which bind the analyte without chemical reaction. The enzymes can be engineered for covalent labeling by insertion of cysteine residues at appropriate locations in the sequence. The glucose induced spectral changes may be larger with other polarity sensitive probes or by the use of FRET between two fluorophores on the protein. In summary, apoenzymes appear to be a valuable source of protein probes for the development of advanced biosensors.

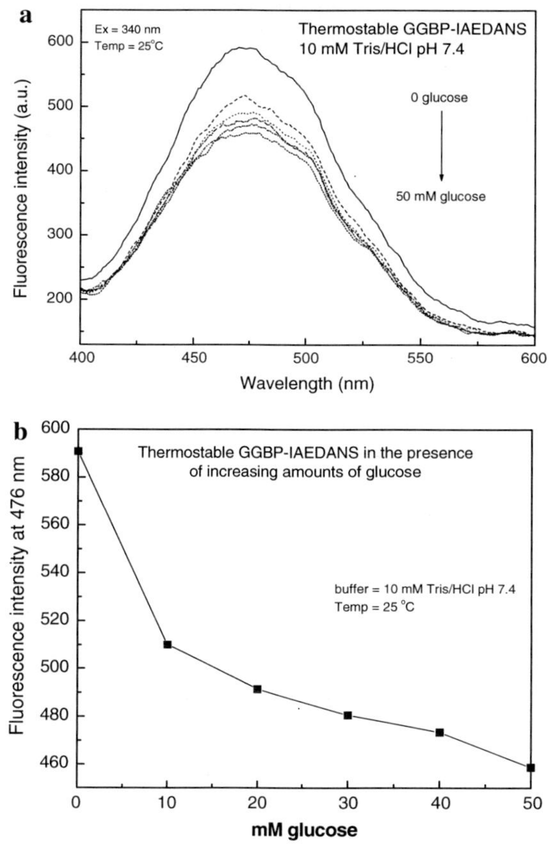


**Fig. 9** (**a**) Fluorescence emission spectra of Ph-SBP upon glucose addition. (**b**) Effect of glucose addition on the fluorescence emission at 476 nm. Ex = 340 nm. The reported values are derived from the spectra showed in (**a**)

**Acknowledgments** This project was realized in the frame of CRdC-ATIBB POR UE-Campania Mis 3.16 activities (S.D., M.R). This work was also realized in the frame of the CNR Commessa "Diagnostica Avanzata ed Alimentazione" (S.D, V.A). This work was supported by a NATO-CNR Advanced Fellowship grant 215.36.S to OV.S and MCB RAN (Program of Molecular and cell biology of the Russian Academy of Science) to OV:S.

## References

Aleshin AE, Zeng C, Bourenkov GP, Bartunik HD, Fromm HJ, Honzatko RB (1998) The mechanism of regulation of hexokinase: new insights from the crystal structure of recombinant human brain hexokinase complexed with glucose and glucose 6-phosphate. Structure 6(1):39–50

Bennet WSJ, Steitz TA (1980) Structure of a complex between yeast hexokinase A and glucose: structure determination and refinement at 3.5 Å resolution. J Mol Biol 140(2):183–209

Boos W, Lucht JM (1995) Periplasmic binding-protein-dependent ABC transports. In: Lin E (ed) *E. coli* and *Salmonella typhimurium*: cellular and molecular biology. American Society for Microbiology, Washington DC, pp 1175–1209

Colacino F, Crichton RR (1997) Enzyme stabilisation: the state of the art. Biotechnol Genet Eng Rev 14:211–277

D'Auria S, Barone R, Rossi M, Nucci R, Barone G, Fessas D, Bertoli E, Tanfani F (1997) Effects of temperature and SDS on the structure of $\beta$-glycosidase from the thermophilic archaeon *Sulfolobus solfataricus*. Biochem J 323:833–840

D'Auria S, Nucci R, Rossi M, Grycznisky I, Malak H, Lakowicz JR (1999a) The $\beta$-glycosidase from the Archaeon Sulfolobus solfataricus: structure and activity in the presence of alcohol. J. Biochemistry 126(3):545–552

D'Auria S, Nucci R, Rossi M, Gryczynski I, Gryczynski Z, Lakowicz JR (1999b) The $\beta$-glycosidase from the hyperthermophilic archaeon *Sulfolobus solfataricus*: enzyme activity and conformational dynamics at temperatures above 100°C. Biophys Chem 81:23–31

D'Auria S, Herman P, Lakowicz JR, Bertoli E, Tanfani F, Rossi M, Manco G (2000) The thermophilic esterase from *Archaeoglobus fulgidus*: structure and conformational dynamics at high temperature. Proteins 38:351–360

D'Auria S, Lakowicz JR (2001) Enzyme fluorescence as a sensing tool: new perspectives in biotechnology. Curr Opin Biotechnol 1:99–104

D'Auria S, Moracci M, Febbraio F, Tanfani F, Nucci R, Rossi M (1998) Structure–function studies on $\beta$-glycosidase from *Sulfolobus solfataricus*. Molecular bases of thermostability. Biochimie 80:949–957

Feldman I (1984) Ionic strength dependence of glucose binding by yeast hexokinase isoenzymes. Biochem J 217(1):335–337

Feldman I, Norton GE (1980) Effects of glucose and magnesium ion on the quenching of yeast hexokinase fluorescence by acrylamide. Biochim Biophys Acta 615(1):132–142

Flocco MM, Mowbray SL (1994) The 1.9 Å X-ray structure of a closed unliganded form of the periplasmic glucose/galactose receptor from *Salmonella typhimurium*. J Biol Chem 269:8931–8936

Gilardi G, Mei G, Rosato N, Finazzi-Agro' AF, Cass AEG (1997) Spectroscopic properties of an engineered maltose-binding protein. Protein Eng 10:479–486

Gilardi G, Zhou LQ, Hibbert L, Cass AEG (1994) Engineered the maltose binding protein for reagentless fluorescence sensing. Anal Chem 66:3840–3847

Goward CR, Scawen MD, Atkinson T (1987) The inhibition of glucokinase and glycerokinase from *Bacillus stearothermophilus* by the triazine dye procion blue MX-3G. Biochem J 246:83–88

Gryczynski I, Lakowicz JR, Gryczynski Z (1999) Polarization sensing with visual detection. Anal Chem 71:1241–1251

Gryczynski Z, Gryczynski I, Lakowicz JR (2000) Simple apparatus for polarization sensing of analytes. Opt Eng 39:2351–2358

Honzatko RB, Aleshin AE, Zeng C, Bartunik HD, Fromm HJ (1998) Regulation of hexokinase I: crystal structure of recombinant human brain hexokinase complexed with glucose and phosphate. J Mol Biol 282(2):345–357

Ishikawa H, Maeda T, Hikita H (1987) Initial-rate studies of a thermophilic glucokinase from *Bacillus stearothermophilus*. Biochem J 248:13–20

Jaenicke R Schuring H Beaucamp N, Ostendorp R (1996) Structure and stability of hyperstable proteins: glycolitic enzymes from hyperthermophilic bacterium *Thermotoga maritima*. Adv Protein Chem 48:181–269

Kramp DC, Feldman I (1978) Tryptophan distribution in yeast hexokinase isoenzyme B. Biochim Biphys Acta 537:406–416

Lakowicz JR (1995) Advances in fluorescence sensing technology II. Proc SPIE 2388:159–170

Lakowicz JR, Szmacinski H (1993) Fluorescence lifetime-based sensing of pH, $Ca^{2+}$, $K^{+}$ and glucose. Sensors Actuators B 11:133–143

Lakowicz JR, Gryczynski I, Gryczynski Z, Dattelbaum JD (1999) Anisotropy-based sensing with reference fluorophores. Anal Biochem 267:397–405

Luck LA, Falke JJ (1991) Open conformation of a substrate binding cleft: 19F NMR studies of cleft angle in the D-galactose chemosensory receptor. Biochemistry 30:6484–6490

Marabotti A, D'Auria S, Rossi M, Facchiano AM (2004) Theoretical model of the three-dimensional structure of a sugar-binding protein from *Pyrococcus horikoshii*: structural analysis and sugar-binding simulation. Biochem J 380:677–684

Marvin JS, Hellinga HW (1998) Engineering biosensors by introducing fluorescent allosteric signal transducers: construction of a novel glucose sensor. J Am Chem Soc 120:7–11

McDonald RC, Steitz TA, Engelman DM (1979) Yeast hexokinase in solution exhibits a large conformational change upon binding glucose or glucose 6-phosphate. Biochemistry 18(2):338–342

Miyawaki A, Llopis J, Heim R, McCaffery JM, Adams JA, Ikura M, Tsien RY (1997) Fluorescent indicators for $Ca^{++}$ based on green fluorescent proteins and calmodulin. Nature 388:882–887

Quiocho FA, Ledvina PS (1996) Atomic structure and specificity of bacterial periplasmic receptors for active transport and chemotaxis: variation of common themes. Mol Microbiol 20:17–25

Romoser VA, Hinkle PM, Persechini A (1997) Detection of living cells of $Ca^{++}$ dependent changes in the fluorescence emission of an indicator composed of two green fluorescent protein variants linked by a calmodulin-binding sequence. J Biol Chem 272:13270–13274

Smith LD, Budgen N, Bungard J, Danson MJ, Hough DV (1989) Purification and characterization of glucose dehydrogenase from the thermoacidophilic archaebacterium *Thermoplasma acidophilum*. Biochem J 261:973–977

Spichiger-Keller UE (1998) Chemical sensors and biosensors for medical and biological applications. Wiley-VCH, New York, pp 313–328

Staiano M, Sapio MR, Scognamiglio V, Marabutti A, Facchiano AM, Bazzicalupo P, Rossi M, D'Auria S (2004) A thermostable sugar-binding protein from the Archaeon *Pyrococcus horikoshii* as a probe for the development of a stable fluorescence biosensor for diabetic patients. Biotechnol Prog 20:1572–1577

Sthal S (1993) Thermostability of enzymes. In: Gupta MN (ed) Springer-Verlag, Berlin, pp 45–74

Sun MC, Tolliday N, Vetriani C, Robb FT, Clark DS (1999) Pressure-induced thermostabilization of glutamate dehydrogenase from the hyperthermophile *Pyrococcus furiosus*. Protein Sci 8:1056–1063

Szmacinski H, Lakowicz JR (1994) Life-time based sensing. Plenum Press, 1991–1994, 233 Spring Str. New York, NY 10013. Top Fluorescence Spectrosc 4:295–334

Tolosa L, Gryczynski I, Eichhorn LR, Dattelbaum JD, Castellano FN, Rao G, Lakowicz JR (1999) Glucose sensor for low-cost lifetime-based sensing using a genetically engineered protein. Anal Biochem 267:114–120

Tomita K, Nagata K, Kondo H, Shiraishi T, Tsubota H, Suzuki H, Ochi H (1990) Thermostable glucokinase from *Bacillus stearothermophilus* and its analytical application. Ann NY Acad Sci 613:421–425

Ureta T, Medina C, Preller A (1987) The evolution of hexokinases. Arch Biol Med Exp 20(3–4):343–357

Wolfbeis OS (2000) Fiber-optic chemical sensors and biosensors. Anal Chem 72:81R–89R

Woolfitt AR, Kellet GL, Hoggett JG (1998) The binding of glucose and nucleotides to hexokinase from *Saccharomyces cerevisiae*. Biochim Biophys Acta 952(2):238–243

Rev Environ Sci Biotechnol (2006) 5:219–231
DOI 10.1007/s11157-006-0008-x

REVIEW PAPER

# Astrobiological significance of minerals on Mars surface environment

**Jesus Martinez-Frias · Gabriel Amaral · Luis Vázquez**

Received: 7 December 2005 / Accepted: 25 May 2006 / Published online: 14 July 2006


**Abstract** Despite the large amount of geomorphological, geodynamic and geophysical data obtained from Mars missions, much is still unknown about Martian mineralogy and paragenetic assemblages, which is fundamental to an understanding of its entire geological history. Minerals are not only indicators of the physical–chemical settings of the different environments and their later changes, but also they could (and do) play a crucial astrobiological role related with the possibility of existence of extinct or extant Martian life. This paper aims: (1) to present a synoptic review of the main water-related Martian minerals (mainly jarosite and other sulfates) discovered up to the present time; (2) to emphasize their significance as environmental geomarkers, on the basis of their geological settings and mineral parageneses on earth (in particular in the context of some selected terrestrial analogues), and (3) to show that their differential UV shielding properties, against the hostile environmental conditions of the Martian surface, are of a great importance for the search for extraterrestrial life.

J. Martinez-Frias (✉) · L. Vázquez
Centro de Astrobiología (CSIC-INTA), 28850
Torrejón de Ardoz, Madrid, Spain
e-mail: martinezfrias@mncn.csic.es

G. Amaral
Departamento de Química Física I, Facultad de Ciencias Químicas, Universidad Complutense, 28040 Madrid, Spain
e-mail: gamaral@quim.ucm.es

L. Vázquez
Departamento de Matemática Aplicada, Facultad de Informática, Universidad Complutense de Madrid, 28040 Madrid, Spain
e-mail: lvazquez@fdi.ucm.es

**Keywords** Mars minerals · Extreme environment · Astrobiology · UV radiation · Jarosite · Gypsum · Sulfates

## Introduction

Over the last half century, Mars has been explored with telescopes, spacecrafts and robotic rovers. All the information obtained from these different sources, along with the results obtained by the study of SNC meteorites and terrestrial analogs, is starting to reveal the geological diversity of the planet and provides data for theorizing about how the different Martian environments evolved. Although it is well known that liquid water is not stable at the surface under today's atmospheric conditions (e.g., Ingersoll 1970; Hecht 2002), there is significant evidence that Mars once had a thicker atmosphere, that liquid water may have been much more abundant on the surface and in the subsurface earlier in Martian history, that it has at least sporadically flowed on the Martian surface, and that it may even still be present in the subsurface today (e.g., Sagan and Mullen 1972; Carr et al. 1977; Cess et al. 1980; Squyres et al. 1992; Mckay and Stoker 1989; Malin and Edgett 2000; Feldman et al. 2002; Boynton et al. 2002; Mitrofanov et al. 2002; Costard et al. 2002; Noe Dobrea et al. 2003; Squyres et al. 2004; Klingelhöfer et al. 2004; Madden et al. 2004; Christensen et al. 2004; Orofino et al. 2005; Glotch and Christensen 2005, among others). However, despite the huge amount of geomorphological, geodynamic and geophysical data obtained: (a) there is a clear ambiguity in interpreting certain geological features of the Martian surface, and (b) much is still unknown about Mars mineralogy and paragenetic assemblages, which is fundamental to an understanding of its whole geological history. Minerals are not only indicators of the physical–chemical settings of the different environments and their later changes, but also they could (and do) play a crucial astrobiological role related with the possibility of existence of extinct or extant Martian life. If thirty years ago Viking landers provided the first elemental analyses of Martian surface materials, the detection of an iron mineral (gray crystalline hematite) by the Mars Global Surveyor Thermal Emission Spectrometer (MGS-TES) (Christensen et al. 2000, 2001; Pearson et al. 2000) led to the selection of Meridiani Planum as one of the landing sites of the two NASA's Mars Exploration Rovers (MERs). In 2004, the Mars Exploration Rover Opportunity's Moessbauer spectrometer obtained new straightforward evidence that, at least in Meridiani Planum, the formation of hematite involved an aqueous mechanism. Hematite at Meridiani Planum consists essentially of spherules interpreted as concretions that have weathered out of a sulfate-rich outcrop. In addition, hematite is also a component of the outcrop matrix material. It also indicated the presence of an iron-bearing mineral called jarosite in the set of rocks dubbed "El Capitan" (Squyres et al. 2004; Klingelhöfer et al. 2004; Madden et al. 2004; Christensen et al. 2004; Glotch and Christensen 2005). "El Capitan" is located within the rock outcrop that lines the inner edge of the small crater where Opportunity landed. The exciting discovery of jarosite indicates the existence of an ancient extreme (acidic) Martian environment and its possible association with other liquid water-related minerals (e.g. gypsum) indubitably stresses its astrobiological interest.

This paper aims: (1) to present a synoptic review of the main water-related Martian minerals discovered till the present; (2) to emphasize their significance as environmental geomarkers, on the basis of their geological settings and mineral parageneses on earth (in particular in the context of some selected terrestrial analogues), and (3) to show that their differential UV shielding properties, against the hostile environmental conditions of the Mars surface, are of a great importance for the search for extraterrestrial life.

## Mineralogy and UV radiation on the surface of Mars

The Martian regolith is made up of an apparently homogenized dust having (broadly) basaltic composition, with admixed local rock components, oxides (e.g. hematite), water-bearing phyllosilicates and salts (mainly sulfates). Quartzofeldspathic materials also have been identified (Bandfield et al. 2004). Information from scientific literature about past Mars missions, together with recent reviews and new findings (see for instance Souza et al. 2004; McSween 2004; Vaniman et al. 2004; Lane et al. 2004; Squyres and Knoll 2005; Clark et al. 2005; Poulet et al. 2005; Yen et al. 2005; Hutchinson et al. 2005) indicate a mineralogical composition of the Martian surface, which displays, in broad terms, the following general distribution: silicates and oxides (mainly olivine ($Mg_2SiO_4$ to $Fe_2SiO_4$), pyroxenes (Ca (Mg, Fe, Al)(Al, Si)$_2$O$_6$) and plagioclases (Na, Ca)(Si, Al)$_4$O$_8$ (87–79%)); hematite, $Fe_2O_3$, goethite, FeO(OH), sulfate salts (jarosite, $KFe_3(SO_4)_2(OH)_6$, kieserite, $MgSO_4 \cdot H_2O$, and very possibly also some polyhydrated sulfates: epsomite, $MgSO_4 \cdot 7H_2O$, hexahydrite, $MgSO_4 \cdot 6H_2O$, pentahydryte, $MgSO_4 \cdot 5H_2O$, starkyite, $MgSO_4 \cdot 4H_2O$, (12%) [Zhu et al. (2006), Bibring et al. (2006), as well as, possibly, szomolnokite and ferricopiapite] (Lane et al. (2004); and carbonates (Banfield et al. 2003) (0–4%), chloride salts (1%), nitrates (0–1%), water (>1%, may be much higher). Poulet (2005) detected the presence of phyllosilicates in the ancient Martian highlands. These authors suggest that Earth-like conditions existed well before 3.5–4 billion years ago. During later martian history, it seems that the surface became more acidic, suppressing the formation of phyllosilicates and carbonates, and leading to the haematite and sulfates spectacularly observed at Meridiani by Opportunity. Very recently, Zhu et al. (2006) suggest the existence of other mineral phases, such as calcopyrite, covellite, garnet (uvarovite, almandine) and thenardite. Marion et al. (2006) developed a model, parameterized for the Na–K–Mg–Ca–Fe–H–Cl–$SO_4$–$NO_3$–OH–$HCO_3$–$CO_3$–$CO_2$–$CH_4$–$H_2O$ system, which includes 81 solid phases. Their simulation suggests the possible existence, among others, of melanterite, rozenite, mirabilite, szomolnokite and schwertmanite.

Infrared observations display evidence for igneous diversity and magmatic evolution on Mars (Christensen et al. 2005).

The very recent MEX-OMEGA results have shed light on the discussion about the mineralogical and petrological characteristics of Mars surface. However, in accordance with Wyatt and McSween (2002) we agree that controversy still remains (and probably it will be necessary to improve "in situ" analysis") about the existence of andesite versus chemically weathered basalts, as the basalts of the northern plains on Mars are more andesitic and weathered than the basalt of the southern highlands. They appear to be well represented by the Bounce Rock at the Meridiani Site, which is dominated by pyroxene (clinopyroxene ~55%, orthpyroxene ~5%) and plagioclase (~20%), and is poor in olivine (~5%). Oxides are accounting for ~10%. The chemical composition of Bounce Rock is more evolved than the basalts in the Gusev crater. It has a high $P_2O_5$ content of 0.95 wt%, a Fe/Mg ratio of 36, a low Mg number (molar MgO/MgO + FeO) of 0.42 and a high Ca/Al ratio of 1.7, a lower FeO (15.6%), and a higher CaO (12.5%) content (Squyres et al. 2004; Klingelhöfer et al. 2004; Squyres and Knoll 2005; Christensen et al. 2005; Clark et al. 2005). Broadly, the basalts in the northern plains are in general rich in sulfur and variably enriched in bromine relative to chlorine, indicating a past interaction with water (Fan and Schulze-Makuch 2005). Generally, Martian basalts are composed of plagioclase, feldspar, clinopyroxene, olivine, plus/minus sheet silicates and occur primarily in the equatorial to mid-latitude southern highlands regions (Banfield 2002). Major surface geological units of the ancient crust consist of pyroxenes and plagioclase, with varying proportions of olivine and alteration minerals. Moreover, Martian (SNC) meteorites display small amounts of secondary minerals (clays, carbonates, halides, sulfates) probably formed by reaction with subsurface fluids.

In accordance with Patel et al. (2004) the study of solar ultraviolet (UV) radiation is of extreme importance in a wide range of scientific disciplines, with UV radiation playing an important role in organic and chemical evolution and also as a major constraint in biological evolution. Unlike Earth, there is a significant amount of UV flux on Mars, mainly due to the influence of the shorter wavelengths UVC (100–280 nm) and UVB (280–315 nm). On the surface of Mars solar radiation which penetrates the thin atmosphere at wavelengths between 200 and 400 nm is capable of interacting directly with biological structures and causing severe damage. Various works on the biological effects of UV radiation (Cockell 1998; Cockell et al. 2000; Rontó et al. 2003; Patel et al. 2003, 2004) have documented that even the current Martian UV flux would not in itself prevent life. Nevertheless, it is a fact that this UV flux contributes, together with the absence of liquid water and extreme low temperatures, to both possible mineralogical alterations (e.g. possible dehydration) and to the biologically harsh nature of the Martian surface. In this sense, UV radiation induced dehydration was already suggested around 30 years ago (Huguenin 1976; Huguenin et al. 1977). According to these works, photons with wavelengths shorter than 280 nm release $H_2O$ (g) from FeO(OH) (goethite) by ejecting OH-ligands which subsequently combine with $H^+$ from nearby sites. Morris and Lauer (1981), however, repeated the experiments and found no UV dehydration effects on goethite (a-FeOOH) or lepidocrocite (g-FeOOH) in exposures equivalent to 10–100 years on the Martian surface.

More recently, Yen et al. (1997) indicated that exposure to the Martian environment over geologic time scales could have removed the initial water content of the hydrated minerals modeled to be present. These authors placed iron oxide samples into an ultra high vacuum (UHV) chamber evacuated to 10–8 torr using an ion pump. One sample was exposed to 254 nm radiation from a mercury vapor lamp through a sapphire window while the other sample was held as a control. The samples were then heated to 500°C, and the evolved water was measured as a function of temperature. The results obtained (Yen et al. 1997) indicated that, although more confirmation is required, ultraviolet radiation was capable of enhancing the rate of dehydration of goethite in high vacuum conditions.

From the astrobiological point of view, it is extremely important to note that the present-day DNA-weighted irradiance on the surface of Mars is similar to the weighted irradiance on the surface of Archean Earth (Cockell 1998). The amount of dust in the atmosphere has a non-trivial effect and its UV shielding role must not be underestimated. Patel et al. (2004) indicate that high wind speeds, dust devils and local/global storms can raise particulate matter from the Mars surface and inject it into the atmosphere, where the dust can remain for long periods of time playing a major role in global circulation and atmospheric dynamics. As indicated by Cockell and Knowland (1999), for 3.8 billion years evolution, the development of strategies to attenuate UV radiation has been an omnipresent issue for life, mainly for photosynthetic organisms that require solar radiation for their energy needs. The authors illustrate a selection of UV shielding methods found on present-day earth which may have been relevant on early earth: iron compounds, sulfur, solid NaCl, water column, sediments, different types of rocks and minerals. Even microbial communities themselves can protect other communities from UV radiation. The surface of microbial mats has been shown to provide protection against UVC radiation for the microbiota beneath (Margulis et al. 1976).

There is a general agreement regarding the exploration and detection of life on Mars: any living organism, as we know it, should have preferentially developed in a particular subsurface microenvironment able to protect it from the harsh conditions on the surface. Thus, the study of the mineralogical and petrologic features of the Martian surface is crucial. Terrestrial endolithic communities that live in the subsurface layers of rock that provide appropriate microenvironments against extreme external conditions have been proposed (Friedmann 1982; McKay 1993; Wynn-Williams and Edwards 2000; Villar et al. 2005) as possible analogs to life on Mars. In this context, Cockell et al. (2003) point out that in natural terrestrial environments, there are a variety of specific substrates (rocks, snow and ice, soils, dust), that can cover microbial communities. Some microbial species inhabit the underside of rocks as "hypolithic" organisms (Broady 1981a, b) or they live in cracks in rocks as "chasmoendoliths" (Broady 1981b). The petrologic and mineralogical composition of the substrate is also important. In fact, where the geological features of the substrate allow, they can inhabit the inside of rocks as "cryptoendolithic" micro-organisms (Friedmann and Ocampo 1976). Therefore, extant Martian life would require strong UV shielding, which, in accordance with some mineralogical studies presented here, could be perfectly accomplished, at the surface, by certain minerals (e.g. sulfate minerals) already discovered on Mars. But if it is important to identify and understand the mineralogical assemblages of the surface, it is also critical to determine the set of geochemical reactions which can modify them, altering the original settings.

With regards to the geochemical processes of Mars' surface, Burns (1987) suggested the possibility of formation of the ferric oxyhydroxysulfate mineral schwertmannite in equatorial regions of Mars, where acidic permafrost melts and is oxidized by the Martian atmosphere and, more recently, Lammer et al. (2003) evaluated the formation of ferric oxy-hydroxides and sulfates. They indicated that the oxidation of iron may schematically be described in terms of the change of the ferrous component of iron-bearing precursor phases into a ferric oxide. Gomez et al. (2003) studied the growth of prokaryotic and eukariotic microorganisms after UV irradiation with and without ferric iron as a protection agent, concluding that ferric iron is an effective protective agent for both cell systems. It is proposed that the ferric oxide on Mars may be hematite and/or maghemite, which are chemically identical and that the oxidation process itself is independent of the transformation of ferric oxides into oxyhydroxides. A significant aspect that they stress is that the formation of sulfates may be as important as rusting, and for the oxidation process itself, the type of sulfate is unimportant. Under the oxidizing conditions on the Martian surface, any sulfur in the soil should be bound in sulfatic weathering phases (Lammer et al. 2003).

## Minerals as environmental geomarkers: astrobiological significance of water-related sulfates

In accordance with Farmer (2004), in defining a site-selection strategy to explore for a Martian fossil record, a key concept is contemporaneous chemical precipitation, or mineralization. On Earth, geological environments where microorganisms are often preserved in this way include, among others: (1) mineralizing systems (subaerial, subaqueous, and shallow subsurface hydrothermal systems, and cold springs of alkaline lakes), (2) ephemeral lacustrine environments (sabkhas), or terminal (evaporative) lake basins, (3) duricrusts and subsoil hard-pan environments formed by the selective leaching and re-precipitation of minerals within soil profiles, and (4) periglacial environments ground ice or permafrost (frozen soils) have captured and cryopreserved microorganisms and associated organic materials. Thus, if we want to identify potential biomarkers regarding the type of microbes which lived (or still live) at the surface of Mars, we will previously need to use the minerals as geomarkers to understand the geological and environmental context.

Sulfates are indeed to be present in the Martian soil as indicated by the sulfur measured and other mineralogical determinations at the Viking, Mars Pathfinder (MPF), and Mars Exploration Rover (MER) (Lane et al. 2005). Sulfates are widespread minerals in nature, mostly linked with different formation mechanisms (O'Connor 2005): (a) alteration of sulfides; (b) genetically related with volcanic or postvolcanic activity (alteration of volcanic rocks in acid fumaroles, or hydrothermal activity with or without implication of bacterial activity, and (c) evaporitic processes.

Some common chemical reactions leading the formation of iron sulfates, which are widespread in the *gossan* areas of many hydrothermal mineral deposits, involve the oxidization of pyrite, marcasite, or other sulfides by the atmosphere and water (Jerz 2002):

$$\mathrm{FeS_{2(s)}(pyrite,\ marcasite)} + 7/2\mathrm{O_{2(aq)}} + \mathrm{H_2O_{(aq)}} = \mathrm{FeSO_{4(s)}} + \mathrm{H_2SO_{4(aq)}}$$

$$\mathrm{Fe_{1-x}S_{(s)}(pyrrhotite)} + 8 - 2\mathrm{x}/4\mathrm{O_{2(aq)}} + \mathrm{xH_2O_{(aq)}} = (1-\mathrm{x})\mathrm{FeSO_{4(s)}} + \mathrm{xH_2SO_{4(aq)}}$$

A third reaction occurs when sulfur is burned and the gas is released, which slowly reacts with free oxygen in humid air to form:

$$\mathrm{S_{(s)}} + \mathrm{O_{2(g)}} + \mathrm{heat} = \mathrm{SO_{2(g)}}$$

$$2\mathrm{SO_{2(g)}} + \mathrm{O_{2(g)}} + 2\mathrm{H_2O_{(g)}} = 2\mathrm{H_2SO_{4(g)}}$$

A well known and very interesting example is represented by the acidic waters of the Rio Tinto, and the associated deposits of hematite, goethite, jarosite and other sulfates, which have been recognized as an important chemical analog to the "Sinus Meridiani" site on Mars (Fernandez-Remolar et al. 2004, 2005; Fairen et al. 2004). The Mars Analog Rio Tinto Experiment (MARTE) (Stoker et al. 2003, 2005, 2006) has been investigating the hypothesis of a subsurface microbial ecosystem based on the metabolism of iron and sulfur minerals. Reduced iron and sulfur might provide electron donors for microbial metabolism while in situ oxidized iron or oxidants entrained in recharge water might provide electron acceptors. The results obtained indicated that geochemical resources are available in the Rio Tinto subsurface to support several kinds of anaerobic chemolithotrophic metabolism (Stoker et al. 2005, 2006).

Likewise, sulfate deposits related with evaporitic processes are also important indicators of their depositional environments, including climate and the hydrochemistry of the water from which the minerals precipitated (Spencer 2000). Spencer and Hardie (1990) calculated the precipitation sequence for the evaporation of modern seawater, and discovered that precipitate minerals would form in the following order: calcite, gypsum, anhydrite ($CaSO_4$), halite, glauberite ($Na_2Ca(SO_4)_2$), halite, polyhalite ($K_2Ca_2Mg(SO_4)_4 \cdot 2H_2O$), epsomite ($MgSO_4 \cdot 7H_2O$), hexahydrite ($MgSO_4 \cdot 6H_2O$), kieserite ($MgSO_4 \cdot H_2O$), carnalite ($KMgCl_3 \cdot 6H_2O$), and bischofite ($MgCl_2 \cdot 6H_2O$) (Spencer 2000). However, caution is needed in the extrapolation of these "precipitation patterns" to possible evaporitic Martian systems as very big differences between modern locations of evaporite mineral deposition and those in the rock record can exist.

Quinn et al. (2005) examined the dry acid deposition and accumulation on the surface of Mars and in the Atacama desert and proposed that the recent discovery of the Martian jarosite, which forms in strongly acidic-sulfate rich environments, increases the importance of understanding the chemical state of the Martian surface material and its behavior in aqueous systems. These authors concluded that the extremely low pH resulting from acid accumulation, combined with limited water availability and high oxidation potential, will result in acid-mediated reactions at the soil surface during low-moisture transient wetting events (i.e. thin films of water) (Quinn et al. 2005). These soil acids are expected to play a significant role in the oxidizing nature of the soils, the formation of mineral surface coatings, and the chemical modification of organics in the surface material. In this context, it is important to stress that Murad and Rojík (2003) found that some sulfate precipitates showed color and mineralogy variations depending on the pH. In initial acidic conditions, which have a pH of about 2.3, the dominant mineral is jarosite. A pH between 3 and 4 yields precipitates that are orange in color, and the most predominant mineral is usually schwertmannite. Ferrihydrite and goethite, brownish-red in color, formed at a more neutral pH between 5 and 7. Jarosite, schwertmannite, ferrihydrite and goethite are all ferric (hydr)oxysulfates, meaning that the minerals all contain $Fe^{3+}$ and OH- in their chemical structures. Murad and Rojík (2003) believe that these changes in mineralogy and the associated color variations are direct indicators of the environments in which the minerals were formed.

As previously defined, some water-related minerals (e.g. gypsum, jarosite) were recently discovered (Squyres et al. 2004; Klingelhöfer et al. 2004; Langevin et al. 2005) on Mars' surface. Both sulfates can be used as idoneous examples which represent very well their environmental and astrobiological applications as potential geomarkers.

Gypsum, $CaSO_4 \cdot 2H_2O$, is a very common terrestrial evaporitic sulfate. It has essentially a layered structure bound by hydrogen bonds. Zig-zag chains of $CaO_8$ polyhedra, running parallel to c, are bound together by similar chains of isolated $(SO_4)^{2-}$ tetrahedra, forming a double sheet perpendicular to (010). Each $Ca^{2+}$ ion is surrounded by six oxygen atoms belonging to the sulfate groups and two oxygen atoms belonging to the $H_2O$ molecules. These $H_2O$ molecules form a layer binding the polyhedral sheets together with weak hydrogen bonds. The $H_2O$ molecules are significantly distorted and are oriented such that the hydrogen bond $H_2 \cdots O_1$ acts almost entirely along b (Schofield et al. 1996). It has been suggested (Moore and Bullock 1999) that evaporite deposits may represent significant sinks of mobile cations (e.g., those of Ca, N, Mg, and Fe) and anions (e.g., those of C, N, S, and Cl) among the materials composing the Martian surface and upper crust.

Some well known gypsum-rich evaporitic areas (e.g. Sorbas area, SE Spain) have been proposed (Martinez-Frias et al. 2001a) as possible Mars analogs to study paleogeographic, paleoclimatic and mineralogical problems associated with catastrophic evaporitic processes. The Sorbas basin contains one of the most complete sedimentary successions of the Mediterranean (gypsum karst) reflecting the increasing salinity during the Messinian salinity crisis (desiccation of the Mediterranean Sea) (Fig. 1) (Martin and Braga 1994; Riding et al. 1998; Krijgsman et al. 1999) and showing a complex paleogeographical evolution, being a signature of its progressive restriction and isolation.

**Fig. 1** Gypsum crystals from the Sorbas evaporitic basin, Almería province, SE Spain. Sulfate-rich layers reflect the increase of salinity during the Messinian crisis (desiccation of the Mediterranean Sea)

Fishbaugh et al. (2006) propose that north polar gypsum deposit of Mars was formed as an evaporite deposit in the unique conditions provided at the north pole. Water from the Chasma Boreale melting event (and possibly a nearby impact into ice) pooled beneath the ice and evaporated, precipitating gypsum. The ice has since retreated, exposing the gypsum source region, allowing gypsum to be eroded from this source by the wind. Sand sized gypsum particles are now saltating and are intimately mixed with the dark, mafic sands.

Recent studies (Parnell et al. 2004) of microbial colonization in impact generated hydrothermal crystalline gypsum deposits in the Haughton Crater, Devon Island, Canadian High Arctic, have demonstrated the presence of cyanobacteria in endolithic habitats up to 50 mm from the crystal margins. The crystalline gypsum was found to exist in the clear selenite form. These authors indicate that the propensity for sulfates to form clear crystals makes them an advantageous habitat for photosynthesisers. In accordance with Parnell et al. (2004) and Edwards et al. (2005), the gypsum colonisation in the Haughton Crater has a particular astrobiological relevance with the recent discoveries of sulfate minerals on Mars. The authors consider interesting to speculate that the colonisation of gypsum deposits on Mars could be a geological niche of microbial activity from periods when there was significant moisture at the Martian surface.

Jarosite is a mineral of the alunite-jarosite family. In accordance with Scott (2000), the alunite–jarosite minerals are defined as having the general formula $AB_3(XO_4)_2(OH)_6$, where A is a large ion in 12-fold coordination (e.g., K, Na, Ca, Pb, and REE), B is usually Fe or Al, and the $XO_4$ anions are usually $SO_4$, $PO_4$ or $AsO_4$. Jarosite was first characterized on Earth, in 1852, in the "Jaroso Ravine" at Sierra Almagrera (Fig. 2), in the Cuevas del Almanzora natural area (Jaroso Hydrothermal System, Almeria province, Spain), which is the world type locality of jarosite (Amar de la Torre 1852; Martinez-Frias 1999). The Jaroso Hydrothermal System (Martinez-Frias et al. 2004) is a volcanism-related multistage hydrothermal episode of Upper Miocene age, which includes oxides and oxy-hydroxides (e.g. hematite, goethite), base- and precious-metal sulfides and different types of sulfosalts. Hydrothermal fluids and sulfuric acid weathering of the ores have generated huge amounts of oxide and sulfate minerals of which jarosite is the most abundant (Martinez-Frias et al. 1992; Martinez-Frias 1998; Rull et al. 2004). Very recently, hallotrichite ($FeSO_4 \cdot Al_2(SO_4)_3 \cdot 22H_2O$) also has been found and characterized at the Jaroso area (Frost et al. 2005). It is important to note that this area of SE Spain had already been proposed as a relevant geodynamic and mineralogical model (Martinez-Frias et al. 2001a, b; Martinez-Frias et al. 2004; Rull et al. 2005; Rull and Martinez-Frias 2006) to follow for the astrobiological exploration of Mars.

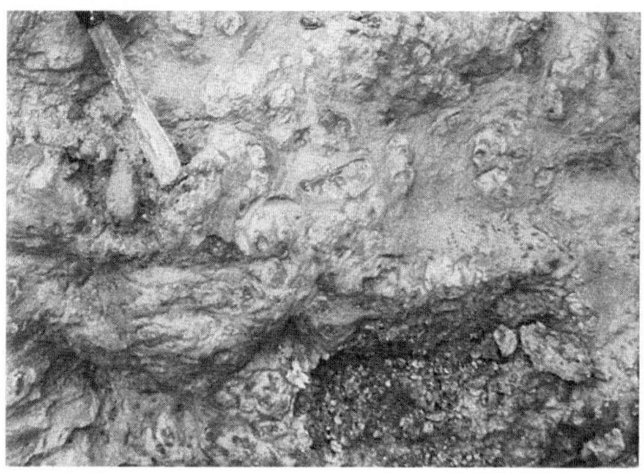

**Fig. 2** Typical alteration crust rich in jarosite at El Jaroso ravine, Sierra Almagrera, Jaroso Hydrothermal System, Cuevas del Almanzora Natural Area, Almería province, SE Spain

Microorganisms typically are involved in the oxidation of sulfides to sulfates in terrestrial acid mine drainage sites. Hence, outcrops on Mars which are rich in acid sulfate minerals (e.g. jarosite) may be a good location to search for evidence of life on that planet (Colmer and Hinkle 1947; Bishop et al. 2005). An astrobiologically significant aspect linked with sulfates was recently described by Aubrey et al. (2005). These authors studied concentrations of organic matter along with amino acids in natural terrestrial sulfate mineral samples. They found that sulfate minerals contain between 0.03 and 0.69% organic carbon as well as high ppb to low ppm abundances of amino acids and their degradation products in samples ranging from 30 million years old to contemporary. Thus amino acids and their amine decarboxylation products are well preserved over long geological time in the sulfate mineral matrices on Earth, and, as suggested by the authors, sulfates should be principal targets in the search for organic compounds, including those of biological origin, on Mars (Aubrey et al. 2005, 2006). Jarosite (Fig. 2) has proven to have a great astrobiological importance, not only for its relation with liquid water, but also because it can act as a sink and source of Fe ions for Fe-related chemolithoautotrophic microorganisms, such as those encountered in numerous extremophilic ecosystems (e.g. Tinto river) (López-Archilla et al. 2001; Gonzalez-Toril et al. 2003; Amaral Zettler et al. 2003; Fernandez-Remolar et al. 2004, 2005).

Considering all previous aspects and the astrobiological relevance of both Martian Ca and Fe sulfates, Amaral et al. (2005) performed UV radiation experiments on jarosite and gypsum samples (Figs. 1 and 2) from Jaroso and Sorbas area, SE Spain (Martinez-Frias et al. 2001b) using a Xe Lamp with an integrated output from 220 to 500 nm of 1.2 $Wm^{-2}$. Samples were flattened to different thicknesses (between 0.1 and 1.6 mm) before being exposed to UV light. The results obtained demonstrated a large difference in the UV protection capabilities of both minerals and also confirmed that the mineralogical composition of the Martian regolith is a crucial shielding factor. In a previous work, Parnell et al. 2004 had determined that a 1 mm thickness of the Haughton selenite gypsum exposed to the environmentally relevant range 290–400 nm, exhibited a mean absorbance of 0.12 (transmission of 0.88). Recent results obtained by Amaral et al. (2005) showed that whereas gypsum showed a much higher transmission percentage, jarosite samples, with a thickness of only 500 µm, prevented transmission. It is well known that, iron and iron-bearing compounds can provide an UV screen for life (Sagan and Pollack 1974; Olsen and Pierson 1986; Pierson et al. 1993; Kumar et al. 1996; Allen et al. 1998; Phoenix et al. 2001; Gomez et al. 2003; among others). The results obtained by Amaral et al. (2005) fit this working hypothesis well and are extremely important for the search for life on Mars as: (a) jarosite typically occurs on Earth as alteration crusts and patinas, and (b) a very thin crust of jarosite on the surface of Mars would be sufficient to shield microorganisms from UV radiation.

## Final remarks

As demonstrated by the MER mission, mineralogy provides the most robust means for discovering ancient aqueous environments and comprises an essential step in selecting the sites that have the best chance for having captured and preserved a record of ancient life or pre-biotic chemistry. A sophisticated spectrometer can accurately identify a specific water-related mineral (e.g. jarosite, gypsum, kieserite, etc.) on Mars; but, what does it mean? We know that the same mineral can be formed in different terrestrial environments; the same sulfate that we can find in a dessert can also be the product of a hydrothermal system. Thus, as previously defined, a previous step to detect possible Martian biomarkers is the utilization of minerals as geomarkers to understand the geological and environmetal context. This implies that if it is essential to determine what minerals and rocks are significant for the search of life on Mars, the appropriate selection and detailed study of the different geological and mineralogenetic terrestrial settings (Mars analogs) in which such minerals occur and evolve, are also of a great interest. Unfortunately the number of minerals unambiguously identified on Mars' surface is still extremely scarce and their textural relationships are not well understood. The interdisciplinary study of potential Mars analogues en Earth (hydrothermal systems, evaporitic areas, acidic rivers, impact craters, (mineralizing) submarine and hydrocarbon vents, etc.) are helping us to recognize the great variety of geological and mineralogical frameworks and the richness of environmental settings useful in astrobiological exploration.

**Acknowledgements** This work was supported by the Spanish Centro de Astrobiologia (CSIC/INTA), associated to the NASA Astrobiology Institute. Thanks to the Rover Environmental Monitoring Station (REMS) project. Maite Fernandez Sampedro, Maria Paz Martín Redondo and Dr Virginia Souza-Egipsy are acknowledged for their assistance with the analyses. Also thanks to three anonymous referees and Dr Alberto G. Fairén for their very helpful comments and remarks that have greatly improved the original manuscript. Special thanks to Dr David Hochberg for the revision of the English version.

## References

Allen CC, Jager KM, Morris RV, Lindstrom DJ, Lindstrom MM, Lockwood JP (1998) Martian soil simulant available for scientific, educational study. EOS 79:405

Amar de la Torre R (1852) Descripcion de los minerales, algunos de ellos nuevos, que constituyen el filón del Barranco Jaroso de Sierra Almagrera, por el caballero profesor el doctor Augusto Breithaupt, de Freiberg. Revista Minera 3:745–754

Amaral G, Martinez-Frias J, Vázquez L (2005) Paragenetic models and influence of UV radiation on selected water-related minerals under simulated Martian conditions. Preliminary experiments. MSL-REMS First Science Team Meeting, Centro de Astrobiologia, CSIC/INTA, associated to the NASA Astrobiology Institute, Torrejon de Ardoz, Madrid, Spain

Amaral Zettler LA, Messerli MA, Laatsch AD, Smith PJS, Sogin ML (2003) From genes to genomes: beyond biodiversity in Spain's Rio Tinto. Biol Bull 204:205–209

Aubrey AD, Cleaves HJ, Chalmers JH, Bada JL (2005) Sulfate minerals as targets for biomolecule detection on Mars Goldschmidt Conference Abstracts 2005 The Robotic Exploration of Mars and Titan, A533

Aubrey AD, Cleaves HJ, Chalmers JH, Skelley AM, Mathies RA, Grunthaner FJ, Ehrenfreund P, Bada JL (2006) Sulfate minerals and organic compounds on Mars. Geology 34:357–360

Bandfield JL (2002) Global mineral distribution on Mars. J Geophys Res107(E6):5042

Bandfield JL, Glotch TD, Christensen PR (2003) Spectroscopic identification of carbonate minerals in the Martian dust. Science 301:1084–1087

Bandfield JL, Hamilton VE, Christensen PR, McSween HY Jr (2004) Identification of quartzofeldspathic materials on Mars. J Geophys Res E100:1–14

Bibring JP, Langevin Y, Mustard JF, Poulet F, Arvidson R, Gendrin A, Gondet B, Mangold N, Pinet P, Forget F, the OMEGA team (2006) Global mineralogical and aqueous Mars history derived from OMEGA/ Mars express data. Science 312:400–404

Bishop JL, Dyar MD, Lane MD, Banfield JF (2005) Spectral identification of hydrated sulfates on Mars and comparison with acidic environments on Earth. Int J Astrobiol 3:275–285

Boynton WV, Feldman WC, Squyres SW, Prettyman TH, Brückner J, Evans LG, Reedy RC, Starr R, Arnold JR, Drake DM, Englert PAJ, Metzger AE, Mitrofanov I, Trombka JI, d'Uston C, Wänke H, Gasnault O, Hamara DK, Janes DM, Marcialis RL, Maurice S, Mikheeva I, Taylor GJ, Tokar R, Shinohara C (2002) Distribution of hydrogen in the near surface of Mars: evidence for subsurface ice deposits. Science 297:81–84

Broady PA (1981a) The ecology of hypolithic terrestrial algae at the Vestfold Hills, Antarctica. Br Phycol J 16:231–240

Broady PA (1981b) The ecology of chasmoendolithic algae at coastal locations of Antarctica. Phycologia 20:259–272

Burns RG (1987) Ferric sulfates on Mars. J Geophys Res 92:570–574

Carr MH, Crumpler LS, Cutts JA, Greeley R, Guest JE, Masursky H (1977) Martian impact craters and emplacement of ejecta by surface flow. J Geophys Res 82:4055–4065

Cess RD, Ramanathan V, Owen T (1980) The martian paleoclimate and enhanced carbon dioxide. Icarus 41:159–165

Christensen PR, Bandfield JL, Clark RN, Edgett KS, Hamilton VE, Hoefen T, Kieffer HH, Kuzmin RO, Lane MD, Malin MC, Morris RV, Pearl JC, Pearson R, Roush TL, Ruff SW, Smith MD (2000) Detection of crystalline hematite mineralization on Mars by the thermal emission spectrometer: evidence for near surface water. J Geophys Res 105 (E4):9623–9642

Christensen PR, Bandfield JL, Clark RN, Edgett KS, Hamilton VE, Hoefen T, Kieffer HH, Kuzmin RO, Lane MD, Malin MC, Morris RV, Pearl JC, Christensen PR, Morris RV, Lane MD, Bandfield JL, Malin MC (2001) Global mapping of martian hematite deposits: remnants of water-driven processes on early Mars. J Geophys Res 106:23,873–23,886

Christensen PR, McSween HY Jr, Bandfield JL, Ruff SW, Rogers AD, Hamilton VE, Gorelick N, Wyatt MB, Jakosky BM, Kieffer HH, Malin MC, Moersch JE (2005) Evidence for magmatic evolution and diversity on Mars from infrared observations. Nature 504–509

Christensen PR, Wyatt MB, Glotch TD, Rogers AD, Arvidson RE, Bandfield JL, Blaney DL, Budney C, Calvin WM, Fergason RL, Graff TG, Hamilton VE, Hayes A, Johnson JR, Knudson AT, McSween HY, Mehall GL, Mehall LK, Moersch JE, Morris RV, Smith MD, Squyres SW, Ruff SW, Wolff MJ (2004)

Initial results from the miniature thermal emission spectrometer experiment at the opportunity landing site on meridiani planum, Science 306:1733–1739

Clark BC, Morris RV, McLennan SM, Gellert R, Jolliff B, Knoll AH, Squyres SW, Lowenstein TK, Ming DW, Tosca NJ, Yen A, Christensen PR, Gorevan S, Bruckner J, Calvin W, Dreibus G, Farrand W, Klingelhoefer G, Waenke H, Zipfel J, Bell JF III, Grotzinger J, McSween HY, Rieder R (2005) Chemistry and mineralogy of outcrops at Meridiani Planum. Earth Planet Sci Lett 240:73–94

Cockell CS (1998) The biological effects of high ultraviolet radiation on early Earth—A theoretical evaluation. J Theor Biol 193:717–729

Cockell CS, Knowland J (1999) Ultraviolet radiation screening compounds. Biol Rev 74:311–345

Cockell CS, Catling D, Davis WL, Kepner RN, Lee PC, Snook K, McKay CP (2000) The ultraviolet environment of Mars: biological implications past, present and future. Icarus 146:343–359

Cockell CS, Rettberg P, Horneck G, Scherer K, Stokes MD (2003) Measurements of microbial protection from ultraviolet radiation in polar terrestrial microhabitats. Polar Biol 26:62–69

Colmer AR, Hinkle ME (1947) The role of microorganisms in acid mine drainage: a preliminary report. Science 106:253–256

Costard F, Forget F, Mangold N, Peulvast JP (2002) Formation of recent martian debris flows by melting of near-surface ground ice at high obliquity. Science 295(5552):110–113

Edwards HGM, Jorge Villar SE, Parnell J, Cockell CS, Lee P (2005) Raman spectroscopic analysis of cyanobacterial gypsum halotrophs and relevance for sulfate deposits on Mars. Analyst 130:917–923

Fairen AG, Fernandez-Remolar D, Dohm JM, Baker VR, Amils R (2004) Inhibition of carbonate synthesis in acidic oceans on early Mars. Nature 431(7007):423–426

Fan Ch, Schulze-Makuch D (2005) Columbia Plateau Basalt as an analog to the basalt of the Martian Northern Plains Goldschmidt Conference Abstracts 2005 The Robotic Exploration of Mars and Titan, A534

Farmer J (2004) Targeting sites for future astrobiological missions to Mars Second Conference on Early Mars, October 11–15, 2004, Jackson Hole, Wyoming, 8088.pdf

Feldman WC, Boynton WV, Tokar RL, Prettyman TH, Gasnault O, Squyres SW, Elphic RC, Lawrence DJ, Lawson SL, Maurice S, McKinney GW, Moore KR, Reedy RC (2002) Global distribution of neutrons from Mars results from Mars Odyssey. Science 297(5578):75–78

Fernandez-Remolar D, Gomez-Elvira J, Gomez F, Sebastian E, Martin J, Manfredi JA, Torres J, Gonzalez Kesler C, Amils R (2004) The Tinto River, an extreme acidic environment under control of iron, as an analog of the Terra Meridiani hematite site of Mars, Planet. Space Sci 52:239–248

Fernandez-Remolar DC, Morris R, Gruener JE, Amils R, Knoll AH (2005) The Río Tinto Basin, Spain: mineralogy, sedimentary geobiology, and implications for interpretation of outcrop rocks at Meridiani Planum, Mars. Earth Planet Sci Lett 240:149–167

Fishbaugh KE, Poulet F, Langevin Y, Chevrier V, Bibring JP (2006) Origin of gypsum in the Mars north polar region. Lunar Planet Sci XXXVII:1642.pdf

Friedmann EI (1982) Endolithic microorganisms in the antarctic cold desert. Science 215:1045–1053

Friedmann EI, Ocampo R (1976) Cryptoendolithic blue-green algae in the dry valleys: primary producers and the Antarctic desert ecosystem. Science 193:1247–1249

Frost RL, Weier ML, Kloprogge JT, Rull F, Martinez-Frias J (2005) Raman spectroscopy of halotrichite from Jaroso, Spain. Spectrochim Acta Part A Mol Biomol Spectrosc 62(1–3):166–180

Glotch TD, Christensen PR (2005) Geologic and mineralogic mapping of Aram Chaos: evidence for a water-rich history. J Geophys Res 110:E09006, doi:10.1029/2004JE002389

Gomez F, Grau A, Vazquez L, Amils R (2003) UV radiation effects over microorganisms and study of protective agents. Proc. Of the III European Workshop on Exo-Astrobiology, Mars, The Search for Life, Madrid, Spain, 18–20 November 2003, 21–26

Gonzalez-Toril E, Llobet-Brossa E, Casamayor EO, Amann R, Amils R (2003) Microbial ecology of an extreme acidic environment, the Tinto River. Appl Environ Microbiol 69:4853–4865

Hecht MH (2002) Metastability of liquid water on Mars. Icarus 156:373–386

Huguenin RL (1976) Chemical evolution of the Martian atmosphere by surface weathering. Proc Colloq Water Planet Regoliths V.I:33–43

Huguenin RL, Prinn RG, Maderazzo M (1977) Mars: photodesorption from mineral surfaces and its effects on atmospheric stability. Icarus 32:270–298

Hutchison L, Mustard JF, Gendrin A, Bibring J-P, Langevin Y, Gondet B, Mangold N, the OMEGA Science Team (2005) Mafic polyhydrated sulfates and Kieserite in Capri Chasma. Lunar Planet Sci XXXVI:1404.pdf

Ingersoll AP (1970) Mars: occurrence of liquid water. Science 168:972–973

Jerz JK (2002) Geochemical reactions in unsaturated mine, wastes. Thesis/dissertation, Blacksburg, Va, University Libraries, Virginia Polytechnic Institute and State University, OCLC:49698307

Klingelhöfer G, Morris RV, Bernhardt B, Schröder C, Rodionov DS Jr, de Souza PA, Yen A, Gellert R, Evlanov EN, Zubkov B, Foh J, Bonnes U, Kankeleit E, Gütlich P, Ming DW, Renz F, Wdowiak T, Squyres SW, Arvidson RE (2004) Jarosite and Hematite at Meridiani Planum from Opportunity's Mössbauer Spectrometer. Science 306(5702):1740–1745

Krijgsman W, Hilgen FJ, Raffi I, Sierro FJ, Wilson DS (1999) Chronology, causes and progression of the Messinian salinity crisis. Nature 400(6745):652–655

Kumar A, Tyagi MB, Srinivas G, Singh N, Kumar HD, Sinha RP, Haeder DP (1996) UVB shielding and the role of $FeCl_3$ and certain cyanobacterial pigments. Photochem Photobiol 63:321–325

Lammer H, Lichtenegger HC, Kolb IM, Ribas I, Guinan EF, Abart R, Bauer SJ (2003) Loss of water from Mars: implications for the oxidation of the soil. Icarus 165:9–25

Lane MD, Bishop JL, Dyar MD, Cloutis E, Forray FL, Hiroi T (2005) Integrated spectroscopic studies of anhydrous sulfate minerals. Lunar Planet Sci XXXVI:1442.pdf

Lane MD, Darby Dyar M, Bishop JL (2004) Spectroscopic evidence for hydrous iron sulfate in the Martian soil. Geophys Res Lett 31:L19702, doi:10.1029/2004GL021231

Langevin Y, Poulet F, Bibring JP, Gondet B (2005) Sulfates in the North polar region of Mars detected by OMEGA/Mars express. Science 307(5715):1584–1586

López-Archilla AI, Marín I, Amils R (2001) Microbial community composition and ecology of an acidic aquatic environment: the Tinto River, Spain. Microb Ecol 41:20–35

Madden MEE, Bodnar RJ, Rimstidt JD (2004) Jarosite as an indicator of water-limited chemical weathering on Mars. Nature 431:821–823

Malin MC, Edgett KS (2000) Evidence for recent groundwater seepage and surface runoff on Mars. Science 288:2330–2335

Margulis L, Walker JCG, Rambler M (1976) Reassessment of roles of oxygen and ultraviolet light in Precambrian evolution. Nature 264:620–624

Marion GM, Kargel JS, Catling DC (2006) Modeling ferrous/ferric iron chemistry with application to Martian surface geochemistry. Lunar Planet Sci XXXVII: 1898.pdf

Martin JM, Braga JC (1994) Messinian events in the Sorbas Basin in southeastern Spain and their implications in the recent history of the Mediterranean. Sediment Geol 90:257–268

Martinez-Frias J, García Guinea J, López Ruiz J, Reynolds GA (1992) Discovery of fossil fumaroles in Spain. Econ Geol 87:444–447

Martinez-Frias J (1998) An Ancient Ba-Sb-Ag-Fe-Hg-bearing hydrothermal system in SE Spain. Episodes 21(4):248–252

Martinez-Frias J (1999) Mining vs. geological heritage: the Cuevas del Almanzora natural area (SE Spain). AMBIO: J Hum Environ 28(2):204–207

Martinez-Frias J, Lunar R, Mangas J, Delgado A, Barragan G, Sanz-Rubio E, Diaz E, Benito R, Boyd T (2001a) Evaporitic and hydrothermal gypsum from SE iberia: geology, geochemistry, and implications for searching for life on Mars, Geological Society of America (GSA) Annual Meeting, Boston, Massachussets, USA

Martinez-Frias J, Lunar R, Rodriguez Losada JA (2001b) Hydrothermal mineralization in SE Spain as possible volcanics-related metallogenetic model for the early Mars, 2nd Astrobiology Minisymposium, CAB (CSIC/INTA), associated to NASA Astrobiology Institute, Madrid, Spain

Martinez-Frias J, Lunar R, Rodríguez-Losada JA, Delgado A (2004) The volcanism-related multistage hydrothermal system of El Jaroso (SE Spain): implications for the exploration of Mars. Earth Planets Space 56:v–viii

McKay CP, Stoker CR (1989) The early environment and it's evolution on Mars: implications for life. Rev Geophys 27:189–214

McKay CP (1993) Relevance of Antarctic microbial ecosystems to exobiology. In: Friedmann EI (ed) Antarctic microbiology. Wiley-Liss, New York, pp 593–601

McSween HY (2004) Mars mineralogy: the view from MER, November 10, 2004. Geological Society of America, Abstracts with Programs 36(5):556

Mitrofanov I, Anfimov D, Kozyrev A, Litvak M, Sanin A, Tret'yakov V, Krylov A, Shvetsov V, Boynton W, Shinohara C, Hamara D, Saunders RS (2002) Maps of subsurface hydrogen from the high energy neutron detector, Mars Odyssey. Science 297(5578):78–81

Moore JM, Bullock MA (1999) Experimental studies of Mars-analog brines. J Geophys Res 104(E9):21,925–21,934

Morris RV, Lauer HV Jr (1981) Stability of goethite (a-FeOOH) and lepidocrocite (g-FeOOH) to dehydration by UV radiation: implications for their occurrence on the Martian surface. J Geophys Res 86:10,893–10,899; J Geophys Res 84:8415–8426

Murad E, Rojík P (2003) Jarosite, schwertmannite, goethite, ferrihydrite and lepidocrocite: the legacy of coal and sulfide ore mining. SuperSoil 2004: 3rd Australian New Zealand Soils Conference, 5–9 December 2004. University of Sydney, Australia, pp 1–8

Noe Dobrea EZ, Bell JF, Wolff MJ, Gordon KD (2003) $H_2O$- and OH-bearing minerals in the Martian regolith: Analysis of 1997 observations from HST/NICMOS. Icarus 163(1):1–20

O'Connor VA (2005) Comparative crystal chemistry of hydrous iron sulfates from different terrestrial environments, Bachelor of Arts with Honors, Department of Geology, Smith College, p 150

Olsen JM, Pierson BK (1986) Photosynthesis 3.5 thousand million years ago. Photosynth Res 9:251–259

Orofino V, Blanco A, D'Elia M, De Carlo F, Fonti S, Marra AC, Marzo GA, Politi R, Verrienti C (2005) Spectroscopic studies of materials relevant for the hydrological evolution of Mars. Mem SAIt Suppl Vol 6:76–81

Parnell J, Lee P, Cockell CS, Osinski GR (2004) Microbial colonization in impact-generated hydrothermal sulphate deposits, Haughton impact structure, and implications for sulphates on Mars. Int J Astrobiol 3:247

Patel MR, Bérces A, Kolb C, Lammer H, Rettberg P, Zarnecki JC, Selsis F (2003) Seasonal and diurnal variations in Martian surface UV irradiation: biological and chemical implications for the Martian regolith. Int J Astrobiol 2:21–34

Patel MR, Berces A, Kerekgyarto T, Ronto Gy, Lammer H, Zarnecki JC (2004) Annual solar UV exposure and biological effective dose rates on the Martian surface. Adv Space Res 33:1247–1252

Pearson R, Roush TL, Ruff SW, Smith MD (2000) Detection of crystalline hematite mineralization on Mars by the Thermal Emission Spectrometer: evidence for near-surface water. J Geophys Res 105:9623–9642

Phoenix VR, Konhauser KO, Adams DG, Bottrell SH (2001) The role of biomineralization as an ultraviolet shield: implications for the Archean. Geology 29:823–826

Pierson BK, Mitchell HK, Ruff-Roberts AL (1993) Chloroflexus aurantiacusand ultraviolet radiation: implications for archean shallow-water stromatolites. Origins Life Evol Biosphere 23:243–260

Poulet F, Bibring JP, Mustard JF, Gendrin A, Mangold N, Langevin Y, Arvidson RE, Gondet B, Gomez C, the Omega Team (2005) Phyllosilicates on Mars and implications for early Martian climate. Nature 438:623–627

Quinn RC, Zent AP, Ehrenfruend P, Taylor CL, McKay CP, Garry JRC, Grunthaner FJ (2005) Dry acid deposition and accumulation on the surface of Mars and in Atacama Desert, Chile. Lunar Planet Sci XXXVI:2282.pdf

Riding R, Braga JC, Martin HM, Sánchez-Almazo IM (1998) Mediterranean Messinian salinity crisis: constraints from a coeval marginal basin, Sorbas, southeastern Spain. Mar Geol 146:1–20

Rontó Gy, Berces A, Lammer H, Cockell CS, Molina-Cuberos GC, Patel MR, Selsis F (2003) Solar UV irradiation conditions on the surface of Mars. Photochem Photobiol 77:34–40

Rull F, Martinez-Frias J, Medina J (2005) Surface mineral analysis from two possible Martian analogs (Rio Tinto and Jaroso Ravine, Spain) using micro-, macro-, and remote laser Raman spectroscopy, European Geosciences Union. Geophys Res Abstr 7:09114

Rull F, Martinez-Frias J, Medina SJ, González-Pastor E (2004) A comparative analysis of mineral sulphates from Rio Tinto and Jaroso Ravin (Spain) using XRD, FTIR and Raman pectroscopy 6th International Conference on Raman Spectroscopy Applied to the Earth and Planetary Sciences June 6–11, 2004, Honolulu, Hawai'i, USA

Rull F, Martinez-Frias J (2006) Raman spectroscopy goes to Mars. Spectrosc Eur 18(1):18–21

Sagan C, Pollack JB (1974) Differential transmission of sunlight on Mars: biological implications. Icarus 21:490–495

Sagan C, Mullen G (1972) Earth and Mars: evolution of atmospheres and surface temperatures. Science 177:52–56

Schofield PF, Knight KS, Stretton IC (1996) Thermal expansion of gypsum investigated by neutron powder diffraction. Am Mineral 81(7–8):847–851

Scott KM (2000) Nomenclature of the alunite supergroup: discussion. Can Mineral 38(5):1295–1297

Souza de PA Jr, Squyres S, Arvidson R, Klingelhöfer G, Morris RV, Schröder C, Rodionov D, Rieder R, Christensen P, Herkenhoff K, Gorevan S, McSween H, Cabrol N, Bernhard B, Gellert R, Calvin W, Renz F, the Athena Science Team (2004) Mars mineralogy: preliminary results from Mars exploration Rovers Spirit and opportunity applied mineralogy. Pecchio et al. (eds) 2004, ICAM-BR, São Paulo, ISBN 85-98656-01-1

Spencer RJ (2000) Sulfate minerals in evaporite deposits. Reviews in mineralogy and geochemistry: sulfate minerals, crystallography, geochemistry, and environmental significance. C. N. Alpers, Jambor, J.L., and Nordstrom, D.K., 40:173–192

Spencer RJ, Hardie LA (1990) Control of seawater composition by mixing of river waters and mid-ocean ridge hydrothermal brines. In: Spencer RJ, I-Ming Chou (eds) Fluid–mineral interactions: a tribute to H.P. Eugster. Geochem Soc Spec Pub 2:409–419

Squyres SW, Clifford SM, Kuzmin RO, Zimbelman JR, Costard FM (1992) Ice in the martian regolith. In: Kieffer HH, Jakosky BM, Snyder CW, Matthews MS (eds) Mars. Univ. of Arizona Press, Tucson, pp 523–554

Squyres SW, Grotzinger JP, Arvidson RE, Bell JF III, Calvin W, Christensen PR, Clark BC, Crisp JA, Farrand WH, Herkenhoff KE, Johnson JR, Klingelhöfer G, Knoll AH, McLennan SM, McSween HY, Morris RV, Rice JW Jr, Rieder R, Soderblom LA (2004) In situ evidence for an ancient aqueous environment at Meridiani Planum, Mars. Science 306(5702):1709–1714

Squyres SW, Knoll AH (2005) Sedimentary rocks at Meridiani Planum: origin, diagenesis, and implications for life on Mars. Earth Planet Sci Lett 240:1–10

Stoker CR, Stevens T, Amils R, Gómez-Elvira J, Rodríguez N, Gómez F, González-Toril E, Aguilera A, Fernández-Remolar D, Dunagan S, Lemke L, Zavaleta J, Sanz JL (2005) Characterization of a subsurface biosphere in a massive sulfide deposit at Rio Tinto, Spain: implications for extant life on Mars. Lunar Planet Sci XXXVI:1534.pdf

Stoker C, Lemke L, Mandell H, Mckay D, George J, Gomez-Elvira J, Amils R, Stevens T, Miller D (2003) Mars analog research and technology experiment (MARTE): a simulated Mars drilling mission to search for subsurface life at the Rio Tinto (Spain). Lunar Planet Sci XXXIV:1076.pdf

Stoker CR, Lemke LG, Cannon H, Glass B, Dunagan S, Zavaleta J, Miller D, Gomez-Elvira J (2006) The search for subsurface life on Mars: results from the MARTE analog drilling experiment in Rio Tinto, Spain. Lunar Planet Sci XXXVII:1537.pdf

Vaniman DT, Bish DL, Chipera SJ, Fialips CI, Carey JW, Feldman WC (2004) Magnesium sulphate salts and the history of water on Mars. Nature 431:663–665

Villar SE, Edwards HG, Cockell CS (2005) Raman spectroscopy of endoliths from Antarctic cold desert environments. Analyst 130(2):156–162

Wyatt MB, McSween HY (2002) Spectral evidence for weathered basalt as an alternative to andesite in the northern lowlands of Mars. Nature 417:263–266

Wynn-Williams DD, Edwards HGM (2000) Laser Raman microspectroscopy of surface microbial communities and protective biomolecules in situ: overview of terrestrial Antarctic habitats and Mars analogs. Icarus 144:486–503

Yen AS, Murray BC, Grunthaner FJ, Rossman GR (1997) Ultraviolet radiation-induced dehydration of minerals: new experiments Lunar Planet Sci XXVIII:1755.pdf

Yen AS, Gellert R, Schroder C, Morris RV, Bell JF III, Knudson AT, Clark BC, Ming DW, Crisp JA, Arvidson RE, Blaney D, Bruckner J, Christensen PR, DesMarais DJ, de Souza PA Jr, Economou TE, Ghosh A, Hahn BC, Herkenhoff KE, Haskin LA, Hurowitz JA, Joliff BL, Johnson JR, Klingelhofer G, Madsen MB, McLennan SM, McSween HY, Richter L, Rieder R, Rodionov D, Soderblom L, Squyres SW, Tosca NJ, Wang A, Wyatt M, Zipfel J (2005) An integrated view of the chemistry and mineralogy of martian soils. vol 436, 7 July 2005, doi:10.1038/nature03637

Zhu M, Xie H, Guan H, Smith RK (2006) Mineral and lithologic mapping of Martian low-albedo regions using OMEGA data. LPSCXXXVII:2173.pdf

Rev Environ Sci Biotechnol (2007) 6:231–259
DOI 10.1007/s11157-006-9117-9

REVIEW PAPER

# Industrial barrens: extreme habitats created by non-ferrous metallurgy

**M. V. Kozlov · E. L. Zvereva**

Received: 6 March 2006 / Accepted: 22 November 2006 / Published online: 21 December 2006


**Abstract** Industrial barrens are bleak open landscapes evolved due to deposition of airborne pollutants, with only small patches of vegetation surrounded by bare land. These extreme environments appeared as a by-product of human activities about a century ago. The comparative analysis of information available from 36 industrial barrens worldwide allowed to identify factors and conditions that are necessary and sufficient for the appearance of these specific habitats. Vast majority of industrial barrens is associated with non-ferrous smelters, located predominantly in mountainous or hilly landscapes. Development of industrial barrens starts from gradual decline of vegetation due to severe pollution impact accompanied by other human-induced disturbances (primarily clearcutting) and is usually concluded by a fire, facilitated by accumulation of woody debris. Since vegetation recovery is hampered by soil toxicity caused by extreme contamination by heavy metals, soils remain bare and suffer from erosion enhanced by altered microclimate. In spite of general reduction in biodiversity, industrial barrens still support a variety of life, including regionally rare and endangered species, as well as populations that evolved specific adaptations to the harsh and toxic environment. Recently, most industrial barrens show some signs of natural recovery due to emission decline or closure of responsible polluters; some of barren sites have been or are being successfully revegetated. The remaining industrial barrens offer unique opportunities for conducting 'basic' ecological research, in particular for testing some general theories in an evolutionary novel stressful environment; some of barren habitats deserve conservation for scientific and educational purposes.

**Keywords** Biodiversity · Clearcutting · Conservation · Contamination · Disturbance · Fire · Heavy metals · Microclimate · Pollution · Recovery · Soil erosion

M. V. Kozlov (✉) · E. L. Zvereva
Section of Ecology, Department of Biology, University of Turku, 20014 Turku, Finland
e-mail: mikoz@utu.fi

## 1 Introduction

Naturalists have always been intrigued by the ability of life to sustain conditions inhospitable to humans. Both scientific and popular literature contains numerous descriptions of biota living 'on the edge'—in deserts, on barren soil of polar islands, under Antarctic ice, in deep waters, and in many other more or less unusual conditions. However, all these habitats exist for millennia, and living beings had sufficient time to evolve biochemical, morphological and behavioural adaptations allowing to live and even flourish in these 'extreme environments'. More astonishing is the diversity of life persisting in industrial barrens—extreme habitats that appeared as a by-product of human activities only about a century ago.

The creation of metal-contaminated, phytotoxic land by atmospheric smelter effluents[1] has occurred in many parts of the world (Vangronsveld et al. 1995; Winterhalder 2000) and attracted agitated public attention: "Bushes shrank and vanished. Grasses died away. Blighted land replaced the forest. All around us dead hills, red, raw, ribbed by erosion, stood stark in the sunshine. Hardly two miles from dense woodland we were in the midst of a moonscape on earth. ... We were in the southeast corner of Tennessee, in the Ducktown Desert of the Copper Basin" (Teale 1951).

Unfortunately, industrial barrens are studied much less than other 'extreme habitats': the scientists were only called to evaluate the damage (e.g. Haywood 1910; Euler 1939; Dean and Swain 1944) or develop rehabilitation measures (e.g. Carter et al. 1977; Pommerening 1977; Winterhalder 2000). Importantly, information on several industrial barrens is reported only in publications describing reclamation measures (e.g. Pancholy et al. 1975; Sopper 1989; Vangronsveld et al. 1995, 1996; Kelley and Tate 1998; Winterhalder 2000). Even the researchers exploring pollution effects on plant communities tend to select the most polluted sites outside industrial barrens, because these habitats seem not comparable with less disturbed sites (e.g. Banásová et al. 1987). Therefore the first goal of our review is to introduce industrial barrens to a wide range of scientists, showing that these easy-to-reach habitats offer an excellent opportunity for a variety of 'basic' ecological and evolutionary studies. From a scientific perspective, unusual anthropogenic ecosystems of denuded barrenland with lifeless lakes can be considered opportunistic macrocosms (unique laboratories) for integrated research on the effects of harsh environmental conditions on ecosystem structure and functions (Nriagu et al. 1998).

Not every strong polluter is surrounded by industrial barrens. Their development seems only possible under specific combination of landscape characteristics, human activities, and co-occurring stressors; however, to our knowledge, no attempt was made to explore this problem by means of comparative analysis. The second goal of our review is therefore to find out both similarities and dissimilarities among the industrial barrens that exist (or did exist) across the Globe and on this basis identify the factors and conditions that are necessary and sufficient for environmental deterioration to reach its nearly final point (not the final point, because some life is always present in industrial barrens).

The barren landscape presents itself not only to ecologists interested in the complex effects of manmade disturbance on ecosystems, but also to those involved with the determination of public policy related to environmental pollution (Anand et al. 2003). It is only rarely appreciated that severely contaminated sites and other post-industrial landscapes may support regionally rare and endangered species (Johnson et al. 1978; Eyre and Luff 1995). Furthermore, industrial barrens clearly show the results of 'gross negligence' typical for early stages of industrial development, and their exploration can serve to ecological education of people. Therefore our third goal is to briefly describe the diversity of life in industrial barrens and call for conservation measures.

Last but not least, we aimed at identification of knowledge gaps that require immediate attention of international scientific community, because most of existing industrial barrens may disappear rather soon due to strict emission control and rehabilitation measures. On the other hand, the serious concern is the potential for developing nations to repeat the environmental mistakes made by the developed world (Padgett and Kee 2004). The knowledge on mechanisms behind the development of industrial barrens may help to prevent habitat deterioration in developing countries, especially in South-Eastern Asia where

[1] Dumps of solid waste, as well as areas that became barren due to mechanical removal of soil (e.g. in the course of open-pit mining) are not considered here.

industrial emissions are predicted to increase drastically (Fowler et al. 1999).

## 2 Terminology

To our knowledge, no attempt was made to unify the terminology associated with description of the habitats severely disturbed by aerial emissions; terminology used by different researchers is summarised in Table 1. McCall et al. (1995, p. 847) defined 'barrens' adjacent to Sudbury smelters as landscapes "characterised by bare and sparsely vegetated land, severely eroded and blackened hilltops and acidic (pH < 4.0) and metal contaminated soils". Later on, McCall et al. (1995, p. 849) introduced a 5% vegetation cover to delineate barrens by using aerial photographs; Eränen and Kozlov (2006) used 10% cover to define industrial barrens. However, these absolute values may appear misleading in landscapes with naturally low vegetation cover, like deserts.

In this paper we classify as industrial barrens the bleak open landscapes evolved around the point sources of industrial pollution due to deposition of airborne pollutants, with small patches of vegetation (cover usually reduced to 10% or less relative to control) surrounded by bare land with illuvial horizon or even the rock exposed due to intensive soil erosion (relict soil cover usually less than 20%) (Figs. 1–3).

Industrial barrens are usually surrounded by strongly modified ecosystems (Figs. 4–6) that have a potential to turn into industrial barrens under some circumstances (see below). In Sudbury, these habitats where "conifers are generally absent and the forest consists of a near monoculture of stunted and coppiced white birch (*Betula papyrifera*)" were defined as 'semi-barren' area (McCall et al. 1995, p. 847) or described as 'savannah woodland' (Courtin 1994). Similar habitats existing around Monchegorsk and Nikel were termed 'birch transitional community' (Kozlov 2001, 2002) or 'birch woodlands' (Kozlov et al. 2001); some Russian authors (e.g. Stepanov and Chernenkova 1989) used an expression 'forests with dead field layer vegetation' for secondary birch regrowth near the Karabash smelter (Fig. 6).

## 3 Documented occurrence of industrial barrens

### 3.1 Responsible industries

We were able to obtain some information on 36 industrial barrens worldwide (Table 1). Nearly all of them (33 of 36) have developed in the impact zones of non-ferrous smelters and refineries, primarily those of copper-, nickel-, zinc- and lead-producing factories. The notable exceptions are the iron sintering plant in Wawa, Canada and the magnesite plants in Satka (Russia) and Lubenik (Slovakia). Industrial barrens are frequently associated with the historical smelting sites: 12 of them occur around industries that started operations in 18–19th centuries (Table 1). The most recent installation that caused development of industrial barrens is Ventanas smelter in Quinteros, Chile (operated since 1964).

Intriguingly, industrial barrens have not been observed around aluminium factories, which emit (or have emitted in the past) large amounts of phytotoxic compounds, primarily fluorine. Only Gilbert (1975, p. 120) mentioned that "very small areas of exposed rocky ground were at the denuding stage with eroding soils sparsely colonised by annuals" around two of three aluminium smelters in Norway. Recently, even the most exposed sites near the Bratsk aluminium factory in Siberia are covered by damaged and dwarfed forest with dense field layer vegetation (pers. obs.). Similarly, environmental impact of power plants, even the largest ones that emit more sulphur dioxide than some of smelters surrounded by barren lands (Pearce 1994), is not strong enough to cause development of industrial barrens.

In general, industrial barrens represent a relatively rare phenomenon: they were recorded around approximately 10% of point polluters which impact on biota was scientifically documented (by 2003, we retrieved the published information on biotic effects of 233 polluters: Kozlov and Zvereva 2003).

**Fig. 1** Sudbury, non-reclaimed industrial barren (2001)

**Fig. 2** Nikel, industrial barren developed from Scots pine forest (2000)

**Fig. 3** Zapolyarnyy, industrial barren developed from birch woodland on a bog (2000)

**Fig. 4** Revda, erosion of bare soils (2002)

**Fig. 5** Krompachy, secondary forest on bare ground behind the smelter (2001)

**Fig. 6** Karabash, secondary birch forest on bare ground (2002)

### 3.2 Geographical distribution and vegetation zones

Nearly a half of identified 36 industrial barrens, including those at incipient stages of development (Table 1), occur in North America, with nine localities in USA and five in Canada. Russia houses 10 industrial barrens, three of which are located in the Kola Peninsula and six in Southern Ural region. The only industrial barrens in the Southern hemisphere are reported from Chile

**Table 1** Documented occurrence of industrial barrens and dying forests on a barren soil (alphabetical order, by location)

| Location: Site | Location: Country | Polluter (establishment-closure) | Primary vegetation | Terminology used to describe the most disturbed landscapes | First record of industrial barrens | Barren/semi-barren area, ha (year of estimation) | Contributing disturbances and secondary stressors | References |
|---|---|---|---|---|---|---|---|---|
| Anaconda | USA | Cu smelter (1884–1980) | Subalpine lodgepole pine and Douglas fir forest | Sites devoid of vegetation; nearly barren | Early 1900s | 4,500/- (1992) | Unfertile soil; extreme climatic fluctuations | Haywood (1910), Galbraith et al. (1995), Marty (2000), Burt et al. (2003) |
| Ashio | Japan | Cu smelter (1877–1988) | Beech and coniferous forest | Biologically destroyed area; devastated area; absolute wasteland | 1893 | 1,395/- (1893) >2,500/- (1970) | Logging; soil erosion | Usui and Suzuki (1973), Shoji and Sugai (1992) |
| Banská Štiavnica | Slovakia | Pb and Cu smelters (1750–1969) | Oak-hornbeam forest | Entirely destroyed plant cover; emission gaps; bare lands | 1880s | <10/- (1970s)/- | – | Kapusta and Múdry (1974), Banásová et al. (1987) |
| Ducktown (Copperhill) | USA | Cu smelter (1854–1912) | Hardwood forests | Barren, eroded landscape; totally bare area/lightly vegetated zone; desert; denuded area; moonscape; blighted land | 1870s(?) | 2,700/6,800 (1910) 13,000/- (1930s) 2,833/19,020 (early 1940s) 930/- (1986) | Logging; fuel procurement; grazing by cattle; fire; soil erosion | Seigworth (1943), Hursh (1948), Teale (1951), Quinn (1989) |
| Flin Flon | Canada | Cu–Zn smelter (1930-) | Mixed boreal forest | Barren soil; a barren tract of rocky hills, parched soil and sparse vegetation | 1970s? | – | – | Hogan and Wotton (1984), Winterhalder (2000, 2003), Lees (2000) |
| Garfield | USA | Cu smelter (1906-) | Coniferous forest and brush community | Extremely denuded areas; highly disturbed zone | Before 1970 | 1,500/- (1970) | Fire; soil erosion | Eastmond (1971) |
| Harjavalta | Finland | Cu–Ni smelter (1945-) | Southern boreal Scots pine forest | Understorey vegetation is almost completely absent; clonal dwarf shrubs have survived in small patches | 1970s? | <50/- (2000) | – | Salemaa et al. 2001; pers. obs. |

**Table 1** continued

| Location | | Polluter (establishment-closure) | Primary vegetation | Terminology used to describe the most disturbed landscapes | First record of industrial barrens | Barren/semi-barren area, ha (year of estimation) | Contributing disturbances and secondary stressors | References |
|---|---|---|---|---|---|---|---|---|
| Site | Country | | | | | | | |
| Henryetta | USA | Zn, Cd roaster/smelter (1916–1968) | Oak forest | Completely denuded area; bare area | 1970s | 400/- (1960s) | – | Pancholy et al. (1975) |
| Karabash | Russia | Cu smelter (1907-) | Southern taiga | Industrial barren, industrial desert | 1930s | <1,000/- (2002) | Soil erosion | Stepanov and Chernenkova (1989), Makunina (2002), E. Vorobeichik (pers. comm.) pers. obs. |
| Kellogg | USA | Pb smelter (1916–1980), Zn plant (1928–1980) | Cedar-hemlock forest | Barren hillsides; denuded area | 1970s | 7,285/- (1977) | Fire | Carter et al. (1977), Hansen and Mitchell (1978), Winterhalder (2000) |
| Kirovgrad | Russia | Cu smelter (1912-) | Southern taiga/ Scots pine forest | Complete destruction of ecosystems; industrial desert | 1930s? | <100/- (1994)/- | – | Vorobeichik et al. (1996), E. Vorobeichik (pers. comm.) |
| Krasno-uralsk | Russia | Cu smelter (1932-) | Southern taiga/ Scots pine forest | Industrial desert | Early 1950s | <300/- (1980s) | – | Makhnev et al. (1990), E. Vorobeichik (pers. comm.) |
| Krompachy | Slova-kia | Cu smelter (1843-) | Mixed forests (Scots pine, beech) | Total degradation; bare soils; emission gaps (*imisne holiny*, Slov.) | 1980s | <50/- (2002) | Soil erosion | Maňkovská (1984), Banásová and Lackovičová (2004), pers. obs. |
| Legnica | Poland | Cu smelter and refinery (1953-) | Arable land; remnants of oak – hornbeam or Scots pine forest | Waste land; bare ground with patchy ruderal vegetation | 1980s | 300–200/- (1990) | Soil erosion | Weber (2002), Lehmann and Rebele (2004) |
| Lubenik | Slovakia | Magnesite plant (1958-) | Broadleaved forest | Seriously damaged zone: no vegetation found; emission gaps | Mid-1970s | 200 (2006) | | Kautz et al. (2001), J. Kulfan and P. Zach (pers. comm.) |
| Maatheide/ Lommel | Belgium | Zn smelter (1904–1974) | Heath developed from agricultural landscape | Bare industrial area; desert-like area | Before 1940 | 450/- (1942) 98/- (1952) 135/- (1990) | – | Vangronsveld et al. (1995, 1996, pers. comm.) |
| Miami-Globe | USA | Two Cu smelters (1890–1924; 1915-) | Desert and desert grassland | | 1970s | <500/- (1976) | Overgrazing | Dawson and Nash (1980) |

**Table 1** continued

| Location | | Polluter (establishment-closure) | Primary vegetation | Terminology used to describe the most disturbed landscapes | First record of industrial barrens | Barren/semi-barren area, ha (year of estimation) | Contributing disturbances and secondary stressors | References |
|---|---|---|---|---|---|---|---|---|
| Site | Country | | | | | | | |
| Mednogorsk | Russia | Cu smelter (1939-) | Steppe | Industrial wasteland; gas-induced barren area | Prior 1980s | <500/- (1976) | Soil erosion | Shilova and Lukjanets (1989) |
| Monche-gorsk | Russia | Ni–Cu smelter (1939–1941, 1947-) | Boreal coniferous forest | Industrial desert; zone of total destruction of ecosystems; industrial barrens; industrially created wasteland | Early 1960s | 21,000/ 44,000 (1990s) | Logging; fire | Doncheva (1978), Kryuchkov (1993), Barcan and Kovnatsky (2002), V. Barcan (pers. comm.) |
| Murgul | Turkey | Cu smelter (1902- mid-1970s) | Coniferous and broadleaved forests | Area with almost no live plants | 1960s | <1,000/- (1967) | Soil erosion; forest pests | Acatay (1968) |
| Nikel | Russia | Ni–Cu smelter (1932-) | Boreal coniferous forest/birch woodland | Industrial barrens | 1960s | 13,600/- (1973) 30,900/- (1999) | Fire | Tømmervik et al. (2003, pers. comm.) |
| Norilsk | Russia | Ni smelter (1939-) | Sparse larch forests, shrubby tundra | Dead forests; lichen desert; de-vegetated zone | Mid-1950s | -/- (total 300,000) (1980s); -/- (total > 400,000) (1990s) | Logging; forest diseases; fire | Filipchuk and Kovalev (1990), Vlasova and Filipchuk (1990), Kharuk (2000) |
| Palmerton | USA | Zn smelter (1898–1980) | Oak-chestnut-white pine forest | Completely barren or sparsely vegetated area; a barren, devastated, biological desert | 1930s | 485/- (1970s) >800/- (1980s) | Logging; fire; soil erosion; desiccation | Jordan (1975), Sopper (1989) |
| Quinteros | Chile | Cu smelter (1964-) | Shrubby grassland | The barrens | Mid-1970s | 2,000/- (2005) | – | Ginocchio (2000, pers. comm.) |
| Queenstown | Aust-ralia | Cu smelter (1895–1969) | Forest of King Billy pine | Bare area; devoid of vegetation/substantially denuded; wasteland; deserted moonscape | By 1900 | 1,500/2,500 (1950s) | Logging; burning of the surface duff | Hodgson et al. (2000), Winterhalder (2000; Anonymous (2005a) |
| Redding | USA | Cu smelter (1860–1907) | Pine forest | All vegetation entirely dead | Early 1900s | 1,000–1,500/- (1900s) | – | Haywood (1905, 1910) |
| Revda | Russia | Cu smelter (1940-) | South taiga/ Scots pine forest | Gas and smoke induced wasteland; industrial barren; industrial desert | Early 1950s | <100/- (1990s) <300/- (2000s) | – | Menstchikov et al. (1997), E. Vorobeichik (pers. comm.) |

**Table 1** continued

| Location | | Polluter (establishment-closure) | Primary vegetation | Terminology used to describe the most disturbed landscapes | First record of industrial barrens | Barren/semi-barren area, ha (year of estimation) | Contributing disturbances and secondary stressors | References |
|---|---|---|---|---|---|---|---|---|
| Site | Country | | | | | | | |
| Satka | Russia | Magnesite plant (1900-) | Scots pine forest | Magnesite desert | 1957 | 2,200/- (1957) 4,000/- (1959) 4,400/- (1963) | Grazing; forest pests; soil erosion | Kulagin (1964), Sokolov (1996) |
| Sudbury | Canada | Ni–Cu smelter (1888-) | Mixed boreal coniferous forest | Barren land/semibarren woodland; micro-desert; totally denuded barrenlands; badly devastated area | 1920? | 19,565/ 83,796 (1970) | Logging; burning of the surface duff; fire, pests, soil erosion, frost | Hazlett et al. (1983), Allum and Dreisinger (1987), Courtin (1994), McCall et al. (1995, Winterhalder (2000), Anand et al. (2003) |
| Superior | USA | Cu smelter (1924–1971) | Shrubby desert | [Some components of vegetation] almost entirely absent | 1960s | – | Grazing | Wood and Nash (1976) |
| Szopienice | Poland | Zn smelter (1834-) | Scots pine forest | Industrial desert | Late 1960s | – | – | Wolak (1970) |
| Trail | Canada | Pb–Zn smelter (1896-) | Mixed forest | Drifting sand dunes | 1930s | – | Logging, fires | Dean and Swain (1944), Winterhalder (2000), Nielsen and Kovats (2004) |
| Wawa | Canada | Iron sintering plant (1939–1998) | Mixed boreal coniferous forest | Very severe damage; total kill/heavy kill areas; fume kill area | 1950s | 13,870/- (1960) 10,850/ 19,100 (1970) | Fire, soil erosion | Gordon and Gorham (1963), Linzon (1975), McGovern (1975), Anonymous (1999) |
| Yellowknife | Canada | Au smelter (1941–1999) | Open sub-arctic woodland | Drastic deterioration | 1970s | <800/- (1970s) | Fires | Hocking et al. (1978) |
| Ykspihlaja | Finland | Zn smelter (1962-), fertiliser and chemical plants (1940-) | Scots pine forest | Almost desert area; no forest floor vegetation; devegetated area | 1970s | 10/40 (1984) | – | Väisänen (1986), pers. obs |
| Zapolyarnyy | Russia | Ni–Cu ore roasting plant (1959-) | Birch woodland | Industrial barrens | Late 1960s | 20,400/- (1973) 37,800/- (1999) | – | Tømmervik et al. (2003, pers. comm.) |

and Tasmania. Industrial barrens have not been described from the tropics, and it can only be guessed whether this reflects geographical distribution of smelting activities, shortage of information or higher resistance of tropical ecosystems to extreme deposition of aerial pollutants.

Most of industrial barrens (27 of 36) occur in forested areas, where they generally replace coniferous or mixed forests. Barrens around Legnica, Poland, and Maatheide, Belgium, evolved from arable lands that were earlier covered by broadleaved forests. Similarly, smelter in Banská Štiavnica was earlier surrounded by oak-hornbeam forest. Industrial barrens near Zapolyarnyy, Norilsk and Yellowknife developed at the northern tree limit from sparse sub-tundra forests, birch woodlands and shrubby tundras. Barrens near Quinteros, Superior and in the Miami-Globe area are surrounded by shrubby grasslands and desert, while barrens at Mednogorsk developed from steppes.

### 3.3 Landscapes

Majority (27 of 36) of industrial barrens occur in the mountainous or hilly areas. However, since ore deposits are often associated with mountains, additional data are necessary to test the hypothesis that mountain landscapes are especially prone to severe disturbance caused by smelter fumes. Industrial barrens in relatively flat regions, like at Legnica in Poland, Harjavalta and Ykspihlaja in Finland, and at Maatheide in Belgium, seems less extensive, possibly due to less intensive soil erosion on planes.

Development of industrial barrens in mountain landscape can be enhanced by local meteorology. Fumes emitted by the lead and zinc smelter at Trail drifted north and south along the relatively narrow, mountain-bordered valley that constitutes a natural channel for the smelter gases (Archibold 1978; Quinn 1989). The prevailing north-westerly winds carry emissions of Palmerton smelter directly toward Blue Mountain, resulting in higher concentrations of pollutants than would be the case in level terrain (Jordan 1975). Similarly, predominance of meridional winds in surroundings of Monchegorsk caused development of extensive industrial barrens to both North and South of the nickel–copper smelter (Doncheva 1978; Kryuchkov 1993); at the same time, nearly undamaged Scots pine (*Pinus sylvestris* L.) forest still occurs some 5 km East of the smelter (pers. obs.).

Mountains can not only channel aerial emissions but also 'trap' the smoke, greatly enhancing vegetation exposure and the resulting damage. For example, the Copper Basin, comprising 3,300 ha, is surrounded by a rim of mountains (Wolt and Lietzke 1982). The Summer Valley near the Anaconda smelter "resembled a bowl, rimmed on three sides by the main range of the Rocky Mountains... During the winter months, and occasionally in the summer, an inversion often placed a lid over the valley and trapped the city's smoke..." (MacMillan 2000, p. 25). Inversion conditions occur frequently near the Kellogg smelter in the fall (Ragaini et al. 1977); similarly, local contamination by nickel–copper smelters at Nikel and Monchegorsk is exacerbated by frequent temperature inversions (Kryuchkov and Makarova 1989).

### 3.4 Topography

Topography has an important effect on spatial pattern of both vegetation decline and recovery (Easmond 1971; Usui and Suzuki 1973; Gilbert 1975). In Sudbury, the line of vegetation discontinuity usually followed elevation contour lines (McCall et al. 1995). Exposed hilltops were particularly prone to fumigation damage in the past (James and Courtin 1985), and are now very slow to recover (McCall et al. 1995). Hilltops in Sudbury remain bare for decades after emission decline, and their sparse soils are still subject to erosion (Dudka et al. 1995). In industrial barrens near Monchegorsk vegetation damage was lower, and vegetation cover and species richness were higher in the bottom of the valley, near the small stream, compared with slopes that were only some 50–80 m above (Kozlov 1997).

## 4 Development of industrial barrens

### 4.1 Effects of pollutants

The largest air pollution problems in industrial barrens are episodic, with high ambient concentrations lasting a few hours to two days (Sivertsen et al. 1994). Near the smelters in the Kola Peninsula these episodes in 1980s occurred during 3–5% of days in winter and 1–2% in summer (Baklanov and Sivertsen 1994). During episodes, hourly concentrations of $SO_2$ reached 1200 μg $m^{-3}$ near Monchegorsk (Baklanov and Rodyushkina 1993), 2500 μg $m^{-3}$ near Nikel (Sivertsen et al. 1994), and 14,800 μg $m^{-3}$ near Norilsk (Savchenko 1998). During the summer of 1994, monthly average concentrations of $SO_2$ in industrial barrens near Monchegorsk were 150–270 μg $m^{-3}$ (Zvereva and Kozlov 2005, and unpublished). These values should be regarded intolerable for local forests, because the proposed $SO_2$ critical levels estimated by different methods range 5–15 μg $m^{-3}$ as a growing season mean (Manninen and Huttunen 1997).

Although acute damage by sulphur dioxide, the main phytotoxic component of smelter fumes, definitely contributed to vegetation decline (e.g. Haywood 1910; Euler 1939; Makhnev et al. 1990), the importance of this damage for the development of industrial barrens remains unclear. Extreme levels of sulphur dioxide, especially near the roastbeds, killed adjacent forests during relatively short time (Hedgcock 1912; Dean and Swain 1944; Gordon and Gorham 1963; Kryuchkov 1993; Hutchinson and Symington 1997). However, absence of industrial barrens around power plants and aluminium smelters hints that development of this kind of landscape is impossible without severe soil contamination by heavy metals (see Sect. 5.1). In particular, near Anaconda smelter, percent bare ground positively correlated with both concentrations of hazardous substances in soil and soil phytotoxicity revealed by laboratory experiments, thus suggesting the leading role of soil contamination in loss of vegetation in the field (Galbraith et al. 1995).

### 4.2 Accompanying disturbances

Smelting industry was usually preceded by or accompanied with other human-induced disturbances, which share the responsibility for the development of industrial barrens. In many historical smelter sites (e.g. Kellogg, Sudbury, Copper Basin, Queenstown, Ashio) deforestation was primarily due to harvesting for mine timber and fuel (Usui and Suzuki 1973; Hansen and Mitchell 1978; Winterhalder 2000). Shortage of fuel for smelters in the Copper Basin was noticed as early as in 1861; by 1878, about 130 $km^2$ had been stripped of vegetation (Anonymous 2005b). Some 3,262,000 $m^3$ of wood were consumed for roasting in Sudbury between 1890 and 1930 (Allum and Dreisinger 1987). Similarly, over 3 million tonnes of timber were cut down around Queenstown between 1896 and 1923 (Anonymous 2005a).

When commercial forests started to die due to pollution impact, the very first reaction of the foresters was to cut down the damaged stands in order to prevent losses of timber. This practice was widely applied in Central and Northern Europe (e.g. Gilbert 1975) until at least the mid-1970s, when it became obvious that the cost of harvested timber is minor in relation to the costs of rehabilitation measures required after clear-cutting. In spite of this knowledge, some local regulations (existing, for example, in Russia) still require immediate felling of pollution-damaged forest in order to make use of the timber (Kozlov 2004). Alternatively, dead trees can be removed to make the landscape more 'attractive' visually; in spite of advice of the local scientists, this selective logging was conducted in spring of 2006 near the Monchegorsk smelter (pers. obs.). However, the polluted forest never become completely dead, and even the dead trees maintain some climatic and biotic stability in the contaminated habitats, in particular by ameliorating microclimate (Wołk 1977) and preventing soil erosion. The old clearcuts under severe pollution impact near both Monchegorsk and Nikel have been rapidly transformed to industrial barrens, while some vegetation in the adjacent uncut areas is still alive (Kozlov 2004).

Industries build in 1920–1940s did not require so much of timber as a fuel. Still forests were cut for building purposes, e.g. around Monchegorsk. Initial deforestation around Norilsk, Siberia, was caused by quite peculiar reasons: since prisoners were used extensively for the construction of the Norilsk industry, a buffer of 3–4 km in width was cut around each of prison camps (Kharuk 2000). Importantly, logging causes soil disturbance that may facilitate erosion.

Forests with unusually open canopies, like one observed near Palmerton smelter in 1970s (Jordan 1975), or forest with stunted trees and dead or nearly dead ground layer vegetation, recently existing e.g. near Krompachy, Harjavalta, Ykspihlaja, Revda and Karabash smelters (pers. obs.), seem to represent a transitional stage between forests and industrial barrens. For example, field layer vegetation cover near Harjavalta is ca. 5%, while cover of Scots pine remains as high as 44% (Salemaa et al. 2001). This kind of forest may gradually turn to industrial barrens; however, it is likely that most, if not all, industrial barrens evolved following a fire that destroyed remnants of forest or shrubby vegetation. In semi-barren and barren sites near both Monchegorsk and Nikel we have observed several highly localised fires that consumed woody debris and destroyed most of remaining vegetation, leading to the extension of the barren area. We suppose that dying forests around the smelters listed above will immediately turn into industrial barrens following a fire, because a natural post-fire regeneration is hampered by soil toxicity and altered microclimate (Jordan 1975; Hansen and Mitchell 1978; Vajda and Venäläinen 2005). The detrimental role of occasional forest fires was considered the primary reason of forest deterioration around Bell Bay aluminium smelter in Australia, mostly due to disruption of nutrient cycles and alteration of soil moisture regime (Mitchell 1982).

Both logging and forest damage by fumes from smelting and roasting increased the amount of woody debris in the impacted areas, making them especially vulnerable to occasional fires. Unusually large amount of woody debris is clearly seen on photographs taken in the vicinity of several smelters (Figs. 1–3). In Sudbury, not only sparks from wood-burning locomotives of the Canadian Pacific Railway started the fires, but also prospectors often burned the remaining vegetation and duff to reveal the bedrock below (Winterhalder et al. 2001). In Palmerton, hillside vegetation behind zinc plant was originally destroyed by fire (Pommerening 1977). In the Copper Basin, in times of dry weather, almost daily fires swept the earth bare of vegetation (Teale 1951). The degradation of the Copper Basin landscape may have been further exacerbated by lowland farmers, who trucked cattle into the Basin for free grazing, and regularly burned the land to encourage growth of a sparse pasture (Clay 1983).

### 4.3 Hampering of natural recovery

All catastrophic events that destroy vegetation, like fires or avalanches, are always followed by natural recovery that, sooner or later, results in restoration of about the same vegetation community. The extant plants growing in industrial barrens may sustain extreme pollution loads and produce viable seeds, sometimes even in larger amounts than in unpolluted sites (Zvereva and Kozlov 2001, 2005; Kozlov and Zvereva 2004). These seeds preserved in the soil seedbank remain viable for decades (Komulainen et al. 1994); still natural regeneration is absent or nearly absent in industrial barrens (Jordan 1975; Kozlov and Haukioja 1999; Rigina and Kozlov 2000). This is most likely due to high concentrations of heavy metals in uppermost soil layers (Table 2) that stunted radicle growth of many plant species (Stavrova 1990; Zeid 2001). Although seeds of several native plant species were capable of germinating in soils of industrial barrens at both Sudbury and Monchegorsk, root growth was so inhibited that seedlings quickly dried off and died completely (Winterhalder et al. 2001; Kozlov 2005). Persistence of some plants in industrial barrens may be transient, being explained not only by higher resistance of the survivors, but also by past phenotypic acclimatisation of mature plants to gradual increase in pollution (Kozlov 2005), and the extent of industrial barrens increases as plants age and die (Zverev and Kozlov, unpublished).

Additional problems for vegetation recovery may be imposed by slow decomposition of litter. This was in particular noticed around the Palmerton smelter, where a considerable portion of the area is covered with a layer of undecomposed tree bark (Sopper 1989) or by thick (6–16 cm) litter (Jordan 1975). Similarly, leaf litter is the major problem of semi-barren communities in Sudbury, because it hinders the establishment of understorey species by seeds (Winterhalder 2000).

**Table 2** Chemistry of humus layer or uppermost soil horizon in industrial barrens

| Location | Extreme pH | Maximum concentrations of contaminants, μg $g^{-1}$ | | | | | | Minimum level of nutrients, mg $g^{-1}$ | | References |
|---|---|---|---|---|---|---|---|---|---|---|
| | | Cu | Cu[a] | Ni | As | Zn | Pb | N | Ca | |
| Anaconda | 3.0 | 2,800 | – | 14 | 408 | 849 | 474 | 2.30 | – | Swain and Harkins (1908), Galbraith et al. (1995), Marty (2000), Redenteet al. (2002), Burt et al. (2003) |
| Ashio | 4.3 | – | – | – | – | – | – | 0.37 | 0.17 | Usui and Suzuki (1973) |
| Banská Štiavnica | 4.5 | 770 | – | – | 130 | 4,000 | 18,000 | 3.22 | 0.56 | Banásová et al. (1987) |
| Ducktown | 3.2 | – | – | – | – | – | – | – | 0.02 | Wolt and Lietzke (1982) |
| Flin Flon | 4.4 | 2,670 | – | 130 | 558 | 7,428 | 1692 | – | – | Henderson et al. (1998) |
| Garfield | 3.7 | – | – | – | – | – | – | – | – | Eastmond (1971) |
| Harjavalta | 3.5 | 49,000 | 7,540 | 913 | 58 | 620 | 204 | – | 1.17 | Helmisaari et al. (1995), Derome and Nieminen (1998), Uhlig et al. (2001), Salemaa et al. (2001), Nieminen et al. (2002), Nieminen (2004) |
| Henryetta | 5.0 | – | 15 | – | – | 19,510 | 2,450 | 0.23 | – | Pancholy et al. (1975), Basta et al. (2001) |
| Karabash | – | 6,744 | – | >10,000 | 1,500 | 10,000 | 3,000 | – | – | Chernenkova (2002), Makunina (2002) |
| Kellogg | 3.9 | >400 | – | >2,000 | 260 | 29,000 | 7,900 | 0.50 | – | Ragaini et al. (1977), Hansen and Mitchell (1978) |
| Kirovgrad | 4.1 | 3,758 | 4,040 | – | 162 | 3,689 | 674 | – | – | Marin (1996), Vorobeichik (2003, 2004) |
| Krasnouralsk | 2.2 | 1,611 | – | 3,216 | – | 1,696 | 1,725 | – | – | Menshchikov et al. (1997) |
| Krompachy | 4.0 | 8,437 | – | 25 | 130 | 1,482 | 3,343 | – | – | Maňkovská (1984), Wilcke et al. (1999) Banásová and Lackovičová (2004) |
| Legnica | 4.4 | 15,443 | 33 | 33 | – | 5,000 | 7,000 | 0.10 | – | Rebele et al. (1993), Rybicka and Jędrzejczyk (1995), Weber (2002), Lehmann and Rebele (2004) |
| Lubenik | 9.2 | 10 | – | 30 | – | 70 | 38 | 0.50 | 0.13 | Šály and Mihálik (1985), Kautz et al. (2001), Tucekova (2001) |
| Maatheide | 5.6 | 1,650 | – | – | – | 11,425 | 1800 | – | – | Vangronsveld et al. (1995, 1996) |
| Mednogorsk | 3.3 | 5,000 | – | – | – | 4200 | 1250 | – | – | Lurie (1986), Shilova et al. (1984), Shilova and Lukjanets (1989) |
| Miami-Globe | 4.0 | 2,183 | 2,250 | – | – | 67 | 90 | – | – | Dawson and Nash (1980) |
| Monchegorsk | 3.9 | 4,622 | 1,268 | 9,288 | 41 | 210 | 88 | 0.15 | 0.02 | Lukina and Nikonov (1996, 1999); Barcan and Kovnatsky (1998), V. Barcan (pers. comm.) |
| Murgul | – | – | – | – | – | – | – | – | – | <No data found> |
| Nikel | 3.7 | 3,489 | 113 | 2,990 | – | 113 | | – | 0.02 | Lukina and Nikonov (1996) |
| Norilsk | 4.4 | 20,600 | – | 7147 | – | | | – | – | Igamberdiev et al. (1994), Kharuk et al. (1996), Gorshkov (1997) |

**Table 2** continued

| Location | Extreme pH | Maximum concentrations of contaminants, μg $g^{-1}$ | | | | | | Minimum level of nutrients, mg $g^{-1}$ | | References |
|---|---|---|---|---|---|---|---|---|---|---|
| | | Cu | Cu[a] | Ni | As | Zn | Pb | N | Ca | |
| Palmerton | 4.5 | 2,390 | 0.69 | 33 | – | 135,000[a] | 5,225 | – | – | Buchauer (1973), Beyer et al. (1985), Sopper (1989), Kelly and Tate (1998) |
| Quinteros | 6.2 | 3,718 | 110 | – | – | 174 | 105 | 0.50 | – | Ginocchio (2000), Ginocchio et al. (2004) |
| Queenstown | – | – | – | – | – | – | – | – | – | <No data found> |
| Redding | – | – | – | – | – | – | – | – | – | <No data found> |
| Revda | 2.9 | 9,585 | 12,120 | – | – | 4,194[a] | 2,348[a] | – | – | Vorobeichik (2003, pers. comm.) |
| Satka | 9.0 | – | – | – | – | – | – | – | – | Sokolov (1996) |
| Sudbury | 2.0 | 9,700 | 900 | 12,300 | – | 336 | 92 | 0.99 | 0.60 | Hutchinson and Whitby (1974), Hazlett et al. (1983), Dudka et al. (1995), Hutchinson and Symington (1997), Winterhalder (2000), Anand et al. (2003) |
| Superior | 5.0 | 9,618 | – | – | – | 299 | 323 | – | – | Wood and Nash (1976) |
| Szopienice | 6.3 | 247 | – | – | – | 24,400 | 3,500 | – | – | Badora et al. (1998) |
| Trail | 3.7 | 106 | – | 67 | 157 | 1,632 | 7,490 | 0.30 | – | Goodarzi et al. (2002), Nielsen and Kovats (2004) |
| Wawa | – | 110 | – | 62 | 935 | 165 | 120 | – | 1.25 | Anonymous (1999) |
| Ykspihlaja | – | – | – | – | – | – | – | – | – | <No data found> |
| Yellowknife | 4.0 | 130 | – | – | 21,213 | 500 | 110 | – | – | Hocking et al. (1978), Hutchinson et al. (1982), Anonymous (2003) |
| Zapolyarnyy | 3.2 | 1,020 | 186 | 2,230 | – | – | – | – | 0.01 | Niskavaara et al. (1996), Reimann et al. (1998), T. Gorbacheva (pers. comm.) |

[a] Plant available forms

[b] In isolated patches of decomposed leaf litter; uppermost soil layer contained 50,000–80,000 μg $g^{-1}$ (Buchauer 1973)

### 4.4 Soil erosion

The soil of developing industrial barrens, having lost its protective vegetation cover, suffers from extensive erosion (Figs. 4, 5). Sparsely vegetated sites near Anaconda smelter are mostly devoid of topsoil (Marty 2000); soil erosion is especially severe on steep slopes (Redente and Richards 1997). The topsoil, and sometimes even the subsoil, have been lost in Ducktown (Seigworth 1943), Queenstown (Anonymous 2005a), Monchegorsk (Kryuchkov 1993), Nikel (Fig. 2), Zapolyarnyy (Fig. 3), Palmerton (Jordan 1975), Karabash (Stepanov and Chernenkova 1989), Ashio (Usui and Suzuki 1973) and several other sites. Even after emission decline, soil is still exposed to water erosion by runoff, summer desiccation, wind erosion and frost heaving (Hazlett et al. 1983; Dudka et al. 1995; McCall et al. 1995; Vangronsveld et al. 1995).

In contrast to dying forests with large amount of litter, erosion of bare ground is exacerbated by the intense frost-heaving and needle ice formation that resulted once the insulating leaf litter was gone (Winterhalder et al. 2001). As the result, soils of industrial barrens around Sudbury (Fig. 1), Monchegorsk, Nikel (Fig. 2), Palmerton, Karabash, and some other polluters have a stony covering (Sopper 1989; Winterhalder 2000; pers. obs.). These eroded lands are unlikely to revegetate naturally (Jordan 1975), and transformation of forests into industrial barrens has sometimes been claimed to be irreversible (Tsvetkov 1991), at least on the time scale of the human life span. In particular, the natural revegetation had not occurred in industrial barrens adjacent to Palmerton smelter for at least 50 years (Sopper 1989). Heavily contaminated sites at Maatheide in Belgium remained bare some 20 years following the closure of zinc smelter (Vangronsveld et al. 1995). Similarly, no regrowth appeared until 2006 in industrial barrens surrounding the Monchegorsk smelter (pers. obs.), although emissions of both sulphur dioxide and heavy metals drastically declined in early 1990s.

### 4.5 Positive feedbacks

In many cases initial (partly pollution-induced) forest disturbance causes secondary effects (like increased snow evaporation, followed by soil freezing and plant damage) that may enhance further disturbance in a positive feedback fashion (Kozlov 2001, 2002). Forest decline results in higher wind speed (see Sect. 5.3) that may enhance environmental stress via changes in snowpack structure. A thin and compact snow layer explains the lower soil temperatures recorded in industrial barrens during the wintertime (Kozlov and Haukioja 1997) that in combination with a pollution-induced decrease in cold-hardiness of conifers (Sutinen et al. 1996) increases the probability of death of extant trees from freezing injury. Importantly, climatic effects of deforestation may hamper recovery of vegetation even in absence of pollution (Arsenault and Payette 1997; Vajda and Venäläinen 2005).

Extant plants in industrial barrens facilitate deterioration of soil quality in their own root-inhabited areas: plant foliage traps contaminants, which then enter the soil (with either rainfall or plant litter) immediately under a plant. As the result, copper concentration in organic soil under *Empetrum nigrum* L. patches near Harjavalta smelter (49,000 μg $g^{-1}$) was much higher than in surrounding barren soils (Uhlig et al. 2001). Similarly, in industrial barrens near Monchegorsk mean concentrations of plant-available nickel and copper in topsoil were higher under dwarf shrub patches than in bare areas (Zvereva and Kozlov 2005, and unpublished), and concentrations of several pollutants were higher under extant trees than in between-tree gaps (Lukina and Nikonov 1996).

Loss of biodiversity and creation of monoculture on semi-barren areas, such as birch woodlands near Sudbury and Monchegorsk, can increase population densities of pests and pathogens (see Sects. 6.1–6.3) which damage or even kill the remaining vegetation, thus contributing to the expansion of industrial barrens. In particular, bronze birch borer (*Agrilus anxius* Gory) killed 60–90% of birch stems near Sudbury by 1990s (Courtin 1994). Death of chestnut from diseases was reported in the impact zone of the Palmerton smelter (Jordan 1975), although it remains unclear whether this disease was facilitated by forest damage by smelter fumes. Forest pests were reported to accelerate dieback of the remaining trees near the polluters at Satka (Kulagin 1964) and Murgul (Acatay 1968).

Due to these positive feedbacks, industrial barrens may be to a certain extent resilient to external impacts, including both emission decline and restoration efforts. By using Monchegorsk smelter as an example, Tarko et al. (1995) predicted extension of the zone of biotic damage during several years following complete ceasing of emissions. In agreement with this prediction, gradual decline of mountain birch populations in industrial barrens surrounding the Monchegorsk smelter continued at least until 2006, some 10–15 year after drastic emission decline (pers. obs.). This kind of resilience is best described by the 'alternative state' models of ecosystems that consider abrupt shifts between two or more ecosystem states and incorporate system thresholds and feedbacks (Suding et al. 2004).

### 4.6 Time frame

Development of industrial barrens may proceed rather fast: although the copper smelter in Queenstown was built in 1895 only, the combination of timber felling, the sulphur fumes and the heavy rainfall in the area (which washed away the top soil) ensured that already by 1900 the whole valley around Queenstown looked like a desert (Anonymous 2005a). Appearance of industrial barrens near Quinteros was first observed some 10 years after the Ventanas smelter started its operations (R. Ginocchio, pers. comm.) But in most of the documented situations, development of industrial barrens takes 20–35 years; this time span was reported for Sudbury, Trail, Monchegorsk, Ducktown, Anaconda and Palmerton smelters (Table 1).

## 5 Environmental conditions in industrial barrens

### 5.1 Soil toxicity

Soils of industrial barrens contain huge amounts of toxic pollutants deposited from aerial emissions. In all documented cases (Table 2), the concentration of at least one pollutant in decomposed litter or in uppermost soil layer exceeded 1,000 μg $g^{-1}$ (i.e. 1 g of pollutant per 1 kg of soil), or at least approached this value (Wawa). The highest concentrations of pollutants exceeded 10,000 μg $g^{-1}$, with an absolute maximum of 135,000 μg $g^{-1}$ of zinc near the Palmerton smelter (Buchauer 1973). These concentrations are much higher than the toxicity limits; however, the large (but highly variable—Table 2) fraction of metals is deposited in insoluble forms, such as oxides, and is therefore not readily available for plants and animals (Kozlov et al. 2000a). In spite of that, total rather than bio-available levels of pollutants are reported in most of studies (Table 2), although soil toxicity can not be judged from these total levels which therefore are only of limited value for ecotoxicology.

Since majority of heavy metals accumulated in soils are in non-soluble forms, their complete leaching from upper soil horizons will take centuries, e.g. 160–270 years for nickel and 100–200 years for copper accumulated in industrial barrens around Monchegorsk (Barcan 2002). It is estimated that mobilisation of metals stored in soils can sustain the high concentrations of copper and nickel in many lakes of the Sudbury basin for well over 1000 years (Nriagu et al. 1998).

Although soil acidification is often mentioned among principal reasons of vegetation damage or even decline, soils of industrial barrens are not always more acidic than soils of surrounding landscapes (Table 2). In some situations, the pattern may be even opposite: soils near Palmerton smelter are less acidic than in the background region, perhaps because of zinc oxide deposition (Jordan 1975). Similarly, soil pH near Vantanas smelter was 6.2 compared with 4.8 in background sites (Ginocchio 2000). Finally, soils around the magnesite plant at Satka are extremely alkaline (pH = 9), and development of industrial barrens was due to formation of thick (up to 10 cm!) cement crust preventing most of plants to grow (Kulagin 1964).

### 5.2 Soil nutritional quality

Although it is widely accepted that soil acidification promotes nutrient leaching from upper soil horizons, the data on soil nutritional quality are available only for some industrial barrens

(Table 2). In particular, exchangeable Ca, Mg and K near Harjavalta smelter are reduced by a factor of 3–5 compared to more distant sites (Mälkönen et al. 1999). Nitrogen near Monchegorsk is reduced by a factor of 10 or more (Lukina and Nikonov 1999). The loss of base cations from the organic layer close to the smelter is primarily due to displacement by copper and nickel (Derome and Lindroos 1998). Since fertilisation generally improved plant performance in industrial barrens (Winterhalder 2000), it is believed that nutritional deficiency is one of the factors adversely affecting plant life in these habitats. Recently, soil nutritional quality was demonstrated to be the leading factor influencing abundance and diversity of microorganism populations in Sudbury pollution gradient (Anand et al. 2003).

### 5.3 Microclimate

Disappearance of vegetation, especially of trees, strongly modifies the climate of industrial barrens. Although this problem is investigated insufficiently, it seems that the most important changes are imposed by altered temperature and wind regime. Even at early stages of pollution-induced forest deterioration air and soil temperatures during the growth season may substantially increase, leading to an increased water loss from upper soil layers (Wołk 1977). Air temperatures in industrial barrens near Ducktown in summer were 1–2° C higher and in winter 0.3–1°C lower than in surrounding undisturbed (forested) sites (Hepting 1971). Similarly, air temperatures in barren sites at Monchegorsk were 1–2° C lower in cool periods (under +7° C) and 1–2° C higher in warm periods compared to undisturbed forests some 20–30 km apart (Kozlov and Haukioja 1997). Daily fluctuations in temperatures of barren soils are much higher than in undisturbed sites where the heat exchange is buffered by vegetation (Hursch 1948; Wołk 1977; Kozlov and Haukioja 1997; Winterhalder 2000). The lack of leaf litter results in enhanced frost action, imposing additional stress on plants (Sahi 1983). Barren soils are overheated in summer (e.g. average soil temperature in barren sites at Ducktown were increased by 11°C; Hepting 1971), and this overheating is accompanied by fast loss of soil moisture (Hursch 1948; Teale 1951; Freedman and Hutchinson 1980; Courtin 1994; Kozlov and Haukioja 1997; Marty 2000; Winterhalder 2000); all these factors obviously influence plant performance. Lower amount of precipitation was reported from Ducktown (Hepting 1971) but not found at Monchegorsk (pers. obs.). Importantly, drought may enhance toxicity of some contaminants, e.g. zinc (Jordan 1975).

Wind speed, measured in industrial barrens adjacent to the copper smelter at Ducktown, was increased by a factor 5–15 (Hepting 1971), while around the nickel–copper smelter at Monchegorsk it was two to three times as high as in nearly unpolluted forests (Kozlov 2002). High winds and subsequent particle movement that can sand blast, bury, and defoliate plants, were reported to impose additional stress on heavily contaminated sites near both Anaconda and Monchegorsk smelters (Marty 2000; Kozlov 2001).

Along with the direct impact of wind on plant performance, higher wind speed is responsible for thin snow layer (about one-third of that in unpolluted forests) in industrial barrens around the Monchegorsk smelter (Kozlov 2001) which, in turn, results in lower soil temperatures during the winter time (Kozlov and Haukioja 1997). Both these factors increase exposure of plants to frost damage. Soil freezing (in autumn) occurred in industrial barrens near Monchegorsk 10–11 weeks earlier and soil thawing (in spring) 3–4 weeks earlier than in unpolluted forests (Kozlov and Haukioja 1997). Furthermore, a lower accumulation of snow in industrial barrens, in combination with higher wind speed, may expose plants to additional drought stress. However, snow cover in industrial barrens near Nikel has about the same depth as in unpolluted forests (Ratkin 1999), and therefore the generality of the effects discovered near Monchegorsk remains unknown.

## 6 Life in industrial barrens

### 6.1 Biodiversity

General loss of biodiversity with the replacement of undisturbed habitats by industrial barrens seems indisputable (e.g. Jordan 1975; Kleinert 1988; Courtin 1994; Koponen and Niemelä 1995;

Cicák et al. 1999); however, the magnitude of this effect had only rarely been documented. In the 'total kill' area around Wawa the ground flora declined to 0 to 1 species per 40 $m^2$ from about 20–40 species in unpolluted sites (Gordon and Gorham 1963). Similarly, grass communities at Garfield contained 1–5 species per 50 $m^2$ plot, compared to 7–22 species at most distant sites (Eastmond 1971). Near the Anaconda smelter, plant species richness in 1992–1994 was about one-third of control (Marty 2000). The number of plant species (observed in five 1-$m^2$ plots) on 10–50 m from the edge of the O'Donnell roast bed (Sudbury, Canada) 66 years after its closure ranged 5–7, compared to 25–31 species at the distances 150–300 m of it (Hutchinson and Symington 1997). In Miami, Arizona, grasses, forbs, and small shrubs were entirely absent at the site proximate, to the coppee smelter, whereas the cover of large shrubs had not been affected (Dawson and Nash 1980). The total floristic diversity near the Quinteros smelter was 17 species compared to 42 species in control sites (Ginocchio 2000).

In industrial barrens near Monchegorsk the number of vascular plant species ranged 0–25% of the number recorded in unpolluted (control) sites if censuses were made by using small plots (1–25 $m^2$). However, if larger plots (100–10,000 $m^2$) were surveyed, the decrease in species richness was around one-third, suggesting that pollution effects were expressed in decline of population densities rather than in selective removal of certain species (Kozlov et al. 1998, and unpublished).

Samples of moths and butterflies, as well as human-biting flies, collected in industrial barrens near Monchegorsk showed nearly the same diversity as samples from primary forests, although abundance of most species was drastically declined (Kozlov 1997; Kozlov et al. 2005a). Maximum species richness of ants in the Monchegorsk pollution gradient in 1993 was discovered in industrial barrens, whereas in 1994 it peaked in birch transitional communities surrounding the barren area (Kozlov 1997). This result is in line with findings by Koponen and Niemelä (1995) who reported higher species richness of ants in a barren site near the Harjavalta smelter than at more distant and less disturbed sites. Species richness of ground beetles (Carabidae) near the Lubenik magnesite plant was reduced to approximately one-third of observed in an unpolluted site (Kleinert 1988); however, rarefaction analysis shows that this difference was mostly due to differences in abundance, and actual species loss was much smaller, about 30% (Kozlov, unpubl.).

Some of insect species clearly prefer barrens to undisturbed habitats. In course of bait-trapping along the Monchegorsk pollution gradient, all the specimens of *Polia conspicua* (B.-H.), a mountain tundra noctuid moth, were collected between the external border of the industrial barrens and the smelter (Kozlov et al. 1996). Moreover, viable population of an extremely rare moth, *Sesia bembeciformis* (Hb.), the species considered extinct in Finland (Rassi et al. 1985), was discovered in an industrial barren south of Monchegorsk (Kozlov 1997). The buprestid beetle *Melanopila formaneki* (Jakobson) was for the first time discovered in Finland when studying Scots pines in a semi-barren site near the Harjavalta smelter (Heliövaara et al. 1990); the same site is also inhabited by several spiders which are missed from unpolluted forests (Koponen and Niemelä 1993). Industrial barrens may favour some life forms at the expense of others: for example, barrens near Lubenik magnesite plant were dominated by smaller ground beetles (Carabidae) with diurnal activity, whereas larger species with nocturnal activity were mostly associated with unpolluted habitats (Kleinert 1988).

In industrial barrens near Monchegorsk species richness of birds was reduced to approximately one third relative to unpolluted sites (Gilyazov 1993; Kozlov et al. 2005b). Near the magnesite factory in Lubenik the reduction in both species richness and abundance of breeding species seem even more pronounced (Cicák et al. 1999), although direct comparison is difficult due to different size of study plots. Among small mammals, only large-toothed redback vole (*Clethrionomus rufocanus* Sund.) and root vole (*Microtus oeconomus* Pall.) were captured near Monchegorsk whereas six species were recorded in undisturbed forests (Kozlov et al. 2005b). Winter censusing of large mammals in industrial barrens revealed tracks of two species, mountain hare (*Lepus timidus* L.) and red fox (*Vulpes vulpes* L.), compared with eight species in unpolluted areas (Kozlov et al. 2005b).

To conclude, species richness of plants and animals in industrial barrens in an average comprises one-third to one-half of the observed in surrounding undisturbed habitats. However, diversity of some groups is not affected, and some species occurring in industrial barrens are not found in undisturbed habitats.

## 6.2 Population density and structure

Microbiota of degraded soils had been shown to lose its resilience to disturbance and become no longer able to perform normal processes of cycling nutrients, assimilating organic residues and maintaining soil structure. As the result, in industrial barrens near Harjavalta accumulated mass loss of Scots pine needle and fine roots was ca. 80% of observed in unpolluted site (McEnroe and Helmisaari 2001), whereas near Sudbury decomposition of birch leaves was reduced to ca. 40% of the control level (Johnson and Hale 2004). This retarded decomposition results in particular in formation of thick matt of litter under extant plants (pers. obs.); this litter seems to mitigate adverse climatic effects on field layer vegetation (Zvereva and Kozlov 2007), but it can also hamper seedling establishment (Jordan 1975).

Although abundance of most species in industrial barrens is extremely low (e.g. Gilyazov 1993; Kozlov 1997; Salemaa et al. 2001; Vorobeichik 2003, 2004; Kozlov et al. 2005a, b), some plants and animals flourish in these habitats. In particular this concern some willow species that are much more abundant in barren sites than in unpolluted forests. Increase in willow density, along with decline of pressure from natural enemies, favours several insect herbivores. Willow-feeding leaf beetle, *Chrysomela lapponica* L., generally not abundant through its distribution range, in some years reached extremely high densities in industrial barrens near both Monchegorsk and Nikel (Zvereva et al. 2002). Also its specialised parasites, scuttle fly *Megaselia opacicornis* Schmitz and tachinid fly *Cleonice nitidiuscula* (Zett.), earlier known from a few specimens only, are abundant in industrial barrens near both Monchegorsk and Nickel (Richter and Zvereva 1996; Zvereva and Kozlov 2000a; Disney et al. 2001). Similarly, densities of birch- and willow-feeding leafrollers, along with some *Eriocrania* leaf-miners, in industrial barrens near both Monchegorsk and Nikel are on an average much higher than in unpolluted forests (Kozlov 1997, Zvereva and Kozlov 2006). The higher mean and peak densities of *Eriocrania* observed in industrial barrens are explained by disturbance of density-dependent feedback with parasitoids (Zvereva and Kozlov 2006).

Population structure of both plants and animals is considerably changed in industrial barrens. Populations of some woody species (mountain birch and willows in Monchegorsk, white birch in Sudbury) become more continuos due to elimination of pollution-sensitive competitors, while ground layer vegetation is highly fragmented (Zvereva and Kozlov 2004, 2007). In Sudbury, the dimension of patches of 'micro-deserts' and 'micro-oases' varies typically between 5 m and 15 m (Courtin 1994). In Monchegorsk, patches of ground layer vegetation are usually smaller, ranging 0.2–2 m in dimension, with barren gaps reaching as much as 20–50 m (pers. obs.). Absence of natural regeneration causes strong shift in age structure of plant populations. The average site-specific age of randomly collected individuals of *Empetrum nigrum* ssp. *hermaphroditum* (Lange ex Hagerup) Böcher in industrial barrens near Monchegorsk and Nikel was 34–38 years, compared with 17–28 years in undisturbed forests (V. Zverev and M. Kozlov, unpublished).

Populations of some insects, like leafmining *Eriocrania*, in industrial barrens became more aggregated than in unpolluted forests, possibly because strong winds force ovipositing females of these tiny moths to walk along a birch twig rather than take a risk of flying to another birch tree (Kozlov 2003; Zvereva and Kozlov 2006).

## 6.3 Growth form of woody plants

The trees which managed to survive in industrial barrens generally demonstrate bush-like or even creeping growth forms (Kryuchkov 1993; Kozlov and Haukioja 1995; Rigina and Kozlov 2000). In the most polluted barren sites near Monchegorsk height of mountain birches was in an average nearly 10 times lower than in healthy forests (Kozlov 2001). Similarly, proportion of the low-stature spruces with abnormalities in crown architecture (dead upper canopies but extensive growth of creeping lowest twigs) increased when approaching the smelter (Kozlov 2001, and unpubl.). The bush-like growth forms were also observed in Scots pine and aspen (*Populus tremula* L.) surviving in industrial barrens near both Monchegorsk and Nikel (Kozlov et al. 1999, unpubl.). Two willow species (*Salix borealis* (Fries.) Nasar. and *S. caprea* L.) in industrial barrens have more epiciormic shoots than in unpolluted forests (Zvereva and Kozlov 2001). The most plausible explanation of all these changes in growth form is the damage of apical (supranival) twigs by wind-driven snow abrasion (Arsenault and Payette 1997), which may be enhanced by pollutants (Alexeyev 1990). Therefore the negative correlations between snow depth and vertical growth of mountain birch, as well as between snow depth and proportion of spruces with abnormal crown architecture, may reflect causal link between these phenomena (Kozlov 2001). An increased light availability (due to forest decline) may also have contributed to higher branching and formation of bush-like crowns in woody plants surviving in barren sites (Zvereva and Kozlov 2001).

### 6.4 Changes in interactions between organisms

In industrial barrens interactions between organisms may differ from those in unpolluted forests. For example, development of delayed inducible resistance (decrease in plant quality next year after intensive damage), one of the important negative feedbacks regulating population dynamics of herbivores, in boreal willow, *Salix borealis* is disturbed in barren sites (Zvereva and Kozlov 2000b). At the same time, winter and spring bud damage, caused by harsh environmental conditions in barrens, may improve willow quality for herbivores (Zvereva and Kozlov 2000c). These changes in host plant–herbivore interactions favour herbivore outbreaks (see Sect. 4.5). Importantly, compensatory responses of boreal willow to herbivore damage are reduced in industrial barrens (Zvereva and Kozlov 2001), thus damaged plants recover slower than in undisturbed habitats.

Plant–plant interactions, which are mostly competitive in favourable habitats, tend to become positive in stressful environments (Brooker and Callaghan 1998). In industrial barrens, where competition among plants is low due to decreased density, the role of facilitation increases. Near Ventanas smelter in Chile shrubs imposed nursery effect on ground layer vegetation, which lead to better plant recruitment under shrub canopies (Ginocchio et al. 2004). Dwarf shrubs grew and reproduced better under birch canopies compared to between tree gaps in barrens near Monchegosk, whereas in undisturbed habitats the effects of trees were negative (Zvereva and Kozlov 2004). These effects may be explained by alleviation of harsh environmental conditions by trees (Zvereva and Kozlov 2007).

### 6.5 Adaptation to life in industrial barrens

Heavy metal tolerance of plants has been studied extensively and provides a well-documented example of rapid evolutionary adaptation (Bradshaw and McNeilly 1981; Macnair 1997). Several grass species, including *Agrostis scabra* Willd., *Deschampsia caespitosa* (L.) Beauv., *Agrostis gigantea* Roth. and *Poa compressa* L., colonising Sudbury's metal-contaminated soils, may have developed local metal-tolerant populations (Winterhalder 2000, and references therein). Although pollution tolerance in populations of long-lived trees has been detected less frequently than in populations of grasses and herbs, recently it had been demonstrated that long-lasting pollution impact increased pollution resistance of mountain birch in industrial barrens near both Monchegorsk and Nikel (Kozlov 2005; Eränen 2006), possibly by elimination of sensitive genotypes (survival selection) from the affected populations. Similarly, we discovered that resistance of non-specific esterases (enzymes degrading various xenobiotics) to heavy metals was higher in populations of leaf beetle, *C. lapponica*, from heavily contaminated barren sites (Zvereva et al. 2003). It should be stressed that metal-tolerant populations of plants and animals developed near point polluters represent unique genetic resource that will be lost (overcompeted) with restoration of natural communities following pollution decline.

Some changes in feeding behaviour of herbivorous insects may also be regarded as adaptations to life in the barren landscape. In particular, feeding niche breadth of the leaf beetle, *C. lapponica*, decreased with increase in pollution: in industrial barrens this species concentrated on the boreal willow that assures the best survival of larvae, whereas in surrounding forests it feeds on other willow species as well (Zvereva et al. 1995). Indirect data suggest that in industrial barrens some other herbivores may also prefer other host plants than in undisturbed forests. In particular, larval density of autumnal moth, *Epirrita autumnata* Bkh., measured on birches (most preferred host plant), strongly declined in industrial barrens near Monchegorsk (Ruohomäki et al. 1996). At the same time, the number of moths attracted by pheromone traps did not decline; the moths collected in industrial barrens were metal-contaminated and thus presumably of local origin (pers. obs.). These data suggest that larvae of autumnal moth near Monchegorsk have been feeding on plants other than mountain birch.

## 7 Significance of industrial barrens

### 7.1 Industrial barrens as refugia of rare and endangered species and populations

The sites of biological significance within severely degraded environments may not be as rare as is commonly thought, and assumption that physically or chemically hostile environments are incapable of attaining biological diversity is far from being true (Johnson et al. 1978). Industrial barrens are rather heterogeneous (Kozlov 1997; Uhlig et al. 2001; Ginocchio et al. 2004), with a range of different substrate types that favour different species. Therefore, in spite of the general loss of biodiversity, these habitats can develop a great richness of unusual and interesting plants and animals, including regionally rare and endangered species (Johnson et al. 1978; Heliövaara et al. 1990; Eyre and Luff 1995; Spalding and Haes 1995; also see above), and the overall site diversity can be high even when each patch is relatively poor. Furthermore, several plant species of low competitive ability benefit from fragmentation of the continuous 'carpet' of vegetation. Invertebrates in industrial barrens may escape from strong enemy pressure, and this 'enemy-free space' phenomenon may explain high abundance of some species that are usually depressed in less disturbed habitats (Zvereva and Kozlov 2000a, 2006).

### 7.2 Conservation of industrial barrens for science, education, and tourism

Although industrial barrens are usually seen as a by-product of human activities that deserves rehabilitation only, and the suggestion to conserve these severely modified landscapes is frequently perceived as a joke, their importance for science, education, and even sightseeing had been sometimes appreciated (Winterhalder 2000). However, conservation of industrial barrens is in line with conservation of historical industrial landscapes which form a part of the World Heritage (UNESCO 2006). Detailed investigation should be conducted prior 're-greening' or rehabilitation of these unusual habitats. Non-selective reclamation works could be destructive to conservation interests, and losses associated with habitat reclamation may easily overcome the benefits gained. In particular, local metal-tolerant populations of plants and animals will definitely be lost.

Prospect of reforestation, that started a more than half of century ago (Seigworth 1943), has created some interest in preserving part of the Copper Basin landscape in its denuded and eroded state for historical purposes (Quinn 1988). Recently, a 200 ha exemplary plot of 'desert' remains as part of the display at the Ducktown Museum (Anonymous 2005c). Also a part of Sudbury industrial barrens (about 600 ha surrounding Alice and Baby lakes just behind the abandoned Coniston smelter) is identified in the city reclamation plan as an 'industrial reserve'. Fortunately or unfortunately, the intention to preserve the extreme level of landscape damage nearly failed: due to natural recovery following pollution decline some 35 years ago the non-reclaimed area is now difficult to distinguish from reclaimed territories (K. Winterhalder and J. Gunn, pers. comm.). Still natural succession on derelict lands may produce unexpected communities of considerable scientific interest (Box 1993). Students and faculty of Laurentian University and other high schools often come to Sudbury to see the 'past and present' of the barren landscape. Industrial barrens at Monchegorsk are frequently visited for 'scientific sightseeing', especially by the researchers from the neighbouring Finland (Ruotsalainen and Markkola 2004); at least once they served a target of an international environmental field course.

As far as we know, only Queenstown had advertised the barren landscapes as tourist attractions. "By any measure Queenstown is one of the wonders of the world. It is a profound reminder of humanity's capacity to destroy and pollute and, in that sense, it deserves to be seen by everyone" (Anonymous 2005a). This attitude, supported by local government, prevented reforestation of the hills around Queenstown, although some locals believe that the rainforest, which characterises the area, should be encouraged to regrow. Industrial barrens around Monchegorsk, although neither advertised nor conserved, still attract attention of foreign tourists: quite frequently, buses travelling to/from Murmansk, stop in the barren site for photographing the landscape.

## 8 Conclusion

### 8.1 Past and future of industrial barrens

Comparative analysis of the existing data demonstrated that industrial barrens have developed due to combined impact of different stressors, among which soil contamination by heavy metals, clearcutting and fires played the leading role. The combined effects are critical, as either stress alone would have caused much less damage (Jordan 1975). Pollution, accompanied by other human-induced disturbances (primarily clearcutting), damages and gradually kills the vegetation; this process is usually concluded by a fire, facilitated by unusually large amount of woody debris accumulated in severely damaged communities. Since vegetation recovery is hampered by soil toxicity due to extreme contamination by heavy metals, soils remain bare for a long time and suffer from extensive wind and water erosion exacerbated by the intense frost-heaving. The destructive processes are enhanced by both disturbance of negative feedbacks regulating e.g. relationships between plants, herbivores, and their natural enemies, and development of positive feedbacks, e.g. increase in climatic severity due to initial vegetation damage enhances decline of the remaining vegetation.

The most extensive industrial barrens appeared before 1970s; some of them have already been (partially) reclaimed or are recovering following closure of polluters (14 of 36 documented situations; Table 1) or rapid emission decline. Natural recovery of the existing barren sites may take long time; the example of Sudbury suggests that some 20 years after pollution decline may be sufficient to initiate this process. On the other hand, human-assisted recovery may proceed much faster (Vangronsveld et al. 1995, 1996; Winterhalder 2000). However, non-selective reclamation works could be destructive to conservation interests, and the care should be taken to preserve at least some of the unusual post-industrial habitats.

### 8.2 Industrial barrens as unintentional experiments

Links between pollution-oriented environmental studies ('applied ecology') and development of ecological theories ('basic ecology') are surprisingly weak (Ormerod et al. 1999). Researchers addressing basic ecological problems only rarely use the results of 'unintentional pollution experiments' (Lee 1998). However, industrial barrens offer unique opportunities for conducting 'basic' ecological research, in particular for testing some theories (assumed to be general) in an evolutionary novel stressful environment (Kozlov and Zvereva 2003; Vorobeichik 2004).

Studies in industrial barrens contributed to the development of the theories on plant growth, compensatory ability and reproductive strategies in stressful environment (Zvereva and Kozlov

2001, 2005), as well as to understanding of the role of host plant quality in regulating population dynamics of herbivores in 'enemy free space' (Zvereva et al. 1997). Exploration of plant–plant interactions in industrial barrens demonstrated that shift from competition to facilitation occurs not only in natural stress gradients, that were used to construct the predictive model (Brooker and Callaghan 1998), but also in human-induced stress gradients (Zvereva and Kozlov 2004, 2007). Importantly, results of 'basic' research on plant–plant facilitation are now suggested for practical application in phytostabilisation of heavily contaminated areas (Frérot et al. 2006).

Although net primary production had not been measured in any of industrial barrens, drastic decline in vegetation cover, along with reported retardation in plant growth (e.g. Kryuchkov 1993; Nöjd et al. 1996), suggest that productivity of plant communities declined by a factor of 10–100. Thus, industrial barrens are unique ecosystems capable to maintain relatively high diversity at very low productivity level, and their exploration is likely to contribute to the long-lasting debate (Hector and Schmid 1999; Tilman 1999) on the relationship between diversity, stability and productivity.

### 8.3 Management of polluted habitats

Recently, most industrial barrens show some signs of natural recovery due to emission decline or closure of responsible polluters; some of them have been or are being successfully revegetated. The latter strategy is in particular justified by the urgent need to prevent dispersion of metals to surrounding areas (e.g. Vandronsveld et al. 1995); it has a strong priority in densely populated sites, like Central Europe, in spite of high scientific and sometimes public interest to unusual post-industrial habitats.

Reclamation strategies used in industrial barrens vary from natural attenuation to extensive (and expensive) revegetation programs. Literature on this problem is rather extensive and had been repeatedly reviewed during the past years (Kozlov et al. 2000b; Winterhalder 2000; Adriano et al. 2004; Kozlov 2004); therefore we provide only a short summary of most relevant findings.

The reduction of emissions is a critical requirement for any reclamation programs. However, until measures to reduce air pollution are fully realised, efforts should be devoted to improving the vitality and maintaining the stability of affected ecosystems. The experience gathered to date suggests that the pollution-induced decline of forests can be slowed down and perhaps even reversed at most stages of degradation. In the case of continuing pollution, its effects can be mitigated to prevent further damage.

Restoration of terrestrial ecosystems after the reduction of emissions is based on the suggestion that some successional stages are not essential and can be bypassed by proper intervention. Thus, in practical terms, the rehabilitation of forests damaged by pollution is an attempt to partially substitute time with money. However, small-scale, short-term experiments still dominate over practical applications. Thus it is reasonably well known how to initiate the early stages of rehabilitation, but there are substantial uncertainties over the longer-term effects (e.g. Eränen and Kozlov 2006). The latter especially concern possible side-effects and ecological risks which may result from human intervention, such as increased leaching of toxic substances and/or nutrients. To reduce these risks, a careful analysis of the need for chemical treatments such as liming or fertilising should be carried out for each specified area. We suggest that whenever possible a 'minimal intervention' approach aimed at promoting natural succession should be used instead of 're-greening' or an artificial re-creation of the desired ecosystem.

**Acknowledgements** We are grateful to V. Barcan, R. Ginocchio, T. Gorbacheva, J. Gunn, J. Kulfan, P. Niemelä, H. Tømmervik, J. Vangronsveld, E. Vorobeichik, P. Zach, V. Zverev and late K. Winterhalder for providing us with both published and unpublished information. The study was financially supported by the Academy of Finland (project 211734), Maj and Tor Nessling Foundation, and by EC through the BALANCE project carried out under contract EVK2–2002–00169.

## References

Acatay A (1968) Damage from injurious gases of Murgul copper plant. Dergisi Istanbul Universitesi Orman Fakultesi 18a(1):1–17

Adriano DC, Wenzel WW, Vangronsveld J, Bolan NS (2004) Role of assisted natural remediation in environmental cleanup. Geoderma 122:121–142

Alexeyev VA (1990) Changes in the growth form of the Scots pine under aerial pollution. In: Tarabrin VP (ed) Industrial botany: the state and perspectives of development. Abstracts of regional scientific conference devoted to the 25th anniversary of Donetsk Botanical Garden, Sept. 1990, Donetsk. Kiev, Ukraine, pp 99–100 (in Russian)

Allum JAE, Dreisinger BR (1987) Remote sensing of vegetation change near Inco's Sudbury mining complexes. Int J Remote Sens 8:399–416

Anand M, Ma KM, Okonski A, Levin S, McCreath D (2003) Characterising biocomplexity and soil microbial dynamics along a smelter-damaged landscape gradient. Sci Tot Environ 311:247–259

Anonymous (1999) Studies of the terrestrial environment in the Wawa Area 1998–1999. http://www.ene.gov.on.-ca/envision/techdocs/3918.pdf

Anonymous (2003) Yellowknife EMAN-North Site: Biotic Monitoring & Studies. http://www.emannorth.ca/ic/ds013/settlement.cfm

Anonymous (2005a) Queenstown. http://www.walk-about.com.au/fairfax/locations/TASQueens-town.shtml

Anonymous (2005b) The environmental impact of past mining in the Copper Basin. http://www.gaminer-al.org/copperbasin.htm

Anononymous (2005c) Ducktown Desert Site. http://ludb.clui.org/ex/i/TN3139

Archibold OW (1978) Vegetation recovery following pollution control at Trail, British Columbia. Can J Bot 56:1625–1637

Arseneault D, Payette S (1997) Landscape change following deforestation at the Arctic tree line in Québec, Canada. Ecology 78:693–706

Badora A, Furrer G, Grünwald A, Schulin R (1998) Immobilization of zinc and cadmium in polluted soils by polynuclear $Al_{13}$ and Al-montmorillonite. J Soil Contamin 7:573–588

Baklanov AA, Rodyushkina IA (1993) Pollution of ambient air by 'Severonikel' smelter complex: observations and modelling. In: Kozlov MV, Haukioja E, Yarmishko VT (eds) Aerial pollution in Kola Peninsula. Proceedings of the international workshop, 14–16 April 1992, St. Petersburg, Russia. Kola Sci. Centre, Apatity, Russia, pp 83–89

Baklanov AA, Sivertsen B (1994) Investigations of local transborder air pollution. NILU F-19/94, Kjeller

Banásová V, Holub Z, Zeleňáková E (1987) The dynamic, structure and heavy metal accumulation in vegetation under long-term influence of Pb and Cu immissions. Ekológia (Bratislava) 6:101–111

Banásová V, Lackovičová A (2004) The disturbance of grasslands in the vicinity of copper plant in the town of Krompachy (Slovenské Rudohorie Mts, NE Slovakia). Bull Slov Bot Spoločn Bratislava 26:153–161

Barcan V (2002) Leaching of nickel and copper from soil contaminated by metallurgical dust. Environ Internat 28:63–68

Barcan VS, Kovnatsky EF (1998) Soil surface geochemical anomaly around the copper–nickel metallurgical smelter. Water Air Soil Pollut 103:197–218

Basta NT, Gradwohl R, Snethen KL, Schroder JL (2001) Chemical immobilization of lead, zinc, and cadmium in smelter-contaminated soils using biosolids and rock phosphate. J Environ Qual 30:1222–1230

Beyer WN, Patee OH, Sileo L, Hoffman DJ, Mulhern BM (1985) Metal contamination in wildlife near two zinc smelters. Environ Pollut 38:63–86

Box J (1993) Conservation or greening? The challenge of post-industrial landscapes. Brit Wildlife 4:273–279

Bradshaw AD, McNeilly T (1981) Evolution and pollution. Edward Arnold, London

Brooker RW, Callaghan TV (1998) The balance between positive and negative plant interactions and its relationship to environmental gradients: a model. Oikos 81:196–207

Buchauer MJ (1973) Contamination of soil and vegetation near a zinc smelter by zink, cadmium, copper, and lead. Environ Sci Technol 7:131–135

Burt R, Wilson MA, Keck TJ, Dougherty BD, Strom DE, Lindhal JA (2003) Trace elements speciation in selected smelter-contaminated soils in Anaconda and Deer Lodge Valley, Montana, USA. Adv Environ Res 8:51–67

Carter DB, Loewenstein H, Pitkin FH (1977) Amelioration and revegetation of smelter-contaminated soils in the Coeur d'Alene mining district of northern Idaho. Paper 13. In: Proceedings of 2nd annual meeting, Canadian Land Reclamation Association, Edmonton, Canada. CLRA, Guelph, Ontario (NV, cited after Winterhalder 2000)

Chernenkova TV (2002) Response of forest vegetation to industrial pollution. Nauka, Moscow (in Russian)

Cicák A, Mihál I, Kulfan J, Šušlík V, Zach P, Krištin A (1999) Health state of forest tree species and selected groups of fungi and animals in surroundings of a magnesium factory (Central Slovakia). Biológia (Bratislava) 18:211–222

Clay G (1983) Copper Basin cover-up. Landscape Architect 94:49–55

Courtin GM (1994) The last 150 years – a history of environmental degradation in Sudbury. Sci Tot Environ 148:99–102

Dawson JL, Nash TH (1980) Effects of air pollution from copper smelters on a desert grassland community. Environ Exptl Bot 20:61–72

Dean RS, Swain RE (1944) Report submitted to the trail smelter arbitral tribunal. US Dept. of the Interior, Bureau of Mines, Bull. 453. US Government Printing Office, Washington

Derome J, Lindroos A-J (1998) Effects of heavy metal contamination on macronutrient availability and acidification parameters in forest soil in the vicinity of the Harjavalta Cu–Ni smelter, SW Finland. Environ Pollut 99:225–232

Derome J, Nieminen T (1998) Metal and macronutrient fluxes in heavy-metal polluted Scots pine ecosystems in SW Finland. Environ Pollut 103:219–228

Disney RHL, Zvereva EL, Mostovsky M (2001) A scuttle fly (Diptera: Phoridae) parasitizing a beetle (Coleoptera: Chrysomelidae) in Russia. Entomol Fenn 12:59–63

Doncheva AV (1978) Landscape in zone of industrial impact. Lesnaya promyshlennost, Moscow

Dudka S, Poncehernandez R, Hutchinson TC (1995) Current level of total element concentrations in the surface layer of Sudbury's soils. Sci Tot Environ 162:161–171

Eastmond RJ (1971) Response of four plant communities to eroded soils and smelter smoke in northern Utah. A Dissertation submitted to the Faculty of the University of Utah. Dept of Biology, Univ. of Utah

Eränen J (2006) Local adaptation of mountain birch to heavy metals in subarctic industrial barrens. Forest Snow Landscape Res 80:161–167

Eränen J, Kozlov MV (2006) Physical sheltering and liming improve survival and performance of mountain birch seedlings: a five-year study in a heavily polluted industrial barren. Restor Ecol 14:77–86

Euler WD (ed) (1939) Effect of sulphur dioxide on vegetation. Nat. Res. Council Canada, Ottawa

Eyre MD, Luff ML (1995) Coleoptera on post-industrial land: a conservation problem? Land Contam Reclam 3:132–134

Filipchuk AN, Kovalev BI (1990) Dynamics of sub-tundra forest decline in Norilsk. In: Anonymous (ed) International Symposium "Northern forests: state, dynamics human impact" July 16–26 1990. The USSR State Forest Committee, Moscow, pp 29–37 (in Russian, English summary)

Fowler D, Cape JN, Coyle M, Flechard C, Kuylenstierna J, Hicks K, Derwent D, Johnson C, Stevenson D (1999) The global exposure of forests to air pollutants. Water Air Soil Pollut 116:5–32

Freedman B, Hutchinson TC (1980) Long-term effects of smelter pollution at Sudbury, Ontario, on forest community composition. Can J Bot 58:2123–2140

Frérot H, Lefèbvre C, Gruber W, Collin C, Dos Santos A, Escarré J (2006) Specific interactions between local metallicolous plants improve the phytostabilization of mine soils. Plant Soil 282:53–65

Galbraith H, Lejeune K, Lipton J (1995) Metal and arsenic impacts to soils, vegetation communities and wildlife habitat in southwest Montana uplands contaminated by smelter emissions. 1. Field evaluation. Environ Toxicol Chem 14:1895–1903

Gilbert OL (1975) Effects of air pollution on landscape and land-use around Norwegian aluminium smelters. Environ Pollut 8:113–121

Gilyazov AS (1993) Air pollution impact on the bird communities of the Lapland biosphere reserve. In: Kozlov MV, Haukioja E, Yarmishko VT (eds) Aerial pollution in Kola Peninsula. Proceedings of the international workshop, 14–16 April 1992, St. Petersburg, Russia. Kola Sci. Centre, Apatity, pp 383–390

Ginocchio R (2000) Effects of a copper smelter on a grassland community in the Puchuncavi Valley, Chile. Chemosphere 41:15–23

Ginocchio R, Carvallo G, Toro I, Bustamante E, Silva Y, Sepulveda N (2004) Micro-spatial variation of soil metal pollution and plant recruitment near a copper smelter in Central Chile. Environ Pollut 127:343–352

Goodarzi F, Sanei H, Labonte M, Duncan WF (2002) Sources of lead and zinc associated with metal smelting activities in the Trail area, British Columbia, Canada. J Environ Monitor 4:400–407

Gordon AG, Gorham E (1963) Ecological aspects of air pollution from an iron-sintering plant at Wawa, Ontario. Can J Bot 41:1063–1078

Gorshkov SP (1997) Landscape geoecological mapping of the Norilsk area of ecological disaster. In: Gennadiev AN, Milanova EB (eds) Changes in the environment. Moscow University, Moscow, pp 148–161

Hansen JE, Mitchell JE (1978) The role of terraces and soil amendments in revegetating steep smelter-affected land. Reclam Rev 1:103–112

Haywood JK (1905) Injury to vegetation by smelter fumes. US Dept. of Agriculture, Bureau of Chemistry, Bull. no. 89. Government Printing Office, Washington

Haywood JK (1910) Injury to vegetation and animal life by smelter wastes. US Dept. of Agriculture, Bureau of Chemistry, Bull. no. 113. Government Printing Office, Washington

Hazlett PW, Rutherford GK, vanLoon GW (1983) Metal contaminants in surface soils and vegetation as a result of nickel/copper smelting at Coniston, Ontario, Canada. Reclam Reveg Res 2:123–137

Hector A, Schmid B (1999) Plant diversity and productivity experiments in European grasslands. Science 286:1123–1127

Hedgcock GG (1912) Winter-killing and smelter-injury in the forests of Montana. Torreya 12:25–30

Heliövaara K, Väisänen R, Mannerkoski I (1990) *Melanophila formaneki* (Jakobson) (Coleoptera, Buprestidae) new to Finland. Ent Fenn 1:221–225

Helmisaari H-S, Derome J, Fritze H, Nieminen T, Palmgren K, Salemaa M, Vanha-Majamaa I (1995) Copper in Scots pine forests around a heavy-metal smelter in south western Finland. Water Air Soil Pollut 85:1727–1732

Henderson PJ, McMartin I, Hall GE, Persival JB, Walker DA (1998) The chemical and physical characteristics of heavy metals in humus and till in the vicinity of the base metal smelter at Flin Flon, Manitoba, Canada. Environ Geol 34:39–58

Hepting GH (1971) Air pollution and trees. In: Matthews WH, Smith FE, Goldberg ED (eds) Man's impact on terrestrial and oceanic ecosystems. MIT Press, Cambridge Mass, pp 116–197

Hocking D, Kuchar P, Plambeck JA, Smith RA (1978) Impact of gold smelter emissions on vegetation and soils of a sub-arctic forest-tundra transition ecosystem. J Air Pollut Control Assoc 28:133–137

Hodgson DA, Vyverman W, Chepstow-Lusty A, Tyler PA (2000) From rainforest to wasteland in 100 years: the limnological legacy of the Queenstown mines, Western Tasmania. Archiv Hydrobiol 149:153–176

Hogan GD, Wotton DL (1984) Pollutant distribution and effects in forests adjacent to smelters. J Environ Qual 13:377–382

Hursh CR (1948) Local climate in Copper Basin of Tennessee as modified by the removal of vegetation. USDA Circ. no. 774. USDA, Washington

Hutchinson TC, Whitbey LM (1974) Heavy metal pollution in the Sudbury mining and smelting region of Canada. I. Soil and vegetation contamination by nickel copper and other metals. Environ Conserv 1:123–132

Hutchinson TC, Symington MS (1997) Persistence of metal stress in a forested ecosystem near Sudbury, 66 years after closure of the O'Donnell roast bed. J Geochem Explor 58:323–330

Hutchinson TC, Aufreiter S, Hancock RGV (1982) Arsenic pollution in the Yellowknife area from gold smelter activities. J Radioanalyt Chem 71:59–73

Igamberdiev VM, Tereshenkov OM, Kutyev HA, Popova EN (1994) Evaluation of the recent state of the environment in the Norilsk industrial area. Econ Komi Repub 3:54–61 (in Russian)

James GI, Courtin GM (1985) Stand structure and growth form of the birch transition community in an industrially damaged ecosystem, Sudbury, Ontario. Can J Forest Res 15:809–817

Johnson D, Hale B (2004) White birch (*Betula papyrifera* Marshall) foliar litter decomposition in relation to trace metal atmospheric inputs at metal-contaminated and uncontaminated sites near Sudbury, Ontario and Rouyn-Noranda, Quebec, Canada. Environ Pollut 127:65–72

Johnson MS, Putwain PD, Holliday RJ (1978) Wildlife conservation value of derelict metalliferous mine workings in Wales. Biol Conserv 14:131–148

Jordan MJ (1975) Effects of zinc smelter emissions and fire on a chestnut-oak woodland. Ecology 56:78–91

Kapusta M & Múdry P (1974) Long-run effect of immissions from smelting works in Banská Štiavnica on vegetation and landscape. Acta Geobot III(8):85–126

Kautz G, Zimmer M, Zach P, Kulfan J, Topp W (2001) Suppression of soil microorganisms by emissions of a magnesite plant in the Slovak Republic. Water Air Soil Pollut 125:121–132

Kelly JJ, Tate RL (1998) Effects of heavy metal contamination and remediation on soil microbial communities in the vicinity of a zinc smelter. J Environ Qual 27:609–617

Kharuk VI (2000) Air pollution impacts on subarctic forests at Noril'sk, Siberia. In: Innes JL, Oleksyn J (eds) Forest dynamics in heavily polluted regions. (IUFRO task force on environmental change. Report no. 1). CAB International, Wallingford, pp 77–86

Kharuk VI, Winterberger K, Tsibulskii GM, Yakhimovich AP, Moroz SN (1996) Technogenic disturbance of sub-tundra forests in the Norilsk valley. Ekologia (Ekaterinburg) 27:406–410 (in Russian)

Kleinert J (1988) Synusiae of ground beetles (Coleoptera: Carabidae) in the environs of Ješlava and Lubenik affected by alkaline emissions. Biológia (Bratislava) 43:159–167

Komulainen M, Vieno M, Yarmishko VT, Daletskaja TD, Maznaja EA (1994) Seedling establishment from seeds and seed banks in forests under long-term pollution stress – a potential for vegetation recovery. Can J Bot 72:143–149

Koponen S, Niemelä P (1993) Ground-living spiders in a polluted pine forest SW Finland. Boll Acc Gioenia Sci Nat 26:221–226

Koponen S, Niemelä P (1995) Ground-living arthropods along pollution gradient in boreal pine forest. Ent Fenn 6:127–131

Kozlov MV (1997) Pollution impact on insect biodiversity in boreal forests: evaluation of effects and perspectives of recovery. In: Crawford RMM (ed) Disturbance and recovery in arctic lands: an ecological perspective. Proceedings of the NATO advanced research workshop on disturbance and recovery of arctic terrestrial ecosystems, Rovaniemi, Finland, 24–30 September 1995 (NATO ASI series. Partnership subseries 2. Environment, vol 25). Kluwer Academic Publishers, Dordrecht, pp 213–250

Kozlov MV (2001) Snowpack changes around a nickel–copper smelter at Monchegorsk northwestern Russia. Can J Forest Res 31:1684–1690

Kozlov MV (2002) Changes in wind regime around a nickel–copper smelter at Monchegorsk northwestern Russia. Int J Biometeorol 46:76–80

Kozlov MV (2003) Density fluctuations of the leafminer *Phyllonorycter strigulatella* (Lepidoptera: Gracillariidae) in the impact zone of a power plant. Environ Pollut 121: 1–10

Kozlov MV (2004) Silviculture in polluted areas. In: Burley J, Evans J, Youngquist JA (eds) Encyclopedia of forest science. Elsevier Academic Press, London, pp 1112–1121

Kozlov MV (2005) Pollution resistance of mountain birch, *Betula pubescens* subsp. *czerepanovii* near the copper–nickel smelter: natural selection or phenotypic acclimation? Chemosphere 59:189–197

Kozlov MV, Haukioja E (1995) Pollution-related environmental gradients around the "Severonickel" smelter complex on Kola Peninsula, Northwestern Russia. In: Munawar M, Luotola M (eds) The contaminants in the Nordic Ecosystem: Dynamics, processes and fate. S.P.B. Academic Publ., The Netherlands, pp 59–69

Kozlov MV, Haukioja E (1997) Microclimate changes along a strong pollution gradient in northern boreal forest zone. In: Uso JL, Brebbia CA, Power H (eds) Ecosystems and sustainable development (Advances in Ecological Sciences vol 1). Computation Mechanics Publ, Southampton, pp 603–614

Kozlov MV, Haukioja E (1999) Performance of birch seedlings replanted in heavily polluted industrial barrens of the Kola Peninsula NW Russia. Restor Ecol 7:145–154

Kozlov MV, Zvereva EL (2003) Impact of industrial polluters on terrestrial ecosystems: a research synthesis. In: Honkanen JO Koponen PS (eds) Sixth Finnish conference of environmental sciences: Proceedings. Finnish Society for Environmental Sciences & University of Joensuu, Joensuu, Finland, pp 72–75

Kozlov MV, Zvereva EL (2004) Reproduction of mountain birch along a strong pollution gradient near Monchegorsk, Northwestern Russia. Environ Pollut 132:443–451

Kozlov MV, Jalava J, Lvovsky AL, Mikkola K (1996) Population densities and diversity of Noctuidae (Lepidoptera) along an air pollution gradient on the Kola Peninsula, Russia. Entomol Fenn 7:9–15

Kozlov MV, Byalt VV, Salo J (1998) Species richness of vascular plants in studies of human-induced disturbances in northern coniferous forests: effect of scale. In: Gorshkov VG, Makarieva AM, Kharkina TG (eds) The role of virgin terrestrial biota in the modern processes of global change: biotic regulation of the environment. Proceedings of international seminar, 12–16 October 1998, Petrozavodsk. St. Petersburg Institute of Nuclear Physics, Gatchina, Russia, pp 305–307

Kozlov MV, Zvereva EL, Niemelä P (1999) Effects of soil quality and air pollution on the rooting and survival of *Salix borealis* cuttings. Boreal Environ Res 4:67–76

Kozlov MV, Haukioja E, Bakhtiarov AV, Stroganov DN, Zimina SN (2000a) Root *vs* canopy uptake of heavy metals by birch in an industrially polluted area: contrasting behaviour of nickel and copper. Environ Pollut 107:413–420

Kozlov MV, Haukioja E, Niemelä P, Zvereva E, Kytö M (2000b) Revitalization and restoration of boreal and temperate forests damaged by aerial pollution. In: Innes JL, Oleksyn J (eds) Forest dynamics in heavily polluted regions (IUFRO task force on environmental change. Report no. 1). CAB International, Wellingford, pp 193–218

Kozlov MV, Zvereva EL, Niemelä P (2001) Shoot fluctuating asymmetry: a new and objective stress index in Norway spruce (Picea abies). Can J Forest Res 31:1289–1291

Kozlov MV, Brodskaya NK, Haarto A, Kuusela K, Schäfer M, Zverev V (2005a) Population densities and diversity of blood-sucking flies (Diptera: Ceratopogonidae, Culicidae, Tabanidae, Simuliidae) around a nickel–copper smelter at Monchegorsk, Northwestern Russia. J Vector Ecol 30:263–271

Kozlov MV, Zvereva EL, Gilyazov AS, Kataev GD (2005b) Contaminated zone around a nickel-copper smelter: a death trap for birds and mammals? In: Burk AR (ed) Trends in biodiversity research. Nova Science Publishers, New York, pp 81–99

Kryuchkov VV (1993) Extreme anthropogenic loads and the northern ecosystem condition. Ecol Appl 3:622–630

Kryuchkov VV, Makarova TA (1989) Impact of industrial air pollution on ecosystems of the Kola North. Kola Sci. Centre, Apatity (in Russian)

Kulagin YZ (1964) Smoke wastes of 'Magnezit' plant and dynamics of green belt forests around the city of Satka (Southern Ural). In: Mamaev SA (ed) Vegetation and industrial pollution (materials of conference). Ural State University, Sverdlovsk, pp 175–187 (in Russian)

Lee JA (1998) Unintentional experiments with terrestrial ecosystems: ecological effects of sulphur and nitrogen pollutants. J Ecol 86:1–12

Lees D (2000) Green rebirth. Can Geogr 120:61–71

Lehmann C, Rebele F (2004) Evaluation of heavy metal tolerance in *Calamagrostis epigejos* and *Elymus repens* revealed copper tolerance in a copper smelter population of *C. epigejos*. Environ Exptl Bot 51:199–213

Linzon SN (1975) Forest ecosystems: the effects of atmospheric sulfur. In: Klein WM, Severson JG (eds) Sulphur in the environment. Missouri Botanical Garden St. Louis, Missouri, pp 112–133

Lukina NV, Nikonov VV (1996) Biogeochemical cycles in the northern forests subjected to air pollution. Institute of North Industrial Ecology Problems, Apatity (in Russian)

Lukina NV, Nikonov VV (1999) Pollution-induced changes in soils subjected to intense air pollution. In: Nikonov VV, Koptsik GN (eds) Acidic deposition and forest soils. Kola Science Centre, Apatity, Russia, pp 79–126 (in Russian)

Lurie NY (1986) Impact of industrial pollution on structure and function of microbial coenoses in the black earth of Southern Ural. PhD. VV Dokuchaev Soil Institute, Moscow (in Russian)

MacMillan D (2000) Smoke wars: Anaconda copper, Montana air pollution and the courts 1890–1924. Montana Historical Society Press, Montana

Macnair MR (1997) The evolution of plants in metal-contaminated environments. In: Bijlsma R, Loeschchke V (eds) Environmental stress adaptation and evolution . Birkhäuser Verlag Basel, Switzerland, pp 3–24

Makhnev AK, Trubina MR, Pryamonosova SA (1990) Forest vegetation in surroundings of non-ferrous plants. In: Shiyatov SG, Shilova A (eds) Natural vegetation of industrial and urban areas of Ural. Ural Division of the Academy of Science of the USSR, Sverdlovsk, Russia, pp 3–40 (in Russian)

Makunina GS (2002) Degradation and chemical properties of soils in the industrially altered area of Karabash. Pochvovedenie [Soil Science, Moscow] 0(3): 368–376 (in Russian, English summary)

Mälkönen E, Derome J, Fritze H, Helmisaari H-S, Kukkola M, Kytö M, Saarsalmi A, Salemaa M (1999) Compensatory fertilization of Scots pine stands polluted by heavy metals. Nutr Cycl Agroecosyst 55:239–268

Maňňkovská B (1984) The effects of atmospheric emissions from the Krompachy, Nižná Slaná, Rudňany iron ore mines on forest vegetation and soils. Ekológia (Bratislava) 3:331–344

Manninen S, Huttunen S (1997) Critical level of $SO_2$ for subarctic Scots pine forests. In: Anonymous (ed) The AMAP international symposium on environmental pollution in the Arctic. 1–5 June 1997. Extended Abstracts. Tromsø, Norway, pp 173–174

Marin YuF (1996) On the impact of Kirovgrad copper smelter on the environment. In: Marin YuF (ed) Problems of reserve management. 25th anniversary of Visim reserve: materials of the scientific conference.

Ekaterinburg Publishers, Ekaterinburg, Russia, pp 87–91 (in Russian)
Marty L (2000) The use of local ecotypes for the revegetation of acid/heavy metal contaminated lands in western Montana. In: Billings land reclamation symposium 2000: striving for restoration, fostering technology and policy for reestablishing ecological function, March 20–24, 2000, Billings, Montana, pp 216–228. http://www.ott.wrcc.osmre.gov/library/proceed/recsym.htm
McCall J, Gunn J, Struik H (1995) Photo interpretive study of recovery of damaged lands near the metal smelters of Sudbury, Canada. Water Air Soil Pollut 85:847–852
McEnroe NA, Helmisaari H-S (2001) Decomposition of coniferous forest litter along a heavy metal pollution gradient south-west Finland. Environ Pollut 113:11–18
McGovern PC (1975) Effects of sulphur dioxide and heavy metals on the Wawa area soils and vegetation (1974). Ontario Minist. Environ., Northeast Region, Sudbury
Menshchikov SL, Vlasenko VE, Estyugin AS (1997) Local monitoring of forest ecosystems under variable types of pollution. In: Makhnev AK (ed) Biological recultivation of disturbed lands. Proceedings of the international congress, 26–29 August 1996. Institute of Forest, Ekaterinburg, pp 184–192 (in Russian)
Mitchell J (1982) Revegetation of lands surrounding the Bell Bay smelter. In: Murray F (ed) Fluoride emissions: their monitoring and effects on vegetation and ecosystems. Academic Press, Sydney, pp 207–215
Nielsen D, Kovats Z (2004) Report on analysis of the effects of smelter emissions on vegetation within Trail regional area. http://www.teckcominco.com/articles/tr-ecorisk/techreports/2004/2004–05-veganalysis.pdf
Nieminen TM (2004) Effects of soil copper and nickel on survival and growth of Scots pine. J Environ Monitoring 6:888–896
Nieminen TM, Ukonmaanaho L, Shotyk W (2002) Enrichment of Cu, Ni, Zn, Pb and As in an ombrotrophic peat bog near a Cu-Ni smelter in southwest Finland. Sci Tot Environ 292:81–89
Niskavaara H, Reimann C, Chekushin V (1996) Distribution and pathways of heavy metals and sulphur in the vicinity of the copper-nickel smelters in Nikel and Zapoljarnij, Kola Peninsula, Russia as revealed by different sample media. Appl Geochem 11:25–34
Nöjd P, Mikkola K & Saranpää P (1996) History of forest damage in Monchegorsk, Kola: a retrospective analysis based on tree rings. Can J Forest Res 26:1805–1812
Nriagu JO, Wong HKT, Lawson G, Daniel P (1998) Saturation of ecosystems with toxic metals in Sudbury basin, Ontario, Canada. Sci Tot Environ 223:99–117
Ormerod SJ, Pienkowski MW, Watkinson AR (1999) Communicating the value of ecology. J Appl Ecol 36:847–855
Padgett PE, Kee SN (2004) Impacts of air pollution on forest ecosystems. In: Burley J, Evans J, Youngquist JA (eds) Encyclopedia of forest science. Elsevier Academic Press, London, pp 132–139
Pancholy SK, Rice EL, Turner JA (1975) Soil factors preventing revegetation of a denuded area near an abandoned zinc smelter in Oklahoma. J Appl Ecol 12:337–342
Pearce F (1994) Europe top 100 polluters. New Scientist 143:9
Pommerening E (1977) Revegetation of Coeur-d'Alene mining district. Mining Congress J 63:20–23
Quinn M-L (1988) Tennessee's Copper Basin: A case for preserving an abused landscape. J Soil Water Conserv 43:140–144
Quinn M-L (1989) Early smelter sites: a neglected chapter in the history and geography of acid rain in the United States. Atmos Environ 23:1281–1292
Ragaini RC, Ralston HR, Roberts N (1977) Environmental trace-metal contamination in Kellogg, Idaho, near a lead smelting complex. Environ Sci Technol 11:773–781
Rassi P, Alanen A, Kemppainen E, Vickholm M, Väisänen R (eds) (1985) Uhanalaisten Eläinten ja Kasvien Suojelutoimikunnan Mietintö. II. Suomen Uhanalaiset Eläimet. Komiteanmietintö 43, Ympäristöministeriö, Helsinki (in Finnish)
Ratkin NE (1999) Application of the calculation method for estimating the pollutants accumulation in snow cover. In: Kalabin GV (ed) Ecological and geographical problems of the Kola North. Institute of North Industrial Ecology Problems, Apatity, pp 42–54 (in Russian)
Rebele F, Surma A, Kuznik C, Bornkamm R, Brej T (1993) Heavymetal contamination of spontaneous vegetation and soil around the copper smelter Legnica. Acta Soc Bot Pol 62:53–57
Redente EF, Zadeh H, Paschke MW (2002) Phytoxicity of smelter-impacted soils in southwest Montana, USA. Environ Toxicol Chem 21:269–274
Redente EF, Richards JL (1997) Effects of lime and fertilizer amendments on plant growth in smelter impacted sols in Montana. Arid Soil Res Rehabil 11:353–366
Reimann C, de Caritat P, Niskavaara H, Finne TE, Kashulina G, Pavlov VA (1998) Comparison of elemental contents in O- and C-horizon soils from the surroundings of Nikel, Kola Peninsula, using different grain size fractions and extractions. Geoderma 84:65–87
Rigina O, Kozlov MV (2000) Pollution impact on sub-Arctic northern taiga forests in the Kola peninsula Russia. In: Innes JL, Oleksyn J (eds) Forest dynamics in heavily polluted regions (IUFRO task force on environmental change. Report no. 1). CAB International, Wallingford, pp 37–65
Richter VA, Zvereva EL (1996) The tachinid species *Cleonice nitidiuscula* Zetterstedt new for fauna of Murmansk Province (Diptera: Tachinidae). Zoosyst Ross 6:202
Ruohomäki K, Kaitaniemi P, Kozlov MV, Tammaru T, Haukioja E, (1996) Density and performance of *Epirrita autumnata* (Lep., Geometridae) along three air pollution gradients in northern Europe. J Appl Ecol 33:773–785

Ruotsalainen AL, Markkola AM (2004) Kuolan niemimaalla kesällä 2003 pieni matkakertomus. Luonnon Tutkija 4:135–143

Rybicka EH, Jędrzejczyk B (1995) Preliminary studies on mobilization of copper and lead from contaminated soils and readsorption on competing sorbents. Appl Clay Sci 10:259–268

Sahi VM (1983) Frost heaving and needle ice formation and their effect upon seedling survival at selected sites in Sudbury. [MSc]. Laurentian University, Ontario

Salemaa M, Vanha-Majamaa I, Derome J (2001) Understorey vegetation along a heavy-metal pollution gradient in SW Finland. Environ Pollut 112:339–350

Šály R, Mihálik A (1985) Influence of magnesite plant immissions on mineral composition of soils. In: Klimo E, Šály R (eds) Air pollution and stability of coniferous forest ecosystems: international symposium. October 1–5, 1984. Institute of Forest Ecology, Brno, Czechoslovakia. University of Agriculture, Brno, pp 99–112

Savchenko VA (1998) Ecological problems of Taymyr. Russian Academy of Engineers, Moscow (in Russian)

Seigworth KL (1943) Ducktown – a postwar challenge. Am Forestry 49:521–523

Shilova II, Lukjanets AI (1989) Successions of steppe vegetation on areas impacted by aerial industrial emissions of non-ferrous metallurgy. In: Makhnev AK, Shilova II (eds) Vegetation of industrial landscapes of Ural. Institute of Plant and Animal Ecology, Sverdlovsk, pp 56–79 (in Russian)

Shilova II, Makhnev AK, Lukjanets AI (1984) Geochemical transformation of soils and vegetation on territories adjacent to non-ferrous metallurgy enterprises. In: Mamaev SA, Shilova II (eds) Ecological aspects of optimisation of industrial landscapes. Institute of Plant and Animal Ecology, Sverdlovsk, pp 14–35 (in Russian)

Shoji K, Sugai M (1992) The Ashio copper mine pollution case: the origins of environmental destruction. In: Ui J (ed) Industrial pollution in Japan. United Nations University Press, Tokyo, pp 18–63

Sivertsen B, Baklanov A, Hagen L, Makarova T (1994) Air pollution in the border areas of Norway and Russia. Summary report 1991–1993. NILU OR 56/94, Kjeller

Sokolov GI (1996) Forest decline near the city of Satka, Chelyabinsk region, due to industrial emissions of 'Magnesit' plant. In: Mozolevskaya EG, Sharapa TV (eds) International scientific conference 'Impact of atmospheric pollution and other human-created and natural factors on destabilization of forests in Central and Eastern Europe'. Abstracts of presentations, vol 1. Moscow State Forest University, Moscow, pp 35–37 (in Russian)

Sopper WE (1989) Revegetation of a contaminated zinc smelter site. Landscape Urban Plan 17:241–250

Spalding A, Haes ECM (1995) Contaminated land – a resource of wildlife: a review and survey of insects on metalliferous mine sites in Cornwall. Land Contam Reclam 3:24–29

Stavrova NI (1990) Regeneration of Scots Pine (*Pinus sylvestris* L.) in Northern Taiga forests under the impact of industrial pollution [PhD]. Komarov Botanical Institute, Leningrad, Russia (in Russian)

Stepanov AM, Chernenkova TV (1989) A study of forest biocenoses in the vicinity of copper plant. Izvestiya Akad Nauk Ser Biol 0(3):356–363 (in Russian)

Suding KN, Gross KL, Houseman GR (2004) Alternative states and positive feedbacks in restoration ecology. Trends Ecol Evol 9:46–53

Sutinen ML, Raitio H, Nivala V, Ollikainen R, Ritari A (1996) Effects of emissions from copper-nickel smelters on the frost hardiness of *Pinus sylvestris* needles in the subarctic region. New Phytol 132:503–512

Swain RE, Harkins WD (1908) Papers on smelter smoke. Arsenic in vegetation exposed to smelter smoke. J Am Chem Soc 30:915–928

Tarko AM, Bykadorov AV, Kryuchkov VV (1995) Modeling of air-pollution impacts on regional forest ecosystems. Doklady Rus Acad Sci 341:571–573 (in Russian)

Teale EW (1951) The murder of a landscape. Nat Hist 60:352–356

Tilman D (1999) The ecological consequences of changes in biodiversity: a search for general principles. Ecology 80:1455–1474

Tømmervik H, Hogda KA, Solheim L (2003) Monitoring vegetation changes in Pasvik (Norway) and Pechenga in Kola Peninsula (Russia) using multitemporal Landsat MSS/TM data. Rem Sens Environ 85:370–388

Tsvetkov VF (1991) State of forests subjected to industrial emission impact in Murmansk region and problems of their preservation. In: Anonymous (ed) Ecological research in forests of the European North. Arkhangelsk Institute of Forestry and Forest Chemistry, Arkhangelsk, pp 125–136 (in Russian)

Tucekova A (2001) Doriesenie umelej obnovy tazko zalesnitelnych ploch v magnezitovej imisnej oblasti Jelsava – Lubenik. [Artificial regeneration of scarcely reforestable area in a polluted area Jelsava – Lubenik]. Štúdia LVÚ, Zvolen (in Slovak)

Uhlig C, Salemaa M, Vanha-Majamaa I, Derome J (2001) Element distribution in *Empetrum nigrum* microsites at heavy metal contaminated sites in Harjavalta, western Finland. Environ Pollut 112:435–442

UNESCO (2006) World Heritage. http://whc.unesco.org/en

Usui H, Suzuki H (1973) Ecological and revegetational problems of the Ashio devastated area. In: Hutnik RJ, Davis G (eds) Ecology and reclamation of devastated land, vol 2. Gordon & Breach, NY, pp 395–406

Väisänen S (1986) Effects of air pollution by metal, chemical and fertilizer plants on forest vegetation at Kokkola, W Finland. Ann Bot Fenn 23:305–315

Vajda A, Venalainen A (2005) Feedback processes between climate, surface and vegetation at the northern climatological tree-line (Finnish Lapland). Boreal Environ Res 10:299–314

Vangronsveld J, Colpaert JV, VanTichelen KK (1996) Reclamation of a bare industrial area contaminated by non-ferrous metals: physico-chemical and biological evaluation of the durability of soil treatment and revegetation. Environ Pollut 94:131–140

Vangronsveld J, Vanassche F, Clijsters H (1995) Reclamation of a bare industrial-area contaminated by nonferrous metals: *in situ* metal immobilization and revegetation. Environ Pollut 87(1):51–59

Vorobeichik EL (2003) Ecological standartization of the toxicity loads on terrestrial ecosystems. [Doct]. Instutute of Plant and Animal Ecology of Ural division of Russian Academy of Scienses, Ekaterinburg (in Russian)

Vorobeichik EL (2004) Ecology of the impacted regions: prospects for basic research. In: Zhuikova TV (ed) Materials of the 6th meetings on populations. State Social and Pedagogical Academy, Nizhnij Tagil, pp 36–45

Vorobeichik EL, Marin YuF, Zhigalskii OA, Goryachev VM, Hantemirova EV, Lukjanov OA, Mikhailova IN, Kaigorodova SYu, Goldberg IL (1996) Prospects of the use of Visim reserve for studies of human impacts on terrestrial ecosystems. In: Marin YuF (ed) Problems of natural reserves. 25th anniversary of the Vitim reserve. Ekaterinburg Publ. House, Ekaterinburg, pp 19–23 (in Russian)

Vlasova TM, Filipchuk AN (1990) Selection of bioindicators for local monitoring of northern forests under aerial industrial pressure. In: Anonymous (ed) International symposium "Northern forests: state, dynamics, human impact", July 16–26, 1990. The USSR State Forest Committee, Moscow, pp 6–17 (in Russian, English summary)

Weber J (2002) Scanning electron microscope approaches to heavy metal polluted soils in the vicinity of the copper smelter. In: International conference on sustainable land use and management 2002, Çanakkale. www.toprak.org.tr/isd/can_52.htm

Wilcke W, Guschker C, Kobza J, Zech W (1999) Heavy metal concentrations, partitioning, and storage in Slovak forest and arable soils along a deposition gradient. J Plant Nutrition Soil Sci 162:223–229

Winterhalder K (2002) Reclamation of smelter-damaged lands. In: Richard I, Barnhisel RI, Darmody RG, Daniels WL (eds) Reclamation of drastically disturbed lands (Agronomy Monograph, no. 41). American Society of Agronomy, Madison, pp 819–853

Winterhalder K, Lautenbach WE, Beckett P (2001) The Sudbury regional land reclamation program: a case study. In: Burley JB (ed) Environmental design for reclaiming surface mines. The Edwin Mellen Press, Lewinston, pp 286–294

Winterhalder K (2003) Minimal Amelioration as appropriate technology in the Green Flin Flon/Creighton. Paper presented at Sudbury 2003 Mining and the Environment, Laurentian University, Sudbury, Ontario, May 25–28, 2003 http://www.ott.wrcc.osmre.gov/library/proceed/sudbury2003/sudbury03/173.pdf

Wolak J (1970) Formation of new ecological systems under the impact of industrial emissions. Sylwan 114(9):35–39 (in Polish)

Wołk A (1977) Microclimate alterations in relation to the extent of forest destruction in vicinity of the nitrogen plant at Puławy. Sylwan 121(7):33–46 (in Polish)

Wolt DJ, Lietzke DA (1982) The influence of anthropogenic sulfur inputs upon soil properties in the Copper Basin region of Tennessee. Soil Sci Soc Am J 46:651–656

Wood CW, Nash TN III (1976) Copper smelter effluent effects on Sonoran desert vegetation. Ecology 57:1311–1316

Zeid IM (2001) Responses of *Phaseolus vulgaris* to chromium and cobalt treatments. Biol Plant 44:111–115

Zvereva EL, Kozlov MV (2000a) Effect of air pollution on natural enemies of the leaf beetle, *Melasoma lapponica*. J Appl Ecol 37:298–308

Zvereva EL, Kozlov MV (2000b) Pollution suppresses delayed inducible resistance in boreal willow *Salix borealis*. Ecol Lett 3:85–89

Zvereva EL, Kozlov MV (2000c) Bud removal alleviates negative effects of pollution on quality of *Salix borealis* for insect herbivore. Basic Appl Ecol 1:171–176

Zvereva EL, Kozlov MV (2001) Effects of pollution induced habitat disturbance on the response of willows to simulated herbivory. J Ecol 89:21–30

Zvereva EL, Kozlov MV (2004) Facilitative effects of top-canopy plants on four dwarf shrub species in habitats severely disturbed by pollution. J Ecol 92:288–296

Zvereva EL, Kozlov MV (2005) Growth and reproduction of dwarf shrubs, *Vaccinium myrtillus* and *V. vitis-idaea*, in a severely polluted area. Basic Appl Ecol 6:261–274

Zvereva EL, Kozlov MV (2006) Top-down effects on population dynamics of *Eriocrania* miners (Lepidoptera) under pollution impact: does enemy-free space exist? Oikos 115:413–426

Zvereva EL, Kozlov MV (2007) Facilitation of bilberry by mountain birch in habitat severely disturbed by pollution: importance of sheltering. Environ. Exptl. Bot. doi:10.1016/j.envexpbot.2006.10.005

Zvereva EL, Kozlov MV, Neuvonen S (1995) Decrease in feeding niche breadth of *Melasoma lapponica* (Coleoptera: Chrysomelidae) with increase in pollution. Oecologia 104:323–329

Zvereva EL, Kozlov MV, Niemelä P, Haukioja E (1997) Delayed induced resistance and increase in leaf fluctuating asymmetry as responses of *Salix borealis* to insect herbivory. Oecologia 102:368–373

Zvereva EL, Kozlov MV, Kruglova OYu (2002) Color polymorphism in relation to population dynamics of the leaf beetle *Chrysomela lapponica*. Evol Ecol 16:523–539

Zvereva EL, Serebrov V, Glupov V, Dubovsky I (2003) Activity and heavy metal resistance of non-specific esterases in leaf beetle *Chrysomela lapponica* from polluted and unpolluted habitats. Comp Biochem Physiol (Part C) 135:383–391

Rev Environ Sci Biotechnol (2006) 5:203–218
DOI 10.1007/s11157-006-0007-y

REVIEW PAPER

# Extremely halophilic archaea and the issue of long-term microbial survival

**Sergiu Fendrihan · Andrea Legat · Marion Pfaffenhuemer · Claudia Gruber · Gerhard Weidler · Friedrich Gerbl · Helga Stan-Lotter**

Received: 6 March 2006 / Accepted: 22 May 2006 / Published online: 19 July 2006


**Abstract** Halophilic archaebacteria (haloarchaea) thrive in environments with salt concentrations approaching saturation, such as natural brines, the Dead Sea, alkaline salt lakes and marine solar salterns; they have also been isolated from rock salt of great geological age (195–250 million years). An overview of their taxonomy, including novel isolates from rock salt, is presented here; in addition, some of their unique characteristics and physiological adaptations to environments of low water activity are reviewed. The issue of extreme long-term microbial survival is considered and its implications for the search for extraterrestrial life. The development of detection methods for subterranean haloarchaea, which might also be applicable to samples from future missions to space, is presented.

**Keywords** Extreme halophiles · Haloarchaea · Life detection · Microbial longevity · Salt mines · Salt sediments · Space missions · Subterranean · Taxonomy of halobacteriaceae

S. Fendrihan · A. Legat · M. Pfaffenhuemer · C. Gruber · G. Weidler · F. Gerbl · H. Stan-Lotter (✉)
Department of Microbiology, University of Salzburg, Billrothstr. 11, A-5020 Salzburg, Austria
e-mail: helga.stan-lotter@sbg.ac.at

## Introduction

The halobacteria are a group of microorganisms with so many unusual features—growth at salt concentrations higher than those used in any food pickling processes, striking pigmentation in red, orange or purple, obligately salt-dependent enzymes, possessors of the first known proton pump, bacteriorhodopsin, which is driven just by sunlight—that early researchers became almost desperate about "the halobacteria's confusion to biology", which was the title of a lecture given by Larsen (1973) and which described what was known at the time about that "life in the borderland of physiological possibilities". While some halobacterial features have turned out to be not completely singular and while the molecular basis for halophilism is being unraveled, there are still many characteristics which are unique, and new ones have been added since—e.g. a square and flat morphology, potential longevity of halophilic microorganisms in salt sediments for millions of years, implications of the discovery of extraterrestrial halite.

Many excellent monographs and books with extensive reviews on halophilic microorganisms exist (Rodriguez-Valera 1988; Javor 1989; Vreeland and Hochstein 1993; Oren 1999, 2002; Ventosa 2004; Gunde-Cimerman et al. 2005). The subjects of this chapter are a brief survey of haloarchaeal taxonomy, a presentation of some of the special properties of extreme halophiles, which are not—or only rarely—present in other microorganisms, and a consideration of the questions about their geological age and the possiblity of halophilic life on other planets or moons.

## Taxonomy and phylogeny

The extremely halophilic archaea (also called haloarchaea or, traditionally, "halobacteria") belong to the order Halobacteriales, which contains one family, the *Halobacteriaceae* (Grant et al. 2001); since the publication of Bergey's Manual of Systematic Bacteriology (2001), which listed 14 recognized haloarchaeal genera, the number has increased to 19 genera, according to the International Committee on Systematics of Prokaryotes (http://www.the-icsp.org). The number of validated species at this time is 57. One taxonomic criterion for the identification and recognition of haloarchaea is the sequence of the 16S rRNA genes; specific signature sequences and signature bases have been described in detail by Kamekura et al. (2004). These authors recommend also the determination of 23S rRNA gene sequences for refined assignment of novel isolates to haloarchaeal genera. The currently recognized genera and species of the family *Halobacteriaceae* are listed in Table 1; shown are also the data bank accession numbers for the 16S rRNA gene sequences, which lead in most cases to the literature for strain isolation and description.

Historically, the composition of membrane polar lipids has long been used as one of the key chemotaxonomic criteria for the differentiation of haloarchaeal genera (Ross et al. 1985; Kamekura and Kates 1999). All haloarchaea examined to date possess ether-linked phosphoglycerides; phosphatidyl glycerol and phosphatidyl glycerol phosphate methyl ester are always present; many strains contain phosphatidyl glycerol sulfate and one or more glycolipids or sulfated glycolipids (Grant et al. 2001); most glycerol ether core lipids contain $C_{20}C_{20}$ (diphytanyl) isoprenoids, although some strains, especially haloalkaliphiles, possesss also $C_{20}C_{25}$ (phytanyl-sesterterpanyl) or $C_{25}C_{25}$ (di-sesterterpanyl) isoprenoid chains. The halobacterial taxonomy based on the polar lipid composition proved remarkably consistent with phylogenetic data deduced from 16S rRNA gene sequence comparisons (Grant et al. 2001). Halobacteria (haloarchaea) are a monophyletic group, with the most distantly related species showing a 16S rRNA gene sequence similarity of 83.2% (Grant et al. 2001). The methanogens, another archaeal group, are their closest relatives, with less than 80% 16S rRNA gene sequence similarity (Olsen et al. 1994). The complete list of required and recommended criteria for the determination and recognition of haloarchaeal species was proposed by Oren et al. (1997). Three genomes of haloarchaea have been sequenced, that of *Halobacterium salinarum* NRC-1 (Ng et al. 2000), *Haloarcula marismortui* (Baliga et al. 2004) and *Natronomonas pharaonis* (Falb et al. 2005).

## Morphology, envelopes and inner structures

The principal morphological types of haloarchaea are rods, cocci and irregular pleomorphic forms, which are mostly rather flat cells. A very unusual shape is exhibited by the well known "square bacterium", which was detected by Walsby (1980, 2005) and belongs to the haloarchaea (Antón et al. 1999): almost perfectly quadratic cells are attached to each other like stamps and can form large thin sheets (Kessel and Cohen 1982); sometimes cell division has occurred, but individual cells have not separated and have grown to sizes of $40 \times 40$ μm or even larger (Bolhuis 2005). The reason for the occurrence of such sheets is unknown; a proposition is that oxygen diffusion might be facilitated by large surface areas in the notoriously oxygen-depleted brines (Grant et al. 2001); another suggestion is that square cells, which are probably also phototrophic due to the presence of bacteriorhodopsin (see below), would reach the water surface passively, without the expense of energy as required by flagellar movement (Walsby 2005).

The morphology of non-coccoid haloarchaea can change, dependent on the salt concentration of the environment. With increasing dilution of salt, club-shaped, swollen and bent rods or spheres appear (Mohr and Larsen 1963; Kushner

**Table 1** The family *Halobacteriaceae*. Genera, species and accession numbers (Source: http://www.the-icsp.org and *Epub Int J Syst Evol Microbiol*)

| Genus | Species | Accession number |
|---|---|---|
| Genus I. *Halobacterium* | *Halobacterium salinarum* (type species) | AJ496185 |
| | *Halobacterium noricense* | AJ548827 |
| Genus II. *Haloarcula* | *Haloarcula vallismortis* (type species) | D50851 |
| | *Haloarcula marismortui* | X61688 (rrnA), X61689 (rrnB) |
| | *Haloarcula hispanica* | U68541 |
| | *Haloarcula japonica* | D28872 |
| | *Haloarcula argentinensis* | D50849 |
| | *Haloarcula quadrata* | AB10964 |
| Genus III. *Halobaculum* | *Halobaculum gomorrense* (type species) | L37444 |
| Genus IV. *Halococcus* | *Halococcus morrhuae* (type species) | D11106 |
| | *Halococcus saccharolyticus* | AB004876 |
| | *Halococcus salifodinae* | AB004877 |
| | *Halococcus dombrowskii* | AJ420376 |
| Genus V. *Haloferax* | *Haloferax volcanii* (type species) | K00421 |
| | *Haloferax gibbonsii* | D13378 |
| | *Haloferax denitrificans* | D14128 |
| | *Haloferax mediterranei* | D11107 |
| | *Haloferax alexandrinus* | AB037474 |
| | *Haloferax lucentensis* | AB081732 |
| | *Haloferax sulfurifontis* | AY458601 |
| Genus VI. *Halogeometricum* | *Halogeometricum borinquense* (type species) | AF002984 |
| Genus VII. *Halorhabdus* | *Halorhabdus utahensis* (type species) | AF071880 |
| Genus VIII. *Halorubrum* | *Halorubrum saccharovorum* (type species) | U17364 |
| | *Halorubrum sodomense* | D13379 |
| | *Halorubrum lacusprofundi* | X82170 |
| | *Halorubrum coriense* | S70839 |
| | *Halorubrum distributum* | D63572 |
| | *Halorubrum vacuolatum* | D87972 |
| | *Halorubrum trapanicum* | X82168 |
| | *Halorubrum tebenquichense* | AJ276887 |
| | *Halorubrum terrestre* | AB090169 |
| | *Halorubrum xinjiangense* | AY510707 |
| | *Halorubrum alkaliphilum* | AY510708 |
| Genus IX. *Haloterrigena* | *Haloterrigena turkmenica* (type species) | AB004878 |
| | *Haloterrigena thermotolerans* | AF115478 |
| Genus X. *Natrialba* | *Natrialba asiatica* (type species) | D14123 |
| | *Natrialba magadii* | X72495 |
| | *Natrialba taiwanensis* | D14124 |
| | *Natrialba aegyptiaca* | AF251941 |
| | *Natrialba hulunbeirensis* | AF262026 |
| | *Natrialba chachannaoensis* | AJ003193 |
| Genus XI. *Natrinema* | *Natrinema pellirubrum* (type species) | AJ002947. |
| | *Natrinema pallidum* | AJ002949 |
| | *Natrinema versiforme* | AB023426 |
| | *Natrinema altunense* | AY277583 |
| Genus XII. *Natronobacterium* | *Natronobacterium gregoryi* (type species) | D87970 |
| Genus XIII. *Natronococcus* | *Natronococcus occultus* (type species) | Z28378 |
| | *Natronococcus amylolyticus* | D43628 |
| Genus XIV. *Natronomonas* | *Natronomonas pharaonis* (type species) | D87971 |
| Genus XV. *Natronorubrum* | *Natronorubrum bangense* (type species) | Y14028 |
| | *Natronorubrum tibetense* | AB005656 |
| Genus *Halomicrobium* | *Halomicrobium mukohataei* (type species) | D50850 |
| Genus *Halobiforma* | *Halobiforma haloterrestris* (type species) | AF333760 |
| | *Halobiforma nitratireducens* | AB045012 |
| | *Halobiforma lacisalsi* | AY277582 |

**Table 1** continued

| Genus | Species | Accession number |
|---|---|---|
| Genus *Halosimplex* | *Halosimplex carlsbadense* (type species) (three 16S rDNA sequences were reported) | AF320478<br>AF320479<br>AF320480 |
| Genus *Halalkalicoccus* | *Halalkalicoccus tibetensis* (type species) | AF435112 |
| Genus, "*Halovivax*" | *Halovivax asiaticus (type species)* Castillo et al. (2006) | AM039978 |

and Bayley 1963). One reason for this behaviour is their cell envelope layer which needs the presence of high concentrations of cations for stability. Rod-shaped and pleomorphic haloarchaeal species possess a surface layer (S-layer), which is composed of a tightly packed hexagonal lattice consisting of one type of glycoprotein (Lechner and Sumper 1987) and is firmly anchored to the plasma membrane by a transmembrane domain; O- and N-glycosylation sites differ from species to species (see Eichler 2003 for a review). The S-layer is thought to have a shape-maintaining function; on lowering the cation concentration, progressively fewer negative charges will be shielded by the positively charge cations, which results in disruption of the envelope layer due to electrostatic repulsion between negative charges on the constituents of the envelope and also of the membranes, with ensuing lysis of cells (Boring et al. 1963). However, other factors play also a role in the process of disintegration, since interference with the biosynthesis of the glycoprotein or its glycosylation produces similarly irregular or spherical cells from rod shaped haloarchaea (Mescher and Strominger 1976). In contrast, a rigid cell wall composed of heteropolysaccharides is present in the coccoid species *Halococcus* and *Natronococcus* (Schleifer et al. 1982; Niemetz et al. 1997). Placing these cells in hypotonic solutions will not cause lysis; cells maintain their viability as was detected by staining with the LIVE/DEAD kit (Leuko et al. 2004, 2005); see also below.

Internal gas vesicles are only produced by prokaryotes; several bacteria and also some halophilic archaea are capable of synthesizing these flotation devices (Walsby 1994). Gas vesicles are filled by diffusion with gases dissolved in the environment; their function is apparently to provide buoyancy and enabling cells to regulate their position in the water. The identification of gas vesicle genes and their regulation is carried out in the laboratories of Pfeifer (Pfeifer et al. 1997; Pfeifer 2004) and DasSarma (Shukla and DasSarma 2004). Walsby (2005) pointed out that without the presence of gas vesicles, which can be detected easily by phase contrast microscopy due to their refractivity, the square haloarchaeal sheets described above would hardly have been recognized as living entities.

Inside the cytoplasm of *Halobacterium salinarum*, fibrillary structures were identified, which apparently consist of a bundle of hollow tubes and were termed "fibrocrystalline bodies" (Cho et al. 1967). Recently, the isolation of these structures was reported and their sensitivity to the drug vincristine (Alba et al. 2001). This feature, together with the appearance of the fibrils, could indicate the presence of a cytoskeleton-like organelle in haloarchaea.

## Biochemistry and bioenergetics

Most enzymes from haloarchaea, with only few exceptions, are optimally active and stable in the presence of 3–4 M KCl or NaCl (for reviews see Lanyi 1974; Eisenberg and Wachtel 1987; Oren 2002). The often observed preference of $K^+$ instead of $Na^+$ is consistent with the unusual intracellular concentration of potassium ions in haloarchaea, which can be as high as 5 M (Christian and Waltho 1962). The stability and solubility of halophilic proteins under conditions, where non-halophilic proteins would precipitate, was puzzling for early researchers; Reistad (1970) was the first to quantitatively analyse the bulk amino acid composition. She found a high excess—more than 10 mole-%—of acidic amino acids (aspartic and glutamic acid) in proteins from *Halobacterium*. The majority of haloarchaeal proteins have isoelectric points around 4.2 and nearly lack basic proteins; the unique acidity of the proteome from *Halobacterium salinarum* NRC-1 was confirmed by analysis of sequencing data and compared with other microbial proteomes in a clear graphic depiction (Kennedy et al. 2001). Another characteristic feature of halophilic proteins is their low content of hydrophobic amino acids and accordingly, low extent of hydrophobic interactions within proteins (Lanyi 1974); thus, high salt is needed to maintain those interactions.

Some haloarchaeal proteins were analysed in detail, notably ferredoxin, malate dehydrogenase, dihydrofolate reductase, and showed in X-ray diffraction studies a tight network of acidic residues on the protein surfaces, where water and hydrated $K^+$ ions are sequestered and form a large number of internal salt bridges (Dym et al. 1995; Frolow et al. 1996; Mevarech et al. 2000; Pieper et al. 1998).

The polar lipids of haloarchaea are used as suitable taxonomic markers (see above); as in polar lipids from all archaea, the core structure is derived from *sn* 2, 3 substituted glycerol, which is a different structure than the polar lipids of most other organisms, where the hydrocarbon chains are connected in a *sn* 1, 2 configuration. Phytanyl ($C_{20}$) and sesterterpanyl ($C_{25}$) chains are present, which represent fixed lengths of isoprenoids (Kates 1993; Kamekura 1993; Kates and Kushwaha 1995). An equivalent to the adaptation to temperature changes by varying chains lengths of fatty acids, as is possible for many bacteria, is therefore precluded for haloarchaea. The hydrocarbon chains are generally fully saturated; in one case, *Halorubrum lacusprofundi* from a lake in Antarctica, which grows at the low temperature of 4°C, unsaturated phytenyl chains were found, which might contribute to the regulation of membrane fluidity (Franzmann et al. 1988).

The intense red, pink or purple pigmentation of haloarchaea is due to carotenoids in their membranes and to the $C_{50}$ compound bacterioruberin or its derivates (Kamekura 1993). These non-polar lipids provide a protection against the strong sunlight which is usually characteristic for the natural environments of haloarchaea, and they may act as membrane reinforcers (Kamekura 1993). The retinal containing protein bacteriorhodopsin is present in some species of haloarchaea (Javor 1989) and has been thoroughly studied, because it is a comparatively simple light driven proton pump, capable of producing a pH gradient across the membrane, which is used for the production of ATP; this was first demonstrated in a famous reconstitution experiment by Racker and Stockenius (1974) and opened the way for modern bioenergetics and Peter Mitchell's chemiosmotic theory. For recent reviews on bacteriorhodopsin and related proteins see Lanyi and Varo (1995), Varo (2000), Essen (2002) and Lanyi (2005). Bacteriorhodopsin is not an obligately halophilic protein, but functions well in the absence of salt. The membrane ATP-synthase of several haloarchaea was isolated and showed structural similarities to the V-type ATPases of plants, animals and some bacteria (Ihara et al. 1991; Stan-Lotter et al. 1991; Steinert et al. 1995), but enzyme activity was also affected by compounds which are inhibitors of the F-type ATPase/ATP-synthase (Hochstein 1992; Hochstein and Lawson 1993). Only the haloarchaeal ATPase moiety has been isolated so far, as well as the membrane-embedded proteolipid, but not the whole ATP synthesizing complex (Ihara et al. 1997). Haloarchaea use protons as coupling ions between their respiratory chains and ATP synthases; this applies also to *Natronomonas pharaonis*, which was rather unexpected, since alkaliphiles had generally been thought to use sodium ions instead (Falb et al. 2005).

## Genetics

The genome of *Halobacterium salinarum* strain NRC-1 was the first halobacterial genome to be completely sequenced (Ng et al. 2000); in 2004 and 2005, the genome sequences of *Haloarcula marismortui* and *Natronomonas pharaonis*, respectively, were published (Baliga et al. 2004; Falb et al. 2005). The genome of *Halobacterium salinarum* NRC-1 consists of a total of 2,571,010 base pairs, which are organised in one large chromosome and two plasmids called pNRC100 (191,346 bp) and pNRC200 (365,425 bp); several smaller minor plasmids may be present, which are deletion derivatives of pNRC100. The plasmids have been called minichromosomes or megaplasmids in earlier work, since it was recognized that halobacterial DNA consists of several fractions; they account for 11–36% of the total DNA (Joshi et al. 1963; Pfeifer 1988) and are found in many members of the *Halobacteriaceae* (Gutierrez et al. 1986). The genome of *Halobacterium salinarum* strain NRC-1 contains a total of 91 insertion sequences, which had been identified and classified previously (Charlebois and DasSarma 1995). The insertion sequences are highly mobile elements and are responsible for the high frequency of spontaneous mutations (Charlebois 1999; Pfeifer et al. 1997). The genomes of *Haloferax volcanii* and *Haloferax mediterranei* contain much less insertion sequences and are therefore more stable (Lopez-Garcia et al. 1995).

The genome of *Haloarcula marismortui* is almost twice the size of that of *Halobacterium salinarum* NRC-1—a total of 4,274,642 bp were reported, which are organised into nine replicons (Baliga et al. 2004). The largest replicon was called chromosome I and consists of 3,131,724 bp. Comparisons of the genome sequences of the two haloarchaea revealed interesting additional capabilities for survival in *Haloarcula marismortui*, such as extra signal transducers, numerous environmental response regulators and extra pathways for amino acid synthesis; the number of identified insertion elements was about 40 (Baliga et al. 2004). The genome of *Natronomonas pharaonis* shows certain similarities to that of *Halobacterium salinarum* NRC-1, including nearly identical transposases, but much less insertion elements, which are often present in multiple copies (a total of about 40). Its size is 2,595,221 bp and it contains one plasmid of 130,989 bp and a second multicopy plasmid of 23,486 bp. High similarities to components of the signal transduction systems for chemo- and phototaxis of *Halobacterium salinarum* were detected, albeit with extensive shuffling of domains; coping with the extreme high pH of its environment appears to be supported by different types of glycoproteins in *Natronomonas pharaonis*, which make up a more complex envelope that the usual S-layer of many haloarchaea (Falb et al. 2005).

Extremely halophilic archaea possess, like other archaea, a range of eukaryote-like features such as multisubunit RNA polymerases, homologues to eukaryotic transcription factors, TATA-box promoters; these features make them distinct from eubacteria at the molecular level (Ng et al. 2000; Kennedy et al. 2001). The transcription of proteins in dependence of medium salinity has begun to be investigated with *Haloferax volcanii*; different salt concentrations triggered a differential transcription (Ferrer et al. 1996); differences in gene expression were identified with restriction patterns from genomic DNA, which originated from cells grown at 12% or 20% NaCl (Juez et al. 1990).

## Viable haloarchaea in rock salt

During several periods in the Earth's history, extensive sedimentation of halite and some other minerals from hypersaline seas took place. An estimated 1.3 million cubic kilometers of salt were deposited in the late Permian and early Triassic periods alone (ca. 240–280 million years ago; Zharkov 1981). The continental land masses were concentrated and formed the supercontinent Pangaea. Salt sediments developed in large basins, which were connected to the open oceans by narrow channels. The paleoclimate was warm and arid in a wide belt around the equator, causing large scale evaporation. About 100 million years ago, Pangaea started to break up; the continents were displaced to the North, and folding of new mountain ranges such as the Alps and Carpathians was underway (Einsele 1992). As a result of these movements driven by plate tectonics, huge salt deposits are found today mainly in the Northern regions of the continents (Fig. 1), e.g. in Siberia, Northern and Central Europe (Zechstein series), South-Eastern Europe (Alps and Carpathian mountains), and the Midcontinent basin in North America (Zharkov 1981).

The formation of most of the Alpine salt sediments and the Zechstein deposits is dated to the Late Permian period, while some Alpine deposits are dated to the Early Triassic period (see also Radax et al. 2001 for a detailed description). No significant salt sedimentation

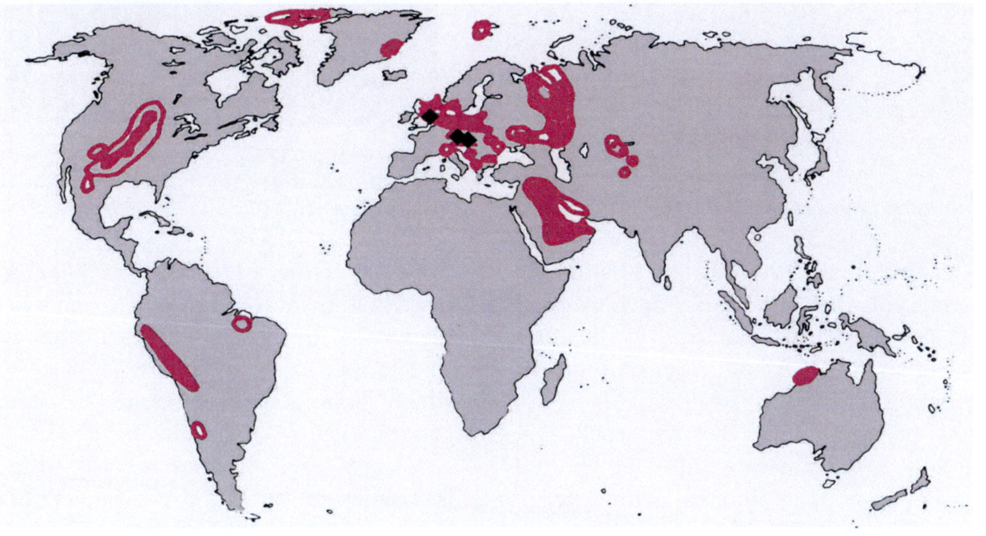

**Fig. 1** Distribution of known and presumed Permian salt sediments (modified from Javor, 1989). Black diamonds indicate locations of rock salt samples (Zechstein deposit in England; alpine deposits in Germany and Austria)

had occurred after that period in the pre-Alpidic regions. This is different from some other salt evaporites in Europe; for instance, waters from the receding Tethys sea in the Eastern parts of Eurasia caused salt sedimentation well into the Miocene (about 20 million years ago). The stratigraphic positions of the evaporites, together with the determinations of sulfur isotope ratios, are indicators for the geological age (Holser and Kaplan 1966). In addition, pollen grains or spores from extinct plants in the sediments, which are often well preserved and exhibit distinct morphological features, have been examined (Klaus 1974). Both methods confirmed the Permo-Triassic origin of the Alpine and also of the Zechstein salt sediments. The Alpine salt deposits are located today at altitudes between 500 and 1200 m; their thickness is between 200 and 500 m, although some deeply buried layers were estimated to be up to 1000 m. The layers of clay and limestone prevented the washing-out of the salt during the heavy precipitation during the ice ages. Many of the salt deposits in Europe have been mined for centuries, and newly opened mine tunnels and shafts, as well as deep drilling operations provide opportunities for obtaining rock salt samples. Figure 2 shows freshly drilled bore cores, which were used for our investigations.

Dombrowski (1963) and Reiser and Tasch (1960) were the first to describe viable microorganisms, which were isolated from ancient rock salt, in the 1960s (reviewed by McGenity et al.

**Fig. 2** Drilled bore cores from the salt mine in Altaussee, Austria, obtained from about 500 m below surface. Pink portions represent halite; greyish portions contain mostly anhydrite and some clay

2000). More recently, Norton et al. (1993) classified isolates from British salt mines of Permian and Triassic age as species of *Haloarcula, Halorubrum*, *Halobacterium* and a variety of new types on the basis of polar lipid composition, and Gemmell et al. (1998) investigated the evolution of the *Haloarcula* representatives by comparing 16S rRNA gene sequences with those of surface isolates. Vreeland et al. (1998) isolated halophilic bacteria, some of them probably haloarchaea, from the Permian Salado formation in the mid-continent basin in the USA, and from brines close to that formation.

Our group isolated from Permian rock salt, which was collected from the salt mine in Bad Ischl, Austria, numerous colonies with intense pigmentation (Fig. 3), which indicated the presence of carotenoids and bacterioruberin. One isolate was a coccus, growing in clusters, which was designated strain BIp (Stan-Lotter et al. 1993). Based upon polyphasic taxonomic data, the strain was recognized as a novel species and named *Halococcus salifodinae* (Denner et al. 1994). This was the first isolate from ancient rock salt, which was formally classified and deposited in several international culture collections. Two independently isolated strains, Br3 (from solution-mined brine in Cheshire, England) and BG2/2 (from a bore core from the mine of Berchtesgaden, Germany) resembled *Halococcus salifodinae* BIp in many properties; in addition, rock salt samples were obtained eight years later

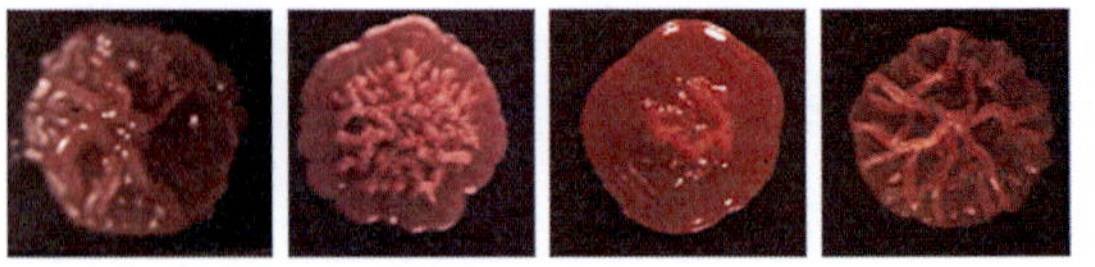

**Fig. 3** Colonies of haloarchaeal cells isolated from alpine rock salt, following growth for 3–4 months on agar with 20% NaCl. Diamenter of individual colonies is about 1 cm

from the same site and several halococci were recovered from these samples, which proved to be identical to strain BIp (Stan-Lotter et al. 1999). The data suggested that viable haloarchaea, which belong to the same species, occur in geographically separated evaporites of similar geological age. Another halococcal isolate from the Bad Ischl salt formation, which differed from the previously described strains, was subsequently identified as a novel species and named *Halococcus dombrowskii* (Stan-Lotter et al. 2002). *Halococcus salifodinae* and *Halococcus dombrowskii* have so far not been found in any hypersaline surface waters, or any location other than salt mines. Several non-coccoid strains were later obtained from a freshly drilled bore core at the salt mine in Altaussee, Austria (about 40 km distance from Bad Ischl), which were similar in their 16S rRNA sequence to *Halobacterium salinarum* NRC-1; however, other properties were different and consequently, a novel species was created, *Halobacterium noricense* (Gruber et al. 2004). Table 2 contains a list of the formally classified isolates from alpine rock salt and a

**Table 2** Haloarchaeal isolates from Permo-Triassic rock salt and salt mine brine

| Organism, strain | Type strain ($^T$), catalogue numbers | Origin | Reference |
|---|---|---|---|
| *Halococcus salifodinae* Bip | DSM8989$^T$<br>ATCC51437$^T$<br>JCM9578$^T$ | Rock salt (lumps), Bad Ischl, Austria | Denner et al. (1994) |
| *Halococcus salifodinae* BG2/2 | DSM13045 | Salt drill core, Berchtesgaden, Germany | Stan-Lotter et al. (1999) |
| *Halococcus salifodinae* Br3 | DSM13046 | Brine in salt mine, Cheshire, England | Stan-Lotter et al. (1999) |
| *Halococcus salifodinae* N1 | DSM13070 | Rock salt (lumps), Bad Ischl, Austria | Stan-Lotter et al. (1999) |
| *Halococcus salifodinae* H2 | DSM13071 | Rock salt (lumps), Bad Ischl, Austria | Stan-Lotter et al. (1999) |
| *Halococcus dombrowskii* H4 | DSM14522$^T$<br>ATCC BAA-364$^T$<br>NCIMB13803$^T$ | Rock salt (lumps), Bad Ischl, Austria | Stan-Lotter et al. (2002) |
| *Halobacterium noricense* A1 | DSM15987$^T$<br>ATCC BAA-852$^T$<br>NCIMB13967$^T$ | Salt drill core, Altaussee, Austria | Gruber et al. (2004) |

strain from a British salt mine. Figure 4 shows the relationships of the haloarchaeal isolates from Permo-Triassic rock salt to several haloarchaeal type species in the form of a phylogenetic tree, based on 16S rRNA gene sequence data.

Another example of an isolate from ancient sediments is a single rod-shaped *Halobacterium* strain from a 97,000-year-old salt formation in the USA (Mormile et al. 2003); the isolate was deemed to resemble *Halobacterium salinarum* NRC-1. The microbial content of ancient rock salt is generally low - estimates range from 1–2 cells/kg of salt from a British mine (Norton et al. 1993) to $1.3 \times 10^5$ colony forming units (CFUs) per kg of alpine rock salt (Stan-Lotter et al. 2000); nevertheless, the reports showed that viable haloarchaeal isolates were obtained reproducibly by several groups around the world.

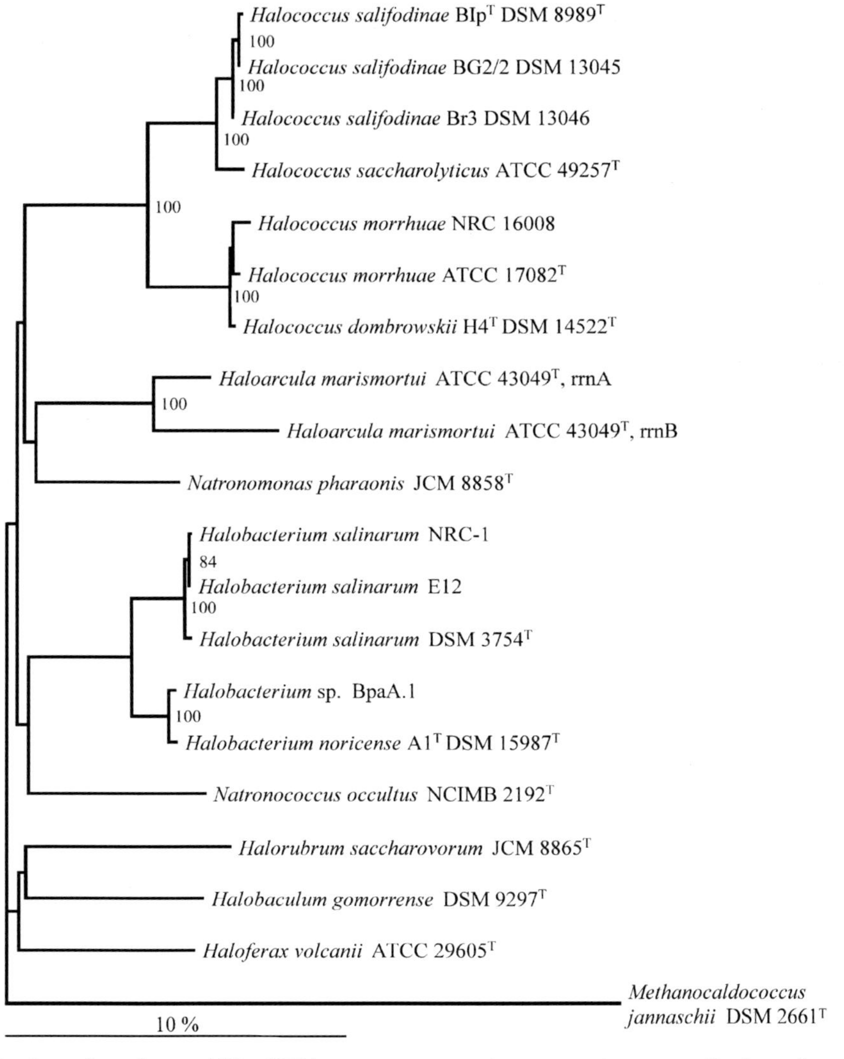


**Fig. 4** Phylogenetic tree based on 16S rRNA gene sequence data indicating the relationship of haloarchaeal species isolated from Permo-Triassic rock salt with established haloarchaeal type species. Catalogue numbers are indicated following type species names. Sequences were aligned with Clustal X (Thompson et al. 1997) and subjected to phylogenetic analysis with distance matrix (Jukes and Cantor 1969), maximum likelihood and maximum parsimony methods, using programs of the PHYLIP package, version 3.5.1.c (Felsenstein 1993). The tree was constructed using the neighbor-joining method (Saitou and Nei 1987). Bootstrap values greater than 70% are indicated at nodes. Scale bar represents 10% sequence difference. *Methanocaldococcus jannaschii* was used as an outgroup

The data support the hypothesis that the halophilic isolates from subterranean salt deposits may be the remnants of populations which once inhabited ancient hypersaline seas; in addition, they provide strong evidence against the notion that the recovered strains could be the result of laboratory contamination, since the isolates were obtained independently from different locations.

Analysis of dissolved alpine rock salt with molecular methods was also performed by extracting DNA and subsequent amplification and sequencing of 16S rRNA genes. The results provided evidence for the occurrence of numerous haloarchaea, which have not yet been cultured (Radax et al. 2001; Fish et al. 2002). Similarities of these 16S rDNA gene sequences were less than 90–95% to known sequences in about 37% of approximately 170 analysed clones (Radax et al. 2001; Stan-Lotter et al. 2004); the remaining clone sequences were 98–99% similar to isolates from rock salts of various ages (McGenity et al. 2000) and to known haloarchaeal genera. These data suggested the presence of a very diverse microbial community in ancient rock salt.

**How old are the cells from rock salt?**

The salt sediments are thought to have been deposited about 280–240 million years ago; while there is no direct proof that viable haloarchaea have been entrapped in rock salt since its deposition, it would also be difficult to prove the opposite, namely that masses of diverse microorganisms entered the evaporites in recent times (see also McGenity et al. 2000). Especially for the Alpine deposits, an influx of meteoric waters containing microorganisms seems rather improbable, because these sediments have been folded up and located at altitudes of 1,000 m or higher for at least the last 100 million years (Einsele 1992), covered by impermeable layers of clay and carbonates (see Radax et al. 2001). In addition, a special property of most haloarchaea (except halococci) is their quick lysis when they are suspended in pure water; therefore, they would not survive for very long in salt-free aquatic environments. If a Permo-Triassic age is postulated for the haloarchaeal isolates, then it becomes necessary to explain the biological mechanisms for such extreme longevity. Grant et al. (1998) discussed several possibilities, such as the formation of resting stages other than spores—since haloarchaea are not known to form spores—or the maintenance of cellular functions with traces of carbon and energy sources within the salt, which would imply an almost infinitely slow metabolism.

What seemed rather inconceivable just a few years ago, is now being considered more seriously: vast numbers of microbes were detected repeatedly in subterranean locations; they may have stayed there for centuries or even millenia, since they were literally found almost everywhere in great depths (the current record is 5,278 m for thermophilic anaerobes; see Pedersen, 2000). Their occurrence is apparently only limited, apart from nutritional support, by the increase in temperature with depth. However, at this time there are no methods available to prove directly a great prokaryotic age, whether it be a bacterium or a haloarchaeon. The mass of an average prokaryotic cell is only about $10^{-12}$ g (picograms); it is composed of about 3,000 different biomolecules, which are present at femtogram levels or less; therefore, no current dating procedures can be applied. Perhaps suitable methods for application to single prokaryotic cells will be available in the future. After all, single atoms can be visualized in the laboratory by element-selective electron spectroscopy (Suenaga et al. 2000), and it is conceivable that certain isotopes may be identifed in a similar way from ancient material.

**Astrobiology**

Mars is a planet where the presence of salts has been demonstrated. Evidence for halite was found in the SNC meteorites (Gooding 1992); their Martian origin has been confirmed independently by several groups (Treiman et al. 2000; Rieder et al. 2004). Elements from Martian soil and rocks, which were recently determined with the alpha-particle X-ray spectrometer, include Na, Mg, Cl, Br (Rieder et al. 2004). Therefore, saturated salt solutions, which would possess greatly depressed freezing points, could be envisaged on Mars. They may not be present as standing pools, but rather could occur in small pore spaces between mineral grains, as suggested by Landis (2001). The apparent longevity of haloarchaeal strains in dry salty environments is of interest for astrobiological studies and the search for life on Mars. On Earth, microorganisms were the first life forms to emerge and were present perhaps as early as 3.8 billion years ago (Schidlowski 1988, 2001). If Mars and Earth had a similar geological past (Schidlowski 2001; Nisbet and Sleep 2001), then microbial life, or the remnants of it, could still be present on Mars. Since halophilic microorganisms appear to survive in dry salt over geological time scales, as our and other studies suggested (Norton et al. 1993; Grant et al. 1998; McGenity et al. 2000), it appears plausible to include specific searches for halophiles in the exploration of extraterrestrial samples or environments. If such searches are planned, several issues should be considered, e.g.

- where are microorganisms located in salt sediments—in fluid inclusions, between grains?
- how can their presence be recognized with some certainty?
- what is the survival potential of microorganisms from salt, under terrestrial and under Martian conditions?

Some answers to these issues can be obtained with the investigation of natural salt samples, e.g. with drill cores and salt lumps from subsurface sediments; some issues would best be approached with laboratory studies, e.g. by embedding halophilic microorganisms in salt crystals, under various settings, in a simulation of extraterrestrial and/or Martian conditions.

## Towards life detection methods

We considered the location and survival of haloarchaea in artificially produced halite and the potential effects of dissolution of salt crystals on embedded haloarchaea. We used analysis of single cells, which was mainly based on the labeling with fluorescent dyes (Leuko et al. 2004; Stan-Lotter et al. 2006) contained in the LIVE/DEAD® *Bac*Light bacterial viability kit (referred to as LIVE/DEAD kit) from Molecular Probes. The kit consists of two nucleic acid stains: SYTO 9, which penetrates most membranes freely, and propidium iodide, which is highly charged and normally cell-impermeant; it will, however, penetrate damaged membranes. Simultaneous application of both dyes can be used for enumeration of active cells: green fluorescence indicates viable cells with an intact membrane, whereas dead cells, due to a compromised membrane, show red fluorescence (Haugland 2002).

It is not known if the microorganisms in sediments survive while embedded in dry salt crystals, or in highly saline fluid inclusions, which occur in halite (Roedder 1984). So far, only one example of a single viable haloarchaeon in a fluid inclusion of a natural salt sample exists: a *Halobacterium* species was isolated by using a microdrill device and subsequently cultured (Mormile et al. 2003). Norton and Grant (1988) noted the preferential enclosure of halobacteria in fluid inclusions upon formation of crystals. We stained haloarchaeal cells with the LIVE/DEAD kit prior to embedding in salt and prepared salt crystals by drying the cellular suspensions (Fendrihan and Stan-Lotter 2004). Figure 5 shows an example of stained *Halobacterium* cells which were examined 2–3 days, following their embedding in artificial halite. At low magnification, the bright green fluorescence of stained haloarchaea was outlining well the morphology of the characteristic rectangular fluid inclusions of halite (Fig. 5, left panel). At higher magnifications, individual cells became visible (Fig. 5, right panel). The data suggested that the localization of halobacterial cells in the salt was probably exclusively in fluid inclusions of artificial halite, which formed during desiccation, and which are present also in natural halite (Roedder 1984; Mormile et al. 2003).

The recovery of viable cells, which were embedded in artificial halite for various times and under different environmental conditions, is being investigated. Figure 6 shows *Halobacterium salinarum* NRC-1 cells, which were carefully resuspended, following storage in halite for 2 weeks at 37°C, and subsequently stained with

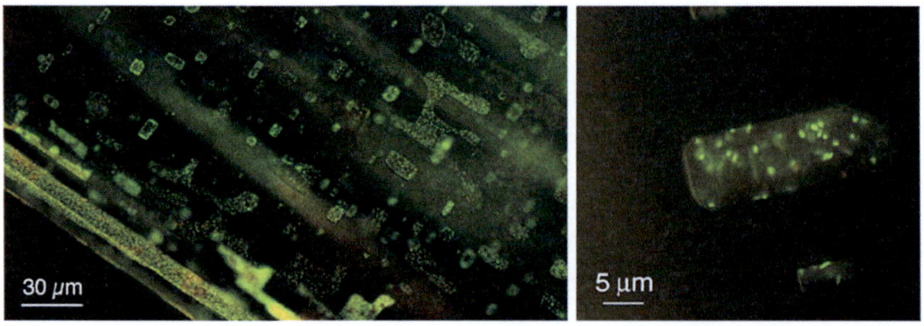


**Fig. 5** Localisation of pre-stained haloarchaea in fluid inclusions. Cells were stained with the LIVE/DEAD *Bac*Light kit prior to embedding in artificial halite. Low (left panel) and high (right panel) magnification of *Halobacterium salinarum* NRC-1 cells, trapped in fluid inclusions for about 2 days. Cells were observed with a Zeiss Axioskope fluorescence microscope

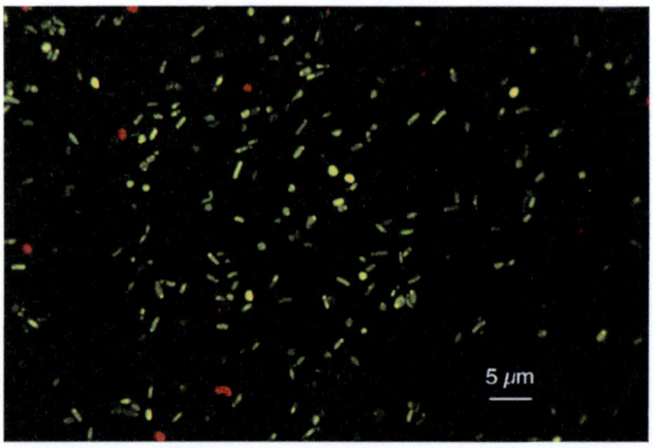


**Fig. 6** Epifluorescence microscopy of *Halobacterium salinarum* NRC-1 cells. Cultures were grown in complex medium, embedded in salt crystals for 14 days at 37°C, resuspended in a salt buffer and stained with the LIVE/DEAD *Bac*Light kit. Pictures were taken on a Aristoplan fluorescence microscope. Green fluorescence indicates intact membranes and thus viable cells, red fluorescence indicates non-viable cells

the LIVE/DEAD dyes. The majority of cells had intact membranes, as indicated by the green fluorescence; viability was confirmed by plating cell suspensions on solidified medium and determination of CFUs.

## Conclusions and outlook

1. Viable haloarchaea were isolated from ancient rock salt; since extraterrestrial halite was detected, a search for halophiles on Mars or in extraterrestrial samples might be plausible. For this purpose, in situ tests, which reveal the presence of cellular entities and provide intense signals, should be considered.

2. Detection of single, preferably viable cells in environmental samples should be improved and developed further. Background signals should be avoided, or their source be clearly identified; studies with different minerals, which are to be expected on Mars, should be done.

3. Embedding studies of haloarchaea in halite should be carried out under Martian or space conditions and the response of haloarchaea to low pressure and a carbon dioxide atmosphere should be examined.

4. The dormancy status of microorganism should be better clarified, preferably on a genetic basis. For this goal, proteomic studies of haloarchaea should be done, which are expected to show differences between protein patterns of actively growing cells and dormant cells, e.g. starved cells, embedded in halite.

**Acknowledgement** This work was supported by the Austrian Science Foundation (FWF), projects P16260-B07 and P18256-B06. We thank Dr. Nikolaus Bresgen, Department of Cell Biology, for access to and help with the Leitz Aristoplan fluorescence microscope, and Michael Mayr, M.Sc., Salinen Austria, for providing rock salt samples.

## References

Alba I, Torreblanca M, Sanchez M, Colom MF, Meseguer I (2001) Isolation of the fibrocrystalline body, a structure present in haloarchaeal species, from *Halobacterium salinarum*. Extremophiles 5: 169–175

Antón J, Llobet-Brossa FE, Rodriguez-Valera F, Amann R (1999) Fluorescence in situ hybridization analysis of the prokaryotic community inhabiting crystallizer ponds. Environ Microbiol 1:517–523

Baliga NS, Bonneau R, Facciotti MT, Pan M, Glusman G, Deutsch EW, Shannon P, Chiu Y, Weng RS, Gan RR, Hung P, Date SV, Marcotte E, Hood L, Ng WV (2004) Genome sequence of *Haloarcula marismortui*: a halophilic archaeon from the Dead Sea. Genome Res 14:2221–2234. Erratum in: Genome Res 2004, 14:2510

Bolhuis H (2005) Walsby's square archaeon. It's hip to be square but even more hip to be culturable. In: Gunde-Cimerman N, Oren A, Plemenitas A (eds) Cellular origins, life in extreme habitats and astrobiology (COLE), Seckbach J (ed) Vol. 9, Adaptations to life at high salt concentrations in archaea, bacteria and eukarya. Springer Verlag, Heidelberg, New York, pp 185–200

Boring J, Kushner DJ, Gibbons NE (1963) Specificity of the salt requirement of *Halobacterium cutirubrum*. Can J Microbiol 9:143–154

Castillo AM, Gutierrez MC, Kamekura M, Ma Y, Cowan DA, Jones BE, Grant WD, Ventosa A (2006) *Halovivax asiaticus* gen. nov., sp. nov., a novel extremely halophilic archaeon isolated from Inner Mongolia, China. Int J Syst Evol Microbiol 56:765–770

Charlebois RL, DasSarma S (1995) Insertion elements of halophiles. In: DasSarma S, Fleischmann EM (eds) Archaea: A laboratory manual—Halophiles. Cold Spring Harbor Laboratory Press, Cold Spring Harbor, New York, pp 253–256

Charlebois RL (1999) Evolutionary origins of the haloarchaeal genome. In: Oren A (eds) Microbiology and biogeochemistry of hypersaline environments. CRC Press, Boca Raton, pp 309–317

Cho KY, Doy CH, Mercer EH (1967) Ultrastructure of the obligate halophilic bacterium *Halobacterium halobium*. J Bacteriol 94:196–201

Christian JHB, Waltho JA (1962) Solute concentration within cells of halophilic and non-halophilic bacteria. Biochim Biophys Acta 65:506–508

Denner EBM, McGenity TJ, Busse H-J, Wanner G, Grant WD, Stan-Lotter H (1994) *Halococcus salifodinae* sp.nov., an archaeal isolate from an Austrian salt mine. Int J Syst Bacteriol 44:774–780

Dombrowski H (1963) Bacteria from Paleozoic salt deposits. Ann NY Acad Sci 108:453–460

Dym O, Mevarech M, Sussman JL (1995) Structural features stabilizing halophilic malate dehydrogenase from an archaebacterium. Science 267: 1344–1346

Eichler J (2003) Facing extremes: archaeal surface-layer (glyco)proteins. Microbiology 149:3347–3351

Einsele G (1992) Sedimentary basins. Evolution, facies and sediment budget. Springer, Berlin, Heidelberg, New York

Eisenberg H, Wachtel EJ (1987) Structural studies of halophilic proteins, ribosomes and organelles of bacteria adapted to extreme salt concentrations. Ann Rev Biophys Biophys Chem 16:69–92

Essen LO (2002) Halorhodopsin: light-driven ion pumping made simple? Curr Opin Struct Biol 12:516–522

Falb M, Pfeiffer F, Palm P, Rodewald K, Hickmann V, Tittor J, Oesterhelt D (2005) Living with two extremes: conclusions from the genome sequence of *Natronomonas pharaonis*. Genome Res 15:1336–1343

Felsenstein J (1993) PHYLIP (phylogenetic interference package) version 3.5.1c (distributed by the author) Department of Genetics. University of Seattle, WA, USA

Fendrihan S, Stan-Lotter H (2004) Survival of halobacteria in fluid inclusions as a model of possible biotic survival in Martian halite. In: Teodorescu HN, Griebel HS (eds) Mars and Planetary Science and Technolgy, selected papers from EMC'04. Performantica Press, Iasi, Romania, pp 9–18

Ferrer C, Mojica FJM, Juez G, Rodriguez-Valera F (1996) Differentially transcribed regions of *Haloferax volcanii* genome depending on the medium salinity. J Bacteriol 178:309–313

Fish SA, Shepherd TJ, McGenity TJ, Grant WD (2002) Recovery of 16S ribosomal RNA gene fragments from ancient halite. Nature 417:432–436. Erratum in: Nature 420 (2002):202

Franzmann PD, Stackebrandt E, Sanderson K, Volkman JK, Cameron DE, Stevenson PL, McMeekin TA, Burton HR (1988) *Halobacterium lacusprofundi* sp. nov., a halophilic bacterium isolated from Deep Lake, Antarctica. Syst Appl Microbiol 11:20–27

Frolow F, Harel M, Sussman JL, Mevarech M, Shoham M (1996) Insights into protein adaptation to a saturated salt environment from the crystal structure of a halophilic 2Fe–2S ferredoxin. Nature Struct Biol 3:452–458

Gemmell RT, McGenity TJ, Grant WD (1998) Use of molecular techniques to investigate possible long-term dormancy of halobacteria in ancient halite deposits. Ancient Biomol 2:125–133

Gooding JL (1992) Soil mineralogy and chemistry on Mars: possible clues from salts and clays in SNC meteorites. Icarus 99:28–41

Grant WD, Gemmell RT, McGenity TJ (1998) Halobacteria: the evidence for longevity. Extremophiles 2:279–287

Grant WD, Kamekura M, McGenity TJ, Ventosa A (2001) Class III. Halobacteria class. nov., In: Boone DR, Castenholz RW, Garrity GM (eds) Bergey's manual of systematic bacteriology, vol. I, 2nd edn. Springer Verlag, New York, pp 294–301

Gruber C, Legat A, Pfaffenhuemer M, Radax C, Weidler G, Busse H-J, Stan-Lotter H (2004) *Halobacterium noricense* sp. nov., an archaeal isolate from a bore core of an alpine Permo-Triassic salt deposit, classification of *Halobacterium* sp. NRC-1 as a strain of *Halobacterium salinarum* and emended description of *Halobacterium salinarum*. Extremophiles 8:431–439

Gunde-Cimerman N, Oren A, Plemenitas A (eds) (2005) Adaptations to life at high salt concentrations in archaea, bacteria and eukarya. Vol 9 of: Cellular origins, life in extreme habitats and astrobiology (COLE), Seckbach J (Series ed). Springer Verlag, Heidelberg, New York

Gutierrez MC, Garcia MT, Ventosa A, Nieto JJ, Ruiz-Berraquero F (1986) Occurrence of megaplasmids in halobacteria. J Appl Bacteriol 61:67–71

Haugland RP (2002) LIVE/DEAD BacLight bacterial viability kits. In: Gregory (eds) Handbook of fluorescent probes and research products, ninth edition. Molecular Probes, Eugene, Oregon, USA, pp 626–628

Hochstein LI (1992) ATP synthesis in *Halobacterium saccharovorum*: evidence that synthesis may be catalysed by an $F_0F_1$-ATP synthase. FEMS Microbiol Lett 97:155–159

Hochstein LI, Lawson D (1993) Is ATP synthesized by a vacuolar-ATPase in the extremely halophilic bacteria? Experientia 49:1059–1063

Holser WT, Kaplan IR (1966) Isotope geochemistry of sedimentary sulfates. Chem Geol 1:93–135

Ihara K, Mukohata Y (1991) The ATP synthase of *Halobacterium salinarium* (*halobium*) is an archaebacterial type as revealed from the amino acid sequences of its two major subunits. Arch Biochem Biophys 286:111–116

Ihara K, Watanabe S, Sugimura K, Katagiri I, Mukohata Y (1997) Identification of proteolipid from an extremely halophilic archaeon *Halobacterium salinarum* as an *N,N'*-dicyclohexyl-carbodiimide binding subunit of ATP synthase. Arch Biochem Biophys 341:267–272

Javor BJ (1989) Hypersaline environments: microbiology and biogeochemistry. Springer Verlag, Berlin, Heidelberg, New York

Joshi JG, Guild WR, Handler P (1963) The presence of two species of DNA in some halobacteria. J Mol Biol 6:34–38

Juez G, Rodriguez-Valera F, Herrero N, Mojica FJM (1990) Evidence for salt-associated restriction pattern modifications in the archaebacterium *Haloferax mediterranei*. J Bacteriol 172:7278–7281

Jukes TH, Cantor CR (1969) Evolution of protein molecules. In: Munro HN (eds) Mammalian Protein Metabolism, vol 3. Academic Press, New York, pp 21–132

Kamekura M (1993) Lipids of extreme halophiles. In: Vreeland RH, Hochstein LI (eds) The biology of halophilic bacteria. CRC Press, Boca Raton, pp 135–161

Kamekura M, Kates M (1999) Structural diversity of membrane lipids in members of *Halobacteriaceae*. Biosci Biotechnol Biochem 63:969–972

Kamekura M, Mizuki T, Usami R, Yoshida Y, Horikoshi K, Vreeland RH (2004) The potential use of signatures bases from 16S rRNA gene sequences to aid the assignment of microbial strains to genera of halobacteria. In: Ventosa A (ed) Halophilic Microorganisms. Springer Verlag, Berlin, Heidelberg, pp 77–87

Kates M (1993) Membrane lipids of extreme halophiles: biosynthesis, function and evolutionary significance. Experientia 49:1027–1036

Kates M, Kushwaha SC (1995) Isoprenoids and polar lipids of extreme halophiles. In: DasSarma S, Fleischmann EM (eds) Archaea, a laboratory manual. Halophiles. Cold Spring Harbor Laboratory Press, Cold Spring Harbor, pp 35–54

Kennedy SP, Ng WV, Salzberg SL, Hood L, DasSarma S (2001) Understanding the adaptation of *Halobacterium* species NRC-1 to its extreme environment through computational analysis of its genome sequence. Genome Res 11:1641–1650

Kessel M, Cohen Y (1982) Ultrastructure of square bacteria from a brine pool in southern Sinai. J Bacteriol 150:851–860

Klaus W (1974) Neue Beiträge zur Datierung von Evaporiten des Oberperm. Carinthia II, 164, Jahrgang 84:79–85

Kushner DJ, Bayley ST (1963) The effect of pH on surface structure and morphology of the extreme halophile, *Halobacterium cutirubrum*. Can J Microbiol 9:53–65

Landis GA (2001) Martian water: are there extant halobacteria on Mars? Astrobiology 1:161–164

Lanyi JK (1974) Salt-dependent properties of proteins from extremely halophilic bacteria. Bacteriol Rev 38:272–290

Lanyi JK, Varo G (1995) The photocycles of bacteriorhodopsin. Isr J Chem 35:365–385

Lanyi JK (2005) Proton transfers in the bacteriorhodopsin photocycle. Biochim Biophys Acta, Dec 9, 2005 [Epub ahead of print]

Larsen H (1973) The halobacteria's confusion to biology. The fourth A. J. Kluyver memorial lecture. Antonie van Leeuwenhoek 39:383–396

Lechner J, Sumper M (1987) The primary structure of a procaryotic glycoprotein. Cloning and sequencing of the cell surface glycoprotein gene of halobacteria. J Biol Chem 262:9724–979

Leuko S, Legat A, Fendrihan S, Stan-Lotter H (2004) Evaluation of the LIVE/DEAD BacLight kit for detection of extremophilic archaea and visualization of environmental hypersaline samples. Appl Environ Microbiol 70:6884–6886

Leuko S, Legat A, Fendrihan S, Wieland H, Radax C, Gruber C, Pfaffenhuemer M, Weidler G, Stan-Lotter H (2005) Isolation of viable haloarchaea from ancient salt deposits and application of fluorescent stains for *in situ* detection of halophiles in hypersaline environmental samples and model fluid inclusions. In: Gunde-Cimerman N, Oren A, Plemenitas A (eds) Cellular origins, life in extreme habitats and astrobiology (COLE), Seckbach J (ed) vol. 9, Adaptations to life at high salt concentrations in archaea, bacteria and eukarya. Springer Verlag, Heidelberg, New York, pp 91–104

Lopez-Garcia P, St Jean A, Amils R, Charlebois RL (1995) Genomic stability in the archaea *Haloferax volcanii* and *Haloferax mediterranei*. J Bacteriol 177:1405–1408

McGenity TJ, Gemmell RT, Grant WD, Stan-Lotter H (2000) Origins of halophilic microorganisms in ancient salt deposits. Environ Microbiol 2:243–250

Mevarech M, Frolow F, Gloss LM (2000) Halophilic enzymes: proteins with a grain of salt. Biophys Chem 86:155–164

Mescher MF, Strominger JL (1976) Structural (shape-maintaining) role of the cell surface glycoprotein of *Halobacterium salinarium*. Proc Natl Acad Sci USA 73:2687–2691

Mohr V, Larsen H (1963) On the structural transformation and lysis of *Halobacterium salinarium* in hypotonic and isotonic solutions. J Gen Microbiol 31:267–280

Mormile MR, Biesen MA, Gutierrez MC, Ventosa A, Pavlovich JB, Onstott TC, Fredrickson JK (2003) Isolation of *Halobacterium salinarum* retrieved directly from halite brine inclusions. Environ Microbiol 5:1094–1102

Ng WV, Kennedy SP, Mahairas GG, Berquist B, Pan M, Shukla HD, Lasky SR, Baliga NS, Thorsson V, Sbrogna J, Swartzell S, Weir D, Hall J, Dahl TA, Welti R, Goo YA, Leithauser B, Keller K, Cruz R, Danson MJ, Hough DW, Maddocks DG, Jablonski PE, Krebs MP, Angevine CM, Dale H, Isenbarger TA, Peck RF, Pohlschroder M, Spudich JL, Jung KW, Alam M, Freitas T, Hou S, Daniels CJ, Dennis PP, Omer AD, Ebhardt H, Lowe TM, Liang P, Riley M, Hood L, DasSarma S (2000) Genome sequence of *Halobacterium* species NRC-1. Proc Natl Acad Sci USA 97:12176–12181

Niemetz R, Karcher U, Kandler O, Tindall BJ, König H (1997) The cell wall polymer of the extremely halophilic archaeon *Natronococcus occultus*. Eur J Biochem 249:905–911

Nisbet EG, Sleep NH (2001) The habitat and nature of early life. Nature 409:1083–1091

Norton CF, Grant WD (1988) Survival of halobacteria within fluid inclusions in salt crystals. J Gen Microbiol 134:1365–1373

Norton CF, McGenity TJ, Grant WD (1993) Archaeal halophiles (halobacteria) from two British salt mines. J Gen Microbiol 139:1077–1081

Olsen GJ, Woese CR, Overbeck R (1994) The winds of (evolutionary) change: breathing new life into microbiology. J Bacteriol 176:1–6

Oren A, Ventosa A, Grant WD (1997) Proposed minimal standards for description of new taxa in the order Halobacteriales. Int J Syst Bacteriol 47:233–236

Oren A (ed) (1999) Microbiology and Biogeochemistry of hypersaline environments. CRC Press, Boca Raton

Oren A (ed) (2002) Halophilic microorganisms and their environments. Kluwer Academic Publishers, Dordrecht

Pedersen K (2000) Mini Review. Exploration of deep intraterrestrial microbial life: current perspectives. FEMS Microbiol Lett 185:9–16

Pfeifer F (1988) Genetics of halobacteria. In: Rodriguez-Valera F (ed) Halophilic bacteria, Vol II. CRC Press, Boca Raton, pp 105–133

Pfeifer F, Krüger K, Röder R, Mayr A, Ziesche S, Offner S (1997) Gas vesicle formation in halophilic Archaea. Arch Microbiol 167:259–268

Pfeifer F (2004) Gas vesicle genes in halophilic archaea and bacteria. In: Ventosa A (ed) Halophilic Microorganisms. Springer Verlag Berlin, Heidelberg, pp 229–241

Pieper U, Kapadia G, Mevarech M, Herzberg O (1998) Structural features of halophilicity derived from the crystal structure of dihydrofolate reductase from the Dead Sea halophilic archaeon, *Haloferax volcanii*. Structure 6:75–88

Racker E, Stoeckenius W (1974) Reconstitution of purple membrane vesicles catalyzing light-driven proton uptake and adenosine triphosphate formation. J Biol Chem 249:662–663

Radax C, Gruber G, Stan-Lotter H (2001) Novel haloarchaeal 16S rRNA gene sequences from Alpine Permo-Triassic rock salt. Extremophiles 5: 221–228

Reiser R, Tasch P (1960) Investigation of the viability of osmophile bacteria of great geological age. Trans Kans Acad Sci 63:31–34

Reistad R (1970) On the composition and nature of the bulk protein of the extremely halophilic bacteria. Arch Microbiol 71:353–360

Rieder R, Gellert R, Anderson RC, Bruckner J, Clark BC, Dreibus G, Economou T, Klingelhofer G, Lugmair GW, Ming DW, Squyres SW, d'Uston C, Wanke H, Yen A, Zipfel J (2004) Chemistry of rocks and soils at Meridiani Planum from the Alpha Particle X-ray Spectrometer. Science 306:1746–1749

Rodriguez-Valera F (ed) (1988) Halophilic bacteria, vol I, vol II. CRC Press, Boca Raton

Roedder E (1984) The fluids in salt. Amer Mineral 69:413–439

Ross HNM, Grant WD, Harris JE (1985) Lipids in archaebacterial taxonomy. In: Goodfellow M, Minnekin DE (eds) Chemical methods in bacterial systematics. Academic Press, London, pp 289–299

Saitou N, Nei M (1987) The neighbor-joining method: a new method for reconstructing phylogenetic trees. Mol Biol Evol 4:406–425

Schidlowski M (1988) A 3,800 million-year old record of life from carbon in sedimentary rocks. Nature 333:313–318

Schidlowski M (2001) Search for morphologigal and biochemical vestiges of fossil life in extraterrestrial settings: utility of terrestrial evidence. In: Horneck G, Baumstark-Khan C (eds) Astrobiology. The Quest for the Conditions of Life. Springer Verlag, Berlin, Heidelberg, New York, pp 373–386

Schleifer KH, Steber J, Mayer H (1982) Chemical composition and structure of the cell wall of *Halococcus morrhuae*. Zentrbl Bakteriol Mikrobiol Hyg I Abt Orig C3:171–178

Shukla HD, DasSarma S (2004) Complexity of gas vesicle biogenesis in *Halobacterium* sp. strain NRC-1: identification of five new proteins. J Bacteriol 186:3182–3186

Stan-Lotter H, Bowman EJ, Hochstein LI (1991) Relationship of the membrane ATPase from *Halobacterium saccharovorum* to vacuolar ATPases. Arch Biochem Biophys 284:116–119

Stan-Lotter H, Sulzner M, Egelseer E, Norton CF, Hochstein LI (1993) Comparison of membrane ATPases from extreme halophiles isolated from ancient salt deposits. Origins Life Evol Biosp 23:53–64

Stan-Lotter H, McGenity TJ, Legat A, Denner EBM, Glaser K, Stetter KO, Wanner G (1999) Very similar strains of *Halococcus salifodinae* are found in geographically separated Permo-Triassic salt deposits. Microbiology 145:3565–3574

Stan-Lotter H, Radax C, Gruber C, McGenity TJ, Legat A, Wanner G, Denner EBM (2000) The distribution of viable microorganisms in Permo-Triassic rock salt. In: Geertman RM (ed) SALT 2000. 8th world salt symposium. Elsevier Science BV, Amsterdam, pp 921–926

Stan-Lotter H, Pfaffenhuemer M, Legat A, Busse H-J, Radax C, Gruber C (2002) *Halococcus dombrowskii* sp. nov., an archaeal isolate from a Permo-Triassic alpine salt deposit. Int J Syst Evol Microbiol 52:1807–1814

Stan-Lotter H, Radax C, McGenity TJ, Legat A, Pfaffenhuemer M, Wieland H, Gruber C, Denner EBM (2004) From intraterrestrials to extraterrestrials – viable haloarchaea in ancient salt deposits. In: Ventosa A (ed) Halophilic microorganisms. Springer Verlag, Berlin, Heidelberg, New York, pp 89–102

Stan-Lotter H, Leuko S, Legat A, Fendrihan S (2006) The assessment of the viability of halophilic microorganisms in natural communities. In: Oren A, Rainey F (eds) Methods in Microbiology. Extremophiles. Elsevier, Oxford (in press)

Steinert K, Kroth-Pancic PG, Bickel-Sandkötter S (1995) Nucleotide sequence of the ATPase A- and B-subunits of the halophilic archaebacterium *Haloferax volcanii* and characterization of the enzyme. Biochim Biophys Acta 1249:137–144

Suenaga K, Tence M, Mory C, Colliex C, Kato H, Okazaki T, Shinohara H, Hirahara K, Bandow S, Iijima S (2000) Element-selective single atom imaging. Science 290:2280–2282

Thompson JD, Gibson TJ, Plewniak F, Jeanmougin F, Higgins DG (1997) The CLUSTAL_X windows interface: flexible strategies for multiple sequence alignment aided by quality analysis tools. Nucleic Acids Res 25:4876–82

Treiman AH, Gleason JD, Bogard DD (2000) The SNC meteorites are from Mars. Planet Space Sci 48:1213–1230

Varo G (2000) Analogies between halorhodopsin and bacteriorhodopsin. Biochim Biophys Acta 1460:220–229

Ventosa A (ed) (2004). Halophilic microorganisms. Springer Verlag, Berlin, Heidelberg, New York

Vreeland RH, Hochstein LI (eds) (1993) The biology of halophilic bacteria. CRC Press, Boca Raton

Vreeland RH, Piselli Jr AF, McDonnough S, Meyers SS (1998) Distribution and diversity of halophilic bacteria in a subsurface salt formation. Extremophiles 2:321–331

Walsby AE (1980) A square bacterium. Nature 283:69–71

Walsby AE (1994) Gas vesicles. Microbiol Rev 58:94–144

Walsby AE (2005) Archaea with square cells. Trends Microbiol 13:193–195

Zharkov MA (1981) History of Paleozoic Salt Accumulation. Springer Verlag, Berlin

Rev Environ Sci Biotechnol (2006) 5:167–185
DOI 10.1007/s11157-006-0003-2

ORIGINAL PAPER

# Planktonic microbial assemblages and the potential effects of metazooplankton predation on the food web of lakes from the maritime Antarctica and sub-Antarctic islands

**Antonio Camacho**

Received: 6 March 2006 / Accepted: 10 May 2006 / Published online: 20 July 2006


**Abstract** Antarctica is the continent with the harshest climate on the Earth. Antarctic lakes, however, usually presents liquid water, at least during part of the year or below the ice cover, especially those from the sub-Antarctic islands and the maritime Antarctic region where climatic conditions are less extreme. Planktonic communities in these lakes are mostly dominated by microorganisms, including bacteria and phototrophic and heterotrophic protists, and by metazooplankton, usually represented by rotifers and calanoid copepods, the latter mainly from the genus *Boeckella*. Here I report and discuss on studies performed during the last decade that show that there is a potential for top–down control of the structure of the planktonic microbial food web in sub-Antarctic and maritime Antarctic lakes. In some of the studied lakes, the effect of copepod grazing on protozoa, either ciliates or flagellates, depending on size of both the predator and the prey, could promote cascade effects that would be transmitted to the bacterioplankton assemblage.

A. Camacho (✉)
Instituto Cavanilles de Biodiversidad y Biología Evolutiva & Departamento de Microbiología y Ecología, Edificio de Investigación "Jeroni Muñoz", Campus de Burjassot, Universitat de València, E-46100 Burjassot (Valencia), Spain
e-mail: antonio.camacho@uv.es

**Keywords** Biotic interactions · Copepods · Sub-Antarctic and maritime Antarctic lakes · Microbial food web · Protozoa

## Introduction

For Biological scientists, Antarctica represents a natural laboratory in which life can be observed at the extreme of a gradient of one of the most important ecological factors, temperature, where this factor acts directly on the organisms and have additional effects on the physical structure of the ecosystems (Simmons et al. 1993). Its geographical isolation and the extreme environmental conditions for human life allowed the maintenance of pristine conditions in many Antarctic ecosystems, which can nowadays be studied as unaltered by anthropogenic impacts. With respect to climate conditions, Antarctica shows the more extreme conditions in the Earth, which restrict chances of most living beings to occupy Antarctic ecosystems (Fountain et al. 1999).

The existence of liquid water is one the most important ecological factors determining the possibility of colonisation of Antarctic ecosystems by aquatic organisms (Kennedy 1993, 1995). This is the reason why most of the known diversity of life in Antarctica is associated to marine environments, which have been extensively studied. Terrestrial aquatic ecosystems, such as streams and lakes, have also been studied in Antarctica, where liquid water can be present at least for a period of time if melting occurs during the austral summer, or even below a layer of ice cover in many lakes during most of the year (Vincent 1988). Antarctic lakes are harsh environments, undergoing large annual variations in levels of photosynthetically available radiation caused by their latitude and by winter ice and snow cover (Vincent 1988), while water temperatures are permanently low and sources of potential colonizing organisms are remote (Butler et al. 2000).

The maritime Antarctic region includes the west coast of the Antarctic Peninsula and the associated islands of the Scotia Arc (Lewis Smith, 1984), and extends from the South Sandwich Islands and Bouvetøya, through the South Orkney and South Shetland Islands and down the western side of the Antarctic Peninsula to approximately 72°S Although Antarctic freshwater lakes are characterized by short food webs dominated by microbes, those from the maritime Antarctica and from sub-Antarctic islands (Frenot et al. 2005) have slightly more complex food webs than continental Antarctic lakes and often include crustacean zooplankton (Ellis-Evans 1996). Freshwater lenitic water bodies in these geographical locations include, among others, both deep and shallow inland lakes and ponds lying on rock substrate (e. g. Toro et al. submitted), epishelf lakes as those described on Alexander Island (e.g. Smith et al. 2006), and coastal lagoons and pools mostly influenced by sea sprays and marine animals (e.g. Hansson et al. 1996; Butler 1999a).

In most lakes from the maritime Antarctic region and from sub-Antarctic islands environmental conditions are less extreme than in continental Antarctica and rapid changes in community composition at certain times of the year can occur despite constant low temperatures (Heywood 1978; Heywood et al. 1980; Ellis-Evans 1996). There, when the snow cover melts and the lake's ice cap thaws in summer (Ellis-Evans et al. 1998), water flows increase nutrient circulation into lakes, which in turn favour the development of planktonic and benthic microbial populations. Among these lakes, special attention have been paid by several researchers to those from South Georgia, South Shetland Islands, South Orkney Islands and the northern part of the Antarctic Peninsula, and there is a disperse bibliography that I have reviewed with special focus on the potential effects of metazooplankton predation on the microbial food web. In this manuscript, I report for several studies that demonstrate the potential for top–down control of the planktonic community structure in lakes from the maritime Antarctica and sub-Antarctic islands, contrasting with the physical control exerted by abiotic ecological factors such as temperature, freezing or light availability, which are usually saw as being the only responsible for structuring these communities. Because fast ecological responses to the quick environmental changes in these regions are occurring (Convey et al. 2002; Quayle et al. 2002), longer productive periods with less restricting abiotic conditions are likely to increase the relevance of such biotic interactions for structuring food webs in these Antarctic and sub-Antarctic lakes. Warming would also probably favour the success of invasive alien species mainly brought by human activity (Frenot et al. 2005), and bigger changes would then be produced in the structure of biotic communities of sub-Antarctic islands and even, perhaps, in those of the maritime Antarctic region.

## Main limnological features of sub-Antarctic and maritime Antarctic lakes

### Physical and chemical features

In Antarctic lakes, during most of the year, the ice cover impedes turbulent mixing and gas exchange between water and the atmosphere, creates a thermal isolation of the lake, and reduces light penetration (Hawes 1983a; Hawes and Schwarz 2001), decreasing the energy input for photosynthesis. When melting occurs these restrictions become less severe, and lake exchanges with its environment increase. Contrastingly to continental Antarctica, most lakes of the sub-Antarctic islands and the maritime Antarctic region loss the ice cover during a certain period in summer, which coincides with the period of higher light supply due to the large seasonal differences in the day length at high latitudes.

A relevant feature to the microbial ecology of lakes, as those from sub-Antarctic islands and maritime Antarctica, is the chemical characteristics of water, especially its salinity and nutrient loads. Salinity is generally determined by sea influence (e.g. Pizarro et al. 2004), with those lakes located close to the sea, although being commonly freshwater lakes (Hansson and Håkansson 1992), presenting higher salt contents than those located further away. Direct flooding by sea intrusion or sublacustrine penetration of sea water can cause salinization of coastal lakes, but sea spray brought by the strong Antarctic winds can also transport salts to inland lakes in zones close to the coast (Gasparon and Burgess 2000), thus influencing their salt content otherwise just depending on the geology of the catchments. When salt accumulation in deep lake waters results in higher water density a crenogenic meromixis would originate, otherwise the lakes would follow an stratification pattern depending on their morphometric features and determined by climatic conditions (Wetzel 2001). The occurrence of meromictic lakes have been mainly reported in zones from Antarctica other than the area of the Antarctic Peninsula, such as those from the Dry Valleys at Southern Victoria Land (Simmons et al. 1993). In the maritime Antarctica, meromictic saline lakes also appear as for instance at the volcanic Deception Island (Drago 1989). For freshwater lakes a cold monomictic regime (Wetzel 2001), with stratification under the ice and summer mixing when ice cover melts, usually occurs.

### Trophic status

A wide range of trophic status can be seen in lakes from the sub-Antarctic islands and the maritime Antarctica (Ellis-Evans 1996). Inorganic nutrient supply, mainly nitrogen and phosphorus compounds, is strongly determined by the influence of seabirds and marine mammals (e.g. Hawes 1983b; Hansson et al. 1996; Mataloni et al. 1998; Izaguirre et al. 2001, Laybourn-Parry et al. 2002). Thus, lakes lacking this influence are generally ultraoligotrophic (Vincent 1988), whereas those influenced by marine animals increase their trophic status even reaching eutrophic conditions that can drive to winter anoxia under the ice (e.g. Butler 1999a). The nutrient impact of animals, either marine mammals or seabirds, affects mainly coastal lakes, whereas the main source of nutrients to inland lakes is runoff from the catchment, although sea sprays can also increase salts and nutrient supply to inland maritime lakes. Biological fixation of nutrients in the catchment by benthic organisms, such as carbon fixation by primary producers (mosses, microalgae and cyanobacteria forming microbial mats and phototrophic biofilms) and nitrogen fixation (mainly by nitrogen-fixing cyanobacteria and other mat-forming bacteria), either by in lake production (Ellis-Evans et al. 1998) or via runoff from the catchment, may considerably increase the nutrient supply to the lakes compared to that becoming from leaching of catchment's mineral substrate, since in the sub-Antarctic islands and the maritime Antarctica these organisms can cover part of the lake catchment area and lake sediments (Moorhead et al. 1997; Vinocur and Pizarro 2000; Toro et al. submitted).

## Composition of the planktonic microbial food web in Antarctic lakes and related ecological factors

Because of the strong environmental conditions derived from the extreme climate, most life forms widely spread through the Earth are not found in Antarctic lakes, where most diversity can be attributed to microorganisms (Ellis-Evans et al. 1998). Bacteria, algae, heterotrophic protists and sometimes metazoan zooplankton can be found both in the planktonic and in the benthic environment of Antarctic lakes, with cyanobacterial mats, which have been studied with detail, being the most important benthic communities (Vincent et al. 1993; Quesada and Vincent 1997; Tang et al. 1997; Fritsen and Priscu 1998; Nadeau and Castenholz 2000; Fernandez-Valiente et al. 2001; Taton et al. 2003).

In lakes from the maritime Antarctica and from some sub-Antarctic islands, planktonic communities usually have a reduced diversity of metazoans, in contrast with a richer microbial assemblage (Tranvik and Hansson 1997; Butler et al. 2005). Although molecular taxonomical studies on these are yet scarce (Peck et al. 2005), this methodological tool is starting to offer some additional results on the biodiversity of the planktonic assemblages from sub-Antarctic and maritime Antarctic lakes (ej. Pearce et al. 2003; Unrein et al. 2005).

## Some key physical factors

As in lakes from lower latitudes, the stratification of the water column, when occurring, becomes relevant as structuring factor of the planktonic community, by creating physical and chemical gradients in which environmental conditions can strongly change, thus influencing the distribution of microorganisms. During winter, ice cover avoids water turbulence allowing an inverse temperature gradient, although summer stratification in lakes of the sub-Antarctic islands and the maritime Antarctica is not likely to occur by thermal stratification because the effect of strong winds would mix the water column when the ice cover disappear, and surface warming is commonly not enough to create a temperature gradient from surface to the bottom. On the other hand, light extinction is higher when the lake is covered by ice (Hawes and Schwarz 2001), whereas light penetration increases after ice melting and favour primary production, together with the increase in the number of light hours per day during the Antarctic summer.

## Phytoplankton and primary production

Among phytoplankton, most studies performed so far have identified the main species using microscopical techniques (e.g. Hansson and Tranvik 1996; Laybourn-Parry et al. 1996; Unrein and Vinocur 1999; Mataloni et al. 2000; Vinocur and Unrein 2000; Allende and Izaguirre 2003; Izaguirre et al. 2003). These studies showed that Chrysophytes (e.g. *Ochromonas* spp., *Chromulina* spp.), Cryptophytes (e.g. *Chroomonas* spp.), Chlorophytes (e.g. *Chlamydomonas* spp., *Monoraphidium* spp., *Elakatothrix* sp.), picocyanobacteria, and diatoms (e.g. *Navicula* spp, *Nitzschia* spp., *Achnantes* spp.), the latter mostly from benthic origin, are usually found dominating the plankton of lakes from sub-Antarctic islands and the maritime Antarctica. However, the dominance of certain algal groups can usually change when the lake shows eutrophic conditions (Izaguirre et al. 2001). Even within each algal group, salinity and/or nutrient levels determined the algal assemblages (Hansson and Håkansson 1992; Jones et al. 1993). Contrastingly with benthic microbial mats, which are usually dominated by filamentous cyanobacteria, among others (Moorhead et al. 1997; Vinocur and Pizarro 2000), filamentous cyanobacteria are rarely important among the phytoplankton of Antarctic lakes. In contrast, high numbers of unicellular picocyanobacteria have been found in some maritime Antarctic lakes, such as Lake Boeckella from the Antarctic Peninsula (Izaguirre et al. 2003), and in other of these lakes their abundance increases in bottom waters (Toro et al. submitted).

Pioneer studies using molecular tools to identify the main nanophytoplankters of lakes from the maritime Antarctica, performed for ten lakes from the Antarctic Peninsula and King George Island, have been recently published by Unrein et al. (2005). These studies confirmed the previously described dominance of Chrysophytes offering up to 5 different DGGE bands corresponding to this taxon, although sequences corresponding to other algal taxa such as Chlorophytes and diatoms were also identified. One of the most remarkable conclusions of this study is the co-occurrence of several common sequences shared by most lakes, corresponding to Chrysophytes, which suggests the existence of well-adapted nanophytoplankton species dispersed throughout the maritime Antarctic lakes. Some sequences, however, appeared exclusively in specific lakes, which was explained by differences in the trophic status and other local conditions of the lakes (Unrein et al. 2005).

Although for benthic species, biogeographical studies on Antarctic diatoms performed using classical microscopical identification techniques have found a high proportion of cosmopolitan species, whereas diversity apparently decreased when moving southwards by comparing sub-Antarctica, maritime Antarctica and the continent (Van de Vijver and Beyens 1999). The supposed diversity decrease at increased latitudes for these algae would follow a pattern similar to that found for organisms of some other groups for large distances along a latitudinal gradient (e.g. Petz 2003) and could probably reflect, if existing, the negative effect of more unfavourable climatic conditions when increasing latitude, as found among many taxa (Begon et al. 2005). However, for most of the taxa, and especially for microbes, there is still a lack of a robust source dataset to test the hypothesis of decreased diversity within the latitudinal gradient, and even some of the pioneer studies using molecular techniques as that of Lawley et al. (2004) failed to support this hypothesis for the maritime Antarctica (60 to 72°S) when studying soil eukaryotic microbial diversity (protists and fungi). Contrarily, these authors even found the higher diversity in the maritime Antarctica at the southernmost site studied in Alexander Island (ca. 72°S), although diversity in maritime Antarctic sites was three to four times higher than in continental Antarctic sites. Thus, the question of the existence of a latitudinal gradient of diversity within concrete Antarctic regions remains unsolved and would be a challenge for further research.

Concerning the studied maritime Antarctic lakes, it has also been reported that their phytoplankton assemblages typically show low diversity with a few species accounting for most of the phytoplankton standing crop (Allende and Izaguirre 2003), although eutrophic conditions sometimes promoted higher species richness (Izaguirre et al. 2001). Hansson and Tranvik (1996), according to their results in manipulative experiments performed in the sub-Antarctic South Georgia suggested that, in addition to the low temperature and short growing season (Priddle et al. 1986), the presence of an unregulated grazer assemblage, such as those communities where metazooplankton are the top consumers, suppress the phytoplankton species richness, and algae adapted to a high grazing pressure would then dominate.

In addition to the availability of inorganic nutrients, which determines primary production rates in any lake, latitude is an important factor for primary production, since in polar ecosystems, the lower is the latitude, the longer the growing season, and less extreme physical conditions are found. Although phytoplankton of lakes of high latitudes is usually considered as well adapted to low temperatures and light availability, (Henshaw and Laybourn-Parry 2002), comparisons should be made by considering lakes of similar latitudes, such as those from the maritime Antarctic zone or those from sub-Antarctic islands. Small geographic areas, such as Byers Peninsula (Livingston Island, South Shetland Islands), present waterbodies with strong differences in phytoplankton abundance, which in terms of chlorophyll means that some ultraoligotrophic lakes show chlorophyll concentrations lower than 0.1 mg Chl-*a* $m^{-3}$, whereas coastal lakes influenced by marine animals easily reach chlorophyll concentrations as high as 40 mg Chl-*a* $m^{-3}$ (Toro et al. submitted). The high chlorophyll-*a* concentrations often found in these eutrophic waterbodies in the maritime Antarctica suggest that, at least during the austral summer, factors other than temperature can be much more important in regulating planktonic communities, since differences in temperature between these compared lakes were not significant. Among these possible factors, the potential role of predation as possible structuring force for the microbial assemblage of these lakes will be further addressed in this manuscript.

## Bacteria and heterotrophic protists

In addition to autotrophic protists and picocyanobacteria, heterotrophic protozoa are another important component of the planktonic community of sub-Antarctic and maritime Antarctic lakes. Petz (2003) studied the diversity of ciliates in polar habitats, and reported the low diversity of ciliated protozoa in planktonic habitats of Antarctica compared to other Antarctic aquatic habitats such as benthic environments, although their abundance in the planktonic habitat was not necessarily low. From a total of 159 different species representing at least 73 genera of ciliated protozoa found in lakes from Livingston Island and Deception Island (South Shetland Islands), only nine species were found in the plankton of an intensively studied lake (Lake Limnopolar) from a total of 29 species found in the lake, and from these, only one was actually an euplanktonic species, with the rest being less abundant benthic species suspended by wind-induced wave action (Petz et al. 2005). Heterotrophic nanoflagellates (HNAN) can also be present at relatively high abundance in the planktonic community of sub-Antarctic and maritime Antarctic lakes (Laybourn-Parry et al. 1996; Tranvik and Hansson 1997; Butler et al. 2000; Unrein et al. 2005), although, as for ciliates, their diversity is commonly higher in the benthic environment (Vincent 1988). In addition to HNAN, mixotrophic phytoflagellates could use phagotrophy as complementary feeding strategy, as already demonstrated for continental Antarctic lakes (Roberts and Laybourn-Parry 1999; Laybourn-Parry 2002; Marshall and Laybourn-Parry 2002; Laybourn-Parry et al. 2005), and its possible impact on bacterial abundance should then be considered when evaluating the role of protozoa in food webs of Antarctic lakes.

The abundance of heterotrophic bacterioplankton in Antarctic lakes show densities usually comprised between $10^4$–$10^6$ cell ml$^{-1}$, even in ultraoligotrophic lakes (Laybourn-Parry et al. 1995, 2001). It has been estimated that bacterial biomass can account for an important part of the planktonic biomass in lakes from continental Antarctica such as in those from Dry Valleys near McMurdo, where they can account for 30–60% of it (Takacs and Priscu 1998). In lakes from the maritime Antarctica, such as those from Byers Peninsula (Livingston Island, South Shetland Islands), bacterial abundance is also relatively high, ranging from $0.59 \times 10^6$ to $6.59 \times 10^6$ cell ml$^{-1}$, with higher bacterial abundance in lakes with higher trophic status (Toro et al. submitted). The relatively high abundance of heterotrophic bacteria could be supported by several carbon sources, additional to phytoplankton, such as, in oligotrophic lakes, microbial mats (Pizarro et al. 2004) and mosses (Light and Heywood 1973; Imura et al. 2003a, 2003b) growing inside the lakes or in the catchment. Mosses and microbial mats covering part of the catchment area can provide organic carbon and other nutrients via runoff in summer. In lakes influenced by marine animals, however, the main origin of organic carbon (Laybourn-Parry et al. 2004) fueling bacterial production could be these animals.

Concerning the community structure of the bacterial assemblage in sub-Antarctic and maritime Antarctic lakes, the first studies using molecular techniques (see review by Pernthaler and Amann 2005 to get a general overview of these techniques) to describe bacterial communities on several lakes from Signy Island have been published in recent years (Pearce 2000, 2003, 2005; Pearce et al. 2003). For instance, in one of the most studied lakes among the sub-Antarctic islands and the maritime Antarctica, the oligotrophic Moss Lake, Pearce (2003), using FISH techniques found a variable dominance of $\beta$-proteobacteria (26.4 to 71.5%), $\alpha$-Proteobacteria (2.3 to 10.6%), $\gamma$-Proteobacteria (0–29.6%) and the Cytophaga-Flavobacterium group (1.8–23.5%). In this study PCR amplification of 16S rRNA gene fragments followed by DGGE, cloning and sequencing, found just a couple of sequences corresponding to previously identified species, whereas most sequences corresponded to uncultured groups. Further studies, in which a mesotrophic (Sombre Lake) and a eutrophic (Heywood Lake) waterbodies were included, showed that both the increased levels of nutrient inputs and the timing and duration of the ice cover leaded to marked changes in the structure and stability of the bacterioplancton community (Pearce 2005).

## Metazoan zooplankton

Contrastingly with most lakes from continental Antarctica, where the presence of crustacean zooplankton, if any, is much less common and with much lower abundance (Bayliss et al. 1997; Ellis-Evans et al. 1998), sub-Antarctic and maritime Antarctic lakes usually present crustacean zooplankton, mainly calanoid copepods (Hansson et al. 1996; Pugh et al. 2002; Dartnall 2005; Toro et al. submitted). In these lakes, the presence of metazooplankton from several groups has been reported, such as planktonic rotifers as *Cephalodella* spp., *Lepadella patella*, *Keratella* spp., *Notholca walterkostei* and other bdelloids (Paggi 1987; Hansson et al. 1996; Izaguirre et al. 2003; Dartnall et al. 2005), fairy shrimps as *Branchinecta gainii* (Paggi 1987, 1996; Toro et al. submitted), and calanoid copepods, mainly

*Boeckella* species, such as *B. michaelseni* and especially *B. poppei*. These copepods are commonly present in freshwater lakes of some sub-Antarctic islands and the maritime Antarctica, as well as in lakes of the Andean region of South America (Paggi 1987; Bayly 1992; Menu-Marque et al. 2000; Bayly et al. 2003), with *B. poppei* being a major species in high latitude lakes of the Southern Hemisphere (Butler et al. 2005). Other fauna, mainly from benthic environments, such as tardigrades, nematodes, and benthic rotifers and crustaceans, can sometimes be linked to the pelagic food-web by benthic-planktonic interactions.

## Predation (grazing) as a potential structuring force of the microbial community

### Potential structuring factors of the planktonic community in Antarctic lakes

Planktonic food webs in Antarctic lakes are of low complexity and essentially composed by microorganisms, lacking vertebrate predators (Ellis-Evans 1996; Hansson et al. 1996; Wynn-Williams 1996; Laybourn-Parry et al. 2001). Consequently, most energy transfer between organisms would there circulate through the so-called microbial loop (Azam et al. 1983; Laybourn-Parry 1997; Roberts et al. 2000). The microbial food web (Pomeroy 1974) implies trophic levels composed by bacteria and protozoa between phytoplankton and crustacean zooplankton (Sommer and Sommer 2006).

Predation (*sensu lato*) and resource availability (Hairston et al. 1960, Currie et al. 1999) are recognized as relevant structuring forces of food webs in biological communities from temperate zones (Matson and Hunter 1992), including aquatic ecosystems (Pace and Funke 1991). Trophic cascades (Paine 1980; Carpenter et al. 1985) can amplify the effect of consumers down in the food web in several types of ecosystems (Pace et al. 1999), impacting entire trophic levels ("community level cascades") or just single species or groups of species ("species level cascades") (Polis et al. 2000). The strength of trophic cascades is usually recognized to be higher in aquatic environments (Strong 1992; Persson 1999; Shurin et al. 2002), depending mainly on metabolic and taxonomic factors (Borer et al. 2005). The extreme weather conditions experienced by the communities living in Antarctica, however, supposes that physical control has been usually considered as the main structuring force of Antarctic communities (Fountain et al. 1999), resulting in very flexible life histories of Antarctic biota favoured by adversity or stress selection (Convey 1996), whereas biotic interactions are usually considered as much less important than in temperate zones. Theoretical considerations assume that biological communities in systems with strong physical control would be far from conditions of biotic equilibrium (Krebs 2001), and consequently the relative importance of biotic interactions as structuring forces would be low. Freezing and ice thaw cycles, rock weathering, sea influence, the physical effect of the ice cover and, when occurring, vertical water stratification, are among others, prominent determinant factors for life in Antarctic lakes (Simmons et al. 1993). However, although these factors could dominate during most of the year, recently published work demonstrates that the role of biotic interactions such as predation (sometimes referred by as grazing because of the organisms involved) could, at least during some periods, have chance to influence the structure of the planktonic microbial community in lakes from sub-Antarctic islands and the maritime Antarctic region, where climatic constrains are less extreme.

### Experimental evidences for potential top–down control

Biological communities of lakes from continental Antarctica are though to be controlled by bottom-up forces, resource availability or physical climate-dependent control by temperature or light availability (Laybourn-Parry and Bayliss 1996; Bell and Laybourn-Parry 1999; Laybourn-Parry et al. 2000, Roberts et al. 2004), although some studies on planktonic copepods in continental Antarctic lakes (e. g. Swadling and Gibson 2000) hypothesize that they could have a considerable impact on lake's carbon cycle. Anyway, the less extreme climatic restrictions in the sub-Antarctic islands and the maritime Antarctica

(Ellis-Evans 1996) would allow, at least during summer periods, higher relevance of biotic interactions. Manipulative microcosms experiments, in spite of their methodological constraints and interpretation limitations (Carpenter, 1996, 1999; Schindler 1998; Drenner and Mazumder 1999; Huston 1999), might reveal possible trends in the functioning of aquatic food webs. These kind of experiments, although less conclusive than ecosystem experiments (Carpenter et al. 1995, 2001; Pace et al. 1998), are useful in pristine habitats such as Antarctic lakes that must be conserved without alteration according to the Antarctic Treaty, which would not allow manipulative whole-ecosystem experiments.

Several researchers (e.g. Hansson 1992; Hansson et al. 1993; Laybourn-Parry et al. 1996; Laybourn-Parry 1997; Mataloni et al. 2000) have argued during recent years that biotic interactions could play a more relevant role as structuring forces of the planktonic food web of sub-Antarctic and maritime Antarctic lakes that what was generally expected. Hansson's experiments at the earliest 90's pointed out this possibility, and further work performed since then, mostly using experimental manipulations in micro or mesocosms, have evidenced that top–down processes could potentially contribute to structure planktonic food webs in sub-Antarctic (Hansson and Tranvik 1996; Tranvik and Hansson 1997) and maritime Antarctic lakes (Almada et al. 2004; Butler et al. 2005; Camacho et al. submitted). Tables 1 and 2 show the main limnological characteristics and the main components of the planktonic community of the lakes in which these experiments have been performed.

Although usually though to be herbivorous, the copepod *Boeckella poppei* can in fact consume heterotrophic protists such as flagellated and ciliated protozoa. This is a general feature of most previously saw as "herbivorous crustaceans", which now are recognized as omnivores (e.g. Stoecker and Capuzzo 1990; Kleppel 1993; Ohman and Runge 1994; Atkinson 1995; Balseiro et al. 2001), with selectivity depending more on individual feeding modes and food size spectra (e.g. Bertilsson et al. 2003) than on the auto- or heterotrophic nature of their prey. Copepods can actively select their prey (DeMott 1988; DeMott and Watson 1991), with prey size, motility and chemical quality being the most important criteria for prey selection (Sommer and Sommer 2006). When size of food items is adequate, they even can select among them (Mitra and Flynn 2005) and mechanoreceptors are then the main way for this selection (Kiørboe et al. 1996). Concerning predation, the role of zooplankton, and especially of copepods, in controlling the abundance of protozoa has been recognized from time ago (e.g. Carrick et al. 1991). In the studied cases with *Boeckella* species in sub-Antarctic and maritime Antarctic lakes, predation of these copepods on protozoa is also size dependent, being determined by both the size of the predator (e.g. life stage or species) and that of the prey (Tranvik and Hansson 1997; Butler et al. 2005; Camacho et al. submitted). The size structure in populations has been recognized as a main factor introducing heterogeneity in food webs (Persson, 1999), which could determine the extent of trophic cascades depending on, for instance, the life stage of the predator. Predation on protozoa, whose main food source are bacteria (Pernthaler and Amann 2005), can make weaker the grazing impact of these protozoa on bacteria, and consequently reduce its possible regulatory effect on the bacterial abundance (Fenchel 1982; Pace 1988; Berninger et al. 1991; Jürgens et al. 1994; Hahn and Höfle 2001; Gasol et al. 2002), as well as modify the effects of grazing selectivity, depending, among others, on bacterial cell morphology (Pernthaler et al. 1996; Pernthaler and Amann 2005). Hence, trophic cascades would also appear through the components of the microbial loop, in which bacteria can transform in particulate biomass the nutritional supply by primary producers, both from benthic (littoral,-Karlsson and Byström 2005-, bottom, or even allochthonous inputs from the catchment) and planktonic (phytoplankton) origin. Bacteria can consequently provide particulate organic matter that can be used by protozoa, but they can be released from the grazing impact of certain protozoa by predation of copepods on these protozoa. Takacs and Priscu (1998), in a study on the productive dynamics and losses by the bacterial assemblage of Antarctic lakes demonstrated that grazing impact can be responsible for most of the losses within bacterial

**Table 1** Location, and values of some physical and chemical features of lakes from maritime Antarctica and sub-Antarctic islands in which trophic relationships within the planktonic community have been experimentally determined by means of manipulative studies

| Lake | Location | Coordinates | Area (Ha) | Maximum depth (m) | Trophic status | Chl-*a* ($\mu g\ l^{-1}$) | P ($\mu M$) | References |
|---|---|---|---|---|---|---|---|---|
| Boeckella | Hope Bay, Antarctic Peninsula. | 63° 24′ S 57° 00′ W | 6.75 | 4 | eutrophic | 1.25–9.44 | n.a | Allende and Izaguirre(2003); Izaguirre et al. (2003); *Almada et al. (2004). |
| Limnopolar | Byers Peninsula, Livingston Island, South Shetland Islands. | 62ª 39′ S 61° 06′ W | 2.22 | 5.5 | oligotrophic | 0.13–0.18 | 0.13 (TP) | Toro et al. (submitted); *Camacho et al. (submitted) |
| Lake 9 | Tønsberg Peninsula, South Georgia Island. | 54° 10′ S 36° 41′ W | 0.95 | 6 | oligo-mesotrophic | 0.86–1.1 | 0.38 (TP) | Hansson et al. (1996); Hansson and Tranvik (*1996, 1997); *Tranvik and Hansson 1997. |
| Lake 11 | Tønsberg Peninsula, South Georgia Island. | 54°10′ S 36° 41′ W | 2.2 | 8 | meso-eutrophic | 4.1–5.77 | 1.51 (TP) | Hansson et al. 1996; Hansson and Tranvik (*1996, 1997); *Tranvik and Hansson (1997). |
| Sombre | Paternoster Valley, Signy Island, South Orkney Islands. | 60°43′ S 45°38′ W | 2.66 | 11 | mesotrophic increasing nutrient input | n.d.–16 | n.d.–2.4 (SRP) | Laybourn-Parry et al. (1996); Butler (1999b); Pearce (2005); *Butler et al. (2005). |

*=reference for the manipulative study performed in such lake. n.a. = data not available in the cited references. n.d. = not detectable

communities, which highlights the possible relevance of top–down effects, when occurring, in structuring microbial food webs in these lakes. Interestingly, a review by Hansen et al. (1997) showed how, among protists, ciliates display maximum ingestion and clearance rates, whereas among metazooplankton, calanoid copepods have usually higher clearance rates. That would mean that ciliates and copepods, which are among the main components of the planktonic food web in maritime Antarctic lakes, could have there a great potential to control food web structure. This coincides with the findings of Camacho et al. (submitted), who in manipulative experiments using microcosms performed in the maritime Antarctic Lake Limnopolar found a significant potential grazing impact of copepods on ciliates, which in turn favoured bacterial abundance (Fig. 1).

Probably the best documented studies on food web interactions among lakes from the maritime Antarctica and sub-Antarctic islands are those performed by Hansson and Tranvik in the sub-Antarctic South Georgia. They performed a series of experiments to study the effect of predation on different trophic levels in two sub-Antarctic lakes, Lakes 9 and 11 (Tables 1 and 2) from South Georgia (Hansson et al. 1996), where two copepods, *Boeckella michaelseni* and *Pseudoboeckella* (*Boeckella*) *poppei*, exerted a substantial pressure on algae. In two of the manuscripts (Hansson and Tranvik 1996; Tranvik and Hansson 1997) they reported that in these lakes the dominant algal species exhibited properties that enabled them to avoid or compensate grazing, such as large size, extruding spines, protective mucilagous, or mechanisms for recruitment from the sediment. This was interpreted as a consequence of the high grazing pressure that can potentially drive to the dominance of inedible or grazing-compensating species (Steiner 2003). On the other hand, the effect of predators, the juvenile stages of the diving beetle *Lancetes angusticollis* (Arnold and Convey 1998) and the copepod *Parabroteas sarsi*, on *Boeckella* spp. populations was low (about 0.4% of removal). Among both *Boeckella* species, *B. michaelseni* (mean length 0.7 mm) grazed mainly on nanoflagellates and favoured bacteria by partly releasing them from grazing pressure by

**Table 2** Main planktonic organisms and bacterioplankton abundance in lakes from maritime Antarctica and sub-Antarctic islands in which trophic relationships among the planktonic community have been experimentally determined by means of manipulative studies

| Lake | HPP abundance (cell $ml^{-1} \times 10^3$) | APP abundance (cell $ml^{-1} \times 10^3$) | Eukaryotic phytoplankton (main genera) | Planktonic heterotrophic protists (main genera or groups) | Metazooplankton | Reference |
|---|---|---|---|---|---|---|
| Boeckella | 89–225 | 50–364 | *Chlamydomonas*<br>*Ochromonas* | HNAN<br>Ciliates | *Boeckella poppei*<br>*Philodinia gregaria*<br>*Notholca walterkostei* | Allende and Izaguirre (2003);<br>Izaguirre et al. (2003);<br>*Almada et al. (2004). |
| Limnopolar | 1229–2167 | 1.2–30 | Chrysophytes<br>Diatoms<br>*Koliella* | *Balanion* | *Boeckella poppei* | *Camacho et al., submitted;<br>Toro et al., submitted;<br>Petz et al. (2005). |
| Lake 9 | 1590–8880 | n.a. | *Mallomonas*<br>*Elakatothrix*<br>*Tribonema* | HNAN<br>Ciliates | *Boeckella michaelseni*<br>*Boeckella poppei*<br>*Parabroteas sarsi*<br>(*Lancetes angusticollis*-diving beetle) | Hansson et al. (1996);<br>Hansson and Tranvik (*1996, 1997, 2003).<br>Tranvik and Hansson (1997). |
| Lake 11 | 1810–4800 | n.a. | *Staurastrum*<br>*Elakatothrix*<br>*Tribonema* | HNAN<br>Ciliates | *Boeckella michaelseni*<br>*Boeckella poppei*<br>*Parabroteas sarsi*<br>(*Lancetes angusticollis*-diving beetle) | Hansson et al. (1996);<br>Hansson and Tranvik (*1996, 1997, 2003);<br>*Tranvik and Hansson (1997). |
| Sombre | 100–3180 | n.a. | *Cryptomonas*<br>*Chlamydomonas*<br>*Ochromonas*<br>*Pseudokephyrion*<br>*Gymnodinium*<br>*Mallomonas*<br>*Ankistrodesmus* | *Monodinium*<br>*Halteria*<br>*Strombidium*<br>*Cinetochilum*<br>*Cyclidium*<br>*Holophrya*<br>*Bicosoeca*<br>*Monosiga*<br>*Bodo*<br>*Paraphysomonas* | *Boeckella poppei* | Laybourn-Parry et al. (1996);<br>Butler (1999b);<br>*Butler et al. (2005). |

*=reference for the manipulative study performed in such lake. n.a. = data not available in the cited references. HPP = Heterotrophic picobacterioplankton. APP = Autotrophic picobacterioplankton. HNAN = Heterotrophic nanoflagellates. Note that data on the composition of the planktonic community correspond to discrete samplings or are averages for a certain period, depending on the source, and would change with time; for more information see the original cited references

nanoflagellates, whereas the larger (mean length 2.4 mm) *Pseudoboeckella* (*Boeckella*) *poppei* grazed on larger food particles and had no significant impact on any studied microbial group (Tranvik and Hansson 1997). Later on, using a stable isotope approach, they investigated the trophic relationships between the different components of the planktonic food web (Hansson and Tranvik 2003), and found that both *B. michaelseni* and *B. poppei* fed on pelagic particulate organic matter, although the latter also fed on benthic algae, whereas both copepods, especially the former, were a prey of the predatory copepod *Parabroteas sarsi*. In the same study they corroborated that *B. poppei* is also a prey of *Lancetes*, but both predators, the copepod and the diving beetle, feed also on the benthic cladoceran *Alona weineckii*.

Zooplankton may affect phytoplankton not only directly by grazing, but may also have effects by regenerating nutrients to the ungrazed part of the algal community (Sterner 1986). In studies performed in a mesotrophic and a eutrophic sub-Antarctic lake, Tranvik and Hansson (1997) found that available phosphorus (orthophosphate) concentrations were lower at high copepod abundances, which was interpreted as inefficient nutrient regeneration by these copepods. Camacho et al (submitted), however, in short-term experiments in a more oligotrophic maritime Antarctic lake (Lake Limnopolar, Tables 1 and 2 and Fig. 1), found that higher abundances of *B. poppei* yielded higher algal abundances and resulted in increased concentrations of available nitrogen, showing that apparently these copepods where providing nitrogen to the algal assemblage within the experimental enclosures in such an oligotrophic environment.

Almada et al. (2004) also performed an study on the trophic interactions among the different components of the planktonic assemblages in Lake Boeckella (Tables 1 and 2), a freshwater lake influenced by nutrient input from a nearby penguin rookery located in Hope Bay (Antarctic Peninsula), near Esperanza Station (Allende and Izaguirre 2003). In this lake dominant phytoplankton were nanoplanktonic volvocales, flagellated crysophyceae and picocyanobacteria, and *Boeckella poppei* was the dominant zooplankter (Izaguirre et al. 2003). In this study, the impact of grazing by different life-stages of this copepod on the dominant fractions of pico-and nanophytoplankton, as well as on benthic algae and on bacterioplankton, was determined. From these experiments they concluded that copepods fed mainly on nanophytoplankton, and used benthic algae as an alternative food source, whereas the smallest picoplancton fraction (both picophytoplankton and heterotrophic bacterioplankton) was not significantly affected.

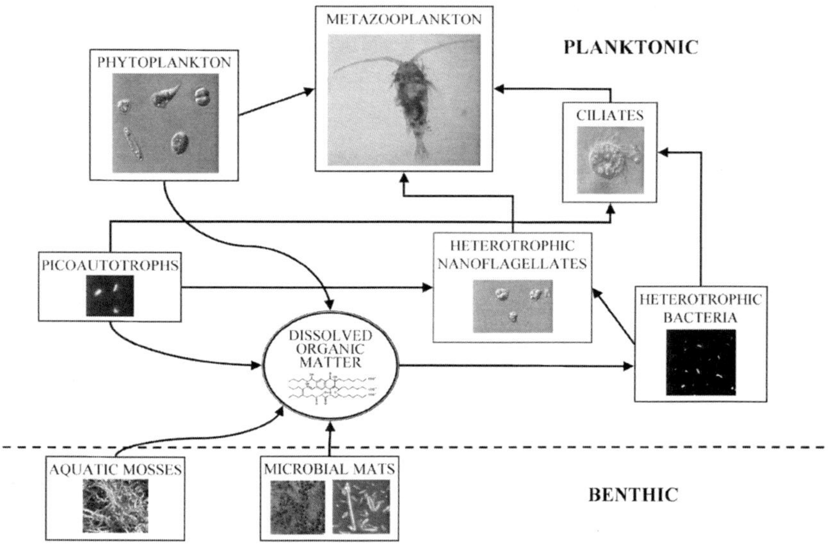


**Fig. 1** Trophic interactions among different compartments of Lake Limnopolar (Byers Peninsula, Livingston Island, South Shetland Islands) according to Camacho et al. (submitted). Arrows go from the donor to the user compartment

Butler et al. (2005) evaluated clearance rates by adults of *B. poppei* from Sombre Lake (Tables 1 and 2), a previously oligotrophic lake with little phytoplankton (Clarke et al. 1989) from South Orkney Islands, which nowadays suffers some animal-induced eutrophication (Jones and Juggins 1995; Butler, 1999b). In this study, Butler et al. (2005) also showed that *B. poppei* was omnivorous, although its diet was based on small phototrophic flagellates that were dominant in the incubation water. Clearance rates were equivalent to those found for freshwater and marine copepods of similar size and at similar temperatures. Grazing impact of this copepod increased with prey size, with stronger impact on planktonic ciliates of about 18 μm (equivalent spherical diameter), with a clearance of 24% of ciliates per day, which shows their potential control on this part of the microbial assemblage. These results can be complemented by those of Laybourn-Parry et al. (1996), who offer some data on the abundance of the different components of this planktonic microbial assemblage and grazing rates on protozoa that show a much stronger impact on bacteria by ciliates (119.3 bact $ind^{-1}$ $h^{-1}$) compared to that of heterotrophic nanoflagellates (HNAN). This shows that there can be a potential influence of ciliates on the abundance of bacteria, usually its main food source (Berninger et al. 1991), but grazing by copepods on ciliates, as described by Butler et al. (2005), would release bacteria for most ciliate grazing, thus allowing high bacterial abundance (up to $3.18 \times 10^6$ cell $ml^{-1}$) and promoting relatively low ciliate abundance compared to neighbour lakes. Interestingly, the study by Laybourn-Parry et al. (1996) showed than in another lake from the same zone, Heywood Lake, ciliate abundance was higher than in Sombre Lake. In comparison specific rates per individual of ciliate grazing on bacteria were, however, lower in Heywood Lake, whereas HNAN were also more abundant, but HNAN grazing on bacteria in Heywood lake was nevertheless curtailed as a result of predation by microcrustacean larvae, another signal of apparent temporal top–down control in the dynamics of the microbial assemblage of these lakes. Since HNAN, as well as PNAN (phototrophic nanoflagellates) as a winter survival strategy (Roberts and Laybourn-Parry 1999; Laybourn-Parry 2002; Marshall and Laybourn-Parry 2002; Laybourn-Parry et al. 2005), can have an impact on bacterial populations, possible effects of grazing either by ciliated protozoa or by metazooplankton grazing on HNAN such as those described by Tranvik and Hansson (1997) would also have chance to cause top down effects through the microbial assemblage.

Experimental evidence on the effect of zooplankton predation on the protists assemblage in pelagic habitats, which in turn could be transmitted to the bacterial community, has been accumulated during the last decade (e.g. Adrian and Schneider-Olt 1999; Zöllner et al. 2003). Although not in Antarctic lakes, the first experiments made with *Boeckella* species by Burns and Schallenberg (1996, 1998) also found a strong negative impact of these copepods on ciliates and in some cases heterotrophic bacteria profiting from the removal of ciliates. The same impact on ciliates was found in other studies with *Boeckella*, such as those of Balseiro et al. (2001) in an ultra-oligotrophic Andean lake, where *B. gracilipes* was sustained by an omnivorous diet based on ciliates and phytoflagellates. Studies with copepods suggest that they can present different prey encounter strategies (Kiørboe et al. 1996) that optimize nutritional gain (Kleppel, 1993), whose selection depended on the prey. In Kiørboe's experimental study, ciliates were captured by the generation of a feeding current that they cannot escape.

Recently, Sommer and Sommer (2006) reviewed the contrasting top–down controls of cladocerans and copepods in planktonic communities and the induced trophic cascades. Concerning copepods, they concluded than they usually had higher impact on large (phyto)plankton, but additionally they released small plankton from grazing pressure by intermediate consumers, such as protozoa, and leave more nutrients than cladocerans for compensatory growth of ungrazed phytoplankton. These authors also reported for a large variety of studies that show the preference of copepod assemblages for protozoans, which was attributed to a better biochemical food quality compared to phytoplankton (Klein-Breteler et al. 1999; Ptacnik et al. 2004). Additionally, they predicted that copepods would favour small phytoplankton by recycling part of the nutrients trapped in bigger plankton. Although nutrient recycling and translocation by zooplankton in sub-Antarctic and maritime Antarctic lakes is not the subject of this review, animals are know to be important in nutrient cycling in freshwater ecosystems, where they can supply nutrients that can support a substantial proportion of the nutrient demands of primary producers (Elser et al. 1988; Atkinson and Whitehouse 2001; Vanni 2002). This is clearly the case of coastal sub-Antarctic and maritime Antarctic lakes influenced by marine animals, but probably also of inland lakes in some polar areas, where copepods can play a role in the transport of nutrients from the richer benthic environment (Bonilla et al. 2005) to the poorer pelagic habitat (Camacho et al. submitted). Interestingly, Sommer and Stibor (2002) and Sommer et al. (2002) have proposed a hypothesis linking the trophic level of marine copepods to nutrient supply and phytoplankton size. A part of this hypothesis predicts that oligotrophic conditions, with the dominance of small phytoplankton, should force copepods to rely on protozoan feeding, thus leading to a trophic level of 3. In some of the reported cases of sub-Antarctic and maritime Antarctic lakes where impact of *Boeckella* on ciliates was strong, these copepods were acting in fact as a trophic level of 3, with less impact on small phytoplankton and bacterioplankton, the latter even been favoured by the release of protozoan grazing.

### Relevance of energy recovery within the microbial loop for planktonic communities of Antarctic lakes

In oligotrophic systems, such as many Antarctic lakes, the amount of energy and materials circulating through the living component of the ecosystem is low compared with nutritionally richer systems. Any energy transfer pathway increasing the transference to upper trophic levels would hypothetically be relevant in these systems, and consequently recovery of energy and materials by bacteria through transformation of allochthonous dissolved compounds in particulate material would increase nutrient availability for phagotrophic organisms, such as heterotrophic protists and metazoan zooplankton (Fig. 1), as a way of energy recovery that could be added to autochthonous primary production as food source for the heterotrophic planktonic organisms.

## Implications of climate change for the functioning of Antarctic lacustrine communities

High latitude lakes are likely to be sensitive indicators of climate change (Vincent et al. 1998). Relatively simple microbial communities, such as those from Antarctic lakes, can be used as indicators of microbial processes and responses to environmental change (Wynn-Williams 1996). Climate warming, which is especially strong in the area of the Antarctic Peninsula (Hansen and Sato 2001; Sun and Hansen 2003), have promoted a rapid retreat of the ice cap and glaciers in sub-Antarctic islands and maritime Antarctica observed during the last recent years (Quayle et al. 2002; Sancho and Pintado 2004, Frenot et al. 2005), as well as fast physical ecosystem changes combined to ecological responses to that change (Quayle et al. 2002). These include longer productive periods favoured by earlier ice and snow melting, and increased nutrient supply to these lakes during the last decades, which in turn could support primary production of these polar systems (Markager et al. 1999; Geider et al. 2001) and affect ecological interactions within the biological community. Therefore, increased knowledge of the functioning of sub-Antarctic and maritime Antarctic lakes seems essential to evaluate future changes in these ecosystems that can act as sentinels of climate change.

## Conclusions

Undoubtfully, climate conditions strongly determine the biological communities in Antarctic lakes. Low temperatures and reduced light availability due to the ice cover are, even during the Antarctic summer, physical constraints that are not easy to cope with by planktonic organisms. However, these climatic constraints are less severe in maritime Antarctica and sub-Antarctic islands, especially in summer when ice melting would occur in most of these lakes and physical constraints become weaker. Under these situations, life goes on, and factors other than temperature or light availability acquire importance to structure biological communities. Even in simple food webs, such as those of Antarctic lakes, trophic interactions are established linking the different components of the community. The referred studies cited in this manuscript show how the effect of top consumers can be somewhat transferred to lower trophic levels in lakes of sub-Antarctic Islands and the maritime Antarctica, at least under some circumstances, and alert us on the potential relevance of predation (*sensu lato*) as structuring force for these communities, contrasting with the general thought looking to physical constraints as the only relevant factors. Quick environmental changes associated to climate change, which are likely to reduce the strength of physical constraints for life in these Antarctic regions, would potentially increase the role of biotic interactions in structuring Antarctic communities, and consequently more attention should be paid to the study of these interactions under the current environmental scenario of planet Earth.

**Acknowledgements** The work on Antarctic lakes performed by the author is nowadays supported by grant CGL2005-06549-CO2-02/ANT from Spanish Ministry of Education and Science to Antonio Camacho, and previous work was supported by grant REN2000-0435 ANT from the same institution to Antonio Quesada. The author is indebted to a number of colleagues and friends from LIMNOPOLAR Project, to Unidad de Tecnologia Marina (UTM-CSIC), to Las Palmas crew (Spanish Navy), and to the Spanish Polar Committee. I also thank Dr. J. Cynan Ellis-Evans for his interesting comments during the ESF workshop on "Life in Extreme Environments" held in San Feliu de Guixols and his work with the edition of the manuscript, to two anonymous reviewers whose comments greatly helped to improve it, and to the European Science Foundation, that invited me to this workshop.

## References

Adrian R, Schneider-Olt B (1999) Top-down effects of crustacean zooplankton on pelagic microorganisms in a mesotrophic lake. J Plankton Res 21:2175–2190

Allende L, Izaguirre I (2003) The role of physical stability on the establishment of steady states in the phytoplankton community of two Maritime Antarctic lakes. Hydrobiologia 502:211–224

Almada P, Allende L, Tell G, Izaguirre I (2004) Experimental evidence of the grazing impact of *Boeckella poppei* on phytoplankton in a maritime Antarctic lake Polar Biol 28:39–46

Arnold RJ, Convey P (1998) The life history of the diving beetle *Lancetes angusticollis* (Curtis) (Coleoptera:Dytiscidae), on sub-Antarctic South Georgia. Polar Biol 20:153–160

Atkinson A (1995) Omnivory and feeding selectivity in 5 copepod species during spring in the Bellingshausen Sea, Antarctica. ICES J Mar Sci 52:385–396

Atkinson A, Whitehouse MJ (2001) Ammonium regeneration by Antarctic mesozooplankton: an allometric approach. Mar Biol 139:301–311

Azam F, Fenchel T, Field JG, Grey JS, Meyer-Reil KA, Thingstad F (1983) The ecological role of water-column microbes in the sea. Mar Ecol Prog Ser 10:257–263

Balseiro EG, Modenutti BE, Queimaliños CP (2001) Feeding of *Boeckella gracilipes* (Copepoda, Calanoida) on ciliates and phytoflagellates in an ultraoligotrophic Andean lake. J Plankton Res 23:849–857

Bayliss P, Ellis-Evans JC, Laybourn-Parry J (1997) Temporal patterns of primary production in a large ultra-oligotrophic Antarctic freshwater lake. Polar Biol 18:363–370

Bayly IAE (1992) Fusion of the genera *Boeckella* and *Pseudoboeckella* (copepoda) and revision of their species from south-America and sub-Antarctic islands. Rev Chil Hist Nat 65:17–63

Bayly IAE, Gibson JAE, Wagner B, Swadling KM (2003) Taxonomy, ecology and zoogeography of two East Antarctic freshwater calanoid copepod species: *Boeckella poppei* and *Gladioferans antarcticus*. Antarctic Sci 15:439–448

Begon M, Harper JL, Townsend CR (2005) Ecology: from individuals to ecosystems. Blackwell Scientific Publishers, Oxford, 1068 pp

Bell EM, Laybourn-Parry J (1999) Annual plankton dynamics in an Antarctic saline lake. Freshwater Biol 41:507–519

Berninger UG, Finlay BJ, Kuuppo-Leinikki P (1991) Protozoan control of bacterial abundances in freshwater. Limnol Oceanogr 36:139–147

Bertilsson S, Hansson LA, Graneli W, Philibert A (2003) Size-selective predation on pelagic microorganisms in Arctic freshwaters. J Plankton Res 25:621–631

Bonilla S, Villeneuve V, Vincent WF (2005) Benthic and planktonic algal communities in a high Arctic lake: pigment structure and contrasting responses to nutrient enrichment. J Phycol 41:1120–1130

Borer ET, Seabloom EW, Shurin JB, Anderson KE, Blanchette CA, Broitman B, Cooper SD, Halpern BS (2005) What determines the strength of a trophic cascade? Ecology 86:528–537

Burns CW, Schallenberg M (1996) Relative importance of cladocerans, copepods and nutrients on the microbial food web of a mesotrophic lake. J Plankton Res 18:683–714

Burns CW, Schallenberg M (1998) Impacts of nutrients and zooplankton on the microbial food web of an ultra-oligotrophic lake. J Plankton Res 20:1501–1525

Butler H (1999a) Seasonal dynamics of the planktonic microbial community in a maritime Antarctic lake undergoing eutrophication. J Plankton Res 21:2393–2419

Butler H (1999b) Temporal plankton dynamics in a maritime Antarctic lake. Arch Hydrobiol 146:311–339

Butler HG, Edworthy MG, Ellis-Evans JC (2000) Temporal plankton dynamics in an oligotrophic maritime Antarctic lake. Freshwater Biol 43:215–230

Butler H, Atkinson A, Gordon M (2005) Omnivory and predation impact of the calanoid copepod *Boeckella poppei* in a maritime Antarctic lake. Polar Biol 28:815–821

Carpenter SR (1996) Microcosm experiments have limited relevance for community and ecosystem ecology. Ecology 77:677–680

Carpenter SR (1999) Microcosm experiments have limited relevance for community and ecosystem ecology: Reply. Ecology 80:1085–1088

Carpenter SA, Kitchell JF, Hodgson J (1985) Cascading trophic interactions and lake productivity. Bioscience 35:634–639

Carpenter SR, Chisholm SW, Krebs CJ, Schindler DW, Wright RF (1995) Ecosystem experiments. Science 269:324–327

Carpenter SR, Cole JJ, Hodgson JR, Kitchell JF, Pace ML, Bade D, Cottingham KL, Essington TE, Houser JN, Schindler DE (2001) Trophic cascades, nutrients, and lake productivity: whole-lake experiments. Ecological Monographs 71:163–186

Carrick HJ, Fahnenstiel GL, Stoezmer EF, Wetzel RG (1991) The importance of zooplankton-protozoan trophic coupling in Lake Michigan. Limnol Oceanogr 36:1335–1345

Clarke A, Ellis-Evans JC, Sanders MW, Holmes LJ (1989) Patterns of energy-storage in *Pseudoboeckella poppei* (Crustacea, Copepoda) from 2 contrasting lakes on Signy island, Antarctica. Hydrobiologia 172:183–191

Convey P (1996) The influence of environmental characteristics on life history attributes on Antarctic terrestrial biota. Biol Rev 71:191–225

Convey P, Pugh PJA, Jackson C, Murray AW, Ruhland CT, Xiong FS, Day TA (2002) Response of Antarctic terrestrial microarthropods to long-term climate manipulations. Ecology 83:3130–3140

Currie DJ, Dilwortth-Christie P, Chapleau F (1999) Assessing the strength of top-down influences on plankton abundance in unmanipulated lakes. Can J Fish Aquat Sci 56:427–436

Dartnall HJG (2005) Freshwater invertebrates of sub-Antarctic South Georgia. J Nat History 39:3321–3342

Dartnall HJG, Hollwedel W, de Paggi JC (2005) The freshwater fauna of Macquarie Island, including a redescription of the endemic water-flea *Daphnia gelida* (Brady) (Anomopoda: Crustacea). Polar Biol 28:922–939

DeMott WR (1988) Discrimination between algae and artificial particles by freshwater and marine copepods. Limnol Oceanogr 33:397–408

DeMott WR, Watson MD (1991) Remote detection of algae by copepods: responses to algal size, odors and motility. J Plankton Res 13:1203–1222

Drago EC (1989) Thermal summer characteristics of lakes and ponds on Deception Island, Antarctica. Hydrobiologia. 184:51–60

Drenner RW, Mazumder A (1999) Microcosm experiments have limited relevance for community and ecosystem ecology: comment. Ecology 80:1081–1085

Elser JJ, Elser MM, MacKay NA, Carpenter SR (1988) Zooplankton-mediated transitions between N-and P-limited algal growth. Limnol Oceanogr 33:1–14

Ellis-Evans JC (1996) Microbial diversity and function in Antarctic freshwater ecosystems. Biodiv Conserv 5:1395–1431

Ellis-Evans JC, Laybourn Parry J, Bayliss PR, Perriss SJ (1998) Physical, chemical and microbial community characteristics of lakes of the Lasermann Hills, Continental Antarctica. Arch Hydrobiol 141:209–230

Fenchel T (1982) Ecology of heterotrophic microflagellates. IV Quantitative occurrence and importance as bacterial consumers. Mar Ecol Prog Ser 9:35–42

Fernandez-Valiente E, Quesada A, Howard-Williams C, Hawes I (2001) $N_2$ fixation in cyanobacterial mats from ponds on the McMurdo Ice Shelf, Antarctica. Microbial Ecol 42:338–349

Fritsen CH, Priscu JC (1998) Cyanobacterial assemblages in permanent ice covers on Antarctic lakes: distribution, growth rate, and temperature response of photosynthesis. J Phycol 10:587–597

Fountain AG, Berry Lyons W, Burkins MB, Dana GL, Doran PT, Lewis KJ, McKnight DM, Moorhead DLL, Parsons AN, Priscu JC, Wall DH, Wharton RA, Ross VA (1999) Physical control on the Taylor Valley Ecosystem, Antarctica. Bioscience 49:961–971

Frenot Y, Chown SL, Whinam J, Selkirk PM, Convey P, Skotnicki M, Bergstrom DM (2005) Biological invasions in the Antarctic: extent, impacts and implications. Biol Rev 80:45–72

Gasol JM, Pedrós-Alio C, Vaque D (2002) Regulation of bacterial assemblages in oligotrophic plankton systems. Antonie van Leeuwenhoek 81:435–452

Gasparon G, Burgess JS (2000) Human impacts in Antarctica trace-element geochemistry of freshwater lakes in the Larsemann Hills, East Antarctica. Environ Geol 39:963–976

Geider RJ, Delucia EH, Falkowski PG, Finzi AC, Grime JP, Grace J, Kana TM, La Roche J, Long SP, Osborne BA, Platt T, Prentice C, Raven JA, Schlesinger WH, Smetacek V, Stuart V, Sathyendranath S, Thomas RB, Vogelmann TC, Williams P, Woodward FI (2001) Primary productivity of planet Earth: biological determinants and physical constraints in terrestrial and aquatic habitats. Global Change Biology 7:849–882

Hahn M, Höfle MG (2001) Grazing of protozoa and its effect on populations of aquatic bacteria. FEMS Microbiol Ecol 35:113–121

Hairston NG, Smith FE, Slobodkin LB (1960) Community structure, population control, and competition. Am Nat 44:421–425

Hansen JE, Sato M (2001) Trends of measured climate forcing agents. PNAS 98:14778–14783
Hansen PJ, Bjorsen PK, Hansen BW (1997) Zooplankton grazing and growth: Sealing within the 2–2000 μm body size range. Limnol Oceanogr 42:687–704
Hansson LA (1992) The role of food-chain composition and nutrient availability in shaping algal biomass development. Ecology 73:241–247
Hansson LA, Håkansson H (1992) Diatom community response along a productivity gradient of shallow Antarctic lakes. Polar Biol 12:463–468
Hansson LA, Tranvik LJ (1996) Quantification of invertebrate predation and hervibory in food chains of low complexity. Oecologia 108:542–551
Hansson LA, Tranvik LJ (1997) Algal species composition and phosphorus recycling at contrasting grazing pressure: An experimental study in sub-Antarctic lakes with two trophic levels. Freshwater Biol 37:45–53
Hansson LA, Tranvik LJ (2003) Food webs in sub-Antarctic lakes: a stable isotope approach. Polar Biol 26:783–788
Hansson LA, Lindell M, Tranvik LJ (1993) Biomass distribution among trophic levels in lakes lacking vertebrate predators. Oikos 66:101–106
Hansson LA, Dartnall HJG, Ellis-Evans JC, MacAlister H, Tranvik LJ (1996) Variation in physical, chemical and biological components in the sub-Antarctic lakes of South Georgia. Ecography 19:393–403
Hawes I (1983a) Light climate and phytoplankton photosynthesis in maritime Antarctic lakes. Hydrobiologia 123:69–79
Hawes I (1983b) Nutrients and their effects on phytoplankton populations in lakes on Signy Island, Antarctica. Polar Biol 2:115–126
Hawes I, Schwarz AMJ (2001) Absorption and utilization of irradiance by cyanobacterial mats in two ice-covered Antarctic lakes with contrasting light climate. J Phycol 37:5–15
Henshaw T, Laybourn-Parry J (2002) The annual patterns of photosynthesis in two large, freshwater, ultra-oligotrophic Antarctic lakes. Polar Biol 25:744–752
Heywood RB (1978) Maritime Antarctic lakes. Verh. Internat. Verein Limnol. 20:1210–1215
Heywood RB, Dartnall HJG, Priddle J (1980) Characteristics and classification of the lakes of Signy Island, South-Orkney Islands, Antarctica. Freshwater Biol 10:47–59
Huston MA (1999) Microcosm experiments have limited relevance for community and ecosystem ecology: synthesis of comments. Ecology 80:1088–1089
Imura S, Bando T, Seto K, Kudoh S, Kanda H (2003a) Distribution of aquatic mosses in the Sôya region, East Antarctica. Polar Biosci 16:1–10
Imura S, Bando T, Saito S, Seto K, Kanda H (2003b) Benthic moss pillars in Antarctic lakes. Polar Biol 22:137–140
Izaguirre I, Mataloni G, Allende L, Vinocur A (2001) Summer fluctuations of microbial planktonic communities in a eutrophic lake -Cierva Point, Antarctica. J Plankton Res 23:1095–1109
Izaguirre I, Allende L, Marinone MC (2003) Comparative study of the planktonic communities of three lakes of contrasting trophic status at Hope Bay (Antarctic Peninsula). J Plankton Res 25:1079–1097
Jones VJ, Juggins S (1995) The construction of a diatom-based chlorophyll a transfer function and its application at three lakes on Signy Island (maritime Antarctic) subject to differing degrees of nutrient enrichment. Freshwater Biol 34:433–445
Jones VJ, Juggins S, Ellis-Evans JC (1993) The relationship between water chemistry and surface sediment diatom assemblages in maritime Antarctic lake. Antarctic Sci 5:339–348
Jürgens K, Arndt H, Rothhaupt KO (1994) Zooplankton-mediated changes of bacterial community structure. Microbial Ecol 27:27–42
Karlsson J, Byström P (2005) Littoral energy mobilization dominates energy supply for top consumers in subarctic lakes. Limnol Oceanogr 50:538–543
Kennedy AD (1993) Water as a limiting factor in the Antarctic terrestrial environment: A biogeographical synthesis. Arctic Alpine Res 25:308–315
Kennedy AD (1995) Antarctic terrestrial ecosystem response to global environmental change. Annu Rev Ecol Syst 26:683–704
Kiørboe T, Saiz E, Viitasalo M (1996) Prey switching behaviour in the planktonic copepod *Acartia tonsa.* Mar Ecol Prog Ser 143:65–75
Klein-Breteler WCM, Schotg N, Baas M, Schouten S, Kraay GW (1999) Trophic upgrading of food quality by protozoans enhancing copepod growth. Mar Biol 135:191–198
Kleppel GS (1993) On the diets of calanoid copepods. Mar Ecol Prog Ser 99:183–195
Krebs CJ (2001) Ecology: the experimental analysis of distribution and abundance. Benjamin Cummings, San Francisco, CA, 695 pp
Lawley B, Ripley S, Bridge P, Convey P (2004) Molecular analysis of geographic patterns of eukaryotic diversity in Antarctic soils. Appl Environ Microbiol 70:5963–5972
Laybourn-Parry J (1997) The microbial loop in Antarctic lakes. In: Howard-Williams C, Lyons WB, Hawes I (eds) Ecosystem processes in Antarctic ice-free landscapes. Balkema, Rotterdam, pp 231–240
Laybourn-Parry J (2002) Survival mechanisms in Antarctic lakes. Phil Trans Roy Soc London Ser B 357:863–869
Laybourn-Parry J, Bayliss P (1996) Seasonal dynamics of the planktonic community in Lake Druzhby, Princess Elisabeth Land, Eastern Antarctica. Freshwater Biol 35:57–67
Laybourn-Parry J, Bayliss P, Ellis-Evans JC (1995) The dynamics of heterotrophic nanoflagellates and bacterioplankton in a large ultra-oligotrophic Antarctic lake. J Plankton Res 17:1834–1850
Laybourn-Parry J, Ellis-Evans JC, Butler H (1996) Microbial dynamics during the summer ice-loss phase in maritime Antarctic lakes. J Plankton Res 18:495–511
Laybourn-Parry J, Bell EM, Roberts EC (2000) Protozoan growth rates in Antarctic lakes. Polar Biol 23:445–451
Laybourn-Parry J, Quayle WC, Henshaw T, Ruddell A, Marchant HJ (2001) Life on the edge: the plankton

and chemistry of Beaver lake, an ultra-oligotrophic epishelf lake, Antarctica. Freshwater Biol 46:1205–1217
Laybourn-Parry J, Quayle W, Henshaw T (2002) The biology and evolution of Antarctic saline lakes in relation to salinity and trophy. Polar Biol 25:542–552
Laybourn-Parry J, Henshaw T, Jones DJ, Quayle W (2004) Bacterioplankton production in freshwater Antarctic lakes. Freshwater Biol 49:735–744
Laybourn-Parry J, Marshall WA, Marchant HJ (2005) Flagellate nutritional versatility as a key to survival in two contrasting Antarctic saline lakes. Freshwater Biol 50:830–838
Lewis Smith RI (1984) Terrestrial plant biology of the sub-Antarctic and Antarctic. In: Law RM (eds) Antarctic Ecology, Vol. 1. Academic Press, London, pp 61–162
Light JJ, Heywood RB (1973) Deep-water mosses in Antarctic lakes. Nature 242:535–536
Markager S, Vincent WF, Tang EPY (1999) Carbon fixation by phytoplankton in high Arctic lakes: implications of low temperature for photosynthesis. Limnol Oceanogr 44:597–607
Marshall W, Laybourn-Parry J (2002) The balance between photosynthesis and grazing in Antarctic mixotrophic cryptophytes during summer. Freshwater Biol 47:2060–2070
Mataloni G, Tesolin G, Tell G (1998) Characterization of a small eutrophic Antarctic lake (Otero Lake, Cierva Point) on the basis of algal assemblages and water chemistry. Polar Biol 19:107–114
Mataloni G, Tesolin G, Sacullo F, Tell G (2000) Factors regulating summer phytoplankton in a highly eutrophic Antarctic lake. Hydrobiologia 432:65–72
Matson PA, Hunter MD (1992) The relative contributions of top-down and bottom-up forces in population and community ecology. Ecology 73:723–723
Menu-Marque S, Morone JJ, de Mitrovich CL (2000) Distributional patterns of the South American species of *Boeckella* (Copepoda: Centropagidae): A track analysis. J Crustacean Biol 20:262–272
Mitra A, Flynn KJ (2005) Predator-prey interactions: is "ecological stoichiometry" sufficient when good food goes bad? J Plankton Res 27:393–399
Moorhead DL, Wolf CF, Wharton RA (1997) Impact of light regimes on productivity patterns of benthic microbial mats in an Antarctic lake: A modelling study. Limnol Oceanogr 42:1561–1569
Nadeau TL, Castenholz RW (2000) Characterization of psychrophilic oscillatorians (Cyanobacteria) from Antarctic meltwater ponds. J Phycol 36: 914–923
Ohman MD, Runge JA (1994) Sustained fecundity when phytoplankton resources are in short supply: omnivory by *Calanus finmarchicus* in the Gulf of St. Lawrence. Limnol Oceanogr 39:21–36
Pace ML (1988) Bacterial mortality and the fate of bacterial production. Hydrobiologia 159:41–49
Pace ML, Funke E (1991) Regulation of planktonic microbial communities by nutrients and herbivores. Ecology 72:904–914
Pace ML, Cole JJ, Carpenter SR (1998) Trophic cascades and compensation: Differential responses of microzooplankton in whole-lake experiments. Ecology 79: 138–152
Pace ML, Cole JJ, Carpenter SR, Kitchell JF (1999) Trophic cascades revealed in diverse ecosystems. Trends Ecol Evol 15:483–488
Paggi JC de (1987) Limnological studies in the Potter Peninsula, 25 de Mayo Island, South Shetland Islands: biomass and spatial distribution of zooplankton. BIOMASS Sci Ser 7:175–191
Paggi JC de (1996) Feeding ecology of *Branchinecta gainii* (Crustacea: Anostraca) in ponds of South Shetland Islands, Antarctica. Polar Biol 16:13–18
Paine RT (1980) Food webs, linkage, interaction, strength and community infrastructure. J Animal Ecol 49:667–685
Pearce DA (2000) A method to study bacterioplankton community structure in Antarctic lakes. Polar Biol 23:352–356
Pearce DA (2003) Bacterioplankton community structure in a maritime Antarctic oligotrophic lake during a period of holomixis, as determined by denaturing gradient gel electrophoresis (DGGE) and fluorescence in situ hybridization (FISH). Microbial Ecol 46:92–105
Pearce DA (2005) The structure and stability of the bacterioplankton community in Antarctic freshwater lakes, subject to extremely rapid environmental change. FEMS Microbiol Ecol 53:61–72
Pearce D, van der Gast CJ, Lawley B, Ellis-Evans JC (2003) Bacterioplankton community diversity in a maritime Antarctic lake, determined by culture-dependent and culture-independent techniques. FEMS Microbiol Ecol 45:59–70
Peck LS, Clark MS, Clarke A, Cockell CS, Convey P, Detrich HW, Fraser KPP, Johnston IA, Methe BA, Murray AE, Romisch K, Rogers AD (2005) Genomics: applications to Antarctic ecosystems. Polar Biol 28:351–365
Pernthaler J, Amann R (2005) Fate of heterotrophic protists in pelagic habitats: Focus in populations. Microbiol Molec Biol Rev 69:440–461
Pernthaler J, Sattler B, Simek K, Schwarzenbacher A, Psenner R (1996) Top-down effects on the size-biomass distribution of a freshwater bacterioplankton community. Aquat Microb Ecol 10:255–263
Persson L (1999) Trophic cascades: abiding heterogeneity and the trophic level concept at the end of the road. Oikos 85:385-397
Petz W (2003) Ciliate biodiversity in Antarctic and Arctic freshwater habitats – a bipolar comparison. Eur J Protistol 39:491–494
Petz W, Valbonesi A, Quesada A (2005) Ciliate biodiversity in freshwater habitats of the maritime and continental Antarctica. Terra Antarctica 11:43–50
Pizarro H, Allende L, Bonaventura SM (2004) Littoral epilithon of lentic water bodies at Hope Bay, Antarctic Peninsula: biomass variables in relation to environmental conditions. Hydrobiologia 529:237–250

Polis GA, Sears ALW, Huxel GR, Strong DR, Maron J (2000) When a trophic cascade is a trophic cascade? Trends Ecol Evol 15:473–475

Pomeroy LR (1974) The ocean's food web, a changing paradigm. Bioscience. 24:499–504

Priddle I, Hawes I, Ellis-Evans JC (1986) Antarctic aquatic systems as habitats for phytoplankton. Biological Rev 61:199–238

Ptacnik R, Sommer U, Hansen T, Martens V (2004) Effects of microzooplankton and mixotrophy in an experimental planktonic food web. Limnol Oceanogr 49:1435–1445

Pugh PJA, Dartnall HJG, McInnes SJ (2002) The non-marine Crustacea of Antarctica and the islands of the Southern Ocean: Biodiversity and biogeography. J Nat History 36:1047–1103

Quayle WC, Peck LS, Peat H, Ellis-Evans JC, Harrigan PR (2002) Extreme responses to climate change in Antarctic lakes. Science 295:645

Quesada A, Vincent WF (1997) Strategies of adaptation by Antarctic cyanobacteria to ultraviolet radiation. Eur J Phycol 32:335–342

Roberts EC, Laybourn-Parry J (1999) Mixotrophic cryptophytes and their predators in the Dry Valley lakes of Antarctica. Freshwater Biol 41:737–746

Roberts EC, Laybourn-Parry J, McKnight DM, Novarino G (2000) Stratification and dynamics of microbial loop communities in Lake Fryxell. Freshwater Biol 44:649–661

Roberts EC, Priscu JC, Wolf C, Lyons WB, Laybourn-Parry J (2004) The distribution of microplankton in the McMurdo Dry Valley lakes, Antarctica: response to ecosystem legacy or present-day climatic controls? Polar Biol 27:238–249

Sancho LG, Pintado A (2004) Evidence of high annual growth rate for lichens in the maritime Antarctic. Polar Biol 27:312–319

Schindler DW (1998) Replication versus realism: The need for ecosystem-scale experiments. Ecosystems 1:323–334

Shurin JB, Borer ET, Seabloom EW, Anderson K, Blanchette CA, Broitman B (2002) A cross-ecosystem comparison of the strength of trophic cascades. Ecol Lett 5:785–791

Simmons GM, Vestal JR, Wharton RA (1993) Environmental regulators of microbial activity in continental Antarctic lakes. In: Friedmann EI (eds) Antarctic microbiology. Wiley-Liss, New York, pp. 491–541

Smith JA, Hodgson DA, Bentley MJ, Verleyen E, Leng MJ, Roberts SJ (2006) Limnology of two Antarctic epishelf lakes and their potential to record periods of ice shelf loss. J Paleolimnol 35:373–394

Sommer U, Sommer F (2006) Cladocerans versus copepods: the cause of contrasting top–down controls on freshwater and marine phytoplankton. Oecologia 147:183–194

Sommer U, Stibor H (2002) Copepoda-Cladocera-Tunicata: the role of three major mesozooplankton groups in pelagic food webs. Ecol Res 17:161–174

Sommer U, Stibor H, Katechakis A, Sommer F, Hansen T (2002) Pelagic food web configurations at different levels of nutrient richness and their implications for the ratio of fish production:primary production. Hydrobiologia 484:11–20

Steiner CF (2003) Keystone predator effects and grazer control of planktonic primary production. Oikos 101:569–577

Sterner R (1986) Hervivores' direct and indirect effects on algal populations. Science 231:605–607

Stoecker DK, Capuzzo SM (1990) Predation on protozoa: its importance to zooplankton. J Plankton Res 12:891–908

Strong DR (1992) Are trophic cascades all wet? Differentiation and donor-control in speciose ecosystems. Ecology. 73:747–754

Sun S, Hansen JE (2003) Climate simulations for 1951–2050 with a coupled atmosphere-ocean model. J Cimate 16:2807–2826

Swadling KM, Gibson JAE (2000) Grazing rates of a calanoid copepod (*Paralabidocera antarctica*) in a continental Antarctic lake. Polar Biol 23:301–308

Takacs CD, Priscu JC (1998) Bacterioplankton dynamics in the McMurdo Dry Valley lakes, Antarctica: Production and biomass loss over four seasons. Microbial Ecol 36:239–250

Tang EPY, Tremblay R, Vincent WF (1997) Cyanobacterial dominance of polar freshwater ecosystems: Are high-latitude mat-formers adapted to low temperature? J Phycol 33:171–181

Taton A, Grubisic S, Brambilla E, de Wit R, Wilmotte A (2003) Cyanobacterial diversity in natural and artificial microbial mats of Lake Fryxell (McMurdo Dry Valleys, Antarctica): a morphological and molecular approach. Appl Environ Microbiol 69:5157–5169

Tranvik LJ, Hansson LA (1997) Predator regulation of aquatic microbial abundance in simple food webs of sub-Antarctic lakes. Oikos 79:347–356

Unrein F, Vinocur A (1999) Phytoplankton structure and dynamics in a turbid Antarctic lake (Potter Peninsula, King George Island). Polar Biol 22:93–101

Unrein F, Izaguirre I, Massana R, Balague V, Gasol JM (2005) Nanoplankton assemblages in maritime Antarctic lakes: characterisation and molecular fingerprinting comparison. Aquat Microb Ecol 40:269–282

Van de Vijver B, Beyens L (1999) Biogeography and ecology of freshwater diatoms in sub-Antarctica: a review. J Biogeogr 26:993–1000

Vanni MJ (2002) Nutrient cycling by animals in freshwater ecosystems. Annu Rev Ecol Syst 33:341–370

Vincent WF (1988) Microbial ecosystems in Antarctica. Cambridge University Press, Cambridge, UK

Vincent WF, Laurion I, Pienitz R (1998) Arctic and Antarctic lakes as optical indicators of global change. Ann Glaciol 27:691–696

Vincent WF, Downes MT, Castenholz RW, Howard-Williams C (1993) Community structure and pigment

organization of cyanobacteria-dominated microbial mats in Antarctica. Eur J Phycol 28:213–221
Vinocur A, Unrein F (2000) Typology of lentic water bodies at Potter Peninsula (King George Island, Antarctica) based on physical-chemical characteristics and phytoplankton communities. Polar Biol 23:858–870
Vinocur A, Pizarro H (2000) Microbial mats of twenty-six lakes from Potter Peninsula, King George Island, Antarctica. Hydrobiologia 437:171–185
Wetzel RG (2001) Limnology. Academic Press, San Diego, CA
Wynn-Williams DD (1996) Antarctic microbial diversity: The basis of polar ecosystem processes. Biodiv Conserv 5:1271–1293
Zöllner E, Santer B, Boersma M, Hoppe HG, Jürgens K (2003) Cascading predation effects of *Daphnia* and copepods on microbial food web components. Freshwater Biol 48:2174–2193

Rev Environ Sci Biotechnol (2007) 6:127–141
DOI 10.1007/s11157-006-9107-y

REVIEWS

# Fungi in Antarctica

**Serena Ruisi · Donatella Barreca · Laura Selbmann · Laura Zucconi · Silvano Onofri**

Received: 23 March 2006 / Accepted: 16 August 2006 / Published online: 29 December 2006


**Abstract** Fungi are generally easily dispersed and are able to colonize a very wide variety of different substrata and to withstand many different environmental conditions. Because of these characteristics they spread all over the world. The Antarctic mycoflora is quite diversified within the different climatic regions of the continent. Most Antarctic microfungi are cosmopolitan; some of them are propagules transported to Antarctica but unable to grow under the Antarctic conditions, while others, termed indigenous, are well adapted and able to grow and reproduce even at low temperatures, mostly as psychrotolerant, or fast sporulating forms, able to conclude their life-cycles in very short time. In the most extreme and isolated areas of the continent, such as the Antarctic Dry Valleys, endemic species showing physiological and morphological adaptations have locally evolved. Most Antarctic fungi, as well as fungi from other dry and cold habitats, are adapted to low temperatures, repeated freeze and thawing cycles, low water availability, osmotic stress, desiccation, low nutrients availability and high UV radiation. Sometimes single strategies are not specific for single stress factors and allow these microorganisms to cope with more than one unfavourable condition.

**Keywords** Adaptation · Antarctica · Fungi · Low temperature · Osmotic stress · UV radiation · Water availability

S. Ruisi · D. Barreca · L. Selbmann · L. Zucconi · S. Onofri (✉)
Dipartimento di Scienze Ambientali, Università degli Studi della Tuscia, Largo dell'Università, 01100 Viterbo, Italy
e-mail: onofri@unitus.it

## 1 Introduction

Antarctica is a remote and inhospitable continent. The climate is the coldest and driest known on Earth; nevertheless it is not uniform across the continent, and different climatic regions can be distinguished (Holdgate 1977; Onofri 1999; Øvstedal and Lewis Smith 2001). The prevailing Antarctic conditions of low temperature, low water availability, frequent freeze–thaw cycles, low annual precipitation, strong winds, high sublimation and evaporation, high incidence of solar and especially ultraviolet radiation together constitute significant limiting factors for plant and animal life. Therefore, the biology of Antarctica, more than other continents, is dominated by microorganisms (Friedmann 1993), with a high level of adaptation and able to withstand extreme conditions.

Abundance and diversity of organisms decrease, along broad latitudinal gradients, from the maritime to the continental Antarctic zone and within the latter with increasing altitude and latitude from the coast to the Ice Slope Region (Pickard and Seppelt 1984; Kappen 1993; Broady 1996), due to increasingly severe environmental conditions and isolation from rich propagules sources such as South America and Australasia.

The harsh climate of the inland Antarctic continent strongly affects microbial settlement on the rock surfaces and microbial life withdraws into microscopic niches inside the rocks, where microorganisms find more stable environmental conditions. In particular, the cold desert of the McMurdo Dry Valleys, or Ross Desert, in Southern Victoria Land, the most extreme ice-free environment known on Earth, was thought to be almost sterile after microbiological analysis of soils by cultural methods (Horowitz et al. 1972). There, the absolute limit for life is reached (Nienow and Friedmann 1993) and microbial communities form a particular cryptoendolithic ecotype (Golubic et al. 1981), consisting of organisms colonizing the interstitial spaces of porous rocks, as a rule under a rock crust, while life on bare rock surfaces is represented by rare epilithic lichens growing in protected niches; thus endolithic microorganisms, living close to the limit of their physiological adaptability, are the dominant life form (Friedmann and Ocampo 1976; Friedmann 1982; Friedmann et al. 1988; Friedmann and Koriem 1989; Kappen 1993; Nienow and Friedmann 1993; Wierzchos and Ascaso 2001, 2002).

Over the last decades the Antarctic regions have been investigated mainly for the presence of psychrophilic bacteria and archea, occasionally for algae and more rarely for fungi (Gunde-Cimerman et al. 2003). Terrestrial microfungal communities in the Antarctic are scarcely investigated (Vincent 1988; Del Frate and Caretta 1990; Wynn-Williams 1990; Vishniac 1993), and there has been little of the physiological information necessary to explain the ecological roles of microfungi in these habitats. Investigations on microorganisms that survive and even thrive in extreme Antarctic environments (extremotolerant) are now providing new insights into the biological mechanisms of tolerance and adaptation (Vincent 2000).

The present paper will examine Antarctic fungi and their adaptive strategies. Papers dealing with morphological and physiological adaptations of Antarctic fungal strains, suggest the lack of specific adaptations, in agreement with those already observed in Antarctic lichens and mosses (Green et al. 1999). In fact, responses of Antarctic fungi to different stresses appear similar to those found even in temperate regions.

Those fungi able to tolerate Antarctic conditions therefore possibly possess pre-existing adaptive capabilities and have been positively selected. Amongst these, those fungi able to grow actively and reproduce in Antarctica could be considered truly indigenous. This situation is exacerbated in black fungi from the Antarctic ice-free desert, which are unable to compete with faster growing cosmopolitan fungi but due to their extremo-tolerance have access to a low competition niche.

Most Antarctic environments continuously receive a rain of microbial propagules from outside the region, as indicated by the high frequency of apparently cosmopolitan species in most habitats but particularly in relative proximity to rich sources such as South America (Kappen 1993). This long-distance dispersal depends on vectors such as winds and animals and it is not easy to discriminate between "indigenous" and "non indigenous" species. For instance, with respect to Antarctic mycoflora, some fungi are represented by mesophilic species, present as viable propagules but unable to reproduce except in rarely favourable climatic conditions. Other species, able to grow actively, at least under Antarctic summer conditions, comprise particular ecotypes of cosmopolitan species showing mesophilic-psychrotolerant behaviour as an adaptation to the cold Antarctic climate (Zucconi et al. 1996). Finally, few of them are true psychrophilic strains; they include some strains of *Thelebolus microsporus* (Berk and Broome) Kimbr. and *Mucor flavus* Bainier, and some cryptoendolithic black fungi strains. Antarctica has been relatively isolated from the rest of the world since the separation from Gondwanaland and later formation of the Southern Ocean polar front. Geographic isolation, combined with environmental stress, makes the Antarctica the first place to look for endemic organisms, in order to shed light on evolutionary processes generating microbial speciation (Vincent 2000; de Hoog et al. 2005; Selbmann et al. 2005). The paradigm of 'everything is everywhere' is widely recognized in microbial ecology as a consequence of large numbers of minute viable cells and the ability to form dormant stages, which together allow dispersal by air and animals. Small organisms tend to have broader geographic distribution than larger ones, but recent molecular data suggest a rather restricted distribution of some Antarctic fungal genera and species (de Hoog et al. 2005; Selbmann et al. 2005).

Despite the few mycological studies and the low number of habitats available for microbial life in the Antarctic continent, a significant number of new taxa of fungi, bacteria, micro-algae and protozoa, have been described. Amongst these, there are some filamentous fungi and yeasts originally described from the Antarctic continent, which have been later isolated elsewhere, as *Penicillium antarcticum* A.D. Hocking & C.F. McRae, also recorded in Denmark and Atlantic Ocean, and *Cryptococcus adeliensis* Scorsetti et al., *Cryptococcus albidosimilis* Vishniac & Kurtzman, *Cryptococcus antarcticus* Vishniac & Kurtzman, and *Mrakia frigida* (Fell et al.) Y. Yamada & Komag. (CBS, website). For at least about 20 of the new taxa described from continental Antarctica, representing 8% of the whole mycoflora of this area (consisting of 251 specific and infraspecific taxa, Onofri et al. 2006), no further records have been reported from elsewhere at present, but given the scant knowledge of microbial diversity and occurrence when compared to higher plants and animals (Foissner 2006) one can't exclude their possible wider distribution. With the advent of molecular techniques, however, our understanding of Antarctic microbial phylogeny, biodiversity and evolution is improving.

## 2 Antarctic fungi

The Antarctic fungal biodiversity has been investigated from floristic (Del Frate and Caretta 1990; Montemartini Corte 1991; Onofri et al. 1991, 1994; Onofri and Tosi 1992; Mercantini et al. 1993; Montemartini Corte et al. 1993) ecophysiological (Caretta et al. 1994; Zucconi et al. 1996; Fenice et al. 1997; Onofri et al. 2000; Tosi et al. 2002), molecular (Vishniac and Onofri 2002) and, most recently, phylogenetic points of view (Selbmann et al. 2005), by isolating and identifying fungal strains from samples collected in different areas of the Antarctic continent (Onofri et al. 2005a). The distribution of fungi in Antarctica is related to the distribution of different substrata such as soils, rocks, bird feathers and dung, vegetation, which consists of plants (largely limited to maritime Antarctica), bryophytes, and lichens (Bridge and Worland 2004; Jumpponen et al. 2003; Leotta et al. 2002; Tosi et al. 2002), and to the distribution of scientific research stations, mainly scattered along the coast of the continent, whose environs have been most extensively investigated. There is a long list of fungal species, including yeasts, colonizing nearly all terrestrial environments occurring in Antarctica (Onofri 1999; Onofri et al. 2006). Most of the filamentous fungi and yeasts are cosmopolitan species; some fungi are psychrophilic, even more are psychrotolerant; whilst the arid conditions on the Antarctic rock surface, usually lead to the dominance of xerotolerant organisms.

Regarding continental Antarctica, Onofri et al. (2005b, 2006) report that a small percentage (0.6%) of known "fungi" species is represented by water moulds (kingdom *Chromista*), while the major proportion (99.4%) is composed of true fungi, including yeasts and filamentous fungi, which together comprise species belonging to phyla of *Chytridiomycota*, *Zygomycota*, *Ascomycota* and *Basidiomycota*. Among these, some species are possibly endemic and some others are indigenous, being able to show active growth and to reproduce in Antarctica; these include psychrophilic or, more commonly, cold-tolerant mesophilic strains, or strains adopting other forms of adaptations. For some species, indigenicity is indicated by the high number of isolations over the years, as in *Cryptococcus vishniacii*, *Geomyces pannorum* and *Thelebolus microsporus*, frequently recorded in Antarctica from different sites and substrata (Onofri et al. 2006). Many microfungal strains, having meristematic growth, mainly found associated with rocks, have also been recorded. Some of them, isolated from the ice-free areas of the Victoria Land, have been described as endemic genera and species (Onofri et al. 1999; Selbmann et al. 2005).

Most of the fungi recorded in the Antarctic continent are anamorphic forms. It seems that fungi gaveup sexual reproduction as this simplification means that life cycles can be concluded in a shorter time and without high metabolic costs. A few exceptions have been reported, such as *Thelebolus* spp. (*Ascomycota*) frequently collected in continental Antarctica. *Thelebolus* species have been reported to reproduce sexually, but the endemic species *T. globosus* Brumm. & de Hoog and *T. ellipsoideus* Brumm. & de Hoog, living in the microbial mat of some Antarctic lakes, show a strong reduction of the ascomata compared with the species living in more permissive conditions and produce anamorphs as adaptations to the extreme conditions of sealed Antarctic lakes (de Hoog et al. 2005).

Cryptoendolithic black fungi living in the prohibitive conditions of the ice-free zones in Victoria Land show an even more extreme simplification of their life-cycles: most of them are unable even to produce anamorphic reproductive structures but propagules are generated directly from disarticulation of toruloid pre-existing hyphae, as observed in *Friedmanniomyces endolithicus* Onofri. Some strains of *Cryomyces* are mostly yeast-like organized and can formally conclude their life-cycles with the production of a single cell being itself a resistant propagule. These high levels of simplification match very well with the prohibitive environment they colonize where the climatic conditions for active life occur just few days a year. Similar situations, recently reported by Kis-Papo et al. (2003) for fungi living in hypersaline environments, support this hypothesis. In fact they found a positive association among ecological stress, genomic diversity and sexual reproduction, and assumed that even if higher levels of genomic diversity provide a higher potential for genetic adaptation, when environmental conditions become closer to the limits for life, and the niche becomes narrower, homogeneous, and extremely stressful, then genomic diversity declines. Thus the harsh conditions experienced by cryptoendolithic black fungi promote a natural selection, which turns from diversification to a selective regime leading to a few highly adapted homozygous clones.

A completely different behaviour has however been discussed by Seymour et al. (2005) for lichens (lichen-forming fungi), which produce abundant sexual structures even in the hostile environments of Antarctica ensuring genetic recombination within populations. In these cases reproductive ascomata require just an initial high metabolic cost, because they persist for several years during which time large numbers of ascospores can be released. But even these organisms, in the very extreme environments of the McMurdo Dry Valleys, give up their typical thallus morphology, epilithic sexual structures disappear and they shift to the cryptoendolithic growth form.

### 2.1 Cryptoendolithic communities

Antarctic cryptoendolithic microorganisms constitute very simple communities comprising only a few species (Nienow and Friedmann 1993). The most common and extensively studied is the "lichen dominated community" found in sandstone (Friedmann 1982). This community colonizes porous sandstones and appears under the rock crust as a conspicuous zone up to 10 mm deep, formed by parallel and differently coloured bands. Typically it consists of a black zone under the crust followed by a white, a green, and sometimes a blue-green zone (Fig. 1). The black and the white zones are formed by filamentous fungi and chlorophycean algae, which together give rise to a cryptoendolithic lichen association. However, they retain the potential of thallus formation and in protected niches, such as small depressions and crevices, characterized by a favourable microclimate, epilithic sexual structures appear. Lichens are pioneer organisms of extreme terrestrial environments (Kappen 1974) showing diverse and strategic growth and distribution (Kappen 1993; Nienow and Friedmann 1993); the transition from epilithic to endolithic morphotype in Antarctica occurs on a gradient that is mainly based on temperature (Friedmann

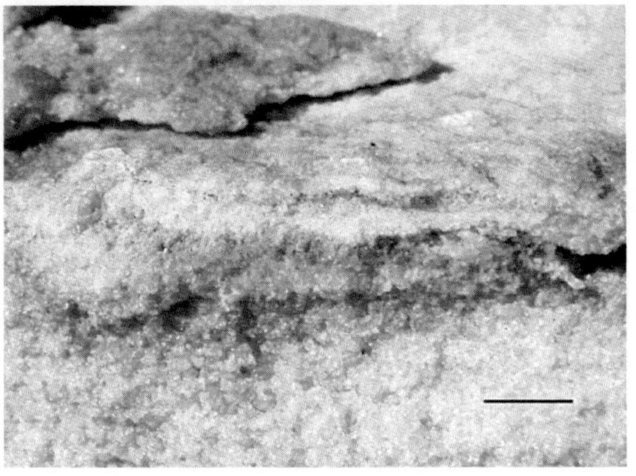

**Fig. 1** Cryptoendolithic lichen dominated community colonizing sandstone in the McMurdo Dry Valleys, Antarctica (bar 10 mm) (Onofri et al. 2006)

et al. 1994). Sandstone outcrops appear to become sterile with increasing distance from the sea (Onofri and Friedmann 1998). The fungal hyphae in the dark zone are melanized, while in the white zone they are colourless; they possibly represent different morphotypes of the same fungal species and the pigmentation in the upper level is supposed to be a response to higher light intensity (Nienow and Friedmann 1993); in fact, a fungal strain isolated from a lichen-dominated community in Southern Victoria Land, producing both pigmented and hyaline hyphae in culture, was demonstrated to be the fungal symbiont by means of laboratory resynthesis experiments (Ahmadjian and Jacobs 1987); later on molecular and microscopy studies demonstrated the ability of a lichen to grow endolitically (de los Rios et al. 2005). In the black zone, thick-walled and dark pigmented non-lichenized fungi, often showing meristematic growth in vitro, grow mixed with the lichen-forming fungi and are isolated as regular members of the lichen dominated cryptoendolithic community (Ocampo-Friedmann and Friedmann 1993; Onofri et al. 1999; Selbmann et al. 2005). In the green zone different species of non-lichenized chlorophycean algae grow, including the endemic species *Hemichloris antarctica*, as well as different extremophilic prokaryotes, such as the cyanobacteria *Chroococcidiopsis* sp. and *Gloeocapsa* sp. (Friedmann 1982). In some cases a further band is observed whenever *Chroococcidiopsis* forms a separate band below *H. antarctica*.

## 2.2 Rock inhabiting black fungi

Rock-inhabiting fungi can be split in two ecological groups: hyphomycetes of soil and of epiphytic origin, and black (melanized) microscopic fungi (Gorbushina et al. 2005). The black melanized microscopic fungi exhibit meristematic growth and form compact restricted microcolonies into rocks (Gorbushina et al. 1993; Sterflinger et al. 1999; Ruibal et al. 2005); they have been named microcolonial fungi (MCF, Staley et al. 1982; Sterflinger 2005) for their growth habit; their presence is ubiquitous and ranges from rock surfaces in the Mediterranean, middle and northern Europe, desert rocks worldwide, bare alpine rock surfaces, hypersaline environments to the Antarctic or as plant, animal and human pathogens. The meristematic fungi are a group of fungi that show very little and slow expansion growth, cauliflower-like colonies, and reproduce by isodiametric enlargement with subdividing cells. From a phylogenetic point of view they represent a quite heterogeneous group of fungi (Sterflinger et al. 1999). Microcolonial fungi show some typical characteristics that are extremely important for their survival in the challenging lithic environments and make them the most stress-tolerant and persistent inhabitants of subaerial rock biofilms on desert rocks (Staley et al. 1982). These fungi express melanized thick cell walls as a stable character, which make them able to withstand dryness, desiccation, and UV exposition. Furthermore, the meristematic growth ensures an optimal surface to volume ratio for the colonies (Wollenzien et al. 1995), a character that was supposed to improve the survival ability under stressful conditions such as high or low temperature, low water availability (Wollenzien et al. 1995), high UV exposition (Urzì et al. 1995), nutrient deficiency (Sterflinger et al. 1999), and high salt concentrations (Zalar et al. 1999c). Interactions between rock inhabiting microorganisms (fungi, algae, cyanobacteria and bacteria) are still poorly understood even if possible contacts between fungi and algal cells in rock crevices penetrated by fungi have been already recognized (Grondona et al. 1997; Turian 1977). Furthermore, in a recent study, Gorbushina et al. (2005) showed interactions between xerophilic photoautotrophic (photobionts of xerophilic lichens) and chemo-organoheterotrophic (free-living microcolonial fungi) rock-inhabiting microorganisms, using axenic cultures. They demonstrated that at least some rock-inhabiting microcolonial fungal strains are capable of forming a very specific contact with algae. Rock-inhabiting black fungi in the Antarctic have been isolated from cryptoendolithic communities. They are fascinating and scarcely known microorganisms, surviving and even thriving in environments more hostile than those of the remaining rock-inhabiting fungi. Among the fungal species reported from the continent, they seem to be the best adapted ones to the harshest conditions typical of this environment (Selbmann et al. 2005). The typical morphology of the rock inhabiting fungi makes them suitable to withstand harsh environmental conditions; they also show the ability to produce extracellular polymeric substances (EPS), possibly polysaccharides, that are thought to be involved specially in the protection against desiccation and repeated freezing and thawing cycles (Selbmann et al. 2002, 2005).

Recently, two new genera and four new species (*Friedmanniomyces endolithicus* Onofri, *F. simplex* Selbmann et al., *Cryomyces minteri* Selbmann et al., *C. antarcticus* Selbmann et al.) have been described (Onofri et al. 1999; Selbmann et al. 2005) by means of morphological and molecular analyses (rDNA ITS and SSU). Among the filamentous hyphomycetes present in endolithic habitats, a strain of *Verticillium* sp. and a non-cultured black pigmented fungus have been recorded from a microbial endolithic community within gypsum crusts at Two Step Cliffs, the closest analogue to continental Dry Valley systems present in the Antarctic Peninsula region (Alexander Island, maritime Antarctica) (Hughes and Lawley 2003).

## 3 Adaptation to stress conditions

Antarctica is characterized by some of the main environmental stress factors, and Antarctic microorganisms withstand them showing different morphological and physiological adaptive strategies. Antarctic microorganisms have to face several stresses simultaneously and they adopt different strategies at the same time to address these stresses. Sometimes single strategies are not specific for single stress factors and allow these microorganisms to cope with more than one unfavourable condition.

### 3.1 Low temperatures

In biology low temperature is usually identified with subzero temperatures with a lower limit of –20°C, below which no life processes persist (Rivkina et al. 2000). Polar terrestrial organisms in particular, must survive long periods of subzero temperatures and quite often daily freeze–thaw events (Montiel 2000). Extremely low temperatures not only restrict microbial enzyme activity and membrane integrity (Russell 1990; Crowe et al. 1992), but also constrain the availability of liquid water for the hydration of biomolecules and as a medium for biochemical processes (Wynn-Williams and Edwards 2000). Several physiological mechanisms of cold-tolerance in fungi have been reported, and it is possible that a combination of these strategies is employed by Antarctic microorganisms: dehydration ability, antifreezes production, high supercooling activity, high chill tolerance, freeze tolerance, selection of micro- and nano-habitats, life in habitats with snow cover, anoxia tolerance (being encased in ice) (Robinson 2001).

It is well established that, among different cold-tolerance mechanisms, a strategy used by microorganisms, in Antarctica as well as in other cold habitats, is the alteration of membrane lipid composition (Russell 1990). Freeze and/or dehydration damage to cells can be caused by transition of membrane lipids from the liquid crystalline to the gel phase at low temperature (Crowe et al. 1987). The temperature at which the transition occurs can be lowered by increasing the degree of fatty acids unsaturation, sustaining cell functions even at very low temperatures. An increase in the linoleic and arachidonic acids production was observed in Antarctic strains of *Mortierella alpina* Peyronel, *Mortierella antarctica* Linnem., and *Cadophora fastigiata* Lagerb. & Melin when cultivated at low temperatures

(Maggi et al. 1991) and changes in fatty acid composition in response to low temperatures were even observed in Antarctic strains of *Geomyces vinaceus* Dal Vesco and *Geomyces pannorum* (Link) Sigler & J.W. Charmich. (Finotti et al. 1993).

Low temperatures induce also dehydration and osmotic stress in fungi and this can result in the synthesis of different compatible solutes with enzyme activity protection roles. Glycerol is supposed to be the most compatible solute (Brown 1978), even if the most efficient ones with cryoprotective ability during desiccation or freezing are sugars, such as trehalose (Weinstein et al. 2000) and mannitol (Feofilova et al. 1994). They can stabilize membranes, maintaining their integrity and function, thus playing a major role for so-called anhydrobiotic organisms capable of surviving complete dehydration (Crowe et al. 1984, 1986), a mechanism that enhances the ability to avoid damages caused by unfavourable environmental conditions. Trehalose is the most widely distributed disaccharide in fungi. It is very common in both vegetative and reproductive stages (Thevelein 1984). It is an important component in fungal spores and it is believed to enhance their resistance under extreme environmental stress conditions, such as low and even high temperatures, dehydration, desiccation and freezing (D'Amore et al. 1991; Gadd et al. 1987; Hottiger et al. 1987; Lewis et al. 1995). Its role as protectant under environmental stresses has also been suggested for yeasts (Thevelein 1984; Van Laere 1989; Wiemken 1990). An accumulation of cryoprotective carbohydrates in response to sub-optimal growth temperature was observed in *Humicola marvinii* M.E. Palm & Weinst. (intracellular trehalose and extracellular glycerol) and *Mortierella elongata* Linnem. (intracellular trehalose), described from fellfield soils in maritime Antarctica (Weinstein et al. 2000). Exopolysaccharides production by fungi represents another response to stress conditions; the mechanism through which these molecules act has still not been fully clarified, but it has been suggested they could protect cells by changing the permeability to $Na^+$ and $K^+$ ions during freezing and thawing and act on the viscosity of extracellular solution avoiding excessive stresses. It seems that they also could alter the structure of water within and around the cells forming a glass structure during freezing which controls water crystallization to a certain extent. Selbmann et al. (2002) demonstrated that the production of exopolysaccharides by an Antarctic fungal strain of *Phoma herbarum* Westend. was related to its ability in preventing damage due to repeated freezing and thawing cycles. Meristematic black fungi from Antarctica indeed (Selbmann et al. 2005) frequently produce extracellular polymeric substances outside the hyphae or surrounding the multicellular conidia, as in the case of *Friedmanniomyces endolithicus* (Onofri et al. 1999). For the Antarctic endolithic microecosystems a similar situation has been reported by de los Rìos et al. (2003), who described a form of microbe-enclosing polymer layer, buffering the physical and chemical conditions at a nano-scale into the rocks. The finding of cold-active enzymes produced by some Antarctic fungal species (Fenice et al. 1998) may also contribute to an understanding of how these fungi thrive at low temperatures; in some other cases Antarctic fungi show wide enzymatic competences, that increase survival chances in unfavourable environments (Fenice et al. 1997).

Some papers have dealt with growth temperature preferences in Antarctic fungal strains. According to van Uden (1984) and Vishniac (1987), fungi having the optimum temperature for growth at about 15°C or lower, a maximum up to 25°C, and still able to grow at 0°C or below, are considered psychrophilic; those capable of growing at 5°C and below, regardless of the optimum temperature, are psychrotolerant. The mycoflora of Antarctica is mainly composed of psychrotolerants (Kerry 1990; Zucconi et al. 1996; Azmi and Seppelt 1997), able to grow in a wide range of temperatures down to 0°C, although psychrophilic strains have been isolated. The preponderance of psychrotolerance is probably a response to the wide temperature fluctuations in Antarctic ice-free microhabitat, allowing microorganisms to survive in unstable environments (Onofri et al. 2004). In fact, as observed by Baross and Morita for bacteria (1978), psychrophiles are almost exclusively isolated from stable cold environments such as the polar ocean, whereas more unstable conditions of the microhabitats of terrestrial Antarctic fungi result in the predominance of mesophilic psychrotolerant strains (Onofri 1999). Psychrophilic fungi are more frequently found in the coldest terrestrial locations. For instance, some Antarctic black fungi isolated from the lichen dominated cryptoendolithic community of the McMurdo Dry Valleys were classified as psychrophilic having a maximum growth temperature below 25°C and being able to grow at 0°C (Selbmann et al. 2005), a specific adaptation that enables them to be metabolically active during the period of highest photosynthetic activity of the community (Selbmann et al. 2005). Soils of the arid highlands of the Dry Valleys provide a unique habitat for a psychrophilic species, *Cryptococcus vishniacii* Vishniac & Hempfling, adapted to this habitat and unknown elsewhere (Vishniac 2006).

The general prevalence of psychrotolerant strains indicates that most of the Antarctic mycoflora is metabolically active whenever a combination of favourable abiotic conditions occurs during the short growing summer season; in protected niches, subject to solar warming, temperatures at the micro- or nano-climatic scale can be far higher than ambient temperatures and facilitate active life. Most of the Antarctic microorganisms overcome the worst environmental conditions in a state of quiescence, and active growth only occurs occasionally, interspersed with periods of prolonged dormancy. High resistance to freezing is due to the presence of specific intracellular compounds, stable and flexible membranes and other adaptations, usually present to facilitate survival during unfavourable times. Freeze-resistance and withstanding unfavourable conditions in a dormant state are not unique features of Antarctic organisms, as microbial growth and activity below the freezing point have also been suggested for a number of fungi from high-Arctic soils (Bergero et al. 1999).

### 3.2 Low water availability

The Antarctic continent as a whole is extremely dry. In fact, although 70% of the Earth's freshwater is in Antarctica, it is almost exclusively locked up in ice, and in some ice-free areas of the continent, such as the McMurdo Dry Valleys, precipitations, represented only by snow, is less than 100 mm a year. There, desiccation can be enhanced by high winds, which promote evaporation, and the sole source of moisture is represented by transient water melted under the influence of the solar heating of the substratum during the austral summer. Larger amounts of liquid water are present in maritime Antarctica where milder conditions exist and, to a lesser extent, along the coasts of continental Antarctica. The bioavailability of liquid water is among the most important factors limiting the distribution and abundance of terrestrial organisms in continental Antarctica and organisms mostly spread along the coasts where melted water more frequently occurs. Organisms capable of growing under conditions of low water availability are commonly referred as xerotolerant, i.e. able to grow under dry conditions, or xerophilic, preferring habitats where little water is available. Freezing leads to cellular dehydration due to reduced water absorption and conduction; thus, the survival of Antarctic microorganisms, in ice-free areas, fundamentally depends on their resistance to dehydration.

In Antarctica there is often a high concentration of salts in shallow ponds and on rock and soil surfaces as a consequence of high evaporation rates (Nishiyama 1977; Des Marais 1995). Since high salinity causes the same effects as freezing, due to osmotic imbalances (Gunde-Cimerman et al. 2003), microorganisms adapted to matrix–water stress, resulting from drought or binding of water into ice, are successfully adapted to osmotic stress as well. Many fungi have been discovered in environments with salinities ranging between 15 and 32%, where it was previously assumed that only bacteria were able to grow (Méjanelle et al. 2001); Gunde-Cimerman et al. (2005) have provided extensive information about fungi adapted to life at high salt concentrations.

Under conditions of desiccation and low water availability microorganisms react by producing both extra- and intra-cellular compounds. Some Antarctic and non-Antarctic fungal species belonging to black meristematic fungi (Urzì and Realini 1998; Selbmann et al. 2005) as well as *Phoma herbarum* from Antarctica (Selbmann et al. 2002) protect themselves from aridity by producing extracellular polysaccharides. Low molecular weight intracellular osmoregulators are accumulated at high concentrations without interferring with enzyme activity and metabolism and these compatible solutes belong to several classes of compounds: polyols and melanin (Bell and Wheeler 1986; Butler and Day 1998; Kogej et al. 2004), mycosporines (Volkmann et al. 2003), sugars and sugar derivatives, such as glycerol, arabitol, mannitol, and trehalose (Grant 2004). Fungi, including yeasts, are well-known for their ability to react to osmotic stresses by intracellular accumulation of compatible solutes, mainly represented by glycerol and arabitol (Brewer 1999; Pascual et al. 2002); the exposition of a strain of *Aspergillus nidulans* (Eidam) G. Winter to high salt concentrations led to the activation of genes whose expression resulted in the synthesis of glycerol and erithritol (Han and Prade 2002). Resistance to drought, high salinity and cold in fungal species was observed by Gunde-Cimerman et al. (2003), who obtained higher halophilic/xerophilic fungal CFU (colony forming units) numbers than previously reported from Arctic sea-ice samples by using selective low water activity media plus low incubation temperatures. They isolated both melanized and non-melanized fungi, most of which known from other geographical regions and environments, including solar salterns of the Mediterranean coast, alpine and tundra soils, permafrost layers and both polar regions. Preliminary growth tests carried out on some strains of the Antarctic rock genus *Cryomyces* highlighted their ability to grow on media containing up to 24% NaCl concentration (Onofri et al. 2006). The rock-inhabiting meristematic fungus *Hortaea werneckii* (Horta) Nishim. & Miyaji isolated in the sea spray area of Delos (Greece), regulates the osmotic potential under NaCl stress by producing intracellular glycerol (Sterflinger 1998). In hypersaline environments a group of truly halophilc melanized meristematic fungi prevails over the other mycoflora (Buchalo et al. 1998; Gunde-Cimerman et al. 2000); among the most frequently recorded species *Hortea wernekii* (Horta) Nishim. & Miyaji, *Phaeotheca triangularis* de Hoog & Beguin, *Trimmatostroma salinum* Zalar, de Hoog & Gunde-Cim., and *Aureobasidium pullulans* (de Bary) G. Arnaud have been reported (de Hoog et al. 1999; Zalar et al. 1999a–c; Sterflinger et al. 1999; Gunde-Cimerman et al. 2000), the latter also isolated from many different mostly dry and/or salty sites in Antarctica (Onofri et al. 2006). The main characteristics of their extremophilic habit are polymorphic life cycles with alterations in hydrophilic and hydrophobic cells, formation of thick, melanized cell walls, meristematic growth, propagation by endoconidiation, adhesion, and production of extracellular polysaccharides (Gunde-Cimerman et al. 2004). It is surmised that fungi that are tolerant to desiccation, cold and high salinity, could potentially grow in Antarctic microenvironments such as thin liquid water films in ice and permafrost (Gunde-Cimerman et al. 2003). In the latter, at the temperature of –16.5°C, 98–99% of the total water volume is represented by ice and the remaining water is in an unfrozen, supercooled vapour state available for microorganisms (Rivkina et al. 2004). In this context it is interesting to note the recording in the Antarctic permafrost sediments and ice samples of several species belonging to the genera *Aspergillus*, *Cladosporium*, *Fusarium*, *Penicillium*, *Thelebolus*, *Geomyces and Cryptococcus* (Abyzov 1993; Kochkina et al. 2001; Gunde-Cimerman et al. 2003), even if the strains isolated possibly came from propagules in a dormant state.

Among several fungi isolated from permafrost (Vishniac 1993; Nienow and Friedmann 1993; Broady 1993), from snow (Abyzov 1993) and from glacier ice (Abyzov 1993, Ma et al. 2000) the presence of both polar and cosmopolitan fungal species has been repeatedly demonstrated (Ma et al. 2000).

### 3.3 High UV radiation

Solar radiation is both a necessary and a damaging environmental factor. Since the 1980s, emissions of chlorofluorocarbons, mainly due to anthropogenic inputs, have led to a depletion of the stratospheric ozone layer (Farman et al. 1985). The consequence has been an increased level of solar UV-B radiation reaching the Earth's surface (Frederick and Snell 1988; Frederick et al. 1989; Blumthaler and Ambach 1990; Kerr and

McElroy 1993). The depletion of the stratospheric ozone over the Antarctica, relative to the values in the 1970s, was estimated to be about 50% in spring, leading to a doubling of UV-B levels at the surface (Madronich et al. 1998); chemicals already present in the atmosphere are expected to continue to deplete ozone for many decades to come. High UV exposure causes damages to DNA, proteins, including cell membrane lipoproteins, and organelles (Karentz 1994) and can affect ecosystems and biological evolution (Cockell and Blaustein 2001).

Terrestrial microorganisms in habitats exposed to high UV-radiation produce different pigments potentially protecting against UV-B damages. They can be located extracellular or inside the cell to protect metabolic critical molecules. Some others are accessory pigments having quenching properties to dissipate excess energy from UV-B, and thus to avoid the generation of toxic single oxygen. UV radiation is one of the most important and widely experienced stress factors in Antarctica, even if few studies have examined the UV radiation consequences on Antarctic fungal species and communities. Among the fungal taxa recorded in Antarctica, some melanized strains are known for their UV resistance: *Alternaria alternata* (Fr.) Keissl. (Ellis 1971; Verona and Firpi 1971; Domsch et al. 1980), *Stachybotrys chartarum* (Ehrenb.) S. Hughes and *Ulocladium consortiale* (Thüm.) E.G. Simmons (Domsch et al. 1980). A brown pigment, most probably melanin, was produced by an Antarctic strain of *Phoma herbarum* within 24 h of exposure to elevated UV-B radiation (Hughes et al., 2003). Similar doses of UV-B inhibited the growth of five Antarctic fungal strains (*Geomyces pannorum*, *Phoma herbarum*, *Verticillum* sp., and *Mortierella parvispora* Linnem.), each isolated from Antarctic terrestrial habitats, and fungi that grew fastest at higher temperatures (25°C) were also the least inhibited by UV-B radiation (Hughes et al. 2003). *Arthrobotrys ferox* Onofri & Tosi, a non-melanized springtail catching Antarctic fungus, seems to be able to produce carotenoids and mycosporines acting as UV protecting agents (Arcangeli et al. 1997; Arcangeli and Cannistraro 2000). Subsequently Zucconi et al. (2002) have demonstrated the higher UV-B tolerance of spores of this Antarctic fungus compared with those of an European strain of *Arthrobotrys oligospora* Fresen., which would seem to relate to their higher germination ability after UV exposure.

Plastic cloches were used at Edmonson Point (continental Antarctica) in manipulation experiments designed to assess differences in soil fungal taxa composition and abundance subjected to altered temperature and/or UV exposure (Tosi et al. 2005). Under the walled cloches, artificial warming led to stress on Antarctic soil fungal assemblages while UV protection led to a higher equilibrium in the assemblage structure; UV radiation was therefore proposed as the most important limiting ecological factor for soil mycobiota in continental Antarctica.

Some microorganisms contain a combination of different pigments, which individually would not effort adequate protection but that together can minimize UV-radiation damage (Wynn-Williams and Edwards 2001). Organisms that can synthesize their own UV-protecting compounds have the opportunity to occupy a greater diversity of habitats, depending of course on the limitations imposed by other environmental factors (Cockell and Knowland 1999).

In the case of rock-inhabiting fungi, mycosporines were detected in the colonies of eight strains from desert and pseudodesert rock surfaces; in addition to their known protective role against excessive UV radiation in the natural environments, these compounds are also linked with survival potential, non-expansive intracolonial growth, and longevity of these fungi (Gorbushina et al. 2003). Further, the meristematic growth of some of these strains represents a protection mechanism against UV radiation, enabling the colonies to expose the lowest possible surface to the sunlight. Therefore the melanized thick walls they express as a stable character are considered a highly effective UV defence strategy (Selbmann et al. 2005; Sterflinger 2005). Hyaline fungi have frequently been recorded in Antarctica; even if they are less protected against UV-radiation compared to melanized fungi, they probably can survive in the Antarctic environment thanks to some other strategies such as fast growth focused on the short period in which temperature, moisture and light supply create favourable conditions and abundant sporulation (Onofri 1999), or growth in available protected microniches. Endolithic habitats offer UV protection; within the rock, or under the soil surface, increasing solar radiation from the beginning of the austral summer increases the rock or soil temperature, so liquid water becomes available and metabolic activity is possible for a short period if the organism is sufficiently responsive.

## 4 Conclusions

Despite the severe conditions of Antarctica, the overall picture shows a relatively rich mycoflora, more diversified than one might expect; the continental Antarctic mycota, is mainly composed of anamorphic fungi (filamentous fungi and yeasts), some *Zygomycota*, and very few species of *Ascomycota*; *Basidiomycota* are represented by basidiomycetous yeasts (Onofri et al. 2006). The knowledge on the Antarctic fungi is the result of more than one century of investigations carried out by isolating in pure culture fungi from several different substrata. The general shared opinion is that not all recorded fungi actively grow under the Antarctic conditions. In Antarctica fungi can be present as occasional propagules or be indigenous, the latter adapted in some extent to the Antarctic conditions. They show a range of morphological and physiological adaptations, similar to those adopted by other taxa from different extreme environments. Under the environmental pressure in some extreme and isolated locations, possibly endemic fungal taxa have evolved, and cryptoendolitic black fungi of the Antarctic desert are one of the most meaningful examples. It has been suggested (Selbmann et al. 2005) that these fungi have been possibly selected among a highly diversified mycoflora, originally present on the Continent when it was located at higher latitudes, during the changing environmental conditions that occurred in the migration of Antarctica to the Pole. These positively selected microfungi are the predominant group in these environments, thanks to their extremotolerance and to the absence of faster growing competitors; they include possibly endemic species, many of which are still waiting to be described.

**Acknowledgements** The authors thank the Italian National Program for Research in Antarctica (PNRA) and the European Commission's Research Infrastructure (SYNTHESYS Project) for financial support.

## References

Abyzov SS (1993) Microorganisms in the Antarctic ice. In: Friedmann EI (ed) Antarctic microbiology. Wiley-Liss, Inc, New York, pp 265–285

Ahmadjian V, Jacobs JB (1987) Studies on the development of synthetic lichens. Bibl Lichenol 25:47–58

Arcangeli C, Cannistraro S (2000) In situ Raman microspectroscopic identification and localization of carotenoids; approach to monitoring of UV-B irradiation stress on Antarctic fungus. Biospectroscopy 57:178–186

Arcangeli C, Zucconi L, Onofri S, Cannistraro S (1997) Fluoroscence study on whole Antarctic fungal spores under enhanced UV irradiation. J Photochem Photobiol B: Biol 39:258–264

Azmi OR, Seppelt RD (1997) Fungi of the Windmill Islands, Continental Antarctica. Effect of temperature, pH and culture medium on the growth of selected microfungi. Polar Biol 18:128–134

Baross JA, Morita RY (1978) Microbial life at low temperatures: ecological aspects. In: Kushner DJ (ed) Microbial life in extreme environments. Academic Press, London, pp 9–71

Bell AA, Wheeler MH (1986) Biosynthesis and functions of fungal melanins. Annu Rev Phytopathol 24: 411–451

Bergero R, Girlanda M, Varese GC, Intili D, Luppi AM (1999) Psychrooligotrophic fungi from Arctic soils of Franz Joseph Land. Polar Biol 21:361–368

Blumthaler M, Ambach W (1990) Indication of increasing solar ultra-violet-B radiation flux in Alpine regions. Science 248:206–208

Brewer MS (1999) Traditional preservatives-sodium chloride. In: Robinson RK, Blatt CA, Patel PD (eds) Encyclopaedia of food microbiology. Academic Press, London, pp 1723–1728

Bridge PD, Worland MR (2004) First report of an entomophthoralean fungus on an arthropod host in Antarctica. Polar Biol 27:190–192

Broady PA (1993) Soils heated by volcanism. In: Friedmann EI (ed) Antarctic microbiology. Wiley-Liss, Inc, New York, pp 413–432

Broady PA (1996) Diversity, distribution and dispersal of Antarctic terrestrial algae. Biodivers Conserv 5:1307–1335

Brown AD (1978) Compatible solutes and extracellular water stress in eukaryotic microorganisms. Adv Microb Physiol 17:181–242

Buchalo AS, Nevo E, Wasser SP, Oren A, Molitoris HP (1998) Fungal life in the extremely hypersaline water

of the Dead Sea: first records. Proc R Soc London B 265:1461–1465
Butler MJ, Day AW (1998) Fungal melanins: a review. Can J Microbiol 44:1115–1113
Caretta G, Del Frate G, Margiarotti AM (1994) A record of *Arthrobotrys tortor* Jarowaja and *Engyodontium album* (Limber) de Hoog from Antarctica. Bol Micol 9:9–13
Cockell CS, Blaustein A (2001) Ecosystems, evolution and UV radiation. Springer, New York
Cockell CS, Knowland J (1999) Ultraviolet radiation screening compounds. Biol Rev 74:311–345
Crowe JH, Crowe LM, Chapman D (1984) Preservation of membranes in anhydrobiotic organisms: the role of trehalose. Science 223:701–703
Crowe LM, Womersley C, Crowe JH, Reid D, Appel L, Rudolph A (1986) Prevention of fusion and leakage in freeze-dried liposomes by carbohydrates. Biochem Biophys Acta 861:131–140
Crowe JH, Crowe LM, Carpenter JF, Wistrom CA (1987) Stabilization of dry phospholipid bilayers and proteins by sugars. Biochem J 242:1–10
Crowe JH, Hoekstra FA, Crowe LM (1992) Anhydrobiosis. Annu Rev Physiol 54:579–599
D'Amore T, Crumplen R, Stewart GG (1991) The involvement of trehalose in yeast stress tolerance. J Ind Microbiol 7:191–196
Del Frate G, Caretta G (1990) Fungi isolated from Antarctic material. Polar Biol 11:1–7
de Hoog GS, Zalar P, Urzì F, de Leo F, Yurlova NA, Sterflinger K (1999) Relationships of dothideaceous black yeasts and meristematic fungi based on 5.8S and ITS2 rDNA sequence comparison. Stud Mycol 43:31–37
de Hoog GS, Göttlich E, Platas G, Genilloud O, Leotta G, van Brummelen J (2005) Evolution, taxonomy and ecology of the genus *Thelebolus* in Antarctica. Stud Mycol 51:33–76
de los Ríos A, Wierzchos J, Sancho LG, Ascaso C (2003) Acid microenvironments in microbial biofilms of Antarctic endolithic microecosystems. Eviron Microbiol 5:231–237
de los Ríos A, Sancho LG, Grube M, Wierzchos J, Ascaso C (2005) Endolithic growth of two *Lecidea* lichens in granite from continental Antarctica detected by molecular and microscopy techniques. New Phytol 165:181–190
Des Marais DJ (1995) The biogeochemistry of hypersaline microbial mats. Adv Microb Ecol 14:251–274
Domsch KH, Gams W, Anderson TH (1980) Compendium of Soil Fungi, Vol 2. Academic Press, London. (reprint IHW – Verlag, Eching 1993)
Ellis MB (1971) Dematiaceous Hyphomycetes. Commonwealth Mycological Institute, Kew, England, 608pp
Farman JC, Gardiner BG, Shanklin JD (1985) Large losses of total ozone in Antarctica reveal seasonal ClOx/NOx interaction. Nature 315:207–210
Fenice M, Selbmann L, Zucconi L, Onofri S (1997) Production of extracellular enzymes by Antarctic fungal strains. Polar Biol 17:275–280
Fenice M, Selbmann L, Di Giambattista R, Federici F (1998) Chitinolytic activity at low temperature of an Antarctic strain (A3) of *Verticillium lecanii*. Res Microbiol 149:289–300
Feofilova EP, Tereshina VM, Gorova IB (1994) Changes in carbohydrate composition of fungi during adaptation to thermostress. Microbiology 63:442–445
Finotti E, Moretto D, Marsella R, Mercantini R (1993) Temperature effects and fatty acid patterns in *Geomyces* species isolated from Antarctic soil. Polar Biol 13:127–130
Foissner W (2006) Biogeography and dispersal of microorganisms: a review emphasizing protists. Acta Protozool 45:111–136
Frederick JE, Snell HE (1988) Ultraviolet radiation levels during the Antarctic spring. Science 241:438–440
Frederick JE, Snell HE, Haywood EK (1989) Solar ultraviolet radiation at the earth's surface. Photochem Photobiol 50:443–450
Friedmann EI (1982) Endolithic microorganisms in the Antarctic cold desert. Science 215:1045–1053
Friedmann EI (1993) Antarctic microbiology. Wiley-Liss, New York, 634pp
Friedmann EI, Koriem AM (1989) Life on Mars: how it disappeared (if it was ever there). Adv Space Res 9:167–172
Friedmann EI, Ocampo R (1976) Endolithic blue-green algae in the dry valleys: primary producers in the Antarctic desert ecosystem. Science 193:1247–1249
Friedmann EI, Hua M, Ocampo-Friedmann R (1988) Cryptoendolithic lichen and cyanobacterial communities of the Ross Desert, Antarctica. Polarforschung 58:251–259
Friedmann EI, Druk AY, McKay CP (1994) Limits if life and microbial extinction in the Antarctic desert. Antarct JUS 29:176–179
Gadd GM, Chalmers K, Reed RH (1987) The role of trehalose in dehydration resistance of *Saccharomyces cerevisiae*. FEMS Microbiol Lett 48:249–254
Golubic S, Friedmann EI, Schneider J (1981) The lithobiontic ecological niche, with special reference to microorganisms. J Sediment Petrol 51:475–478
Gorbushina AA, Krumbein WE, Panina L, Soukharjevsky S, Wollenzien U, Hamann KH (1993) On the role of black fungi in colour change and biodeterioration of antique marbles. Geomicrobiol J 11:205–221
Gorbushina AA, Whitehead K, Dornieden T, Niesse A, Schulte A, Hedges JI (2003) Black fungal colonies as units of survival: hyphal mycosporines synthesized by rock-dwelling microcolonial fungi. Can J Bot 81: 131–138
Gorbushina AA, Beck A, Schulte A (2005) Microcolonial rock inhabiting fungi and lichen photobionts: evidence for mutualistic interactions. Mycol Res 109(11): 1288–1296
Grant WD (2004) Life at low water activity. Phil Trans R Soc London B 359:1249–1267
Green TGA, Schroeter B, Sancho LG (1999) Plant life in Antarctica. In: Pugnaire FI, Valladares F (eds) Handbook of functional plant ecology. Marcel Dekker, Inc, New York, pp 495–543
Grondona I, Monte E, Rives V, Vicente MA (1997) Lichenized association between *Septonema tormes* sp.

nov., a coccoid cyanobacterium, and a green alga with an unforeseen biopreservation effect of Villamayor sandstone at 'Casa lis' of Salamanca, Spain. Mycol Res 101:1489–1495

Gunde-Cimerman N, Zalar P, de Hoog S, Plemenitaš A (2000) Hypersaline waters in salterns-natural ecological niches for halophilic black yeast. FEMS Microbiol Ecol 32:235–240

Gunde-Cimerman N, Sonjak S, Zalar P, Frisvad JC, Diderichsen B, Plemenitaš A (2003) Extremophilic fungi in arctic ice: a relationship between adaptation to low temperature and water activity. Phys Chem Earth 28:1273–1278

Gunde-Cimerman N, Zalar P, Petrovič U, Turk M, Kogej T, de Hoog GS, Plemenitaš A (2004) Fungi in salterns. In: Ventosa A (Eds) Halophilic microorganisms. Springer-Verlag, Berlin pp 103–113

Gunde-Cimerman N, Oren A, Plemenitaš A (2005) Adaptation to life at high salt concentrations in Archaea, Bacteria, and Eukarya. Springer, The Netherlands, 577pp

Han KH, Prade RA (2002) Osmotic stress-coupled maintenance of polar growth in *Aspergillus nidulans*. Mol Microbiol 43(5): 1065–1078

Holdgate MV (1977) Terrestrial ecosystems in the Antarctic. Philos T Roy Soc B 279:5–25

Horowitz NH, Cameron RE, Hubbard JS (1972) Microbiology of the Dry Valleys of Antarctica. Science 193:242–245

Hottiger T, Boller T, Wiemken A (1987) Rapid changes of heat and desiccation tolerance correlated with changes of trehalose content in *Saccharomyces cerevisiae* cells subjected to temperature shifts. FEBS Lett 220:113–115

Hughes KA, Lawley B (2003) A novel Antarctic microbial endolithic community within gypsum crusts. Environ Microbiol 69:1488–1491

Hughes KA, Lawley B, Newsham KK (2003) Solar UV-B radiation inhibits the growth of Antarctic terrestrial fungi. Appl Environ Microbiol 69:1488–1491

Jumpponen A, Newsham KK, Neises DJ (2003) Filamentous ascomycetes inhabiting the rhizoid environment of the liverwort *Cephaloziella varians* in Antarctica are assessed by direct PCR and cloning. Mycologia 95:457–466

Kappen L (1974) Response to extreme environments. In: Ahmadjian V, Hale ME (eds) The lichens. Academic Press, New York, pp 311–380

Kappen L (1993) Lichens in the Antarctic region. In: Friedmann EI (ed) Antarctic microbiology. Wiley-Liss, New York, pp 433–490

Karentz D (1994) Ultraviolet tolerance mechanisms in Antarctic marine organisms. In: Weiler CS, Penhale PA (eds) Ultraviolet radiation in Antarctica: measurements and biological effects. American Geophysical Union, Washington, DC, pp 92–110

Kerr JB, McElroy CT (1993) Evidence for large upward trends of ultraviolet-B radiation linked to ozone depletion. Science 262:1032–1034

Kerry E (1990) Effects of temperature on growth rates of fungi from subantarctic Macquarie Island and Casey, Antarctica. Polar Biol 10:293–299

Kis-Papo T, Kirzhner V, Wasser SP, Nevo E (2003) Evolution of genomic diversity and sex at extreme environments: fungal life under hypersaline Dead Sea stress. PNAS 100:14970–14975

Kochkina GA, Ivanushkina NE, Karasev SG, Gavrish E Yu, Gurina LV, Evtushenko LI, Spirina EV, Vorob'eva EA, Gilichinskii DA, Ozerskaya SM (2001) Survival of micromycetes and actinobacteria under conditions of long term natural cryopreservation. Microbiology 70:356–364

Kogej T, Wheeler MH, Rižner TL, Gunde-Cimerman N (2004) Evidence for 1,8–dihydroxynaphthalene melanin in three halophilic black yeasts grown under saline and non-saline conditions. FEMS Microbiol Lett 232:203–209

Leotta GA, Paré JA, Sigler L, Montalti D, Vigo G, Petruccelli M, Reinoso EH (2002) *Thelebolus microsporus* mycelial mats in the trachea of wild brown skua (*Catharacta antarctica lonnbergi*) and South Polar skua (*C. maccormicki*) carcasses. J Wildlife Dis 38(2): 443–447

Lewis JG, Learmonth RP, Watson K (1995) Induction of heat, freezing and salt tolerance by heat and salt shock in *Saccharomyces cerevisiae*. Microbiology 141: 687–694

Ma L, Rogers SO, Catranis CM, Starmer TS (2000) Detection and charaterization of ancient fungi entrapped in glacial ice. Mycologia 92:286–295

Madronich S, McKenzie RL, Björn LO, Caldwell MM (1998) Changes in biologically active ultraviolet radiation reaching the Earth's surface. J Photochem Photobiol B: Biol 46:5–19

Maggi O, Persiani AM, Fabbri AA, De Luca C, Lunghini D, Quadraccia L, Fanelli C (1991) Differenze nella composizione fosfolipidica di diversi taxa fungini. Giorn Bot Ital 125(3):259

Méjanelle L, Lòpez JF, Gunde-Cimerman N, Grimalt JO (2001) Ergosterol biosynthesis in novel melanized fungi from hypersaline environments. J Lipid Res 42:352–358

Mercantini R, Marsella R, Moretto D, Finotti E (1993) Keratinophilic fungi in the Antarctic environment. Mycopathalogia 122:169–175

Montemartini Corte A (1991) Funghi di ambienti acquatici. In: Battaglia B, Bisol PM, Varotto V (eds) Proceedings of the 1st Meeting on 'Biology in Antarctica' (English summaries). Roma CNR 22–23 June 1989, Scienza e Cultura, Edizioni Universitarie Patavine, pp 67–76

Montemartini Corte A, Caretta G, Del Frate G (1993) Notes on *Thelebolus microsporus* isolated in Antarctica. Mycotaxon 48:343–358

Montiel PO (2000) Solubile carbohydrates (trehalose in particular) and cryoprotection in polar biota. Cryo-Lett 21:83–90

Nienow JA, Friedmann EI (1993) Terrestrial lithophytic (rock) communities. In: Friedmann EI (ed) Antarctic microbiology. Wiley-Liss, New York, pp 343–412

Nishiyama T (1977) Studies on evaporite minerals from Dry Valley, Victoria Land, Antarctica. Antarct Rec 58:171–185

Ocampo-Friedmann R, Friedmann EI (1993). Biologically active substances produced by Antarctic cryptoendolithic fungi. Ant JUS 28:252–254

Onofri S (1999) Antarctic microfungi. In: Seckbach J (ed) Enigmatic microorganisms and life in extreme environments. Kluwer Academic publishers, Dordrecht, Boston, London, pp 323–336

Onofri S, Friedmann EI (1998) Cryptoendolithic microorganisms in sandstone and pegmatite in the northern Victoria Land. In: Tamburrini M, D'Avino R (eds), Newsletter of the italian biological research in Antartica. Camerino University Press, Camerino, pp 45–51

Onofri S, Tosi S (1992) *Arthrobotrys ferox* sp. nov. a springtail-capturing hyphomycete from Continental Antarctica. Mycotaxon 44:445–451

Onofri S, Rambelli A, Maggi O, Persiani AM, Riess S, Tosi S, Grasselli E (1991) Micologia del suolo. In: Battaglia B, Bisol PM, Varotto V (eds) Proceedings of the 1st Meeting on 'Biology in Antarctica' (English summaries). Roma CNR 22–23 June 1989. Scienza e Cultura, Edizioni Universitarie Patavine, pp 55–65

Onofri S, Tosi S, Persiani AM, Maggi O, Riess S, Zucconi L (1994) Mycological reserches in Victoria Land terrestrial ecosystem. Proceedings of the Second Meeting on "Antarctic Biology", Padova, 26–28 February 1992. Scienza e cultura, Edizioni Universitarie Patavine, Padova, pp 19–32

Onofri S, Pagano S, Zucconi L, Tosi S (1999) *Friedmanniomyces endolithicus* (Fungi, Hyphomycetes), anam.-gen. and sp. nov., from continental Antartica. Nova Hedwigia 68:175–181

Onofri S, Fenice M, Cicalini AR, Tosi S, Magrino A, Pagano S, Selbmann L, Zucconi L, Vishniac HS, Ocampo-Friedmann R, Friedmann EI (2000) Ecology and biology of microfungi from Antarctic rocks and soil. Ital J Zool 67(suppl. 1):163–168

Onofri S, Selbmann L, Zucconi L, Pagano S (2004) Antarctic microfungi as models for exobiology. Planet Space Sci 52:229–237

Onofri S, Selbmann L, Zucconi L, Tosi S, Fenice M, Barreca D, Ruisi S (2005a) Studies on Antarctic fungi. Polarnet Tech Rep.1:49–52

Onofri S, Selbmann L, Zucconi L, Tosi S, de Hoog GS (2005b) The Mycota of Continental Antarctica. Terra Antart Rep 11:37–42

Onofri S, Zucconi L, Tosi S (2006) Continental Antarctic Fungi. IHW-Verlag, Eching (in press)

Øvstedal DO, Lewis Smith RI (2001) Lichens of Antarctica and South Georgia. In: Øvstedal DO, Lewis Smith RI (eds) A guide to their identification and ecology. Studies in Polar Research, University of Cambridge, pp 4–5

Pascual S, Melgarejo P, Magan N (2002) Water availability affects the growth, accumulation of compatible solutes and the viability of the biocontrol agent *Epicoccum nigrum*. Mycopathologia 156:93–100

Pickard J, Seppelt RD (1984) Phytogeography of Antarctica. J Biogeogr 11:83–102

Rivkina E, Friedmann EI, McKay CP, Gilichinsky DA (2000) Metabolic activity of permafrost bacteria below the freezing point. Appl Environ Microbiol 66:3230–3233

Rivkina E, Laurinavichius K, McGrath J, Tiedje J, Shcherbakova V, Gilichinsky D (2004) Microbial life in permafrost. Adv Space Res 33:1215–1221

Robinson CH (2001) Cold adaptation in Arctic and Antarctic fungi. New Phytol 151:341–353

Ruibal C, Platas G, Bills GF (2005) Isolation and characterization of melanized fungi from limestone in Mallorca. Mycol Progr 4:23–38

Russell NJ (1990) Cold adaptation of microorganisms. Philos Trans Roy Soc London Ser B 326:595–611

Selbmann L, Onofri S, Fenice M, Federici F, Petruccioli M (2002) Production and structural characterization of the exopolysaccharide of the Antarctic fungus *Phoma herbarum* CCFEE 5080. Res Microbiol 153:585–592

Selbmann L, de Hoog GS, Mazzaglia A, Friedmann EI, Onofri S (2005) Fungi at the edge of life: cryptoendolithic black fungi from Antarctic desert. Stud Mycol 51:1–32

Seymour FA, Crittenden PD, Dyer PS (2005) Sex in the extremes: lichen-forming fungi. Mycologist 19:51–58

Staley JT, Palmer FE, Adams JB (1982) Microcolonial fungi: common inhabitants on desert rocks? Science 215:1093–1095

Sterflinger K (1998) Temperature and NaCl-tolerance of rock inhabiting meristematic fungi. Antonie van Leeuwenhoek 74:271–281

Sterflinger K (2005) Black yeasts and meristematic fungi: ecology, diversity and identification. In: Rosa C, Gabor P (eds) Yeast handbook: biodiversity and ecophysiology of yeasts. Springer, New York, pp 505–518

Sterflinger K, de Hoog GS, Haase G (1999) Phylogeny and ecology of meristematic ascomycetes. Stud Mycol 43:5–22

Thevelein JM (1984) Regulation of trehalose mobilization in fungi. Microbiol Rev 48:42–59

Tosi S, Begoña C, Gerdol R, Caretta G (2002) Fungi isolated from Antarctic mosses. Polar Biol 25:262–268

Tosi S, Onofri S, Brusoni M, Zucconi L, Vishniac H (2005) Response of Antarctic soil fungal assemblages to experimental warming and reduction of UV radiation. Polar Biol 28:470–482

Turian G (1977) *Coniosporium aeroalgicolum* sp. nov. – a dematiaceous fungus living in balanced parasitism with aerial algae. B Soc Bot Suisse 87:19–24

Urzì C, Realini M (1998) Colours changes of Noto's calcareous sandstone as related to its colonization by microorganisms. Int Biodeter Biodegr 42:45–54

Urzì C, Wollenzien U, Criseo G, Krumbein WE (1995) Biodiversity of the rock inhabiting microflora with special reference to black fungi and black yeasts. In: Allsopp D, Colwell RR, Hawksworth DL (eds) Microbial diversity and ecosystem function. CAB International, Wallingford, UK, pp 289–302

van Laere A (1989) Trehalose, reserve and/or stress metabolite? FEMS Microbiol Rev 63:201–210

van Uden N (1984) Temperature profiles of yeasts. Adv Microb Physiol 25:195–251

Verona O, Firpi M (1971) Sui micromiceti carticoli dei generi *Alternaria*, *Ulocladium*, *Stemphylium*. Cell Carta 8:55–72

Vincent WF (1988) Microbial ecosystems of Antartica. Cambridge University Press, Cambridge, 303pp

Vincent WF (2000) Evolutionary origins of Antarctic microbiota: invasion, selection and endemism. Antarct Sci 12(3): 374–385

Vishniac HS (1987) Psychrophily and systematics of yeast-like fungi. In: de Hoog GS, Smith MTh, Weijman ACM (eds) The expanding realm of yeast-like fungi. Stud. Mycol. 30:389–402

Vishniac HS (1993) The microbiology of Antarctic soils. In: Friedmann EI (ed) Antarctic microbiology. Wiley-Liss, New York, pp 433–490

Vishniac HS (2006) Yeast biodiversity in the Antarctic. In: Rosa CA, Gabor P (eds) Biodiversity and ecophysiology of Yeasts, Springer, pp 419–440

Vishniac HS, Onofri S (2002) *Cryptococcus antarcticus* var. *circumpolaris* var. nov., a basidiomycetous yeast from Antarctica. Antonie van Leeuwenhoek 83:233–235

Volkmann M, Whitehead K, Rüttgers H, Rullkötter J, Gorbushina AA (2003) Mycosporine-glutamicol-glucoside: a natural UV absorbing secondary metabolite of rock-inhabiting microcolonial fungi. Rapid Comm Mass Spectr 17:897–902

Weinstein RN, Montiel PO, Johnstone K (2000) Influence of growth temperature on lipid and soluble carbohydrate synthesis by fungi isolated from fellfield soil in the maritime Antarctic. Mycologia 92:222–229

Wiemken A (1990) Trehalose in yeast, stress protectant rather than reserve carbohydrate. Antonie van Leeuwenhoek 58:209–217

Wierzchos J, Ascaso C (2001) Life, decay and fossilisation of endolithic microorganisms from the Ross Desert, Antarctica: suggestions for *in situ* further research. Polar Biol 24:863–868

Wierzchos J, Ascaso C (2002) Microbial fossil record of rocks from the Ross Desert, Antarctica: implications in the search for past life on Mars. Int J Astrobiol 1:51–59

Wollenzien U, de Hoog GS, Krumbein WE, Urzì C (1995) On the isolation of microcolonial fungi occurring on and in marble and other calcareous rocks. Sci Total Environ 167:287–294

Wynn-Williams DD (1990) Ecological aspects of Antarctic microbiology. Adv Microb Ecol 11:71–146

Wynn-Williams DD, Edwards HGM (2000) Antarctic ecosystems as models for extraterrestrial surface habitats. Planet Space Sci 48:1065–1075

Wynn-Williams DD, Edwards HGM (2001) In: Horneck G, Baumstark-Khan C (eds) Environmental UV radiation: biological strategies for protection and avoidance, in astrobiology: the quest for the conditions of life. Springer-Verlag, Berlin, pp 244–259

Zalar P, de Hoog GS, Gunde-Cimerman N (1999a) Ecology of halotolerant dothideaceous black yeasts. Stud Mycol 43:38–48

Zalar P, de Hoog GS, Gunde-Cimerman N (1999b) Taxonomy of the endoconidial black yeast genera *Phaeotheca* and *Hyphospora*. Stud Mycol 43:49–56

Zalar P, de Hoog GS, Gunde-Cimerman N (1999c) *Trimmatostroma salinum*, a new species from hypersaline water. Stud Mycol 43:57–62

Zucconi L, Pagano S, Fenice M, Selbmann L, Tosi S, Onofri S (1996) Growth temperature preferences of fungal strains from Victoria Land, Antarctica. Polar Biol 16:53–61

Zucconi L, Ripa C, Selbmann L, Onofri S (2002) Effects of UV on the spores of the fungal species *Arthrobotrys oligospora* and *A. ferox*. Polar Biol 25:500–505

Rev Environ Sci Biotechnol (2006) 5:323–331
DOI 10.1007/s11157-006-9105-0

REVIEW PAPER

# Ecology and molecular adaptations of the halophilic black yeast *Hortaea werneckii*

**Nina Gunde-Cimerman · Ana Plemenitaš**

Received: 4 March 2006 / Accepted: 3 August 2006 / Published online: 30 August 2006


**Abstract** Molecular studies on halophilic adaptations have focused on prokaryotic microorganisms due to a lack of known appropriate eukaryotic halophilic microorganisms. However, the black yeast *Hortaea werneckii* has been identified as the dominant fungal species in hypersaline waters on three continents. It represents a new model organism for studying the mechanisms of salt tolerance in eukaryotes. Ultrastructural studies of the *H. werneckii* cell wall have shown that it synthesizes dihydroxynaphthalene (DHN) melanin under both saline and non-saline growth conditions. However, melanin granules in the cell walls are organized in a salt-dependent way, implying the potential osmoprotectant role of melanin. At the level of membrane structure, *H. werneckii* maintains a sterol-to-phospholipid ratio significantly lower than the salt-sensitive *Saccharomyces cerevisiae*. Accordingly, membranes of *H. werneckii* are more fluid over a wide range of NaCl concentrations, indicating high intrinsic salt stress tolerance. Even *H. werneckii* grown in high NaCl concentrations maintains very low intracellular amounts of potassium and sodium, demonstrating the sodium-excluder character of this organism. The salt-dependent expressions of two *HwENA* genes suggest roles for them in the adaptation to changing salt concentrations. The high similarity of these ENA ATPases to other fungal ENA ATPases involved in $Na^+/K^+$ transport indicates their potential importance in *H. werneckii* ion homeostasis. Glycerol is the main compatible solute which accumulates in the cytoplasm of *H. werneckii* at high salinity, although it seems that mycosporines may also act as supplementary compatible solutes. Salt dependent increase in glycerol synthesis is supported by the identification of two copies of a gene putatively coding for glycerol-3-phosphate-dehydrogenase. Expression of only one of these genes is salt dependent.

**Keywords** Compatible solutes · Ecology · Halophiles · *Hortaea werneckii* · Hypersaline water · Ions · Melanin · Membrane · Mycosporines · Salterns

N. Gunde-Cimerman (✉)
Department of Biology, Biotechnical Faculty, University of Ljubljana, Večna POT 111, 1000 Ljubljana, Slovenia
e-mail: nina.gunde-cimerman@bf.uni-lj.si

A. Plemenitaš
Faculty of Medicine, Institute of Biochemistry, University of Ljubljana, Vrazov trg 2, 1000 Ljubljana, Slovenia
e-mail: ana.plemenitas@mf.uni-lj.si

## 1 Ecology of *Hortaea werneckii*

*Bacteria* and *Archaea* dominate natural environments characterized by extreme physico-chemical properties. Nevertheless, eukaryotic microorganisms such as the "black yeasts" (de Hoog and Hermanides-Nijhof 1977), microcolonial fungi (Staley et al. 1982) and meristematic ascomycetes (Sterflinger et al. 1999) are also remarkably successful in adapting to certain extreme environments (Staley et al. 1982; Gorbushina et al. 1993; Nienow and Friedmann 1993; Wollenzien et al. 1995; Gunde-Cimerman et al. 2000). Although they have been known since the end of the 19th century (de Hoog et al. 1999), difficulties in their morphological identification together with their slow growth and low competitive ability frequently hindered the isolation and identification of these fungi. The representatives of black yeasts belong to the ascomycetous orders *Chaetothyriales*, *Dothideales* and *Pleosporales* (Sterflinger et al. 1999). Several genera and species of black yeasts from the order *Dothideales* represent a group of rare halophilic eukaryotic microorganisms, highly appropriate for studying the mechanisms of salt tolerance in eukaryotes (Petrovič et al. 1999); (Petrovič et al. 2002); (Turk and Plemenitaš 2002). So far little is known about eukaryotic halophilic microorganisms, let alone the mechanisms of their adaptation to such conditions (Petrovič et al. 2002).

Microscopy and histological studies of black yeasts have revealed that their morphological ecotype is important for their survival in various extreme environments. Black yeasts are characterized by melanized slowly expanding colonies, with reproduction by isodiametric enlargement of subdividing cells (Sterflinger et al. 1999; Wollenzien et al. 1995). Their distinctive features are polymorphism, meristematic growth, endoconidiation or sarcinic conidiogenesis, frequently muriform cells which develop by conversion from undifferentiated hyphae, and thick, melanized cell walls (de Hoog 1993; Zalar et al. 1999; Sterflinger et al. 1999; Wollenzien et al. 1995).

*Hortaea werneckii*, the best described eukaryotic halophilic model organism to date, is a characteristic black yeast. In the past, its identification was based solely on morphological characteristics and has thus received many designations: *Cladosporium werneckii*, *Exophiala werneckii*, *Pullularia werneckii*, *Aureobasidium werneckii*, *A. mansonii*, *Sarcinomyces crustaceus* and *Phaeoannellomyces werneckii*. Nowadays, the identification is additionally based on nutritional physiology and molecular methods. The physiological tests include metabolism of different C and N sources with or without NaCl in high concentrations, as well as temperature-tolerance tests. Molecular differentiation is based on the sequencing of the ITS rDNA region and RFLP markers from SSU rDNA and ITS rDNA regions (de Hoog et al. 1999).

Until recently, the genus *Hortaea* contained only a single species, *H. werneckii* (Horta) Nishimura and Miyaji, with no known sexual stage (Zalar et al. 1999). In 2004, another species was described and named *H. acidophila* (Holker et al. 2004).

*Hortaea werneckii* was long known primarily as the etiological agent of human tinea nigra, a superficial infection of the human hand, particularly frequent in warmer areas of the world. Investigations have revealed that the fungus is strictly limited to the dead surface of the skin (stratum corneum), in particular to the grease on the skin. Since *H. werneckii* does not show any keratin-degrading activity, it does not invade the living tissue below and so the infection is only a cosmetic problem (Göttlich et al. 1995). Besides its involvement in tinea nigra, *H. werneckii* was also known as one of the few species of fungi capable of contaminating food preserved with high concentrations of NaCl (Mok and Barreto da Silva 1981), without showing any obligate requirement for NaCl (Andrews and Pitt 1987). In addition to human skin and salty food, the fungus has been isolated from seawater (Iwatsu and Udagawa 1988), marine fish (Todaro et al. 1983), beach soil (de Hoog and Guého 1998) and arid inorganic and organic surfaces (Krumbein et al. 1996). On the basis of *H. werneckii* random isolations from different low water activity substrates and in vitro ecophysiological studies it was suggested that salt might be the decisive factor in its ecology and therefore in the etiology of tinea nigra. Furthermore, it was thus speculated that *H. werneckii* might grow in drying salty ponds at the seaside (de Hoog and Gerrits van den Ende 1992). However, the primary environmental ecological niche of *H. werneckii* remained unknown until we investigated salterns along the Slovenian

Adriatic coast for the potential presence of halophilic fungi. Their distribution was followed throughout two successive years in five different evaporitic ponds, covering the entire salinity range (3–32% NaCl) (Gunde-Cimerman et al. 2000; Butinar et al. 2005). This study revealed that hypersaline waters of the salterns harbour different species of melanized fungi from the order *Dothideales*. They appeared in three peaks, at the water salinities 5–8%, 10–20% and 18–25% NaCl, which correlated primarily with high environmental nitrogen values. At the highest environmental salinities, melanized fungi represented 85–100% of the total isolated mycobiota, but they were partially replaced by non-melanized fungi with lowering salinities and they were detected only occasionally at the end of the season, when NaCl concentrations were below 5%. *H. werneckii* was the dominant black-yeast species in the Adriatic salterns during the season of salt production (Gunde-Cimerman et al. 2000). Initially it was isolated from the crystallization ponds of the Adriatic salterns, but it was later also identified in hypersaline waters of six salterns on three continents (Butinar et al. 2005).

Besides being the dominant black yeast species in hypersaline water at salinities above 20%, *H. werneckii* was also isolated from various microniches within the salterns: the surface and interior of wood submerged in brine, from biofilms on the surface of hypersaline water, and from dry ponds and microbial mats (Butinar et al. 2005; Zalar et al. 2005). However, it appears that *H. werneckii* survives in eutrophic thalassohaline waters of salterns in temperate climatic zones, since it was only occasionally retrieved from salterns in Puerto Rico, but never from the oligotrophic salterns in Eilat at the Red Sea in Israel, or the athalassohaline waters of the Dead Sea, or those of Salt Lake, Utah (Gunde-Cimerman et al. 2005).

In contrast to most prokaryotic halophiles, which display better growth in the presence of NaCl, the growth salinity range for *H. werneckii*, defined in vitro, is from 0% to saturation (32%) NaCl, with a broad optimum from 6% to 14% NaCl. Its complex polymorphic life cycle enables *H. werneckii* to show versatile ecotype adaptations in response to changing environmental concentrations of NaCl, UV intensity, nutrients and water availability. If sufficient nutrients are available, the hydrophilic yeast phase rapidly colonizes hypersaline water. At salinites above 15% NaCl, yeast cells begin to differentiate into meristematic budding cells, and at the highest salinities they form dormant meristematic sclerotial bodies with endogenous conidiation. Clustered growth allows sheltering of interior cells and minimizes the number of cells directly in contact with the hostile environment. Under conditions of drought, with no water in the ponds, the fungus changes into an aerophilic, hydrophobic hyphal stage, producing conidia which can be dispersed by air currents. These conidia can germinate in saline water, giving rise to actively propagating yeast cells (Butinar et al. 2005).

## 2 Molecular adaptations of *Hortaea werneckii*

### 2.1 Membranes and cell-wall pigmentation

As the *H. werneckii* cell is in contact with its environment via the plasma membrane, the adaptability and flexibility of this membrane is of vital importance to the survival of the cell. The cell should be able to modify the lipid composition and consequently properties of its membranes if the conditions in the environment change. The fluidity of biological membranes is largely influenced by the length, branching and degree of saturation of the fatty acids, the amount of sterols and the nature of the phospholipids (Russell 1989a, b).

The influence of salt stress on lipid composition and membrane properties has been studied in bacteria (Russell et al. 1995) and in yeasts, including the salt-sensitive *Saccharomyces cerevisiae* (Sharma et al. 1996; Tunblad-Johansson and Adler 1987) and several halotolerant yeasts. These organisms showed different responses to salt stress. While *Zygosaccharomyces rouxii* showed increased amounts of free ergosterol, decreased amounts of unsaturated fatty acids and decreased membrane fluidity when grown with NaCl (Hosono 1992; Yoshikawa et al. 1995), high salinity did not induce significant changes in the unsaturation of fatty acids in *Yarrowia lipolitica*

(Andreishcheva et al. 1999; Yoshikawa et al. 1995) but caused a decrease in phospholipid and sterol contents. In contrast, *Candida membranefaciens* grown at high NaCl concentrations exhibited increased unsaturation in fatty acids and an increase in the contents of phosphatidylinositol (PI) and phosphatidylethanolamine (PE), resulting in slightly higher membrane fluidity (Khaware et al. 1995). The plasma membrane of the marine yeast *Debaryomyces hansenii* adapts to stress conditions by decreasing fluidity and increasing the sterol-to-phospholipid ratio in the presence of salt (Turk et al., submitted). Our studies have shown that salt stress does not significantly influence the total sterol content in halophilic *H. werneckii*, but does cause an increase in the phospholipid content. The most abundant fatty acids in phospholipids contained $C_{16}$ and $C_{18}$ chain lengths with a high percentage of $C18{:}2^{\Delta 9,12}$. Salt stress also caused an increase in the fatty acid unsaturation. Halophilic fungi maintained their sterol-to-phospholipid ratio significantly lower than the salt-sensitive *S. cerevisiae*. Additionally, EPR measurements showed that the membranes of *H. werneckii* were significantly more fluid in comparison with the membranes of the above mentioned fungi, of the halotolerant black yeast *A. pullulans* and also the salt-sensitive *S. cerevisiae* (Turk et al. 2004). Additionally, it was demonstrated that halophilic *H. werneckii* maintained this very high fluidity over a wide range of NaCl concentrations, indicating high intrinsic salt-stress tolerance. Results were in good agreement with eco-physiological data and the dominance of *H. werneckii* in hypersaline waters of salterns. As the membrane fluidity of the related halotolerant black yeast *A. pullulans* was different from that of halophilic *H. werneckii* and resembled more that of salt-sensitive *S. cerevisiae* (Turk et al. 2004), membrane fluidity appears to be a good indicator of the degree of salt tolerance.

It has been reported that hyperosmotic shock induces changes in the organization of cell wall, most probably as a result of displacement of periplasmic and cell wall matrix material into invaginations of the plasma membrane (Slaninova et al. 2000). No significant invaginations occurred in *H. werneckii* at low salinities, but such changes in its cell-wall structure were apparent at high salinities (Kogej 2006). However, we also observed salt-dependent changes of the cell wall structure previously unobserved in fungi, which was caused by the redistribution of the melanin which is responsible for the dark colour of *H. werneckii* (Kogej 2006). We demonstrated that *H. werneckii* synthesizes 1,8-dihydroxynaphthalene (DHN) melanin under saline as well as non-saline growth conditions (Kogej et al. 2004). While the biosynthesis was not salt-dependent, the ultrastructural studies of the cell wall of *H. werneckii* showed for the first time that the melanin granules in the outer part of the cell walls are loosely organized in the medium without salt, and are more densely packed as the salt concentration in the medium increases (Kogej 2006; Plemenitaš and Gunde-Cimerman 2005). Since melanins are known to confer protection to UV-irradiation, temperature extremes (Bell and Wheeler 1986), desiccation (Zhdanova et al. 1990; Zhdanova and Pokhodenko 1973) and have an osmotic role (Elliot and Henson 2001), it is reasonable to suggest a potential osmoprotectant role of melanin in the cell wall of halophilic *H. werneckii*.

### 2.2 Sodium and potassium in the cells of *Hortaea werneckii*

Cells living in natural saline systems, where high salt concentrations cause high osmotic pressure, must maintain lower water potential than their surroundings in order to survive and proliferate. At the same time they have to keep the intracellular concentrations of sodium ions bellow the toxic level for the cells. Halophilic microorganisms have developed different strategies for counterbalancing osmotic pressure. Extremely halophilic *Archaea* accumulate KCl up to molar concentrations when exposed to high external salinity (Oren 1999). In contrast, eukaryotic microorganisms cannot tolerate high intracellular ion concentrations. Mechanisms of salt tolerance have been studied on salt-sensitive *S. cerevisiae* (Blomberg 2000) and a few halotolerant fungi such as the filamentous yeasts *Debaryomyces hansenii*, *Candida versatilis*, *Rhodotorula mucilaginosa* and *Pichia guillermondii* (Andre et al.

1988; Ramos 1999, 2005; Almagro et al. 2000; Silva-Graça and Lucas 2003; Prista et al. 2005). Data on these fungi show that the maintenance of positive turgor pressure at high salinity is mainly due to an increased production and accumulation of glycerol, trehalose and other organic compatible solutes. However, it is also known that in *D. hansenii* osmotic adjustments of the major intracellular cations also occurs in response to osmotic stress (Blomberg and Adler 1992; Ramos 2005). It was shown that *D. hansenii* keeps relatively high amounts of internal sodium when grown under salt stress and so this yeast has been defined as a $Na^+$-includer organism (Ramos 2005).

In contrast we demonstrated that *H. werneckii* keeps very low intracellular amounts of potassium and sodium even when grown in the presence of 25% NaCl, which prevents the growth of other investigated fungi, thus indicating the efficient Na-excluding character of this organism. Interestingly, in *H. werneckii* the amounts of $K^+$ and $Na^+$ were the lowest in the cells grown at 17% NaCl. At this salinity of the medium *H. werneckii* still grows well, but most probably this salinity represents a turning point, shown in restricted colony size, slower growth rate and characteristic changes of physiological behaviour (Plemenitaš and Gunde-Cimerman 2005). When cells of *H. werneckii* were exposed to hyperosmotic shock, both non-adapted cells grown without NaCl and salt-adapted cells grown at 10% NaCl reacted by an immediate drop in the amount of $K^+$. On the other hand, the amount of $Na^+$ in non-adapted cells remained almost unchanged after hyperosmotic shock, while the salt-adapted cells showed increased values and fluctuating values of sodium. The observed pattern of ion fluctuations after hyperosmotic shock is in accordance with the growth characteristics of *H. werneckii*. Twenty percent NaCl slowed the growth rate of *H. werneckii*, indicating the sensitivity to increased intracellular sodium. Our results nevertheless indicate that due to their low concentrations, cations probably do not contribute significantly to osmoadaptation in *H. werneckii* and that *H. werneckii* possesses a very efficient export system for $Na^+$ and $K^+$ (Kogej et al. 2005).

### 2.3 Sodium and potassium efflux through the membrane of *Hortaea werneckii*

When exposed to a hypersaline environment, many organisms exclude $Na^+$ ions from the cytoplasm, due to the potentially toxic effects of these ions. Many fungi use ENA P-type ATPases as one of the mechanisms for $Na^+$ and/or $K^+$ export. However, most of the data on ENA ATPases are derived from studies on salt-sensitive and moderately halotolerant fungi. In *S. cerevisiae*, several ENA ATPases mediate sodium efflux processes (Garciadeblas et al. 1993), while most other fungi have only one ENA ATPase, like in *Schizosaccharomyces pombe* (Benito et al. 2002), *Zygosaccharomyces rouxi* (Watanabe et al. 1999) and *Neurospora crassa* (Benito et al. 2002) or two ENA ATPases, as found in *Schwanniomyces occidentalis* (Banuelos and Rodriguez-Navarro 1998) and *D. hansenii* (Almagro et al. 2000, 2001). All of these ENA ATPases are plasma membrane proteins. Most of them are equally effective in suppressing the sensitivity of cells to $K^+$ or $Na^+$, while *NcENA1* from *N. crassa* was reported to be more effective for $Na^+$.

We have recently isolated and characterized two *HwENA* genes coding for Ena ATPases from *H. werneckii* (Gorjan and Plemenitaš, submitted). Their protein sequences, deduced from the cDNA sequences of the respective genes, revealed that both of the amino acid sequences contain conserved domains significant for cation-transporting ATPases. Since that both *HwENA* genes were seen to be located on the same chromosome, we speculate that they are arranged in tandem, similarly to those in *S. cerevisiae*, where *ENA* genes constitute a tandem array of 4–5 genes on one chromosome.

We also found that the expressions of both *HwENA* genes, *HwENA1* and *HwENA2*, were highly salt responsive. In adapted cells, the expression levels of both *HwENA* genes were relatively low below 17% NaCl, but became strongly induced in a hyper saline environment (25% NaCl), and this was particularly marked with the expression of the *HwENA2* gene. In contrast, the expression profile of *HwENA* genes from the cells which were exposed to salt stress by sudden increase in NaCl concentration in the medium, revealed that the level of *HwENA2* mRNA was lower then *HwENA1* mRNA. We also found that mRNA expression of both genes was induced only after 90 min. These results suggest that *HwENA* genes are involved in the late response of the cells to salt stress, with *HwENA1* being more responsive to sudden changes in the salt concentrations, and *HwENA2* playing a more important role in the mechanism of maintaining low cation content in adapted cells. It was found that adapted cells of *H. werneckii* have low sodium and potassium intracellular contents, which does not vary much with increasing extracellular salt concentrations (Kogej et al. 2005). We assume that the adapted cells employ a variety of mechanisms to maintain low intracellular cation contents over a wide range of salinity, and we suggest that HwENA ATPases contribute to these mechanisms only at extremely high NaCl concentrations. A high concentration of $Na^+$ ions in the environment probably causes the entry of more $Na^+$ ions into the cells, so more $Na^+$ exporters are needed. The higher demand for those exporters presumably triggers the additional transcription of *HwENA* genes.

The high similarity of ENA ATPases between *H. werneckii* and other fungal ENA ATPases involved in $Na^+/K^+$ transport, together with salt-responsive gene expressions described above, indicate the potential importance of this system in ion homeostasis in the halophilic black yeast *H. werneckii*. Phylogenetically, HwENA ATPases belong to a separate group of fungal alkali cation P-ATPases, evolving alongside other fungal P-type ATPases from a common ancestor (Gorjan and Plemenitaš, submitted). While the function of most of these P-type ATPases has not yet been determined, it was speculated that NcENA2 from *N. crassa* is involved in $K^+$ metabolism (Benito et al. 2002). It is believed that genes encoding fungal $K^+$- or $Na^+$-ATPases (ENA P-type ATPases) have most probably evolved from an ancestral $K^+$-ATPase through the processes of gene duplication and that the capacity of ENA ATPases to pump $Na^+$ has evolved as an adaptation mechanism to increased salinity (Benito et al. 2002).

### 2.4 Compatible solutes

Halophilic microorganisms living in hypersaline conditions adapt to high osmotic pressure and avoid associated water loss by accumulation of osmolytes—either cations or various organic solutes in the cytoplasm. In eukaryotic species, exemplified by the salt-sensitive yeast *S. cerevisiae* and the green alga *Dunaliella salina*, osmoadaptation is predominantly achieved through accumulation of organic solutes, also termed compatible solutes. Glycerol is the predominant solute used for this purpose, increasing the intracellular osmotic potential and simultaneously protecting cellular structures from adverse conditions (Blomberg and Adler 1992; Oren 1999).

Our first measurements (Petrovič et al. 2002) showed that the intracellular glycerol concentration in halophilic *H. werneckii* cells grown at different salinities steadily increases from 0% to 10% NaCl. At higher salinities, the intracellular glycerol concentration remains virtually unchanged. On the other hand, the extracellular glycerol concentrations are seen to be low and independent of salt concentration between 0% and 17% NaCl and start to increase above this salinity. We hypothesized that synthesized glycerol is efficiently kept inside the cells; although we could not exclude the possibility that glycerol crosses the plasma membrane and is then actively taken up into the cells (Petrovič et al. 2002). This assumption is in accordance with previous reports from other microorganisms (Nevoigt and Stahl 1997; Blomberg 2000). The increase in glycerol synthesis in *H. werneckii* was supported with the identification of two genes coding for putative glycerol-3-phosphate dehydrogenase; *HwGPD1* and *HwGPD2*. According to the expression profile obtained by RT-PCR, only *HwGPD2* is expressed differentially at increased salinity (Petrovič et al. 2002). These results suggest that glycerol biosynthesis in *H. werneckii* is regulated at the level of transcription of glycerol-3-phosphate-dehydrogenase, like that of *S. cerevisiae*.

Although our results show that glycerol is the most important compatible solute, we investigated the presence of other compatible solute(s) in this halophilic black yeast. Indeed, *H. werneckii* accumulates other higher polyols besides glycerol in a salt-dependent manner (Kogej 2006). This is in accordance with the findings in other fungi, where polyols such as erythritol, inositol, arabinitol, xylitol and mannitol, were also found to be used for osmoadaptation (Pfyffer et al. 1986; Blomberg and Adler 1992). Furthermore, other compatible solutes besides polyols are also produced in response to salt stress in microorganisms, such as nitrogen-containing compounds like glycine betaine and free amino acids (Galinski 1995), while mycosporine-like amino acids (MAAs) were suggested as osmoprotectants in halophilic cyanobacteria (Oren 1997). MAAs preferentially contain an aminocyclohexenimine unit bound to an amino acid or amino alcohol group and absorb maximally in the range 310–360 nm (Badaranayake 1998; Libkind et al. 2004). Mycosporines are similar substances with an aminocyclohexenone unit bound to an amino acid or amino alcohol group (Bandaranayake 1998), which have been described in fungi as UV sunscreens (absorbing 310–320 nm) and as being involved in morphogenesis and sporulation (Leach 1965; Trione et al. 1966). We determined the mycosporine and MAA content in fungi grown in hypersaline conditions (Kogej et al. 2006). Mycosporine-glutaminol-glucoside and mycosporine-glutamicol-glucoside were two mycosporines detected in the extracts of *H. werneckii*. These mycosporines were previously identified as metabolites of microcolonial fungi (MCF) inhabiting UV-exposed rocks in arid and semi-arid regions. The amount of mycosporine-glutaminol-glucoside was clearly salt-dependent, whereas the amount of mycosporine-glutamicol-glucoside decreased as salt increased. It seems that mycosporine-glutaminol-glucoside might act as supplementary compatible solute in *H. werneckii* exposed to variations in water activity or to hypersaline conditions. Further studies are needed to confirm its role in osmoadaptation (Kogej et al. 2006).

In conclusion, *H. werneckii* uses a unique set of compatible solutes for osmoregulation. Whereas *D. hansenii* uses $K^+$ (Blomberg and Adler 1992; Ramos 2005) and glycerol (Almagro et al. 2000) to achieve this purpose, *H. werneckii* uses no cations for osmoadaptation (Kogej et al. 2005), but rather accumulates compatible solutes including glycerol (Petrovič et al. 2002), higher polyols and mycosporine-glutaminol-glucoside (Kogej 2006).

## 3 *Hortaea werneckii*—a eukaryotic model organism for halophilic adaptations

The halophilic black yeast *H. werneckii* is one of the most salt tolerant eukaryotic organisms so far described. Our studies on its adaptations have revealed some new mechanisms that enable *H. werneckii* to thrive at extremely high NaCl concentrations and also to quickly adapt to a wide range of NaCl concentrations. This adaptive halophilic behaviour and complex polymorphic life cycle ensure *H. werneckii* dominance amongst fungi inhabiting the hypersaline waters of eutrophic salterns. Differences have been observed on the level of cell-wall pigmentation and structure, membrane fluidity and compatible solutes. Preliminary global studies on the transcriptome of *H.werneckii* allow us to speculate on the existence of double sets of genes involved in the mechanism of adaptation to life in hypersaline environments, such as the genes *HwENA* and *HwGDP*, which are expressed differentially at different salinities (Petrovič et al. 2002; Gorjan and Plemenitaš, submitted). This seems a plausible explanation for the successful euryhaline strategy employed by *H. werneckii*.

**Acknowledgements** This work was supported by the Ministry of Higher Education and Technology of the Republic of Slovenia.

## References

Almagro A, Prista C, Quintas C, Madeira Lopes A, Ramos J, Loureiro-Dias MC (2000) Effects of salts on *Debaryomyces hansenii* and *Saccharomyces cerevisiae* under stress conditions. Int J Food Microbiol 56:191–197

Almagro A, Prista C, Benito B, Loureiro-Dias MC, Ramos J (2001) Cloning and expression of two genes coding for sodium pumps in the salt-tolerant yeast *Debaryomyces hansenii*. J Bacteriol 183(10):3251–3255

Andre L, Nillsson A, Adler L (1988) The role of glycerol in osmotolerance of the yeast *Debaromyces hansenii*. J Gen Microbiol 134:669–677

Andreishcheva EN, Isakova EP, Sidorov NN, Abramova NB, Ushakova NA, Shaposhnikov GL, Soares MIM, Zvyagilskaya RA (1999) Adaptation to salt stress in a salt-tolerant strain of the yeast *Yarrowia lipolytica*. Biochemistry 64(9):1061–1067 (Moscow)

Andrews S, Pitt JI (1987) Further studies on the water relations of xerophilic fungi, including some halophiles. J Gen Microbiol 133:233–238

Bandaranayake WM (1998) Mycosporines: are they nature's sunscreens? Nat Prod Rep 15(2):159–172

Banuelos MA, Rodriguez-Navarro A (1998) P-type ATPases mediate sodium and potassium effluxes in *Schwanniomyces occidentalis*. J Biol Chem 273(3):1640–1646

Bell AA, Wheeler MH (1986) Biosynthesis and functions of fungal melanins. Annu Rev Phytopathol 24:411–451

Benito B, Garciadeblas B, Rodriguez-Navarro A (2002) Potassium- or sodium-efflux ATPase, a key enzyme in the evolution of fungi. Microbiology 148(Pt 4):933–941

Blomberg A, Adler L (1992) Physiology of osmotolerance in fungi. Adv Microb Physiol 33:145–212

Blomberg A (2000) Metabolic surprises in *Saccharomyces cerevisiae* during adaptation to saline conditions: questions, some answers and a model. FEMS Microbiol Lett 182(1):1–8

Butinar L, Santos S, Spencer-Martins I, Oren A, Gunde-Cimerman N (2005) Yeast diversity in hypersaline habitats. FEMS Microbiol Lett 244(2):229–234

Elliot ML, Henson JM (2001) Effect of osmotic stress on growth of *Gaeumannomyces graminis* strains differing in hyphal pigmentation. Mycologia 93(4):617–625

Galinski EA (1995) Osmoadaptation in bacteria. Adv Microb Physiol 37:272–328

Garciadeblas B, Rubio F, Quintero FJ, Banuelos MA, Haro R, Rodriguez-Navarro A (1993) Differential expression of two genes encoding isoforms of the ATPase involved in sodium efflux in *Saccharomyces cerevisiae*. Mol Gen Genet 236(2–3):363–368

Gorbushina AA, Krumbein WE, Hamann CH, Panina LK, Soukharjevski SM, Wollenzien U (1993) Role of black fungi in color change and biodeterioration of antique marbles. Geomicrobiol J 11:205–211

Göttlich E, de Hoog GS, Yoshida S, Takeo K, Nishimura K, Miyaji M (1995) Cell surface hydrophobicity and lipolysis as essential factors in human tinea nigra. Mycoses 38:489–494

Gunde-Cimerman N, Frisvad JC, Zalar P, Plemenitaš A (2005) Halotolerant and halophilic fungi. Oxford & IBH Publishing Co. Pvt. Ltd.

Gunde-Cimerman N, Zalar P, de Hoog GS, Plemenitaš A (2000) Hypersaline waters in salterns – natural ecological niches for halophilic black yeasts. FEMS Microbiol Ecol 32(3):235–240

Holker U, Bend J, Pracht R, Tetsch L, Muller T, Hofer M, de Hoog GS (2004) *Hortaea acidophila*, a new acid-tolerant black yeast from lignite. Anton Van Leeuwen 86(4):287–294

de Hoog GS (1993) Evolution of black yeasts: possible adaptation to the human host. Anton Van Leeuwen 63:105–109

de Hoog G, Hermanides-Nijhof E (1977) Survey of black yeasts and allied fungi. Stud Mycol 15:178–221

de Hoog GS, Gerrits van den Ende AHG (1992) Nutritional pattern and eco-physiology of *Hortaea werneckii*, agent of human tinea nigra. Anton Van Leeuwen 62:321–329

de Hoog GS, Guého E (1998) Agents of white piedra, black piedra and tinea nigra. In: Asello L, Hay RJ (eds) Topley and Wilsons microbiology and microbial infections, 3rd edn., vol 4. Arnold, London, pp 1–15

de Hoog GS, Zalar P, Urzi C, de Leo F, Yurlova NA, Sterflinger K (1999) Relationships of dothideaceous black yeasts and meristematic fungi based on 5.8S and ITS2 rDNA sequence comparison. Stud Mycol 43:33–40

Hosono K (1992) Effect of salt stress on lipid composition and membrane fluidity of the salt-tolerant yeast *Zygosaccharomyces rouxii*. J Gen Microbiol 138:91–96

Iwatsu T, Udagawa S (1988) *Hortaea werneckii* isolated from sea-water. Jpn J Med Mycol 29(2):142–145

Khaware RK, Koul A, Prasad R (1995) High membrane fluidity is related to NaCl stress in *Candida membranefaciens*. Biochem Mol Biol Int 35(4):875–880

Kogej T (2006) Physiological adaptations of halophilic black yeast *Hortaea werneckii* to growth at saline conditions on the levels of cell wall and accumulation of compatible solutes. Doctoral Thesis, Ljubljana, 198 pp

Kogej T, Gostinčar C, Volkmann M, Gorbushina AA, Gunde-Cimerman N (2006) Mycosporines in extremophilic fungi – novel complementary osmolytes? Environ Chem 3(2):105–110

Kogej T, Ramos J, Plemenitaš A, Gunde-Cimerman N (2005) The halophilic fungus *Hortaea werneckii* and the halotolerant fungus *Aureobasidium pullulans* maintain low intracellular cation concentrations in hypersaline environments. Appl Environ Microbiol 71(11):6600–6605

Kogej T, Wheeler MH, Lanišnik Rižner T, Gunde-Cimerman N (2004) Evidence for 1,8-dihydroxynaphthalene melanin in three halophilic black yeasts grown under saline and non-saline conditions. FEMS Microbiol Lett 232(2):203–209

Krumbein WE, Gorbushina AA, Sterflinger K, Haroska U, Kunert U, Drewello R, Weißmann R (1996) Biodeterioration of historical window panels of the former Cistercian Monastery church of Haina (Hessen, Germany). DECHEMA Monographs 133:417–424

Leach CM (1965) Ultraviolet-absorbing substances associated with light-induced sporulation in fungi. Can J Bot 43:185–200

Libkind D, Perez P, Sommaruga R, Dieguez Mdel C, Ferraro M, Brizzio S, Zagarese H, van Broock M (2004) Constitutive and UV-inducible synthesis of photoprotective compounds (carotenoids and mycosporines) by freshwater yeasts. Photochem Photobiol Sci 3(3):281–286

Mok WYC, Barreto da Silva MS (1981) Occurrence of *Exophiala werneckii* on salted freshwater fish *Osteoglossum bicirrhosum*. J Food Technol 16:505–512

Nevoigt E, Stahl U (1997) Osmoregulation and glycerol metabolism in the yeast *Saccharomyces cerevisiae*. FEMS Microbiol Rev 21(3):231–241

Nienow JA, Friedman EI (1993) Terrestrial lithophytic (rock) communities. In: Friedmann EI (ed) Antarctic Microbiology (Wiley Series in Ecological and Applied Microbiology), pp 342–412. Wiley-Liss, 644 pp

Oren A (1997) Mycosporine-like amino acids as osmotic solutes in a community of halophilic cyanobacteria. Geomicrobiol J 14:231–240

Oren A (1999) Bioenergetic aspects of halophilism. Microbiol Mol Biol Rev 63(2):334–348

Petrovič U, Gunde-Cimerman N, Plemenitaš A (1999) Salt stress affects sterol biosynthesis in the halophilic black yeast *Hortaea werneckii*. FEMS Microbiol Lett 180(2):325–330

Petrovič U, Gunde-Cimerman N, Plemenitaš A (2002) Cellular responses to environmental salinity in the halophilic black yeast *Hortaea werneckii*. Mol Microbiol 45(3):665–672

Pfyffer GE, Pfyffer BU, Rast DM (1986) The polyol pattern, chemotaxonomy, and phylogeny of the fungi. Sydowia 39:160–201

Plemenitaš A, Gunde-Cimerman N (2005) Cellular reponses in the halophilic black yeast *Hortaea weneckii* to high environmental salinity. In: Gunde-Cimerman N, Oren A, Plemenitaš A (eds) Adaptation to life at high salt concentrations in Archea, Bacteria and Eukarya. Springer, Dordrecht, The Netherlands, pp 455–470

Prista C, Loureiro-Dias MC, Montiel V, García R, Ramos J (2005) Mechanisms underlying the halotolerant way of *Debaryomyces hansenii*. FEMS Yeast Res 5:693–701

Ramos J (1999) Contrasting salt tolerance mechanisms in *Saccharomyces cerevisiae* and *Debaryomyces hansenii*. In: Pandalai SG (ed) Recent research developments in microbiology, vol 3. Research Signpost, Trivandrum, India, pp 377–390

Ramos J (2005) Introducing *Debaryomyces hansenii*, a salt-loving yeast. In: Gunde-Cimerman N, Oren A, Plemenitaš A (eds) Adaptation to life at high salt concentrations in Archea, Bacteria and Eukarya. Springer, Dordrecht, The Netherlands, pp 441–451

Russell NJ (1989a) Adaptive modifications in membranes of halotolerant and halophilic microorganisms. J Bioenerg Biomembr 21(1):93–113

Russell NJ (1989b) Structural and functional role of lipids. In: Ratledge C, Wilkinson SG (eds) Microbial lipids. Academic Press, New York, pp 279–349

Russell NJ, Evans, RI, ter Steeg PF, Hellemons J, Verheul A, Abee T (1995) Membranes as a target for stress adaptation. Int J Food Microbiol 28(2):255–261

Sharma SC, Raj D, Forouzandeh M, Bansal MP (1996) Salt-induced changes in lipid composition and ethanol tolerance in *Saccharomyces cerevisiae*. Appl Biochem Biotechnol 56(2):189–195

Silva-Graça M, Lucas C (2003) Physiological studies on long-term adaptation to salt stress in the extremely halotolerant yeast *Candida versatilis* CBS 4019 (syn *C. halophila*). FEMS Yeast Res 3(3):247–260

Slaninova I, Sestak S, Svoboda A, Farkas V (2000) Cell wall and cytoskeleton reorganization as the response to hyperosmotic shock in *Saccharomyces cerevisiae*. Arch Microbiol 173(4):245–252

Staley JT, Palmer F, Adams JB (1982) Microcolonial fungi: common inhabitants on desert rocks? Science 215:1093–1095

Sterflinger K, de Hoog GS, Haase G (1999) Phylogeny and ecology of meristematic ascomycetes. Stud Mycol 43:5–22

Todaro F, Berdar A, Cavaliere A, Criseo G, Pernice L (1983) Gasophtalmus in black sea bream (*Spodyliosoma cantharus*) caused by *Sarcynomyces crustaceus* Lindner. Mycopathologia 81:95–97

Trione EJ, Leach CM, Mutch JT (1966) Sporogenic substances isolated from fungi. Nature 212:163–164

Tunblad-Johansson I, Adler L (1987) Effect of sodium chloride concentration on phospholipid fatty acid composition of yeasts differing in osmotolerance. FEMS Microbiol Lett 43:275–278

Turk M, Plemenitaš A (2002) The HOG pathway in the halophilic black yeast *Hortaea werneckii*: isolation of the HOG1 homolog gene and activation of HwHog1p. FEMS Microbiol Lett 216(2):193–199

Turk M, Mejanelle L, Sentjurc M, Grimalt JO, Gunde-Cimerman N, Plemenitaš A (2004) Salt-induced changes in lipid composition and membrane fluidity of halophilic yeast-like melanized fungi. Extremophiles 8(1):53–61

Watanabe Y, Iwaki T, Shimono Y, Ichimiya A, Nagaoka Y, Tamai Y (1999) Characterization of the $Na^+$-ATPase gene (ZENA1) from the salt-tolerant yeast *Zygosaccharomyces rouxii*. J Biosci Bioeng 88(2):136–142

Wollenzien U, de Hoog GS, Krumbein WE, Urzì C (1995) On the isolation of microcolonial fungi occurring on and in marble and other calcareous rocks. Sci Tot Env 167:287–294

Yoshikawa S, Mitsui N, Chikara KI, Hashimoto H, Shimosaka M, Okazaki M (1995) Effect of salt stress on plasma membrane permeability and lipid saturation in the salt-tolerant yeast *Zygosaccharomyces rouxii*. J Ferment Bioener 80(2):131–135

Zalar P, de Hoog GS, Gunde-Cimerman N (1999) Ecology of halotolerant dothideaceous black yeasts. Stud Mycol 43:38–48

Zalar P, Kocuvan MA, Plemenitaš A, Gunde-Cimerman N (2005) Halophilic black yeast colonize wood immersed in hypersaline water. Bot Mar 48:323–326

Zhdanova NN, Pokhodenko VD (1973) The possible participation of a melanin pigment in the protection of the fungus cell from desiccation. Microbiology 42:753–757

Zhdanova NN, Borisyuk LG, Artzatbanov VY (1990) Ocurrence of the K-type of life strategy in some melanin-containing fungi under experimental conditions. Folia Microbiol 35:423–430

Rev Environ Sci Biotechnol (2006) 5:141–166
DOI 10.1007/s11157-006-0002-3

VIEW PAPER

# Ultraviolet radiation shapes seaweed communities

**Kai Bischof · Ivan Gómez · Markus Molis · Dieter Hanelt · Ulf Karsten · Ulrike Lüder · Michael Y. Roleda · Katharina Zacher · Christian Wiencke**

Received: 10 March 2006 / Accepted: 8 May 2006 / Published online: 15 July 2006


**Abstract** Stratospheric ozone depletion and the concomitant increase in irradiance of ultraviolet-B radiation (UVB) at the earth's surface represent major threats to terrestrial and aquatic ecosystems. In costal rocky shore environments, seaweeds constitute a group of organisms of particular significance to ecosystem function. Thus, impairment of seaweed performance by UVB-exposure may result in severe changes in the functioning of coastal ecosystems. Here we present our view on how UVB radiation affects seaweed physiology and ecology and, thus, shapes the coastal environment by affecting the spatial, species and functional structure of seaweed communities.

**Keywords** Acclimation · Photosynthesis · Seaweeds · Ultraviolet radiation · Vertical zonation

K. Bischof (✉)
Institute for Polar Ecology, University of Kiel, Wischhofstr. 1-3, Building 12, 24148 Kiel, Germany
e-mail: kbischof@ipoe.uni-kiel.de

I. Gómez
Instituto de Biologia Marina, Universidad Austral de Chile, Casilla 567, Valdivia, Chile

M. Molis · M. Y. Roleda
Alfred Wegener Institute for Polar and Marine Research, Marine Station, 27498 Helgoland, Germany

D. Hanelt
Biocenter Klein-Flottbek, University of Hamburg, 22609 Hamburg, Germany

U. Karsten
Institute of Biological Sciences, Applied Ecology, University of Rostock, 18059 Rostock, Germany

U. Lüder · K. Zacher · C. Wiencke
Alfred Wegener Institute for Polar and Marine Research, 27570 Bremerhaven, Germany

## Introduction to seaweed communities

Seaweeds (also referred to as "marine macroalgae") represent key components within coastal ecosystems (Lüning 1990). As primary producers these green, brown or red coloured marine plants serve a multitude of ecosystem functions. They may grow from just a few millimetres in size up to 60 m or even more. At their growth site they often form submersible forests (i.e., "kelp forests" in cold-temperate oceans) characterized by high primary productivity. For instance, along the coasts of cold-temperate regions, the communities dominated by the brown algal genus *Laminaria* have annual productivity rates of about 2 kg carbon per $m^2$ (Thomas 2002). Seaweeds are globally distributed from the Tropics to the Polar Regions and primarily settle on hard bottom substrate, such as rocky shorelines. Here, seaweed communities typically show distinct vertical zonation patterns exhibiting a characteristic sequence of species with increasing depth (Stephenson and Stephenson 1972). Along with the depth gradient, the environmental parameters seaweeds are exposed to change drastically. Species settling in the supralittoral zone (the fringe above the high tide level) are exposed to drought, high solar radiation and also atmospheric changes in temperature. The intertidal fringe (the eulittoral zone) is moreover characterised by regular and extreme changes in abiotic conditions, based on tidal influence (Davison and Pearson 1996). During low tide, organisms are exposed to high solar irradiance, drought, atmospheric temperatures, and, depending on current weather conditions, sometimes extreme changes in salinity (Schonbeck and Norton 1978; Davison and Pearson 1996). Furthermore, mechanical stress due to tidal currents and wave exposure has a strong impact on intertidal communities. Underneath, in the sublittoral zone extending below the low tide level, seaweeds usually encounter a more stable habitat, as the water column above buffers against strong changes in abiotic parameters. As outlined below, it is also the respective characteristics of the water column, which determines exposure of sublittoral organisms towards UVB exposure.

Seaweeds interact with and play an existential role for many marine animals (Lubchenco and Gaines 1981; Hay and Fenical 1992). The most important functions of seaweeds in this context are the supply of food and habitat to and their effects on the recruitment and dispersal of animals. Being at the basis of marine food webs, seaweeds are directly consumed by a diversified suite of micro- and macrograzers, which in turn can alter structure and species composition of seaweed communities (Duffy and Hay 2000). In addition, algal exudates might fuel the microbial loop if they are used by free-living and alga-associated bacteria. However, algae can also adversely affect animals. For instance, decomposing algal mats produce anoxic conditions affecting survival of covered animals.

Seaweeds serve many species as habitat, to which sessile forms attach directly, and may host motile animals by provision of shelter from predators (Lippert et al. 2001). Open space is a key resource for sessile marine species and in this way, seaweeds ameliorate competition for space. Seaweeds are also among the first multi-cellular colonizers and thus precondition the substratum for later successional recruits. Subsequent settlement of spores and larvae may be facilitated, tolerated, or inhibited in dependency of the already existing seaweed community (Sousa and Connell 1992). Furthermore, larger seaweeds serve as egg-deposition sites of predatory fin and shell fish, thereby supporting indirectly top-down control in coastal habitats. Furthermore, dislodged seaweeds serve as dispersal vectors for species, accelerating the transport of invasive species and the recolonisation of defaunated habitats (Thiel and Gutow 2005).

Due to their enormous importance within coastal ecosystems a decrease in seaweed abundance related to environmental change e.g., under increased UVB irradiance will thus have dramatic consequences for the sum of organisms associated. Even under the present radiation conditions, UVB represents a crucial environmental factor organisms have to deal with. The knowledge on the physiological and ecological effects of present UVB irradiances is a precondition in order to be able to estimate future consequences of ozone depletion and increased UVB levels. Thus, in the following we present our view on how UVB radiation may shape seaweed dominated coastal ecosystems under the present radiation conditions and how the spatial and functional structure of seaweed communities might be affected in the future.

## Enhanced UVB radiation

Solar radiation is the most important prerequisite for life on earth. In the process of photosynthesis, photoautotrophic organisms (like seaweeds) convert light energy into chemically bound energy, which is used for biomass production; as a side effect, molecular oxygen is generated as a basis for all heterotrophic organisms. Changes in irradiance and light quality can either promote photosynthesis, but can also inhibit many biological processes if radiation becomes excessive (Barber and Andersson 1992), or if short wavelength radiation with high energy content, such as UVB radiation, is absorbed by biomolecules (Vass 1997). Consequently, damage to important components in plant metabolism results in reduced photosynthetic and general metabolic activity and, hence, lead to a decrease in biomass production.

Ever since the discovery of stratospheric ozone depletion in the Antarctic in the 1970s (Farman et al. 1985), serious concerns have arisen about the impacts of increasing UVB radiation on the biosphere (Madronich et al. 1998; Björn et al. 1999). Ozone is predominantly generated in the low latitudes, by photolysis of molecular oxygen. In the stratosphere, ozone molecules are subject to UV-mediated photolysis and may also be degraded due to the reaction within catalytic cycles with NO, Cl or Br serving as catalysts (Lary 1997; Langer 1999). The concentration of these compounds in the atmosphere increases mainly due to anthropogenic emissions, thus leading to ozone depletion.

Ultraviolet radiation includes the wavelengths below those visible to the human eye. This spectral range is according to the CIE definition (Commission Internationale de l'Eclairage 1935) divided into three wavebands: 315–400 nm UVA, 280–315 nm UVB, and 190–280 nm UVC, which does not reach the earth's surface, as it is completely absorbed on its way through the atmosphere. Due to the optical characteristics of ozone, it is the UVB range, which is likely to increase at the earth's surface, as a consequence of a decrease in stratospheric ozone concentration. Calculations based on the absorption characteristics of $O_3$ indicate that a 10% decline in column ozone would result in an approx. 5% increase of surface irradiance at 320 nm while the same decline would be accompanied by a 100% increase at 300 nm (Frederick et al. 1989).

## UVB in the aquatic environment

The effects of UVB on aquatic ecosystems are strongly dependent on the optical properties of the water body (Holm-Hansen et al. 1993; Hanelt et al. 2001). Therefore, it is necessary to estimate the penetration of UV radiation into the water column. The UV irradiance reaching the water surface is influenced by various *atmospheric* factors, such as latitude and altitude, elevation of the sun coinciding to the season and the time of day, weather conditions (clouds and fog), ozone and aerosol concentrations. The *underwater* light field is even more variable: In coastal sea waters, UV radiation and blue light are strongly attenuated due to dissolved organic matter (DOM; Kirk 1994), and this depends largely on the input of this material from the terrestrial ecosystem. According to Kirk (1994), the diffuse vertical attenuation coefficient of downward irradiance ($K_d$) in the water column is determined by the following formula:

$$K_d = \ln(\mathrm{Ed}_{(z_2)}/\mathrm{Ed}_{(z_1)})^*(z_1 - z_2)^{-1}$$

with $\mathrm{Ed}_{(z_1)}$ and $\mathrm{Ed}_{(z_2)}$ as the respective irradiance in depth $z_1$ and $z_2$. The value of the $K_d$ in the UV-waveband is naturally much higher than for the PAR range (Jerlov 1976) and typically increases tremendously during the summer season in polar coastal waters (Hanelt et al. 2001). The reason is that the turbidity of the coastal water in polar regions rises due to rainfall or melt water from snow layers and glaciers. A high discharge of turbid fresh water into the coastal zone carries fine terrigenous sediments into the seawater (Svendsen et al. 2002). In the Arctic the variation in $K_d$-values showed that from middle of June, UVB transparency decreased strongly due to the input of turbid melt waters on Spitsbergen (Hanelt et al. 2001) or in the Hudson bay (Vincent and Belzile 2003). While the UVB transmittance decreased only by about 22% per meter in clear waters during spring, the attenuation increased to about 53% per meter in summer, so that UVB was almost fully absorbed within the first 3 m of the water body at Spitsbergen. In spring and late autumn at low air temperatures the water conditions are relatively clear. Then, UV radiation penetrates deeply into the water column and the threshold irradiance of UVB with the potential to affect primary plant productivity negatively is still at about 5–6 m depth. In contrast to the situation in sheltered bays and fjords, at open coastlines with strong current, melt water can be replaced much faster with clearer oceanic water, which diminishes the observed turbidity effects on light penetration. Climate-induced changes in planktonic or allochthonous sources of DOM, either through changes in vegetation cover, decomposition, or glacial meltwater may have a higher impact on the underwater UVB regime than ozone depletion would produce (Franklin et al. 2003). Vincent and Belzile (2003) found a close correlation between UV attenuation and seston concentration in the Antarctic region. In the south western Ross Sea a high particulate absorption in the UVB waveband was observed during a phytoplankton bloom which was rather caused by the absorption of chromophoric dissolved organic matter (Arrigo and Brown 1996). Arctic and Antarctic Oceans both experience increased biological UV exposure resulting from stratospheric ozone depletion. However, climate related sea ice melting in the Arctic may potentially result in greater change in underwater UVB exposure than the increase caused by recent ozone depletion in Antarctica (Vincent and Belzile 2003). In summary, the respective optical characteristics of the water column determine the under water light climate and thus also UVB absorption in the aquatic environment. As UVB exposure is decreasing with increasing depth, the attenuation of UV radiation in the water column represents a major structuring frame for seaweed communities.

## Biological effects of UVB

The effects of UVB exposure on biological systems are manifold, and reach from the molecular to the organism level, thereby affecting growth and production, and, consequently, ecosystem structure and function. A prerequisite for UVB induced damage is the absorption by biomolecules. Potential UV-chromophores in plants mainly include nucleic acids (such as DNA, RNA) and proteins (Vass 1997). DNA is one of the most UV-sensitive molecules and UV-induced damage occurs directly by the absorption of UVB quanta by aromatic residues. The results are direct structural alterations such as formation of cyclobutane dimers (Lois and Buchanan 1994), but can also be indirectly mediated due to the presence of free oxygen radicals, generated by the electron transfer from chromophore molecules, excited by UV absorption (Mitchell and Karentz 1993). UV-induced damage to the DNA represents a serious effect, as photoproducts can inhibit replication or even cause mutations, thereby affecting gene expression. UVB absorbing aromatic residues are also present in certain amino acids (e.g., tyrosine, phenylalanine, tryptophan) and, therefore, in proteins. Consequently, damage to protein molecules is a major effect of UVB in organisms. Furthermore, disulphide bonds between cysteine residues in the protein can be cleaved by UVB radiation (Vass 1997). These bonds have an important role in protein folding, and thus, are essential for proper functioning of the protein. Lipids, a major compound in all biological membranes, may be destroyed by UVB in the presence of oxygen. This peroxidation of unsaturated fatty acids has a direct effect on membrane structure and the generation of lipid peroxy radicals can induce further damage by participating in free radical cascades (Murphy 1983). In plants, pigments of the photosynthetic apparatus can also be destroyed by UV exposure (Strid et al. 1990), with the phycobilins being the most sensitive, and carotenoids generally being less affected than the chlorophylls (Teramura 1983; Häder and Häder 1989). As a consequence of a number of molecular effects, several physiological processes are impaired, such as photosynthesis (Bornman 1989; Strid et al. 1990; Nogues and Baker 1995; Allen et al. 1997), and nutrient uptake (Döhler 1985, 1992; Flores-Moya et al. 1998; Gómez et al. 1998), while others, e.g., respiration, appear to be less affected (Larkum and Wood 1993; Aguilera et al. 1999).

Photosynthesis is probably the most intensively studied process in plant biology. Due to its central role in plant metabolism, as well as its importance for all oxygen dependent life on earth, studies on adverse effects on photosynthesis, in the context of a globally changing environment are of particular interest. Due to numerous effects of UVB radiation to the respective molecules involved in photosynthesis, the effects of UV-exposure are also multiple (see Vass 1997 for review). The common consequences on photosynthetic function are decreased $CO_2$-fixation and oxygen evolution (Renger et al. 1986; Allen et al. 1997). This could be caused by several molecular events: While most studies have found that photosystem I (PS I) is only minimally affected by UVB (by inhibiting PS I-mediated cyclic photophosphorylation; Iwanzik et al. 1983; Renger et al. 1986), photosystem II (PS II) seems to be a more important target (Bornman 1989). It is likely that UVB causes an inhibition of energy transfer within the PS II reaction centre by blocking electron flow. Furthermore, the function of the $D_1$ protein may be impaired by the UVB induced fragmentation of the protein (Renger et al. 1986; Vass 1997). On the oxidising side of PS II, the oxygen evolving system (water splitting complex) is another sensitive target of UVB (Renger et al. 1986). Furthermore, it has been suggested that UVB may affect the light-harvesting complex (LHC) by its functional disconnection from the photosystem, resulting in an impairment of energy transfer to the reaction centre (Renger et al. 1986; Lorenz et al. 1997). A decrease in photosynthetic activity may also be due to the photodestruction of pigments; within the chlorophylls, Chl *a* has been observed to be more affected than Chl *b* (Teramura 1983; Strid et al. 1990).

The $CO_2$ fixing enzyme RubisCO has been shown to be another critical component in UVB-induced inhibition of photosynthesis. The UVB-induced decline in its activity is related to the decreasing amount of both subunits as well as the corresponding mRNA levels (Strid et al. 1990; Jordan et al. 1992; Bischof et al. 2000a, 2002a). Another effect of UVB on reactions related to photosynthesis represents the inactivation of chloroplast ATPase (Strid et al. 1990). Impairment of any of the components mentioned above contributes to lower the photosynthetic activity during and following UV-exposure.

The physiological effects are also reflected on the ultrastructural level. UVB radiation can lead to dramatic changes of the fine structure of chloroplasts and mitochondria. Mild UV stress leads to a wrinkled appearance of the thylakoids, lumen dilatations and damage of the outer membranes. In the mitochondria a swelling of the cristae is often observed (Poppe et al. 2003; Holzinger et al. 2004). After strong UVB exposure the formation of 'inside-out' vesicles from thylakoids was demonstrated in four red algal species. In *Palmaria decipiens* the fine structural changes are reversible indicating acclimation to UV stress (Poppe et al. 2002, 2003).

On the organism level, the effects mentioned above can result in reduced growth and production, as shown in higher plants, seaweeds, phytoplankton and ice algae (Caldwell 1971; Worrest 1983; Ekelund 1990; Karentz et al. 1991a, b; Holm-Hansen 1993; McMinn et al. 1999, Han 1996a, b; Makarov 1999; Aguilera et al. 2000; Altamirano et al. 2000a, b). Other effects include the impairment of reproductive success or may even bear lethal consequences. Consequently, all aspects mentioned may also affect ecosystem structures (Holm-Hansen et al. 1993; Johanson et al. 1995; Caldwell et al. 1998).

## Seaweed responses to UVB

Seaweeds became a prominent group of organisms in UVB research for two reasons. Firstly, seaweeds represent a crucial component for coastal ecosystems. Thus, UVB related damage to these organisms might have drastic consequences to the entire ecosystem. Secondly, in a range of field and laboratory studies, seaweeds were proven to be in general rather sensitive to UVB exposure (at least compared to terrestrial plants) but also it was shown in some species growing over a wide depth or even geographical range, that seaweeds do apply a vast variety of acclimation mechanisms. Thus they became very suitable model systems also in basic stress physiology.

### Adaptation versus acclimation

The respective reaction of a species towards UVB exposure is determined by the interplay of genetically fixed adaptation and physiological acclimation. Generally spoken, adaptation is setting the frame in which acclimation to changing environmental conditions might occur.

In a laboratory study on Antarctic seaweeds, which were isolated in the field decades ago and subsequently kept in stock cultures it was shown that these specimens still exhibit distinct species-specific differences in UV-tolerance once they are grown to macrothalli and exposed to identical culture and experimental UV conditions (Bischof et al. 1998a). Due to the cultivation of sporophytes under low-light conditions and UV-exclusion no acclimation to UV radiation had occurred prior to the experiments. The two shallow-water green algae *Enteromorpha bulbosa* and *Acrosiphonia arcta* were least affected by UVB radiation. Photosynthesis in the brown algae *Desmarestia antarctica* and *D. anceps* and the red alga *Gymnogongrus antarcticus*, inhabiting slightly deeper waters, was inhibited to a similar and intermediate extent. However, two other red algal species from the lower subtidal, *Phycodrys austrogeorgica* and *Delesseria lancifolia*, responded extremely sensitively towards UVB-exposure. In the case of *Delesseria sanguinea*, a deep sublittoral species, growth can be also strongly impaired when the alga is exposed to surface solar radiation (Pang et al. 2001) indicating these plants may lack all protecting mechanisms against excessive radiation. In the field protective mechanisms against UV radiation might not be necessary because they live in the shade of the canopy algae and/or in great depths. Similarly, zoospores of the deep-water species *L. saccharina* and *L. hyperborea* are more strongly photoinhibited after exposure to UV radiation than zoospores from the shallow water species *L. digitata* (Roleda et al. 2005a). In addition, recovery of PS II activity is high in *L. digitata*, low in *L. saccharina* and lowest in *L. hyperborea*. Spores of the eulittoral *M. stellatus* and *C. crispus* are photoinhibited after UV exposure but recover quickly after exposure to dim white light (Bischof et al. 2000b; Roleda et al. 2004). The first attempt to study kinetics of photoinhibition and recovery in zoospores of Arctic Laminariales showed that zoospores of the lower sublittoral *L. saccharina* were more sensitive to PAR- and UV-induced photoinhibition than upper- to midsublittoral *S. dermatodea, A. esculenta* and *L. digitata*. Kinetics of recovery in zoospores showed a fast phase in *S. dermatodea* which indicates a reduction of the photoprotective process while a slow phase in *L. saccharina* indicates recovery from severe photodamage (Roleda et al. 2006d). These experiments were focussed on short-term effects, thus the preadaptive setting of species could be revealed, but not the respective potential of acclimation, also determined by the genetic features of species.

In another study, six different red algal species from cold-temperate regions and with different zonation patterns were cultivated under identical culture conditions and exposed to similar irradiance of UVB (van de Poll et al. 2001). The inhibition of growth became stronger in accordance with the position on the shore these algae usually take in the field. The two species from the upper sublittoral or even lower eulittoral zone (*Palmaria palmata* and *Chondrus crispus*) did hardly exhibit inhibition in growth, whereas growth in the species from the middle sublittoral zone (*Phyllophora pseudoceranoides* and *Rhodymenia pseudopalmata*) was inhibited up to 50%. In the deep-water algae *Phycodrys rubens* and *Polyneura hilliae*, growth was inhibited almost completely. Curiously, pronounced accumulation of damaged DNA, expressed as thymine dimer formation, was only found in these two species. Apparently these true deep-water algae do lack mechanisms to shield the DNA from UVB exposure or to repair already damaged DNA e.g., by the activity of repair-enzymes (as e.g., DNA phytolyase). This study illustrates how genetic pre-adaptation is setting the frame in which acclimation may occur. The fact that not any acclimation to UVB was shown by these species points to a strong degree of adaptation to low irradiance environments.

Modulation of ecophysiological reactions towards variation in abiotic factors is conditioned by genetic adaptation. This is also visible in two red algal species from Spitsbergen with slightly different vertical zonation preferences (Karsten et al. 1999). *Devaleraea ramentacea* as a species from shallow waters is permanently equipped with high activities of reactive oxygen scavenging superoxide dismutase (SOD). This high but static activity is reasonable for a species from shallow waters, where usually strong variation in abiotic conditions, and thus the onset of stressfull conditions to photosynthesis resulting in increased ROS production, is more likely than in more stable deeper waters. However, maintaining a protective systems on such a high level throughout the year is probably energetically costly. Thus, species which are not permanently exposed to stressfull conditions, e.g., in deeper waters, may favour the strategy to respond to abiotic stress and to increase protective strategies, like SOD activity, only when they are needed during times of e.g. high UV irradiance. *Palmaria palmata* inhabiting slightly deeper waters than *D. ramentacea* is applying this strategy (Karsten et al. 1999). In the case of the estuarine red alga *Gracilaria chilensis*, which is subject to extremely changing light conditions both during tidal cycles and seasonally, exposures to surface UVB irradiances induce marked reductions in photosynthesis (Gómez et al. 2005a). Although the species can display rapid acclimation mechanisms, the constitutively high pigment contents and low concentrations of sunscreen substances (e.g., MAAs), clearly suggest that the *Gracilaria* retains its shade adapted characteristics, probably as a consequence of the normally turbid waters at the estuary.

*Acclimation of photosynthesis to UV radiation*

In seaweed species inhabiting a flexible environment, i.e., the shallow water zones down to approx. 15 m depth, acclimation to changing abiotic conditions is important, to adjust photosynthetic performance in order to maintain energy supply for growth, but also to prevail under periods of stressfull conditions. The ability for fast acclimation to increased UV irradiance has been demonstrated in the Arctic/cold-temperate kelp *Alaria esculenta* from Spitsbergen (Bischof et al. 1999). It was shown that its macrothalli are able to adjust photosynthetic performance to changes in irradiance at their respective growth site. This capability may represent one prerequisite for this species to establish over a wide depth range and also to endure the seasonal variation of radiation conditions (Chapman and Lindley 1980; Falkowski and LaRoche 1991; Klöser et al. 1996; Bischof et al. 1998b, 1999).

Within the brown algae studied so far, two different responses were observed in the process of acclimation of photosynthetic activity to changing radiation conditions. Firstly, the rate of recovery from UV-induced photoinhibition increases. Secondly, the degree of inhibition becomes smaller (Bischof et al. 1998b, 1999). Increases in the rate of recovery may result from an activation of different repair mechanisms, counteracting the impact of UV-exposure by a faster replacement of damaged molecules. The molecular mechanism responsible may be the same as discovered in the cyanobacteria *Synechocystis* sp. and *Synechococcus* sp. In both species, it was found that exposure to moderate doses of UVB results in an increased turnover rate of the $D_1$ and $D_2$ reaction centre subunits of PS II, thus, rapidly replacing damaged protein by newly synthesised polypeptides (Campbell et al. 1998; Máté et al. 1998). The latter authors found that UVB induces the transcription of psbA genes, which encode the $D_1$ reaction centre protein of PS II. Although comparable studies are lacking for macroalgae, it may be that a similar response provides an explanation for the increasing rate of recovery in the studied brown algal species. However, this mechanism may only be successful as long as UVB exposure does not induce stronger damage to DNA, thus impairing gene expression. Results also showed, that in algae previously acclimated to high PAR, additional UV-exposure rather results in a delay of the recovery process than in a further inhibition of photosynthesis (Bischof et al. 1999). These findings support data from field experiments on *Fucus distichus* from Spitsbergen, indicating that at their natural growth site in the eulittoral zone photoinhibition is mainly caused by high irradiances of PAR and natural UVB causing a delay in recovery (Hanelt et al. 1997a). The observed delay in recovery is indicative for damage to the $D_1$ protein (Aro et al. 1993). Under UVB exclusion the rate of $D_1$ degradation mediated by solar radiation was found to be as much as 30% slower than under full sunlight (Greenberg et al. 1989), thus supporting those results for high light acclimated algae.

In contrast to subtidal species, intertidal brown algae, have to cope with highly changing solar radiation scenarios on a short term basis. In this sense, high PAR irradiances become as ecological important as UV radiation. Flores-Moya et al. (1999) observed a significant delay in recovery from photoinhibition in the brown alga *Dictyota dichotoma* from Southern Spain, when samples were exposed to solar radiation depleted from the UVB range and subsequently transferred to dim light conditions. Recovery in samples receiving either the whole solar spectrum or PAR only, recovered at the same rate. This indicates the presence of complex synergistic effects involved in the inhibition of photosynthesis in the field, which need to be studied further. On the other hand, it must be emphasized that brown algae include many large, perennial species, which exhibit complex responses to UV-exposure. For example, in the southern kelp *Lessonia nigrescens*, the photosynthetic responses to seasonally changing UV conditions form part of the suite of adaptations along with its ontogenic development, i.e., the alga has a complex UVB exposure history characterized by high levels of UVB in summer and low levels in winter (Huovinen et al. 2006). Moreover, gradients in UV tolerance have been reported along of the massive thallus (Gómez et al. 2005b). Thus, the morpho-functional factors involved in UV photobiology of brown algae are important and have to be considered in further studies in order to evaluate more accurately the effects of enhanced UVB on coastal primary productivity.

A common response observed in the brown algal species during acclimation to UV radiation is the reduction in the degree of photoinhibition. This effect may be explained either by the activation of the antioxidative response, increased activity of repair and recovery mechanisms counteracting the inhibitory effects (see above), or by the formation of UV-screening compounds (Lesser 1996).

*Mycosporine-like amino acids (MAA)*

One of the most important physiochemical acclimation mechanism against biologically harmful UV radiation involves the biosynthesis and accumulation of UV-screening substances. Typically absorbing in the UVA and UVB range, these biomolecules were invoked to function as passive shielding solutes by dissipating the absorbed short wavelength radiation energy in form of harmless heat without generating photochemical reactions (Bandaranayake 1998). The most common photoprotective sunscreens in Antarctic macroalgae are the mycosporine-like amino acids (MAAs), a suite of chemically closely related, colourless, water-soluble, polar and at cellular pH uncharged or zwitterionic amino acid derivatives. MAAs exhibit a high molar absorptivity for UVA and UVB, and have been reported as photochemically stable molecules, which are prerequisites for their sunscreen function (Conde et al. 2000). While MAAs have been mainly observed in numerous Antarctic (Karentz et al. 1991a; Hoyer et al. 2001) and cold-temperate Rhodophyta (Huovinen et al. 2004), Phaeophyta and most Chlorophyta typically lack these compounds, except the green alga *Prasiola crispa* ssp. *antarctica* which contains high concentrations of an unique MAA with an absorption maximum at 324 nm (Hoyer et al. 2001; Karsten et al. 2005). Many Phaeophyta synthesise photoprotective phlorotannins under UV exposure (Pavia et al. 1997; Schoenwaelder 2002b; Schoenwaelder et al. 2003), this strategy will be reviewed in detail below.

The function of MAAs as intracellular screening agents has been inferred from a decrease in concentration with increasing depth (Hoyer et al. 2001, 2003). Supra- and eulittoral Antarctic red algal species experience the strongest insolation, and consequently synthesise and accumulate very high MAA contents, which generally are positively correlated with the natural UV doses (Karsten et al. 1998a; Huovinen et al. 2004). In contrast, many taxa growing in the sublittoral are physiologically not capable to produce MAAs, which well explains their strong sensitivity, for example, of photosynthesis against ambient solar radiation. These Rhodophyta avoid any UV exposure, and, hence, there is no physiological need to synthesise and accumulate metabolically expensive nitrogen-containing MAAs. This in turn would save energy to better support other essential pathways such as light-harvesting phycobilisomes to guarantee sufficient PAR absorption under the prevailing low-light conditions.

While juvenile lateral fronds of the red alga *Palmaria decipiens* collected in Antarctic winter contained low concentrations of UV-absorbing compounds, mature plants in late spring and summer exhibited significantly higher values indicating strong seasonal effects (Post and Larkum 1993), which may be related to the changing daylengths and radiation conditions. Based on the MAA concentrations and the induction patterns after exposure to different radiation conditions Antarctic Rhodophyta can be physiologically classified in 3 categories (Hoyer et al. 2001): Type I—no MAAs at all; Type II—MAAs inducible in variable concentrations, and Type III—permanently high MAA values. While Type I represents deep-water red algae, Type II and III species are growing in the supra- and eulittoral zone. Experiments with Antarctic Rhodophyta under defined radiation sources indicate that the induction, biosynthesis and accumulation of MAAs is a very flexible and species-specific process. While some taxa synthesise MAAs particularly under UVB, others prefer UVA or higher PAR only (Hoyer et al. 2003). Although experimental evidence for a particular trigger mechanism as well as details for the biosynthetic pathway for individual MAAs are still missing, it is reasonable to assume that a signal transduction pathway must be involved in the overall process leading to high MAA concentrations. Due to the different types of MAA induction patterns the presence of various photoreceptors, most probably between the blue light and UVB wavelengths, have to be taken into consideration (Kräbs et al. 2002).

Not the whole red algal thallus is uniformly responding to the ambient solar conditions, but especially young apical or marginal zones, i.e., growing cells synthesise and accumulate MAAs leading to cross sectional and longitudinal concentration gradients (Hoyer et al. 2001). Older tissue regions exhibit thicker cell walls and a leathery texture, and are therefore optically well protected. In contrast, higher MAA concentrations in the most exposed outer cortex are essential to guarantee protection of the delicate meristematic cells.

Besides the stimulating effect of increasing solar radiation on the biosynthesis and accumulation of MAAs in macroalgae other environmental factors may also act as controlling parameter. Particularly lower temperatures have been experimentally proven to stimulate the MAA concentration of Antarctic Rhodophyta (Hoyer, unpublished data). Nutrient availability may also affect the MAA contents (Korbee et al. 2005). Some MAAs also exhibit antioxidative activity (Dunlap and Yamamoto 1995). However, further functional abilities of MAAs are unexplored in macroalgae.

Also in this example, the genetically determined ability to synthesise UV-screening MAAs in the different species of seaweeds is closely related to the spatial structure of the algal community in the field. In summary, the comparison of the species-dependent ability to form UV-screening compounds under laboratory and field conditions provide strong indications for differential genetic preadaptations to the potentially harmful radiation at the natural growth site.

*Phenolic compounds*

A special class of polyphenolic compounds are phlorotannins, which are exclusively found in brown seaweeds (Ragan and Glombitza 1986). Phlorotannins are secondary metabolites and occur in tissue concentration of up to 20% of the algal dry weight. Several functions are commonly accepted, including a role in adhesion and a strengthening role in cell walls (Schoenwaelder 2002b). Phlorotannins absorb UV radiation, mainly UVC and partly UVB, with maxima at 195 nm and 265 nm (Ragan and Glombitza 1986; Pavia et al. 1997; Henry and Van Alstyne 2004). As tannins of higher plants, phlorotannins possess a high antioxidant activity. Thus, phlorotannins are important for scavenging cell toxic reactive oxygen species (ROS), such as superoxide anion radicals produced by harmful UVB radiation. Therefore, phlorotannins are proposed to function in protecting against excess irradiance, in particular ultraviolet radiation, by screening UV radiation and/or by being an antioxidant.

We suggest four strategies to consider phlorotannins as UV-protecting compounds (1) a generally high tissue concentration of phlorotannins that absorb harmful radiation and prevent cell damages, (2) an induction of phlorotannins stimulated by harmful radiation, (3) an exudation of phlorotannins in the surrounding medium shielding harmful radiation, or (4) an excess inclusion of phlorotannins in cell walls shielding harmful radiation.

High concentrations of phlorotannins in the outer cell layers protect *Hormosira banksii* from sunburn during Australian summer

(Schoenwaelder 2002a). While the outer cell layers are damaged and disrupted by sunlight, phlorotannins are released from the phlorotannin containing vesicles, the physodes, into the cytoplasma where they cause oxidative burn. Oxidised phlorotannins become brownish and form a dark brownish protective cell layer for the underlying photosynthetic tissue. Similarly, the high density of physodes at the periphery of egg and zygote cells of *Fucus spiralis* is most probably responsible for the high tolerance to UV-exposure, in contrast to *F. serratus*, in which physodes are less abundant (Schoenwaelder and Wiencke 2000; Schoenwaelder et al. 2003).

An induction of phlorotannins after a 2-week exposure of artificial UVB radiation was first described in *Ascophyllum nodosum* (Pavia et al. 1997). A weaker response was found in *A. nodosum* when exposed to natural UVB radiation (Pavia and Brock 2000). In that study significant differences in the phlorotannin content of UVB treated and untreated individuals were found after 7 weeks of exposure to natural UVB radiation, while only slightly increased phlorotannin concentrations were measured in the UVB treated individuals at week 2 and week 4. An induction of phlorotannins due to UVB and UVA radiation was also described for *Macrocycstis integrifolia* (Swanson and Druehl 2002). In contrast, no induction of phlorotannins was found in juveniles and embryos of *Fucus gardneri* after 3-week exposure to UV radiation, while growth of embryos was inhibited and growth of juveniles was not affected by UVB radiation (Henry and Van Alstyne 2004). An increase in the size of physodes was observed in various Laminariales from Spitsbergen after UVB exposure indicating an induction of phlorotannin synthesis (Wiencke et al. 2004a). This has recently been verified in the UV tolerant species *Alaria esculenta* and *Saccorhiza dermatodea*. The absorbance of zoospore suspensions from these species increased considerably after UVB exposure, whereas the absorbance of spore suspensions of the UVB sensitive species *L. digitata* did not change at all (Roleda et al. 2006c).

An exudation of phlorotannins as response to artificial UVB radiation was observed in *Macrocyctis integrifolia* (Swanson and Druehl 2002). As a result UVB transmission through the water column was reduced, thereby protecting germinating meiospores of *Laminaria groenlandica* against harmful UVB radiation. Similarly, biofilters containing phloroglucinol (the monomer of phlorotannins) mitigated the harmful effect of UV-exposure on developing zygotes and embryos of *Fucus serratus* (Schoenwaelder et al. 2003). Biofilters made of UV transparent acrylic sheet, filled with zoospore suspensions of *S. dermatodea*, *A. esculenta*, *L. digitata* or phloroglucinol showed a varying capacity to protect zoospore cultures from the lethal effects of UVB (Roleda et al. 2006c). Generally, high phlorotannin content and high exudation rates might reflect an adaptation of seaweeds to enhanced UV radiation. Induction and variable exudation rates of phlorotannins reflect the acclimation potential to environmental changes.

Do phlorotannins play a role in the determination of depth zonation of brown algae? Evidence was found in zygotes and embryos of *Fucus* species (Schoenwaelder and Wiencke 2000; Schoenwaelder et al. 2003). *F. spiralis* from the high intertidal zone is less sensitive to UV-exposure and contains much more physodes than *F. serratus*, which is found further down the shore. Similarly, zoospores of *Laminaria digitata*, growing on Helgoland in the upper sublittoral, exhibit a strong absorbance below 300 nm, indicative of phlorotannins, compared to zoospores of the mid and lower sublittoral *Laminaria* species (Roleda et al. 2005a). Phlorotannins may also play a role for the UV-sensitivity of zoospores of Arctic Laminariales from different depths (Wiencke et al. 2004a). Similarly, higher phlorotannin concentrations were found in *Desmarestia anceps* collected in shallow waters compared to deep water samples, which might indicate an induction of phlorotannins due to higher irradiances and/or UV exposure (Fairhead et al. 2005). Conversely, *D. menziesii* shows no differences in phlorotannin concentrations in samples from different collecting depth and even exhibits lower concentrations than *D. anceps*, which usually occurs in deeper waters (Fairhead et al. 2005).

In contrast to the data sets available on potential UV-screens in red and brown algal species information on specific UV-absorbing compounds in marine green algae is still very limited. Very recent studies demonstrated the solely UVB-induced increase in thallus absorption in *Ulva pertusa* with a maximum at 295 nm (Han and Han 2005), however, the responsible screening compound is still unidentified.

*DNA damage and repair*

As outlined above, a particularly hazardous event of UV-exposure is the induction of DNA damage. Thus strategies in order to prevent damage or to efficiently repair existing damage represent acclimation mechanisms of crucial importance. Differential induction and accumulation of UVB-induced cyclobutane-pyrimidine dimers (CPD) were measured in several red macroalgae from Brittany and Spitsbergen (van de Poll et al. 2001, 2002). In the course of 15 days repeated exposure to artificial UV radiation, no accumulation of DNA damage but rather a decrease in CPD concentration was observed in the temperate littoral species *Palmaria palmata* and *Chondrus crispus*. Conversely, an approximately 6-fold increase in the amount of CPD was observed in sublittoral species *Phycodrys rubens* and *Polyneura hilliae* (van de Poll et al. 2001). In Arctic red macrophytes, the amount of solar radiation-induced CPD concentrations in *Devaleraea ramentacea*, *Palmaria palmata*, *Odonthalia dentata*, *Coccotylus truncatus* and *Phycodrys rubens* (van de Poll et al. 2002) is related to their upper depth distribution limit in Kongsfjorden described by Wiencke et al. (2004b). In shallow coastal waters, blooms of floating *Ulva* are exposed to the full solar radiation. The amount of thallus DNA damage is, however, relatively low ranging from 1.0 to 1.88 CPD $Mb^{-6}$ depending on solar exposure of the investigated canopy layer (Bischof et al. 2002b) compared to those reported in red macrophytes (van de Poll et al. 2001, 2002). These data suggest that the low sublittoral habitat of the Ceramiales (*P. hilliae*, *C. truncatus* and *P. rubens*) is primarily due to their lack of tolerance to UV radiation and that UV protection mechanisms are not sufficient to prevent the accumulation of DNA-damage in these species.

Obviously, UV-susceptibility of DNA damage is highly depending on the respective developmental stage of the species under investigation. The impact of UVB-induced DNA damage on the early life stages of macroalgae is important in shaping up community structure and zonation pattern. DNA damage in carpospores of eulittoral *Mastocarpus stellatus* and *Chondrus crispus* (Roleda et al. 2004) was lower compared to zoospores of the sublittoral Laminariales in Helgoland (*Laminaria digitata*, *L. saccharina* and *L. hyperborea*; Roleda et al. 2005a) and Spitsbergen (*Saccorhiza dermatodea*, *Alaria esculenta* and *L. digitata*; Roleda et al. unpublished data). The UV-sensitivity of the carpospores and zoospores were related to the depth distribution of the foliose gametophyte and adult sporophytes, respectively. In the subsequent life history stage investigated, young gametophytes and sporophytes were less susceptible to UVB-induced DNA damage compared to spores. No detectable CPDs were observed in the young foliose gametophyte stages of the eulittoral *M. stellatus* and *C. crispus* (Roleda et al. 2004). Conversely, the remaining tissue DNA damage among juvenile Laminariales sporophytes was observed to be dependent on the thallus thickness and optical property (Roleda et al. 2005b, 2006a, 2006b). Increasing thallus thickness and opacity (in relation to available cell-bound UV-absorbing compounds) minimise UV-effects where outer phlorotannin-rich cortical layers can selectively filter short UV-wavelength from reaching the UV-sensitive targets (i.e., chloroplasts).

Less genetic damage was incurred in diploid carpospores compared to haploid zoospores (Roleda 2006). Zoospores were, however, found to be more efficient in DNA damage repair. In a sexual organism, the advantage of both ploidy states can be combined by spending much of the life cycle in the haploid state, then fusing to become diploid. During the diploid state DNA damage can be repaired, since there are two copies of the gene in the cell and one copy is presumed to be undamaged (Long and Michod 1995). In the life history of Laminariales, haploid zoospores were more sensitive to DNA damage compared to the diploid young sporophytes.

Between the investigated Gigartinales, the lower DNA damage and effective DNA damage repair mechanism in carpospores of *M. stellatus* is responsible to its recruitment success and colonization of the upper eulittoral zone effectively changing the rocky intertidal biotope of Helgoland (Roleda et al. 2004).

### *Acclimation via morphogenetic variation*

Several morphogenetic effects have been described for higher plants grown under UVB irradiation. Compared to white light, UVB exposed plants exhibit reduced leaf area and stem growth, but increased leaf thickness (Tevini and Teramura 1989; Mepsted et al. 1996). Information on UV-induced morphogenetic effects on the thalli of marine macroalgae is very limited. Studies on the brown alga *Alaria esculenta* have shown that UVB radiation results in reduced growth in length and a significant increase in fresh weight, indicative for increasing thallus thickness (Michler et al. 2002). In previous studies on various Laminariales and seagrasses it was shown that thicker thalli generally exhibit a higher UV-tolerance than thin thalli (Dawson and Dennison 1996; Dring et al. 1996a; Hanelt et al. 1997b). On the other hand, blades of the southern kelp *Lessonia nigrescens* are more UV sensitive than stipes and holdfasts, which has been though as a strategy to minimise mortality. In this species, the higher UV damage is concentrated in the phylloids, which are transient and with high turnover rate, whereas the massive stipes and holdfast, which are supporting structures, are less exposed to UV radiation (Gómez et al. 2005b). In general, the individual effects of UVB radiation on algal thalli with differences in morphological features are largely unexplored, thus we lack information on how UVB might influence the individual thallus structure of the individual seaweed.

## Vertical versus latitudinal distribution

### *Depth zonation*

Ecophysiological studies indicate a general correlation between stress tolerance and vertical distribution of seaweeds (Davison and Pearson 1996; Hanelt 1998). Hitherto, there is agreement that the species sensitivity to solar radiation stress is a function of depth distribution (Dring et al. 1996b; Larkum and Wood 1993; Hanelt et al. 1997a, c; Hanelt 1998; Bischof et al. 1998a; Yakovleva et al. 1998). Moreover, some authors regard solar UVB as one of the most important factors controlling the upper distribution limit of seaweeds in the field (Maegawa et al. 1993). Therefore, it is reasonable to assume that increased UVB, penetrating deeper into the water column, may result in a shift of the upper distribution limit of single species to greater water depths.

Larkum and Wood (1993) were the first who have stressed the correlation between UV-tolerance and the depth-zonation of marine macroalgae. For the vertical distribution of tropical seagrasses UV radiation was also proven to be an important factor (Dawson and Dennison 1996). Dring et al. (1996b) showed that sensitivity to UV in red algae growing around the island of Helgoland (Germany) varies with species and depth of collection. As for UVB radiation, investigations on the photoinhibition induced by high levels of PAR also shows a correlation between depth-zonation and the ability for dynamic photoinhibition of macroalgae both from sublittoral (Hanelt 1992, 1998; Hanelt et al. 1994, 1997a) and intertidal populations (Gómez et al. 2004)

In support of this concept the green algae *Enteromorpha bulbosa* and *Acrosiphonia arcta* which occur in the middle and lower eulittoral at the Antarctic Peninsula (Wiencke and Clayton 2002) do almost show no negative UV-effects on photosynthesis and are able to acclimate to even further elevated UV-exposure within hours or days (Bischof et al. 1998a). In contrast, the brown algae *Desmarestia antarctica* and *D. anceps* are described for the middle sublittoral zone off the Antarctic Peninsula (Klöser et al. 1996) but grow mostly in greater depths and occur only occasionally in depths shallower than 17 m. In these depths biologically relevant doses of UVB radiation occur only in very transparent waters, under clear skies and at a high solar declination (Karentz 1989). This might explain why *D. anceps* is quite sensitive to UVB radiation. The red alga

*Gymnogongrus antarcticus* occurs from the upper sublittoral zone down to 20 m (Klöser et al. 1996), i.e., the upper distribution limit is similar to that of the brown algae *D. anceps* and *Himantothallus grandifolius* and upon UVB exposure these three species show a similar inhibition rate of photosynthesis. *H. grandifolius* is found at 5 m (Lamb and Zimmermann 1977) with the lowest depth at 90 m (Zielinski 1990). This zonation pattern is in line with its high sensitivity. Finally, the red algae *Phycodrys austrogeorgica* and *Delesseria lancifolia* are described for the middle sublittoral zone (Delaca and Lipps 1976; Zielinski 1990; Klöser et al. 1996), but they grow under canopy plants such as *D. anceps* and *H. grandifolius*. This explains their extreme sensitivity to UV radiation. These plants may lack all protecting mechanisms against excessive radiation, because recovery from UV-exposure is poor in both species. The observed differences in species dependent UV-sensitivity are as outlined above genetically determined.

The decisive role of UVB radiation in determining vertical zonation patterns in seaweeds was recently evidenced in studies on the UV effects on the particularly sensitive developmental and reproductive stages. Only few studies were conducted using the unicellular propagation units of seaweeds, although it is widely recognised that the early developmental stages are the most susceptible to a variety of anthropogenic stresses (Coelho et al. 2000).

Seaweed propagules, spores, gametes and zygotes, are the unicellular products of asexual and sexual reproduction and have essential functions in the life-history of seaweeds with respect to dispersal, settlement, attachment, survival and recruitment (Clayton 1992; Norton 1992). They are naked cells, bounded by a plasma membrane, and in some species covered by a layer of mucilage. Their size ranges from 2 μm to >250 μm in diameter and determines the sinking rate in the water column (Okuda and Neushul 1981). Brown and green algal spores and gametes are flagellated and usually small, between 3 μm and 10 μm in diameter (Henry and Cole 1982; Clayton 1992). Zygotes of the brown algal order Fucales have diameters between 60 μm and 250 μm (Clayton 1992; Schoenwaelder et al. 2003). Spores of red algae are between about 10–100 μm in diameter (Clayton 1992). A positive phototaxis is found in spores of the brown algal genus *Ectocarpus* (Müller 1977) and the green algal genus *Ulva* (Evans and Cristie 1970), a negative phototaxis is found in some primitive Laminariales including *Saccorhiza dermatodea*. However, the capability for active swimming over long distances is rather limited. Even motile propagules have little control over their fate. The range of dispersal is at least 200 m and is driven mainly by currents and water motion (Norton 1992; Frederiksen et al. 1995). The number of chloroplasts per propagule can be very low. The zoospores e.g. of the more advanced families of the Laminariales, the Alariaceae, Laminariaceae and Lessoniaceae only contain one small chloroplast (Henry and Cole 1982), with comparatively low photosynthetic activity (Amsler and Neushul 1991; Wiencke et al. 2000). Storage lipids are certainly the main energy source of spores, supporting swimming and potentially germination processes (Brezezinski et al. 1993; Reed et al. 1999).

Among the different stages in the life-cycle of seaweeds the unicellular propagules are clearly the stages most susceptible to UV radiation. Germination of *Ectocarpus rhodochondroides* spores is inhibited by UVB, while adult specimens survive (Santas et al. 1998). Similarly, the photosynthetic efficiencies ($F_v/F_m$) of sporophytes and zoospores of *Laminaria digitata* differ strongly when exposed to PAR, PAR + UVA or PAR + UVA + UVB (Wiencke et al. 2000). Irrespective of the radiation treatment large sporophytes show always considerably higher $F_v/F_m$ values compared to zoospores. Motile zoospores of *Lessonia nigrescens* and *L. trabeculata* from Chile are more UV susceptible than settled spores, gametophytes and young sporophytes (Véliz et al. 2006). The macrothalli of red algae *Mastocarpus stellatus* and *Chondrus crispus* are relatively UV tolerant whereas their carpospores are not (Bischof et al. 2000b; Roleda et al. 2004). Similarly, photosynthetic efficiency of zoospores of the green alga *Ulva intestinalis* exhibits an up to 6-fold higher UVB sensitivity compared to the mature thalli (Cordi et al. 2001). Interestingly, the UV susceptibility of the gametes is even greater.

The UV susceptibility of the photosynthesis and of the DNA in spores is unequivocally related to the depth distribution of the macrothalli as outlined above (see chapters 5.1, 5.1.1.3; Bischof et al. 2000a; Wiencke et al. 2000; Roleda et al. 2004, 2005a). Other negative effects of UV radiation especially on macroalgal spores presumably important for determination of the upper distribution limit concern the cytoskeleton. The phototactic response of zoids of the brown algae *Scytosiphon lomentaria* and *Petalonia fascia* is negatively influenced by UVB (Flores-Moya et al. 2002). Moreover, the motility of zoospores of *L. saccharina* is affected by UVB depending on the actual UV doses (Makarov and Voskoboinikov 2001). This might be related to the observation by Huovinen et al. (2000) who reported that the nuclear division of spores of *Macrocystis pyrifera* is inhibited after UVB exposure. Both nuclear division and the activity of the flagellar apparatus depend on a functional microtubular system, which might be damaged by UV. This would explain also the drastic affects on zygotes observed in *Fucus serratus* and *F. distichus* (Schoenwaelder et al. 2003). No polarisation has been observed in UV exposed zygotes, rather, they remain spherical and there is no further development. Similarly, amphibian and fish zygotes remain undifferentiated and this has been related to UV effects on the microtubuli (Scharf and Gerhard 1983; Strahle and Jesuthasan 1993). The actin cytoskeleton may be affected as well as studies on other fucoids suggest. Actin inhibitors prevent polarisation, cross wall formation and vesicle movement (Schoenwaelder and Clayton 1999), the same effect as after UV treatment of zygotes of the two *Fucus* species (Schoenwaelder et al. 2003).

In spores the balance between the damaging effects of UV radiation and the various repair and protective mechanisms is indicated by the integrative parameter germination. If germination is not inhibited after UV exposure the repair and protective mechanisms are strong enough to outweigh the damaging effects of UV. Spore vitality after UV exposure in *M. stellatus* is higher compared to *C. crispus*, probably due to the higher capability for DNA repair in *Mastocarpus* (Roleda et al. 2004). In this context the size of the algal propagules may be of importance as the UV susceptibility of zoospores of various Laminariales from the eastern Pacific depends on the size of the spores (Swanson and Druehl 2000). The spores most tolerant to UV stress come from shallow water species, whereas the progeny of kelps occupying low-level UV-environments exhibit a lower germination capacity after UV stress. These results are similar to studies of phytoplankton species, in which larger organisms are more tolerant to UV exposure (Karentz et al. 1991b).

A clear dependence of the UV susceptibility of germination and growth of sporelings of coralline algae on the radiation conditions in the habitat of the various species has been demonstrated by Bañares et al. (2002). Spores from species in sun-exposed habitats in the eulittoral were more UV tolerant than spores from a species growing in shaded crevices. Similarly, zygotes of *F. serratus* from the mid eulittoral on Helgoland show an abnormal development after UV exposure, whereas zygotes of *F. spiralis*, a species from the upper eulittoral much stronger exposed to UV, is not affected by the same UV treatment (Schoenwaelder et al. 2003). A clear dependence of the upper depth distribution limit of seaweeds on the UV susceptibility of their spores has also been proven for the three *Laminaria* species from Helgoland mainly due to the different DNA repair capacities and the different content of UV absorbing compounds (Roleda et al. 2005a). Similarly, the UV tolerance of different brown algae from Spitsbergen determines the upper depth distribution limit as indicated by experiments in the laboratory (Wiencke et al. 2000, 2004) and in the field (Wiencke et al. 2006).

*Microscale gradients in UVB exposure*

Pronounced acclimation of photosynthesis to UV exposure does also take place along microscale gradients as for example in algal mats: Green algal mats do frequently occur as a result of eutrophication in sheltered coastal lagoons. Within these mats usually a steep gradient of solar radiation occurs (Vergara et al. 1997): Top layers are exposed to high surface irradiance, whereas bottom layers are permanently exposed to very low light conditions or even stay in darkness. It has been show for the two bloom-forming green algal species *Ulva rotundata* and *Chaetomorpha linum* that UVB irradiance substantially contributes in structuring these mat-like canopies (Bischof et al. 2002b, 2003, 2006). Inside the mats, strong gradients are visible for solar (UVB) radiation, but also for physiological responses: loss in photosynthetic pigments and proteins is strongly pronounced in top layers and is diminished with increasing depth inside the mat. Photosynthetic activity is inhibited in the top layers, performs at maximum rates in intermediate thallus layers and, due to light limitation, is again strongly reduced in bottom layers (Krause-Jensen et al. 1996; Vergara et al. 1997; Bischof et al. 2002b). A specific response to the degree of UVB exposure inside these mat-like structures is the specific activity of superoxide dismutase (SOD), an enzyme responsible for the scavenging of superoxide anions generated in photosynthesis under stress full conditions (Bischof et al. 2003, 2006). Depending on the degree of UVB exposure and inhibition of photosynthesis along the depth gradient, the activity of SOD becomes stimulated. Two mechanisms might be involved in this response: either the enzyme becomes activated due to a light effect (due to exposure to short wavelength radiation) or it is the presence of previously generated oxygen radicals (ROS) which act as trigger for increasing SOD activity (Bischof et al. 2006). For higher plants, it was shown that ROS play an important signalling role in signal transduction pathways, in order to respond to UVB exposure (Mackerness et al. 1999).

*From the tropics to the poles*

As the problem of ozone depletion is rather restricted to the polar and temperate regions, organisms inhabiting the (shallow water zone of) tropical regions are unlikely to be affected by seasonally *enhanced* UVB irradiance, but are of course permanently exposed to much higher UV levels than encountered in polar regions (even under ozone depleted conditions). Thus, tropical species must have developed particularly effective protection strategies in response to high UVB irradiance. Unfortunately, data on UVB protection in tropical species as well as comparative studies on related species from different geographical regions are extremely scarce and the limited information available is not conclusive. In a study on different species of the green alga genus *Cladophora* no hints were found for a particular higher UVB tolerance of a true tropical species (*C. zolingerii* from the Philippines) compared to its cold-temperate congener *C. rupestris* (Bischof and van de Poll unpublished data). Preliminary data on UVB absorption characteristics within the cell wall of both species did not provide any hint for a more effective UVB shielding in the tropical species. Moreover, also physiological parameters tested did not point to a particularly higher UVB tolerance of the tropical species compared to the congener from temperate waters.

Like at the species level, studies on latitudinal UV-patterns at the community level are extremely scarce. To our knowledge, only the study by Wahl et al. (2004) investigated UV-effects on shallow-water macro-epibenthic assemblages at different bio-geographical regions. Their study revealed a consistent pattern of UV-effects at both hemispheres. Species richness and community biomass were negatively affected by UV-treatments. Surprisingly, effects by UVA were more detrimental than those of UVB. UV-effects were transient, disappearing at 80% of all sites within 2–3 months, but persisted at one polar (Norway) and one tropical (China) site, suggesting lack of latitudinal patterns of UV-effects at the community level.

## Ecological implications—UVB and the structure of seaweed communities

The consequences of enhanced UVB exposure to ecosystem function are still largely unexplored and hypothetical. However, based on the studies accomplished so far the following assumptions can be made.

### Potential effects on primary productivity

Throughout the previous studies, it is shown that all species, which have to withstand UVB in the field (i.e., the species inhabiting the intertidal and the upper sublittoral zone) possess different and largely efficient mechanisms for acclimation to respond to changes in light climate. However, formation of screening compounds as well as the development of further protective mechanisms require additional energy costs, which may result in reduced growth and primary productivity (Roleda et al. 2006a). This problem is still largely unaddressed but an important field of future investigation. As long as the energy costs for applying protective mechanisms remain unknown a reduction of seaweed productivity in response to increased UVB levels in future cannot be excluded.

### Reduced reproductive success and shifts in age structure

The information available on UV-susceptibility of different developmental stages indicates the unicellular spores and zygotes as being most sensitive. Other information on how UV-exposure may affect reproduction and the timing of developmental cycles is hardly available. It is obvious that increased UV-induced spore mortality will result in impaired reproductive success, but may also affect the age structure of seaweed populations (Wiencke et al. 2000, 2004a). Due to high interannual variations in light climate particularly in polar areas (Hanelt et al. 2001) high UV-irradiance may affect zoospore survival in shallow waters in 1 year, as in another year it may not. Thus new recruits are only likely to develop to adult sporophytes if they are protected from high UVB exposure. In certain years with particularly high fluences of UVB reaching the benthic communities a new generation of recruits might therefore fail to develop, while larger age and size classes of sporophytes were rather unaffected due to the already accomplished acclimation on a physiological and morphological base (Altamirano et al. 2003a, b).

### Downward shifts in depth distribution

As outlined above there is now evidence, that the species sensitivity to solar radiation stress is a function of depth distribution (Larkum and Wood 1993; Dring et al. 1996b; Hanelt et al. 1997a, c; Hanelt 1998; Bischof et al. 1998a; Wiencke et al. 2006). For several physiological parameters a strong correlation of UVB sensitivity and vertical position on the shore was shown for numerous species (van de Poll et al. 2001, 2002; Bischof et al. 1998a, Gómez et al. 2004, Wiencke et al. 2006). Therefore, it is reasonable to argue that increased UVB, penetrating deeper into the water column, results in a shift of the upper distribution limit of single species to greater water depths. However, at least in the macrothalli several acclimation processes can counteract radiation stress, but the particular sensitivity of spores may be the decisive aspect in this scenario, thus preventing recruitment in shallow waters with high UVB irradiances. Through this process, elevated UVB may result in a shift of seaweed communities to deeper waters (Wiencke et al. 2006).

## Succession

Species do not exist in isolation. Rather, multiple species form complex, interacting habitat-specific communities. Results from physiological studies at the species level might not reflect biological responses when repeated at the community level due to e.g., indirect or synergistic effects. Consequently, it seems reasonable that an ultimate assessment of UV-effects should be inferred at the community level (Bothwell et al. 1994). Open space represents a key resource for many species of seaweeds and its colonisation is characterised by a series of species replacements (Sousa and Connell 1992). Three alternative successional models have been proposed by Connell and Slatyer (1977). The facilitation model suggests that initially the propagules of a few species will recruit on open space and their existence will modify the substratum for the settlement of other species. The tolerance model predicts a neutral effect of early on later colonizers with the latter replacing early colonizers due to more efficient resource use by competitive exclusion. Under the inhibition model, early settlers pre-empt the substratum, hindering invasion of subsequent species. The predictions of these models may be altered by overlaying patterns of climatic or ecological factors (e.g., grazing Farrell 1991; Sousa and Connell 1992).

Only a few studies assessed UV-effects during species succession of shallow-water seaweed-dominated communities (Santas et al. 1998; Lotze et al. 2002; Molis et al. 2003; Molis and Wahl 2004; Wahl et al. 2004; Dobretsov et al. 2005). Surprisingly in all but one study (Dobretsov et al. 2005), UV-effects were only apparent during the early phase of succession. In the study by Dobretsov et al. (2005) it seemed very unlikely that later successional species could cope with UV-effects by acclimatizing at the same time to the UV-regime, e.g., by induction of UV-screening substances. Alternatively and more likely, transient UV-effects during community assemblage might result from intra- and interspecific differences in the susceptibility to UV-exposure for which experimental evidence exists. For instance, differential UV-sensitivities of carpospores influenced recruitment success of two competing Gigartinales-species, partly explaining vertical zonation patterns of both species (Roleda et al. 2004). Similarly, spores from deeper dwelling species exhibit higher mortality rates compared to spores from shallow water species (Wiencke et al. 2000, 2004a, 2006). Moreover, sorus of *Laminaria digitata* is completely opaque, resulting in a sudden and drastic change in UV-exposure ('UV-shock') for released zoospores (Gruber, pers. communication). Finally, several seaweed species, e.g. green filamentous forms, seem particularly well adapted to recruit at UV-exposed sites in the upper eulittoral, i.e., where UV-irradiance is strongest on the shore, and it was experimentally demonstrated that their abundance correlates positively with UV-exposure time (Molis et al. 2003). The invasion of sites high on the intertidal shore might be a selective advantage, as this guarantees a relative shorter exposure to grazers compared to sites lower on the shore. Thus, macroalgal propagules may show species-specific differential UV-sensitivities, which favour initial colonisation of empty space by UV-resistant species. Wahl et al. (2004) suggested that these early colonizers facilitate the recruitment of later successional seaweed and invertebrate species at the majority of their study sites by amelioration of UV-regimes due to protective shading. Similarly Vinebrooke and Leavitt (1999) concluded that diatom mats can precondition the substratum for macroalgal spores by provision of UV-free micro-climates.

The effect of UVB exposure on the early succession of macroalgae was also studied at a rocky intertidal platform at King George Island, Antarctica, revealing a significant reduction of species diversity based on the effects of UVB (Zacher et al. unpublished data). Species recruitment and dry weight was monitored over a period of 15 weeks on artificial substrates exposed in the intertidal zone. Natural UV-exposure was found to affect the density of the green alga *Monostroma hariotii* in the first 10 weeks of the experiment, whereas the density of red algal recruits decreased significantly due to UV after 8 and 15 weeks, respectively. Shannon diversity *H'* dropped as succession proceeded in the PAR + UVA + UVB treatment, increased slightly in the PAR + UVA treatment, and increased in the PAR treatment from the beginning until the end of the experiment. Furthermore, the treatment with PAR alone resulted in a significantly higher diversity at the end of the experiment than the treatment including the total UV-range. After 15 weeks the community excluded from UV radiation showed a significantly higher diversity, evenness and number of red algal germlings and species than communities exposed to the whole solar spectrum. This led to a significant dissimilarity in species composition between these two communities. Diversity was negatively influenced by both UVA and UVB radiation. The results show that Antarctic macroalgal recruits are particularly sensitive to UV-exposure during their first month of development, but that effects change during succession.

## Competition

Very few examples of UV-mediated changes in competitive abilities of macroalgae exist (Bischof et al. 2000b; Roleda et al. 2004). However, UV-exposure may quite commonly affect the competitive ability of a macroalgae, if UV-induced changes, e.g., production of MAAs, represent a metabolic cost for the alga. Moreover, the competitive ability of seaweeds may be negatively affected, if UVB has detrimental effects on growth. As a result, the algae will be shaded by more UV-resistant species and experience further reduced growth due to limited PAR-regimes, similar to what has been predicted for terrestrial plants (Caldwell et al. 1989).

Several species of algae are known to affect the community structure of nearby developing benthic assemblages (Wahl 2001). One possible mechanism of this biogenic neighbourhood effect is the exudation of metabolites that influence settlement. Little is known about UV-induced changes in chemical composition of macroalgal exudates and the possible indirect effects of UV-exposure on the structure and species composition of communities that establish in the vicinity of macroalgae. For example, tri-hydroxycoumarins (phenolics) excreted by the green alga *Dasycladus vermicularis* in response to enhanced radiation conditions protects other macroalgal species from UV-exposure (Pérez-Rodriguez 2000). UV-resistant seaweeds may have a selective advantage over UV-sensitive species by indirectly reducing herbivory. Given that two species of macroalgae have identical physiological properties, except with respect to UV-susceptibility, one would expect the less sensitive species to occur higher on the shore, i.e., where UV-regimes are more severe than lower on the shore, than the UV-sensitive counterpart. Consequently, the UV-sensitive species would be exposed over longer periods to motile consumers, e.g., isopods, and other important herbivores, e.g. sea urchins.

## Alga-herbivore interactions

The consumption of plant biomass represents the most fundamental plant-animal interaction, affecting biomass accrual and community composition of photoautotrophs (Sousa and Connell 1992; Duffy and Hay 2000). Consequently, herbivory is one key factor controlling the central ecosystem services of primary producers. Grazing intensity is considered to be lower on terrestrial plants than on aquatic macrophytes (Cyr and Pace 1993), due to a higher availability of seaweed biomass to herbivore attacks, the larger proportion of digestible algal biomass (Hay 1991), and higher mass-specific consumption rates of aquatic herbivores (Cyr and Pace 1993). As a result, aquatic plants may be more strongly top-down controlled than terrestrial counterparts and, thus, trophic interactions in aquatic habitats may be of stronger ecological outreach, e.g., in driving algal recruitment (Diaz-Pulido and McCook 2003), community succession (Farrell 1991) or mediating stability-diversity-productivity relationships of communities (Worm and Duffy 2003).

Seaweeds actively participate in the interaction with herbivores by tolerating consumption with compensatory growth (Cronin 2001), escape from herbivory (Hay et al. 1988), or the defence of grazing (e.g., Cronin 2001). Anti-herbivory defences in seaweeds may either be permanently expressed (constitutive) or induced in response to herbivory (Amsler and Fairhead 2006). Experimental evidence suggests that the induction of anti-herbivory defences is of selective advantage in variable grazing regimes (Karban and Nagasaka 2004), while constitutive defences are more beneficial to algae under a constant herbivory load (Karban et al. 1999). Induced chemical anti-herbivory defences can result in reduced consumption of previously attacked tissues (Ceh et al. 2005), increased feeding dispersal with a concomitant higher risk of herbivores becoming visible to predators (Borell et al. 2004) or both (Borell et al. 2004). The induction of anti-herbivory defences in seaweeds is known to be tissue- and (Sotka et al. 2002; Taylor et al. 2002) grazer-specific (Pavia and Toth 2000; Molis et al. 2006) and that at least some of this specificity is seasonally variable (Molis et al. 2006). Thus, phenotypic plasticity of algal responses to herbivory adds complexity to the trophic interactions between algae and their consumers. Furthermore, seasonality in inducible defences suggests that top-down forces may vary under variable environmental conditions, e.g., UV-regimes.

Cronin and Hay (1996) reported that UV-exposure reduced chemical defences in *Dictyota ciliolata*, making this brown seaweed more palatable to sea urchins and to a lesser extend also to amphipods. The selected UV-exposure time represents common emergence periods for intertidal algae. Consequently, UV-exposure seems to have the potential to increase algal susceptibility to consumers, especially during summer, i.e., at maximum UV-levels. However, algae form multi-layered piles during low tide, suggesting that only top-layered individuals are fully affected by UV, while specimens positioned deeper in the pile are protected to some extend from UV-exposure. Similar indirect UV-protective effects are known from cyanobacterial mats (Karsten et al. 1998b) and macroalgal dominated subtidal macrobenthic communities (Wahl et al. 2004). Furthermore, UV-effects on algae-animal interactions were absent in the study of Macaya et al. (2005) who did not detect any changes in response patterns between UV-exposed and -shielded *Macrocystis* individuals. This suggests that UV-exposure did not alter the ability to induce anti-herbivory defences in this brown macroalga, at least in the case of some invertebrates. Lotze and Worm (2002) investigated the interactive effects of UV-exposure and grazing on early life stages of a green alga. Their study revealed that the influence of ecological controls, i.e., grazers, on *Enteromorpha*-recruitment were stronger than climatic controls, i.e., UV and temperature, and of opposite sign.

Linking UV-effects on macroalgae with trophic interactions seems to be an interesting, but as yet not fully accounted venue for further exploiting the effects of climatic and ecological drivers on the performance and fitness of seaweeds and the species composition and productivity of seaweed communities. Presently, too few studies tested the interactive effects of UV-exposure and herbivory on alga-consumer interactions to draw general conclusions about the influence of UV-radiation on herbivory. It seems likely that UV-induced chemicals may indirectly change the susceptibility of algae to consumers by altering the function of existing chemicals, which could in- or decrease algal palatability. Alternatively, UV-induced compounds may have multiple functions. Schmitt et al. (1995) demonstrated that herbivore-deterrent chemicals displayed also anti-fouling activities. To our knowledge, only the study by Pavia et al. (1997) suggests multiple functions in UV-induced phlorotannins. The concentration of phlorotannins was lower in controls than in UV-exposed *Ascophyllum nodosum*, with the latter stimulating *Idotea* grazing. As UV-exposure is known to induce the production of many secondary metabolites (e.g., MAAs) it would be interesting to study the direct and indirect effects of these compounds on algae-herbivore interactions. For instance, UV-induced secondary metabolites may affect the anti-fouling ability of a macroalgae and altered epibiontic communities may change algal susceptibility to consumers, resulting in shared doom scenarios or associational resistance (*sensu* Wahl and Hay 1995).

Besides UV-induced effects on macroalgae, patterns of herbivory can also be altered by UV-effects on consumers. McNamara and Hill (1999) showed differential susceptibility of consumers to UVB, suggesting shifts in grazing regimes at sites of high UVB-exposure. Experimental evidence comes from studies on periphyton communities. DeNicola and Hoagland (1996) showed that herbivore density under UV-exposure was on average 50% reduced compared to PAR-treatments at the end of a 28 d long experiment. Bothwell et al. (1994) reported that, in the presence of herbivores, biomass accrual in UVB-exposed periphyton communities was higher than in PAR-irradiated communities. This suggests stronger negative effects at the consumer than at the producer level and, consequently, an indirect positive UVB-effect on periphytic biomass production. Future studies should focus on UV-effects on herbivores to elicit the relative harmfulness of UV-exposure on seaweeds and their consumers.

## Synthesis

Under present radiation conditions, UVB radiation has already to be regarded as a common abiotic factor, which influences the physiology of individual seaweeds but does also contribute in structuring seaweed communities in various ways. However, its potential threat to biologic processes is counteracted by several adaptive strategies activating different protective and repair mechanisms. These mechanisms (e.g., formation of screening compounds, antioxidative systems, regulation of enzyme activity and gene expression, DNA repair mechanisms) serve as a physiological filter to reduce the adverse effect of the impinging solar UVB, in addition canopy algae also protect more susceptible subcanopy organisms from solar exposure. Based on the differential adaptation and acclimation capabilities realised in the different species UVB radiation may, even under non-depleted ozone conditions, substantially affect the structure of seaweed communities. As outlined above, UVB exposure may modulate productivity, reproduction, vertical distribution, species diversity and succession, competition and alga-herbivore interactions. However, the effect extent of each of the aspects under ozone-depleted conditions is largely unknown. To improve our knowledge on how UVB may shape seaweed communities and thus rocky coastal ecosystems in the future, further research should predominantly focus on the following three directions: (1.) UVB is not the only factor potentially causing abiotic stress to seaweed photosynthesis. Particularly the intertidal zone is characterized by large variation of abiotic parameters, furthermore anthropogenically caused increases in seawater temperature and nutrient loads also affect seaweed performance in the field. Physiological studies on the interaction of a multitude of abiotic (stress) factors are indispensable. (2.) The molecular mechanisms triggering acclimation strategies are largely unknown. The application of molecular tools also in seaweed physiology will be a major step forward in this respect. (3.) Certainly, the most serious lack of information is still present regarding the ecological interactions modulated by UVB exposure. As outlined above, only fragmentary data sets are available so far addressing how UVB is presently interfering with seaweed succession or in interactions with seaweed-associated consumers. Field studies on a large spatial and time scale would be required including the generation of UV-free, ambient UV and increased UV irradiation treatments. Such as kind of experiment is logistically ambitious but would substantially broaden our knowledge on how UVB will alter the structure of seaweed communities in the future.

**Acknowledgements** A large part of the studies reported here were funded by the following institutions: German Academic Exchange Service (DAAD), CONICYT Chile, German Research Foundation (DFG), European Union, Helmholtz Association of German Research Centres (HGF), Alexander von Humboldt Foundation (AvH)

## References

Aguilera J, Karsten U, Lippert H, Vögele B, Philipp E, Hanelt D, Wiencke C (1999) Effects of solar radiation on growth, photosynthesis and respiration of marine macroalgae from the Arctic. Mar Ecol Prog Ser 191:109–119

Aguilera J, Gordillo FJL, Karsten U, Figueroa FL, Niell FX (2000) Light quality effect on photosynthesis and efficiency of carbon assimilation in the red alga *Porphyra leucosticta*. J Plant Physiol 157:86–92

Allen DJ, McKee IF, Farage PK, Baker NR (1997) Analysis of limitations to $CO_2$ assimilation on exposure of leaves of two *Brassica napus* cultivars to UV-B. Plant Cell Envir 20:633–640

Altamirano M, Flores-Moya A, Figueroa FL (2000a) Long-term effects of natural sunlight under various ultraviolet radiation conditions on growth and photosynthesis of intertidal *Ulva rigida* (Chlorophyceae) cultivated in situ. Bot Mar 43:19–126

Altamirano M, Flores-Moya A, Figueroa FL (2000b) Growth seasonality, photosynthetic pigments, and C and N content in relation to environmental factors: a field study on *Ulva olivascens* (Ulvales, Chlorophyta). Phycologia 39:50–58

Altamirano M, Flores-Moya A, Figueroa FL (2003a) Effects of UV radiation and temperature on growth of germlings of three species of *Fucus* (Phaeophyceae). Aquatic Bot 75:9–20

Altamirano M, Flores-Moya A, Kuhlenkamp R, Figueroa FL (2003b) Stage-dependent sensitivity to ultraviolet radiation in zygotes of the brown alga *Fucus serratus*. Zygote 11:101–106

Amsler CD, Neushul M (1991) Photosynthetic physiology and chemical composition of spores of the kelps *Macrocystis pyrifera, Nereocystis luetkana, Laminaria farlowii* and *Pterigophora californica* (Phaeophyceae). J Phycol 27:26–34

Amsler CD, Fairhead VA (2006) Defensive and sensory chemical ecology of brown algae. Adv Bot Res 43:1–91

Arrigo KR, Brown CW (1996) Impact of chromophoric dissolved organic matter on UV inhibition of primary productivity in the sea. Mar Ecol Prog Ser 140: 207–216

Aro EM, Virgin I, Andersson B (1993) Photoinhibition of photosystem II. Inactivation, protein damage and turnover. Biochim Biophys Acta 1143:113–134

Bandaranayake WM (1998) Mycosporines: are they nature's sunscreens? Nat Prod Rep 15:159–172

Bañares E, Altamirano M, Figueroa LF, Flores-Moya A (2002) Influence of UV radiation on growth of sporelings of three non-geniculate coralline red algae from Southern Iberian Peninsula. Phycol Res 50:23–30

Barber J, Andersson B (1992) Too much of a good thing: light can be bad for photosynthesis. Trend Biochem Sci 17:61–66

Bischof K, Hanelt D, Wiencke C (1998a) UV-radiation can affect depth-zonation of Antarctic macroalgae. Mar Biol 131:597–605

Bischof K, Hanelt D, Tüg H, Karsten U, Brouwer PEM, Wiencke C (1998b) Acclimation of brown algal photosynthesis to ultraviolet radiation in Arctic coastal waters (Spitsbergen, Norway). Pol Biol 20:388–395

Bischof K, Hanelt D, Wiencke C (1999) Acclimation of maximal quantum yield of photosynthesis in the brown alga *Alaria esculenta* under high light and UV radiation. Plant Biol 1:435–444

Bischof K, Hanelt D, Wiencke C (2000a) Effects of ultraviolet radiation on photosynthesis and related enzyme reactions of marine macroalgae. Planta 211:555–562

Bischof K, Janknegt PJ, Buma AGJ, Rijstenbil JW, Peralta G, Breeman AM (2003) Oxidative stress and enzymatic scavenging of superoxide radicals induced by solar UV-B radiation in *Ulva* canopies from southern Spain. Sci Mar 67:353–359

Bischof K, Kräbs G, Hanelt D, Wiencke C (2000b) Photosynthetic characteristics and mycosporine-like amino acids under UV-radiation: a competitive advantage of *Mastocarpus stellatus* over *Chondrus crispus* at the Helgoland shoreline? Helgol Mar Res 54:47–52

Bischof K, Kräbs G, Wiencke C, Hanelt D (2002a) Solar ultraviolet radiation affects the activity of ribulose-1,5-bisphosphate carboxylase-oxygenase and the composition of photosynthetic and xanthophyll cycle pigments in the intertidal green alga *Ulva lactuca* L. Planta 215:502–509

Bischof K, Peralta G, Kräbs G, van de Poll WH, Perez-Llorens JL, Breeman AM (2002b) Effects of solar UV-B radiation on canopy structure of *Ulva* communities from Southern Spain. J Exp Bot 53:2411–2421

Bischof K, Rautenberger R, Brey L, Perez-Llorens JL (2006) Physiological acclimation along gradients of solar irradiance within mats of the filamentous green macroalga *Chaetomorpha linum* from southern Spain. Mar Ecol Prog Ser 306:165–175

Björn LO, Callaghan TV, Gehrke C, Johanson U, Sonesson M (1999) Ozone depletion, ultraviolet radiation and plant life. Chemosphere: Global Change Sci 1:449–454

Borell EM, Foggo A, Coleman RA (2004) Induced resistance in intertidal macroalgae modifies feeding behaviour of herbivorous snails. Oecologia 140:328–334

Bornman JF (1989) Target sites of UV-radiation in photosynthesis of higher plants. J Photochem Photobiol B Biol 4:145–158

Bothwell ML, Sherbot DMJ, Pollock CM (1994) Ecosystem response to solar ultraviolet-B radiation: influence of trophic level interactions. Science 265:97–100

Brezezinski MA, Reed DC, Amsler CD (1993) Neutral lipids as major storage products in zoospores of the giant kelp *Macrocystis pyrifera* (Phaeophyceae). J Phycol 29:16–23

Caldwell MM (1971) Solar UV irradiation and the growth and development of higher plants. In: Giese AC (ed) Photophysiology, vol 6. Academic Press, New York, pp 131–177

Caldwell MM, Björn LO, Bornman JF, Flint SD, Kulandaivelu G, Teramura AH, Tevini M (1998) Effects of increased solar ultraviolet radiation on terrestrial ecosystems. J Photochem Photobiol B Biol 46:40–52

Caldwell MM, Teramura AH, Tevini M (1989) The changing solar ultraviolet climate and the ecological consequences for higher plants. Trend Ecol Evol 4:363–367

Campbell D, Eriksson MJ, Öquist G, Gustafsson P, Clarke AK (1998) The cyanobacterium *Synechococcus* resists UV-B by exchanging photosystem II reaction-center $D_1$ proteins. Proc Nat Acad Sci 95:364–369

Chapman ARO, Lindley JE (1980) Seasonal growth of *Laminaria solidungula* in the Canadian high Arctic in relation to irradiance and dissolved nutrient concentrations. Mar Biol 57:1–5

Ceh J, Molis M, Dzeha TM, Wahl M (2005) Induction and reduction of anti-herbivore defenses in brown and red macroalgae off the Kenyan coast. J Phycol 41:726–731

Clayton MN (1992) Propagules of marine macroalgae: structure and development. Br Phycol J 27:219–232

Coehlo SM, Rijstenbil JW, Brown MT (2000) Impacts of anthropogenic stresses on the early development stages of seaweeds. J Aquat Ecosyst Stress Rec 7:317–333

Conde FR, Churio MS, Previtali CM (2000) The photoprotector mechanism of mycosporine-like amino acids. Excited-state properties and photostability of porphyra-334 in aqueous solution. J Photochem Photobiol B Biol 56:139–144

Connell JH, Slatyer RO (1977) Mechanisms of succession in natural communities and their role in community stability and organisation. Am Nat 111:1119–1144

Cordi B, Donkin ME, Peloquin J, Price DN, Depledge MH (2001) The influence of UV-B radiation on the reproductive cells of the intertidal macroalga, *Enteromorpha intestinalis*. Aquat Toxicol 56:1–11

Cronin G (2001) Resource allocation in seaweeds and marine invertebrates: chemical defense patterns in relation to defense theories. In: McClintock JB, Baker BJ (eds) Marine chemical ecology. CRC Press, New York, pp 325–353

Cronin G, Hay ME (1996) Susceptibility to herbivores depends on recent history of both the plant and animal. Ecology 77:1531–1543

Cyr H, Pace ML (1993) Magnitude and patterns of herbivory in aquatic and terrestrial ecosystems. Nature 361:148–150

Davison IR, Pearson GA (1996) Stress tolerance in intertidal seaweeds. J Phycol 32:197–211

Dawson SP, Dennison WC (1996) Effects of ultraviolet and photosynthetically active radiation on five seagrass species. Mar Biol 125:629–638

DeLaca TE, Lipps JH (1976) Shallow-water marine associations. Antarctic Peninsula, Antarctic J US 11:12–20

DeNicola DM, Hoagland KD (1996) Effects of solar spectral irradiance (visible to UV) on a prairie stream epilithic community. J N Am Benthol Soc 15:155–169

Diaz-Pulido G, McCook LJ (2003) Relative roles of herbivory and nutrients in the recruitment of coral-reef seaweeds. Ecology 84:2026–2033

Dobretsov SV, Qian PY, Wahl M (2005) Effect of solar ultraviolet radiation on the formation of shallow,

early successional biofouling communities in Hong Kong. Mar Ecol Prog Ser 290:55–65
Döhler G (1985) Effect of UV-B radiation (290–320 nm) on the nitrogen metabolism of several marine diatoms. J Plant Physiol 118:391–400
Döhler G (1992) Impact of UV-B radiation on uptake of $^{15}$N-ammonia and $^{15}$N-nitrate by phytoplankton of the Wadden Sea. Mar Biol 112:485–489
Dring MJ, Makarov V, Schoschina E, Lorenz M, Lüning K (1996a) Influence of ultraviolet-radiation on chlorophyll fluorescence and growth in different life history stages of three species of *Laminaria*. Mar Biol 126:183–191
Dring MJ, Wagner A, Boeskov J, Lüning K (1996b) Sensitivity of intertidal and subtidal red algae to UVA and UVB radiation, as monitored by chlorophyll fluorescence measurements: influence of collection depth and season, and length of irradiation. Eur J Phycol 31:293–302
Duffy JE, Hay ME (2000) Strong impacts of grazing amphipods on the organization of a benthic community. Ecol Monogr 70:237–263
Dunlap WC, Yamamoto Y (1995) Small-molecule antioxidants in marine organisms: antioxidant activity of mycosporine-glycine. Comp Biochem Physiol 112B:105–114
Ekelund NGA (1990) Effects of UV-B radiation on growth and motility of four phytoplankton species. Physiol Plant 78:590–594
Evans LV, Christie AO (1970) Studies on the ship-fouling alga *Enteromorpha* I. Aspects of the fine-structure and biochemistry of swimming and newly settled zoospores. Ann Bot 34:451–466
Fairhead VA, Amsler CD, McClintock JB, Baker BJ (2005) Variation in phlorotannin content within two species of brown macroalgae (*Desmarestia anceps* and *D. menziesii*) from the Western Antarctic Peninsula. Polar Biol 28:680–686
Falkowski PG, LaRoche J (1991) Acclimation to spectral irradiance in algae. J Phycol 27:8–14
Farman JC, Gardiner BG, Shanklin JD (1985) Large losses of total ozone in Antarctica reveal seasonal $ClO_x/NO_x$ interaction. Nature 315:207–210
Farrell TM (1991) Models and mechanisms of succession: an example from a rocky intertidal community. Ecol Monogr 61:95–113
Flores-Moya A, Gómez I, Viñegla B, Altamirano M, Peréz-Rodríguez E, Maestre C, Caballero RM, Figueroa FL (1998) Effects of solar radiation on the endemic Mediterranean alga *Rissoella verruculosa*: photosynthetic performance, pigment content and activities of enzymes related to nutrient uptake. New Phytol 139:673–683
Flores-Moya A, Hanelt D, Figueroa FL, Altamirano M, Viñegla B, Salles S (1999) Involvement of solar UV-B radiation in recovery of inhibited photosynthesis in the brown alga *Dictyota dichotoma* (Hudson) Lamouroux. J Photochem Photobiol B Biol 49:129–135
Flores-Moya A, Posudin YI, Fernández JA, Figueroa FL, Kawai H (2002) Photomovement of the swarmers of the brown algae *Scytosiphon lomentaria* and *Petalonia fascia*: effect of photon irradiance, spectral composition and UV dose. J Photochem Photobiol B Biol 66:134–140
Franklin LA, Osmond CB, Larkum AWD (2003) UV-B and algal photosynthesis. In: Larkum AWD, Douglas SE, Raven JA (eds) Photosynthesis in algae, advances in photosynthesis and respiration, vol 14. Kluwer Academics Publishers, Dordrecht, pp 351–384
Frederick JE, Snell HE, Haywood EK (1989) Solar ultraviolet radiation at the earth's surface. Photochem Photobiol 50:443–450
Frederiksen S, Sjotun K, Lein EL, Rueness J (1995) Spore dispersal in *Laminaria hyperborea* (Laminariales, Phaeophyceae). Sarsia 80:47–54
Gómez I, Figueroa FL, Huovinen P, Ulloa N, Morales V (2005a) Photosynthesis of the red alga *Gracilaria chilensis* under natural solar radiation in an estuary in southern Chile. Aquaculture 244:369–382
Gómez I, Figueroa FL, Ulloa N, Morales V, Lovengreen C, Huovinen P, Hess S (2004) Patterns of photosynthesis in 18 species of intertidal macroalgae from southern Chile. Mar Ecol Progr Ser 270:103–116
Gómez I, Pérez-Rodríguez E, Viñegla B, Figueroa FL, Karsten U (1998) Effects of solar radiation on photosynthesis, UV-absorbing compounds and enzyme activities of the green alga *Dasycladus vermicularis* from southern Spain. J Photochem Photobiol B Biol 47:46–57
Gómez I, Ulloa N, Orostegui M (2005b). Morpho-functional patterns of photosynthesis and UV sensitivity in the kelp *Lessonia nigrescens* (Laminaria, Phaeophyta). Mar Biol 148:231–240
Greenberg BM, Gaba V, Canaani O, Malkin S, Mattoo AK, Edelman M (1989) Seperate photosensitizers mediate degradation of the 32-kDa photosystem II reaction center protein in the visible and UV spectral regions. Proc Nat Acad Sci USA 86:6617–6620
Häder DP, Häder M (1989) Effects of solar and artificial radiation on motility and pigmentation in *Cyanophora paradoxa*. Arch Microbiol 152:453–457
Han YS, Han T (2005) UV-B induction of UV-B protection in *Ulva pertusa* (Chlorophyta). J Phycol 41:523–530
Han T (1996a) Effect of ultraviolet-B radiation on *Ulva pertusa* Kjellman (Chlorophyta) I. Growth and pigment content. Algae Kor J Phycol 11:155–159
Han T (1996b) Far-UV action on growth, pigmentation and photosynthesis of *Ulva lactuca* (Chlorophyta). Algae Kor J Phycol 11:203–206
Hanelt D, Tüg H, Bischof K, Gross C, Lippert H, Sawall T, Karsten U, Wiencke C (2001) Light regime in an Arctic fjord: a study related to stratospheric ozone depletion as a basis for determination of UV effects on algal growth. Mar Biol 138:649–658
Hanelt D (1992) Photoinhibition of photosynthesis in marine macrophytes of the South China Sea. Mar Ecol Prog Ser 82:199–206
Hanelt D (1998) Capability of dynamic photoinhibition in Arctic macroalgae is related to their depth distribution. Mar Biol 131:361–369
Hanelt D, Li J, Nultsch W (1994) Tidal dependence of photoinhibition of photosynthesis in marine macrophytes of the South China Sea. Bot Acta 107:66–72

Hanelt D, Wiencke C, Nultsch W (1997a) Influence of UV radiation on the photosynthesis of Arctic macroalgae in the field. J Photochem Photobiol B Biol 38:40–47

Hanelt D, Wiencke C, Karsten U, Nultsch W (1997b) Photoinhibition and recovery after high light stress in different developmental and life-history stages of *Laminaria saccharina* (Phaeophyta). J Phycol 33: 387–395

Hanelt D, Melchersmann B, Wiencke C, Nultsch W (1997c) Effects of high light stress on photosynthesis of polar macroalgae in relation to depth distribution. Mar Ecol Prog Ser 149:255–266

Hay ME (1991) Marine-terrestrial contrasts in the ecology of plant chemical defenses against herbivores. Trend Ecol Evol 6:362–365

Hay ME, Fenical W (1992) Chemical mediation of seaweed-herbivore interactions. In: John DM, Hawkins SJ, Price JH (eds) Plant-animal interactions in the marine benthos 46: 319–337

Hay ME, Paul VJ, Lewis SM, Gustafson K, Tucker J, Trindell RN (1988) Con tropical seaweeds reduce herbivory by growing at night? Diel patterns of growth, nitrogen content, herbivory, and chemical versus morphological defenses. Oecologia 75:233–245

Henry BE, Van Alstyne KL (2004) Effects of UV radiation on growth and phlorotannins in *Fucus gardneri* (Phaeophyceae) juveniles and embryos. J Phycol 40:527–533

Henry EC, Cole KM (1982) Ultrastructure of swarmers in the Laminariales (Phaeophyceae). I. Zoospores. J Phycol 18:550–569

Holm-Hansen O, Lubin D, Helbling EW (1993) Ultraviolet radiation and its effects on organisms in aquatic environments. In: Young AR et al (eds) Environmental UV photobiology. Plenum Press, New York, pp 379–425

Holzinger A, Lütz C, Karsten U, Wiencke C (2004) The effect of ujltraviolet radiation on ultrastructure and photosynthesis in the red macroalgae *Palmaria palmata* and *Odonthalia dentata* from Arctic waters. Plant Biol 6:568–577

Hoyer K, Karsten U, Wiencke C (2003) Inventory of UV-absorbing mycosporine-like amino acids in polar macroalgae and factors controlling their concentration. In: Huiskes AHL, Gieskes WWC, Rozema J, Schorno RML, Va der Vies SM, Wolff WJ (eds) Proceedings of the VIII SCAR meeting. Backhuys Publ Leiden, The Netherlands, pp 56–62

Hoyer K, Karsten U, Sawall T, Wiencke C (2001) Photoprotective substances in Antarctic macroalgae and their variation with respect to depth distribution, different tissues and developmental stages. Mar Ecol Prog Ser 211:117–129

Huovinen P, Gómez I, Lovengreen C (2006) A five-year study of solar ultraviolet radiation in southern Chile (39 S): potential impact on physiology of coastal marine algae? Photochem Photobiol 82:515–522

Huovinen P, Gómez I, Figueroa FL, Ulloa N, Morales V, Lovengreen C (2004) Ultraviolet-absorbing mycosporine-like amino acids in red macroalgae from Chile. Bot Mar 47:21–29

Huovinen PS, Oikari AOJ, Soimasuo MR, Cherr GN (2000) Impact of UV radiation on the early development of the giant kelp (*Macrocystis pyrifera*) gametophytes. Photochem Photobiol 72:308–313

Iwanzik W, Tevini M, Dohnt G, Voss M, Weiss W, Gräber O, Renger G (1983) Action of UV-B radiation on photosynthetic primary reactions in spinach chloroplasts. Physiol Plant 58:401–407

Jerlov NG (1976) Marine optics. Elsevier, Amsterdam

Johanson U, Gehrke C, Björn LO, Callaghan TV, Sonesson M (1995) The effects of enhanced UV-B radiation on a subarctic heath ecosystem. Ambio 24:106–111

Jordan BB, He J, Chow WS, Anderson JM (1992) Changes in mRNA levels and polypeptide subunits of ribulose-1,5-bisphosphate carboxylase in response to supplementary ultraviolet-B radiation. Plant Cell Envir 15:91–98

Karban R, Agrawal AA, Thaler JS, Adler LS (1999) Induced plant responses and information content about risk of herbivory. Trend Ecol Evol 14: 443–447

Karban R, Nagasaka K (2004) Are defenses of wild radish populations well matched with variability and predictablity of herbivory? Evol Ecol 18:283–301

Karentz D (1989) Report on studies related to the ecological implications of ozone depletion on the Antarctic environment. Antarct J US 24:175–176

Karentz D, Cleaver JE, Mitchell D (1991b) Cell survival characteristics and molecular responses of Antarctic phytoplankton to ultraviolet-B radiation. J Phycol 27:326–341

Karentz D, McEuen FS, Land MC, Dunlap WC (1991a) Survey of mycosporine-like amino acid compounds in Antarctic organisms: potential protection from ultraviolet exposure. Mar Biol 108:157–166

Karsten U, Bischof K, Hanelt D, Tüg H, Wiencke C (1999) The effect of UV-radiation on photosynthesis and UV-absorbing substances in the endemic Arctic macroalga *Devaleraea ramentacea* (Rhodophyta). Physiol Plant 105:58–66

Karsten U, Friedl T, Schumann R, Hoyer K, Lembcke S (2005) Mycosporine-like amino acids (MAAs) and phylogenies in green algae: *Prasiola* and its relatives from the Trebouxiophyceae (Chlorophyta). J Phycol 41:557–566

Karsten U, Maier J, Garcia-Pichel F (1998b) Seasonality in UV-absorbing compounds of cyanobacterial mat communities from an intertidal mangrove flat. Aquat Microb Ecol 16:37–44

Karsten U, Sawall T, Hanelt D, Bischof K, Figueroa FL, Flores-Moya A, Wiencke C (1998a) An inventory of UV-absorbing mycosporine-like amino acids in macroalgae from polar to warm-temperate regions. Bot Mar 41:443–453

Kirk JTO (1994) Light and photosynthesis in aquatic ecosystems. Cambridge University Press, Cambridge

Klöser H, Quartino ML, Wiencke C (1996) Distribution of macroalgae and macroalgal communities in gradients of physical conditions in Potter Cove, King George Island, Antarctica. Hydrobiologia 333:1–17

Korbee N, Huovinen P, Figueroa FL, Aguilera J, Karsten U (2005) Availability of ammonium influences photosynthesis and the accumulation of MAAs in two *Porphyra* species (Bangiales, Rhodophyta). Mar Biol 146:645–654

Kräbs G, Bischof K, Hanelt D, Karsten U, Wiencke C (2002) Wavelength-dependent induction of UV-absorbing mycosporine-like amino acids in the red alga *Chondrus crispus*. J Exp Mar Biol Ecol 268:69–82

Krause-Jensen D, McGlathery K, Rysgaard S, Christensen PB (1996) Production within dense mats of the filamentous macroalga *Chaetomorpha linum* in relation to light and nutrient availability. Mar Ecol Prog Ser 134:207–216

Lamb IM, Zimmermann MH (1977) Benthic marine algae of the Antarctic Peninsula. Antarctic Res Ser 23:129–229

Langer J (1999) Messungen des arktischen stratosphärischen Ozons: Vergleich der Ozonmessungen in Ny-Ålesund, Spitzbergen 1997 und 1998. Ber Polarforsch 322:1–222

Larkum AWD, Wood WF (1993) The effect of UV-B radiation on photosynthesis and respiration of phytoplankton, benthic macroalgae and seagrasses. Photosyn Res 36:17–23

Lary DJ (1997) Catalytic destruction of stratospheric ozone. J Geophys Res 102:21515–21526

Lesser MP (1996) Acclimation of phytoplankton to UV-B radiation: oxidative stress and photoinhibition of photosynthesis are not prevented by UV-absorbing compounds in the dinoflagellate *Prorocentrum micans*. Mar Ecol Prog Ser 132:287–297

Lippert H, Iken K, Rachor E, Wiencke C (2001) Macrofauna associated with macroalgae in the Kongsfjord (Spitsbergen). Polar Biol 24:512–522

Lois R, Buchanan BBN (1994) Severe sensitivity to ulraviolet radiation in an *Arabidopsis* mutant deficient in flavonoid accumulation: II. Mechanisms of UV-resistance in *Arabidopsis*. Planta 194:504–509

Lorenz M, Schubert H, Forster RM (1997) *In vitro*- and *in vivo*-effects of ultraviolet-B radiation on the energy transfer in phycobilisomes. Photosynthetica 33:517–527

Long A, Michod RE (1995) Origin of sex for error repair. 1. Sex, diploidy, and haploidy. Theroret Pop Biol 47:18–55

Lotze HK, Worm B (2002) Complex interactions of climatic and ecological controls on macroalgal recruitment. Limnol Oceanogr 47:1734–1741

Lotze HK, Worm B, Molis M, Wahl M (2002) Effects of UV radiation and consumers on recruitment and succession of a marine macrobenthic community. Mar Ecol Prog Ser 243:57–66

Lubchenco J, Gaines SD (1981). A unified approach to marine plant-herbivore interactions. I. Poluations and communities. Annu Rev Syst 12:405–437

Lüning K (1990) Seaweeds. Their environment, biogeography, and ecophysiology. Wiley & Sons Inc., New York, p 527

Macaya EC, Rothausler E, Thiel M, Molis M, Wahl M (2005) Induction of defenses and within-alga variation of palatability in two brown algae from the northern-central coast of Chile: effects of mesograzers and UV radiation. J Exp Mar Biol Ecol 325:214–227

Mackerness SAH, Jordan BR, Thomas B (1999) Reactive oxygen species in the regulation of photosynthetic genes by ultraviolet radiation (UV-B: 280–320 nm) in green and etiolated buds of pea (*Pisum sativum* L). J Photochem Photobiol B Biol 48:180–188

Madronich S, McKenzie RL, Björn LO, Caldwell MM (1998) Changes in photobiologically active ultraviolet radiation reaching the earth's surface. J Photochem Photobiol B Biol 46:5–19

Maegawa M, Kunieda M, Kida W (1993) The influence of ultraviolet radiation on the photosynthetic activity of several red algae from different depths. Jpn J Phycol 41:207–214

Makarov M (1999) Influence of ultraviolet radiation on the growth of the dominant macroalgae of the Barents Sea. Chemosphere: Global Change Sci 1:461–467

Makarov MV, Voskoboinikov GM (2001) The influence of Ultraviolet-B radiation on spore release and growth of the kelp *Laminaria saccharina*. Bot Mar 44:89–94

Máté Z, Sass L, Szekeres M, Vass I, Nagy F (1998) UV-B induced differential transcription of psbA genes encoding the D1 protein of photosystem II in the cyanobacterium *Synechocystis* 6803. J Biol Chem 273:17439–17444

McMinn A, Ashworth C, Ryan K (1999) Growth and productivity of Antarctic sea ice algae under PAR and UV irradiances. Bot Mar 42:401–407

McNamara AE, Hill WR (1999) Effects of UV-B dose and irradiance: comparison among grazers. J North Am Benthol Soc 18:370–380

Mepsted R, Paul ND, Stephen J, Corlett JE, Nogués S, Jones NR, Ayres PG (1996) Effects of enhanced UV-B radiation on pea (*Pisum sativum* L.) grown under field conditions in the UK. Global Change Biol 2:325–334

Michler T, Aguilera J, Hanelt D, Bischof K, Wiencke C (2002) Long term effects of ultraviolet radiation on growth and photosynthetic performance of polar and cold-temperate marcoalgae. Mar Biol 140:1117–1127

Mitchell DL, Karentz D (1993) The induction and repair of DNA photodamage in the environment. In: Young AR, Björn LO, Moan J, Nultsch W (eds) Environmental UV photobiology. Plenum Press, New York, pp. 345–377

Molis M, Körner J, Ko YW, Kim JH, Wahl M (2006) Inducible responses in the brown seaweed *Ecklonia cava*: the role of grazer identity and season. J Ecol 94:243–249

Molis M, Lenz M, Wahl M (2003) Radiation effects along a UV-B gradient on species composition and diversity of a shallow-water macrobenthic community in the western Baltic. Mar Ecol Prog Ser 263:113–125

Molis M, Wahl M (2004) Transient effects of solar ultraviolet radiation on the diversity and structure of a field-grown epibenthic community at Luderitz, Namibia. J Exp Mar Biol Ecol 302:51–62

Müller DG (1977) Sexual reproduction in British *Ectocarpus siliculosus*. Br Phycol J 12:131–136

Murphy TM (1983) Membranes as targets of ultraviolet radiation. Physiol Plant 58:381–388

Nogués S, Baker NR (1995) Evaluation of the role of damage to photosystem II in the inhibition of $CO_2$ assimilation in pea leaves on exposure to UV-B radiation. Plant Cell Envir 18:781–787

Norton TA (1992) Dispersal by macroalgae. Br Phycol J 27:293–301

Okuda T, Neushul M (1981) Sedimentation studies of red algal spores. J Phycol 17:113–118

Pang S, Gómez I, Lüning K (2001) The red macroalga *Delesseria sanguinea* as a UVB-sensitive model organism: selective growth reduction by UVB in outdoor experiments and rapid recording of growth rate during and after pulses. Eur J Phycol 36:297–216

Pavia H, Brock E (2000) Extrinsic factors influencing phlorotannin production in the brown alga *Ascophyllum nodosum*. Mar Ecol Prog Ser 193:285–294

Pavia H, Cervin G, Lindgren A, Aberg P (1997) Effects of UV-B radiation and simulated herbivory on phlorotannins in the brown alga *Ascophyllum nodosum*. Mar Ecol Prog Ser 157:139–146

Pavia H, Toth GB (2000) Inducible chemical resistance to herbivory in the brown seaweed *Ascophyllum nodosum*. Ecology 81:3212–3225

Pérez-Rodriguez E (2000) Caracterización bio-optica de aguas oceánicas y costeras. Fotosíntesis, fotoinhibición y fotoprotección bajo radiación ultravioleta y fotosintética en macrofitos marinos. PhD Thesis, Universidad de Málaga, p 356

Poppe F, Hanelt D, Wiencke C (2002) Changes in ultrastructure, photosynthetic activity and pigments in the Antarctic red alga *Palmaria decipiens* during acclimation to UV radiation. Bot Mar 45:253–261

Poppe F, Schmidt RAM, Hanelt D, Wiencke C (2003) Effects of UV radiation on the ultrastructure of several red algae. Phycol Res 51:11–19

Post A, Larkum AW (1993) UV-absorbing pigments, photosynthesis and UV exposure in Antarctica: comparison of terrestrial and marine algae. Aquat Bot 45:231–243

Ragan MA, Glombitza KW (1986) Phlorotannins, brown algal polyphenols. In: Round FE, Chapman DJ (eds) Progress in phycolocical research 4. Biopress Ltd, Bristol, pp 129–241

Reed DC, Brezezinski MA, Coury DA, Graham WM, Petty RL (1999) Neutral lipids in macroalgal spores and their role in swimming. Mar Biol 133:737–479

Renger G, Voss M, Gräber P, Schulze A (1986) Effect of UV irradiation on differential partial reactions of the primary processes of photosynthesis. In: Worrest RC, Caldwell MM (eds) Stratospheric ozone reduction, solar ultraviolet radiation and plant life. NATO ASI Series, vol G8. Springer, Heidelberg, pp. 171–184

Roleda MY (2006) Effects of ultraviolet radiation on early life stages of cold temperate and Arctic macroalgae: implications for recruitment and vertical depth distribution. Ber Polar Meeresforsch 526:1–176

Roleda MY, Clayton MN, Wiencke C (2006c) Screening capacity of UV-absorbing compounds in spores of Arctic Laminariales. J Exper Mar Biol Ecol (subm.)

Roleda MY, Hanelt D, Wiencke C (2005b) Growth kinetics related to physiological parameters in young *Saccorhiza dermatodea* and *Alaria esculenta* sporophytes exposed to UV radiation. Pol Biol 28:466–479

Roleda MY, Hanelt D, Wiencke C (2006b) Growth and DNA damage in young *Laminaria* sporophytes exposed to ultraviolet radiation: implication for depth zonation of kelps on Helgoland (North Sea). Mar Biol 148:1201–1211

Roleda MY, Hanelt D, Wiencke C (2006d) Exposure to ultraviolet radiation delays photosynthetic recovery in Arctic kelp zoospores, Photosynth Res DOI 10.1007/s11120-006-9055-y

Roleda MY, van de Poll WH, Hanelt D, Wiencke C (2004) PAR and UVBR effects on photosynthesis, viability growth and DNA in different life stages of two coexisting Gigartinales: implications for recruitment and zonation pattern. Mar Ecol Progr Ser 281:37–50

Roleda MY, Wiencke C, Hanelt D (2006a) Thallus morphology and optical characteristics affect growth and DNA damage by UV radiation in juvenile Arctic *Laminaria* sporophytes, Planta 223:407–417

Roleda MY, Wiencke C, Hanelt D, van de Poll WH, Gruber A (2005a) Sensitivity of Laminariales zoospores from Helgoland (North Sea) to ultraviolet and photosynthetically active radiation: implications for depth distribution and seasonal reproduction. Plant Cell Environ 28:466–479

Santas R, Korda A, Lianou C, Santas P (1998) Community response to UV radiation. I. Enhanced UVB effects on biomass and community structure of filamentous algal assemblages growing in a coral reef mesocosm. Mar Biol 131:153–162

Scharf ER, Gerhard JC (1983) Axis determination in eggs of *Xenopus laevis*: a critical period before first cleavage. Identified by the common effects of cold, pressure and untraviolet radiation. Developm Biol 99:75–87

Schmitt TM, Hay ME, Lindquist N (1995) Constraints on chemically mediated coevolution: multiple functions for seaweed secondary metabolites. Ecology 76:107–123

Schoenwaelder MAE, Wiencke C, Clayton MN, Glombitza KW (2003) The effect of elevated UV radiation on *Fucus* spp. (Fucales, Phaeophyta) zygote and embryo development. Plant Biol 5:366–377

Schoenwaelder MEA, Clayton MN (1999) The role of the cytoskeleton in brown algal physode movement. Eur J Phycol 34:223–229

Schoenwaelder MEA, Wiencke C (2000) Phenolic compounds in the embryo development of several northern hemisphere fucoids. Plant Biol 2:24–33

Schoenwaelder MEA (2002a) Physode distribution and the effects of 'Thallus Sunburn' in *Hormosira banksii* (Fucales, Phaeophyceae). Bot Mar 45:262–266

Schoenwaelder MEA (2002b) The occurence and cellular significance of physodes in brown algae. Phycologia 41:125–139

Schoembeck M, Norton TA (1978) Factors controlling the upper limits of fucoid algae on the shore. J Exp Mar Biol Ecol 31:303–313

Sotka EE, Taylor RB, Hay ME (2002) Tissue-specific induction of resistance to herbivores in a brown seaweed: the importance of direct grazing versus waterborne signals from grazed neighbors. J Exp Mar Biol Ecol 277:1–12

Sousa WP, Connell JH (1992) Grazing and succession in marine algae. In: John DM, Hawkins S, Price JH (eds) Plant-animal interactions in the marine benthos. Clarendon Press, Oxford, pp 425–441

Stephenson TA, Stephenson A (1972) Life between tidemarks on rocky shores. Freeman, San Francisco, p 425

Strahle S, Jesuthasan U (1993) Ultraviolet irradiation impairs epiboly in zebrafish embryos: evidence for a microtubule-dependent mechanism of epiboly. Development 119:909–919

Strid A, Chow WS, Anderson JM (1990) Effects of supplementary ultraviolet-B radiation on photosynthesis in *Pisum sativum*. Biochim Biophys Acta 1020:260–268

Svendsen H, Beszczynska-Møller A, Lefauconnier B, Tverberg V, Gerland S, Hagen JO, Ørbæk JB, Bischof K, Papucci C, Zajaczkowski M, Azzolini R, Bruland O, Wiencke C, Winther JG, Hodson A, Mumford P (2002) The physical environment of Kongsfjorden–Krossfjorden—an Arctic fjord system in Svalbard. Pol Res 21:133–166

Swanson AK, Druehl LD (2000) Differential meiospore size and tolerance of ultraviolet light stress within and among kelp species along a depth gradient. Mar Biol 136:657–664

Swanson AK, Druehl LD (2002) Induction, exudation and the UV protective role of kelp phlorotannins. Aquat Bot 73:241–253

Taylor RB, Sotka E, Hay ME (2002) Tissue-specific induction of herbivore resistance: seaweed response to amphipod grazing. Oecologia 132:68–76

Teramura AH (1983) Effects of ultraviolet-B radiation on the growth and yield of crop plants. Physiol Plant 58:415–427

Tevini M, Teramura AH (1989) UV-B effects in terrestrial plants. Photochem Photobiol 50:479–487

Thiel M, Gutow L (2005) The ecology of rafting in the marine environment. II. The rafting organisms and community. Oceanogr Mar Biol - Ann Rev 43:279–418

Thomas D (2002) Seaweeds. The natural Hostory Museum London, Smithsonian Institution Press, p 96

van de Poll WH, Eggert A, Buma AGJ, Breeman AM (2001) Effects of UV-B-induced DNA damage and photoinhibition on growth of temperate marine red macrophytes: habitat-related differences in UV-B tolerance. J Phycol 37:30–37

van de Poll WH, Hanelt D, Hoyer K, Buma AGJ, Breeman AM (2002) Ultraviolet-B-induced cyclobutane-pyrimidine dimer formation and repair in arctic marine macrophytes. Photochem Photobiol 76:493–500

Vass I (1997) Adverse effects of UV-B light on the structure and function of the photosynthetic apparatus. In: Pessarakli M (ed) Handbook of photosynthesis. Marcel Dekker Inc., New York, pp 931–949

Véliz K, Edding M, Tala F, Gómez I (2006) Effects of ultraviolet radiation on different life cycle stages of the Pacific kelps, *Lessonia nigrescens* and *Lessonia trabeculata* (Laminariales, Phaeophyceae). Mar Biol DOI 10.1007/s00227-006-0301-9

Vergara JJ, Pérez-Lloréns JL, Peralta G, Hernández I (1997) Seasonal variation of photosynthetic performance and light attenuation in Ulva canopies from Palmones river estuary. J Phycol 33:773–779

Vincent WF, Belzile C (2003) Biological UV exposure in the polar oceans: Arctic–Antarctic comparisions. In: Huiskes AHL et al (eds) Antarctic biology in a global context. Backhuys Publishers, Leiden Netherland, pp 176–181

Vinebrooke RD, Leavitt PR (1999) Differential responses of littoral communities to ultraviolet radiation in an alpine lake. Ecology 80:223–237

Wahl M (2001) Small scale variability of benthic assemblages: biogenic neighbourhood effects. J Exp Mar Biol Ecol 258:101–114

Wahl M, Hay ME (1995) Associational resistance and shared doom: effects of epibiosis on herbivory. Oecologia 102:329–340

Wahl M, Molis M, Davis A, Dobretsov S, Durr ST, Johansson J, Kinley J, Kirugara D, Langer M, Lotze HK, Thiel M, Thomason JC, Worm B, Ben-Yosef DZ (2004) UV effects that come and go: a global comparison of marine benthic community level impacts. Global Change Biol 10:1962–1972

Wiencke C, Clayton MN (2002) Antarctic seaweeds. Synopses of the Antarctic benthos, vol 9. A.R.G. Gantner Verlag KG, Ruggell, p 239

Wiencke C, Clayton MN, Schoenwaelder M (2004a) Sensitivity and acclimation to UV radiation of zoospores from five species of Laminariales from the Arctic. Mar Biol 145:31–39S

Wiencke C, Gómez I, Pakker H, Flores-Moya A, Altamirano M, Hanelt D, Bischof K, Figueroa FL (2000) Impact of UV-radiation on viability, photosynthetic characteristics and DNA on brown algal zoospores: implications for depth zonation. Mar Ecol Progr Ser 197:217–229

Wiencke C, Roleda MY, Gruber A, Clayton M, Bischof K (2006) Susceptibility of zoospores to UV radiation determines upper depth distribution limit of Arctic kelps: evidence through field experiments. J Ecol 94:455–463

Wiencke C, Vögele B, Kovaltchouk NA, Hop H (2004b) Species composition and zonation of marine benthic macroalgae at Hansneset in Kongsfjorden, Svalbard. Ber Polar Meeresforsch 492:55–62

Worm B, Duffy JE (2003) Biodiversity, productivity and stability in real food webs. Trend Ecol Evol 18:628–632

Worrest RC (1983) Impact of solar ultraviolet-B radiation (290–320 nm) upon marine microalgae. Physiol Plant 58:428–434

Yakovleva IM, Dring MJ, Titlyanov EA (1998) Tolerance of North Sea algae to UV and visible radiation. Russ J Plant Physiol 45:45–54

Zielinski K (1990) Bottom macroalgae of the Admirality Bay (King George Island, South Shetlands, Antarctica). Pol Polar Res 11:95–131

Rev Environ Sci Biotechnol (2007) 6:95–126
DOI 10.1007/s11157-006-9106-z

REVIEW

# Life strategy, ecophysiology and ecology of seaweeds in polar waters

C. Wiencke · M. N. Clayton · I. Gómez · K. Iken · U. H. Lüder · C. D. Amsler · U. Karsten · D. Hanelt · K. Bischof · K. Dunton

Received: 7 March 2006 / Accepted: 8 May 2006 / Published online: 19 August 2006


**Abstract** Polar seaweeds are strongly adapted to the low temperatures of their environment, Antarctic species more strongly than Arctic species due to the longer cold water history of the Antarctic region. By reason of the strong isolation of the Southern Ocean the Antarctic marine flora is characterized by a high degree of endemism, whereas in the Arctic only few endemic species have been found so far. All polar species are strongly shade adapted and their phenology is finely tuned to the strong seasonal changes of the light conditions. The paper summarises the present knowledge of seaweeds from both polar regions with regard to the following topics: the history of seaweed research in polar regions; the environment of seaweeds in polar waters; biodiversity, biogeographical relationships and vertical distribution of Arctic and Antarctic seaweeds; life histories and physiological thallus anatomy; temperature demands and geographical distribution; light demands and depth zonation; the effect of salinity, temperature and desiccation on supra- and eulittoral seaweeds; seasonality of reproduction and the physiological characteristics of

C. Wiencke (✉) · U. H. Lüder
Alfred Wegener Institute for Polar and Marine Research, Am Handelshafen 12, D-27570 Bremerhaven, Germany
e-mail: cwiencke@awi-bremerhaven.de

M. N. Clayton
School of Biological Sciences, Monash University, P.O Box 18, Melbourne, Victoria 3800, Australia

I. Gómez
Instituto de Biología Marina, Universidad Austral de Chile, Casilla 567, Valdivia, Chile

K. Iken
Institute of Marine Science, University of Alaska Fairbanks, O'Neill Bldg., Fairbanks, AK 99775, USA

C. D. Amsler
Department of Biology, University of Alabama at Birmingham, Birmingham, AL 35294-1170, USA

U. Karsten
Institute of Biological Sciences, Applied Ecology, University of Rostock, Albert-Einstein-Strasse 3, D-18051 Rostock, Germany

D. Hanelt
Biocenter Klein Flottbek, University of Hamburg, Ohnhorst-Str. 18, D-22609 Hamburg, Germany

K. Bischof
Institute for Polar Ecology, University of Kiel, Wischhofstr. 1–3, D-24148 Kiel, Germany

K. Dunton
Marine Science Institute, University of Texas at Austin, 750 Channel View Drive, Port Aransas, TX 78373, USA

microscopic developmental stages; seasonal growth and photosynthesis; elemental and nutritional contents and chemical and physical defences against herbivory. We present evidence to show that specific characteristics and adaptations in polar seaweeds help to explain their ecological success under environmentally extreme conditions. In conclusion, as a perspective and guide for future research we draw attention to many remaining gaps in knowledge.

**Keywords** Chemical ecology · Freezing · Growth · Light · Phenology · Photosynthesis · Polar algae · Salinity · Seaweeds · Temperature

## History of seaweed research in the polar regions

The first phycological studies in the Arctic and the Antarctic regions took place in the first half of the 19th century. Studies on Antarctic seaweeds started in 1817 with the work of Gaudichaud, Bory, Montagne, Hooker and Harvey (Godley 1965). Later, between 1882 and 1909, intensive studies were carried out by Skottsberg, Reinsch, Gain, Hariot, Kylin, Hylmö, Foslie and others (Papenfuss 1964). After this period of mostly taxonomic and biogeographical investigations, which culminated in the publication of the catalogue by Papenfuss (1964), numerous diving investigations were conducted in West Antarctica by, among others, Neushul (1965), Délépine et al. (1966), Lamb and Zimmermann (1977), Moe and DeLaca (1976), Richardson (1979), Klöser et al. (1996) and in East Antarctica by Zaneveld (1968) and Cormaci et al. (1997). These studies gave greater insight into the depth distribution of Antarctic seaweeds and allowed more detailed studies on the life histories, ecophysiology and ecology of seaweeds from Antarctica. The highly fragmented information on the seaweeds of Antarctica was summarized recently in a synopsis by Wiencke and Clayton (2002).

Early algal research in the Arctic started in 1849 in the Canadian Arctic with Harvey and Dickie (Lee 1980). The Russian phycologist Kjellman worked in many places in the Arctic, especially in the Russian Arctic, the northern Bering Sea and Spitsbergen. His book "The algae of the Arctic Sea" (Kjellman 1883) is a taxonomic and biogeographic baseline study. Detailed studies on seaweeds from Greenland were made by Rosenvinge (1898), and in the 20th century by Lund (1951, 1959a, b) and Pedersen (1976). Taxonomic information concerning seaweeds from Spitsbergen and the Russian Arctic is also available in publications by Svendsen (1959), Zinova (1953; 1955), Vinogradova (1995) and many others. A very valuable contribution on seaweeds from the north-eastern coast of North America is the book by Taylor (1966; first published 1937). The first extensive diving studies were performed by Wilce (1963), Chapman and Lindley (1980) and by Dunton et al. (1982) in the Canadian and Alaskan Arctic. As in Antarctica, diving is an essential prerequisite for detailed studies on all aspects of seaweed biology and this technique has since been widely applied in several regions of the Arctic. Since 1991 there has been intensive research on the ecology and ecophysiology of Arctic seaweeds on Spitsbergen (Hop et al. 2002; Wiencke 2004).

## The environment of seaweeds in polar waters

The polar regions are characterized by pronounced seasonal variations of the light regime, low temperatures, and long periods of snow and ice cover. Littoral seaweed communities are often strongly impacted by icebergs and sea ice especially in areas with high wave action (Klöser et al. 1994). Icebergs can run aground in the Antarctic down to 600 m depth (Gutt 2001). First-year sea ice has a thickness of up to 2 m, and in the Arctic multi-year ice can reach depths of about 40 m (Gutt 2001). Anchor ice forms on the seabed in shallow waters and can enclose seaweeds especially in East Antarctica (Miller and Pearse 1991).

The seasons in the sublittoral are characterized in polar regions by short periods of favourable light conditions and extended periods of darkness due to the polar night and sea ice cover in winter (Kirst and Wiencke 1995). At the poleward distribution limits of seaweeds, at 77° S and 80° N, respectively, the annual solar radiation is 30–50% less than in temperate to tropical regions, and the polar night lasts for about 4 months

(Lüning 1990). At lower latitudes, e.g. close to the northern limit of the Antarctic region, in the South Shetland Islands, daylength varies between 5 h in winter and 20 h in summer (Wiencke 1990a). This extreme light regime has strong implications for primary productivity in general and for the seasonal development of seaweeds. In addition the long periods of darkness are further extended due to the formation of sea ice. If the ice is also covered by snow, irradiance can be diminished to less than 2% of the surface value. Consequently, seaweeds may be exposed up to about 10 months of darkness or very low light conditions (Chapman and Lindley 1980; Dunton 1990; Miller and Pearse 1991; Drew and Hastings 1992; Klöser et al. 1993).

After sea ice break-up in spring, light penetrates deeply into the clear water. On King George Island (South Shetland Islands) at this time average photon fluence rates are 70 μmol $m^{-2}$ $s^{-1}$ at midday in 30 m depth (Gómez et al. 1997). On Signy Island (South Orkney Islands) the euphotic zone, i.e. the 1% depth for photosynthetically active radiation (PAR), was determined at 29 m (Brouwer 1996). UV-radiation (UVR) and blue light are, however, more strongly attenuated. The 1% depth for UVB radiation, representing more or less the threshold irradiance of UV-B with the potential to affect primary plant productivity negatively, is located between 4 and 8 m on Spitsbergen (Hanelt et al. 2001; van de Poll et al. 2002). Later, in summer, coastal waters in polar regions become more turbid due to the development of phytoplankton blooms and the inflow of melt water carrying fine sediments and detritus that strongly affect light transmission. With increasing turbidity the wavelengths shift from the blue to the green waveband in deep waters (Jerlov 1976). Consequently, sublittoral seaweeds are exposed only to low irradiances even though the sun altitude is relatively high (e.g. about 34° elevation in July compared to only 14° in April at 79°North at noon).

In a canopy of kelps or other overstory brown algae, PAR is strongly attenuated, and understorey species are exposed to even lower irradiances (Hanelt et al. 2003). The light quality also changes. Below the canopy the spectrum is enriched in green and in far red light, probably affecting photosynthesis as well as the photomorphogenetic development of the understorey (Salles et al. 1996).

The possible impact of enhanced UVB radiation due to stratospheric ozone depletion on primary production of seaweeds is higher during the spring season, as organisms in the eulittoral and upper sublittoral zone are already affected by UVB radiation throughout the polar day and turbid melt water input occurs only during the summer season. The strong increase of turbidity due to a discharge of sediments by melt water and glaciers applies especially to Arctic shorelines in half-open fjord systems where the water exchange with the clearer oceanic water is retarded (Svendsen et al. 2002). On open coastlines the melt water is exchanged much faster with clear oceanic water so that seaweeds are exposed to biologically effective UVR for longer time periods.

The inflow of melt water in summer has considerable effects on the salinity and temperature regime in inshore waters. During times of calm weather, there is a strong stratification in the water body with a layer of fresh water above a layer of denser sea water. However, due to vertical water mixing by wave action and wind the deeper water zones also become affected and a salinity decrease may occur down to about 20 m depth (Hanelt et al. 2001).

In contrast to the strong seasonal variation in the radiation conditions, water temperatures in the sublittoral mostly change only little between –1.8°C in winter and +2.0°C in summer, e.g. in the Antarctic Peninsula region (Drew and Hastings 1992; Klöser et al. 1993). At the boundary with the temperate region, maximum summer temperatures can go up to about 5°C in the Antarctic and up to about 8–10°C in the Arctic (Wiencke and tom Dieck 1989; Lüning 1990; Svendsen et al. 2002).

Macronutrient concentrations are high and not limiting for seaweeds at any time of the year in the Antarctic (Deacon 1937; Drew and Hastings 1992). However, iron concentrations are low in the Antarctic and inhibit phytoplankton growth (de Baar 1994; de Baar et al. 1995). Whether iron deficiency also affects seaweeds is unclear. High macronutrient concentrations are also present in the area of Spitsbergen, which obtains nutrient-rich water from the south during parts of the year (the so-called Spitsbergen current). This is one reason why the seas of the European Arctic belong to the most productive seas in the world (Orheim et al. 1995). In general, there is a considerable seasonal fluctuation of macronutrients. Nitrogen and phosphorus levels are relatively high during the winter months but both macronutrients are almost fully depleted in summer (Chapman and Lindley 1980). This applies also to the area close to Spitsbergen (Aguilera et al. 2002).

The tidal amplitudes on King George Island and on Spitsbergen vary between 120 and 150 cm (Schöne et al. 1998; Svendsen et al. 2002). Like terrestrial vegetation, supra-and eulittoral seaweeds are exposed to strongly changing light conditions and temperatures, but also frequently to desiccation and low or high salinities due to the tidal regime and the weather. Photon fluence rates close to terrestrial cryptogams in the maritime Antarctic vary between 0.1 mol $m^{-2}$ $d^{-1}$ in winter and more than 30 mol $m^{-2}$ $d^{-1}$ with maxima around 1600 µmol $m^{-2}$ $s^{-1}$ in summer (Schroeter et al. 1995; Winkler et al. 2000). The annual variation of thallus temperatures of a lichen growing on a coastal rock on King George Island (Antarctica) was between +27.4°C in summer and –27.3°C in winter (Schroeter et al. 1995). However, smaller annual amplitudes of only 10°C also occur depending on the degree of exposure and snow cover (Winkler et al. 2000).

With increasing temperature in spring the supralittoral is flushed by meltwater, followed by a period of desiccation in summer, which can be interrupted by rehydration during rain-or snowfall (Davey 1989; Schroeter et al. 1995) or during high tides. Salinity changes strongly in the supralittoral due to the effects of tides, salt spray, desiccation, overflow with melt water and precipitation. In the eulittoral salinity rises are less strong and less frequent than in the supralittoral. On the other hand, low salinities are quite common during low tides, precipitation and meltwater discharge. Klöser (1994) determined salinities between 27 and 41 PSU in the eulittoral on King George Island. At low tide temperatures in tide pools may rise far above the coastal water temperatures up to almost 14°C in summer (Klöser 1994; Moore et al. 1995). Winter temperatures may be very low and depend, apart from the air temperatures, on the thickness of the ice cover in winter.

Despite the remoteness of the polar regions pollution is an issue both in the Arctic and the Antarctic. For example, radionuclide contaminants such as Technetium-99 ($^{99}$Tc) are present around the Arctic archipelago of Svalbard through the long-range oceanic transport of discharges of radioactive effluents from nuclear reprocessing plants in Europe and global fallout (Gerland et al. 2003; Gwynn et al. 2004). Another problem are oil spills. When the Argentinean ship Bahia Paraiso ran aground near Anvers Island at the Antarctic Peninsula more than 150,000 gallons of petroleum were released to the surrounding bays (Kennicutt II et al. 1990). Such oil spills can potentially cause servere harm depending on the amount and type of material released.

## Biodiversity, biogeographical relationships and vertical distribution of Arctic and Antarctic seaweeds

The Arctic and the Antarctic Oceans differ considerably with respect to their genesis and their cold water history. The Southern Ocean has had no land bridge to temperate regions since the late Mesozoic and has been further separated from the neighbouring southern continents by the Antarctic Circumpolar Current since 26 Ma (Hempel 1987; Lüning 1990; Kirst and Wiencke 1995). During the first major glaciation in East Antarctica 14 Ma ago water temperatures decreased and they have been low in the Southern Ocean since then (Crame 1993). In contrast, the broad shelf areas of the Arctic Ocean have had a continuous connection to the temperate coasts of America and Eurasia, and a perennial ice cover did not develop before 0.7–2.0 Ma ago (Clarke 1990). These distinctions are the major reasons for the differences in seaweed biodiversity of both polar regions.

The strong isolation of the benthic seaweed flora of the Southern Ocean has resulted in a high degree of endemism in Antarctica. 33% of all seaweed species are endemic to the Antarctic region. Within the Heterokontophyta 44% of the species are endemic, within the Rhodophyta 32% and within the Chlorophyta 18% (Wiencke and Clayton 2002). There is one endemic order, the kelp-like brown algal order Ascoseirales, and there are several endemic genera: among the brown algae, *Himantothallus*, *Cystosphaera* and *Phaeurus*, among the red algae *Gainia*, *Notophycus* and *Antarcticothamnion*, and among the green algae *Lambia* and *Lola* (Wiencke and Clayton 2002). Well-known Antarctic endemic species include the brown algae *Himantothallus grandifolius*, *Cystosphaera jacquinotii*, the red alga *Porphyra endiviifolium* and the green alga *Lambia antarctica*. The Antarctic region is the only region in the world devoid of kelps, brown algae of the order Laminariales (Moe and Silva 1977). This order is ecologically replaced by the Desmarestiales. According to phylogenetic and morphological analyses and studies on ecophysiological traits this order is considered to have originated in the Southern Ocean and subsequently radiated into the Northern Hemisphere (Peters et al. 1997). A conspicuous character of the Antarctic seaweed flora is the scarcity of small macroalgal epiphytes compared to temperate regions. In fact, these epiphytes are not absent, they rather occur as endophytes inside larger seaweeds, well protected against herbivory by mesograzers (Peters 2003). This suggests that mesoherbivory can be quite intense in Antarctica.

In contrast to the strong endemism in Antarctic macroalgae only very few endemic Arctic species have been found so far. About half a dozen species are restricted to the Arctic including the brown algae *Punctaria glacialis*, *Platysiphon verticillatus* and the red alga *Petrocelis polygyna* (Wilce 1990). Most species have a distribution that extends well into the temperate region, e.g. the kelp *Laminaria solidungula* and the red algae *Devaleraea ramentacea*, *Turnerella pennyi*, *Neodilsea integra* and *Pantoneura baerii* (Lüning 1990).

Compared to rich seaweed floras like those in temperate southern Australia with about 1155 species (Womersley 1991), low species richness is characteristic for both polar seaweed floras. In the Antarctic, 119 species have been recorded so far (Wiencke and Clayton 2002) and in the Arctic there are about 150 species (Wilce 1994). These numbers are, however, likely to be underestimated due to the infrequency of scientific collections, the extreme remoteness and logistic difficulties. In Antarctica most species occur in the Antarctic Peninsula region and only very few species are recorded at the southernmost distribution limit in the Ross Sea. A similar decrease in species richness has been detected in East Greenland between Scoresby Sound, Franz Josephs Fjord and Jörgen Brönlunds Fjord (70, 74 and 82° N, respectively; Lund 1951, 1959a, b). On the panarctic level, macroalgal species richness seems to dramatically decrease from the western (Atlantic) sector to the eastern (Pacific) sector. While about 70 species have been recorded from the Svalbard region (Weslawski et al. 1993, 1997; Vinogradova 1995), only about 10 species are known from the rocky littoral regions in the Alaskan Beaufort Sea (Dunton and Schonberg 2000).

As a result of the strong effect of the Antarctic Circumpolar Current on the dispersal of seaweed propagules (Lüning 1990) many non-endemic species of the Antarctic seaweed flora have a circumpolar distribution. Among the species also occurring on sub-Antarctic islands and Tierra del Fuego are the red alga *Iridaea cordata*, the brown alga *Geminocarpus geminatus* and the green alga *Monostroma hariotii*. Some species, e.g. the red alga *Ballia callitricha* and the brown alga *Adenocystis utricularis*, even occur in New Zealand and Australia. A similar connection between the Antarctic and the cold temperate region of South America has also been recorded for marine invertebrates (Clarke et al. 2005). At least 20 algal species from the Antarctic are cosmopolitan, e.g. the red alga *Plocamium cartilagineum*, the brown alga *Petalonia fascia* and the green alga *Ulothrix flacca* (Papenfuss 1964; Wiencke and Clayton 2002). It is possible that some such species may be recent invaders from temperate regions (Clayton et al. 1997).

A few seaweed species have a disjunct amphi-equatorial distribution and occur both in the Antarctic and the Arctic, e. g. *Acrosiphonia arcta* and *Desmarestia viridis/confervoides*. Molecular studies indicate that the biogeographic disjunctions of these species are recent and probably date back to the maximum of the Würm/Wisconsin glaciation 18,00 years ago (van

Oppen et al. 1993). The units of dispersal were certainly the microscopic stages in the life cycle because they are more resistant to high temperatures allowing the possibility of a passage through the tropics during times of lowered water temperatures (Peters and Breeman 1992; Wiencke et al. 1994; Bischoff and Wiencke 1995a).

Apart from the few species endemic to the Arctic and various cosmopolitan species such as the red alga *Audouinella purpurea*, the brown alga *Scytosiphon lomentaria* and the green alga *Blidingia minima*, the Arctic seaweed flora has affinities to the Atlantic, the Pacific and the Indo-Pacific region (Wilce 1990). An example of an invader from the Pacific may be *Acrosiphonia arcta*. Populations of this species from the Arctic and the North Atlantic seem to originate from the Pacific according to molecular studies (van Oppen et al. 1994). But over 90% of the species in the Arctic originate from Atlantic populations (Dunton 1992). This is particularly obvious in regions with strong influx of Atlantic waters, e.g. in Svalbard (Svendsen et al. 2002, Hop et al. 2002). However, even in the coastal areas of the Beaufort Sea there is a strong Atlantic influence, marked by the kelps *Laminaria solidungula* and *Alaria esculenta* (Dunton 1992) while the Canadian Arctic also has a high proportion of macroalgae of Pacific origin (Cross et al. 1987).

In both polar regions seaweeds are almost entirely subtidal. However, there are some specialized species with a bipolar distribution that occur exclusively in the supralittoral, i.e. in the spray zone. These are the green alga *Prasiola crispa*, which also grows more inland close to seabird rockeries under conditions of low pH and high nutrient concentrations (Knebel 1936) and the red alga *Bangia atropurpurea* (Bird and Mc Lachlan 1992; Clayton et al. 1997). The green alga *Urospora penicilliformis* and species of the genus *Ulothrix* grow just around the high tide level both in the Antarctic and the Arctic. *Acrosiphonia arcta* is a species typical for the lower eulittoral in both polar regions.

A conspicuous species in the upper eulittoral in the Antarctic is the red alga *Porphyra endiviifolium* and in the lower eulittoral/upper sublittoral the green alga *Enteromorpha bulbosa* and the brown alga *Adenocystis utricularis* are common (Westermeier et al. 1992; Wiencke and Clayton 2002). Tide pools and crevices in the lower sublittoral are often colonized by upper sublittoral species such as the red algae *Palmaria decipiens* and *Iridaea cordata*. The upper 5–15 m of the sublittoral are exposed to ice floes and are often devoid of large, perennial algae. Only crustose species or developmental stages can persist here. Below this zone, large brown algae dominate the sublittoral in West Antarctica: *Ascoseira mirabilis* and *Desmarestia menziesii* occur in the upper sublittoral, *D. anceps* in the mid sublittoral and *Himantothallus grandifolius* grows in the lower sublittoral (DeLaca and Lipps 1976; Lamb and Zimmermann 1977; Amsler et al. 1995; Klöser et al. 1996; Quartino et al. 2001, 2005). At higher latitudes in East Antarctica only few species occur, in particular *Palmaria decipiens*, *Phyllophora antarctica* and *I. cordata* (Zaneveld 1968; Miller and Pearse 1991; Cormaci et al. 1992).

In the Arctic the barren zone heavily exposed to floating ice extends down to about 2 m water depth. The upper sublittoral on Spitsbergen is characterized by the brown algae *Fucus distichus*, *Pylaiella littoralis*, *Chordaria flagelliformis*, the kelp *Saccorhiza dermatodea*, the red alga *Devaleraea ramentacea* and green algae of the genus *Acrosiphonia*. The zone below is dominated by the kelps *Alaria esculenta*, *Laminaria digitata* and *L. saccharina*. Characteristic for the lower sublittoral are the red algae *Phycodrys rubens* and *Ptilota gunneri*. The Arctic kelp, *L. solidungula*, occurs in this zone as well but only in the inner part of the fjords (Svendsen 1959; Hop et al. 2002; Wiencke et al. 2004). The seaweed vegetation in the Canadian Arctic (Baffin Island) is similar to that in Spitsbergen with a shallow *F. distichus* zone, the upper sublittoral characterized by *Stictyosiphon tortilis*, *P. littoralis* and *Dictyosiphon foeniculaceus* (Cross et al. 1987). Deeper communities are dominated by the kelps *L. saccharina*, *L. solidungula*, *A. esculenta* and *Agarum cribrosum*, interspersed with the red algae *Dilsea integra*, *D. ramentacea* and *Rhodomela confervoides*. In contrast, no supralittoral or intertidal algae have been recorded from the Alaskan Beaufort Sea. Here, algae are restricted to scattered rocky habitats in shallow waters (about

5–10 m), which are protected by barrier islands from iceberg scouring. *Laminaria solidungula* is the dominant kelp species, although *Alaria esculenta* and *L. saccharina* can also be found regularly (Dunton and Schonberg 2000). Characteristic red algal species are *Phycodrys rubens*, *Odonthalia dentata*, *Dilsea integra*, *Phyllophora truncata* and *Rhodomela subfusca*. Exposed rock is usually covered by encrusting coralline red algae, *Lithothamnion* spp. (Konar and Iken 2005).

## Reproductive and physiological thallus anatomy

The enigma of reproduction and life histories of endemic Antarctic Desmarestiales and of *Ascoseira mirabilis* has been clarified. The sorus structure of *H. grandifolius* (Moe and Silva 1981; Wiencke and Clayton 1990), *D. anceps* (Moe and Silva 1981; Wiencke et al. 1996) and *D. antarctica* (Moe and Silva 1989; Wiencke et al. 1991) is similar, indicating an evolutionary relationship. Unilocular sporangia are cylindrical, borne terminally on 2–4 celled stalks and are interspersed with paraphyses composed of 2–4 cells with many physodes. With respect to sorus morphology, *D. menziesii* exhibits a closer similarity to the Arctic cold-temperate species *D. aculeata* (Wiencke et al. 1995), whereas *Phaeurus antarcticus* resembles the North Atlantic-Mediterranean species *Arthrocladia villosa* (Clayton and Wiencke 1990). In *P. antarcticus*, sporangia are catenate and develop in rows in adjacent cells as filamentous outgrowths of the cortex, interspersed with club-shaped sterile hairs. These phylogenetic relationships have been confirmed meanwhile by molecular data (Peters et al. 1997). *Ascoseira mirabilis*, whose exact phylogenetic relationships are still unresolved, has a fucalean type of life history (Clayton 1987). There is one free-living diploid generation with conceptacles scattered over both surfaces. These contain densely packed chains of gametangia that release biflagellate isomorphic gametes. Zygotes develop into new individuals.

A striking feature of several Antarctic Desmarestiales and *A. mirabilis* is their anatomical complexity resembling that of *Laminaria* species from the northern Hemisphere. As far as we know today these taxa are characterized by a highly diversified structure and complex morpho-functional processes including a temporal synchronization between photosynthesis and growth as well as the long-distance transport of organic substances (Lüning et al. 1973; Schmitz 1981; Clayton and Ashburner 1990; Drew and Hastings 1992; Gómez et al. 1996). Such characteristics represent strategies to cope with the marked seasonality in polar and cold-temperate regions (Chapman and Lindley 1980; Gómez and Wiencke 1998; Gómez and Lüning 2001).

As in other species of *Laminaria* the blade of *L. solidungula* has a basally located meristem, which is active only in winter and forms—comparable with *L. hyperborea*—a new blade at the distal end (Taylor 1966; Dunton and Jodwalis 1988). Up to three generations of blades may be found in one individual. As in species of *Laminaria*, blades of *Himantothallus grandifolius* grow in their lower part (Dieckmann et al. 1985; Drew and Hastings 1992). Punch-hole experiments with *A. mirabilis* indicate that the blade elongates longitudinally as in *Laminaria solidungula* through a seasonally active intercalary meristem (Gómez et al. 1995a).

Carbon fixation rates in kelps and kelp-like brown algae differ between blade parts (Arnold and Manley 1985; Gómez et al. 1995a, b; Cabello-Pasini and Alberte 2001). During the growth phase in late winter-spring, net photosynthetic rates (net $P_{max}$) in *A. mirabilis* are slightly higher in the middle region compared to the basal and distal regions (Gómez et al. 1995a). In comparison, in *L. solidungula*, C-fixation rates are lowest in meristematic tissue, highest in first year tissue and intermediate in second-blade tissue (Dunton and Jodwalis 1988). This pattern is related to the ontogeny within the blade, i.e. photosynthetic activity increases with age of the tissues reaching a maximum but then decreases with further aging. In apical regions, erosion and senescence take place.

Over the years, the blade in *Ascoseira* becomes thicker and more complex as new tissue is formed each growth season, resulting in modifications not only of the photosynthetic gradients but also of the photosynthetic capacity and efficiency along the thallus. Two age-components must be considered here: The age of the different blade tissues and the age of the whole plant. In this context, 2-year old plants of *Ascoseira* are characterized by net $P_{max}$ and $\alpha$-values almost twice that of 3-year old individuals. However, other parameters such as dark respiration, saturation ($E_k$) and compensation ($E_c$) points do not show obvious differences (Gómez et al. 1996). The photosynthetic variability correlated with tissue age may also be seen in terms of the development of the photosynthetic apparatus with ontogeny. In both age classes, middle regions have a higher net $P_{max}$ than basal and distal regions, suggesting that photosynthesis is low due to the presence of a non-developed photosynthetic apparatus in the basal region, and decreases in the oldest distal tissues due to senescence processes. Similar results were obtained in young and adult plants of *L. solidungula* (Dunton and Jodwalis 1988).

Küppers and Kremer (1978) associated the increased $^{14}C$-assmilation in the middle and distal regions of *Laminaria* species with a higher activity of the Calvin cycle enzyme ribulose 1,5-bisphosphate carboxylase-oxygenase (RUBISCO). Moreover, these authors demonstrated longitudinal profiles in light independent C-fixation in these species coupled to a high activity of the enzyme phosphoenolpyruvate carboxykinase (PEP-CK) in the meristematic region ($\beta$-carboxylation). The activities of these carboxylating enzymes apparently respond to the growth characteristics of *Laminaria*. In fact, *Laminaria* species from cold-temperate and Arctic regions grow in winter or in dim light and it is likely that such an alternative carboxylating mechanism is advantageous for these species (Küppers and Kremer 1978). $\beta$-carboxylation is also linked to the growth requirements of these algae. For example, light-independent carbon fixation provides C-skeletons (preferentially amino-acids) for both biosynthesis and anabolic processes, thus replenishing some carbon intermediates in e.g. the Krebs cycle (Kremer 1981b; Falkowski and Raven 1997). Particularly, in the meristematic region of *Laminaria*, PEP-CK uses $CO_2$ lost in glycolysis of mannitol translocated from the distal region to the meristem (Kremer 1981a; Kerby and Evans 1983).

The patterns of C-fixation in *Ascoseira* exhibit also intra-blade variations (Gómez et al. 1995a) although some differences with respect to patterns reported in *Laminaria* are observed. Both light dependent and light independent C-fixation increase with tissue age reaching a maximum in the middle/distal parts of the blade. In contrast, members of the Laminariales studied to date show highest rates of light dependent carbon fixation in distal and highest light independent C fixation in the basal parts (Küppers and Kremer 1978). The different thallus allocation of light independent C fixation in *A. mirabilis* may be related to the high dark respiration rates observed in distal blade regions (Gómez et al. 1995a, 1996). It is not clear, however, whether light independent C-fixation may compensate for C losses due to respiration as reported by Kremer (1981b). Thomas and Wiencke (1991) could not conclusively demonstrate a relationship between light independent C-fixation and dark respiration in Antarctic marine algae. In general, dark C-fixation varied between 4.9% and 31% of dark respiration in five brown algae and one red alga. In species such as *H. grandifolius* and *D. anceps*, low dark C-assimilation rates were coupled to high respiration rates as in *Ascophyllum nodosum* indicating a net C loss due to respiration in the dark (Johnston and Raven 1986).

In kelps sugars are synthesized and stored as polysaccharides (e.g. laminaran) in the middle and distal assimilatory tissues of the algae, from which they are then translocated as low-molecular-weight compounds (e.g. mannitol) in trumpet hyphae to the growth region (Kremer 1981a, b). The trumpet hyphae of Laminariales are elongated, longitudinally oriented cells in the medulla with trumpet like ends. The cross walls are perforated by numerous pits resembling the sieve plates of sieve tubes in higher plants (Schmitz 1981, 1990; Buggeln 1983). Similar trumpet hyphae are present in the medulla of *H. grandifolius* (Moe and Silva 1981) and in the other Desmarestiales *Phaeurus antarcticus* (Clayton and Wiencke 1990), *Desmarestia antarctica* (Moe and Silva 1989; Wiencke et al. 1991) and *D. menziesii* (Scrosati 1992; Wiencke et al. 1995). The trumpet hyphae can be interconnected by sieve plates and are always surrounded by numerous small cells that probably function as transfer cells. In *A. mirabilis* longitudinally arranged "conducting channels" comparable to the trumpet cells found in Laminariales are present in the medulla (Clayton and Ashburner 1990). The channels are aseptate, multinucleate structures with ensheathing filaments often interconnected through sieve-like wall perforations. Although no conclusive evidence for a possible transport function of the conducting channels in *A. mirabilis* is available, differential allocation of reserve carbohydrates associated to meristem activity strongly suggest the involvement of these structures in long distance transport (Gómez and Wiencke 1998).

Overall, the findings that photosynthesis and carbon allocation vary as a function of blade development add new evidence to a convergent morpho-functional evolution between large Antarctic brown algae and Laminariales of the northern Hemisphere. In this case not only morphological organization, but also the metabolic differentiation along the blade, are common characteristics of the taxa. Such characteristics certainly reflect adaptive mechanisms to withstand resource limitation in seasonally changing environments, which have ultimately led to the ecological success of these algae in polar regions.

## Temperature demands and geographical distribution

Photosynthesis of polar seaweeds shows a considerable adaptation to the low temperatures of the environment. Maximum photosynthetic rates of endemic Antarctic species are at a temperature of 0°C similar to values from temperate species measured at higher temperatures (Drew 1977; Thomas and Wiencke 1991; Wiencke et al. 1993; Weykam et al. 1996; Eggert and Wiencke 2000). Moreover, the temperature optima for photosynthesis at least in some species from the Antarctic are well below values determined in temperate species. The lowest temperature optima have been determined in the brown algae *Ascoseira mirabilis* (1–10°C) and *Himantothallus grandifolius* (10–15°C; Drew 1977; Wiencke et al. 1993). The red algae *Ballia callitricha* and *Gigartina skottsbergii* also exhibit values between 10 and 15°C, whereas *Kallymenia antarctica*, *Gymnogongrus antarcticus* and *Phyllophora ahnfeltioides* exhibit broad optima between 10 and 20 (to 25)°C (Eggert and Wiencke 2000). Studies on the few Arctic seaweeds show a temperature optimum at 20°C (Healey 1972). For comparison, in cold-and warm-temperate species optimum values of 20–25°C and 25–35°C were determined, respectively (Lüning 1990). The temperature optima for respiration are clearly above the optima for photosynthesis but temperatures higher than 30°C have never been tested in Antarctic seaweeds (Drew 1977; Wiencke et al. 1993; Eggert and Wiencke 2000). Photosynthesis:respiration ratios are highest at the lowest tested temperature, 0°C, and decrease with increasing temperatures due to different Q10 values for photosynthesis (1.4–3.5) and respiration (2.5–5.1).

The high P:R ratios at low temperatures are the major reason for the high growth rates of Antarctic species at low temperatures. In particular red and brown algal species from both polar regions exhibit temperature growth optima at 0–5°C or even at –2°C (Wiencke and tom Dieck 1989, 1990; Novaczek et al. 1990; Bischoff and Wiencke 1993; Bischoff-Bäsmann and Wiencke 1996; Eggert and Wiencke 2000; McKamey and Amsler 2006). Some species like *Georgiella confluens*, *Gigartina skottsbergii* and *Plocamium cartilagineum* from the Antarctic grow only at 0°C, the lowest temperature tested, but not at 5°C (Bischoff-Bäsmann and Wiencke 1996). The upper survival temperatures (USTs), determined after 2-week-exposures to different temperatures, are as low as 7–11°C. Other Antarctic red algae grow up to 5°C or 10°C and have USTs of ≤19°C (Bischoff-Bäsmann and Wiencke 1996). The sporophytes of endemic Antarctic Desmarestiales grow up to 5°C and exhibit USTs of 11–13°C. In contrast, their gametophytes grow up to 10°C or 15°C and have USTs between 15°C and 18°C (Wiencke and tom Dieck 1989, 1990). The upper limit for gametogenesis (ULG) in *D. antarctica* is 5°C (Wiencke et al. 1991). Antarctic cold-temperate species (especially from the eulittoral) have higher temperature requirements. They grow up to 10–15 (or 20)°C and show USTs between 13.5°C and 19°C. Microthalli of Antarctic cold-temperate brown algae exhibit

USTs between 21°C and 25°C and ULGs between 13°C and 15°C (Wiencke and tom Dieck 1990; Peters and Breeman 1993).

As far as we know today the temperature demands for growth of Arctic seaweeds are somewhat higher than endemic Antarctic species. It must, however, be pointed out that the temperature demands of truly endemic Arctic species (see chapter 3) have not been investigated so far. The sporophyte of the kelp *Laminaria solidungula*, whose distribution extends into the cold-temperate region as far as Newfoundland, grows at temperatures up to 15°C with an optimum at 5–10°C and USTs of 16°C (tom Dieck 1992). The male and female gametophytes of this species exhibit an UST of 18°C and lower survival temperatures (LSTs) of ≤–1.5°C (Bolton and Lüning 1982; tom Dieck 1993). The red alga *Devaleraea ramentacea*, which is distributed from the Arctic to the cold temperate North Atlantic region, grows at up to 10°C with an optimum at 0°C and exhibits an UST of 18–20°C and an LST of ≤–5°C. The ULG of this species is 8°C (Novaczek et al. 1990; Bischoff and Wiencke 1993). Macrothalli of species with a prominent distribution in both, the Arctic and the cold-temperate region, grow at up to 15 or 20 (to 25)°C with optima between 5 and 15 (to 20)°C and exhibit USTs between 17 and 25 (to 26)°C. The LSTs are ≤–1.5 or 2°C (Wiencke et al. 1994). The microscopic gametophytes of Arctic cold-temperate Laminariales grow at up to 20°C and have USTs between 22 and 25 (to 28)°C. Gametophytes of *Alaria esculenta* and *Agarum cribrosum*, however, exhibit USTs of 19–21°C, as low as those of the more Arctic species *Laminaria solidungula*. The LSTs are either 0, –1.5 or –2°C, the few data on ULGs vary between 10°C and 17°C (Wiencke et al. 1994).

The northern distribution of endemic Antarctic species is often limited by the temperature demands for growth. This applies especially to members of the endemic Antarctic species of the order Desmarestiales. In these taxa the northern distribution limit is determined by the temperature requirements for growth of the sporophytes (Wiencke and tom Dieck 1989), which occur only south of the Antarctic convergence in areas with maximum temperatures ≤5°C allowing growth of their sporophytes. The USTs of their sporophytes and gametophytes as well as the temperature demands for growth of the gametophytes are irrelevant for the explanation of the geographic distribution of these species.

The distribution of Arctic North Atlantic species is often limited both by the USTs and the ULGs (van den Hoek 1982a, b; Breeman 1988; Lüning 1990). In such taxa, distribution limits in the West Atlantic are determined by lethal, high summer temperatures, whereas they are determined in the East Atlantic by high winter temperatures inhibiting reproduction. Examples for species from this group are *Laminaria digitata*, *Chorda filum* and *Halosiphon tomentosus*.

During the ice ages both polar regions were inhospitable for seaweeds. In the Southern Hemisphere the sub-Antarctic islands probably served as refugia. The South American Archipelago may also have hosted species escaping from the coasts of the Antarctic continent (Skottsberg 1964). At the maximum of the last ice age the 5°C summer isotherm, the boundary of the Antarctic region just touched the southern tip of South America (CLIMAP project members 1981). Migration of species from the Antarctic continent to South America and vice versa probably occurred along the Scotia Arc, a well-recognized route for animals (Knox and Lowry 1978). In the Northern Hemisphere cold water seaweeds of the Atlantic with ULGs and upper temperature limits for growth at 10 or 15°C experienced an extreme reduction in the distribution area during the ice ages (van den Hoek and Breeman 1989). The distribution of these species was compressed by the glaciers in the North and the 10–15°C winter isotherm, their southern reproduction boundary. This is the likely explanation for the present depauperate flora in the North West Atlantic. Comparable conditions did not exist in the North Pacific, probably a major reason for the richness of the cold North Pacific.

During periods of lowered temperatures taxa with relatively high temperature demands were able to extend their distribution limits towards the equator. Species with an amphiequatorial distribution e.g. *Acrosiphonia arcta* (Bischoff and Wiencke 1995a), *Urospora penicilliformis* (Bischoff and Wiencke 1995b) and *Desmarestia viridis/confervoides* (Peters and Breeman 1992) have probably crossed the equator during the Pleistocene lowering of the water temperatures in the tropics (Wiencke et al. 1994). The USTs of these taxa are with 25–27°C slightly above the minimum tropical sea surface temperatures of 23–25°C during the last glaciation (CLIMAP Project members 1981). These results are supported by molecular studies indicating an almost complete sequence identity in amphiequatorial populations of *A. arcta* and *D. viridis/confervoides* in the fast-evolving internal transcribed spacer regions of the ribosomal DNA (van Oppen et al. 1993).

## Light demands and depth zonation

At high latitudes, the radiation regime imposes severe constraints not only in terms of seasonal light availability but also in regard to the vertical distribution of benthic algae (Gómez et al. 1997). Because of the extreme environmental conditions in the eulittoral, polar algae are mainly sublittoral and thus low light demands and tolerance to darkness are a pre-requisite for occurrence down to great depths (Arnoud 1974; Zielinski 1981; Richardson 1979; Klöser et al. 1993; Amsler et al. 1995). An important feature is the dark tolerance of the microscopic developmental stages. Various Antarctic and Arctic seaweeds tolerate a dark period of up to 18 months (tom Dieck 1993; Wiencke 1988, 1990a). Further evaluation of the relation between photosynthetic characteristics and algal zonation in 36 macroalgal species from King George Island (Weykam et al. 1996, Gómez et al. 1997) indicates a high degree of shade adaptation: (a) photosynthetic efficiency ($\alpha$) is high in all the studied species, reflecting a clear shade adaptation over a broad range of depth; (b) seaweeds growing at depths >10 m exhibit low saturation points for photosynthesis ($E_k$; < 40 μmol $m^{-2}$ $s^{-1}$) irrespective of their taxonomic position; (c) the highest $E_k$ values (>50 μmol $m^{-2}$ $s^{-1}$) are found in species common in upper sublittoral or eulittoral; (d) shallow water species have higher photosynthetic capacity ($P_{max}$) than species from deeper waters. These data obtained using field material are supported also by results obtained in studies using specimens grown from unialgal cultures from the same study area (Wiencke et al. 1993; Eggert and Wiencke 2000). The highest degree of shade adaptation with average $E_k$ values around 3 μmol $m^{-2}$ $s^{-1}$ has been demonstrated in coralline algae from the Ross Sea (Schwarz et al. 2005).

The vertical distribution of dominant Antarctic seaweeds such as *Desmarestia* or *Himantothallus* can be extremely wide in contrast to the algal zonation patterns from cold-temperate and temperate regions, where zonation patterns are characterized by narrow belts of species (Lüning 1990). However, no evidence of an acclimation to the prevailing light conditions at different depths could be demonstrated in five species of seaweeds along a depth gradient between 10 m and 30 m (Gómez et al. 1997). Apparently, the intrinsic low light requirements for photosynthesis account for these patterns. In the Antarctic spring-summer, due to high water transparency irradiances of PAR around 20 μmol $m^{-2}$ $s^{-1}$ reach the depth of 30 m, with 1% of the surface irradiance present at depths >40 m (Gómez et al. 1997). Although these levels are clearly lower than average midday irradiances (30–325 μmol photon $m^{-2}$ $s^{-1}$) at 30 m in some clear temperate and tropical waters (Peckol and Ramus 1988), they exceed reported saturation and compensation points of photosynthesis in many Antarctic seaweeds (Weykam et al. 1996).

As for photosynthesis, the light demands for growth are similarly very low. In microscopic developmental stages of Antarctic seaweeds growth is light saturated already at 4–12 μmol $m^{-2}$ $s^{-1}$ (Wiencke 1990a). In young sporophytes of Antarctic Desmarestiales growth is saturated at 15–20 μmol $m^{-2}$ $s^{-1}$ (Wiencke and Fischer 1990).

In the case of Arctic species, low light requirements for photosynthesis have also been found in nine species from Spitsbergen (Latala 1990) and in crustose coralline red algae from the high Arctic (Roberts et al. 2002). Meristematic tissue of both juvenile and adult individuals of *Laminaria solidungula* from the Alaskan high Arctic exhibits $E_k$ values between 20 and 30 μmol $m^{-2}$ $s^{-1}$, while the average $E_k$ value of vegetative blade tissue in adult plants is 38 μmol $m^{-2}$ $s^{-1}$ (Dunton and Jodwalis 1988). Photosynthetic efficiency ($\alpha$) is higher than in Laminariales from the temperate region. Both the low $E_k$ and $\alpha$ values indicate strong adaptation to the long period of exposure to darkness in winter under ice cover and in summer during times of high water turbidity. Another feature characterising the dark adaptation of *L. solidungula* is the low compensation point ($E_c$) for growth close to 0.6 μmol $m^{-2}$ $s^{-1}$ (Chapman and Lindley 1980).

Seaweeds able to grow in deep waters have developed metabolic strategies to maximize carbon fixation by avoiding excessive carbon losses due to respiration. Because at great depths, $E_c$ for photosynthesis exceed $E_c$ values for growth and available irradiances are normally below the levels required for saturation of photosynthesis, carbon assimilation may be just compensating dark respiration. In Antarctic brown algae, dark respiration has a strong seasonal component and during the growth period, respiratory activity may account for a considerable proportion of the gross photosynthesis (Gómez et al. 1995a, b). If one relates the daily light course of the irradiance to the $E_k$ value, then the daily period for which plants are exposed to irradiances >$E_k$ (denominated $H_{sat}$) can be estimated. The obtained metabolic daily carbon balance can be regarded as a physiological indicator for the ability to grow in deep waters (Gómez et al. 1997).

Polar seaweeds exposed to marked seasonal changes in daylength exhibit generally $H_{sat}$ values >0 h only during the short open water season. *Laminaria solidungula* in a kelp bed in the Alaskan Beaufort Sea at 70 °N was exposed during August to September 1986 to total $H_{sat}$ periods of up to 148 h (Dunton 1990). This value corresponds to an average daily $H_{sat}$ of 3 h. However, depending on the year, the average daily $H_{sat}$ may decrease down to <0.5 h. Year-round underwater light measurements have yielded annual quantum budgets of 45 mol $m^{-2}$ $yr^{-1}$, the lowest ever documented for kelp populations globally (Dunton 1990). A similar low value, 49 mol $m^{-2}$ $yr^{-1}$, was already obtained earlier by Chapman and Lindley (1980), values considerably lower than in the temperate *L. hyperborea*, which receives 71 mol $m^{-2}$ $yr^{-1}$ (Lüning and Dring 1979).

For Antarctic seaweeds, $H_{sat}$ measured during optimum light conditions in spring in five brown and red algae decreases with depth from values close to 14 h at 10 m to values between 7 h and 12 h at 30 m depth (Gómez et al. 1997). For the red algae *Palmaria decipiens*, *Kallymenia antarctica* and *Gigartina skottsbergii* a metabolic carbon balance between 0.6 and 0.8 mg C $g^{-1}$ FW $d^{-1}$ sets the limits for growth at >30 m. For *Himantothallus grandifolius*, which dominates depths below 15 m, the daily carbon balance was low but relatively similar over a range between 10 m and 30 m, indicating that this species can potentially grow even considerably deeper. Only in *Desmarestia anceps* is growth clearly limited at 30 m due to its negative carbon balance at this depth (–1.9 mg C $g^{-1}$ FW $d^{-1}$; Gómez et al. 1997). The lower depth distribution limit of the red alga *Myriogramme mangini* from the South Orkney Islands has been predicted by photosynthetic measurements to be at 23 m water depth (Brouwer 1996).

Overall, polar seaweeds by virtue of their low light demands are potentially able to grow over large ranges of depths. Depending on the water characteristics the lower depth distribution limit of seaweeds both in the Antarctic and the Arctic is located between 30 m and 90 m (Wilce 1994; Wiencke and Clayton 2002). However, other constraints such as competition also control algal zonation. For example, red algae are metabolically able to live at great depths, however, they can be outcompeted by the large canopy brown alga *Himantothallus* (Klöser et al. 1996). Furthermore, ice abrasion and grazers, e.g. some invertebrate and demersal fishes, have a strong influence on zonation patterns (Iken 1996; Iken et al. 1997).

Antarctic seaweeds are not only strongly shade-adapted but can also cope with high light conditions in summer due to their ability for dynamic photoinhibition, a photoprotective mechanism by which excessive energy absorbed is rendered harmless by thermal dissipation (Krause and Weis 1991; Demmig-Adams and Adams III 1992). This capability is best expressed in species from the eulittoral. Upper and mid sublittoral species show a more or less pronounced decrease in photosynthetic activity during high light stress and full recovery during subsequent exposure to dim light (Hanelt et al. 1997). In contrast, deep water and understory species recover only slightly and slowly, indicating photodamage. Similar findings as these on Antarctic seaweeds have been demonstrated also in a detailed analysis of seaweeds from Spitsbergen (Hanelt 1998). Similarly, UV radiation is now also regarded as a key factor affecting the depth zonation of seaweed assemblages. This topic will be treated in detail in a separate paper in this issue (Bischof et al. in press).

## Effect of salinity, temperature and desiccation on supra-and eulittoral seaweeds

Until now few ecophysiological studies have been undertaken on the salinity acclimation in polar seaweeds (Karsten et al. 1991a, b; Jacob et al. 1991, 1992a). These papers indicate that cells of eulittoral Chlorophyta from the Antarctic survive salinities between 7 psu and 102 psu with a low rate of mortality, and that most taxa grow, photosynthesize and respire optimally under normal seawater conditions with rather broad tolerances between 7 psu and 68 psu. The supralittoral *Prasiola crispa* ssp. *antarctica* even grows between 0.3 psu and 105 psu. Consequently, eulittoral and supralittoral macroalgae from Antarctica can be characterized as euryhaline organisms.

Osmotic acclimation in response to salinity changes is a fundamental mechanism of salinity tolerance that conserves intracellular homeostasis (Kirst 1990). The acclimation process in Antarctic Chlorophyta involves the metabolic control of the cellular concentrations of osmolytes. The major inorganic osmolytes are potassium, sodium and chloride (Karsten et al. 1991b; Jacob et al. 1991), the cellular concentrations of which can be rapidly adjusted with low metabolic energy costs, especially compared to the cost of organic osmolyte biosynthesis or degradation (Kirst 1990). However, protein and organelle function, enzyme activity and membrane integrity in seaweeds are adversely affected by increased electrolyte concentration. Hence, the biosynthesis and accumulation of organic osmolytes in the cytoplasm permits the generation of low water potentials without incurring metabolic damage. For these organic compounds that are tolerated by the metabolism even at high intracellular concentrations, the term 'compatible solute' was introduced by Brown and Simpson (1972).

All Antarctic Chlorophyta studied possess as main organic osmolytes the carbohydrate sucrose and the imino acid proline, and—with the exception of *Prasiola*—a third compound, $\beta$-dimethylsulphoniumpropionate (DMSP). *Prasiola* is able to synthesize polyols such as sorbitol and ribitol instead of DMSP. The concentrations of all osmolytes are actively regulated as a function of the external salinity. Because of its physicochemical properties proline is one of the most potent organic osmolytes, which not only balances salinity stress, but also may stimulate enzymatic activity. This imino acid is the most important osmolyte in ice-algae too (Thomas and Dieckmann 2002). Sucrose is also a well-known osmotically active compound in many higher and lower plants, and also exhibits a cryoprotective function. The osmotic function of DMSP seems to be unique to Antarctic Chlorophyta in comparison with temperate ones, because of very high intracellular concentrations and the strong biosynthesis and accumulation with increasing salinities. In addition, DMSP acts as compatible solute and cryoprotectant (Karsten et al. 1996a). Polyols such as sorbitol are the most water-like molecules, and therefore represent not only ideal osmolytes and compatible solutes, but exert a function as rapidly available respiratory substrate (e.g. under desiccation) and as antioxidant (Karsten et al. 1996b). The presence of high concentrations of three organic osmolytes in Antarctic Chlorophyta may be related to the extremely cold habitat suggesting that, besides their osmotic role, these solutes are playing multiple functions in metabolism such as cryoprotectants. In the case of *P. crispa* ssp. *antarctica* the broad salinity tolerance is also assisted by the thickness and mechanical properties of the cell walls (Jacob et al. 1992a). Another ultrastructual feature of *Prasiola* cells is the absence of vacuoles under hyposmotic conditions up to full seawater and the formation of vacuoles at high salinities. Under the latter conditions, they may serve as compartments accumulating inorganic ions.

Compared to the eulittoral Chlorophyta, much less is known of the ecophysiology of the red algae *Bangia atropurpurea* and *Porphyra endiviifolium* from Antarctica. The few data on temperature requirements for growth and survival indicate eurythermal characteristics for both Rhodophyta, i.e. they survive temperatures up to 21–22°C (Bischoff-Bäsmann and Wiencke 1996). Studies on closely related species of both genera from other biogeographic regions clearly indicate an extremely broad tolerance of all environmental factors, which is based on a high metabolic flexibility (Karsten et al. 1993; Karsten and West 2000). *Bangia* and *Porphyra* synthesize and accumulate three isomeric heterosides with different physiological functions as osmolytes, compatible solutes and soluble carbon reserve.

Besides hypersaline conditions decreasing temperatures also strongly stimulate the biosynthesis and accumulation of DMSP in Antarctic Chlorophyta (Karsten et al. 1996a). In addition, DMSP not only stabilises the cold-labile model enzyme lactate dehydrogenase and malate dehydrogenase extracted from *Acrosiphonia arcta* during several freezing and thawing cycles, but also stimulates both enzyme activities at in situ concentrations. Consequently, DMSP acts as an effective cryoprotectant (Nishiguchi and Somero 1992). Recent studies on Antarctic *Prasiola* and sea ice diatoms indicate the presence of ice-binding proteins (IBP) that modify the shape of growing intracellular ice crystals during freezing (Raymond and Fritsen 2001). IBPs do not lower the freezing point, they rather seem to prevent damage to membranes by the inhibition of the recrystallization of ice (Raymond and Knight 2003). During recrystallization small ice particles typically grow to large grains of ice, which may physically injure cell membranes. Consequently, IBPs act as effective structural cryoprotectants.

In the eulittoral and supralittoral zone, low temperatures combined with high irradiance represent a particular challenge to algal physiology. At low temperature, and, thus, reduced enzyme activities and decreased turn-over velocity of D1 reaction centre protein in photosystem II (Davison 1991; Andersson et al. 1992; Aro et al. 1993), a persisting high irradiance of PAR may result in increased electron pressure in photosynthesis. This may ultimately result in the generation of reactive oxygen species within the Mehler reaction (Polle 1996) and increase oxidative stress (Asada and Takahashi 1987). The consequences of increased oxidative stress are chronic photoinhibition/photoinactivation, bleaching of photosynthetic pigments, peroxidation of membrane lipids and enhanced degradation of D1 protein (Aro et al. 1993; Osmond 1994). Hitherto this phenomenon is hardly studied in Antarctic macroalgae (Hanelt et al. 1994, 1997). However, the limited information available suggests that polar algae from the eulittoral, such as the brown alga *Adenocystis utricularis*, may overcome radiation stress at low temperatures by their ability for dynamic photoinhibition/photoprotection (Osmond 1994), which proceeds much faster than in sublittoral algae (Hanelt et al. 1994, 1997). The rate of inhibition seems to be independent of temperature. The underlying mechanism is still unexplored, but possible explanations include the adaptation of enzymes involved in the D1 repair cycle at low temperatures, and the insignificant role of D1 in photoinhibition and recovery of photosynthesis in Antarctic macroalgae (Hanelt et al. 1994). The ability for fast photoinhibition and to completely halt photosynthesis under steadily decreasing temperatures at constant irradiance levels was also found in eulittoral *Fucus distichus* from Arctic Spitsbergen (Bischof and Walter, unpublished data). Here, a critical temperature level to initiate pronounced photoinhibition was found to be –3°C. As soon as temperatures fall below this threshold, photosynthesis becomes rapidly inhibited, and photosynthetic quantum yield completely ceases at –15°C. Recovery from freezing temperature proceeds almost in reverse, with a rapid increase in quantum yield as soon as temperatures climb above –3°C. By their ability to reduce photosynthetic quantum yield, polar eulittoral algae seem to be able to avoid increasing oxidative stress under low temperature conditions. However, some species seem to apply other strategies of photoprotection under low temperature, since Becker (1982) reported for Antarctic *Prasiola* photosynthetic activity down to –15°C. The capacity to photosynthesize as efficiently in air (when hydrated) as in water, in combination with a high affinity for inorganic carbon, has been described for a temperate species of this genus (Raven and Johnston 1991).

Desiccation is a strong stress parameter in supralittoral species. Air exposed thalli of

*Prasiola crispa* loose 75% of the cellular water during the first 6 h of desiccation (Jacob et al. 1992b). A water loss of more than 90% leads to irreversible damage. Growth rates after reimmersion in seawater depend on the thallus water content and the length of the desiccation period. Few ultrastructural changes were found after desiccation. As after salinity stress, the very thick cell walls of the species and the absence of vacuoles are regarded as prerequisites to survive periods of desiccation (Jacob et al. 1992b).

Although there is a lack of knowledge of the ecophysiological performance of many eulittoral macroalgae from Antarctica, the known data clearly indicate broad tolerances against the prevailing environmental, often extreme parameters. In particular, the biochemical capability to synthesize and accumulate various protective compounds seems to be the main prerequisite for long-term survival.

The response of polar seaweeds to inorganic and organic pollutants has only randomly been investigated so far. Especially brown algae take up the radionuclide contaminant $^{99}$Tc very strongly (Topcuoglu and Fowler 1984; Gwynn et al. 2004). The uptake mechanisms and the actual effect of $^{99}$Tc are, however, not yet investigated. With respect to the oil spill of the Bahia Paraiso near Anvers Island, Antarctic Peninsula (Kennicutt II et al. 1990), there were no observable differences between oiled and non-oiled sites with respect to the percentage cover of either sublittoral macroalgal overstory or crustose coralline algae (Amsler et al. 1990). Similarly, Dunton et al. (1990) could not determine significant effects on photosynthesis in *Porphyra endiviifolium* and *Palmaria decipiens*. On the other hand, Stockton (1990) reported that the principal intertidal alga, *Urospora* sp. turned brown soon after the spill. The degree of damage certainly depends on the amount and type of material released.

## Seasonality of reproduction and the physiological characteristics of microscopic developmental stages

The heteromorphic life history of large brown algae is characterized by the development of perennial sporophytes and a reduction of the gametophyte generation (Clayton 1988). The ultimate step in this evolutionary trend is observed in members of the Fucales and Ascoseirales, which lack free-living gametophytes. In Laminariales and Desmarestiales, free-living gametophytes are still present; however, they are morphologically inconspicuous, generally consisting of microscopic filaments and they are important mainly in sexual reproduction. It has been suggested that the dissimilar reproductive phase expression in large kelps may be primarily associated with a differential response to wave action (Neushul 1972), herbivory (Slocum 1980; Lubchenco and Cubit 1980; Dethier 1981) or physical factors such as temperature and light (Kain 1964; Lüning and Neushul 1978; Lüning 1980b; Fain and Murray 1982; Novaczek 1984; tom Dieck 1993). Early developmental stages of seaweeds (zoospores, gametophytes and small sporophytes) are shade-adapted organisms unlike the adult sporophytes (Kain 1964; Amsler and Neushul 1991; Gómez and Wiencke 1996a; Dring et al. 1996). The different physiological performance of the heteromorphic phases has implications for algal ecology: reproduction, metabolic performance (e.g. photosynthesis) and growth are seasonally synchronized improving survival under changing environmental conditions (Lüning 1990). Thus, light adaptation is an important functional character connecting the different stages of life in these species.

In Antarctic Desmarestiales reproduction and further development of gametophytes and sporophytes are governed by the photoperiod. Culture studies carried out on all members of the Antarctic Desmarestiales indicate that life history depends strongly on the seasonal variation of daylengths. In general, the development of gametangia, fertilization (oogamy) and the formation of the early stages of sporophytes take place in winter in all Antarctic species of Desmarestiales, whereas growth of sporophytes begins with increasing daylengths in late winter-spring. In *Himantothallus grandifolius*, *Desmarestia anceps* and *D. menziesii*, gametophytes become fertile under short day conditions only (Wiencke 1990a; Wiencke and Clayton 1990; Wiencke et al. 1995, 1996). In *D. antarctica* and *Phaeurus antarcticus* gametogenesis occurs both in short and long days. In these species seasonality is controlled by the sporophytic stage. Sporophytes become fertile in culture at daylengths between 6 h and 8 h light per day and the developing gametophytes formed gametangia soon after germination (Clayton and Wiencke 1990; Wiencke 1990a; Wiencke et al. 1991). Similarly, in the Arctic kelp, *Laminaria solidungula*, sori are produced in winter (November) just before the growth of the lamina begins, but spore release does not occur until the following spring (Hooper 1984). Gametophytes become fertile in this species under short day conditions only, not under long days and short days combined with a night-break regime (Bartsch, pers. comm.). Irradiance levels $<5$ µmol $m^{-2}$ $s^{-1}$ are necessary to induce gametogenesis and fertilization in Antarctic Desmarestiales (Wiencke 1990a; Wiencke and Clayton 1990; Wiencke et al. 1995, 1996). The upper temperature limit for gametogenesis is 5°C in *D. antarctica* (Wiencke et al. 1991), a feature probably common in Desmarestiales from Antarctica.

Interestingly, all of the Antarctic Desmarestiales studied to date show in situ fertilization, i. e. the juvenile sporophytes remain attached to the gametophytes (Wiencke et al. 1995, 1996). This adds new evidence for the importance of the gametophytic generation on the early stages of the large sporophytes. In terms of ecological significance, one may speculate that the recruitment of sporophytes and consequently the observed dominance of these species may be conditioned by the survival (or mortality) of the gametophytes.

The development of gametophytes or at least their reproductive capacity appears to be constrained by high light conditions suggesting that the fitness associated with the winter development of gametophytes lies partly in a differentiation of light requirements for photosynthesis. For example, in *Desmarestia menziesii* early stages of sporophytes and gametophytes are better suited to live under low light conditions than adult sporophytes: photosynthetic efficiency ($\alpha$) measured in these phases is five times higher than in adult sporophytes (Gómez and Wiencke 1996a). The differences in pigment allocation, cross section pathlength of radiation and general development between the different stages account for the light harvesting efficiency at low irradiance (Ramus 1981). In filamentous or thin sheet-like thalli, pigment-based photosynthesis follows a linear curve, whereas in thick morphs, characterized by several cell layers and low ratios of photosynthetic to non-photosynthetic tissues, photosynthesis becomes uncoupled from the pigment content due to a greater self-shading (Ramus 1978).

In general, early life stages have higher dark respiration rates than adult ones, with consequences for light compensation points ($E_c$). Because of their high respiration rates, $E_c$ in gametophytes and small sporophytes are not significantly lower than in adult sporophytes, whereas the photon fluence rate required for saturation ($E_k$) is considerably higher in adult sporophytes due to lower $\alpha$ values. Photosynthesis in adult sporophytes is saturated at significantly higher irradiances (30 µmol $m^{-2}$ $s^{-1}$) than in gametophytes or juvenile sporophytes (16 µmol $m^{-2}$ $s^{-1}$). Despite their very high respiratory activities sporophytes and gametophytes from several Antarctic Desmarestiales show light saturation of growth at irradiances close to 10 µmol $m^{-2}$ $s^{-1}$ (Wiencke 1990a; Wiencke and Fischer 1990) indicating that growth is not constrained by such low irradiances (Gómez and Wiencke 1996b).

The capacity of gametophytes and small sporophytes to grow and photosynthesize under low light conditions may be regarded as adaptive and allows the algae to survive under the seasonally changing light environment in polar regions. During winter, when incident irradiance is low and daylength is short, high photosynthetic rates and growth of gametophytes and juvenile sporophytes are favoured in virtue of their higher surface-area/volume ratio, higher pigment content and more efficient light use.

## Seasonal growth and photosynthesis

In polar regions, seasons are characterized by short periods of favourable light conditions and extended periods of darkness due to polar nights and sea ice covering during winter (see chapter 2). The seasonal development of macroalgae is generally triggered by light, temperature and/or nutrient conditions (Chapman and Craigie 1977; Lüning 1980a; Lüning and tom Dieck 1989). As temperatures and nutrient levels show only a small variation over the year in Antarctic waters (see chapter 2), seasonality of Antarctic seaweeds depends mainly on variable light conditions and, especially, daylengths. Thus, assessment of the seasonal development of Antarctic species in long-term culture experiments is possible simply by mimicking the seasonally fluctuating daylengths at the collecting site (Wiencke 1990a, b). Seasonality of growth and photosynthesis as well as seasonal formation of gametes/spores can be monitored much more closely than possible in the field. The results of these studies complement the available fragmentary field observations and indicate that the phenology of Antarctic macroalgae can be controlled in the laboratory. Data on Arctic species are so far based on field studies only.

Antarctic seaweeds follow two different growth strategies in order to cope with the strong seasonality of the light regime (Wiencke 1990a, b; Dummermuth and Wiencke 2003). One group, mainly endemics, begins to grow under short day conditions in late winter-spring, even under sea ice, and reaches maximal growth in spring. Some species even reproduce in winter. This group has been classified as season anticipators sensu Kain (1989). In these species annual growth and reproduction is most probably based on photoperiodism and circannual rhythms, triggered or synchronized by daylength as shown for species from other phytogeographic regions (Lüning and tom Dieck 1989; tom Dieck 1991; Lüning and Kadel 1993; Schaffelke and Lüning 1994). To this group belong the brown algae *Desmarestia menziesii*, *D. antarctica*, *D. anceps*, *Himantothallus grandifolius*, *Phaeurus antarcticus*, *Ascoseira mirabilis* (Wiencke 1990a; Drew and Hastings 1992; Gómez et al. 1995a, 1996b; Gómez and Wiencke 1997) and the red algae *Palmaria decipiens*, *Delesseria salicifolia*, *Gymnogongrus antarcticus*, *G. turquetii*, *Hymenocladiopsis crustigena*, *Kallymenia antarctica*, *Phyllophora ahnfeltioides* (Gain 1912; Wiencke 1990b; Weykam et al. 1997; Dummermuth and Wiencke 2003). The second group, mainly composed of Antarctic-cold temperate species, are season responders sensu Kain (1989). Species in this group start growth later coinciding with favourable light conditions in spring and summer. They react directly to changing environmental conditions and show an opportunistic life strategy. To this group belong the brown alga *Adenocystis utricularis* (Wiencke 1990a), the red algae *Iridaea cordata* (Weykam et al. 1997) and *Gigartina skottsbergii* (Wiencke 1990b), and the green algae *Enteromorpha bulbosa* and *Acrosiphonia arcta* (Wiencke 1990b).

As for growth, a strong seasonal pattern of photosynthetic performance was found in long-term culture studies (Weykam and Wiencke 1996; Weykam et al. 1997; Gómez and Wiencke 1997; Lüder et al. 2001a, 2002,2003) and in field experiments (Gutkowski and Maleszewski 1989; Drew and Hastings 1992; Gómez et al. 1995b, 1997). In the brown algal season anticipators optimal photosynthetic rates are highest in late winter (September) for *A. mirabilis* (Gómez et al. 1995b) or in spring (November) for *Himantothallus grandifolius* (Drew and Hastings 1992) and *D. menziesii* (Gómez et al. 1997). In these species, morpho-functional processes strongly regulate photosynthetic seasonality: In spring respiration rates increase in all thallus parts, indicating growth activity in the basal meristem powered by remobilization of carbohydrates from the distal thallus part (Gómez et al. 1995b) as in *Laminaria* species (Lüning et al. 1973; Schmitz 1981). This is supported by the decrease of the storage carbohydrate laminaran in late winter and spring in *A. mirabilis* (Gómez and Wiencke 1998). The low molecular weight compound, mannitol, is presumably used as substrate for respiration during the period of active growth. This is reflected by its low levels during the main growth period. In late summer, the mannitol content increases significantly in the basal and middle region. In the distal region it may serve as substrate for light independent carbon fixation or is stored as laminaran which attains its highest content in the distal thallus part (Gómez and Wiencke 1998). Similar results have been obtained in *D. menziesii* (Gómez and Wiencke 1997; Gómez et al. 1998). In contrast, the season responder *Adenocystis utricularis* (Gutkowski and

Maleszewski 1989) features lowest photosynthetic capacity in spring and highest in autumn (no data in summer available).

In red algae, seasonality has been studied in particular with respect to the acclimation of light harvesting capacities. The light requirements for growth and photosynthesis of the Antarctic season anticipator *Palmaria decipiens* and in the Antarctic season responder *Iridaea cordata* are very low (Wiencke 1990b; Wiencke et al. 1993; Weykam et al. 1997). *P. decipiens* develops new blades during Antarctic late winter/spring (Wiencke 1990b) and starts growth in Antarctic winter (July) with optimum in spring (Wiencke 1990b). The photosynthetic capacity ($P_{max}$ and $ETR_{max}$) and photosynthetic efficiency ($\alpha$ and $F_v/F_m$) are, as growth, maximal in Antarctic spring (Weykam and Wiencke 1996; Lüder et al. 2001a). A strong relationship between the seasonal patterns of photosynthesis and pigment contents has been demonstrated in *P. decipiens* (Lüder et al. 2001a). Photosynthetic capacity and pigment contents increase in parallel continuously during the entire mid autumn, winter and spring. There is a positive correlation between phycobiliproteins and photosynthetic capacity, but a weaker correlation between Chl *a* and photosynthetic capacity. Phycobiliproteins are used to increase number and size of phycobilisomes, the main light harvesting antennae of red algae (Lüder 2003). Photosynthetic efficiency is optimal in autumn, winter and spring. During Antarctic summer, *P. decipiens* reduces its photosynthetic apparatus to a minimum: Photosynthetic efficiency, photosynthetic capacity, and the contents of phycobiliproteins and Chl *a* tissue are low (Lüder et al. 2001a). The existence of two phycobilisome forms with different aggregation states in *P. decipiens* might be considered as an advantage for a rapid acclimation to changes in environmental light conditions (Lüder et al. 2001b). Studies on seasonal assembly of the light harvesting antennae in *P. decipiens* have shown that both phycobilisome forms are extremely variable in size and number during the entire year (Lüder 2003).

The effect of darkness on the seasonal growth pattern in two Antarctic red macroalgae has been examined in long-term culture experiments by Weykam et al. (1997). Growth rates in the Antarctic season anticipator *P. decipiens* are low or even negative during darkness but increase strongly after re-illumination with maximal growth in early spring. The content of floridean starch decreases gradually in the dark and drops suddenly powering the formation of new blades in winter (early August) after 6 months exposure to darkness (Weykam et al. 1997). This feature supports the theory that seasonal growth is controlled in season anticipators either by photoperiodism or cirannual rhythms. In contrast, the Antarctic season responder *Iridaea cordata* develops no blades during darkness and starts growth more slowly after re-illumination with maximal growth rates in late spring.

Photosynthetic capacity and efficiency in *P. decipiens* is almost unaffected by 2 months in darkness albeit an initial increase in pigments (Weykam et al. 1997; Lüder et al. 2002; Lüder 2003). After 4 months in darkness the light harvesting antennae, the phycobilisomes, are degraded. One of the two forms of phycobilisomes in *P. decipiens* disappears (Lüder 2003). This is followed later by a degradation of the chl *a*-containing inner antennae (Lüder et al. 2002). At the end of the 6 months dark period, *P. decipiens* has lost its ability to photosynthesize (Lüder et al. 2002; Weykam et al. 1997). After re-exposure to light, pigments are synthesized rapidly and enhance photosynthetic performance to normal values within a week (Lüder et al. 2002). The season responder *I. cordata*, in contrast, maintains its photosynthetic apparatus functional during exposure to darkness (Weykam et al. 1997). This species is therefore better able to grow in regions with less predictable light conditions.

In contrast to the Antarctic there is a great seasonal variation in the Arctic not only with respect to daylength but also with respect to the macronutrient levels, which are high in winter and low in summer (see chapter 2). Nevertheless, the seasonal growth of the few seaweeds from the Arctic studied so far is triggered primarily by daylength and is supported by high nutrient levels in winter, as discussed below. The Arctic kelp *Laminaria solidungula* and the Arctic-cold temperate *L. saccharina* are clear season anticipators. In *L. solidungula* new blade formation is initiated in autumn under conditions of decreasing daylengths. Maximal growth rates occur in late-winter to early spring under thick ice and decline in summer and autumn in the ice-free period (Chapman and Lindley 1980; Dunton 1985). In the Alaskan Beaufort Sea, *L. solidungula* fronds continue to grow under a turbid ice canopy that produces conditions of nearly complete darkness, even under continuous daylight (Dunton 1990). In comparison, *L. saccharina* appears to delay nearly all of its annual growth to a brief period in late spring when light first starts to penetrate the water column during sea ice break-up (Dunton 1985).

The energy required to survive darkness and to start blade formation in winter is provided by mobilization of stored carbon reserves, accumulated during the previous summer, when inorganic nitrogen becomes available similar as in *L. longicruris* (Hatcher et al. 1977; Gagne et al. 1982; Chapman and Craigie 1978; Dunton and Dayton 1995). In *L. solidungula*, mobilization of these reserves occurs during the dark 9-month ice covered period, when the plant completes over 90% of its annual linear growth, and results in a carbon deficit where up to 30% of its original total carbon content is depleted before photosynthetic production begins in early summer (Dunton and Schell 1986). During the summer open-water period (July to October), when the concentration of inorganic-N is nearly undetectable in coastal waters, high rates of photosynthetic carbon fixation result in carbon storage and not in blade elongation in *L. solidungula* (Dunton and Schell 1986).

The utilization of stored photosynthate for growth was first reported by Lüning (1971), who found that *Laminaria hyperborea* in northern Europe accumulated reserves during its period of slow growth in summer to support the formation and growth of a new blade the following spring. Subsequent work by Lüning (1979) on three *Laminaria* species in the North Sea confirmed the importance of these reserves in *L. hyperborea*, but he also suggested that a circannual rhythm was involved in the regulation of seasonal growth since no one particular factor appeared responsible for triggering the onset of new growth. Lüning's hypothesis was later proved correct; the onset of short daylengths in autumn, combined with lower temperatures and normally low levels of inorganic-N were important cues that led to both sorus formation (Lüning, 1988) and new blade formation in several species of kelp (Lüning and tom Dieck 1989). But of all these cues, elaborate experimental studies demonstrated that daylength was most important in the setting of internal clocks in species that possess circannual rhythms which control the periodicity in linear growth (Lüning 1991; Lüning and Kadel 1993).

## Elemental and nutritional contents

The most commonly used indicator of nutrient limitation in marine algae is the carbon to nitrogen ratio (C:N), with low values indicative of nitrogen replete growth conditions. To date, C:N data have been reported from 54 species of macroalgae in Antarctica including 8 species of green algae, 36 species of red algae, and 10 species of brown algae (Dhargalkar et al. 1987; Weykam et al. 1996; Dunton 2001; Peters et al. 2005). Reported C:N ratios (on a weight:weight basis) range from a minimum of 5.0 to a maximum of 23.6 with an overall mean of 10.2, indicating that most if not all of the macroalgal species are not nitrogen limited (Peters et al. 2005). Likewise, tissue nitrogen levels are relatively high and almost always above the 1.5% level thought to be indicative of nitrogen replete growth conditions (Rakusa-Suszczewski and Zielinski 1993; Weykam et al. 1996; Peters et al. 2005). Generally, C:N values are much lower than most reported outside of Antarctica (a mean C:N value for temperate and tropical marine plants is approximately 19; Atkinson and Smith 1983) and presumably result from the high nitrate concentrations present in coastal Antarctic waters throughout the year (Weykam et al. 1996; Peters et al. 2005), although morpho-functional processes such as storage and remobilization of C and N in response to growth requirements are also important (Gómez and Wiencke 1998). Along the Antarctic Peninsula, there is a slight trend for somewhat lower C:N values at King George Island (62°14′ S, 58°40′ W) compared to Anvers Island (64°46′ S, 64°03′ W), suggesting a possibility of minor geographically based variation, but there was no apparent variation between early and late growing seasons (Weykam et al. 1996; Dunton 2001; Peters et al. 2005).

We are aware of only one published report of C:N ratios or nitrogen contents in macroalgae from the Arctic. Henley and Dunton (1995) reported on seasonal and within-thallus variation in the kelps *Laminaria solidungula* and *L. saccharina* from the Alaskan Arctic. C:N ratios were relatively low (approximately 11.6–13.5, converted from atom:atom to weight:weight basis) and tissue nitrogen levels relatively high (1.9–3.0%) in second year blades of both species regardless of season. In first year blades, however, C:N ratios increased significantly from early to late growing seasons in both species (from approx. 9.4 to 18.3 in *L. solidungula* and 9.8 to 24.8 in *L. saccharina*). This correlated with significant decreases in tissue nitrogen (Henley and Dunton 1995) but laboratory studies suggest important roles for both light and nitrogen availability in driving these patterns (Henley and Dunton 1997).

Macroalgae are a potentially important nutritional source for a variety of Antarctic animals (Iken et al. 1998; Dunton 2001) and macroalgal nutritional parameters, particularly protein content, can be important determinants of palatability to consumers (e.g., Horn and Neighbors 1984; Duffy and Paul 1992; Bolser and Hay 1996). Although we are unaware of any published reports of protein or other nutritional parameters in macroalgae from the Arctic, protein and other nutritional content data are available from 44 species of Antarctic macroalgae including 7 species of green algae, 28 species of red algae, and 9 species of brown algae (Czerpak et al. 1981; Dhargalkar et al. 1987; Rakusa-Suszczewski and Zielinski 1993; Gómez and Westermeier 1995; Gómez and Wiencke 1998; Peters et al. 2005). Protein levels are usually considerably higher when compared to temperate or tropical macroalgae in which protein contents were determined with similar methodology (Peters et al. 2005) although this is not always true (Gómez and Westermeier 1995). By itself, this would suggest that Antarctic macroalgae should be relatively palatable to consumers and the fact that this is not so for a majority of species (Amsler et al. 2005a, b) is probably related to chemical defenses as described in chapter 11. Peters et al. (2005) reported that there was no significant correlation between protein and nitrogen contents across all 40 species they examined, brown algae alone, or red algae late in the growing season but that there was a significant, positive correlation for red algae early in the growing season. In a more detailed study focused on the endemic brown alga *Ascoseira mirabilis*, Gómez and Wiencke (1998) reported a significant, positive correlation between protein and nitrogen content in the distal portion of the thallus but not in middle or basal portions. The general lack of correlation between protein and nitrogen content indicates that protein production is not constrained by nitrogen availability and is probably another reflection of the high nutrient environment present in coastal antarctic waters. This general lack of correlation also indicates that older methods which estimate protein concentrations in macroalgae based on nitrogen contents are less appropriate for use with Antarctic macroalgae.

## Chemical and physical defences against herbivory and fouling

Defence mechanisms against grazing pressure, and fouling organisms, including bacteria and pathogens can be critical to the success and survival of polar seaweeds. Seaweeds constitute a potential food and carbon source for many kinds of heterotrophic organisms, e.g. amphipods, gastropods, sea urchins or fishes. Tropical and temperate algal species often display physical defensive mechanisms such as structural toughness, physiological defensive mechanisms such as low nutritional content (see chapter 10 of this review) or chemical defensive mechanisms such as a high content or an induction of allelopathic chemicals. Allelopathy refers to the usually detrimental effects that seaweeds have on other interacting organisms through the production of secondary metabolites.

The incidence of herbivory in Arctic as well as Antarctic shallow water systems is low but several organisms depend fully or in part on seaweeds as a food source in the Antarctic (Brand 1974; Richardson 1977; Iken et al. 1997, 1999; Iken 1999; Dunton 2001; Graeve et al. 2001) and the

Arctic (Lippert et al. 2001; Wessels et al. 2006; Dunton and Schell 1987). A comprehensive investigation of palatability in 35 fleshy seaweeds from the Antarctic Peninsula (Anvers Island) against a sympatric (occurring within the same geographical region) fish and sea star algal consumer showed that about 60% and 80%, respectively, of the algae were unpalatable (Amsler et al. 2005a). These included all dominant overstory brown algae and a variety of abundant red algal species of the region. Unpalatable algal species were extracted and the palatability of organic extracts was tested after incorporation into artificial foods (Amsler et al. 2005a). Again, the majority of species were unpalatable as at least one extract type to the fish, sea star and an amphipod consumer. No overall correlation in feeding patterns could be established with tissue toughness or nutritional content of the algae (Amsler et al. 2005a; Peters et al. 2005) although toughness could play a role in the unpalatability of some particular macroalgal species to amphipods (Huang et al. 2006).

Only two fleshy red macroalgae, *Iridaea cordata* and *Phyllophora antarctica*, occur in McMurdo Sound, the southernmost location of open ocean conditions (Miller and Pearse 1991). Thallus fragments and paper disks laced with organic extracts of these species were tested in a phagostimulation assay where the retention time of a test or control food was measured over the mouth of the sympatric sea urchin, *Sterechinus neumayeri*. Algal treatments were retained for significantly shorter times compared to controls (blank paper disks and disks with a feeding stimulant), indicating the presence of deterrent chemicals in both red algal species (Amsler et al. 1998).

Along the Antarctic Peninsula, within-thallus variation of chemical and physical defences was tested for the ecologically important brown macroalgae *Desmarestia anceps* and *D. menziesii* (Fairhead et al. 2005a). This study found that overall chemical defences are prevalent, but some tissue types depend more on physical than chemical protection, such as holdfasts. In the highly differentiated *D. anceps*, high allocation of chemical and physical defences to holdfast and stipe tissue was consistent with predictions of optimal defence theory (Rhoades 1979), which assumes that defences are costly and therefore should be allocated to the most valuable and vulnerable tissue types.

A recent study of palatability using 19 abundant macroalgal species from Spitsbergen (Arctic) showed that most species were at least moderately palatable to a sympatric amphipod and sea urchin (Wessels et al. 2006). Consumption by both herbivores differed among algal species and preference patterns for algae differed for both herbivores, indicating species-specific preference patterns. Tissue-specific palatability of stipe and blade tissues of the kelp species differed for both herbivores with no preference in sea urchins and a preference for *Laminaria* blades and *Alaria* stipes in the amphipod (Wessels et al. 2006). Although not explicitly discussed by the authors, higher defences in kelp stipes compared to blades could be explained by resource allocation considerations as mentioned above for Antarctic brown algae. Generally, however, and in contrast to the findings for Antarctic seaweeds, unpalatability in Arctic seaweeds seems to be more related to structural than chemical properties of the algae (Wessels et al. 2006). Only one red algal species, *Ptilota gunneri*, showed indications of being chemically defended against the two grazers.

In addition to deterring grazers, organic extracts of all Antarctic macroalgal species tested (2 from McMurdo, 22 from Anvers Island) also were toxic to sympatric diatoms (Amsler et al. 2005b). To the best of our knowledge, no comparable information is available on antifouling properties in Arctic seaweeds.

Little is known about the particular compounds that may be active in chemical defence against herbivores and epiphytes in either of the two polar regions. Phlorotannins are a class of polyphenolic compounds found exclusively in brown algae. Their ecological functions, especially as anti-herbivore agents for temperate and tropical species have frequently been discussed (Targett and Arnold 2001; Amsler and Fairhead 2006). Phlorotannins were shown to be present in Antarctic brown algae (Iken 1996; Iken et al. 2001; Fairhead et al. 2005b), but their ecological significance remains ambiguous (Iken et al. 2001;

Iken unpubl. data). Microscopic studies in the Arctic brown seaweed, *Laminaria solidungula*, have shown that phlorotannins play an important role in wound healing (Lüder unpubl. data) as in the temperate Australasian *Ecklonia radiata* (Lüder and Clayton 2004). An induction of phlorotannin production in the cells of the wound area was evident after mechanical wounding simulating a grazer attack. Phlorotannins at the wound area probably deter bacteria or other microbes and may prevent the algae from being eaten further. In addition, phlorotannins bound in cell walls (Schoenwaelder 2002) may form a barrier for grazers or fouling organisms in *L. solidungula* (Lüder unpubl. data).

A variety of other secondary metabolites such as terpenes, furanones and acetogenins have been isolated and structurally resolved in Antarctic seaweeds (esp. red algae, for review see Amsler et al. 2001; Ankisetty et al. 2004), although the ecological significance of most of these compounds in grazer or fouler deterrence remains to be established. Chemical screening of organic extracts of Antarctic seaweeds revealed that no chemicals containing nitrogen are present (Amsler et al. 2005a) even though it had been hypothesized that the nitrogen-replete conditions in Antarctic coastal waters should promote the synthesis of nitrogenous compounds.

Another group of compounds shown to be present in Arctic and Antarctic seaweeds are volatile halogenated organic compounds (VHOC) (Laturnus 1996, 2001; Laturnus et al. 1996). These compounds are released continuously in substantial amounts by Arctic brown algae, while red algae release only trace amounts. The deterrent effect of VHOC on grazers is often assumed but so far has been detected only at exceptionally high and probably ecologically irrelevant concentrations (Amsler and Fairhead 2006).

Chemical defences in polar seaweeds may also be employed for protection against deleterious environmental conditions. In polar regions, UVB radiation due to the thinning ozone layer can be particularly harmful. Red algae from both polar regions have been shown to produce mycosporine-like amino acids (MAA) as photoprotective substances (e.g., McClintock and Karentz 1997; Hoyer et al. 2001; Aguilera et al. 2002). For detailed review on MAAs in polar seaweeds see Bischof et al. in press.

## Future perspectives

Although considerable progress has been achieved in recent years, our present knowledge of seaweeds from polar regions is still fragmentary. Due to the extreme remoteness of the polar regions and the infrequency of scientific studies even the basic features of the polar seaweed floras such as the numbers of recorded species are not firmly established. More studies are therefore needed to precisely document the biodiversity of Arctic and Antarctic seaweeds. If such studies are combined with molecular investigations it should be possible to demonstrate, for example, the presence of gene flow between populations of the same species in the Antarctic/Arctic and the adjacent cold-temperate regions. Alternatively, where gene flow is absent, divergence times between the populations can be estimated. Furthermore, molecular studies could definitively demonstrate seaweed dispersal routes and give a much clearer pattern of the biogeographic relationships between the polar and cold-temperate regions than is presently available.

More culture studies are necessary especially on seaweeds from the Arctic. None of the truly endemic Arctic species is in culture today and no data are available on the ecophysiological properties of these species. In this context it would be very interesting to know whether the relatively short cold water history of the Arctic could have promoted the evolution of species with temperature demands as low as in the Antarctic region. Further, culture studies have been proven to give pertinent insights into the life strategies of polar seaweeds not only with respect to the temperature requirements and geographic distribution but also with respect to the light requirements and depth zonation. Even the phenology of seaweeds could be monitored in culture studies under simulated polar light conditions.

Such studies should ideally be combined with field experiments. Data on the underwater radiation regime would allow an estimate of benthic primary production in different water depths through the determination of the metabolic carbon balance. Apart from studies with the Arctic *Laminaria solidungula*, the physiological performance of seaweeds in winter is obscure. Only baseline field studies are available together with a few papers based on laboratory studies. Reproduction, growth and photosynthetic rates, the content of pigments and cryoprotective substances, the elemental and nutritional composition as well as enzyme activities must be monitored over the entire year to understand the life strategy of these species in more detail.

With respect to elemental and nutritional content the major questions are, first, whether the greater availability of nitrogen in Antarctic waters and the apparent uncoupling of nitrogen availability and protein content alter our understanding about resource allocation in seaweeds that come from studies in other systems. Secondly, if the higher overall protein concentrations in Antarctic macroalgae indeed makes them more nutritious to herbivores compared to algae from other regions (which remains to be experimentally tested), does the response of herbivores to the (often few) palatable Antarctic seaweeds alter ideas about herbivore feeding ecology that are based on lower latitude systems? In this respect the functioning of the internal long distance transport system in members of the Desmarestiales and Ascoseirales in comparison with members of the Laminariales is obscure and must be investigated with respect to the allocation of storage compounds.

The few data available on community ecology make further studies in this area very urgent. In this respect, the interesting hypothesis that the relative lack of filamentous epiphytic seaweeds in Antarctica could be due to a relatively high level of mesoherbivory needs to be tested.

In this context more studies are needed on herbivory and chemical defence of seaweeds. Chemical deterrence should be tested against a larger set of potential herbivores so that the importance of chemical defence can be interpreted on a community-wide scale; this also requires a better understanding of herbivore abundances and grazing rates. Spatial coverage of chemical defences against herbivores should be extended to obtain a more comprehensive understanding of their importance in seaweeds on an ocean-wide basis. Few compounds have been identified so far as being active in chemical defence. Detailed knowledge will allow better comparisons with what is known from temperate and tropical systems. A question of interest concerning seaweeds from all regions is whether chemical defenses are constitutive, i.e. consistently present in the alga, or induced by the attack of herbivores? This may be particularly useful for answering the question of the yet unsolved roles of phlorotannins in brown algae: constitutive levels did not show clear patterns in herbivore deterrence, but microscopic studies show that wounding can lead to high local accumulation of phlorotannins. Whether this has broader impact needs to be addressed.

The ecophysiology of supra- and eulittoral species needs to be investigated in more detail in field studies. To survive in the intertidal zone, algae need to protect their metabolism and particularly their photosynthetic apparatus against temporary light stress conditions, heat stress, cold, freezing and also against pollution. Virtually nothing is known about protective mechanisms in polar seaweeds. Therefore, a detailed analysis of such mechanisms ensuring homeostasis and functional integrity of the photosynthetic apparatus in supra-and eulittoral polar seaweeds is necessary. Changes in the expression of photosynthetic key proteins (D1, RubisCO, ATP-synthase), and the synthesis of specific protective proteins (Heat Shock Proteins, Chaperonins, Cryoproteins) and protective metabolites (osmolytes, cryoprotectants, antioxidants, sunscreens etc.) in relation to stress exposure should be studied in the field, mesocosms and controlled laboratory experiments. So far, gene expression, biosynthetic pathways and metabolic regulation, for example, are scientific fields almost unexplored in polar seaweeds. In particular, the quest for potential cryoproteins represents a novel aspect in plant research in general. The results of such studies will substantially increase the understanding of biochemical adaptation to plant life in extreme environments.

All these studies need to be complemented by a detailed monitoring of the environmental conditions especially in the supra-and eulittoral. Our knowledge in this area is also still fragmentary.

**Acknowledgements** Many results described in this paper would not have been possible without the generous financial support by the Deutsche Forschungsgemeinschaft, the German Academic Exchange Service, the European Commission, the Alexander von Humboldt Foundation, the Australian Research Council and the US National Science Foundation. This help is greatly acknowledged here. Finally, we thank Akira Peters and another unknown referee for constructive comments on the manuscript.

## References

Aguilera J, Bischof K, Karsten U, Hanelt D, Wiencke C (2002) Seasonal variation in ecological patterns in macroalgae from an Arctic fjord. II. Pigment accumulation and biochemical defence systems against high light stress. Mar Biol 140:1087–1095

Amsler CD, Fairhead VA (2006) Defensive and sensory chemical ecology of brown algae. Adv Bot Res 43:1–91

Amsler CD, Neushul M (1991) Photosynthetic physiology and chemical composition of spores of the kelps *Macrocystis pyrifera*, *Nereocystis luetkeana*, *Laminaria farlowii*, and *Pterygophora californica* (Phaeophyceae). J Phycol 27:26–34

Amsler CD, Laur DR, Quetin LB, Rowley RJ, Ross R, Neushul M. (1990) Quantitative analysis of the vertical distribution of overstory macroalgae near Anvers Island. Antarctica. Antarctic J US 25:201–202

Amsler CD, Rowley RJ, Laur DR, Quetin LB, Ross RM (1995) Vertical distribution of Antarctic Peninsular macroalgae: cover, biomass and species composition. Phycologia 34:424–430

Amsler CD, McClintock JB, Baker BJ (1998) Chemical defenses against herbivory in the Antarctic marine macroalgae *Iridaea cordata* and *Phyllophora antarctica* (Rhodophyceae). J Phycol 34:53–59

Amsler CD, Iken KB, McClintock JB, Baker BJ (2001) Secondary metabolites from Antarctic organisms and their ecological implications. In: McClintock JB, Baker BJ (eds) Marine chemical ecology. CRC Press, Boca Raton, pp 267–300

Amsler CD, Okogbue IN, Landry DM, Amsler MO, McClintock JB, Baker BJ (2005a) Potential chemical defenses against diatom fouling in Antarctic macroalgae. Bot Mar 48:318–322

Amsler CD, Iken K, McClintock JB, Amsler MO, Peters KJ, Hubbard JM, Furrow FB, Baker BJ (2005b) Comprehensive evaluation of the palatability and chemical defenses of subtidal macroalgae from the Antarctic Pensinsula. Mar Ecol Prog Ser 294:141–159

Andersson B, Salter AH, Virgin I, Vass I, Styring S (1992) Photodamage to photosystem II-primary and secondary events. J Photochem Photobiol B: Biol 15:15–31

Ankisetty S, Nandiraju S, Win H, Park YC, Amsler CD, McClintock JB, Baker JA, Diyabalanage TK, Pasaribu A, Singh MP, Maiese WM, Walsh RD, Zaworotko MJ, Baker BJ (2004) Chemical investigation of predator-deterred macroalgae from the Antarctic Peninsula. J Nat Prod 67:1295–1302

Arnoud PM (1974) Contributions à la binomie marine bentique des régions Antarctiques et subantarctiques. Thethys 6:465–653

Arnold KE, Manley SL (1985) Carbon allocation in *Macrocystis pyrifera* (Phaeophyta): intrinsic variability in photosynthesis and respiration. J Phycol 21:147–167

Aro EM, Virgin I, Andersson B (1993) Photoinhibition of photosystem II. Inactivation, protein damage and turnover. Biochim Biophys Acta 1143:113–134

Asada K, Takahashi M (1987) Production and scavenging of active oxygen in photosynthesis. In: Kyle DJ, Osmond CB, Arntzen CJ (eds) Photoinhibition. Topics in photosynthesis, vol 9. Elsevier Science Publishers, Amsterdam, pp 89–109

Atkinson M, Smith S (1983) C:N:P ratios of benthic marine plants. Limnol Oceanogr 28:568–574

de Baar HWJ (1994) Von Liebig's law of the minimum and plankton ecology. Progr Oceoanogr 33:347–386

de Baar HJW, de Jong JTM, Bakker DCE, Löscher BM, Veth C, Bathmann U, Smetacek, V (1995) Importance of Iron for Plankton Blooms and Carbon Dioxide Drawdown in the Southern Ocean. Nature 373:412–415

Becker EW (1982) Physiological studies on Antarctic *Prasiola crispa* and *Nostoc commune* at low temperatures. Polar Biol 1:99–104

Bird CJ, Mc Lachlan JL (1992) Seaweed flora of the Maritimes, I. Rhodophyta-The red algae. Biopress Ltd. Bristol

Bischof K, Gómez J, Molis M, Hanelt D, Karsten U, Lüder, UH, Roleda M, Zacher K, Wiencke C (2006) UV radiation shapes seaweed communities. Rev Environ Sci Biotechnol in press

Bischoff B, Wiencke C (1993) Temperature requirements for growth and survival of macroalgae from Disko-Island (Greenland). Helgol Mar Res 47:167–191

Bischoff B, Wiencke C (1995a) Temperature ecotypes and biogeography of Acrosiphoniales (Chlorophyta) with Arctic-Antarctic disjunct and Arctic/cold-temperate distributions. Eur J Phycol 30:19–27

Bischoff B, Wiencke C (1995b) Temperature adaptation in strains of the amphi-equatorial green alga *Urospora penicilliformis* (Acrosiphoniales) – biogeographical implications. Mar Biol 122:681–688

Bischoff-Bäsmann B, Wiencke C (1996) Temperature requirements for growth and survival of Antarctic Rhodophyta. J Phycol 32:525–535

Bolser RC, Hay ME (1996) Are tropical plants better defended? Palatability and defenses of temperate vs tropical seaweeds. Ecology 77:2269–2286

Bolton JJ, Lüning K (1982) Optimal growth and maximal survival temperatures of Atlantic *Laminaria* species (Phaeophyta) in culture. Mar Biol 66:89–94

Brand TE (1974) Trophic interactions and community ecology of the shallow-water marine benthos along the Antarctic Peninsula. PhD Dissertation, University of California, Davis, pp 1–220

Breeman AM (1988) Relative importance of temperatures and other factors in determining geographic boundaries of seaweeds: experimental and phenological evidence. Helgoländer Meeresunters 42:199–241

Brouwer PEM (1996) *In situ* photosynthesis and estimated annual production of the red macroalga *Myriogramme mangini* in relation to underwater irradiance at Signy Island (Antarctica). Antarctic Sci 8:245–252

Brown AD, Simpson JR (1972) Water relations of sugar-tolerant yeasts: the role of intracellular polyols. J General Microbiol 72:589–591

Buggeln RG (1983) Photoassimilate translocation in brown algae. Progr Phycol Res 2:283–332

Cabello-Pasini A, Alberte RS (2001) Expression of carboxylating enzymes in *Laminaria setchelli* (Phaeophyceae). Phycologia 40:351–358

Chapman ARO, Craigie JS (1977) Seasonal growth in *Laminaria longicruris*: relation with dissolved inorganic nutrients and internal reserves of nitrogen. Mar Biol 40:197–205

Chapman ARO, Craigie JS (1978) Seasonal growth in *Laminaria longicruris*: relations with reserve carbohydrate storage and production. Mar Biol 46:209–213

Chapman ARO, Lindley JE (1980). Seasonal growth of *Laminaria longicruris* in the High Arctic in relation to irradiance and dissolved nutrient concentration. Mar Biol 57:1–5

Clarke DL (1990) Arctic Ocean ice cover; geologic history and climatic significance. In: Grantz A, Johnson L, Sweeney JL (eds) The Arctic Ocean region. Geol. Soc. America, Boulder Colo, pp 53–62

Clarke A, Barnes KA, Hodgson DA (2005) How isolated is Antarctica. Trends Ecol Evol 20:1–3

Clayton MN (1987) Isogamy and a fucalean type of life history in the Antarctic brown alga Ascoseira mirabilis (Ascoseirales, Phaeophyta). Bot Mar 30:447–455

Clayton, MN (1988) Evolution and life histories of brown algae. Bot Mar 31:379–387

Clayton MN, Ashburner CM (1990) The anatomy and ultrastructure of "conducting channels" in *Ascoseira mirabilis* (Ascoseirales, Phaeophyceae). Bot Mar 33:63–70

Clayton MN, Wiencke C (1990) The anatomy, life history and development of the Antarctic brown alga *Phaeurus antarcticus* (Desmarestiales, Phaeophyceae). Phycologia 29:303–315

Clayton MN, Wiencke C, Klöser H (1997) New records of temperate and sub-Antarctic marine benthic macroalgae from Antarctica. Polar Biol 17:141–149

CLIMAP Project Members (1981) Seasonal re constructions of the earths surface at the last glacial masimum. The Geol. Soc. of America Map and Chart Service MC-36, Washington DC

Cormaci M, Furnari G, Scammacca B (1992) The benthic algal flora of Terra Nova Bay (Ross Sea, Antarctica). Bot Mar 35:541–552

Crame JA (1993) Latitudinal range fluctuations in the marine realm through geological time. Trends Ecol Evol 8:162–266

Cross WE, Wilce RT, Fabijan MF (1987). Effects of experimental releases of oil and dispersed oil on Arctic nearshore macrobenthos. III, Macroalgae. Arctic 40(Suppl. 1):211–219

Czerpak R, Mical A, Gutkowski R, Siegien I (1981) Chemism of some species of Antarctic macroalgae of the genera *Adenocystis*, *Himantothallus*, *Leptosomia*, and *Monostroma*. Pol Polar Res 2:95–107

Davey MC (1989) The Effects of freezing and desiccation on photosynthesis and survival of terrestrial Antarctic algae and cyanobacteria. Polar Biol 10:29–36

Davison IR (1991) Environmental effects on algal photosynthesis: temperature. J Phycol 27:2–8

Deacon GER (1937) The hydrology of the Southern Ocean. Discovery Rep 15:125–152

DeLaca TE, Lipps JH (1976) Shallow-water marine associations, Antarctic Peninsula. Antarctic J 11:12–20

Delépine R, Lamb JM, Zimmermann MH (1966) Preliminary report on the vegetation of the Antarctic Peninsula. Proc 5th Int Seaweed Symp, pp 107–116

Demmig-Adams B, Adams III WW (1992) Photoprotection and other responses of plants to high light stress. Annu Rev Plant Physiol Plant Mol Biol 43:599–626

Dethier MN (1981) Heteromorphic algal life histories: the seasonal pattern and response to herbivory of the brown crust, *Ralfsia californica*. Oecologia (Berl.) 49:333–339

Dhargalkar V, Reddy C, Deshmukhe G, Unatawale A (1987) Biochemical composition of some benthic marine algae of the Vestfold Hills. Antarctica Indian J Mar Sci 16:269–271

tom Dieck I (Bartsch) (1991) Circannual growth rhythm and photoperiodic sorus induction in the kelp *Laminaria setchellii* (Phaeophyta). J Phycol 27:341–350

tom Dieck I (1992) North Pacific and North Atlantic digitate *Laminaria* species (Phaeophyta): hybridisation experiments and temperature responses. Phycologia 31:147–163

tom Dieck I (1993) Temperature tolerance and survival in darkness of kelp gametophytes (Laminariales, Phaeophyta): ecological and biogeographical implications. Mar Ecol Prog Ser 100:253–264

Dieckmann G, Reichardt W, Zielinski K (1985) Growth and production of the seaweed, *Himantothallus grandifolius*, at King George Island. In: Siegfried WR, Condy PR, Laws RM (eds) Antarctic nutrient cycles and food webs. Springer-Verlag, Berlin Heidelberg, pp 104–108

Drew EA (1977) The physiology of photosynthesis and respiration in some Antarctic marine algae. Br Antarct Surv Bull 46:59–76

Drew EA, Hastings RM (1992) A year-round ecophysiological study of *Himantothallus grandifolius* (Desmarestiales, Phaeophyta) at Signy Island. Antarctica. Phycologia 31:262–277

Dring MJ, Makarov V, Schoschina E, Lorenz M, Lüning K (1996) Influence of ultraviolet radiation on chlorophyll fluorescence and growth in different life history stages of three species of *Laminaria* (Phaeophyta). Mar Biol 126:183–191

Duffy J, Paul V (1992) Prey nutritional quality and effectiveness of chemical defenses against tropical reef fishes. Oecologia 90:333–339

Dummermuth AL, Wiencke C (2003) Experimental investigation of seasonal development in six Antarctic red macroalgae. Antarct. Sci. 15:449–457

Dunton KH (1985) Growth of dark-exposed *Laminaria saccharina* (L.) Lamour and *Laminaria solidungula* J. Ag (Laminariales: Phaeophyta) in the Alaskan Beaufort Sea. J Exp Mar Biol Ecol 94:181–189

Dunton KH (1990) Growth and production in *Laminaria solidungula*: relation to continuous underwater light levels in the Alaskan high Arctic. Mar Biol 106:297–304

Dunton KH (1992) Arctic biogeography: the paradox of the marine benthic fauna and flora. Trends Ecol Evol 7:183–189

Dunton KH (2001) $\delta^{15}$N and $\delta^{13}$C measurements of Antarctic Peninsula fauna: trophic relationships and assimilation of benthic seaweeds. Am Zool 41:99–112

Dunton, KH, Dayton, PK (1995) The biology of high latitude kelp. In: Skjoldal HR (ed) Ecology of fjords and coastal waters: proceedings of the Mare Nor symposium on the ecology of fjords and coastal waters, Tromso, Norway, 5–9 December 1994. Elsevier, Amsterdam, pp 499–507

Dunton KH, Jodwalis CM (1988) Photosynthetic performance of *Laminaria solidungula* measured *in situ* in the Alaskan High Arctic. Mar Biol 98:277–285

Dunton KH, Schell DM (1986) Seasonal carbon budget and growth of *Laminaria solidungula* in the Alaskan High Arctic. Mar Ecol Prog Ser 31:57–66

Dunton KH, Schell DM (1987) Dependence of consumers on macroalgal (*Laminaria solidungula*) carbon in an arctic kelp community: $\delta^{13}$C evidence. Mar Biol 93:615–625

Dunton KH, Schonberg SV (2000) The benthic faunal assemblage of the Boulder Patch kelp community. In: Johnson SR, Truett JC (eds) The natural history of an Arctic oil field. Academic Press, San Diego, pp 372–397

Dunton KH, Reimnitz E, Schonberg S (1982) An arctic kelp community in the Alaskan Beaufort Sea. Arctic 35:465–484

Eggert A, Wiencke C (2000) Adaptation and acclimation of growth and photosynthesis of five Antarctic red algae to low temperatures. Polar Biol 23:609–618

Fain SR, Murray SN (1982) Effects of light and temperature on net photosynthesis and dark respiration of gametophytes and embryonic sporophytes of *Macrocystis pyrifera*. J Phycol 18:92–98

Fairhead VA, Amsler CD, McClintock JB, Baker BJ (2005a) Within-thallus variation in chemical and physical defences in two species of ecologically dominant brown macroalgae from the Antarctic Peninsula. J Exp Mar Biol Ecol 322:1–12

Fairhead VA, Amsler CD, McClintock JB, Baker BJ (2005b) Variation in phlorotannin content within two species of brown macroalgae (*Desmarestia anceps* and *D. menziesii*) from the Western Antarctic Peninsula. Polar Biol 28:680–686

Falkowski PG, Raven J (1997) Aquatic photosynthesis. Blackwell Science

Gagne J, Mann K,, Chapman ARO (1982) Seasonal patterns of growth and storage in *Laminaria longicruris* in relation to differing patterns of availability of nitrogen in the water. Mar Biol 69:91–101

Gain L (1912) La flore algologique des régions antarctiques et subantarctiques. In: Charcot J (ed) Deuxième expédition Antarctique française (1908–1910). Sciences naturelles: Documents scientifiques, tome 8. Masson et Companie, Paris, pp 1–218

Gerland S, Lind B, Dowdall M, Karcher M, Kolstad AK (2003) $^{99}$Tc in seawater in the West Spitsbergen Current and adjacent areas. J Environ Radioactivity 69:119–127

Godley EJ (1965) Botany of the southern zone. Exploration to 1843, Tuatara 13:140–181

Gómez I, Lüning K (2001) Constant short-day treatment of outdoor-cultivated *Laminaria digitata* prevents summer drop in growth rate. Eur J Phycol 36:391–395

Gómez I, Westermeier R (1995) Energy contents and organic constituents in Antarctic and south Chilean marine brown algae. Polar Biol 15:597–602

Gómez I, Wiencke C (1996a) Photosynthesis, dark respiration and pigment contents of gametophytes and sporophytes of the Antarctic brown alga *Desmarestia menziesii*. Bot Mar 39:149–157

Gómez, Wiencke C (1996b) Seasonal growth and photosynthetic performance of the Antarctic macroalga *Desmarestia menziesii* (Phaeophyceae) cultivated under fluctuating Antarctic daylengths, Bot Acta 110:25–31

Gómez I, Wiencke C (1997) Seasonal growth and photosynthetic performance of the Antarctic macroalga *Desmarestia menziesii* (Phaeophyceae) cultivated under fluctuating Antarctic daylengths. Bot Acta 110:25–31

Gómez I, Wiencke C (1998) Seasonal changes in C, N, and major organic compounds and their significance to morphofunctional processes in the endemic Antarctic brown alga *Ascoseira mirabilis*. Polar Biol 19:115–124

Gómez I, Thomas DN, Wiencke C (1995a) Longitudinal profiles of growth, photosynthesis and light independent carbon fixation in the Antarctic brown alga *Ascoseira mirabilis*. Bot Mar 38:157–164

Gómez I, Wiencke C, Weykam G (1995b) Seasonal photosynthetic characteristics of the brown alga *Ascoseira mirabilis* from King George Island (Antarctica). Mar Biol 123:167–172

Gómez I, Wiencke C, Thomas DN (1996) Variations in photosynthetic characteristics of the Antarctic marine brown alga *Ascoseira mirabilis* in relation to thallus age and size. Eur J Phycol 31:167–172

Gómez I, Weykam G, Wiencke C (1998) Seasonal photosynthetic metabolism and major organic compounds in the marine brown alga *Desmarestia menziesii* from King George Island (Antarctica). Aquat Bot 60:105–118

Gómez I, Weykam G, Klöser H, Wiencke C (1997) Photosynthetic light requirements, daily carbon balance and zonation of sublittoral macroalgae from King George Island (Antarctica). Mar Ecol Progr Ser 148:281–293

Graeve M, Dauby P, Scailteur Y (2001) Combined lipid, fatty acid and digestive tract content analyses: a penetrating approach to estimate feeding modes of Antarctic amphipods. Polar Biol 24:853–862

Gutkowski R, Maleszewski S (1989) Seasonal changes of the photosynthetic capacity of the Antarctic macroalga *Adenocystis utricularis* (Bory) Skottsberg. Polar Biol 10:145–148

Gutt J (2001) On the direct impact of ice on Marine Benthic communities, a review. Polar Biol 24:553–564

Gwynn JP, Dowdall M, Gerland S, Selnæs ØG, Wiencke C (2004) Technetium-99 in Arctic marine algae from Kongsfjorden. Svalbard Ber Polarforsch Meeresforsch 492:35–45

Hanelt D (1998) Capability of dynamic photoinhibition in Arctic macroalgae is related to their depth distribution. Mar biol 131:361–369

Hanelt D, Jaramillo J, Nultsch W, Senger S, Westermeier R (1994) Photoinhibitioin as a regulative mechanism of photosynthesis in marine algae of Antarctica. Ser Cient INACH 44:67–77

Hanelt D, Melchersmann B, Wiencke C, Nultsch W (1997) Effects of high light stress on photosynthesis of polar macroalgae in relation to depth distribution. Mar Ecol Prog Ser 149:255–266

Hanelt D, Tüg H, Bischof K, Gross C, Lippert H, Sawall T, Karsten U, Wiencke C (2001) Light regime in an Arctic fjord: a study related to Stratospheric Ozone depletion as a basis for determination of UV effects on algal growth. Mar Biol 138:649–658

Hanelt D, Wiencke C, Bischof K (2003) Photosynthesis in marine Macroalgae. In: Larkum AW, Douglas SE, Raven JA (eds) Photosynthesis in algae. Kluwer Academic Publisher, Dordrecht, pp 413–435

Hatcher BG, Chapman ARO, Mann KH (1977) An annual carbon budget for the kelp *Laminaria longicruris*. Mar Biol 44:85–96

Healey FP (1972) Photosynthesis and respiration of some Arctic seaweeds. Phycologia 11:267–271

Hempel G (1987) Die Polarmeere – ein biologischer Vergleich. Polarforsch 57:173–189

Henley WJ, Dunton KH (1995) A seasonal comparison of carbon, nitrogen, and pigment content in *Laminaria solidungula* and *L. saccharina* (Phaeophyta) in the Alaskan Arctic. J Phycol 31:325–331

Henley WJ, Dunton KH (1997) Effects of nitrogen supply and continuous darkness on growth and photosynthesis of the arctic kelp *Laminaria solidungula*. Limnol Oceanogr 42:209–216

van den Hoek C (1982a) Phytogeographic distribution groups of benthic marine algae in the North Atlantic Ocean. A review of experimental evidence from life history studies. Helgoländer Meeresunters 35:153–214

van den Hoek C (1982b) The distribution of benthic marine algae in relation to the temperature regulation of their life histories. Biol J Linn Soc 18:1–144

van den Hoek C, Breeman AM (1989) Seaweed biogeography in the North Atlantic: where are we now? In: Garbary DJ, South GR (eds) Evolutionary biogeography of the marine algae of the North Atlantic. NATO ASI Series, Vol. G 22, Springer-Verlag, Berlin, Heidelberg, pp 57–86

Hooper RG (1984) Functional adaptations to the polar environment by the arctic kelp, *Laminaria solidungula*. Br Phycol J 19:194

Hop H, Pearson T, Hegseth EN, Kovacs KM, Wiencke C, Kwasniewski S, Eiane K, Mehlum F, Gulliksen B, Wlodarska-Kowalczuk M, Lydersen C, Weslawski JM, Cochrane S, Gabrielsen GW, Leakey RJG, Lønne OJ, Zajaczkowski M, Falk-Petersen S, Kendall M, Wängberg S-Å, Bischof K, Voronkov AY, Kovaltchouk NA, Wiktor J, Poltermann M, di Prisco G, Papucci C, Gerland S (2002) The ecosystem of Kongsfjorden, Svalbard. Polar Res 21:167–208

Horn M, Neighbors M (1984) Protein and nitrogen assimilation as a factor in predicting the seasonal macroalgal diet of the monkeyface prickleback. Trans Am Fish Soc 113:388–396

Hoyer K, Karsten U, Sawall T, Wiencke C (2001) Photoprotective substances in Antarctic macroalgae and their variation with respect to depth distribution, different tissues and developmental stages. Mar Ecol Prog Ser 211:117–129

Huang YM, McClintock JB, Amsler CD, Peters KJ, Baker BJ (2006) Feeding rates of common Antarctic gammarid amphipods on ecologically important sympatric macroalgae. J Exp Mar Biol Ecol 329:55–65

Iken K (1996) Trophic relations between macroalgae and herbivores in Potter Cove (King George Island, Antarctica). Ber Polarforsch Meeresforsch 201:1–206. (in German)

Iken K (1999) Feeding ecology of the Antarctic herbivorous gastropod *Laevilacunaria Antarctica*. J Exp Mar Biol Ecol 236:133–148

Iken K, Barrera-Oro ER, Quartino ML, Casaux RJ, Brey T (1997) Grazing in the Antarctic fish *Notothenia coriiceps*: evidence for selective feeding on macroalgae. Ant Sci 9:386–391

Iken K, Quartino M, Barrera Oro E, Palermo J, Wiencke C, Brey T (1998) Trophic relations between macroalgae and herbivores. Ber Polarforsch Meeresforsch 299:258–262

Iken K, Quartino ML, Wiencke C (1999) Histological identification of macroalgae from stomach contents of the Antarctic fish *Notothenia coriiceps* gives new insights in its feeding ecology. Mar Ecol 20:11–18

Iken KB, Amsler CD, Hubbard JM, McClintock JB, Baker BJ (2001) Preliminary results on secondary metabolites from Antarctic brown algae and their ecological relevance. J Phycol 37:25–26. (Suppl.)

Jacob A, Kirst GO, Wiencke C, Lehmann H (1991) Physiological responses of the Antarctic green alga *Prasiola crispa* ssp. *antarctica* to salinity stress. J Plant Physiol 139:57–62

Jacob A, Lehmann H, Kirst GO, Wiencke C (1992a) Changes in the ultrastructure of *Prasiola crispa* ssp. *antarctica* under salinity stress. Bot Acta 105:41–46

Jacob A, Wiencke C, Lehmann H, Kirst GO (1992b) Physiology and ultrastructure of desiccation in the green alga *Prasiola crispa* from Antarctica. Bot Mar 35:297–303

Jerlov NG (1976) Marine optics. Elsevier, Amsterdam

Johnston AM, Raven J (1986) Dark carbon fixation studies on the intertidal macroalga *Ascophyllum nodosum* (Phaeophyta). J Phycol 22:483–485

Kain (Jones) JM (1964) Aspects of the biology of *Laminaria hyperborea*. III. Survival and growth of gametophytes. J Mar Biol Assoc UK 44:415–433

Kain JM (1989) The seasons in the subtidal. Br Phycol J 24:203–215

Karsten U, West JA (2000) Living in the intertidal zone – seasonal effects on heterosides and sun-screen compounds in the red alga *Bangia atropurpurea* (Bangiales). J Exp Mar Biol Ecol 254:221–234

Karsten U, Wiencke C, Kirst GO (1991a) The effect of salinity changes upon physiology of eulittoral green macroalgae from Antarctica and Southern Chile. I. Cell viability, growth, photosynthesis and dark respiration. J Plant Physiol 138:667–673

Karsten U, Wiencke C, Kirst GO (1991b) The effect of salinity changes upon physiology of eulittoral green macroalgae from Antarctica and Southern Chile. II. Inorganic ions and organic compounds. J Exp Bot 42:1533–1539

Karsten U, Barrow KD, King RJ (1993) Floridoside, L-isofloridoside and D-isofloridoside in the red alga *Porphyra columbina*: seasonal and osmotic effects. Plant Physiol 103:485–491

Karsten U, Kück K, Vogt C, Kirst GO (1996a) Dimethylsulfoniopropionate production in phototrophic organisms and its physiological function as a cryoprotectant. In: Kiene RP (ed) Biological and environmental chemistry of DMSP and related sulfonium compounds. Plenum Press, New York, pp 143–153

Karsten U, Barrow KD, Nixdorf O, King RJ (1996b) The compatibility of unusual organic osmolytes from mangrove red algae with enzyme activity. Aust J Plant Physiol 23:577–582

Kennicutt II MC, et al (1990) Oil spillage in Antarctica. Environ Sci Technol 24:620–624

Kerby WN, Evans LV (1983) Phosphoenolpyruvate carboxykinase activity in *Ascophyllum nodosum* (Phaeophyceae). J Phycol 19:1–3

Kirst GO (1990) Salinity tolerance of eukaryotic marine algae. Annu Rev Plant Physiol Plant Mol Biol 41:21–53

Kirst GO, Wiencke C (1995) Ecophysiology of polar algae. J Phycol 31:181–199

Kjellman FR (1883) The algae of the Arctic Sea. Boktryckeriet, Stockholm

Klöser, H (1994) Descripción Basica de la Caleta Potter y Costas Abiertas Adyacentes. Dirección Nacional del Antártico; reporte de datos: Estructura y Dinamica de un Ecosistema Costero Antártico, Estación Cientifica "Teniente Jubany" en la Isla 25 de Mayo (King George Island), Islas Shetland del Sur. Contributión 419B

Klöser H, Ferreyra G, Schloss I, Mercuri G, Laturnus F, Curtosi A (1993) Seasonal variation of algal growth conditions in sheltered Antarctic bays: the example of Potter Cove (King Goerge Island, South Shetlands). J Mar Systems 4:289–301

Klöser H, Quartino ML, Wiencke C (1996) Distribution of macroalgae and macroalgal communities in gradients of physical conditions in Potter Cove, King George Island, Antarctica. Hydrobiologia 333:1–17

Knebel G (1936) Monographie der Algenreihe Prasiolales, insbesonder von *Prasiola crispa*. Hedwigia 75:1–120

Knox GA, Lowry JK (1978) A comparison between the benthos of the Southern Ocean and the North Polar Ocean with special reference to the Amphipoda and the Polychaeta. In: Dunbar JM (ed) 'Polar Oceans' Proc. SCOR/SCAR polar ocean conference, Montreal 1974, pp 423–462

Konar B, Iken K (2005) Competitive dominance among sessile marine organisms in a high Arctic boulder community. Polar Biol 29:61–64

Krause GH, Weis E (1991) Chlorophyll fluorescence and photosynthesis: the basics. Annu Rev Plant Physiol Plant Mol Biol 42:313–349

Kremer BP (1981a) Metabolic implications of non-photosynthetic carbon fixation in brown macroalgae. Phycologia 20:242–250

Kremer BP (1981b) Aspects of carbon metabolism in marine macroalgae. Oceanogr Mar Biol Annu Rev 19:41–94

Küppers U, Kremer BP (1978) Longitudinal profiles of carbon dioxide capacities in marine macroalgae. Plant Physiol 62:49–53

Lamb IM, Zimmermann MH (1977) Benthic marine algae of the Antarctic Peninsula, Antarctic research series 23, Biology of the Antarctic Seas V, Paper 4, pp 129–229

Latala A (1990) Photosynthesis and respiration of some marine benthic algae from Spitsbergen. Polar Res 8:303–308

Laturnus F (1996) Volatile halocarbons released from Arctic macroalgae. Mar Chem 55:359–366

Laturnus F (2001) Marine macroalgae in polar regions as natural sources for volatile organohalogens. Environ Sci Pollut Res 8:103–108

Laturnus F, Wiencke C, Klöser H (1996) Antarctic macroalgae – sources of volatile halogenated organic compounds. Mar Environ Res 41:169–181

Lee RKS (1980) A catalogue of the marine algae of the Canadian Arctic. Natl Mus Can Nat Sci Publ Bot 9:1–83

Lippert H, Iken K, Rachor E, Wiencke C (2001) Macrofauna associated with macroalgae at Kongsfjord (Spitsbergen) – species composition and distribution on abundant macroalgal species. Polar Biol 24:512–522

Lubchenco J, Cubit J (1980) Heteromorphic life histories of certain marine algae as adaptations to variations in herbivory. Ecology 61:676–687

Lüder UH (2003) Acclimation of the photosynthetic apparatus of the endemic Antarctic red macroalga *Palmaria decipiens* to seasonally changing light conditions. Ber Polarforsch Meeresforsch 469:141 pp

Lüder UH, Clayton MN (2004) Induction of phlorotannins in the brown macroalga *Ecklonia radiata* (Laminariales, Phaeophyta) in response to simulated herbivory – the first microscopic study. Planta 218:928–937

Lüder UH, Knoetzel J, Wiencke C (2001a) Acclimation of photosynthesis and pigments to seasonally changing light conditions in the endemic Antarctic red macroalga *Palmaria decipiens*. Polar Biol 24:598–603

Lüder UH, Knoetzel J, Wiencke C (2001b) Two forms of phycobilisomes in the Antarctic red macroalga *Palmaria decipiens* (Palmariales, Florideophyceae). Physiol Plant 112:572–581

Lüder UH, Knoetzel J, Wiencke C (2002) Acclimation of photosynthesis and pigments during and after six months of darkness in *Palmaria decipiens* (Rhodophyta) – a study to simulate Antarctic winter sea ice cover. J Phycol 38:904–913

Lund S (1951) Marine algae from Jörgen Brönlunds fjord in eastern North Greenland. Meddr Grønland 128:1–26

Lund S (1959a) The marine algae from East Greenland. I. Taxonomical Part. Meddr Grønland 156(1):1–247

Lund S (1959b) The marine algae from East Greenland. II. Geographic distribution. Meddr Grønland 156(2): 1–70

Lüning K (1971) Seasonal growth of *Laminaria hyperborea* under recorded underwater light conditions near Helgoland. In: Crisp DJ (ed) Proc 4th European Mar Biol Symp. University Press, Cambridge, pp 347–361

Lüning K (1979) Growth strategies of three *Laminaria* species (Phaeophyceae) inhabiting different depth zones in the sublittoral region of Helgoland (North Sea). Mar Ecol Prog Ser 1:195–207

Lüning K (1980a) Control of algal-life-history by daylength and temperature. In: Price JH, Irvine DEG, Farnham WF (eds) The shore environment, vol 2. Ecosystems. Academic Press, New York, pp 915–945

Lüning K (1980b) Critical levels of light and temperature regulating the gametogenesis of three *Laminaria* species (Phaeophyceae). J Phycol 16:1–15

Lüning K (1988) Photoperiodic control of sorus formation in the brown alga *Laminaria saccharina*. Mar Ecol Prog Ser 45:137–144

Lüning K (1990) Seaweeds. Their environment, biogeography and ecophysiology. John Wiley, Sons, Inc., New York, Chichester, Brisbane, Toronto, Singapore, 527 pp

Lüning K (1991) Circannual growth rhythm in a brown alga, *Pterygophora californica*. Bot Acta 104:157–162

Lüning K, Dring MJ (1979) Continuous underwater light measurement near Helgoland (North Sea) and its significance for characteristic light limits in the sublittoral region. Helgol Meeresunters 32:403–424

Lüning K, Kadel P (1993) Daylength range for circannual rhythmicity in *Pterygophora californica* (Alariaceae, Phaeophyta) and synchronisation of seasonal growth by daylength cycles in several other brown algae. Phycologia 32:379–387

Lüning K, Neushul M (1978) Light and temperature demands for growth and reproduction of Laminarian gametophytes in southern and central California. Mar Biol 45:297–309

Lüning K, tom Dieck I (1989) Environmental triggers in algal Seasonality. Bot Mar 32:389–397

Lüning K, Schmitz K, Willenbrink J (1973) $CO_2$ fixation and translocation in benthic marine algae. III. Rates and ecological significance of translocation in *Laminaria hyperborea* and *L. saccharina*. Mar Biol 23:275–281

McClintock JB, Karentz D (1997) Mycosporine-like amino acids in 38 species of subtidal marine organisms from McMurdo Sound. Antarctica. Ant Sci 9:392–398

McKamey KA, Amsler CD (2006) Effects of temperature and light on growth of the Antarctic algae *Geminocarpus geminatus* (Ectocarpales: Phaeophyceae) and *Cladophora repens* (Cladophorales: Cladophorophyceae) in culture. Phycologia 45:225–232

Miller AK, Pearse JS (1991) Ecological studies of seaweeds in McMurdo sound. Antarctica Am Zool 31:35–48

Moe RL, Silva PC (1977) Antarctic marine flora: uniquely devoid of kelps. Science 196:1206–1208

Moe RL, Silva PC (1981) Morphology and taxonomy of *Himantothallus* (including *Phaeoglossum* and *Phyllogigas*), an Antarctic member of the Desmarestiales (Phaeophyceae). J Phycol 17:15–29

Moe RL, Silva PC (1989) *Desmarestia antarctica* (Desmarestiales, Phaeophyceae), a new ligulate Antarctic species with an endophytic gametophyte. Plant Syst Evol 164:273–283

Moe RL, DeLaca TE (1976) Occurrence of macroscopic algae along the Antarctic Peninsula. Antarctic J 11:20–24

Moore PG, MacAlister HE, Taylor AC (1995) the environmental tolerances and behavioural ecology of the sub-Antarctic Beach-Hopper "*Orchestia*" *scutigerula* Dana (Crustacea: Amphipoda) from Husvik, South Georgia. J Exper Mar Biol Ecol 189:159–182

Neushul M (1965) Diving observations of subtidal Antarctic marine vegetation. Bot Mar 8:234–243

Neushul M (1972) Functional interpretation of benthic marine algal morphology. In: Abbott IS, Kurogi M (eds) Contributions to the systematic of benthic marine algae of the North Pacific. Japanese Society Phycology, Kobe, pp 47–71

Nishiguchi MK, Somero GN (1992) Temperature-and concentration-dependence of compatibility of the organic osmolyte $\beta$-dimethylsulfoniopropionate. Cryobiology 29:118–124

Novaczek I (1984) Response of gametophytes of *Ecklonia radiata* (Laminariales) to temperature in saturating light. Mar Biol 82:241–245

Novaczek I, Lubbers GW, Breeman AM (1990) Thermal ecotypes in amphi-Atlantic algae. I. Algae of Arctic to cold-temperate distribution (*Chaetomorpha melagonium*, *Devaleraea ramentacea* and *Phycodrys rubens*). Helgoländer Meeresunters 44:459–474

Orheim O, Allegrini I, Boissonnas J, Drewry D, Gascard JC, Hedberg D, Müller-Wille L, Prestrud P, Sors A, Tilzer M (1995) European research in the Arctic – looking ahead. Norsk Polarinstitutt, Oslo

Osmond CB (1994) What is photoinhibition? Some insights from comparisons of shade and sun plants. In: Baker NR, Bowyer NR (eds) Photoinhibition of photosynthesis, from the molecular mechanisms to the field. BIOS Scientific Publ., Oxford, pp 1–24

Papenfuss GF (1964) Catalogue and bibliography of Antarctic and Subantarctic benthic marine algae. Am Geophys Union, Antarctic Res Ser 1:1–76

Pedersen PM (1976) Marine, benthic algae from southernmost Greenland. Meddr Grønland 199(3):1–79

Peters AF (2003) Molecular identification, taxonomy and distribution of brown algal endophytes, with emphasis on species from Antarctica. In: Chapman ARO, Anderson RJ, Vreeland V, Davison IF (eds) Proceedings of the 17th international seaweed symposium. Oxford University Press, New York, pp 293–302

Peters AF, Breeman AM (1992) Temperature responses of disjunct temperate brown algae indicate long-distance dispersal of microthalli across the tropics. J Phycol 28:428–438

Peters AF, Breeman AM (1993) Temperature tolerance and latitudinal range of brown algae from temperate Pacific South America. Mar Biol 115:143–150

Peters AF, van Oppen MJH, Wiencke C, Stam WT, Olsen JL (1997) Phylogeny and historical ecology of the Desmarestiaceae (Phaeophyceae) support a Southern Hemisphere origin. J Phycol 33:294–309

Peters KJ, Amsler CD, Amsler MO, McClintock JB, Dunbar RB, Baker BJ (2005) A comparative analysis of the nutritional and elemental composition of macroalgae from the western Antarctic Peninsula. Phycologia 44:453–463

Polle A (1996) Mehler reaction: friend or foe in photosynthesis. Bot Acta 109:84–89

van de Poll WH, Eggert E, Buma AGJ, Breeman AM (2002) Temperature dependence of UV radiation effects in arctic and temperate isolates of three red macrophytes. Eur J Phycol 37:59–68

Quartino ML, Klöser H, Schloss IR, Wiencke C (2001): Biomass and associations of benthic marine macroalgae from the inner Potter Cove (King George Island, Antarctica) related to depth and Substrate. Polar Biol 24:349–355

Quartino ML, Zaixso HE, Boraso de Zaixso AL (2005) Biological and environmental characterization of marine macroalgal assemblages in Potter cove, South Shetland Islands, Antarctica. Bot Mar 48:187–197

Rakusa-Suszczewski S, Zieliński K (1993) Macrophytobenthos. In: Rakusa-Suszczewski S (ed) The Maritime Antarctic Coastal ecosystem of Admiralty Bay. Polish Academy of Sciences, Warsaw, pp 57–60

Ramus J (1978) Seaweed anatomy and photosynthetic performance: the ecological significance of light guides, heterogenous absorption and multiple scatter. J Phycol 14:352–362

Ramus J (1981) The capture and transduction of light energy. In: Lobban CS, Wynne MJ (eds) The biology of seaweeds. University of California Press, Berkeley, pp 458–492

Raven JA, Johnston AM (1991) Photosynthetic inorganic carbon assimilation by *Prasiola stipitata* (Prasiolales, Chlorophyta) under emersed and submersed conditions: relationship to the taxonomy of *Prasiola*. Br Phycol J 26:247–257

Raymond JA, Fritsen CH (2001) Semipurification and ice recrystallization inhibition activity of ice-active substances associated with Antarctic photosynthetic organisms. Cryobiology 43:63–70

Raymond JA, Knight CA (2003) Ice binding, recrystallization inhibition, and cryoprotective properties of ice-active substances associated with Antarctic sea ice diatoms. Cryobiology 46:174–181

Rhoades D (1979) Evolution of plant chemical defenses against herbivores. In: Rosenthal G, Janzed D (eds) Herbivores. Academic Press, New York, pp 4–54

Richardson M (1977) The ecology including physiological aspects of selected Antarctic marine invertebrates associated with inshore macrophytes. PhD Dissertation, University of Durham, pp 1–165 (plus references and appendix)

Richardson MG (1979) The distribution of the Antarctic marine macro-algae related to depth and substrate. Br Antarctic Bull 49:1–13

Roberts RD, Kühl M, Glud RN, Rysgard S (2002) Primary production of crustose coralline red algaein a high Arctic fjord. J Phycol 38:273–283

Rosenvinge LK (1898) Deuxième mémoire sur les algues marines du Groenland. Meddr Grønland 20:1–125

Salles S, Aguilera J, Figueroa FL (1996) Light field in algal canopies: changes in spectral light ratios and growth of *Porphyra leucosticta*. Thur. in Le Jol. Sci Mar 60:29–38

Schaffelke B, Lüning K (1994) A circannual rhythm controls seasonal growth in the kelp *Laminaria hyperborea* and *L. digitata* from Helgoland (North Sea). Eur J Phycol 29:49–56

Schmitz K (1981) Translocation. In: Lobban, CS, Wynne M J. (eds) The biology of seaweeds, Botanical monographs, vol 17. University of California Press, Berkeley, Los Angeles, pp 534–558

Schmitz K (1990) Algae. In: Behnke H-D, Sjolund RD (eds) Sieve elements. Comparative structure, induction and development. Springer Verlag, Berlin, Heidelberg New York, pp 1–18

Schoene T, Pohl M, Zakrajsek AF, Schenke HW (1998) Tide gauge measurements – a contribution for the long term monitoring of the sea level. Ber Polarforsch Meeresforsch 299:12–14

Schoenwaelder MEA (2002) The occurrence and cellular significance of physodes in brown algae. Phycologia 41:125–139

Schroeter B, Olech M, Kappen L, Heitland W (1995) Ecophysiological investigations of *Usnea antarctica* in the Maritime Antarctic. I. Annual microclimatic conditions and potential primary production. Antarct Sci 7:251–260

Schwarz A-M, Hawes I, Andrew N, Mercer S, Cummings V, Trush S (2005) Primary production potential of non-geniculate coralline algae at Cape Evans, Ross Sea, Antarctica. Mar Ecol Prog Ser 294:131–140

Scrosati RA (1992) Estudio anatomico de *Desmarestia ligulata* de Argentina y *D. menziesii* de Antártida (Phaeophyceae). Physis A 113:89–98
Skottsberg CJF (1964) Antarctic phycology. In: Carrick R (ed) Comptes rendues premier symposium biologie Antarctique, Paris 1962. Hermann, Paris, pp 147–154
Slocum CJ (1980) Differential susceptibility to grazers in two phases of an intertidal alga: advantages of heteromorphic generations. J Exp Mar Biol Ecol 46: 99–110
Stockton W (1990) The intertidal zone at Palmer Station, Anvers Island, in the wake of the *Bahia Paraiso* spill. Antarctic J US 25:203
Svendsen P (1959) The algal vegetation of Spitsbergen. A survey of the marine algal flora of the outer part of Isfjorden. Norsk Polarinst Skrifter 116:1–49
Svendsen H, Beszczynska-Møller A, Hagen JO, Lefauconnier B, Tverberg V, Gerland S, Bischof K, Papucci C, Ørbæk JB, Zajaczkowski M, Azzolini R, Bruland O, Wiencke C, Winther JG, Dallmann W (2002) The physical environment of Kongsfjorden-Krossfjorden, an Arctic fjord system in Svalbard. Polar Res 21:133–166
Targett NM, Arnold TM (2001) Effects on secondary metabolites on digestion in marine herbivores. In: McClintock JB, Baker BJ (eds) Marine chemical ecology. CRC Press, Boca Raton, pp 391–412
Taylor WR (1966) Marine algae of the northeastern coast of North America. The University of Michigan Press, Ann Arbor
Thomas DN, Dieckmann GS (2002) Antarctic sea ice – a habitat for extremophiles. Science 295:641–644
Thomas DN, Wiencke C (1991) Photosynthesis, dark respiration and light independent carbon fixation of endemic Antarctic macroalgae. Polar Biol 11:329–337
Topcuoglu S., Fowler SW (1984) Factors affecting the biokinetics of Technetium ($^{95m}$Tc) in marine macroalgae. Mar Environ Res 12:25–43
van Oppen MJH, Olsen JL, Stam W, van den Hoek C, Wiencke C (1993) Arctic–Antarctic disjunctions in the benthic seaweeds *Acrosiphonia arcta* (Chlorophyta) and *Desmarestia viridis/willii* (Phaeophyta) are of recent origin. Mar Biol 115:381–386
van Oppen MJH, Diekmann OE, Wiencke C, Stam WT, Olsen JL (1994) Tracking dispersal routes: phylogeography of Arctic–Antarctic disjunct seaweed *Acrosiphonia arcta* (Chlorophyta). J Phycol 30:67–80
Vinogradova KL (1995) The checklist of the marine algae from Spitsbergen. Bot J 80:50–61
Weslawski JM, Wiktor J, Zajaczkowski M, Swerpel S (1993) Intertidal zone of Svalbard. 1. Macroorganism distribution and biomass. Polar Biol 13:73–79
Weslawski JM, Zajaczkowski M, Wiktor J, Szymelfenig M (1997) Intertidal zone of Svalbard 3. Littoral of a subarctic, oceanic island: Bjornoya. Polar Biol 18:45–52
Wessels H, Hagen W, Molis M, Wiencke C, Karsten U (2006) Intra- and interspecific differences in palatability of Arctic macroalgae from Kongsfjorden (Spitzbergen) for two benthic sympatric invertebrates. J Exp Mar Biol Ecol 329:20–33
Westermeier B, Gomez I, Rivera PJ, Müller DG (1992) Antarctic marine macroalgae: distribution, abundance and necromass at King George Island, South Shetland. Antarctica. Ser Cient INACH 42:21–34
Weykam G, Wiencke C (1996) Seasonal photosynthetic performance of the endemic Antarctic alga *Palmaria decipiens* (Reinsch) Ricker. Polar Biol 16:357–361
Weykam G, Gómez I, Wiencke C, Iken K, Klöser H (1996) Photosynthetic characteristics and C:N ratios of macroalgae from King George Island (Antarctica). J Exp Mar Biol Ecol 204:1–22
Weykam G, Thomas DN, Wiencke C (1997) Growth and photosynthesis of the Antarctic red algae *Palmaria decipiens* (Palmariales) and *Iridaea cordata* (Gigartinales) during and following extended periods of darkness. Phycologia 36:395–405
Wiencke C (1988) Notes on the development of some benthic marine macroalgae of King George Island (Antarctica). Ser Cient INACH 37:23–47
Wiencke C (1990a) Seasonality of brown macroalgae from Antarctica – a long-term culture study under fluctuating Antarctic daylengths. Polar Biol 10:589–600
Wiencke C (1990b) Seasonality of red and green macroalgae from Antarctica – a long-term culture study under fluctuating Antarctic daylengths. Polar Biol 10:601–607
Wiencke C (2004) The coastal ecosystem of Kongsfjorden, Svalbard. Synopsis of biological research performed at the Koldewey Station in the years 1991–2003. Ber Polarforsch.Meeresforsch. 492:1–244
Wiencke C, Clayton MN (1990) Sexual reproduction, life history, and early development in culture of the Antarctic brown alga *Himantothallus grandifolius* (Desmarestiales, Phaeophyceae). Phycologia 29:9–18
Wiencke C, Clayton MN (2002) Antarctic seaweeds. Synopses of the Antarctic Benthos (Wägele JW, Sieg J (eds)), vol 9. Gantner, Ruggell
Wiencke C, Fischer G (1990) Growth and stable carbon isotope composition of cold-water macroalgae in relation to light and temperature. Mar Ecol Prog Ser 65:283–292
Wiencke C, tom Dieck I (1989) Temperature requirements for growth and temperature tolerance of macroalgae endemic to the Antarctic region. Mar Ecol Progr Ser 54:189–197
Wiencke C, tom Dieck I (1990) Temperature requirements for growth and survival of macroalgae from Antarctica and Southern Chile. Mar Ecol Prog Ser 59:157–170
Wiencke C, Stolpe U, Lehmann H (1991) Morphogenesis of the brown alga *Desmarestia antarctica* cultivated under seasonally fluctuating Antarctic daylengths. Ser Cient INACH 41:65–78
Wiencke C, Rahmel J, Karsten U, Weykam G, Kirst GO (1993) Photosynthesis of marine macroalgae from Antarctica: light and temperature requirements. Bot Acta 106:78–87
Wiencke C, Bartsch I, Bischoff B, Peters AF, Breeman AM (1994) Temperature requirements and biogeography of Antarctic, Arctic and amphiequatorial seaweeds. Bot Mar 37:247–259

Wiencke C, Clayton MN, Schulz D (1995) Life history, reproductive morphology and development of the Antarctic brown alga *Desmarestia menziesii* J. Agardh. Bot Acta 108:201–208

Wiencke C, Clayton MN, Langreder C (1996) Life history and seasonal morphogenesis of the endemic Antarctic brown alga *Desmarestia anceps* Montagne. Bot Mar 39:435–444

Wiencke C, Vögele B, Kovaltchouk NA, Hop H (2004). Species composition and zonation of marine benthic macroalgae at Hansneset in Kongsfjorden, Svalbard. Ber Polarforsch Meeresforsch 492:55–62

Wilce RT (1963) Studies on benthic marine algae in north-west Greenland. Proc Int Seaweed Symp 4:280–287

Wilce RT (1990) Role of the Arctic Ocean as a bridge between the Atlantic and Pacific Oceans: fact and hypothesis. In: Garbary DJ, South GR (eds) Evolutionary biogeography of the marina algae of the North Atlantic. Springer-Verlag Berlin, Heidelberg, New York, London, Paris, Tokyo, Hong Kong, Barcelona, pp 323–348

Wilce RT (1994) The Arctic subtidal as habitat for macrophytes. In: Lobban CS, Harrison PJ (eds) Seaweed ecology and physiology. Cambridge University Press, Cambridge, pp 89–92

Winkler JB, Kappen L, Schulz F (2000) Snow and ice as an important ecological factor fort the cryptogams in the maritime Antarctic. In: Davison W, Howard-Williams C, Broady P (eds) Antarctic ecosystems: models for wider ecological understanding. New Zealand Natural Sciences, University of Canterbury, pp 220–224

Womersley HBS (1991) Biogeography of Australasian marine macroalgae. In: Clayton MN, King RJ (eds) Biology of marine plants. Longman Cheshire, Melbourne, pp 367–381

Zaneveld JS (1968) Benthic marine algae, Ross Island to Balleny Islands. Antarctic Map Folio Series. Am Geograph Soc NY, Folio 10:1–12

Zielinski K (1981) Benthic macroalgae of Admiralty Bay (King Goerge Island, Antarctica) and circulation of algal matter between the water and the shore. Pol Polar Res 2:71–94.

Zinova AD (1953) Brown algae of the northern seas of the U.S.S.R. Izdatel'Akademii Nauk SSSR, Moscow, Leningrad

Zinova AD (1955) Red algae of the northern seas of the U.S.S.R. Izdatel'Akademii Nauk SSSR, Moscow, Leningrad

Rev Environ Sci Biotechnol (2006) 5:187–191
DOI 10.1007/s11157-006-9102-3

REVIEW PAPER

# Life expansion in Sørkapp Land, Spitsbergen, under the current climate warming

**Wieslaw Ziaja**

Received: 27 February 2006 / Accepted: 25 July 2006 / Published online: 7 September 2006


**Abstract** Terrestrial life response to climate warming in Sørkapp Land after the Little Ice Age is described. Plant succession starts in the areas abandoned by glaciers and continues in the areas which have been outside glaciers. Continuous tundra becomes denser in many places of the west and south. A denser tundra attracts animal species feeding on plants. Vegetation of non-glaciated areas in the north-east is very sparse until now in spite of persisting development. Scarce fauna of the NE Sørkapp Land consists mainly of several bird species feeding on the sea. They are of great importance in plant and soil development, delivering organic matter to the land by fertilizing. Persisting warming leads an increase of the landscape heterogenity.

**Keywords** Climate warming · Glacial recession · Plant and animal succession · Spitsbergen

## 1 Introduction

Sørkapp Land is the southern peninsula of Spitsbergen, which narrows towards the south, situated between two open seas – the Barents Sea in the east and the Greenland Sea in the west (Fig. 1). More than 60% of the Sørkapp Land territory is still covered by glaciers in spite of climate warming since the beginning of the 20th century, after a definite end of the Little Ice Age in the 1890s. The areas situated near the extent of glaciers there, including nunataks, belong surely to the most extreme terrestrial environments for life due to sever climatic and edaphic conditions connected with intensive denudation processes. The similar extremity is a basic feature of all the unglaciated areas in the north and east of the peninsula. Only two parts of the peninsula, the big western one and the small southernmost one, are overgrown with a comparatively dense (but low) Arctic tundra covering more than 50% of the ground. Such a vegetation (Dubiel and Olech 1990, 1991, 1993) is an indicator of a really severe but not extreme environment due to the better local climate conditions (under the influence of warmer Atlantic waters and foehn winds) there. A characteristic animal world is connected with the tundra. Those two areas persist during all the Holocene and take 9–10% of the peninsula's territory. They compose plant refuges, especially during the colder climatic fluctuations, from which plant succession (colonization) spreads out in warmer periods (Ziaja 1999). Outside those two areas, soil, vegetation and even the fauna distribution are extremely irregular in Sørkapp

W. Ziaja (✉)
Institute of Geography and Spatial Management, Jagiellonian University, 7 Gronostajowa Str., 30-387 Cracow, Poland
e-mail: w.ziaja@uj.edu.pl

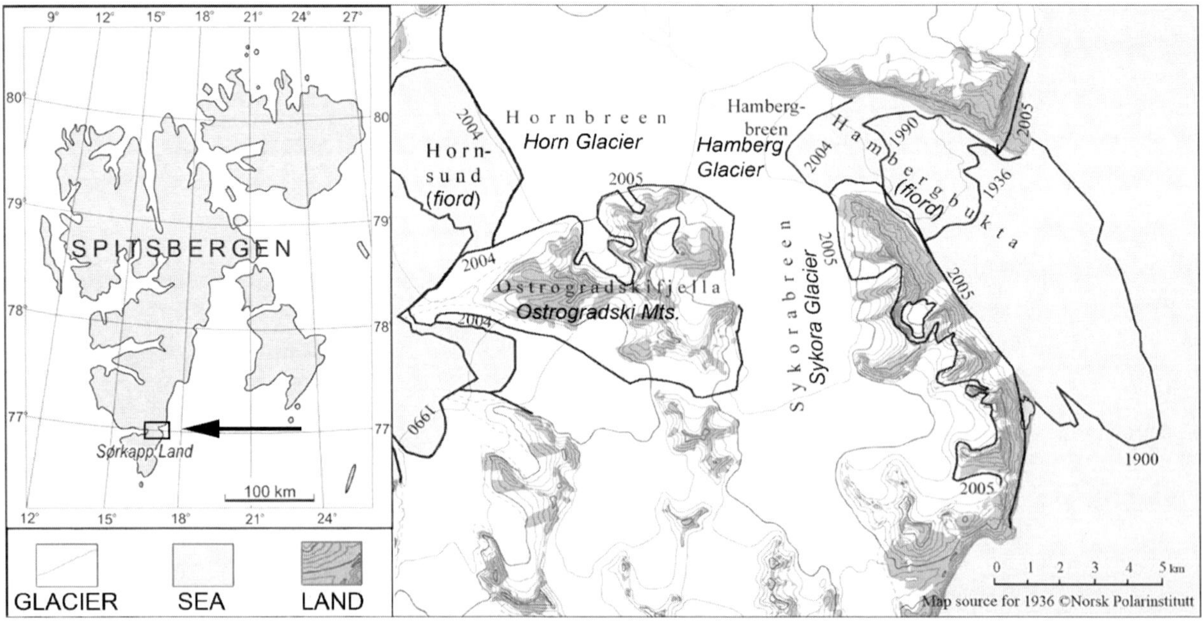


**Fig. 1** Situation of Sørkapp Land in Spitsbergen and changes of the extent of glaciers and marine coastline in the NE Sørkapp Land and the land pass between Sørkapp Land and the rest of Spitsbergen, i.e. between two fjords: Hornsund and Hambergbukta, since 1900 (Lefauconnier and Hagen 1991; Birkenmajer et al. 1992, satellite pictures from NASA, the author's field investigations)

Land, occuring in isolated pockets throughout the landscape (Ziaja 2004). The glaciated interior, with numerous nunataks and glaciers falling into the sea, separates three parts of the peninsula, described below, from each other.

The aim of the paper is to outline the knowledge on the terrestrial plant and animal life response to climate warming in Sørkapp Land after the Little Ice Age.

## 2 Plant succession

### 2.1 General remarks

In areas abandoned by glaciers, the succession starts with single plants just after deglaciation, long before the total ablation of the dead ice, forming and stabilizing the new relief and water net, developing soils, etc. The plants belong to the most resistant, i.e. with wide life amplitudes. Vascular plants and mosses usually appear earlier than lichens after deglaciation (Dubiel and Ziaja 1993).

### 2.2 Western Sørkapp Land

Continuous tundra (growing on the majority of the ground) becomes even denser under the climate warming in many places of the western part of the peninsula. Vascular plant development can also be observed in some areas of the former polar desert there, e.g. in Kulmstranda (Ziaja 2002).

### 2.3 Southern Sørkapp Land

Increasing density of the continuous tundra takes place in the southernmost Sørkapp Land too. However, the plant succession is most rapid in the lowest and most stabilized marginal zones, abandoned by glaciers, with climatologically optimum exposures. For example numerous and comparatively large clumps of plants, taking a few percent of the ground surface, have grown in places near sea level in the marginal zone of Keilhaubreen (SE Sørkapp Land), covered by a glacier as late as 1936. That is the only marginal zone with visible vegetation on infrared air photos at the scale 1:50 000 from 1990. However, just nearby, only small and sparse clumps of a few vascular plants and mosses grew that time in the marginal zone of Randbreen, which is located at higher elevation, more shaded and was free of ice about 20 years later (Ziaja 1999, 2004).

### 2.4 North-eastern Sørkapp Land

Extensive areas on the north-eastern coast of Sørkapp Land were abandoned by glaciers in the 20th century and the beginning of the 21st century (Fig. 1). There is good evidence of extents of glaciers there since 1900 in old cartographic materials, aerial photographs and satellite pictures. Both abiotic and biotic landscape components have been transformed intensively due to glacial recession under the influence of climate warming after the Little Ice Age, i.e. since 1900 (Fig. 2). These transformations were investigated in the field in August 2005 (Ziaja and Ostafin 2005; Ziaja and Maciejowski 2005). The following vascular plants were evidenced in areas abandoned recently by glaciers or newly formed coastal plains: *Oxyria digyna*, *Sagina nivalis*, *Cerastium arcticum*, *Cerastium regelii*, *Papaver dahlianum*, *Cochlearia groenlandica*, *Draba* sp. (*arctica*?), *Saxifraga cespitosa*, *Saxifraga cernua*, *Saxifraga hyperborea*, *Saxifraga rivularis*, *Luzula confusa*, *Poa alpina var. vivipara*, *Arctophila fulva*, *Colopodium vahlianum* (15 taxa). Three of them (*Cerastium arcticum*, *Saxifraga cernua*, *Luzula confusa*) are commonly dispersed in the whole NE Sørkapp Land. The next three (*Oxyria digyna*, *Papaver dahlianum*, *Colopodium vahlianum*) are limited to the areas fertilized by birds. *Cochlearia groenlandica* grows only near the coastline, probably due to the optimal thermal-humidity conditions. The other vascular taxa do not grow at numerous sites, and because of that it is difficult to designate their habitat requirements, apart from that they occur in the lower altitudes only. Two species of bryophytes and 9–10 species of lichens were found also. The plants which are expanding, mainly in the lower parts of the area (below 200 m a.s.l.), grow singly or in very dispersed small clumps. Usually the vegetation covers no more than 2–3% of the ground surface. However, there are a few dozen older patches of dense tundra, from ca. 1 $m^2$ to ca. 30 $m^2$, fertilized by small bird colonies on steep mountain slopes on the sea (apart from the biggest one at Daudbjørnpynten taking several hundred $m^2$). Some of the new streams and small lakes (with humid sites nearby) in deglaciated areas have been colonized by algae.

## 3 Animal colonization

### 3.1 General remark

A denser tundra has attracted animal species feeding on plants, e.g. reindeer *Rangifer tarandus*,

**Fig. 2** The new coastal plain formed on the southern shore of Hambergbukta after 1990. The retreating front of Hambergbreen in the background, seen from SE. The photo was taken by the author in 2005

ptarmigan *Lagopus mutus hyperboreus*, or some species of geese (Ziaja 2004).

### 3.2 Western Sørkapp Land

According to the summer field observations, the birth-rate of reindeer increased greatly in the west Sørkapp Land between 1982–1986 (1 or 2 animals met during a summer) and 2000 (more than 100 animals, with a large part of the young ones, met during 2 summer weeks). That is a really impressive manifold population increase (Ziaja 2004) which influences environment both in the areas with well developed tundra and outside them, supporting the plant succession to the areas with the extreme environment.

### 3.3 Southern Sørkapp Land

The constant presence of reindeer in south Sørkapp Land in the 1990s is a new phenomenon, not observed there in the 1980s (Ziaja 1999, 2004). The same occured to some bird species like ptarmigan *Lagopus mutus hyperboreus* and wild geese *Branta leucopsis* not observed before (Norderhaug 1989). Both reindeer and ptarmigan depend closely on plants.

### 3.4 North-eastern Sørkapp Land

The general remark from the beginning of the chapter does not refer to the NE Sørkapp Land because of scarcity of vegetation. Avifauna of the NE Sørkapp Land is quite scanty: 14 species have been observed and only six of them breed there (*Stercorarius parasiticus*, *Larus hyperboreus*, *Rissa tridactyla*, *Uria lomvia*, *Cepphus grylle*, *Alle alle*). The last three of them are of great importance in delivering organic matter from the sea to the land fertilizing the rocky walls and their foot. The Arctic skua *Stercorarius parasiticus* and kittiwake*Rissa tridactyla* perform a similar function on the coastal plains. The single Arctic fox *Alopex lagopus* observed (the only land mammal observed in NE Sørkapp Land), feeds on the birds during the summer. The vegetation is too poor to feed geese or ptarmigans. Hence, some geese which are typical for tundra (*Anser brachyrhynchus*, *Branta leucopsis*, *Branta bernicla*) can only be observed during flight. No eiders have been observed there, may be because of the low plankton count in the cold waters of the East Spitsbergen Current washing the coast from the north (Ziaja and Maciejowski 2005).

## 4 Discussion

According to Wüthrich (1991), the expansion of richer plant communities in areas previously occupied by poorer ones is possible after a long-term warming by 2–4°C. He formulates a prognosis of that expansion in the form of a modification of the old map showing the extents of four plant communities (*Inner Fjord Zone*, *Cassiope Zone*, *Dryas Zone*, *Barren Zone*) in Spitsbergen (Summerhayes and Elton 1928). However, in his new, modified map, showing the state of vegetation after a warming by 2–4°C, the entire coast of Sørkapp Land remains within the *Barren Zone*, i.e. with a very sparse vegetation which covers much less than 50% of the ground. Hence, his prognosis is pointless because the main part of the western and southern coasts of the peninsula has already been covered with continuous tundra in 1980s. Moreover, the present rate of the colonization suggests a future development of the vegetation (Ziaja 2004). A starting temperature cannot be considered in his prognosis due to lack of data for Sørkapp Land. However, it is sure that the temperature has been lower in Sørkapp Land than in central-west Spitsbergen, on the base of fragmentary meteorological measurements cited by Ziaja (2004) and observations of indirect environmental factors (e.g. influencing Sørkapp Land by the cold sea current).

## 5 Conclusions

The passage from the Little Ice Age to the current warmer period led an increase of the landscape heterogenity.

During the cold period of the LIA, the development of glaciation was followed by a diminished significance of the relief and all of the dependent landscape components (relief, waters, vegetation, fauna, soils) resulting in a relatively homogenous landscape.

Inversely, the contemporary warmer period, with a fairly high rate of deglaciation, generated a quick superficial development of the five not-glacial landscape components mentioned above and thus the increase of the heterogenity. This development will surely intensify in the event of climatic stabilization or further warming.

However, the biotic components (animals, vegetation and soils) play a relatively weak part in the Sørkapp Land environment yet, especially when compared to their efficiency in temperate environments, due to their sparse distribution in the High Arctic regions (Ziaja 2004).

Sørkapp Land is the Artctic region really predisposed to terrestrial studies on the ecological results of the contemporary climate change.

**Acknowledgments** Part of this research was financed by the Polish national Project PBZ-KBN-108/P04/2004 titled *Structure, evolution and dynamics of lithosphere, cryosphere and biosphere in the European Arctic and Antarctic* managed by Institute of Geophysics of the Polish Academy of Sciences.

## References

Birkenmajer K, Nagy J, Dallmann WK (1992) Geological Map, Svalbard, 1:100000, Markhambreen. In: Norsk Polarinstitutt Temakart 22

Dubiel E, Olech M (1990) Plant communities of NW Sörkapp Land (Spitsbergen). Zesz Nauk UJ Pr Bot 21:35–74

Dubiel E, Olech M (1991) Phytosociological map of NW Sörkapp Land (Spitsbergen). Zesz Nauk UJ Pr Bot 22:47–54

Dubiel E, Olech M (1993) State of studies on the flora and vegetation of NW Sørkapp Land (Spitsbergen). Zesz Nauk UJ Pr Geogr 94:105–110

Dubiel E, Ziaja W (1993) Contribution to the vascular plant flora of Sörkapp Land (Spitsbergen). Polish Bot Stud 5:49–51

Lefauconnier B, Hagen JO (1991) Surging and calving glaciers in eastern Svalbard. Norsk Polarinstitutt Meddelelser 116:130

Norderhaug M (1989) Svalbards fugler. Dreyer, Oslo (in Norwegian)

Summerhayes VS, Elton CS (1928) Further contributions to the ecology of Spitsbergen. J Ecol 16:193–268

Wüthrich C (1991) Landschaftökologische Umweltforschung: Beiträge zu den Wechselwirkungen zwischen biotischen und abiotischen Faktoren im hocharktischen Ökosystem (Spitzbergen). Die Erde 122:335–352 (in German)

Ziaja W (1999) Rozwoj geosystemu Sørkapplandu, Svalbard, Wydawnictwo UJ, Krakow (in Polish)

Ziaja W (2002) Changes in the landscape structure of Sørkappland. In: Ziaja W, Skiba S (eds) Sørkappland landscape structure and functioning (Spitsbergen, Svalbard). Wydawnictwo Uniwersytetu Jagiellońskiego, Krakow, pp 18–50

Ziaja W (2004) Spitsbergen Landscape under 20th Century Climate Change: Sørkapp Land. AMBIO 33(6):295–299

Ziaja W, Maciejowski W (2005) Life expansion in NE Sorkapp Land, Spitsbergen, under the contempoary climate warming. In: Investigating life in extreme environments. Book of Abstracts. On line at: http://www.esf.org/conferences/ILEE. European Science Foundation, Sant Feliu de Guixols, pp 61–62

Ziaja W, Ostafin K (2005) Dynamika krajobrazu przesmyku ladowego miedzy Sørkapplandem a reszta Spitsbergenu, Polish Polar Studies: 218–225 (in Polish)

Rev Environ Sci Biotechnol (2007) 6:33–45
DOI 10.1007/s11157-006-0014-z

REVIEW PAPER

# Some views on plants in polar and alpine regions

**F. E. Wielgolaski · S. R. Karlsen**

Received: 7 March 2006 / Accepted: 9 June 2006 / Published online: 6 October 2006


**Abstract** Many plants growing in polar and alpine regions clearly solve serious problems of life under extreme climatic conditions, as low temperatures, strong winds, unstable soils and in the North partly 24-h of light.

**Keywords** Adaptation · Alpine · Arctic · Climate change · Light · Polar · Precipitation · Snow · Temperature · Wind

## Characteristics of the polar and alpine biomes

The most important environmental factor to restrict plant life in polar and alpine regions, both having a short growing season, is low temperature (Billings 1973). Permafrost in the ground, normally with a shallow active layer on top in summer, is common in most polar regions. It may also occur in some alpine regions but this is less frequent. Alpine areas are found in all parts of the world, in both tropical and polar regions, in oceanic as well as in continental and often dry areas. In the mountains of the equatorial zone there is often an enormous diurnal temperature variation. It may be said to be winter every night and summer every day (Hedberg 1957). In polar regions, the monthly mean temperatures in winter may be below – 40°C (e.g. Obasi 1996; Chernov and Matveyeva 1997), even lower in some of the most continental areas, while commonly well above 0°C during summer months in areas where snow melts. Particularly because of these differences in the temperature conditions between polar and alpine regions they are often said to have greater dissimilarities than similarities (e.g. Körner 1995), although there are also several similarities (Billings 1973).

However, in both biomes few plants only can stand the harsh temperature and other extreme environmental conditions, e.g. strong wind, often found there. Normally trees are missing both in alpine and polar regions in the Northern Hemisphere, and the absence of trees is often used as a definition of these biomes. In some districts, however, of both northern Russia and North America, the melted soil top layer in summer is deep enough to foster tree growth. Then, we get a so-called Forest Tundra in areas with a thick frozen layer further down in the ground. Detailed division of the Arctic in bioclimatic/biogeographichal zones is presented by Bliss (1981) and later defined by the CAVM Team (2003), while a brief presentation of the Alpine division and the

F. E. Wielgolaski
Department of Biology, University of Oslo, 1045, 0316 Blindern, Oslo, Norway
e-mail: f.e.wielgolaski@bio.uio.no

S. R. Karlsen (✉)
NORUT IT, 6434, 9294 Tromsø, Norway
e-mail: stein-rune.karlsen@itek.norut.no

previous Arctic divisions is given in Wielgolaski (1997a).

In temperate regions the summer temperature in particular, is the limiting factor for tree growth (Dahl 1986). Generally, the tree line is higher in continental areas with large mountain massifs than in more oceanic areas at the same latitude (Fig. 1). It is found to be as high as above 4000 m elevation in tropical areas of South America and in continental Asia (Troll 1973a), while at 0 m elevation e.g. along the coast of northern Norway (e.g. Aas and Faarlund 2001; Karlsen et al. 2005). In polar regions, soil may be unstable even on relatively gentle slopes, also at low elevation, but it might also be found at high elevations in lower latitudes. Such soil movement (solifluction) easily breaks plant roots, which, therefore, normally prevents tree growth, and has been used as one criterion to define alpine belts (Troll 1973b). Strong wind under extreme environmental conditions may also stop tree growth. Near the upper tree line, groups of "krumholz" trees, particularly of *Picea abies* (Fig. 2), are well known (e.g. Wielgolaski 1997b; Kullman 1998). The outer stems protect the inner ones against the most extreme conditions, e.g. strong winds, in such a way that the inner protected ones normally are considerably taller than the border stems. Small stands of trees at the tree line often show "flagging" i.e. most the branches lacking at the wind side, except near the ground where the buds and branches are protected by snow during winter.

In alpine areas there are often strong regional variations in the amount of precipitation. The actual snow line in the tropical Puna of South America is above 6000 m elevation, mainly as a result of extreme aridity. Low precipitation and deserts are also common at high altitudes in many continental regions of Asia (Troll 1972). However, precipitation in many, normally somewhat more humid regions of the world increases with elevation, e.g. in the Rocky Mountains (Kittel et al. 2002) and in western Fennoscandia (Aune 1993; Førland 1993), and then may cause a thick cover of snow in winter. The melting of the snow shows great local variations depending on the topography, normally of greatest importance on steep mountains, but always influencing the plant growing season (e.g. Sonesson et al. 1975; Inouye and Wielgolaski 2003).

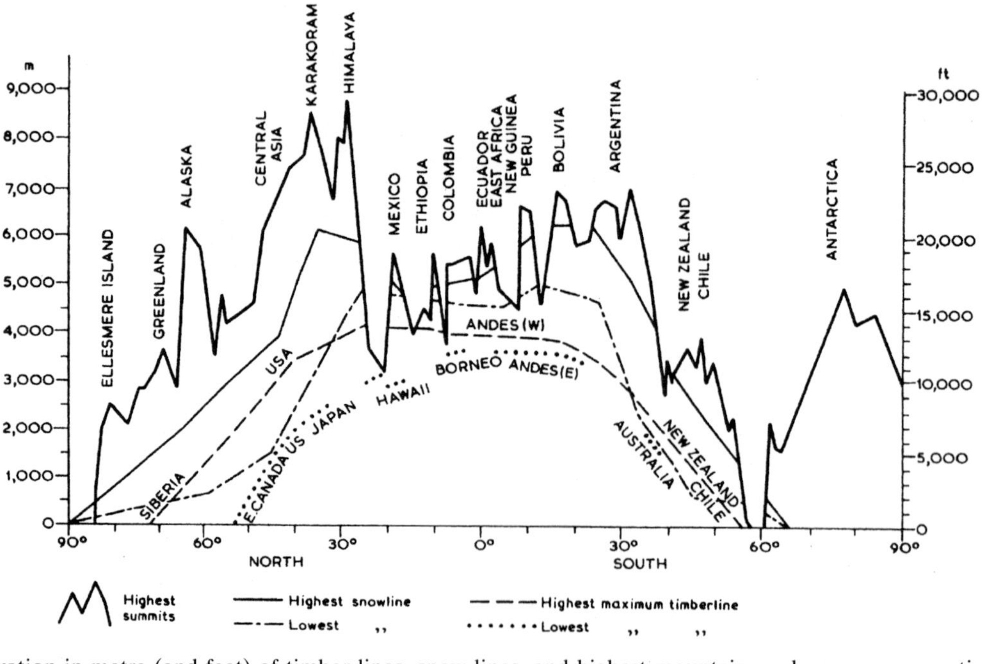


**Fig. 1** Elevation in metre (and feet) of timber lines, snow lines, and highest mountain peaks on a cross-section of global alpine regions (after Swan 1967)

**Fig. 2** "Krumholz " of Norway spruce (*Picea abies*) at the climatic tree line in south-eastern Norway, wind dominating from the left (west) of the picture. Most branches near the snow cover. (Photo: F.E Wielgolaski)

Often, polar regions show a low annual precipitation, commonly <100–400 mm (Bliss 1997; Chernov and Matveyeva 1997). Due to the long period of low temperature, most of it comes as snow, which further delays the length of growing season. The time for disappearance of snow cover in spring is normally the primary factor for growth start in cold regions, particularly in continental regions (Inouye and Wielgolaski 2003). However, growth may start under some snow, if the ground in more humid districts is insulated by a heavy cover and then nearly free of frost in the upper soil (Resvoll 1917; Bliss 1971; Wielgolaski et al. 2004). Due to the generally low temperatures at high latitudes also in summer, the evapotranspiration is also often low but most often approximating precipitation (Brown 1981). Therefore, wet areas are often commonly tundra, e.g. in the lowlands of polar regions (Rydén 1981).

In some cases permafrost in the tundra causes part of the nearly flat areas to build up mounds with ice-cores. This is common e.g. in more continental northern Europe (palsas), but may be even more spectacular in other parts of the world as in Polar parts of North America, where conical ice-core hills may be several meters tall, the so-called "Pingos" (Fig. 3). Under such conditions

**Fig. 3** Several metre tall vegetated "Pingo", built up by an ice core in North American wetlands (Photo: F.E. Wielgolaski)

the roots of plants growing on the surface easily break, which only few species will tolerate, of course.

Particularly the absence of trees both in polar and alpine biomes can very often give a similar visual impression (Wielgolaski 1986), and in both biomes there are variations from wet sedge-moss communities to dry dwarf-shrub heaths and rocky fell-fields. Solifluction, mass-flow and patterned ground, as e.g. polygons (Bliss 1997), are common both in polar communities (Fig. 4) and in high alpine or nival belts (also in the temperate region), which of course strongly influences the sparse vegetation. The broad comparability in the physiognomy of polar and alpine vegetation is clearly stressed by Barry and Ives (1974). However, they also say that the extensive wet tundra

**Fig. 4** Stone-open soil polygons in a nearly flat ground in High Arctic at Spitsbergen, made by frequent ice formations in the soil, with vegetation in between. (Photo: F.E. Wielgolaski)

commonly found in the Arctic is largely lacking in alpine areas, where a complex mosaic of more mesic and xeric communities occur due to the dissected nature of the terrain.

A very important dissimilarity between polar regions and alpine regions further south is the light regime. Whereas there are possibilities for 24 h of sunshine during at least parts of the growing season in polar regions (Fig. 5), this is not the case in temperate and tropical montaneous regions. This often causes different ecotypes or provenances to develop at various altitudes and latitudes (Mooney and Billings 1961). It is also an important factor to find the best fitted cultivar of agricultural species for various districts of polar regions.

## Adaptation

Plants in polar region as well as in alpine tundra are adapted by very rapid growth to the normally short growing season (Savile 1972; Wielgolaski 1997c). Heide (1985) stated that the more severe the environment, the more important survival adaptations seemed to be, while biological competition tended to be less important to the vegetation. Some tundra plant species will under extreme environmental conditions require several years to finish their life cycle, while only one season under better conditions (e.g. Sørensen 1941). A typical example of such adaptation is

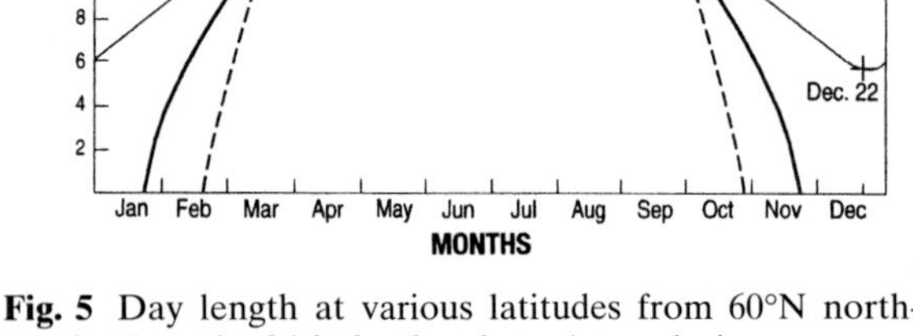


**Fig. 5** Day length at various latitudes from 60°N northwards. Note the 24 h day for about 4 months in summer at 79°N (e.g. at Spitsbergen), and a similar period with sun always below the horizon in winter

found in species of the genus *Salix*. Dahl (1986) stresses that in prostrate tundra species within the genus (*S. herbacea, S. polaris* and *S. reticulata*) it may take four growing seasons for development of catkins, while in taller lowland species it normally takes only one year.

Transplantation of trees originating in a temperate lowland area to a subarctic-subalpine region or between extreme environments of different types (e.g. from a subalpine temperate region to a high latitude area with 24 h daylight in parts of the growing season) has been carried out in many years. Often this has shown strong influence in productivity and phenology of the trees (Hagem 1931; Kalela 1938; Beuker 1994; Wielgolaski and Inouye 2003; Ovaska et al. 2005). Planting trees of a southern origin (provenance or ecotype) in the north, often means that the trees continue their growth late in summer and autumn in their new growth place as they are adapted normally to do on their southern district of origin. This again may cause the trees to have a weak hardening of their new shoots before an early start of low temperature and winter in the more extreme environment. Often that leads to frost damage of the buds during winter or the new shoots in spring.

If a southern provenance is transplanted to northern latitude regions with 24 h daylight during mid summer, that may also lead to growth problems due to plant physiological responses by the changes in spectral composition of the light. The plants originating from northern latitudes are for instance normally adapted to a low red to far red ratio particularly at the end of an Arctic summer (Nilsen 1986) compared to more southern provenances. This may be one factor to induce growth cessation of woody plants adapted to northern latitudes, and is also found to influence the shoot elongation (Håbjørg 1972a, b; 1978). However, temperature variations e.g. day and night may also induce senescence (Marchand et al. 2004), and is therefore important in the plant adaptation by a continued climate change (see next section).

Normally, the well hardened buds of coniferous trees (Beuker 1994) and leaf buds of mountain birch (Ovaska et al. 2005) from the northernmost ecotypes have an earlier break in spring than buds from trees of a more southern ecotype. Many agricultural plants are found to be adapted to use fewer days between various growth and flowering stages when grown in long days (Skjelvåg 1998). As temperature normally decreases to the north, this also means that many specific cultivars of agricultural plants can be grown further north than otherwise would have been possible.

The phenological transplant studies referred to above clearly indicate that the latitude, i.e. the light regime, is of strong importance for development of plants in polar regions. Adaptation to the growing condition at the place of origin in a species is also clearly stated by studies in controlled climate (e.g. in *Betula* by Myking 1999). Plants originally growing in districts with mild and unstable winters (as in the south and along the coast) showed later bud burst, particularly after only a short period of chilling (November) than more northern and inland ecotypes with cold and stable winters (Fig. 6). Differences in the date of bud burst of mountain birch originating from different elevations but approximately the same latitude, indicate that also adaptation to certain temperature regimes is of importance for start of the growing season (Ovaska et al. 2005), which is as expected.

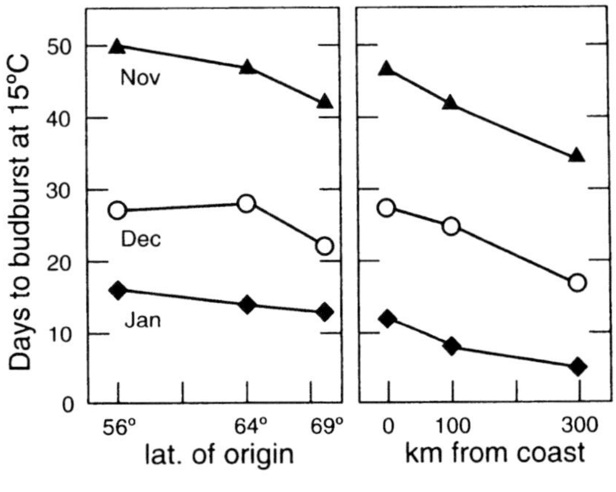


**Fig. 6** Days to bud burst at 15°C in a short-day period (8 h), after different periods of chilling at 5°C, in three *Betula pubescens* ecotypes along two gradients, in **a** (left) from three different latitudes of origin and in **b** originating from various distances from the coast (reprinted from Myking 1999)

The normally very rapid growth found in plants of a short growing season, demands an adaptation to a rapid and good supply in spring of both energy and nutrients from winter storage organs (Wielgolaski 1984). The storage organs of polar and alpine plants are commonly situated below-ground, close to the growing points, which normally are just above, but also at the ground itself, chamaephytes and hemicryptophytes according to the life form classification by Raunkiaer (1934), plant types dominating in these extreme climatic conditions (Table 1). That is a good adaptation to a short translocation distance to and from the buds where the stored material is used for growth in the short and very active growing season. Particularly in monocotyledons of wet sedge communities but also in forbs of mesic low alpine and low Arctic tundra, the storage organs are mainly found in the roots (Mooney and Billings 1960), resulting in a higher proportion of root mass compared to shoot mass than in most other plant communities even in tundra regions (Table 2).

In some species of tundra forbs, sedges and grasses parts of the leaves even stay green during winter to speed up growth start in spring (but then sometimes they die in early summer) (Wielgolaski 1997c). In evergreen dwarf shrubs, however, there is a late bud break and a slow translocation rate but in these species generally having a low nutrient level (Wielgolaski et al. 1975), the life form is probably more an adaptation to grow on nutrient-poor soil than a direct adaptation to live in climatic extreme conditions.

Generally, shrubs are close to their limit of survival in tundra. The plants have to be creeping to be protected by a snow cover against the lowest temperatures and the strongest wind. Close to the permanent snow cover of alpine regions and in the High Arctic the conditions are too extreme for formation of above-ground reserves in buds and woody above-ground parts, and even dwarf-shrubs are missing in these regions. Dahl (1986), however, stressed that the hardiest plants are pulvinate chamaephytes with buds just above the soil. There, the close neighbouring between several, densely tufted leaves makes the cushion to exchange heat more like a single organ of the size of the total cushion. It is normally important for

**Table 1** Number of vascular plant species and percentage of different life forms, based on the classification by Raunkiaer (1934), at various alpine altitudes in southern Norway

| Altitudinal limit (m) | Number of species | Life form | | | | | | |
|---|---|---|---|---|---|---|---|---|
| | | Phanerophyte (%) | Chamaephyte (%) | Hemicrytophyte (%) | Geophyte (%) | Helophyte (%) | Hydrophyte (%) | Therophyte (%) |
| > 2000 | 29 | | 44.8 | 55.2 | | | | |
| 1800–1999 | 39 | | 25.6 | 66.7 | 2.6 | 2.6 | | 2.6 |
| 1600–1799 | 75 | 8.0 | 32.0 | 48.0 | 4.0 | 5.3 | | 2.7 |
| 1400–1599 | 63 | 4.8 | 9.5 | 65.1 | 11.1 | 1.6 | 3.2 | 4.8 |
| 1200–1399 | 138 | 6.5 | 10.1 | 51.4 | 10.1 | 10.1 | 2.9 | 8.7 |
| 1000–1199 | 109 | 2.8 | 5.5 | 55.0 | 10.1 | 11.9 | 5.5 | 9.2 |

**Table 2** Different average ratios between plant parts of some tundra vegetation types (based on Wielgolaski et al. 2001)

| Region | Root/shoot (live) | Non-green/green live: vascular plants | Dead/live above-ground: vascular plants |
|---|---|---|---|
| Desert and semi-desert | 0.9 | 2.3 | 1.9 |
| Wet sedge meadows | 21 | 23 | 1.6 |
| Mesic dry meadows | 5.0 | 7.7 | 0.8 |
| Dwarf-shrub tundra | 3.1 | 12 | 0.6 |
| Low shrub tundra | 2.0 | 19 | 0.2 |
| Forest tundra | 0.8 | 15 | 0.1 |

the polar and alpine plants to absorb as much heat of the incoming radiation as possible via the soil close to the plants or by the plants themselves. This is also reflected by the plant types found and their morphological adaptations.

Very often tundra plants are covered by hairs to protect against cooling wind and evapotranspiration, and in this way also directly absorb heat for the hairy organs. They may be relatively dark in colour as often in *Draba* spp. and *Erigeron* spp. and the early emerging sepals of some polar/alpine *Ranunculus (R. glacialis, R. nivalis, R. sulphurous)*. In other plants the hairs are light coloured, silky or woolly, and light-transparent to the dark and heat-absorbing organs inside the hairs or bracts, as e.g. in cotton-grass, *Eriophorum* spp. (Dahl 1986). In some willows (*Salix* spp.), Krog (1955) observed a 15–25°C higher temperature in the catkins than in the ambient air. In these willows solar radiation penetrates the silky hairs of the dark catkin scales, which are heated also because of lower outward heat transport due to the hairs, a really effective adaptation to life in extreme environments.

It is also observed that both the orientation of many plant leaves and flowers in polar/alpine regions may be adaptations to the environmental conditions of the biomes. The leaves e.g. of many cushion plants at high altitudes (even at lower latitudes) often have a nearly vertical position to get the best reflection of incoming light to hit other leaves instead of being reflected back to the atmosphere (Dahl 1986). Better known is probably the parabolic form of many tundra flowers, which increases the absorbed radiation by reflection to the centre of the flower (Kevan 1975; Wielgolaski 1987). This is favourable to increase the temperature for pollination of the flowers and for development of the ovary. This adaptation seems to be most common in plants with a high reflection coefficient for incoming radiation by having light coloured flowers e.g. white or yellow (for instance in *Dryas* spp., Fig. 7, *Papaver* spp. and *Ranunculus* spp.). Often the flowers also change their direction during the day always to be open towards the sun (Kevan 1975; Wielgolaski 1997c). Sometimes also the colour of the flowers are adapted to change to be darker in colour after

**Fig. 7** Parabolic flowers of near white colour as *Dryas octopetala* collect maximum of incoming heat to the ovary in the centre of the flower. (Photo: F.E. Wielgolaski)

pollination by insects, which are attracted by white colours, to absorb even more heat, e.g. *Ranunculus glacialis*.

## Climate change

Plants living in the extreme environments of polar and alpine biomes are thus adapted to the conditions there in many ways, but growth is always modified, as in other climatic zones of the world, by the nutrient availability. Although, normally temperature is the dominant factor for growth in cold regions, light is of course also very important, particularly at increasing latitude and at the end of the growing season (see e.g. references in Arft et al. 1999). There will always be changes in the climate during a period. Generally, it is estimated that for the Northern Hemisphere annual temperature during the last 150 years has increased by 0.055° per decade (Overpeck et al. 1997; Jones and Moberg 2003). Both in northern and southern Norway Klaveness and Wielgolaski (1996) have observed earlier first flowering of most plant species in the mid 20th century than about 100 years earlier. However, warming has accelerated in recent decades in the Northern Hemisphere (IPCC 2001; ACIA 2004), which is of extreme importance for plant growth particularly in polar and alpine regions, especially if it continues also in the future. Most important is the increase in temperature but also in precipitation, both calculated to be stronger during winter than summer and in models predicted to be strongest in northern latitudes (Dickinson 1986; Maxwell 1997; ACIA 2004; Schwartz et al. 2006). In Europe the positive phase of North Atlantic Oscillation (NAO) has increased clearly in the period February–April during the last decades, leading to prevailing westerly winds and, therefore, higher temperatures and a more humid climate, particularly since the end of 1980s (Post and Stenseth 1999; Chmielewski and Rötzer 2001; Wanner et al. 2001). In northern Norway it is observed that *Cornus suecica* avoiding the most continental districts recently has been common more to the east, indicating a more humid climate there than before (Tømmervik et al. 2004). In high latitudes and altitudes of the Northern Hemisphere the recent climate change has caused higher winter precipitation as snow. Anyhow, in most lowland areas, and particularly in coastal regions, the increased temperature has caused earlier snowmelt (Maxwell 1992), the growing season to be longer (e.g. Bliss and Matveyeva 1992) and the plant population rates to be higher (Carlsson and Callaghan 1994). In extreme cold climates, however, as at the highest elevations and in the continental parts of northernmost Europe, winter temperatures have not increased enough through the last decades of the 1900s to foster earlier snowmelt. However, by continued increase of the winter temperature during the coming years, it is expected that this will change (Shutova et al. 2005; 2006).

In autumn, on the other hand, it may be speculated that somewhat lower global radiation by more cloudy weather and then also partly slightly decreasing temperatures, will continue the earlier end of the growing season calculated for parts of the more continental northernmost Europe by satellite images during the last two decades of 1900 as well as by 40 years of field phenological studies of birch leaf yellowing at Kola Peninsula (Shutova et al. 2005; 2006). The day length has been found to be important for the time of end of growing season, particularly at extreme high latitudes. Håbjørg (1972a, b) found in controlled climate (phytotron) studies that plants from northern latitudes were stronger dependent in cessation of growth by reduction in light than more southern provenances. He concluded that this might be an adaptation in the northern provenances to the markedly lowered red to far red ratio at the end of the Arctic summer.

Low autumn temperatures and then particularly low minimum values have for a long time also been suggested to be triggering yellowing and senescence of plants (Galakhoff 1938). In the International Tundra Experiments (ITEX), the temperature of intact ecosystems in the field have been manipulated at 13 circumarctic and alpine sites by using transparent open top chambers through several years. The plants within these 1–3°C warmer chambers, generally, showed somewhat later senescence than outside, but significantly only at one alpine site (Arft et al. 1999). This is in contrast to a significantly earlier start of the growing season (Henry and Molau 1997) and also of flowering at all sites by the warming. However, Marchand et al. (2004) found in experiments at Northeast Greenland, that plants heated in the field during the growing season with infrared radiation by about 2.5°C, showed a higher maximum percentage of vascular plant cover and a delay in yellowing in the autumn by approximately 15 days (Fig. 8) compared to the surrounding vegetation in NDVI analyses. They concluded that also in High Arctic tundra, higher temperatures postpone the senescence process. This may cause longer growing seasons if temperatures increases in the future. On the other hand, many plant species may be less matured for the winter, and also less fitted for the Polar light regime.

In a paper summarising the ITEX results (Walker et al. 2006), it is stressed that plant diversity is reduced within the open top chambers. However, the height and cover of deciduous shrubs and monocotyledons have increased during the experimental period, while the cover of mosses and lichens has been decreasing (Fig. 9). The changes are stronger in the Low than in the High Arctic, maybe as an indication of less nutrients being released by increased temperature at the coldest sites. It is suggested in several papers (e.g. Shaver et al. 2000) that if "soil organic matter turnover is increased due to warming, there is a high potential for redistribution of nitrogen from soils to vegetation". This results in higher production at least for a certain period.

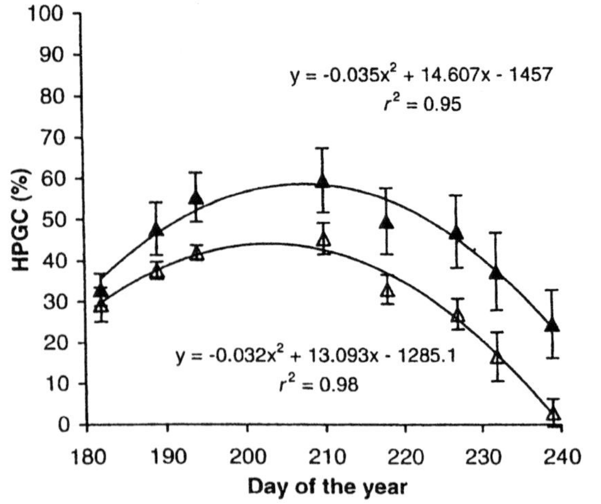


**Fig. 8** Percentage higher-plant cover at unheated (lower curve) and heated in the field with infrared radiation (upper curve) at Northeast Greenland. Note that senescence is clearly delayed in the heated vegetation as studied in NDVI analyses (Marchand et al. 2004)

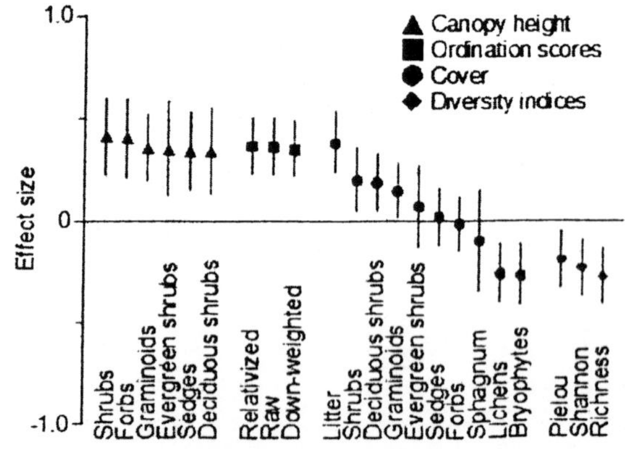


**Fig. 9** Response in several tundra plant forms on heating in the field by keeping the vegetation within open top transparent chambers. Mean of results from several ITEX sites in the response on various community variables, as e.g. on the species composition (ordination scores) (Walker et al. 2006)

Near the tree line in Fennoscandia this is observed in the field particularly in mountain birch growth (own unpubl. obs.). It is also obvious that the tree line in the region has increased by 100–150 m since the mid 1980s as a result of higher temperatures (e.g. Kullman 2002), but for e.g. *Betula pubescens* ssp. *tortuosa* in Fennoscandia also by reduced browsing by livestock. The same author (Kullman 1998) has also found that trees increasing their growth after earlier temperature rises (e.g. in the 1920–1930s) at the tree line locally died in the tops when there came a decrease in winter temperature again afterwards.

The reduced cover of lichens in the ITEX-experiments was found to be significant after 3 years, while the increased cover of vascular plants was significant only after 4–6 years of warming (Walker et al. 2006). The reduction in lichen cover may be seen as a result of competition with increasing biomass of vascular plants (Cornellisen et al. 2001) or of snow melt and refreezing several times during the winter season. Such conditions will cause an ice cover influencing the $O_2/CO_2$ conditions for the lichens under the ice. In transplant studies of vegetation mats in an alpine region from a nearly snow free lichen heath to a near by *Vaccinium myrtillus* snow bed, one of the authors (Wielgolaski 2001) found that already after 1 year the lichen *Cetraria nivalis* had changed colour to yellow-brown, and after 4 years nearly all arbuscular lichens were dead, probably because of lack of oxygen under the ice cover. Lichen death due to climate change will be a serious problem for the survival of reindeers in many northern areas, where lichens are the main diet, particularly during winter. However, overgrazing during the last decades also has caused a dramatic decline of lichen cover in parts of northernmost Norway (Johansen and Karlsen 2005).

Remote sensing has the potential to monitor and give evidence of the ongoing climate change in Arctic vegetation at a variety of spatial and temporal scales (e.g. Stow et al. 2004). It can for instance identify changes in the above-ground production, structure and cover, in the phenological cycle, and changes in ecotone boundaries. NOAA Advanced Very High Radiometer (AVHRR) imagery is particularly valuable for land cover studies at decadal time scales since these data are available from the early 1980s to the present. Studies based on the Normalised Difference Vegetation Index (NDVI) from the AVHRR instrument have found an extensive greening trend at higher northern latitudes (e.g. Myneni et al. 1997; Bogaert et al. 2002). In arctic tundra and boreal forest there is a close relationship between temperature and NDVI (e.g. Suzuki et al. 2001; Karlsen et al. 2006; Walker et al. 2003), and the increased greenness is associated with increased air-temperature (Buermann et al. 2003; Gong and Ho 2003; Walker et al. 2003). However, the greening trend shows large regional differences. Generally, in the 1990s the North American greening trend was higher than the Eurasian trend (Slayback et al. 2003). Locally, a trend of slightly shorter length of the growing season for the period 1982 to 1998 was found in the most continental parts of northern Fennoscandia and Kola Peninsula in northwestern Russia (Høgda et al. 2001) (Fig. 10), mostly as a result of later onset of spring. A trend of changes in the short Arctic growing season length over decadal time scales is of key interest, since it could be the first indication of a shift in the above-ground production and cover. The time-integrated NDVI during the growing season is associated with the above-ground plant biomass (e.g. Walker et al. 2003). Increased time-integrated NDVI is found in northern Alaska the last decades (Jia et al. 2003; Stow et al. 2003), and the greenness increases most rapidly in areas of moist non-acidic tundra (Jia et al. 2003).

The ongoing climate change with strongly increasing temperatures particularly in the North of the Northern Hemisphere may according to

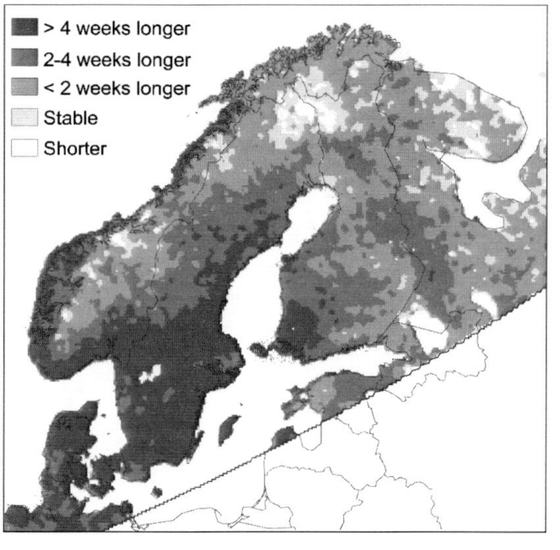


**Fig. 10** Changes in length of the growing season in northernmost Europe during the period 1982–1998 as analysed by the GIMMS-NDVI satellite dataset (reprinted from Høgda et al. 2001 with permission from NORUT IT)

modelling scenarios cause a 40% reduction of the current tundra by an expected temperature increase of 4–7°C over the next 100 years (ACIA 2004). In many ways, this may cause less extreme environments for plants with replacement of the tundra by trees. However, many of these plants will not be adapted to grow in 24 h daylight. Such an adaptation takes many years, and in the meantime many of the invading plant species to the North probably will suffer seriously.

## Concluding remarks

The examples given here clearly demonstrate the serious problems of life for many plants growing in Polar and Alpine Regions under extreme climatic conditions, but other problems may also occur. Low temperatures both during winter and often also summer strongly limit the diversity of plant species. Also high wind speed (Fig. 11) and unstable soil are important for many species. Both extremely low and high precipitation is observed, often in combination with temperature changes. In polar regions also the light regime is special. Having 24 h of light during parts of the growing season clearly causes a need for adaptation in plants to such conditions, both for growth, development and survival. In a changing climate with an expected much higher temperature, many of the plant species will not be mature by the end of the growing season and will die back or freeze to death during winter.

**Fig. 11** Mountain birch exposed to strong wind mainly coming from the left (west) of the picture. Note also the much darker upper parts of the tree stems. That is due to growth mainly above the general snow line in winter by in particular the foliose dark brown lichen *Parmelia olivacea* (Sonesson 2001). It can be hypothesised that ice cover on snow covered trunks is one reason for lack of this lichen by reducing the $O_2$ availability as suggested earlier in the paper for fruticose lichens in snow beds (Photo: F.E. Wielgolaski, from a coastal mountain in Norway)

## References


Aas B, Faarlund T (2001) The holocene history of the nordic mountain birch belt. In: Wielgolaski FE (ed) Nordic mountain birch ecosystems. UNESCO, Paris and The Parthenon Publ. Group, New York, London, pp 5–22

ACIA (2004) Impacts of a warming arctic. Arctic climate impact assessment. Cambridge University Press, Cambridge, UK

Arft AM, Walker MD, Gurevitch J, Alatalo JM, Bret-Harte MS, Dale M, Diemer M, Gugerli F, Henry GHR, Jones MH, Hollister RD, Jónsdóttir IS, Laine K, Lévesque E, Marion GM, Molau U, Mølgaard P, Nordenhäll U, Raszivin V, Robinson CH, Starr G, Stenström M, Totland Ø, Turner PL, Walker LJ, Webber PJ, Welker JM, Wookey PA (1999) Responses of tundra plants to experimental warming: meta-analysis of the international tundra experiment. Ecol Monogr 69:491–511

Aune B (1993) Climate (Engl. Summ. pp 56–61). Nasjonalatlas for Norge. Statens Kartverk. Hønefoss, Norway

Barry RG, Ives JD (1974) Introduction. In: Ives JD, Barry RG (eds) Arctic and alpine environments. Methuen & Co Ltd., London, pp 1–13

Beuker E (1994) Adaptation to climatic changes of the timing of bud burst in populations of *Pinus sylvestris* L. and *Picea abies* (L.) karst. Tree Physiol 14:961–970

Billings WD (1973) Arctic and alpine vegetations: similarities, differences, and susceptibility to disturbance. Bioscience 23:697–704

Bliss LC (1971) Arctic and alpine life cycles. Ann Rev Ecol Syst 2:405–438

Bliss LC (1981) North American and Scandinavian tundras and polar deserts. In: Bliss LC, Heal OW, Moore JJ (eds) Tundra ecosystems: a comparative analysis. Cambridge Univ. Press, Cambridge, London, New York, Sydney, pp 8–34

Bliss LC (1997) Arctic ecosystems of North America. In: Wielgolaski FE (ed) Polar and alpine tundra. Ecosystems of the world, vol. 3. Elsevier, Amsterdam, New York, Oxford, Tokyo, pp 551–683

Bliss LC, Matveyeva NV (1992) Circumpolar arctic vegetation. In: Chapin III FS, Jefferies RL, Reynolds JF, Shaver GR, Svoboda J (eds) Arctic ecosystems in a changing climate. Academic Press, New York, pp 59–89

Bogaert J, Zhou L, Tucker CJ, Myneni RB, Ceulemans R (2002) Evidence for a persistent and extensive greening trend in Eurasia inferred from satellite vegetation index data. J Geophys Res 107(D11), doi:10.1029/2001JD001075

Brown J (1981) Introduction. In: Bliss LC, Heal OW, Moore JJ (eds) Tundra ecosystems: a comparative analysis. Cambridge Univ. Press, Cambridge, London, New York, Sydney, p 79

Buermann W, Anderson B, Tucker CJ, Dickinson RE, Lucht W, Potter CS, Myneni RB (2003) Interannual covariability in northern hemisphere air temperatures and greenness associated with El Niño-southern oscillation and the arctic oscillation. J Geophys Res 108(D13):4396, doi:10.1029/2002JD002630,2003

CAVM Team (2003) Circumpolar arctic vegetation map scale 1:7,500000. Conservation of arctic flora and fauna (CAFF) Map No. 1. US Fish and Wildlife Service, Anchorage, Alaska

Carlsson BA, Callaghan TV (1994) Impact of climatic change factors on the clonal sedge *Carex bigelowii*: implication for population growth and spread. Ecography 17:321–330

Chernov YI, Matveyeva NV (1997) Arctic ecosystems in Russia. In: Wielgolaski FE (ed) Polar and alpine tundra. Ecosystems of the world, vol. 3. Elsevier, Amsterdam, New York, Oxford, Tokyo, pp 361–507

Chmielewski F-M, Rötzer (2001) Response of tree phenology to climate change across Europe. Agr For Meteorol 108:101–112

Cornelissen JH, Callaghan TV, Alatalo JM, Michelsen A, Graglia E, Hartley AE, Hik DS, Hobbie SE, Press MC, Robinson CH, Henry GHR, Shaver GR, Phoenix GK, Gwynn Jones D, Jonasson S, Chapin III FS, Molau U, Neill C, Lee JA, Melillo JM, Sveinbjörnsson B, Aerts R (2001) Global change and arctic ecosystems: is lichen decline a function of increases in vascular plant biomass? J Ecol 89:984–994

Dahl E (1986) Zonation in arctic and alpine tundra and fellfield ecobiomes. In: Bliss LC, Heal OW, Moore JJ (eds) Tundra ecosystems: a comparative analysis. Cambridge Univ. Press, Cambridge, London, New York, Sydney, pp 35–62

Dickinson R (1986) The climate system and modelling of future climate. In: Bolin B, Doos B, Jager J, Warrick RA (eds) The greenhouse effect, climate change and ecosystems. Wiley, Chichester, UK, pp 207–270

Førland E (1993) Annual precipitation. Det Norske Meteorologiske Institutt, Kartblad (Map) 3.1.1. Nasjonalatlas for Norge. Statens Kartverk. Hønefoss, Norway

Galakhoff NN (1938) The autumnal hue and shedding of foliage (in Russ.). Bot J USSR 23:193–208

Gong D-Y, Ho C-H (2003) Detection of large-scale climate signals in spring vegetation index (normalized difference vegetation index) over the northern hemisphere. J Geophys Res 108(D16):4498, doi: 10.1029/2002JD002300

Håbjørg A (1972a) Effects of photoperiod and temperature on growth and development of three latitudinal and three altitudinal populations of *Betula pubescens* Ehrh. Meld N Landbr Høgsk 51(2):1–27

Håbjørg A (1972b) Effects of light quality, light intensty and night temperature on growth and development of three latitudinal populations of *Betula pubescens* Ehrh. Meld N Landbr Høgsk 51(26):1–17

Håbjørg A (1978) Photoperiodic Ecotypes in Scandinavian Trees and Shrubs. Meld N Landbr Høgsk 57(33):1–20

Hagem O (1931) Forsøk med vestamerikanske treslag (in Norw., German summ.). Medd Vestl Forstl Forsk 12:1–217

Hedberg O (1957) Afro-alpine vascular plants: a taxonomic revision. Symb Bot Upsal 15:1–411

Heide OM (1985) Physiological aspects of climatic adaptation in plants with special references to high-latitude environments. In: Kaurin A, Juntila O, Nilsen J (eds) Plant production in the North. Norw. Univ. Press, Tromsø, Norway, pp 1–22

Henry GHR, Molau U (1997) Tundra plants and climate change: the international tundra experiment (ITEX). Global Change Biol 3(Suppl. 1):1–9

Høgda KA, Karlsen SR, Solheim I (2001) Climatic change impact on growing season in Fennoscandia studied by a time series of NOAA AVHRR NDVI data. Proceedings IGARSS 2001 (ISBN 0–7803–7536-X) (pp 1–3). Sydney, Australia

Inouye DW, Wielgolaski FE (2003) High altitude climates. In: Schwartz MD (ed) Phenology: an integrative environmental science. Kluwer Acad. Publ., Dordrecht, Boston, London, pp 195–214

IPCC (Intergovernmental Panel on Climate Change) (2001) Summary for policymakers, third assessment report. In: Watson RT, the Core Writing Team (eds) Synthesis report. Cambridge Univ. Press, Cambridge, London, New York, Sydney, pp 1–34

Jia GJ, Epstein HE, Walker DA (2003) Greening of arctic Alaska, 1981–2001. Geophys Res Lett 30(20):2067, doi:10.1029/2003GL018268

Johansen B, Karlsen SR (2005) Monitoring vegetation changes on Finnmarksvidda, Northern Norway, using Landsat MSS and Landsat TM/ETM+ satellite images. Phytocoenologia 35:969–984

Jones PD, Moberg A (2003) Hemispheric and large-scale surface air temperature variations: an extensive revision and an update to 2001. J Clim 16:206–223

Kalela A (1938) Zur Synthese der experimentellen Untersuchungen über Klimarassen der Holzarten. Comm Inst For Fenn 26:1–445

Karlsen SR, Elvebakk A, Johansen B (2005) A vegetation-based method to map climatic variation in the arctic-boreal transition area of Finnmark, north-easternmost Norway. J Biogeogr 32:1161–1186

Karlsen SR, Elvebakk A, Høgda KA, Johansen B (2006) Satellite based mapping of the growing season and bioclimatic zones in Fennoscandia. Global Ecol Biogeogr 15:416–430

Kevan PG (1975) Sun-tracking solar furnaces in high arctic flowers: significance for pollination and insects. Science 189:723–726

Kittel TGF, Thornton PE, Royle JA, Chase TN (2002) Climates of the rocky mountains: historical and future patterns. In: Baron J, Fagre D, Hauer R (eds) Rocky

Mountain Futures: An Ecological Perspective. Island Press, Covelo, California, pp 59–82
Klaveness D, Wielgolaski FE (1996) Plant phenology in Norway—a summary of past and present first flowering dates (FFDs) with emphasis on conditions within three different areas. Phenol Season 1:47–61
Körner C (1995) Alpine plant diversity: a global survey and functional interpretations. In: Chapin III FS, Körner C (eds) Arctic and alpine biodiversity. Springer-Verlag, Berlin, Heidelberg, pp 45–62
Krog J (1955) Notes on temperature measurements indicative on special organization in arctic and subarctic plants for utilization of radiated heat from the sun. Phys Plant 8:836–839
Kullman L (1998) Tree-limits and Montane forests in the Swedish scandes: sensitive biomonitors of climate change and variability. Ambio 27:312–321
Kullman L (2002) Rapid recent range-margin rise of tree and shrub species in the Swedish scandes. J Ecol 90:68–77
Marchand FF, Nijs I, Heuer M, Mertens S, Kockelbergh F, Pontailler J-Y, Impens I, Beyens L (2004) Climate warming postpones senescence in high arctic tundra. Arctic Antarctic Alp Res 36:390–394
Maxwell B (1992) Arctic climate: potential for change under global warming. In: Chapin III FS, Jefferies RL, Reynolds JF, Shaver GR, Svoboda J (eds) Arctic ecosystems in a changing climate. Academic Press, New York, pp 11–34
Maxwell B (1997) Recent climate patterns in the arctic. In: Oechel WC, Callaghan T, Gilmanov T, Holten JI, Maxwell B, Molau U, Sveinbjörnsson B (eds) Global change and arctic terrestrial ecosystems. Springer-Verlag, New York, pp 21–46
Mooney HA, Billings WD (1960) The annual carbohydrate cycle of alpine plants as related to growth. Am J Bot 47:594–598
Mooney HA, Billings WD (1961) Comparative physiological ecology of arctic and alpine populations of *Oxyria digyna*. Ecol Monogr 31:1–29
Myking T (1999) Winter dormancy release and budburst in *Betula pendula* ROTH and *B. pubescens* EHRH. Ecotypes Phyton (Austria) 39:139–145
Myneni RB, Keeling CD, Tucker CJ, Asrar G, Nemani RR (1997) Increased plant growth in the northern high latitudes from 1981 to 1991. Nature 386:698–702
Nilsen J (1986) Plantenes reaksjon på daglengde og spektral energifordeling (in Norw.). Ottar: 30–38
Obasi GOP (1996) Climatological normals (Clino) for the period 1961–1990. WMO/OMM-No 847
Ovaska JA, Nilsen J, Wielgolaski FE, Kauhanen H, Partanen R, Neuvonen S, Kapari L, Skre O, Laine K (2005) Phenology and performance of mountain birch provenances in transplant gardens: latitudinal, altitudinal and oceanity–continentality gradients. In: Wielgolaski FE (ed) Plant ecology, herbivory, and human impact in nordic mountain birch forests. Springer-Verlag, Berlin, Heidelberg, New York, pp 99–115
Overpeck J, Hughen K, Hardy D, Bradley R, Case R, Douglas M, Finney B, Gajewski K, Jacoby G, Jennings A, Lamoureux S, Lasca A, MacDonald G, Moore J, Retelle M, Smith S, Wolfe A, Zielenski G (1997) Arctic environmental change of the last four centuries. Science 278:1251–1256
Post E, Stenseth NC (1999) Climatic variability, plant phenology, and northern ungulates. Ecology 80:1322–1339
Raunkiaer C (1934) The life forms of plants and statistical plant geography. Clarendon Press, Oxford
Resvoll TR (1917) Om Planter som Passer til Kort og Kold Sommer (in Norw.). Archiv for Matematik og Naturvidenskab 35:1–224
Rydén BE (1981) Hydrology of Northern tundra. In: Bliss LC, Heal OW, Moore JJ (eds) Tundra ecosystems: a comparative analysis. Cambridge Univ. Press, Cambridge, London, New York, Sydney, pp 115–137
Savile DBO (1972) Arctic adaptations in plants. Can Dept Agric Monogr 6:1–81
Schwartz MD, Ahas R, Aasa A (2006) Onset of spring starting earlier across the northern hemisphere. Global Change Biol 12:343–331
Shaver GR, Canadell J, Chapin FS III, Gurevitch J, Harte J, Henry G, Ineson P, Jonasson S, Melillo J, Pitelka L, Rustad L (2000) Global warming and terrestrial ecosystems: a conceptual framework for analysis. BioSci 50:871–882
Shutova E, Wielgolaski FE, Karlsen SR, Makarova O, Haraldsson E, Aspholm PE, Berlina N, Filimonova T, Flø, L, Høgda KA (2005) Phenology of nordic mountain birch in relation to climate change at Kola Peninsula and the trans-boundary pasvik-enare region. Proceedings of the 17th international congress of biometeorology ICB September 2005, Garmisch-Partenkirchen, Germany. Deutscher Wetterdienst, Annal Meteorol 41(2):524–527
Shutova E, Wielgolaski FE, Karlsen SR, Makarova O, Berlina N, Filimonova T, Haraldsson E, Aspholm PE, Flø L, Høgda KA (2006) Growing seasons in nordic mountain birch in northernmost Europe as indicated by long-term field studies and analyses of satellite images. Int J Biometeorol DOI: 10.1007/s 00484-006-0042-y
Skjelvåg AO (1998) Climatic conditions for crop production in nordic countries. Agr Food Sci Finl 7:149–160
Slayback DA, Pinzon JE, Los SO, Tucker CJ (2003) Northern hemisphere photosynthetic trends 1982–99. Global Change Biol 9:1–15
Sørensen T (1941) Temperature relations and phenology of northeast greenland flowering plants. Medd Grønl 125:1–305
Sonesson M (2001) Ecology of some epiphytic lichens on the mountain birch. In: Wielgolaski FE (ed) Nordic mountain birch ecosystems. UNESCO, Paris and The Parthenon Publ. Group, New York, London, pp 63–70
Sonesson M, Wielgolaski FE, Kallio P (1975) Description of Fennoscandian tundra ecosystems. In: Wielgolaski FE (ed) Fennoscandian tundra ecosystems, part 1 plants and microorganisms. Springer-Verlag, Berlin, Heidelberg, New York, pp 3–28

Stow DA, Daeschner S, Hope A, Douglas D, Petersen A, Myneni R, Zhou L, Oechel W (2003) Variability of the seasonally integrated normalized difference vegetation index across the north slope of Alaska in the 1990s. Int J Rem Sens 24:1111–1117

Stow DA, Hope A, McGuire D, Verbyla D, Gamon J, Huemmrich F, Houston S, Racine C, Sturm M, Tape K, Hinzman L, Yoshikawa K, Tweedie C, Noyle B, Silapaswan C, Douglas D, Griffith B, Jia G, Epstein H, Walker D, Daeschner S, Petersen A, Zhou L, Myneni R (2004) Remote sensing of vegetation and land-cover change in arctic tundra ecosystems. Rem Sens Environ 89:281–308

Suzuki R, Nomaki T, Yasunari T (2001) Spatial distribution and its seasonality of satellite-derived vegetation iindex (NDVI) and climate in Siberia. Int J Climat 21:1321–1335

Swan LW (1967) Alpine and aeolian regions of the world. In: Wright HE Jr, Osbura WS Jr (eds) Arctic and alpine environments. Indiana Univ. Press, Bloomington, pp 29–54

Tømmervik H, Johansen B, Tombre I, Thannheiser D, Høgda KA, Gaare E, Wielgolaski FE (2004) Vegetation changes in the nordic mountain birch forests: the influence of grazing and climate change. Arctic Antarctic Alpine Res 36:323–332

Troll C (1972) Geoecology and the world-wide differentiation of high-mountain ecosystems. In: Troll C (ed) Geoecology of the high-mountain regions of Eurasia. Franz Steiner Verlag, Wiesbaden, pp 1–13

Troll C (1973a) The upper timberlines in different climatic zones. Arctic Alp Res A3–A18

Troll C (1973b) High mountain belts between the polar caps and the equator: their definition and lower limit. Arctic Alp Res A19–A27

Walker DA, Epstein HE, Jia GJ, Balser A, Copass C, Edwards EJ, Gould WA, Hollingsworth J, Knudson J, Maier HA, Moody A, Raynolds MK (2003) Phytomass, LAI, and NDVI in northern Alaska: relationship to summer warmth, soil pH, plant functional types, and extrapolation to the circumpolar arctic. J Geophys Res 108(D2):8169, doi: 10.1029/2001JD000986

Walker MD, Wahren CH, Hollister RD, Henry GHR, Ahlquist LE, Alatalo JM, Bret-Harte MS, Calef MP, Callaghan TV, Carroll AB, Epstein HE, Jónsdóttir IS, Klein JA, Magnússon B, Molau U, Oberbauer SF, Reva SP, Robinson CH, Shaver GR, Suding KN, Thompson CC, Tolvanen A, Totland Ø, Turner PL, Tweedie CE, Webber PJ, Wookey PA (2006) Plant community responses to experimental warming across the tundra biome. PNAS 103:1342–1346

Wanner H, Brönnimann S, Casty C, Gyalistras D, Luterbacher J, Schmutz C, Stephenson DB, Xoplaki E (2001) North Atlantic oscillation–concepts and studies. Surv Geophys 22:321–382

Wielgolaski FE (1984) Functioning of tundra vegetation in relation to life forms. In: Margaris NS, Arianoustou-Farragitaki M, Oechel WC (eds) Being alive on land. Dr W Junk Publ., The Hague, pp 13–19

Wielgolaski FE (1986) Production-ecology and functioning of polar and alpine ecosystems. In: Polunin N (ed) Ecosystem theory and application. John Wiley & Sons, Chichester, pp 12–34

Wielgolaski FE (1987) Floral processes. In: Nelson JG, Needham R, Norton L (eds) Arctic heritage. IUCN, Ottawa, pp 80–94

Wielgolaski FE (1997a) Introduction. In: Wielgolaski FE (ed) Polar and alpine tundra. Ecosystems of the world, vol. 3. Elsevier, Amsterdam, New York, Oxford, Tokyo, pp 1–5

Wielgolaski FE (1997b) Fennoscandian tundra. In: Wielgolaski FE (ed) Polar and alpine tundra. Ecosystems of the world, vol. 3, Elsevier, Amsterdam, New York, Oxford, Tokyo, pp 27–83

Wielgolaski FE (1997c) Adadaptation in plants. In: Wielgolaski FE (ed) Polar and alpine tundra. Ecosystems of the world, vol. 3, Elsevier, Amsterdam, New York, Oxford, Tokyo, pp 7–10

Wielgolaski FE (2001) Field-and bottom layer vegetation mat transplantation: a method to simulate possible effects of climate change? Skogræktarritid Iceland 2001:167–170

Wielgolaski FE, Kjelvik S, Kallio P (1975) Mineral content of tundra and forest tundra plants in Fennoscandia. In: Wielgolaski FE (ed) Fennoscandian tundra ecosystems, part 1 plants and microorganisms. Springer-Verlag, Berlin, Heidelberg, New York, pp 316–332

Wielgolaski FE, Bliss LC, Svoboda J, Doyle G (2001) Primary production of tundra. In: Bliss LC, Heal OW, Moore JJ (eds) Tundra ecosystems: a comparative analysis. Cambridge Univ. Press, Cambridge, London, New York, Sydney, pp 187–225

Wielgolaski FE, Inouye DW (2003) High Latitude Climates. In: Schwartz MD (ed) Phenology: an integrative environmental science. Kluwer Acad. Publ., Dordrecht, Boston, London, pp 175–194

Wielgolaski FE, Karlsen SR, Høgda KA (2004) Plant phenology in Norway related to climate change and latitude. In: van Vliet AJH (ed) Proceedings of the international conference "Challenging Times", April 2003. Wageningen, The Netherlands, pp 96–102

Rev Environ Sci Biotechnol (2006) 5:269–279
DOI 10.1007/s11157-006-0015-y

REVIEW PAPER

# Desiccation-tolerant plants in dry environments

T.-N. Le · S. J. McQueen-Mason

Received: 24 February 2006 / Accepted: 13 June 2006 / Published online: 14 July 2006


**Abstract** The majority of terrestrial plants are unable to survive in very dry environments. However, a small group of plants, called 'resurrection' plants, are extremely desiccation-tolerant and are capable of losing more than 90% of the cellular water in vegetative tissues. Resurrection plants can remain dried in an anabiotic state for several years and, upon rehydration, are able to resume normal growth and metabolism within 24 h. Vegetative desiccation tolerance is thought to have evolved independently several times within the plant kingdom from mechanisms that allow reproductive organs to survive air-dryness. Resurrection plants synthesise a range of compounds, either constitutively or in response to dehydration, that protect various components of the cell wall from damage during desiccation and/or rehydration. These include sugars and late embryogenesis abundant (LEA) proteins that are thought to act as osmoprotectants, and free radical-scavenging enzymes that limit the oxidative damage during dehydration. Changes in the cell wall composition during drying reduce the mechanical damage caused by the loss of water and the subsequent shrinking of the vacuole. These include an increase in expansin or cell wall-loosening activity during desiccation that enhances wall flexibility and promotes folding.

T.-N. Le (✉) · S. J. McQueen-Mason
CNAP, Department of Biology, University of York, P.O. Box 373, York, YO10 5YW, UK
e-mail: ntl500@york.ac.uk

**Keywords** Cell wall · Desiccation tolerance · Drought stress · Dry environments · Expansins · Resurrection plants · Water deficit

## Introduction

Plants exhibit several strategies to deal with life in extremely dry environments; namely avoidance, resistance, or tolerance to desiccation (Levitt 1980). Some plants (e.g. annuals) complete their life cycles during the part of the year when water is plentiful and growth conditions are favourable, and avoid times during which desiccation is frequent (e.g. summer). Desiccation avoidance may enable plants to achieve maximal growth and productivity, but it also confines them to areas or periods with favourable conditions. Other plants, for example perennials such as cacti, have modified morphological structures that allow them to retain most of their cellular water and resist equilibration with the air. These adaptations allow desiccation-resistant plants to extend the range of conditions and habitats in which they can remain metabolically active throughout the year. The main drawback for plants with a desiccation survival strategy geared towards resistance is that growth often proceeds very slowly. A small group of plants, the so-called 'resurrection plants' (Gaff 1971), are extremely desiccation-tolerant, and can survive almost complete desiccation, usually considered as drying to $< 0.1$ g $H_2O$ $g^{-1}$ dry mass or to 10% absolute water content or less (Alpert 2005).

Desiccation tolerance has been defined as the ability of an organism to equilibrate its internal water potential with that of moderately dry air, and then resume normal function after rehydration (Alpert 2000). Air-dryness at a relative humidity of 50% at 28°C would equate to a water potential of $-100$ MPa (Gaff 1997), a water deficit that would be lethal to the majority of modern day flowering plants. However, a few resurrection plants possess vegetative tissues that can tolerate loss of greater than 90% of their cellular water, and some can tolerate drying to water potentials as low as $-650$ MPa without injury (Gaff 1997). Resurrection plants are usually found in habitats of sporadic rainfall, including rocky outcrops, and arid zones within tropical and subtropical areas (Rascio and La Rocca 2005). It is estimated that there are about 300 angiosperms which possess vegetative desiccation tolerance (Porembski and Barthlott 2000). Desiccation-tolerant plants have been found in all continents except Antarctica (Alpert 2000), and in all groups of seed plants except for the gymnosperms (Oliver 1996). Resurrection plants, like their desiccation-tolerant animal counterparts, are generally small in size. The absence of vegetative desiccation tolerance within the gymnosperms, or within any vascular plant more than 3 m in height, is thought to be due to the physical constraint of re-establishing water flow in the xylem during rehydration after it has been interrupted during desiccation due to cavitation (Alpert 2005).

This review will focus on the ability of vegetative tissues of resurrection plants to survive extreme desiccation. First we examine how desiccation tolerance of vegetative tissues evolved in vascular plants. Then we investigate the different mechanisms used by resurrection plants to protect cell membranes and organelles during desiccation, and to repair dehydration-induced damage upon rehydration. This includes examining the roles of key proteins (e.g. LEAs and transcription factors), sugars and other compatible solutes in the protection and maintenance of cellular integrity. We also review some adaptations of resurrection plants for surviving desiccation including changes in the cell wall and lipid composition, and the establishment of anti-oxidant systems.

## Evolution of desiccation tolerance

In order to colonize the land, the earliest terrestrial plants must have been desiccation-tolerant at every stage of the life cycle (Oliver et al. 2005). With the evolution of more complex vascular plants, desiccation tolerance was lost in vegetative tissues but was retained in reproductive tissues (Oliver et al. 2000). The acquisition of desiccation tolerance is part of a maturation program during normal seed development. The majority of terrestrial plants today are capable of producing desiccation-tolerant structures such as spores, seeds and pollen which can remain viable in the desiccated state for decades, or centuries, in the case of the ancient *Nelumbo nucifera* (sacred lotus) seed from China (Shen-Miller et al. 1995).

The structural genes required for desiccation tolerance are not unique to resurrection plants, but are also present in the genomes of desiccation-sensitive plants (Bartels and Salamini 2001). However, resurrection plants may have enlisted genes normally expressed in seed tissue for expression in vegetative tissue to allow these plants to withstand desiccation (Illing et al. 2005). For example, genes encoding late embryogenesis abundant (LEA) proteins (see below) that are normally expressed in the seeds of desiccation-sensitive plants during embryo maturation (Close 1996) have been isolated from drought-stressed vegetative tissues of resurrection plants such as *Sporobolus stapfianus* (Neale et al. 2000) and *Craterostigma plantagineum* (Bartels 2005). Recent phylogenetic evidence suggests that these vascular plants gained the ability to withstand desiccation of their vegetative tissues from a mechanism present first in spores, and that this evolution (or re-evolution) has occurred on at least ten independent occasions within the angiosperms (Oliver et al. 2005).

## Plant strategies to survive desiccation

Several different strategies are utilised by resurrection plants to survive desiccation. Some plants have evolved mechanisms that protect cellular membranes and organelles during desiccation. Other plants employ a constitutive repair system which allows them to rapidly mobilize repair mechanisms in the damaged cell upon rehydration. In general, both the protection and repair mechanisms are used by resurrection plants, with mosses generally utilizing a repair mechanism and vascular plants using a protective mechanism (Oliver 1996). Regardless of the strategy used by a particular resurrection plant, three factors have been identified as being crucial for surviving desiccation: maintaining life in the dried state, limiting the extent of damage to existing tissues so that repair is unnecessary or manageable, and mobilizing repair systems upon rehydration (Bewley 1979).

Tolerance to near-complete desiccation of vegetative organs is a widespread capability in bryophytes (Rascio and La Rocca 2005), a group that includes mosses and liverworts. The desiccation-tolerant moss *Tortula ruralis* is a cosmopolitan species that is found in many parts of the world (Oliver 1996). *T. ruralis*, whose natural habitat includes fallen trees and rocks, has evolved tolerance mechanisms that allow the plant to survive rapid drying rates. These mechanisms include a constitutive cellular protection component, and a rehydration-inducible recovery process designed to repair cellular damage experienced during rapid desiccation and subsequent rehydration (Oliver et al. 2000). This strategy allows for frequent and rapid desiccation to be tolerated; often desiccation can occur within 1 h (Bohnert 2000). The main drawbacks of the constitutive repair strategy for desiccation survival are low metabolic rates, possibly due to the cost associated with constitutive expression of repair genes, that (at least in vascular plants) result in slow growth and a limit in size (Bohnert 2000).

The capacity of *T. ruralis* to repair damaged membranes and organelles can be found in ribonucleoprotein particles which contain a complex of proteins and mRNAs that can be rapidly translated into repair enzymes upon imbibition (Wood and Oliver 1999). Proteins whose synthesis is initiated or increased during rehydration after a desiccation event are termed rehydrins (Scott and Oliver 1994). A recent investigation of a *T. ruralis* rehydration cDNA library has found that many of the 10,368 expressed sequence tags (ESTs) encode gene products that are involved in protein synthesis, ion and metabolite transport, oxidative stress metabolism, and membrane biosynthesis and repair (Oliver et al. 2004).

Desiccation-tolerant angiosperms such as the dicot *C. plantagineum* and the monocot *S. stapfianus* utilise a protection-based strategy for surviving desiccation. These plants can persist in the air-dried state for months, and revive from the desiccated state even after several years. Desiccation tolerance in *C. plantagineum* is induced in vegetative tissues upon slow drying of the whole plant, and plants do not survive if dehydration occurs too rapidly (Bartels and Salamini 2001). Similarly, the desiccation-tolerant *C. wilmsii* also exhibits substantial ultrastructural damage upon rapid drying (Cooper and Farrant 2002). This suggests that desiccation-tolerant flowering plants requires a slow drying time in order for mechanisms that protect membranes and organelles during desiccation to be established.

## Effects of hormones on desiccation tolerance

Abscisic acid (ABA) is known to be involved in embryo maturation and germination, and in the response of vegetative tissues to osmotic stress. ABA has been proposed to be an essential mediator in triggering plant responses to various adverse environmental conditions such as drought, high salinity, and cold. Endogenous ABA levels have been reported to increase as a result of water stress, and ABA induces stomatal closure in guard cells by activating $Ca^{2+}$, potassium ($K^+$), and anion channels (Leung and Giraudat 1998). ABA has also been shown to play a role in the desiccation tolerance of *C. plantagineum* vegetative tissues. Undifferentiated callus tissue from *C. plantagineum*, although intrinsically not desiccation-tolerant, acquires tolerance after it has been cultured on medium containing ABA (Bartels and Salamini 2001). Treatment of *C. plantagineum* callus with ABA also induces the expression of a similar set of genes to that activated in vegetative tissues upon drying. In *S. stapfianus*, ABA level increases 6-fold during strong drought stress, and reaches a maximum during severe desiccation (20–10% relative water content). However, studies with *S. stapfianus* protoplasts showed that exogenously applied ABA only promotes protoplasmic drought tolerance (PDT) to a degree that is very much less than is required for desiccation tolerance (Ghasempour et al. 2001). Brassinosteroids, methyl jasmonate, and ethylene promoted PDT in *S. stapfianus* to a higher degree than ABA, but not to a sufficient level to allow the plant to survive desiccation. Of the hormones normally associated with cell differentiation and growth, auxins had no effect, while cytokinins had a deleterious effect on PDT (Ghasempour et al. 2001).

## Molecular approaches to the study of water deficit

Apart from *T. ruralis* (Oliver et al. 2005), *S. stapfianus* (Neale et al. 2000) and *C. plantagineum* (Bartels 2005), the molecular basis of desiccation tolerance has been studied in relatively few resurrection plants. Recently, the mechanisms of desiccation tolerance have been studied in several other resurrection plants including the ferns *Polypodium virginianum* (Reynolds and Bewley 1993) and *P. polypodioides* (Helseth and Fischer 2005), and desiccation-tolerant monocots such as *Xerophyta viscosa* (Sherwin and Farrant 1998) and *X. umilis* (Collett et al. 2004). Genes that are expressed in response to drought stress are classified into two main types: those that encode for products with putative protective functions, and those that encode for regulatory proteins such as transcription factors.

### *Compatible solutes*

Upon dehydration, many plants accumulate non-toxic or 'compatible' solutes such as proline, mannitol, and glycine betaine (Chen and Murata 2002). The net result of compatible solute accumulation is an increase in cellular osmolarity which, in turn, leads to an influx of water into, or at least a reduced efflux from, cells (Hare et al. 1998). Compatible solutes have been proposed to protect cells during drying through the stabilisation of cytoplasmic constituents, ion sequestration, and increased water retention. Transgenic non-resurrection plants over-producing various types of compatible solutes, either individually or in combination, show improved tolerance to drought as well as other osmotic stresses [for a review, see Chen and Murata (2002)].

### *Sugar contents*

Water deficit results in a reduction of photosynthesis and starch content in leaf tissues. There is a rapid conversion of starch into sugars, the most common of which are sucrose and trehalose (Crowe et al. 1998). Conversion of starch, which is usually stored in plastids, into sucrose is attributed primarily to activation of sucrose phosphate synthase by reversible protein phosphorylation upon perception of osmotic stress. The activation of sucrose synthesis may also be associated with a concurrent inhibition of starch synthesis. Sucrose biosynthetic genes have been shown to be induced by both desiccation and ABA in the resurrection plant *C. plantagineum* (Kleines et al. 1999).

Sucrose and trehalose have been proposed to contribute towards the maintenance of turgor during stress, and the prevention of protein denaturation and membrane fusions in the cell (Crowe et al. 1998). Both sugars are capable of forming biological glasses (a process called vitrification) within the dried cell. Vitrification of the cytoplasm may not be due to the effects of sugars only, but probably results from the interaction of sugars with other molecules, most likely proteins (Hoekstra 2005). The formation of an intracellular glass phase is believed to be indispensable to survival during desiccation, and cells of many desiccation-tolerant plants and animals undergo vitrification upon drying to protect organelles from damage (Buitink and Leprince 2004). Vitrification is thought to limit the production of free radicals through the slowing down of chemical reaction rates and molecular diffusion in the cytoplasm (Hoekstra 2005). It may also help to preserve protein structure. Thus, vitrification may be an important protective mechanism for resurrection plants against oxidative damage during desiccation.

In fully hydrated *S. stapfianus* leaves, glucose, fructose, and galactose are present in large amounts. During dehydration, sucrose increases to high levels in air-dry leaves to become the predominant sugar (Ghasempour et al. 1998). Other sugars such as raffinose and trehalose are also detected, and they may supplement the role of sucrose as osmotic protectants during drying. The sugar changes observed in *S. stapfianus* leaves were quantitatively similar to those reported for other resurrection plants where the common trend was the conversion of monosaccharides present in fresh leaves into sucrose in dried leaves, and vice versa in rehydrated leaves (Murelli et al. 1996). In *C. plantagineum*, the unusual eight-carbon sugar, 2-octulose, is the predominant sugar in fully-hydrated leaves. During dehydration, 2-octulose is converted to sucrose, with the reverse process being observed during rehydration (Bartels and Salamini 2001). Sucrose also accumulates in the roots of drying *C. plantagineum* plants, suggesting that it may play a protective role within the root systems of resurrection plants during desiccation (Norwood et al. 2003). The importance of sugars in desiccation tolerance has been implicated in several studies. Transgenic rice plants expressing two genes encoding trehalose biosynthetic enzymes from *Escherichia coli* showed high levels of tolerance to drought stress, as well as salt and cold stress (Garg et al. 2002).

### *LEA proteins*

The ability of a plant to survive desiccation appears to depend upon the accumulation of a large set of proteins with putative protective functions. These proteins include the late embryogenesis abundant (LEA) proteins which, as their name suggests, increase markedly in abundance during the latter stages of embryo development as the seed dries (Dure 1993). Some *LEA* genes are also induced in vegetative tissues in response to various abiotic stresses, including drought, cold, and salt, as well as to ABA. *LEA* transcripts are also abundant in *T. ruralis* during rehydration, suggesting that LEAs may also play a role in recovery from desiccation when water is reintroduced into dried tissues (Oliver et al. 2004).

LEA proteins have been classified into eighteen different groups based on sequence homology (Dure 1993). A common feature of LEA proteins is that they are extremely hydrophilic, and are soluble at high (80°C) temperatures. LEA proteins do not possess any apparent catalytic activity or structural domains, and most of them lack cysteine and tryptophan residues (Close 1996). The predicted functions of LEA proteins include the unwinding or repair of DNA, forming cytoskeletal filaments to counteract the physical stresses imposed by desiccation, and acting as molecular chaperones (Wise and Tunnacliffe 2004). It has also been suggested that LEA proteins, possibly in combination with compatible solutes, could replace water and thus maintain the hydration shell of proteins and other molecules during desiccation (Bartels 2005). In rehydrating *T. ruralis* gametophytes, LEA proteins may function in stabilizing membranes, or perhaps in the transport of lipids for reconstitution of damaged membranes (Oliver et al. 2005). Although many *LEA* genes have been isolated, a functional role in desiccation tolerance has been demonstrated for only few. The expression of the barley LEA gene, *HVA1*, increased drought tolerance in transgenic wheat plants (Sivamani et al. 2000). In rice, over-expression of *HVA1* enhances tolerance to high-salt as well as drought stress (Rohila et al. 2002). Recently, it has been shown that two LEAs of distinct phylogenetic origins (one from wheat and the other from a nematode) can prevent protein aggregation during desiccation, and this protective effect is synergistically enhanced when trehalose is added (Goyal et al. 2005).

### *Transcription factors*

Many genes that are responsive to cold and drought stress in the desiccation-sensitive model plant *Arabidopsis thaliana* contain DRE/CRT (dehydration-responsive element/C-repeat) elements within their 5′ regulatory regions, and are regulated by DREB/CBF (DRE-binding protein/ C-repeat binding factor) transcription factors (Stockinger et al. 1997; Liu et al. 1998). The DREB/CBF proteins belong to the AP2/EREBP (ethylene responsive-element binding protein) family of transcription factors which are important regulators of flower development and plant responses to abiotic stresses (Riechmann and Meyerowitz 1998; Shigyo et al. 2006). Homologues of the *Arabidopsis DREB/CBF* genes have also been isolated from other non-resurrection plants where they have been shown to function in both the drought and cold response pathways (Jaglo et al. 2001). However, only one such AP2/EREBP transcription factor has been reported as a regulator of the drought response pathway in a resurrection plant (Le 2005).

Nearly all of the drought-induced transcription factors isolated thus far from resurrection plants have come from *C. plantagineum* and they belong to either the Myb or the homeodomain-leucine zipper (HD-Zip) family (Iturriaga et al. 1996; Frank et al. 1998). Two Myb-related genes, *cpm10* and *cpm7*, show differential expression and regulation in response to desiccation and ABA in *C. plantagineum* (Iturriaga et al., 1996). *Cpm10* is expressed only in undifferentiated callus tissue, and is up-regulated by ABA, while *cpm7* is induced only by dehydration in the roots. Transgenic *Arabidopsis* plants over-expressing *cpm10* displayed increased tolerance to drought and salt stress (Villalobos et al. 2004). Interestingly, these plants also showed ABA hypersensitivity and glucose-insensitivity, suggesting that *cpm10* is involved in mediating ABA and glucose signaling responses in *Arabidopsis* as well as the response to drought stress. Two HD-Zip genes, *CPHB-1* and *CPHB-2*, are induced by dehydration in leaves and roots of *C. plantagineum*, but show different responses to exogenously applied ABA (Frank et al. 1998). ABA treatment induces the transcription of *CPHB-2*, but not that of *CPHB-1*. Five other HD-Zip genes have been isolated from *C. plantagineum*, two of which were induced by ABA in undifferentiated callus, while the other three were not (Deng et al. 2002). This suggests that these HD-Zip genes act in different pathways of the dehydration response; some mediated by ABA, while others are independent of ABA (Bartels and Salamini 2001). Many studies have shown that the expression of drought-responsive genes in both resurrection and non-resurrection plants is mediated by ABA-independent as well as ABA-dependent signal transduction pathways (Bartels 2005).

## Morphological adaptations

Water deficit induces many morphological changes in desiccation-tolerant vascular plants, the most obvious of which is leaf folding. The folding of leaves during drying is not unique to resurrection plants and also occurs in desiccation-sensitive plants. Leaves of the desiccation-tolerant dicot *C. wilmsii*, which are fully expanded when watered, progressively curl inward during drying and become tightly folded so that only the abaxial surfaces of the older leaves in the outer whorl are exposed to the sun (Sherwin and Farrant 1998). Leaf folding is thought to limit oxidative stress damage from UV radiation, and is an important morphological adaptation for surviving desiccation. Indeed, *C. wilmsii* plants do not survive desiccation in sunlight if the leaves are mechanically prevented from folding (Farrant et al. 2003).

The leaf blades of the desiccation-tolerant monocot *X. humilis* fold in half along the midrib upon dehydration, leaving only the abaxial surfaced exposed to the light (Sherwin and Farrant 1998). In the desiccation-tolerant grass *S. stapfianus*, the leaf adaxial side, which is most exposed to sun radiation, is very rich in epicuticular waxes, whose main function is probably to reflect light, and to limit irradiation and heating of leaf tissues (Dalla Vecchia et al. 1998). During dehydration, this cuticular wax covering, together with the closure of stomata, helps to decrease the rate of water loss. This may also be an important protective mechanism for the thylakoid membranes, which are maintained in the chloroplasts and are particularly sensitive to light damage in water stress (Dalla Vecchia et al. 1998).

## Cell wall changes during desiccation

When water availability to roots decreases, plants tend to limit water loss from transpiration by closing the stomata, and thereby reducing the water flux through the plant, and by reducing leaf growth, which results in a smaller transpiring leaf area (Tardieu 2005). The reduction of growth under water deficit is accomplished by a reduction in cell division rate and an increase in cell wall stiffening, which inhibits cell wall expansion (Cosgrove 2000). This stiffening of the cell wall poses a problem for drying cells. In order to tolerate desiccation, any cell with a large, water-filled vacuole must overcome or limit the mechanical stress caused by its shrinking during drying. The cell walls of some resurrection plants have special adaptations that promote folding to reduce the mechanical stress caused by desiccation. In *C. wilmsii*, a significant increase in xyloglucans and unesterified pectins is observed in the cell wall during drying (Vicre et al. 1999). Dehydration also induces a considerable reduction of glucose in the hemicellulosic fraction of *C. wilmsii* cell walls (Vicre et al. 2004). These changes are thought to enhance the tensile strength of the *C. wilmsii* cell wall, allowing it to contract and to fold without collapsing in the dried tissue. Thus, cell wall flexibility is an important factor in tolerance to injury caused by desiccation. In *C. plantagineum*, an increase in expansin activity during desiccation is associated with a rise in cell wall flexibility and folding (Jones and McQueen-Mason 2004). Similarly, desiccation-sensitive maize plants adapted to low water potentials are able to maintain root growth by increasing the extensibility of their cell walls (Wu et al. 1996). This increase in cell wall flexibility is correlated with an increase in expansin activity and transcript accumulation (Wu et al. 2001). Expansins were first identified as cell wall-loosening factors in acid-induced cell expansion (McQueen-Mason et al. 1992). They are thought to act by disrupting the hydrogen bonds between cellulose and hemicellulose polymers in the cell wall (McQueen-Mason and Cosgrove 1995).

## Lipid composition of cellular membranes

In the desiccation-tolerant mosses *T. ruralis* and *Selaginella lepidophylla*, the shrinking of cells caused by dehydration results in cells with highly convoluted walls and membranes that are similar in appearance to cells in dry seeds (Platt et al. 1994). The plasma membrane during desiccation contains numerous tightly associated lipid droplets and shows a normal lipid bilayer organization. In desiccation-tolerant vascular plants, dehydration generally causes a general decrease in total lipids as well the unsaturation level of individual phospholipids (Quartacci et al. 2002). However, the opposite trend is observed in the resurrection plant *Boea hygroscopica* where increased unsaturation of fatty acids was observed in all lipid classes upon dehydration regardless of whether it was slow or rapid (Navari-Izzo et al. 1995). It is generally known that a high degree of poly-unsaturation in phospholipids results in greater membrane fluidity. As discussed earlier, increased cell wall and membrane fluidity may be an important protective mechanism for the survival of resurrection plants during desiccation. In *S. stapfianus* during desiccation, phospholipid content increased in leaves dried attached to the parent, but decreased in leaves that have been dried detached (Quartacci et al. 1997). The fact that attached dried leaves develop desiccation tolerance, while detached dried leaves do not (Neale et al. 2000), may in part be due to the level of polyunsaturated lipids within the plasma membrane. Upon rehydration in *S. stapfianus*, leaves desiccated on the plant regained almost all of the lipid content, whereas detached dried leaves suffered a complete lipid degradation with the loss of polyunsaturated fatty acids (Quartacci et al. 1997).

## Anti-oxidant systems

Recovery of a resurrection plant correlates with its capacity to establish a number of anti-oxidant protective mechanisms during dehydration, and to maintain these systems upon rehydration (Kranner et al. 2002). Desiccation results in the production of reactive oxygen species (ROS) that can damage membrane lipids and proteins. An important protective mechanism in *S. stapfianus* is the induction of free radical-scavenging enzymes such as glutathione reductase, ascorbate peroxidase, and dehydroascorbate reductase to remove the ROS (Sgherri et al. 1994a). Antioxidant activity also increases in other resurrection plants during desiccation (Sgherri et al. 1994b; Sherwin and Farrant 1998). It has been shown that more damage occurs during rehydration than during desiccation because of intensified oxidative stress during the recovery phase (Sgherri et al. 1994a).

In desiccation-sensitive *S. stapfianus* leaves dried detached, however, ascorbate peroxidase activity decreased during desiccation, resulting in reduced antioxidant capacity (Sgherri et al. 1994a). This may be another reason for the failure of detached dried leaves of *S. stapfianus* to recover from desiccation.

Anti-oxidant enzymes such as superoxide dismutase, glutathione reductase, and ascorbate peroxidase are induced in response to various abiotic stresses in both desiccation-tolerant as well as desiccation-sensitive organisms, and are considered as general 'housekeeping' protectants (Illing et al. 2005). However, only in desiccation-tolerant tissues can the activities of these enzymes remain elevated during dehydration. This may be a consequence of mechanisms that protect and maintain the anti-oxidant enzymes in their native states, which results in increased activity and/or half-life of the proteins, rather than a unique desiccation tolerance mechanism (Illing et al. 2005).

Resurrection plants can persist in the desiccated state for several years. However, they cannot remain indefinitely in this anabiotic state because, in part, oxidative damage increases with duration of desiccation and results in the gradual loss of viability. Failure of the anti-oxidant system during long-term desiccation triggers programmed cell death, causing ageing and eventual death of the plant (Illing et al. 2005). The acyl chains of the membrane polar lipids, which may contain unsaturated double bonds, are particularly sensitive to free radical attack. The leaves of resurrection plants contain a relatively high number of double bonds in their polar lipids, which is a general characteristic of chloroplasts (Hoekstra, 2005). Hoekstra (2005) has shown that there is a negative correlation between the longevity of desiccation-tolerant tissues and the number of double bonds in the polar lipids of membranes, and that the lifespan of resurrection plants during prolonged desiccation is generally limited to several years maximally.

### Photosynthesis and photosystem (PS)II activity

Some homoiochlorophyllous resurrection plants retain all of their chlorophyll during desiccation, whereas others lose their chlorophyll and are thus termed poikilochlorophyllous (Oliver et al. 2000). It has been hypothesised that certain desiccation-tolerant monocots evolved the strategy of poikilochlorophylly to survive and compete in marginal habitats where light availability is variable (Oliver et al. 2000). In contrast, the photosynthetic machinery in desiccation-tolerant bryophytes appears to be constitutively protected during drying such that photosynthetic activity recovers quickly following rehydration (Oliver et al. 2005). The desiccation-tolerant grass *S. stapfianus* is partially poikilochlorophyllous, and retains most of its chlorophyll content during desiccation (Quartacci et al. 1997). After a cycle of dehydration to air-dryness and rehydration to full turgor, *S. stapfianus* regained all of its photosynthetic capability within 24 h. Like all other plants, *S. stapfianus* lowered its PSII during dehydration. In general, water deficit causes a reduction in the photosynthesis rate, resulting in the decline in the photochemical efficiency of PSII and electron transport rate in desiccation-tolerant as well as desiccation-sensitive plants (Ekmekci et al. 2005). The decline in PSII activity could represent a protective mechanism from toxic oxygen production in order to maintain membrane integrity and to ensure protoplast survival (Di Blasi et al. 1998). However, only proteins within the thylakoid membranes of resurrection plants remain stable during desiccation and rehydration, whereas those of desiccation-sensitive plants are completely destroyed after a short-term desiccation event (Deng et al. 2003).

## Conclusion

The ability of vegetative tissues to survive near complete desiccation is shared by about 400 species of resurrection plants, and appears to have evolved from reproductive tissues at least 10 times independently during the evolution of vascular plants. Genes involved in desiccation tolerance may be constitutively expressed, as in mosses, or induced during dehydration stress, as in most resurrection angiosperms, a process that may or may not involve abscisic acid as a signal. In order to survive in dry environments, plants must limit damage from desiccation and/or rehydration to a minimum, maintain cellular integrity in the dried state, and activate repair mechanisms upon rehydration. This is accomplished by the synthesis of compatible solutes including sugars (sucrose in most plants), and proteins with putative protective functions such as LEAs. Sugars such as sucrose and trehalose are particularly important in vitrification of the cell cytoplasm during desiccation, which may decrease chemical reaction rates and molecular diffusion, and limit oxidative damage. Oxidative stress damage is further reduced by the induction of anti-oxidant enzymes such as glutathione reductase, and the inhibition of photosynthesis, with some resurrection plants losing all of their chlorophyll during desiccation altogether. Dehydrating plant cells must also prevent or minimize mechanical damage to the cell wall and membranes due to the shrinking of the vacuole. The increase in cell wall flexibility that promotes wall folding may be due to the increase in activity and transcript accumulation of cell wall-loosening proteins such as expansins. Thus, the ability of resurrection plants to survive in extremely dry environments is a multigenic trait that relies on systems of cellular protection during dehydration, and recovery and repair upon rehydration (Rascio and La Rocca 2005).

## References


Alpert P (2000) The discovery, scope, and puzzle of desiccation tolerance in plants. Plant Ecol 151:5–17

Alpert P (2005) The limits and frontiers of desiccation-tolerant life. Integr Comp Biol 45:685–695

Bartels D (2005) Desiccation tolerance studied in the resurrection plant *Craterostigma plantagineum*. Integrative and Comparative Biology 45:696–701

Bartels D, Salamini F (2001) Desiccation tolerance in the resurrection plant *Craterostigma plantagineum*. A contribution to the study of drought tolerance at the molecular level. Plant Physiol 127:1346–1353

Bewley JD (1979) Physiological aspects of desiccation tolerance. Ann Rev Plant Physiol 30:195–238

Bohnert H (2000) What makes desiccation tolerable? Genome Biol 1:1010.1011–1010.1014

Buitink J, Leprince O (2004) Glass formation in plant anhydrobiotes: survival in the dry state. Cryobiology 48:215–228

Chen THH, Murata N (2002) Enhancement of tolerance of abiotic stress by metabolic engineering of betaines and other compatible solutes. Curr Opin Plant Biol 5:250–257

Close TJ (1996) Dehydrins—emergence of a biochemical role of a family of plant dehydration proteins. Physiol Plantarum 97:795–803

Collett H, Shen A, Gardner M, Farrant JM, Denby KJ, Illing N (2004) Towards transcript profiling of desiccation tolerance in *Xerophyta humilis*: construction of a normalized 11 k *X. humilis* cDNA set and microarray expression analysis of 424 cDNAs in response to dehydration. Physiol Plantarum 122:39–53

Cooper K, Farrant JM (2002) Recovery of the resurrection plant *Craterostigma wilmsii* from desiccation: protection versus repair. J Exp Bot 53:1805–1813

Cosgrove DJ (2000) Expansive growth of plant cell walls. Plant Physiol Biochem 38:109–124

Crowe JH, Carpenter JF, Crowe LM (1998) The role of vitrification in anhydrobiosis. Annu Rev Physiol 60:73–103

Dalla Vecchia F, Elasmar T, Calamassi R, Rascio N, Vazzana C (1998) Morphological and ultrastructural aspects of dehydration and rehydration in leaves of *Sporobolus stapfianus*. Plant Growth Regul 24:219–228

Deng X, Phillips J, Meijer AH, Salamini F, Bartels D (2002) Characterization of five novel dehydration-responsive homeodomain leucine zipper genes from the resurrection plant *Craterostigma plantagineum*. Plant Mol Biol 49:601–610

Deng X, Hu Z-A, Wang H-X, Wen X-G, Kuang T-Y (2003) A comparison of photosynthetic apparatus of the detached leaves of the resurrection plant *Boea hygrometrica* with its non-tolerant relative *Chirita heterotrichia* in response to dehydration and rehydration. Plant Sci 165:851–861

Di Blasi S, Puliga S, Losi L, Vazzana C (1998) *S. stapfianus* and *E. curvula* cv. Consol in vivo photosynthesis, PSII activity and ABA content during dehydration. Plant Growth Regul 25:97–104

Dure L (1993) Structural motifs of LEA proteins in higher plants. In: Close TJ, Bray EA (eds) Responses of plants to cellular dehydration during environmental stress. American Society for Plant Physiology, Rockville, pp 104–118

Ekmekci Y, Bohms A, Thomson JA, Mundree SG (2005) Photochemical and antioxidant responses in the leaves of *Xerophyta viscosa* Baker and *Digitaria sanguinalis* L. under water deficit. Z Naturforschung C-A J Biosci 60:435–443

Farrant JM, van der Willigen C, Loffell DA, Bartsch S, Whittaker A (2003) An investigation into the role of light during desiccation of three angiosperm resurrection plants. Plant Cell Environ 26:1275–1286

Frank W, Phillips J, Salamini F, Bartels D (1998) Two dehydration-inducible transcripts from the resurrection plant *Craterostigma plantagineum* encode interacting homeodomain-leucine zipper proteins. Plant J 15:413–421

Gaff DF (1971) Desiccation tolerant plants in southern Africa. Science 174:1033–1034

Gaff DF (1997) Mechanisms of desiccation tolerance in resurrection vascular plants. In: Basra AS, Basra RK (eds) Mechanisms of environmental stress resistance in plants. Harwood Academic Publishers, Amsterdam, pp 43–58

Garg AK, Kim J-K, Owens TG, Ranwala AP, Choi YD, Kochian LV, Wu RJ (2002) Trehalose accumulation in rice plants confers high tolerance levels to different abiotic stresses. PNAS 99:15898–15903

Ghasempour HR, Anderson EM, Gaff DF (2001) Effects of growth substances on the protoplasmic drought tolerance of leaf cells of the resurrection grass, *Sporobolus stapfianus*. Aust J Plant Physiol 28:1115–1120

Ghasempour HR, Gaff DF, Williams RPW, Gianello RD (1998) Contents of sugars in leaves of drying desiccation tolerant flowering plants, particularly grasses. Plant Growth Regul 24:185–191

Goyal K, Walton LJ, Tunnacliffe A (2005) LEA proteins prevent protein aggregation due to water stress. Biochem J 388:151–157

Hare PD, Cress WA, van Staden J (1998) Dissecting the roles of osmolyte accumulation during stress. Plant Cell Environ 21:535–553

Helseth LE, Fischer TM (2005) Physical mechanisms of rehydration in *Polypodium polypodioides*, a resurrection plant. Phys Rev E 71(6):061903

Hoekstra FA (2005) Differential longevities in desiccated anhydrobiotic plant systems. Integr Comp Biol 45:725–733

Illing N, Denby KJ, Collett H, Shen A, Farrant JM (2005) The signature of seeds in resurrection plants: a molecular and physiological comparison of desiccation tolerance in seeds and vegetative tissues. Integr Comp Biol 45:771–787

Iturriaga G, Leyns L, Villegas A, Gharaibeh R, Salamini F, Bartels D (1996) A family of novel myb-related genes from the resurrection plant *Craterostigma plantagineum* are specifically expressed in callus and roots in response to ABA or desiccation. Plant Mol Biol 32:707–716

Jaglo KR, Kleff S, Amundsen KL, Zhang X, Haake V, Zhang JZ, Deits T, Thomashow MF (2001) Components of the *Arabidopsis*C-repeat/dehydration-responsive element binding factor cold-response pathway are conserved in *Brassica napus* and other plant species. Plant Physiol 127:910–917

Jones L, McQueen-Mason S (2004) A role for expansins in dehydration and rehydration of the resurrection plant *Craterostigma plantagineum*. FEBS Lett 559:61–65

Kleines M, Elster RC, Rodrigo MJ, Blervacq AS, Salamini F, Bartels D (1999) Isolation and expression analysis of two stress-responsive sucrose-synthase genes from the resurrection plant *Craterostigma plantagineum* (Hochst.). Planta 209:13–24

Kranner I, Beckett RP, Wornik S, Zorn M, Pfeihofer HW (2002) Revival of a resurrection plant correlates with its antioxidant status. Plant J 31:13–24

Le T-N (2005) Genetics of desiccation tolerance in the resurrection plant *Sporobolus stapfianus*. PhD thesis. Monash University, Melbourne, Australia

Leung J, Giraudat J (1998) Abscisic acid signal transduction. Ann Rev Plant Physiol Plant Mol Biol 49:199–222

Levitt J (1980) Responses of plants to environmental stresses, 2nd edn. Academic Press, New York

Liu Q, Kasuga M, Sakuma Y, Abe H, Miura S, Yamaguchi-Shinozaki K, Shinozaki K (1998) Two transcription factors, DREB1 and DREB2, with an EREBP/AP2 DNA-binding domain separate two cellular signal transduction pathways in drought- and low-temperature-responsive gene expression, respectively, in *Arabidopsis*. Plant Cell 10:1391–1406

McQueen-Mason SJ, Cosgrove DJ (1995) Expansin mode of action on cell walls—analysis of wall hydrolysis, stress-relaxation, and binding. Plant Physiol 107:87–100

McQueen-Mason SJ, Durachko DM, Cosgrove DJ (1992) Two endogenous proteins that induce cell wall expansion in plants. Plant Cell 4:1425–1433

Murelli C, Adamo V, Finzi PV, Albini FM, Bochicchio A, Picco AM (1996) Sugar biotransformations by fungi on leaves of the resurrection plant *Sporobolus stapfianus*. Phytochemistry 43:741–745

Navari-Izzo F, Ricci F, Vazzana C, Quartacci MF (1995) Unusual composition of thylakoid membranes of the resurrection plant *Boea hygroscopica*—changes in lipids upon dehydration and rehydration. Physiol Plantarum 94:135–142

Neale AD, Blomstedt CK, Bronson P, Le T-N, Guthridge K, Evans J, Gaff DF, Hamill JD (2000) The isolation of genes from the resurrection grass *Sporobolus stapfianus* which are induced during severe drought stress. Plant Cell Environ 23:265–277

Norwood M, Toldi O, Richter A, Scott P (2003) Investigation into the ability of roots of the poikilohydric plant *Craterostigma plantagineum* to survive dehydration stress. J Exp Bot 54:2313–2321

Oliver M, Dowd S, Zaragoza J, Mauget S, Payton P (2004) The rehydration transcriptome of the desiccation-tolerant bryophyte *Tortula ruralis*: transcript classification and analysis. BMC Genomics 5:89

Oliver MJ (1996) Desiccation tolerance in vegetative plant cells. Physiol Plantarum 97:779–787

Oliver MJ, Tuba Z, Mishler BD (2000) The evolution of vegetative desiccation tolerance in land plants. Plant Ecol 151:85–100

Oliver MJ, Velten J, Mishler BD (2005) Desiccation tolerance in bryophytes: a reflection of the primitive strategy for plant survival in dehydrating habitats? Integr Comp Biol 45:788–799

Platt KA, Oliver MJ, Thomson WW (1994) Membranes and organelles of dehydrated *Selaginella* and *Tortula* retain their normal configuration and structural integrity—freeze fracture evidence. Protoplasma 178:57–65

Porembski S, Barthlott W (2000) Granitic and gneissic outcrops (inselbergs) as centers of diversity for desiccation-tolerant vascular plants. Plant Ecol 151:19–28

Quartacci MF, Forli M, Rascio N, Dallavecchia F, Bochicchio A, Navari-Izzo F (1997) Desiccation-tolerant *Sporobolus stapfianus*: lipid composition and cellular ultrastructure during dehydration and rehydration. J Exp Bot 48:1269–1279

Quartacci MF, Glisic O, Stevanovic B, Navari-Izzo F (2002) Plasma membrane lipids in the resurrection plant *Ramonda serbica* following dehydration and rehydration. Journal of Experimental Botany 53:2159–2166

Rascio N, La Rocca N (2005) Resurrection plants: the puzzle of surviving extreme vegetative desiccation. Crit Rev Plant Sci 24:209–225

Reynolds T, Bewley J (1993) Characterisation of protein synthetic changes in a desiccation tolerant fern *Polypodium virginianum*: comparison of the effects of drying, rehydration and abscisic acid. J Exp Bot 44:921–928

Riechmann JL, Meyerowitz EM (1998) The AP2/EREBP family of plant transcription factors. Biol Chem 379:633–646

Rohila JS, Jain RK, Wu R (2002) Genetic improvement of Basmati rice for salt and drought tolerance by regulated expression of a barley Hva1 cDNA. Plant Sci 163:525–532

Scott HB, Oliver MJ (1994) Accumulation and polysomal recruitment of transcripts in response to desiccation and rehydration of the moss *Tortula ruralis*. J Exp Bot 45:577–583

Sgherri CLM, Loggini B, Puliga S, Navari-Izzo F (1994a) Antioxidant system in *Sporobolus stapfianus*—changes in response to desiccation and rehydration. Phytochemistry 35:561–565

Sgherri CLM, Loggini B, Bochicchio A, Navari-Izzo F (1994b) Antioxidant system in *Boea hygroscopica*: changes in response to desiccation and rehydration. Phytochemistry 37:377–381

Shen-Miller J, Mudgett MB, Schopf JW, Clarke S, Berger R (1995) Exceptional seed longevity and robust growth—ancient sacred lotus from China. Am J Bot 82:1367–1380

Sherwin HW, Farrant JM (1998) Protection mechanisms against excess light in the resurrection plants *Craterostigma wilmsii* and *Xerophyta viscosa*. Plant Growth Regul 24:203–210

Shigyo M, Hasebe M, Ito M (2006) Molecular evolution of the AP2 subfamily. Gene 366:256–265

Sivamani E, Bahieldin A, Wraith JM, Al-Niemi T, Dyer WE, Ho DT-H, Qu R (2000) Improved biomass productivity and water use efficiency under water deficit conditions in transgenic wheat constitutively expressing the barley *HVA1* gene. Plant Sci 155:1–9

Stockinger EJ, Gilmour SJ, Thomashow MF (1997) *Arabidopsis thaliana* CBF1 encodes an AP2 domain-containing transcriptional activator that binds to the C-repeat/DRE, a cis-acting DNA regulatory element that stimulates transcription in response to low temperature and water deficit. PNAS 94:1035–1040

Tardieu F (2005) Plant tolerance to water deficit: physical limits and possibilities for progress. C R Geosci 337:57–67

Vicre M, Lerouxel O, Farrant J, Lerouge P, Driouich A (2004) Composition and desiccation-induced alterations of the cell wall in the resurrection plant *Craterostigma wilmsii*. Physiol Plantarum 120:229–239

Vicre M, Sherwin HW, Driouich A, Jaffer MA, Farrant JM (1999) Cell wall characteristics and structure of hydrated and dry leaves of the resurrection plant *Craterostigma wilmsii*, a microscopical study. J Plant Physiol 155:719–726

Villalobos MA, Bartels D, Iturriaga G (2004) Stress tolerance and glucose insensitive phenotypes in *Arabidopsis* overexpressing the *CpMYB10* transcription factor gene. Plant Physiol 135:309–324

Wise MJ, Tunnacliffe A (2004) POPP the question: what do LEA proteins do? Trends Plant Sci 9:13–17

Wood AJ, Oliver MJ (1999) Translational control in plant stress: the formation of messenger ribonucleoprotein particles (mRNPs) in response to desiccation of *Tortula ruralis* gametophytes. Plant J 18:359–370

Wu YJ, Sharp RE, Durachko DM, Cosgrove DJ (1996) Growth maintenance of the maize primary root at low water potentials involves increases in cell-wall extension properties, expansin activity, and wall susceptibility to expansins. Plant Physiol 111:765–772

Wu YJ, Thorne ET, Sharp RE, Cosgrove DJ (2001) Modification of expansin transcript levels in the maize primary root at low water potentials. Plant Physiol 126:1471–1479

Rev Environ Sci Biotechnol (2006) 5:243–251
DOI 10.1007/s11157-006-0012-1

REVIEW PAPER

# Energy dependant plant stress acclimation

**Cristina Dobrota**

Received: 23 February 2006 / Accepted: 6 June 2006 / Published online: 25 July 2006


**Abstract** Plants may live and grow under suboptimal environmental conditions having certain biochemical and metabolic adaptations that facilitate their survival. Plant "metabolic flexibility" consists of the accomplishment of the same step in a metabolic pathway in a variety of different ways. Pyrophosphate which works as an energy donor when cellular ATP pools become diminished during stresses, alternative glycolytic reactions that bypass ATP-requiring steps, additional pathways for electron transport in plant mithocondria and the salvage pathways are some of the aspects related to "energetic flexibility". This key feature that plays an important role in plant acclimation to stress can be an important target for engineering enhanced stress tolerance in crop plants.

**Keywords** ATP · Energy pool · Extreme environment · Metabolic engineering · Stress

## Introduction

Organisms in extreme environments are adapted to the conditions of their surroundings, multiple abiotic stresses such as dry habitats and thermophilic (cells in hot springs and undersea thermal vents up to 121°C), psychrophilic (cryophiles) and halophilic (high salt concentrations) niches, which are among the harshest conditions found on Earth where life is frequently detected. Some organisms grow in extreme pH conditions (acidity versus alkalinity) or under hydrostatic pressure in the deep sea.

Many of the absolute extreme environments are not lifeless but are inhabited by organisms living "on the edge" near the absolute limits of their physiological potential. Under these conditions, even a slight deterioration in the environment may result in death and extinction. Such absolute extreme environments are desert regions such as Antarctica, the Arctic, Atacama (Chile), Negev (Israel) or Gobi (Mongolia). Many microbial inhabitants (algae, fungi, cyanobacteria, heterotrophic bacteria) as well as many animals and plants manage to survive in these environments.

An important ecological limiting factor is represented by low temperatures below the freezing point. In the mountains, the rock surfaces are almost lifeless, but rich communities of microorganisms exist under the surface, colonizing the air spaces inside porous sandstone rocks. Bacteria in Arctic and Antarctic permafrost (frozen soil) maintain their metabolic activity at temperatures as low as − 20°C. *Chlamydomonas nivalis*, a snow

C. Dobrota (✉)
Babes-Bolyai University, 1 Kogalniceanu St., Cluj, Romania
e-mail: crisd@bioge.ubbcluj.ro

algae, has adaptations for efficient capture of both light and $CO_2$ that allow photosynthesis to occur while submerged in snow. This snow microenvironment may be one of the most extreme for photosynthesis in the plant kingdom.

For plants, "extreme environments" have narrower limits than for other organisms. Plants may live and grow under suboptimal environmental conditions having certain biochemical and metabolic adaptations that facilitate their survival. Plants frequently alter their growth and development patterns in an attempt to alleviate some of the unfavorable changes in their environment that they may have been exposed to (Plaxton 1996).

There are certain groups of plants better adapted to extreme conditions than others are. Lichen-dominated vegetation accounts for about 8 percent of land surface. Lichens are extremely effective colonizers in environments such as surfaces that are newly exposed after volcanic eruptions or other environmental disasters, shingle beaches, moorlands and areas where glaciers have retreated. The basic physiology of lichens is what determines their efficiency at growing in extreme environments (Purvis 2000). The interchange of substances between the two components of the symbiosis ensures that lichens have the extraordinary ability to maintain themselves during very long dry periods, when they hardly metabolize at all. In the Antarctic, many lichens can function under very low temperature conditions and actually photosynthesize under the ice to protect themselves under high levels of ultraviolet radiation.

The huge chemical arsenal that lichens possess helps ensure their survival under extreme conditions. Some species contain up to 30 percent dry weight of organic compounds, which act as 'stress metabolites'. Others have antibiotic activity, and this acts as a deterrent to other organisms that may prey on the slow-growing lichens. The same compounds also act as detoxifiers where toxic levels of metals are present, and play an important role in keeping some lichen tissues dry to allow gaseous exchange and carbon fixation (Brodo et al. 2001).

Seed plants are considered the most successful terrestrial plants even though they are less tolerant to extreme environments. Seed plants have been able to adapt to an extraordinary range of habitats. The embryo, protected and nourished inside the seed, is able to survive in a dormant state during unfavorable growing conditions such as drought. Seed dispersal also facilitates the dissemination of the embryos away from the parent plant. Another important feature of seed plants is their mode of fertilization. Pollination is independent of external water, a distinct advantage in terrestrial environments. All the crops with economic importance are from this group and the lab experiments are mainly focused on them.

Experimental studies concerning plant life in extreme environments suffers from the lack of enough well-established model systems and from the infancy of implementation of recent developments in the field of molecular biology, in situ measurements, laboratory simulation techniques and systems biology modeling.

Plant acclimation to any change causing genotypic and phenotypic alterations represents the temporal integration of short-term and long-term fluctuations in an environmental signal. In nature, acclimation is usually initiated by a short-term fluctuation in an environmental factor which perturbs metabolic homeostasis and induces a stress response. Depending on the amplitude, frequency and duration of the stress, acclimation is the result of some long-term adjustment reflecting the development of a new homeostatic state. Tolerant species are able to acclimate to different stresses. Mechanisms to overcome the constraints of short-term and long-term exposure to stress include changes in energy absorption and photochemical transformation through energy partitioning and concomitant changes in chloroplastic carbon metabolism, allocation and partitioning.

Photosynthesis is the principal mechanism to transform light energy into biochemical usable chemical potential energy (ATP) and redox potential energy (NADPH). ATP and NADPH are consumed in the reduction of $CO_2$ to triose phosphate and the continuous regeneration of ribulose1,5-bisphosphate (RuBP). The energy represented by the fixed carbon is then used to maintain cellular homeostasis, cell division and expansion. The mechanisms involved in photosynthetic acclimation to stresses range from modifications within the thylakoid membrane system, affecting photosynthetic electron transport to post-transcriptional activation and increased expression of enzymes for sucrose synthesis, changed expression of Calvin cycle enzymes, changes in leaf protein content, as well as the signals that trigger these processes (Ensminger et al. 2006).

The balance of energy flow between the photophysical and photochemical processes that transform light and the metabolic sinks that consume the energy is called photostasis (Öquist and Huner 2003). Balancing the energy flow through the process of photosynthesis is a challenge, because the extremely rapid, temperature-independent photochemical reactions are integrated with relatively slow, temperature-dependent biochemical reactions. Because the biochemical reactions limit the rate of energy flow, the thermodynamic constraints imposed by low temperatures exacerbate a potential energy imbalance by differentially affecting energy consumption by metabolism.

Adjustment to imbalances in the energy flow from source to sink via electron transport requires a mechanism to sense cellular energy status. This is achieved in phototrophic cells through a redox sensor within the photosynthetic electron transport chain, the PQ pool (Pfannschmidt 2003). As a primary energy sensor, the PQ pool transduces the electron transport signal into biochemical signals that regulate transcription of genes involved in cold acclimation in both the chloroplast and the nucleus. To maintain an energy balance, regulation of photosynthetic genes reflects either the adjustment of the source and hence primary photosynthetic reactions and redox components of the photosynthetic electron transport chain or the adjustment of the sink capacity and hence enzymes involved in chloroplastic and cytosolic carbon metabolism (Ensminger et al. 2006).

It seems that there is complex cross-talk between the photosynthetic redox state of the cell and sugar-signalling pathways to regulate overall plant acclimation to low temperature. In addition, there are overlapping relationships between cold acclimation and responses to dehydration and salt stress in terms of the biochemical changes and in the transcriptional regulation of the responses (Kacperska 2004).

## ATP the universal energy donor

Adenosine triphosphate (ATP) is a vital molecule used by living organisms as a universal source of energy required for intracellular biochemical reactions. ATP functions as a carrier of energy in all living organisms from bacteria and fungi to plants and animals, including humans. ATP captures the chemical energy released by the combustion of nutrients and transfers it to reactions that require energy, e.g., the building up of cell components, muscle contraction, transmission of nerve messages and many other functions.

ATP consists of the nucleoside adenosine linked to three phosphate groups. On removal of the outermost phosphate group, adenosine diphosphate (ADP) is formed; at the same time the energy released can be employed for other reactions (Cohn 2001). Conversely, with the help of energy, an inorganic phosphate group can be bound to ADP and form ATP. The phosphorylation processes occurring in particular circumstances are referred to as substrate level phosphorylation, oxidative phosphorylation and photophosphorylation (Dobrotă 2004). Energy released in some spontaneous (exergonic) reactions may be captured in ATP for use at a later time or in another part of the cell.

Stresses such as anoxia (Huang et al. 2005), extreme temperature (Stupnikova et al. 2006), low pH (Messerli et al. 2005) and nutritional Pi starvation (Plaxton 2004) deregulate the physiology of the plant cell and cause ATP overconsumption.

Recent researches showed that besides its functions inside the cell, ATP may be released to the extracellular milieu, where it functions as an important signaling molecule of a diverse range of physiological processes (Demidchik et al. 2003).

In animal cells, extracellular ATP seems to be a peripheral mediator of pain (Hamilton et al. 2000), serving as a ubiquitous signaling substance in neuronal and non-neuronal systems (Sperlágh and Vizi 1996). In the central nervous system, the stimulation-dependent release and inactivation of extracellular ATP in response to neuronal activity and its origin has been demonstrated from different brain regions. After the release, an extracellular interconversion of ATP into adenosine occurs. Adenosine is a well known neuromodulator, which acts as a neuroprotective substance under various pathological conditions such as hypoxia and ischemia. Because large amounts of extracellular ATP and its degradation products can accumulate under any kind of energy deficit and as a result of cell death, interest was turned to the study of the mechanisms of the release and extracellular metabolism of ATP under pathological conditions, such as those in experimental models of neurological diseases.

Although recent findings revealed that intact plant tissues release ATP as well, there is little known about the physiological function of extracellular ATP in plants. It seems that extracellular ATP is essential for maintaining plant cell viability. Its removal by the cell-impermeant traps (glucose–hexokinase and apyrase) triggered death in both cell cultures and whole plants. Experimental studies suggest that ATP suppresses a default death pathway in plants and that some forms of pathogen-induced cell death are mediated by the depletion of extracellular ATP (Chivasaa et al. 2005).

## Plant adaptations to energy deficit

In certain conditions plants have the ability to respond to environmental changes by altering the expression of complex gene networks through sensing environmental cues, signal transduction, and modification of biochemical pathways. These transcriptional changes can result in successful adaptations leading to tolerance.

There is growing recognition that the abundance of mRNA transcripts is not always representative of cognate protein levels and that mechanisms of post-translational regulation must also play an important role for mediating and integrating many physiological responses of plants to their stressful environment (Renaut et al. 2006). Post-translational modifications of proteins greatly increase protein complexity and dynamics, co-ordinating the intricate regulation of biological events (Kwon et al. 2006).

Plant "metabolic flexibility" consists of the accomplishment of the same step in a metabolic pathway in a variety of different ways. This metabolic flexibility allows the preferential utilization of inorganic pyrophosphate (PPi) as an energy donor, particularly when cellular ATP pools become diminished during stresses such as anoxia and nutritional Pi starvation. Pyrophosphate permits plants and microbes to conserve ATP (Plaxton 2002).

PPi has been assumed to be a byproduct of secondary metabolism and anabolism, therefore use of PPi as an energy donor helps to conserve limited ATP pools (Duff et al. 1989). Plant cytosol contains PPi in concentrations of up to about 0.5 mM (Stitt 1998). These large amounts may be employed to enhance the energetic efficiency of several cellular processes. PPi levels of the plant cytosol are remarkably insensitive to abiotic stresses, which elicit significant reductions in cellular ATP pools (Plaxton 1996, 1999; Stitt 1998). As a consequence of the marked decline (up to fiftyfold) in cytoplasm Pi levels that follows severe Pi stress, large (up to 80 percent) reductions in intracellular levels of ATP and related nucleoside phosphates also occur.

During stress conditions, PPi would continue to be generated (albeit at a lower rate) as a byproduct of the synthesis of essential macromolecules. PPi may be either recycled from the starch biosynthesis pathway where PPi is produced in the amyloplast by ADP-Glc pyrophosphorylase (AGPase), or may be provided by a cycling process. In the latter case, there are two possibilities: the PPi-dependent reaction of pyrophosphate: Fru-6-P phosphotransferase (PFP) and the tonoplast pyrophosphatase (vPPase), both of which would produce PPi by coupling to a parallel ATP consuming process (ATP-dependent tonoplast proton pump or phosphofructokinase).

PPi-powered processes may be a crucial facet of the metabolic adaptations of stress-tolerant plant species to environmental extremes that cause depressed ATP (but not PPi) pools (Palma et al. 2000; Plaxton 1996, 1999).

Davies et al. (1993) computed the standard free energy changes for PPi and ATP hydrolysis in order to assess the relative importance of PPi versus ATP as an energy donor in the plant cytosol. The results indicated that PPi would be particularly favored as a phosphoryl donor, relative to ATP, under cytosolic conditions known to accompany stresses such as anoxia (Huang et al. 2005; Mustroph et al. 2005) or nutritional Pi deprivation.

Alternative PPi-dependent cytosolic processes represent a considerable bioenergetic benefit that may extend the survival time of ATP-depleted plant cells during different stresses. The flexibility of plant bioenergetics helps plants to acclimate to environmental stresses such as anoxia and Pi starvation. Pyrophosphate may be considered as an alternate cytosolic energy 'transmitter' to ATP.

Alternative glycolytic reactions (Fig. 1) can bypass Pi-or ATP-requiring steps of glycolysis under environmental stress conditions (Duff et al. 1989; Theodorou et al. 1992). One alternative glycolytic pathway is catalyzed by a $PP_i$-dependent phosphofructokinase (PFP) that, under P deficiency, can bypass the ATP-dependent phosphofructokinase (PFK), generating fructose 1,6-bisphosphate (Plaxton and Carswell 1999). Other processes that might use PPi are the cleavage of sucrose by a PPi-dependent sucrose synthase pathway and the active transport of protons into the vacuole by a PPi-dependent $H^+$ pump in the tonoplast (Plaxton and Carswell 1999).

Another alternative glycolytic pathway known in plants is catalyzed by the action of a non-phosphorylating NADP-dependent glyceraldehyde-3P dehydrogenase (NADP-G3PDH) that bypasses $P_i$-dependent NAD-G3PDH and phosphoglycerate kinase (Duff et al. 1989; Theodorou et al. 1992). The third bypass of the glycolytic pathway can be catalyzed by the combined activities of PEPC, MDH and NAD-malic enzyme (Theodorou and Plaxton 1996). Phosphorus stress can severely limit the activity of pyruvate kinase (PK), an enzyme requiring Pi and ADP. The PEPC, MDH and NAD-malic enzymes can bypass PK and thus maintain the flow of carbon from glycolysis to the TCA cycle by avoiding the use of ADP but generating free Pi (Plaxton and Carswell 1999). The described metabolic adaptations to Pi stress appear to be plant specific (Vance et al. 2003).

A very unique and important aspect of plant bioenergetic flexibility is the presence of both phosphorylating and non-energy conserving (non-ATP producing) pathways of the mitochondrial electron transport chain. Plant mitochondria have several additional pathways for electron transport such as the non proton pumping NAD(P)H dehydrogenase 'bypass' to complex I, and the non proton pumping alternative oxidase bypass to complex IV.

The presence of the alternative oxidase (AOX) is one of the features that sets plant (and fungal) mitochondria apart from mammalian mitochondria. The AOX does not pump protons, so when electrons from NADH oxidation flow through the AOX, complexes III and IV are bypassed. Therefore, energy conservation in the form of ATP is much smaller when the AOX is active. The alternative oxidase is regulated by the rate of electron transport and the reduction level of quinone, but is independent of the protonmotive force.

The alternative oxidase may also work as a bypass to oxidize NADH and $FADH_2$ under ADP-limiting conditions under which the cytochrome oxidase pathway is restricted. In this way, the citric acid cycle and the glycolytic pathway can continue to run and provide the cell with biosynthetic precursors.

The induction of AOX expression by inhibitors of the cytochrome pathway allows the mitochondria to maintain electron flux to oxygen when electron flux through the phosphorylating cytochrome pathway is restricted by high adenylate energy charge, or under a number of stresses such as low temperature or phosphate starvation. By 'draining' excess electrons the alternative oxidase helps to minimize the production of reactive oxygen species (ROS) that otherwise accumulate when ubiquinone becomes overreduced. ROS can cause damage to proteins, lipids, and DNA and the cell must therefore limit their formation.

Maxwell et al. (1999) demonstrated the importance of the AOX for limiting ROS formation in living cells. They measured the ROS level in a wild-type tobacco cell culture and in cells in which the amount of AOX was either overexpressed or reduced by antisensing. Overexpression of AOX lowered the steady state level of ROS, whereas underexpression caused a five-fold increase in cellular ROS.

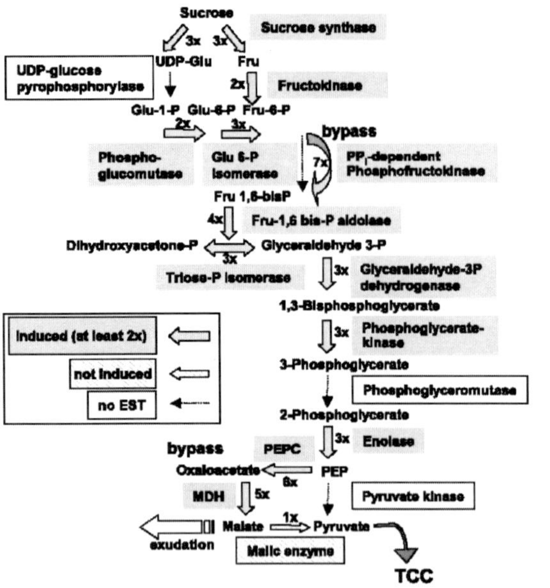


**Fig. 1** Schematic representation of the glycolytic pathway. Expressed sequence tags (ESTs) are represented in gray boxes. The average of gene induction, as determined by two independent macroarrays is indicated at the corresponding arrows (e.g. 2 ×). Enzymes of the glycolytic pathway that were not found in a collection of 1250 ESTs are shown in white boxes and are represented by dotted arrows (after Vance et al. 2003, with permission from Blackwell Publishing)

Another interesting mechanism related to the energy state of the cell is represented by the salvage pathways. These salvage pathways are active in every cell of the organisms and help keep the energy pools from being used up too quickly.

De novo production of purines in which nucleotides are made starting with ribose is an energy-intensive process (for the synthesis of a single purine ring are consumed 5 molecules ATP). Purines and their nucleosides may be converted directly to nucleotides by a salvage pathway. Purine salvage pathway provides up to 80% of the purines required for intermediary metabolism, DNA and RNA synthesis, and energy supplies. Salvage reactions requires less energy than does de novo synthesis. As nucleic acids are degraded, the free bases are scavenged and converted back to nucleotides using 5-phosphoribosyl-1-pyrophosphate (PRPP) which is a source of ribose-5-phosphate. PRPP is used in de novo synthesis of nucleotides such as adenosine, and inosine. PRPP is also an essential participant in the salvage pathways for ATP regeneration (Fig. 2).

In the salvage pathway of ATP formation, ribose enables the cells to quickly and efficiently recycle (i.e., salvage) the end products formed by the breakdown of ATP to form new ATP molecules.

Nucleotides are essential energy sources for basic metabolic reactions and play important roles in protein, glycogen and nucleic acid synthesis (ribonucleotides and deoxyribonucleotides), cyclic nucleotide metabolism, and energy transfer reactions.

Whereas ATP serves as the universal currency of cellular energy, NAD and NADP have specialized roles in carrying high-energy electrons, after becoming NADH and NADPH. NADPH operates with enzymes that synthesize energy-rich molecules or fuel the anti-oxidant defense of the cell. NADH by contrast, has mainly a role as an intermediate in the system of reactions that generate ATP or steer transcription-repair events in the nucleus. NAD and NADH are converted into each other many times along the pathway of the energy cycles.

NAD is an ubiquitous molecule that participates in many metabolic reactions. It also plays important roles in transcriptional regulation, longevity, calorie-restriction-mediated life-span extension and age-associated diseases. In yeast, it has been shown that NAD affects longevity and transcriptional silencing through the regulation of the Sir2p family of NAD-dependent deacetylases (Lin et al. 2004). NAD is synthesized via two major pathways in both prokaryotic and eukaryotic systems: de

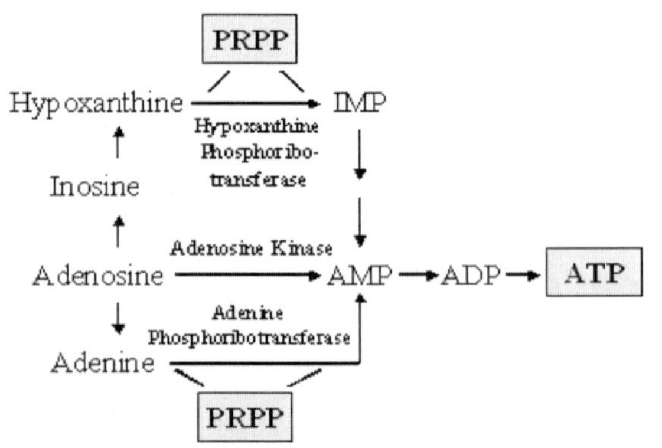


**Fig. 2** ATP salvage pathway

novo pathway and the salvage pathway where the nicotinic acid moiety of NAD is generated by recycling degraded NAD products.

Low calorie intake is considered to be a mild stress, which invokes a general stress response. Calorie restriction (CR) extends life span in a wide spectrum of organisms and for decades was the only regimen known to promote longevity in mammals (Roth et al. 2001). CR has also been shown to delay the onset or reduce the incidence of many age-related diseases including cancer and diabetes (Roth et al. 2001). Although it has been suggested that CR may work by reducing the levels of reactive oxygen species through a slowing in metabolism, the mechanism by which CR extends life span is still uncertain. A switch to oxidative metabolism during CR increases the NAD/NADH ratio by decreasing NADH levels (Lin et al. 2004).

## Engineering high energy-use efficient plants tolerant to stress

All living organisms must acclimate to environmental stresses that can potentially damage cellular processes. Studies of plant metabolic responses to extreme environments have revealed some remarkably adaptive mechanisms that may serve to limit the deleterious consequences of stress.

Plant 'bioenergetic flexibility' that likely plays an important role in plant acclimation to stress can be an important target for engineering enhanced stress tolerance in crop plants. A better understanding of the extent to which changes in flux through alternative enzymes and pathways influences plant stress tolerance is of significant practical interest. This knowledge is relevant to the ongoing efforts of agricultural biotechnologists to produce transgenic crops having improved resistance to environmental extremes via the process of metabolic engineering (DeBlock et al. 2005).

Crop yields are frequently lowered by biotic and abiotic stresses, and one of the most effective strategies to improve agricultural output is to breed or engineer plants tolerant of or resistant to stress. Molecular understanding of the stress perception, signal transduction and transcriptional regulation of stress responsive genes may help to engineer tolerance for multiple stresses.

When a plant is exposed to stress, survival mechanisms turn on to reduce damage. Under stress conditions, reactive oxygen species (ROS) highly toxic occur (Mittler et al. 2005). It is considered that breeding or engineering for high energy use efficiency under stress conditions is a valuable approach to enhance overall stress tolerance of crops.

Marc DeBlock et al. (2005) report a strategy of improving stress tolerance in plants by maintaining the plants energy homeostasis under stress. Since it is not very practical to completely avoid growing crops in conditions that are stressful, scientists focus on engineering plants to be indifferent to stress.

The work on engineering of tolerance to abiotic stresses started within a decade of the molecular understanding of pathways induced in response to one or more of the abiotic stresses. In most of the cases the transgenes expressed faithfully but only a limited level of tolerance was provided under stress conditions as compared to the non-transformed wild type plants. In many cases the transgenic plants had morphological abnormalities and slower growth under non-stressed environment.

When plants are exposed to a stress signal, they expend a lot of energy, in certain cases exhibiting enhanced respiration rates. This is partially due to a breakdown in the $NAD^+$ pool caused by the enhanced activity of poly(ADP-ribose) polymerase (PARP), which uses $NAD^+$ as a substrate to synthesize polymers of ADP-ribose. Stress-induced depletion of $NAD^+$ results in a similar depletion of energy, since ATP molecules are required to resynthesize the depleted $NAD^+$. It seems that plants with lowered poly(ADP ribosyl)ation activity appear tolerant to multiple stresses. Inhibiting PARP activity prevents energy overconsumption under stress, allowing normal mitochondrial respiration.

Another candidate for engineering increased stress resistance in plant cells, especially in organelles such as the peroxisome, where ROS are produced at high levels, is glutathione. Since the ascorbate/glutathione scavenging pathway is present in peroxisomes, chloroplasts, and mitochondria, it is likely that glutathione levels are high in all these subcellular compartments. Transport of glutathione across the peroxisomal and mitochondrial bounding membranes is one possible focus for engineering increased stress resistance. Engineering increased flux through the cytosolic glutathione biosynthetic pathway is another possibility; a strategy that has proved to provide increased protection to photosynthesis (Green 2002).

Another foreseeable application consists in exploitation of economic potential of extremophiles. Extremophilic plants can survive under conditions toxic or otherwise harmful to crop plants. Therefore there is the potential to transfer by molecular cloning, some of these abilities from extremophiles to crop plants with the aim to enhance the tolerance to various stresses, e.g., extreme temperatures, salt, heavy metal, drought or UV (Gurley 2000).

Besides the discussion of economic aspects, the study of plant reactions in extreme environments offers an important potential for predictions. Understanding what limits plant life in extreme conditions may allow drawing conclusions on future developments of ecosystems in environments altered by human activity or as a consequence of natural climatic changes.

The use of multiple tolerance mechanisms for one or more of the abiotic stresses through stepwise or co-transformation may help to achieve high levels of tolerance for commercial exploitation. Understanding the molecular mechanism for providing protection against biotic and abiotic stresses may lead to a generalized master mechanism for stress tolerance. Optimum homeostasis is always a key to living organisms for adjusted environments. Thus, abiotic stress accompanying a number of biological phenomena must be precisely investigated by consideration of plant homeostasis. new opportunities for manipulating plant chemistry for improvement of plant traits such as disease and stress resistance and nutritional qualities.

## References


Brodo I, Sharnoff Duran S, Sharnoff S (2001) Lichens of North America. Yale University Press, New Haven

Chivasaa S, Bongani K, Ndimbab W, Simonc J, Lindseyc K, Slabasc A (2005) Extracellular ATP functions as an endogenous external metabolite regulating plant cell viability. Plant Cell 17:3019–3034

Cohn M (2001) Adenosine triphosphate. In: Encyclopedia of life science. Nature Publishing Group

Davies JM, Poole RJ, Sanders D (1993) The computed free energy changes of hydrolysis of inorganic pyrophosphate and ATP: apparent significance of inorganic-pyrophosphate-driven reactions of intermediary metabolism. Biochim Biophys Acta 1141:29–36

DeBlock M, Verduyn C, De Brouwer D, Cornelissen M (2005) Poly(ADP-ribose) polymerase in plants affects energy homeostasis, cell death and stress tolerance. Plant J 41:95–106

Demidchik V, Nichols C, Oliynyk M, Dark A, Glover BJ, Davies JM (2003) Is ATP a signaling agent in plants? Plant Physiol 133:456–461

Dobrotă C (2004) The biology of phosphorus. In: Valsamy-Jones E, Gray R (eds) Phosphorus in environmental technology: principles and applications. IWA Publishers, London, pp 51–77

Duff SMG, Moorhead GBG, Lefebvre DD, Plaxton WC (1989) Phosphate starvation inducible 'bypasses' of adenylate and phosphate dependent glycolytic enzymes in *Brassica nigra* suspension cells. Plant Physiol 90:1275–1278

Ensminger I, Busch F, Huner NPA (2006) Photostasis and cold acclimation: sensing low temperature through photosynthesis. Physiol Plantarum 126(1):28–44

Greene R (2002) Oxidative stress and acclimation mechanisms in plants. The Arabidopsis Book: Vol. 49, No. 1 pp. 1–20, BioOne Publishers, Washington

Gurley WB (2000) HSP101: a key component for the acquisition of thermotolerance in plants. Plant Cell 12:457–460

Hamilton SG, Warburton J, Bhattacharjee A, Ward J, McMahon S (2000) ATP in human skin elicits a dose-related pain response which is potentiated under conditions of hyperalgesia. Brain 123(6):1238–1246

Huang S, Greenway H, Colmerm TD, Millar AH (2005) Protein synthesis by rice coleoptiles during prolonged anoxia: Implications for glycolysis, growth and energy utilization. Ann Bot 96:703–715

Kacperska A (2004) Sensor types in signal transduction pathways in plant cells responding to abiotic stressors: do they depend on stress intensity? Physiol Plant 122:159–168

Kwon SJ, Choi EY, Choi YJ, Ahn JH, Park O (2006) Proteomics studies of post-translational modifications in plants. J Exp Bot 57(7):1547–1551

Lin SJ, Ford E, Haigis M, Liszt G, Guarente G (2004) Calorie restriction extends yeast life span by lowering the level of NADH. Genes Dev 18:12–16

Maxwell DP, Wang Y, McIntosh L (1999) The alternative oxidase lowers mitochondrial reactive oxygen production in plant cells. Proc Natl Acad Sci USA 96:8271–8276

Messerli MA, Amaral-Zettler LA, Zettler E, Jung SK, Smith PJS, Sogin ML (2005) Life at acidic pH impose an increase energetic cost for eukaryotic acidophile. J Exp Biol 208:2569–2579

Mittler R, Vanderauwera S, Gollery M, Van Breusegem F (2005) Reactive oxygen gene network of plants. Trends Plant Sci 9(10):490–498

Mustroph A, Albrecht G, Hajirezaei M, Grimm B, Biemelt S (2005) Low levels of pyrophosphate in transgenic potato plants expressing *E.coli* pyrophosphatase lead to decreased vitality under oxygen deficiency. Ann Bot 96:717–726

Öquist G, Huner NPA (2003) Photosynthesis of overwintering evergreen plants. Annu Rev Plant Biol 54:329–355

Palma DA, Blumwald E, Plaxton WC (2000) Upregulation of vacuolar $H^+$-translocating pyrophosphatase by phosphate starvation of *Brassica napus* (rapeseed) suspension cell cultures. FEBS Lett 486:155–158

Pfannschmidt T (2003) Chloroplast redox signals: how photosynthesis controls its own genes. Trends Plant Sci 8:33–41

Plaxton WC (2004) Plant response to stress: Biochemical adaptations to phosphate deficiency. In: Goodman R (ed) Encyclopedia of plant and crop science. Marcel Dekker, Inc., NY

Plaxton WC (2002) Metabolic flexibility helps plants to survive stress. Web-essay (www.plantphys.net) supplement the 3rd edition of Plant Physiology (textbook by Taiz & Zeiger)

Plaxton WC (1999) Metabolic aspects of phosphate starvation in plants. In: Deikman J, Lynch J (eds) Phosphorus in plant biology: regulatory roles in molecular, cellular, organismic, and ecosystem processes. American Society of Plant Physiologists, Rockville, pp 164–176

Plaxton WC, Carswell MC (1999). Metabolic aspects of the phosphate starvation response in plants. In: Lerner HR (ed) Plant responses to environmental stress: from phytohormones to genome reorganization. Marcel-Dekker, New York, NY, USA, pp 350–372

Plaxton WC (1996) The organization and regulation of plant glycolysis. Annu Rev Plant Physiol Plant Mol Biol 47:185–214

Purvis W (2000) Lichens natural history museum. Natural World Series, London

Renaut J, Hausman J-F, Wisniewski ME (2006) Proteomics and low-temperature studies: bridging the gap between gene expression and metabolism. Physiol Plantarum 126(1):97–109

Roth GS, Ingram DK, Lane MA (2001) Caloric restriction in primates and relevance to humans. Ann NY Acad Sci 928:305–315

Sperlágh B, Vizi ES (1996) Neuronal synthesis, storage and release of ATP. Semin Neurosci 8:175–186

Stitt M (1998) Pyrophosphate as an energy donor in the cytosol of plant cells: an enigmatic alternative to ATP. Bot Acta 111:167–175

Stupnikova I, Benamar A, Tolleter D, Grelet J, Borovskii G, Dorne A, Macherel D (2006) Pea seed mitochondria are endowed with a remarkable tolerance to extreme physiological temperatures. Plant Physiol 140:326–335

Theodorou ME, Plaxton WC (1996). Purification and characterization of pyrophosphate-dependent phosphofructokinase from phosphate-starved *Brassica nigra* suspension cells. Plant Physiol 112:343–351

Theodorou ME, Cornel FA, Duff SM, Plaxton WC (1992) Phosphate starvation-inducible synthesis of the alpha-subunit of the pyrophosphate-dependent phosphofructokinase in black mustard suspension cells. J Biol Chem 267:21901–21905

Vance CP, Uhde-Stone C, Allan DL (2003) Phosphorus acquisition and use: critical adaptations by plants for securing a nonrenewable resource. New Phytol 157(3):423–424

Rev Environ Sci Biotechnol (2007) 6:47–60
DOI 10.1007/s11157-006-9101-4

REVIEW PAPER

# Hypometabolic induced state: a potential tool in biomedicine and space exploration

**Manuela Malatesta · Marco Biggiogera · Carlo Zancanaro**

Received: 1 March 2006 / Accepted: 25 July 2006 / Published online: 29 September 2006


**Abstract** This paper will first review the issue of hypometabolism in mammals with a focus on the strategies these animals evolved to cope with life challenge in hostile environments (e.g., cold weather and/or shortage of food). The different types of natural hypometabolism (hibernation, torpor, winter sleep) will be briefly described as well as major adaptations in body temperature, and energy and cell metabolism. In the second part of this review the issue of inducing a hypometabolic state in mammals will be afforded with special attention paid to changes in body temperature and metabolism, regulation of gene expression and the possible role of hibernation inducing factors. Finally, an overview of the potential of inducing a hypometabolic state in the human as related to the broad field of biomedicine will be given.

M. Malatesta
Istituto di Istologia e Analisi di Laboratorio, Università di Urbino, Urbino, Italy

M. Biggiogera
Dipartimento di Biologia Animale, Università di Pavia, Pavia, Italy

C. Zancanaro (✉)
Anatomy & Histology Section, Dipartimento di Scienze Morfologico-Biomediche, Università di Verona, Strada Le Grazie, 8, I-37134 Verona, Italy
e-mail: carlo.zancanaro@univr.it

**Keywords** Body temperature · Extreme environments · Hibernation · Human · Mammals · Metabolism · Transplantation · Surgery

## 1 Natural hypometabolism in mammals

### 1.1 Hibernation, torpor, winter sleep

Mammals have evolved the ability to produce endogenous heat to maintain a high and constant body temperature over a wide range of ambient temperatures, thus achieving a high degree of independence from the restrictions imposed by the environment. However, in many mammals, especially the small ones, the cost for the thermoregulatory heat production can exceed the energy supply when the environmental conditions become adverse (low ambient temperature, water or food scarcity). In order to survive, these animals have evolved the capacity of entering a state of torpor. This is a natural hypometabolic state characterized by a drastic reduction of body temperature, metabolic activity, heart rate, and energy demand as well, which may last from a few hours to some weeks, thus facilitating survival (reviews in Lyman et al. (1982), French (1988).

Unlike ectotherms (e.g. frogs and snakes), hypometabolic mammals are able to leave this depressed metabolic state at any time, using the endogenously produced heat to restore a normal body temperature.

The natural hypometabolism of mammals varies depending on the timing, duration and depth.

Aestivation and hibernation are considered as seasonal torpors, being both characterized by subsequent torpor bouts. In fact, the seasonal torpor never spans the entire hibernating season, but it is interrupted by periodic arousals and brief normothermic periods (see Sect. 1.2.5). Some common groups characterized by seasonal torpor include the hedgehogs, marmots and woodchucks, ground squirrels and bats. In these animals, during the pre-hibernating phase, a drastic, rapid weight gain occurs, and exposure to low ambient temperature results in hibernation with body temperature decreasing to near 0°C.

However, in some species a hypometabolic state can be induced at any time of the year by proper environmental stimuli. A typical form of such non-seasonal torpor is the daily torpor, with a duration of less than 24 h and a body temperature ranging 10–25°C (considerably higher that that found during hibernation). Moreover, the daily torpor appears to be integrated into the normal circadian rhythm of activity and rest, although the torpor is not restricted only to the normal rest phase of the animal (Körtner and Geiser 2000). Marsupials, insectivores, chiropterans, primates and rodents have members which exhibit daily torpor. The Syrian (golden) hamster (*Mesocricetus auratus*), can enter deep torpor (body temperature decreasing to about 5°C and lasting a few days), but requires a long preliminary period of cold exposure.

A particular type of natural hypometabolism is represented by the winter sleep in bears. During winter dormancy, bears do not eat, drink, defecate or urinate and they use fat exclusively as their energy source. The bear hibernates at a near normal body temperature (31–35°C), and its metabolic depression is much less than that found even in the daily torpor. In addition, unlike torpid animals, when disturbed the bear is easily aroused into a mobile, reactive state.

### 1.2 General aspects of natural hypometabolism

Despite the different torpor types, the patterns of metabolism, heart rate, respiratory rate and body temperature of the natural hypometabolism are basically similar, differing only in their quantitative aspects. Thus, there is a physiological convergence in achieving a hypometabolic state for energy conservation.

Typically, a torpor bout consists of entry into, maintenance of and arousal from hypometabolism. During entry into the hypometabolic state, a progressive inhibition of heart and respiratory rate and a fall in oxygen consumption and body temperature occur. During the hypometabolic bout, which may last from hours to weeks, all physiological functions are kept at a minimum: heart rate decreases to 1/30 or less, oxygen consumption to 1/100 of their euthermic levels; prolonged apnoea (40–150 min) as well as Cheyne–Stoke breathing (apnoea followed by bursts of breathing) can occur; body temperature is maintained near ambient temperature (sometimes near 0°C). Arousal from torpor is a chain-reaction event requiring from 20 min to 30 min in small hibernators to a few hours in larger animals: substrates are mobilized for energy production, the cardiovascular system is stimulated for tissue perfusion and the non-shivering thermogenesis in the brown adipose tissue starts.

Unfortunately, in spite of extensive studies on hibernation, the basic mechanisms of the natural hypometabolism are still unknown.

Whatever the mechanism utilized by the hibernator, the final goal is to reach a hypometabolic state in which energy consumption is reduced to a minimum. That means that a novel regulation of metabolic and physiological pathways is needed to allow survival.

#### *1.2.1 Temperature*

During torpor, body temperature is still controlled by a CNS regulator: thermosensitive neurons within the hypothalamic region induce cyclic bursts of metabolic heat production if cooled below a certain set-point (Buck and Barnes 2000). Photoperiod duration would play also a role in the modulation of body temperature (Heldmaier et al. 1989).

The tissue mainly responsible for non-shivering thermogenesis is the brown adipose tissue (BAT), a unique lipid storing tissue which grows substantially during the pre-hibernation period. During hibernation, BAT is apparently quiescent, although the brown adipocyte nuclei maintain an "active" configuration (Zancanaro et al. 1993), but it switches to vigorous thermogenesis in the early phase of arousal (Horwitz et al. 1985): fatty acids are oxidized in specialized mitochondria where an extensive uncoupling of oxidative phosphorylation occurs with a consequent heat production (Himms-Hagen 1986).

### *1.2.2 Homeostasis*

The extraordinary capacity of varying body temperature in hibernators implies structural and molecular adaptation of cell and tissue components to maintain homeostasis.

Cell membranes are particularly sensitive to temperature modifications: as the temperature is lowered, the microviscosity of the lipid environment increases, probably influencing the activity of enzymes therein during hibernation. Particularly interesting is the ability of a hibernator's cells to maintain the intracellular $Ca^{++}$ homeostasis by down-regulating the activity of the specific channels in the cell membranes (Wang et al. 2002).

Ventilation is decreased during hibernation and entrance into hibernation is also accompanied by a decrease in the respiratory exchange rate, possibly by a retention of carbon dioxide in body fluids.

A dramatic reduction of heart rate and cardiac output takes place during hibernation, whereas stroke volume increases. A key phenomenon to be explained is the hibernator's heart ability to cope with lowering body temperature skipping fatal arrhythmia (Johansson 1996). Blood clotting times are dramatically increased during hibernation, thereby alleviating the risk of thrombus formation when blood flow is reduced. The mechanism involved could be a reduction in platelets blood count or an increase of the liver $\alpha$ 2-macroglobulin, a universal protease inhibitor (Srere et al. 1995).

The kidney activity is markedly reduced during hibernation (Zatzman 1984) in association with lower kidney blood flow and reduction or cessation of glomerular filtration rate; despite this, the fine structure of the organ is well preserved (Zancanaro et al. 1999).

During hibernation, animals show a remarkable suppression of responsiveness to stimuli; however, handling, re-warming, or exposure to daylight are able to induce arousal: it is apparent that hibernation, unlike coma, results from a highly regulated process rather than a loss of function. Several brain areas seem to be involved in regulating metabolic depression during hibernation. The hippocampus, septum, and hypothalamus retain periodic electroencephalographic (EEG) activity at temperatures below which EEG in other structures becomes isoelectric (Pakotin et al. 1993). The hypothalamic nuclei—the suprachiasmatic nucleus (containing pacemaker cells for circadian rhythms), the lateral septal nucleus, the medial septum—as well as the supraoptic and paraventricular nuclei show modifications related to the hibernating state (O'Hara et al. 1999;). Moreover, the preoptic/anterior hypothalamic area might play a role in hibernation due to the presence of sleep promoting and temperature sensitive neurons therein (Heller 1979).

### *1.2.3 Energy*

The primary energy source utilized during hibernation is the lipid accumulated in the white adipose tissue, while the utilization of carbohydrate is drastically reduced, even in daily torpor.

During the summer and fall months hibernators became hyperphagic and prepare for hibernation drastically increasing their spring body weight, probably because of the combined action of environmental factors (Körtner and Geiser 2000), enzymes responsible for lipogenesis (Mostafa et al. 1993) and specific neuropeptides able to regulate food intake (Boswell et al. 1993).

During hibernation mitochondria preferentially utilize lipid reserves to produce ATP, probably due to the activity of leptin, responsible for mobilization of fat (Rousseau et al. 2003), and of enzymes responsible for fatty acid degradation (Kabine et al. 2003). Accordingly, the enzymes responsible for fatty acid utilization increase (Kabine et al. 2003); many mitochondria-encoded genes are up-regulated to facilitate the transport rate of fatty acids (Hittel and Storey 2002) and the thermogenic activity, especially in BAT (Liu et al. 2001). However, the mitochondrial proton conductance is unchanged during hibernation (Barger et al. 2003), supporting the idea that the reduced metabolism in hibernators is a partial consequence of tissue-specific depression of substrate oxidation (Martin et al. 1999).

However, some tissues, such as brain, need carbohydrates for their metabolic functions. Since the level of glycolytic activity remains low during hibernation and torpor (Heldmaier et al. 1999), the gluconeogenesis (the resynthesis, mainly in liver and kidney, of glucose from amino acids, lactate, glycerol) is the only means of regeneration of carbohydrate reserves during hibernation. Accordingly, the adrenocortical cells, involved in regulation of metabolic processes such as protein and lipid catabolism for gluconeogenesis, maintain an "active" arrangement (Malatesta et al. 1995; Zancanaro et al. 1997).

### *1.2.4 Cell metabolism*

The hibernator cells undergo striking seasonal morpho-functional modifications, due to the drastic reduction of metabolic activities. During hibernation, the proteosynthetic apparatus (RER and Golgi complex) undergoes reduction in various cell types (e.g. Malatesta et al. 1998, 2001, 2002), consistently with the drastic reduction in protein synthesis rate (Hittel and Storey 2002; van Breukelen and Martin 2002a). However, some proteins become more abundant during hibernation than in euthermia; for example the myoglobin of the skeletal muscle (Postnikova et al. 1999) as well as the intestinal stress protein GRP75 (Carey et al. 2000) are overexpressed in hibernating ground squirrels and the phosphoprotein named pp98 is specifically expressed in the brain only during hibernation (Ohtsuki et al. 1998).

The transcriptional activity too is severely inhibited during hibernation (van Breukelen and Martin 2002b). During hibernation a differential expression of several genes in various tissues occurs (see van Breukelen and Martin 2002a). Some genes can be up-regulated in some tissues while in others are down-regulated (e.g. Eddy and Storey 2003), or genes generally silenced in a tissue can be activated during hibernation (Andrews et al. 1998). Accordingly, cell nuclei of hibernating mammals undergo important structural reorganization during the hypometabolic period: various nuclear bodies containing transcription and splicing factors and even regulatory factors of the circadian rhythm (Zancanaro et al. 1993; Malatesta et al. 1994a, b, 1995, 1999, 2003; Tamburini et al. 1996) form during the lethargic period and disappear upon arousal (Malatesta et al. 1994a, 2001). In addition, chromatin organization changes suggesting a decrease in biosynthetic activities (Kolaeva et al. 1980; Giacometti et al. 1989).

### *1.2.5 Periodic arousals*

No mammalian hibernator remains continuously at low body temperature all winter long, but all rewarm and then briefly maintain euthermia at periodic intervals throughout the dormant season. The duration of these euthermic episodes depends on the animal size and the frequency of arousals is related to changes in ambient temperature (Strijkstra and Daan 1997).

The reason for such periodic arousals is still unknown, and the hypotheses are numerous and heterogeneous. Some authors suggested as a trigger for periodic arousal the accumulation of ketone bodies in the blood, due to a shutdown in mobilization and utilization of carbohydrates (Baumber et al. 1971); however, other authors found that elevated ketone levels are associated with increased survival time in hypoxia (D'Alecy et al. 1990). It has been suggested that periodic euthermic episodes may serve to replenish carbohydrate reserves (Galster and Morrison 1975) or to allow "normal" sleep (Daan et al. 1991). Other authors suggested that the periodic rewarming may be related to replacement of mRNA lost during torpor due to degradation processes

(Knight et al. 2000). It has been found that periodic arousals allow for degradation of ubiquitylated proteins accumulated during hibernation in the gut (van Breukelen and Carey 2002), thus restoring protein pools, as well as for the restoration of the host-defense mechanisms which appear to be downregulated during hibernation (Prendergast et al. 2002).

### 1.3 Hibernation triggers

It is evident that survival of natural hibernators under hypometabolic conditions implies the coordination of multiple new regulatory pathways involving different districts of the organism. The whole phenomenon probably occurs through serial metabolic events, but what are the factors responsible for the initiation of the cascade? Unfortunately, this fundamental question remains unanswered.

Among the different mechanisms reputed to be involved in hibernation (e.g. resetting of the temperature set-point, gene expression regulation, new metabolic balance), the potential role of several neuropeptides and transmitters can be regarded as a candidate for control systems of hibernation. In particular, endogenous opioids seem to play a basic role in regulating the hibernation cycle. In fact, it has been found that neuronal complexes immunoreactive for endogeneous opiates, especially enkephalin, and also for vasopressin, somatostatin, substance P, corticotropin-releasing factor and serotonin, are involved in the neuroendocrine control of hibernation (review in Nurnberger 1995). Opioids have been shown to cause changes similar to those observed during the annual cycle of mammalian hibernators, such as an increase in feeding at low doses and anorexia at high doses (Nizielski et al. 1986), bradicardia and hypotension (Kunos et al. 1988), respiratory suppression (Yeadon and Klitchen 1989), and lowering of set-point in thermoregulation (Burks 1991). Moreover, it has been shown that blockade of endogenous opioid activity by exogenous opioid antagonists shortens hibernation bouts or induces premature arousal (Wang 1993). Finally, opioids belonging to all of the three opioid families (proopiomelanocortin, proenkephalin and prodynorphin) have been found in the brain of ground squirrels (Cui et al 1996), showing specific increase in selected brain districts (hippocampus, septum, hypothalamus) during hibernation. All these data strongly suggest an involvement of endogenous opioids in regulating the hibernation cycle.

A natural proto-opioid whose chemical identity is not yet completely clarified has been found in blood and urine of lethargic—but not euthermic—hibernators like bats, brown bears and woodchucks (Bolling et al. 1997a). It is considered a powerful metabolic inhibitor since it can induce hypometabolic effects in hibernators when administered to active animals in summer, probably via the interaction with the peripheral and central opioid receptors (Benedict et al. 1999). Because of these capability, it has been called Hibernation Induction Trigger (HIT) and is at present the most interesting candidate as regulatory factor of natural hypometabolism.

## 2 Induced hypometabolism in mammals

Due to the lack of complete scientific data on mammalian natural hypometabolism, a large amount of speculation must be used in planning experimental designs to induce a reversible hypometabolic state in non-hibernating species. However, on the basis of the available knowledge, some selected mechanisms could conceivably lead to induce hypometabolism in mammals.

### 2.1 Changing body temperature

Lowering of ambient temperature could help inducing hibernation since, in general, decrease in body temperature parallels the fall in metabolic rate in hibernators (Geiser 2004). The decrease in body temperature is apparently a key factor to reduce the rate of metabolic and enzymatic activities on a purely thermodynamic drive; however, during entry into torpor metabolic rates drop rapidly even before a significant decrease in body temperature (Ortmann and Heldmaier 2000) suggesting that low temperature alone cannot explain this phenomenon. Moreover, when a non-hibernator is exposed to low environmental temperatures, body temperature begins to fall and hypothermia ensues; the homeostatic mechanism of shivering fails, the heart fibrillates and ventilation stops. Neuroprotective adaptations existing in hibernators have been reviewed by Drew et al. (2001).

Hibernating animals retain the ability to sense and defend body temperature; however, when they enter torpor their hypothalamic set-point for body temperature regulation is gradually lowered (Heller 1979). Therefore, resetting of the body temperature set point should help getting a hypometabolic state. It is generally accepted that the set-point is reset during fever, therefore it is regulatable. Unfortunately, the mechanism by which the set-point is determined is still a mystery (Cooper 2002). Possible targets are "cold" and "warm" thermosensitive neurones in the hypothalamus; their firing rate could be affected by thermoregulatory neurotransmitters such as catecholamines, serotonin, dopamine, GABA, glutamate, acetylcholine, or nitric oxide (Gertsberger 1999); however, all these molecules have many important effects on different body systems other than affecting temperature.

Maybe an ischemic preconditioning could help the induction of a hypometabolic state. In fact, recent research has shown that before entering the true hibernation phase, animals go through a number of cycles where metabolic rate and body temperature drop briefly. This could be a natural form of hypothermia preconditioning. Interestingly, studies in non-hibernators have shown that one or more short periods of ischaemia substantially improve the ability of cells and organs to tolerate a subsequent, longer period of ischaemia. For instance, repeated myocardial stunning in dogs (Di Carli et al. 2000) proved to be at the basis of prolonged and reversible reduction in systolic functions. In the same model, suspended animation by hypothermia can allow survival after a 60 min cardiac arrest (Nozari et al. 2004). In addition, recent reports indicate that induced hypothermia in swine does not affect learning and memory (Alam et al. 2002); moreover, it has a direct effect on the post shock immune/inflammatory system (Chen et al. 2005).

Lowering temperature could imply freezing process. However, some hibernators (nonmammals and mammals) are able to survive body temperature as low as −3°C by adopting the strategy of super cooling to resist freezing (Lee and Costanzo 1998). Frogs are able to produce and introduce in the bloodstream low molecular weight, cryoprotective carbohydrates (especially glycerol) before cooling (they are echtotherm animals). On the other hand, super cooling can be dangerous because it is a metastable condition where ice nucleation starts with ease. Interestingly, it has been found (Lee et al. 2000) that in rats infusion of the long-lasting volume expander hetastach was able to improve survival during hypothermia (22°C).

## 2.2 Changing metabolism

During torpor bouts, hibernating species are able to down regulate their cellular metabolic activity to a new hypometabolic steady state (down to 1/100th of basal metabolic rate) without damage during the prolonged cold exposure as well as during the transitions between warm and cold, and vice-versa.

Basically, a new balance between the ATP demand and the ATP supply is established (Boutilier 2001). Consequently, during natural hibernation mitochondrial activity drastically decreases and lipids become the preferential energy source (Geiser et al. 1994). On this basis, it could be hypothesized that a shift from carbohydrate to lipid utilization could be useful to promote a hypometabolic state. To this aim, hormones such as leptin, responsible for fatty acid mobilization (Rousseau et al. 2003), could represent a key factor. Moreover, during hibernation, several mitochondrial functions undergo modifications, in association with increased activity of the enzymes responsible for fatty acid transport and utilization (e.g. acyl-CoA oxidase, pyruvate dehydrogenase kinase) (Kabine et al. 2003; Hittel and Storey 2002). Changes in mitochondrial enzyme activity similar to those reported in hibernating animals have been observed in non-hibernators under conditions favouring a shift from carbohydrate to fatty acid oxidation. For example, pyruvate dehydrogenase kinase levels increase in humans fed on an isocaloric, high-fat diet (Peters et al. 2001), and acetyl-CoA carboxylase decreases during starvation, when lipid becomes the main fuel source. Therefore, the change in metabolic fuelling could facilitate torpor entrance. However, the activation/deactivation of mitochondrial functions in hibernators appears as a quite complex phenomenon in which numerous and various factors seem to be involved. In fact, in addition to mitochondrial enzyme activities modifications, calcium ions are responsible for the inactivation of the intramitochondrial ATPase, thereby preventing the exhaustion of cellular ATP in de-energized mitochondria (Bronnikov et al. 1990) and the intracellular pH also plays a regulatory role (Malan et al. 1988). Moreover, the drastic reduction in ATP production should be accompanied by a contemporary counterbalancing decrease in ATP consumption. In hibernating animals, all metabolic activities undergo a more or less profound depression: and the mechanisms responsible for the natural hypometabolic state identified so far are mostly based on the reversible phosphorylation of several regulatory enzymes (MacDonald and Storey 1999, Arendt et al. 2003) as well as on the differential enzyme control at different body temperatures (van Breukelen and Martin 2001). These mechanisms appear again very complex and based on the coordinated functional modifications of different factors.

Recent studies (Blackstone et al. 2005) have demonstrated that hydrogen sulphide ($H_2S$) administration is able to induce suspended animation by inhibition of the cytochrome C oxydase. Mice exposed to up to 80 ppm $H_2S$ had a 90% drop in their metabolic rate without showing any behavioural or functional damage at arousal.

### 2.3 Regulating gene expression

Some seasonal hibernators can enter the hibernation condition even in the absence of external input from environmental cues; this provides evidence that the ability to hibernate is driven by a molecular genetic mechanism rather than being an acute response to e.g., low ambient temperature. Therefore, interventions at the gene expression/transcription level could be envisaged in non-hibernators in order to achieve a hypometabolic state. However, when a gene modification is planned, any cascade effects at the systems level must be taken into account.

For instance, the expression of a particular factor involved in the induction of a hypometabolic state could be forced, or, more simply, fuel storage could be increased. In hibernators, lipid accumulation is achieved by a careful balance of increased plasma insulin in the presence of unaltered glucagon levels. According to the "sliding set-point" hypothesis (Mrosovsky and Fisher 1970), adipose tissue mass is controlled by a hypothalamic "lipostat" that senses body lipid content and initiates compensatory changes in appetite and energy expenditure to maintain a seasonally appropriate level of adiposity, leading to fat gain during late summer and autumn and loss during winter. Molecular support for a lipostat has come from the recent cloning of the *leptin* (*lep*) gene, which was first identified as the defective gene in obese ob/ob mice (Zhang et al. 1994). In humans, mice, and rats, blood concentrations of leptin, a 16 kD protein, are proportional to total body fat, and leptin production is hypothesized to function as a peripheral signalling component of the lipostat, with high levels causing decreased food intake and increased energy expenditure and low levels resulting in greater hunger and energy conservation (Stephens and Caro 1998).

Another interesting opportunity would be to induce the regulated expression $\alpha$2-macroglobulin, a protein playing an important role in preventing blood clotting (Srere et al. 1995). During induced torpor, in fact, blood circulation would be severely slowed down, and the risk of impairing microcirculation is high. The presence of this protein has been shown to enhance survival in hibernators.

Nevertheless, the simplest and more attractive possibility for inducing a hypometabolic state in non-hibernators would be represented by the modulated expression of HIT, the elusive protein factor present in the blood and urine of natural hibernators; HIT is reputed to initiate the cascade effect finally leading to hypometabolic state. The rationale for this intervention will be discussed in the next section.

## 2.4 Triggering hibernation

An accessible strategy to get an induced hypometabolic state in non-hibernators could be based on interference with primary cell functions responsible for the regulation of the whole cell metabolism.

There is clear evidence for depression in the cold and reactivation during arousal for protein translation and protein synthesis in natural hibernators, indicating a role for temperature in depressing transcriptional and translational activities. However, hibernators employ mechanisms to preserve mRNA pools that could aid in the resumption of gene expression during the interbout arousal when protein pools are replenished. Moreover, in hibernating animals cell nuclei undergo modifications of their constituents involved in RNA transcription and splicing (Zancanaro et al. 1993; Malatesta et al. 1994a, b, 1995, 1999, 2001, 2003; Tamburini et al. 1996), probably facilitating the transitions involved in the euthermia-hibernation-arousal cycle. Such a deep modification of cellular activities in hibernation needs a complex coordination, and HIT or other natural opioids can be regarded as candidates for such a role (see Sect. 1.3). HIT has never been completely characterized, and there are also contradictory results as to its efficacy. Nevertheless, in principle, a HIT-like molecule (i.e. a trigger) must exist. A recent report suggests that woodchuck plasma (containing HIT) is effective in protecting skeletal muscle from ischemia/reperfusion in non-hibernators (Hong et al. 2005).

Some recent papers have focussed attention on [D-Ala$^2$, D-Leu$^5$] enkephalin (DADLE), a synthetic delta opioid. DADLE is able to mimic the effects of HIT (Bolling et al. 1997a, b) as well as induce a hypometabolic hibernation-like state in hibernators (Oeltgen et al. 1988; Malatesta et al. 2001); moreover, it has yielded promising results in non-hibernators, also. The effects of DADLE have been investigated in in vitro systems, where the proliferation rate of different cell lines derived from hibernating (woodchuck) and non-hibernating (rat, human) species was significantly slowed down (Kampa et al. 1997). DADLE reduces both RNA transcription and export to cytoplasm; moreover, it probably provokes a cascade effect thus affecting other cellular functions such as protein synthesis and cell proliferation (Vecchio et al. 2006; Baldelli et al. 2004). Interestingly, no cytotoxic effects have been observed, according to previous studies demonstrating a certain antiapoptotic activity of DADLE (Tsao and Su 2001). Finally, after removing DADLE from the culture medium, the effects disappear quite rapidly (Vecchio et al. 2006).

It has been reported that DADLE activity is associated to binding delta opioid receptors (Benedict et al. 1999; Bolling et al. 1998); however, DADLE can be internalized in cells both exhibiting opioid receptors (not only of delta type) and devoid of such receptors (Baldelli et al. 2004, 2006; Vecchio et al. 2006). It should be noted that DADLE is a small, partially lipophylic molecule; moreover, the opioid peptides enkephalins interact with different subtypes of opioid receptors located in the plasma membrane and in the cell nucleus, as well as with polyspecific membrane transporters (see e.g. Hu et al. 2003); consequently, it is likely that specific receptors are not needed for DADLE crossing the cell membrane.

At the organ level, DADLE showed promising results in the experimental preservation of explanted organs: their survival time increased, and their morphological and functional preservation improved (Oeltgen et al. 1996; Bolling et al. 1997a, b; Su 2000). When injected intravenously in a multiorgan block preparation (heart, liver, lung and kidney), DADLE increases the survival time from 8 h to 46 h (Chien et al. 1994). In addition, DADLE is able to increase the survival time of single organs, such as heart and lung, which are quite delicate and difficult to preserve singly (Wu et al. 1996; Bolling et al. 1997a). This peptide can also ameliorate the functional recovery of heart tissue after prolonged ischemia induced in non-hibernating mammals (Bolling et al. 1997a, b). In this context, there are data in the literature showing that DADLE administration helps preserving cell ATP, thereby increasing myocardial tolerance to ischemia (Bolling et al. 1998). DADLE can also promote the survival of neurons in the CNS; this effect is achieved by e.g., contrasting the effects of methamphetamine (which is responsible of the destruction of dopaminergic terminals) through the modulated expression of necrosis tumour factor p53 and c-fos (Su 2000). Further, DADLE increases the survival time of the rat brain after hypoxia (Mayfield and D'Alecy 1994).

At the organism level, some evidence exists of a DADLE-induced hypometabolic hibernation-like state in natural hibernators during summertime (Oeltgen et al. 1988). In addition, DADLE administration can induce a short hypothermic effect in cold-exposed rabbits (Vybiral and Jansky 1997) and rats (Biggiogera et al. 2006), suggesting a possible hibernation mimicry in non-hibernating species.

Notwithstanding the promising results obtained with DADLE, the mechanisms of its action at the cell level as well on the whole organisms are still unclear and its effects as well as the activity of analogous molecules are currently under investigation.

## 3 Induced hypometabolism in the human

### 3.1 Hypometabolism in the human

Can we learn from mammalian hibernation how to achieve controlled metabolic depression/hypometabolism in the human? Metabolic depression in hibernation seems to be a robust evolutionary, entropy-slowing strategy permitting several species to survive hostile environments. Non-hibernating animals, inclusive of the human, do not present such ability, albeit they possess organs which are diverse as to ischaemia/hypoxia tolerance; for example, brain and heart are significantly damaged after minutes of oxygen lack whereas the gut and skeletal muscle tolerate well much longer periods of ischaemia (Hochachka et al. 1996). To exploit natural hibernation mechanisms in the human we should assume that hibernators and man share genes common to all mammals which are differently expressed in response to ambient and/or internal stimuli; this is actually the case according to some investigators (Srere et al. 1992; Carey et al. 2003). Interestingly enough, recent evidence showed the ability of a primate, the Madagascan fat-tailed dwarf lemur, *Cheirogaleus medius*, to suspend endogenous thermogenesis and hibernate even at 30°C ambient temperature (Dausmann et al. 2004).

The potential of a controlled, artificially induced hypometabolic state in the human is obviously enormous; it should be underlined that a deep hibernating state similar to that of small mammals (very low body temperature, a few heart beats per minute, striking reduction of brain electrical activity) is not needed to get consistent benefits in the human; indeed, a controlled mild hypometabolic state would pay enough in terms of energy saving. Several possible applications of induced hypometabolism are envisaged in relation with human life in extreme environments/conditions e.g., space exploration; of course, medical and military applications are foreseen, also.

### 3.2 Potential applications of hypometabolism in space exploration

During manned space missions, astronauts are faced with prolonged low gravity and reduced physical activity; this has negative effects on the musculoskeletal apparatus, leading to reduced muscle and bone mass. Skeletal muscle adaptations include changes in the expression of metabolic, structural, and contractile proteins (Fitts et al. 2001); moreover, weightlessness during space flight results in calcium, vitamin D, and vitamin K deficiency, increased urinary calcium excretion, decreased intestinal calcium absorption, and increased serum calcium level, with decreased levels of serum parathyroid hormone and calcitriol. Bone resorption is increased, whereas bone formation is decreased (Oganov 2004). Induced hypometabolism could be an effective countermeasure, as large hibernating mammals like bears are able to prevent disuse atrophy in the bone (Donahue et al. 2005) and the skeletal muscle (Tinker et al. 1998) A further advantage would be reducing the overall energy (food) requirement by the crew, thereby increasing the payload. Moreover, it is well known that interpersonal problems have been a recurring problem for long-duration space flights, and even after completion of the space mission, intense psychological effects are reported (Collins 2003). Therefore, psychosocial elements of behaviour and performance are likely to have a significant impact on the outcome of long-duration missions in space (Palinkas 2001). Maintaining part of the crew under hypometabolic conditions i.e., unconscious, for long time would help with the stress associated with long-term missions. Of course, several medical and technical problems must be addressed and solved before this option come of time (Ayre et al. 2004).

### 3.3 Potential applications of hypometabolism in biomedicine

Inducing a controlled, reversible hypometabolic state would have profound impact on human medicine. A few of the more relevant issues are indicated below.

- Improving cadaver/organ preservation for transplant. A key factor to improve transplantation success rates is extending storage times before transplantation. Currently, refrigeration and perfusion with preservation solutions are used, but viability of donor organs is limited to hours. Inducing a hypometabolic state in the intact donor would obviously have impact on this limitation; there is available evidence that livers from hibernating ground squirrels tolerate cold ischemia in current preservation solution followed by warm reperfusion for longer times and with better quality than livers from rats or summer squirrels (Lindell et al. 2005); therefore, applied treatments are envisaged that could prolong the viability of excised organs in cold storage and/or improve recovery of function after implantation (Storey 2004). Hibernation-related opioids are good candidates (see Sect. 2.4) in this setting.
- Neuro/cardioprotection following ischaemic accidents. Brain and heart ischaemia is a major cause of morbidity and mortality (stroke, myocardial infarction). Hibernators are tolerant to ischaemia as they can survive for long periods at a greatly reduced blood flow in the brain with no evidence of neurological damage; further, they show an ability to protect brain tissue against traumatic injury (Drew et al. 2001) during dormancy. Therefore, understanding and transferring the mechanism(s) turning down the brain's requirement for oxygen delivery and blood flow, and/or mimicking hibernation would allow for extra time to lyse a clot after a stroke or reducing neuronal damage in the "penumbra" region surrounding brain haemorrhage. Currently, the only therapies proven to improve stroke outcome involve reversing the arterial blockage. Chemical neuroprotection revealed to be not effective and exogenous hypothermia yielded conflicting results (Clifton et al. 2001; Hypothermia after cardiac arrest study group 2002).
- Improving major surgical procedures requiring hypothermia. Moderate to profound hypothermia is a major current procedure in open-heart surgery, repairs of the aorta, and coronary bypass grafting; however, these procedures are associated with up a 30% incidence of post-operative neurocognitive impairment and up to 9% incidence of stroke (Ahonen and Salmenpera 2004). It should be underlined that hypothermia is induced in these patients essentially by cooling i.e., by means of a physical, non-physiologically controlled mechanism. Cooling (and rewarming) of patients can be cumbersome. Vasoconstriction can significantly reduce the efficacy of surface cooling and the administration of anaesthetics or alpha-2 adrenergic agonists to address this problem may lead to hypotension. Rewarming takes time and may not always be achievable within appropriate time frames. Moreover, shivering is likely to accompany hypothermia; while shivering can be prevented by drugs e.g., meperidine (Giesbrecht et al. 1997), such a treatment will add significant risk to the whole procedure. Therefore, induction of a controlled, physiologically driven hypothermia would add much to hypothermic surgery. Lessons learned from how hibernators obtain reversible lowering of core body temperature and especially their ability to manage calcium homeostasis at low temperature (Wang et al. 2002) would help designing strategies to improve hypothermia tolerance at the cell and organism level.

– Improving war surgery. In principle, induced hypometabolism would help military surgeons with battlefield emergencies involving profuse bleeding and/or requiring major surgery in the zone behind the front by reducing haemorrhage and "stabilizing" the patient until arrival to the hospital.

It is apparent from the above discussion that clarifying and/or mimicking the mechanisms of mammalian hibernation in order to reproduce a controlled hypometabolism state in the human is of potential great interest in several settings involving human life under extreme conditions.

**Note added in proof** In a very recent paper (Kondo et al. 2006), a hibernation-specific protein (HP) has been identified as a candidate hormone for hibernation. This protein occurs in the blood of active chipmunks but, during hibernation, it accumulates in the brain, where it probably plays an essential regulatory role.

## References

Ahonen J, Salmenpera M (2004) Brain injury after adult cardiac surgery. Acta Anaesthesiol Scand 48:4–19

Alam HB, Bowyer MW, Koustova E, Gushchin V, Anderson D, Stanton K, Kreishman P, Cryer CM, Hancock T, Rhee P (2002) Learning and memory is preserved after induced asanguineous hyperkalemic hypothermic arrest in a swine model of traumatic exsanguination. Surgery 132:278–288

Andrews MT, Squire TL, Bowen CM, Rollins MB (1998) Low-temperature carbon utilization is regulated by novel gene activity in the heart of a hibernating mammal. Proc Natl Acad Sci USA 95:8392–8397

Arendt T, Stieler J, Strijkstra AM, Hut RA, Rudiger J, Van der Zee EA, Harkany T, Holzer M, Hartig W (2003) Reversible paired helical filament-like phosphotylation of tau is an adaptive process associated with neuronal plasticity in hibernating animals. J Neurosci 23:6972–6981

Ayre M, Zancanaro C, Malatesta M (2004) Morpheus-hypometabolic stasis for long-term spaceflight. J Brit Interplan Soc 57:325–339

Baldelli B, Vecchio L, Biggiogera M, Vittoria E, Muzzonigro G, Gazzanelli G, Malatesta M (2004) Ultrastructural and immunocytochemical analyses of opioid treatment effects on PC3 prostatic cancer cells. Microsc Res Tech 64:243–249

Baldelli B, Vecchio L, Bottone MG, Muzzonigro G, Biggiogera M, Malatesta M (2006) The effect of the enkephalin DADLE on transcription does not depend on opioid receptors. Histochem Cell Biol. In press

Barger JL, Brand MD, Barnes BM, Boyer BB (2003) Tissue-specific depression of mitochondrial proton leak and substrate oxidation in hibernating arctic ground squirrels. Am J Physiol Regul Integr Comp Physiol 284:R1306–R1313

Baumber J, South FE, Ferren L, Zatznan ML (1971) A possible basis for periodic arousals during hibernation: accumulation of ketone bodies. Life Sci 10:462–467

Benedict PE, Benedict MB, Su TP, Bolling SF (1999) Opiate drugs and delta-receptor-mediated myocardial protection. Circulation 100:357–360

Biggiogera M, Fabene P, Zancanaro C (2006) DADLE: a cue to human hibernation? J Brit Interplan Soc 59:115–118

Blackstone E, Morrison M, Roth MB (2005) H2S induces a suspended animation-like state in mice. Science 308:518

Bolling K, Kronon M, Allen BS, Wang T, Ramon S, Feinberg H (1997a) Myocardial protection in normal and hypoxically stressed neonatal hearts: the superiority of blood versus crystalloid cardioplegia. J Thorac Cardiovasc Surg 113:994–1003

Bolling K, Halldorsson A, Allen BS, Rahman S, Wang T, Kronon M, Feinberg H (1997b) Prevention of the hypoxic reoxygenation injury with the use of a leukocyte-depleting filter. J Thorac Cardiovasc Surg 113:1081–1089

Bolling SF, Benedict MB, Tramontini NL, Kilgore KS, Harlow HH, Su TP, Oeltgen PR (1998) Hibernation triggers and myocardial protection. Circulation 98:220–223

Boswell T, Richardson RD, Schwartz MW, D'Alessio DA, Woods SC, Sipols AJ, Baskin DG, Kenagy GJ (1993) NPY and galanin in a hibernator: hypothalamic gene expression and effects on feeding. Brain Res Bull 32:379–384

Boutilier RG (2001) Mechanisms of cell survival in hypoxia and hypothermia. J Exp Biol 204:3171–3181

Bronnikov GE, Vinogradova SO, Chernyak BV (1990) Regulation of ATP hydrolysis in liver mitochondria from ground squirrel. FEBS 266:8386

Buck CL, Barnes BM (2000) Effects of ambient temperature on metabolic rate, respiratory quotient, and torpor in an arctic hibernator. Am J Physiol Regul Integr Comp Physiol 279:R255–R262

Burks TF (1991) Opioid and opioid receptors in thermoregulation. In: Thermoregulation:pathology, pharmacology, biosynthesis and analysis. Pergamon Press, NY, pp 489–508

Carey HV, Frank CL, Seifert JP (2000) Hibernation induces oxidative stress and activation of NK-kappaB in ground squirrel intestine. J Comp Physiol 170B:551–559

Carey HV, Andrews MT, Martin SL (2003) Mammalian hibernation: cellular and molecular responses to depressed metabolism and low temperature. Physiol Rev 83:1153–1181

Chen Z, Chen H, Rhee P, Koustova E, Ayuste EC, Honma K, Nadel A, Alam HB (2005) Induction of profound hypothermia modulates the immune/inflammatory response in a swine model of lethal hemorrhage. Resuscitation 66:209–216

Chien SF, Oeltgen PR, Diana JN, Salley RK, Su TP (1994) Extension of tissuesurvival time in multiorgan block preparation using a delta opioid DADLE. J Thorac Cardiovasc Surg 107:964–967

Clifton GL, Miller ER, Choi SC, Levin HS, McCauley S, Smith KR Jr, Muizelaar JP, Wagner FC Jr, Marion DW, Luerssen TG, Chesnut RM, Schwartz M (2001) Lack of effect of induction of hypothermia after acute brain injury. N Engl J Med 344:556–563

Collins DL (2003) Psychological issues relevant to astronaut selection for long-duration space flight: a review of the literature. Hum Perf Extrem Environ Spring 7:43–67

Cooper KE (2002) Molecular biology of thermoregulation – some historical perspectives on thermoregulation. J Appl Physiol 92:1717–1724

Cui Y, Lee TF, Wang LCH (1996) State-dependent changes of brain endogenous opioids in mammalian hibernation. Brain Res Bull 40:129–133

D'Alecy LG, Lundy EF, Kluger MJ, Harker CT, LeMay DR, Shlafer M (1990) Beta-hydroxybutyrate and response to hypoxia in the ground squirrel, *Spermophilus tridecimlineatus*. Comp Biochem Physiol 96B:189–193

Daan S, Bernes BM, Strijkstra AM (1991) Warming up for sleep? Ground squirrels sleep during arousal from hibernation. Neurosci Lett 128:265–268

Dausmann KH, Glos J, Ganzhorn JU, Heldmaier G (2004) Hibernation in a tropical primate. Nature 429:825–826

Di Carli MF, Prcevski P, Singh TP, Janisse J, Ager J, Muzik O, Vander Heide R (2000) Myocardial blood flow, function, and metabolism in repetitive stunning. J Nucl Med 41:1227–1234

Donahue SW, McGee ME, Harvey KB, Vaughan MR, Robbins CT (2005) Hibernating bears as a model for preventing disuse osteoporosis. J Biomech doi:10.1016/j.jbiomech.2005.03.030

Drew KL, Rice ME, Kuhn TB, Smith MA (2001) Neuroprotective adaptations in hibernation: therapeutic implications for ischemia-reperfusion, traumatic brain injury and neurodegenerative diseases. Free Radic Biol Med 31:563–573

Eddy SF, Storey KB (2003) Differential expression of Akt, PPARgamma, and PGC-1 during hibernation in bats. Biochem Cell Biol 81:269–274

Fitts RH, Riley DR, Widrick JJ (2001) Functional and structural adaptations of skeletal muscle to microgravity. J Exp Biol 204:3201–3208

French AR (1988) The patterns of mammalian hibernation. Am Sci 76:569–575

Galster W, Morrison PR (1975) Gluconeogenesis in arctic squirrels between periods of hibernation. Am J Physiol 228:325–330

Geiser F (2004) Metabolic rate and body temperature reduction during hibernation and daily torpor. Annu Rev Physiol 66:239–274

Geiser F, McAllan BM, Kenagy GJ (1994) The degree of dietary fatty acid unsaturation affects torpor patterns and lipid composition of a hibernator. J Comp Physiol 164:299–305

Gertsberger R (1999) Nitric oxide and body temperature control. News Physiol Sci 14:30–36

Giesbrecht GG, Goheen MS, Johnston CE, Kenny GP, Bristow GK, Hayward JS (1997) Inhibition of shivering increases core temperature afterdrop and attenuates rewarming in hypothermic humans. J Appl Physiol 83:1630–1634

Giacometti S, Scherini E, Bernocchi G (1989) Seasonal changes in the nucleoli of Purkinje cells of the hedgehog cerebellum. Brain Res 488:365–368

Heldmaier G, Steinlechner S, Ruf T, Wiesinger H, Klingenspor M (1989) Photoperiod and thermoregulation in vertebrates: body temperature rhythms and thermogenic acclimation. J Biol Rhytms 4:251–265

Heldmaier G, Klingenspor M, Werneyer M, Lampi BJ, Brooks SP, Storey KB (1999) Metabolic adjustements during daily torpor in the Djungarian hamster. Am J Physiol 276:E896–E906

Heller HC (1979) Hibernation: neural aspects. Annu Rev Physiol 41:305–321

Himms-Hagen J (1986) Brown adipose tissue and cold acclimatation. In: Trayhurn P, Nicholls DG (eds) Brown adipose tissue. Edward Arnold, London, pp 214–268

Hittel DS, Storey KB (2002) The translation state of differentially expressed mRNAs in the hibernating 13-lined ground squirrel (*Spermophilus tridecemlineatus*). Arch Biochem Biophys 401:244–254

Hochachka PW, Buck L, Doll C, Laand S (1996) Unifying theory of hypoxia tolerance: defense and rescue mechanisms for surviving oxygen lack. Proc Natl Acad Sci USA 93:9493–9498

Hong J, Sigg DC, Coles JA Jr, Oeltgen PR, Harlow HJ, Soule CL, Iaizzo PA (2005) Hibernation induction trigger reduces hypoxic damage of swine skeletal muscle. Muscle Nerve 32:200–207

Horwitz BA, Hamilton JS, Kott KS (1985) GDP binding to hamster brown fat mitochondria is reduced during hibernation. Am J Physiol 249:R689–R693

Hu H, Miyauchi S, Bridges CC, Smith SB, Ganapathy V (2003) Identification of a novel $Na^+$-and $Cl^-$-coupled transport system for endogenous opioid peptides in retinal pigment epithelium and induction of the transport system by HIV-1 Tat. Biochem J 375:17–22

Hypothermia after Cardiac Arrest Study Group (2002) Mild therapeutic hypothermia to improve the neurologic outcome after cardiac arrest. N Engl J Med 346:549–56. Erratum in: N Engl J Med 2002 346:1756

Johansson BW (1996) The hibernator heart – nature's model of resistance to ventricular fibrillation. Cardiov Res 31:826–832

Kabine M, Clemencet MC, Bride J, El Kebbaj M-S, Latruffe N, Cherkaoui-Malki M (2003) Changes of peroxisomal fatty acid metabolism during cold acclimatization in hibernating jerboa (*Jaculus orientalis*). Biochimie 85:707–714

Kampa M, Bakogeorgou E, Hatzoglou A, Damianaki A, Martin P-M, Castanas E (1997) Opioid alkaloids and casomorphin peptides decrease the proliferation of prostatic cancer cell lines (LNCaP, PC3 and DU145) through a partial interaction with opioid receptors. Eur J Pharmacol 335:255–265

Knight JE, Narus EN, Martin SL, Jacobson A, Barnes BM, Boyer BB (2000) mRNA stability and polysome loss in hibernating Arctic ground squirrels (*Spermophilus parryii*). Mol Cell Biol 20:6374–6379

Kolaeva SG, Kramarova LI, Ilyasova EN, Ilyasova FE (1980) The kinetics and metabolism of the cells of hibernating animals during hibernation. Int Rev Cytol 66:148–169

Kondo N, Sekijima T, Kondo J, Takamatsu N, Tohya K, Ohtsu T (2006) Circannual control of hibernation by HP complex in the brain. Cell 125:161–172

Körtner G, Geiser F (2000) The temporal organization of daily torpor and hibernation: circadian and circannual rhythms. Chronobiol Int 17:103–109

Kunos G, Mosqueda-Garcia R, Mastrianni JA (1988) Endophinergic neurons in the brainstem and the control of blood pressure and heart rate. In: Illes P, Farsang C (eds) Regulatory roles of opioid peptides. VCH, Weinheim, pp 460–470

Lee RE Jr, Costanzo JP (1998) Biological ice nucleation and ice distribution in cold-hardy ectothermic animals. Annu Rev Physiol 60:55–72

Lee TF, Westly J, Wang LCH (2000) Effects of hetastarch and mannitol on prolonging survival in stable hypothermia in rats. Am J Physiol Reg Integr Comp Physiol 278:R1040–R1047

Lindell SL, Klahn SL, Piazza TM, Mangino MJ, Torrealba JR Southard JH, Carey HV (2005) Natural resistance to liver cold ischemia-reperfusion injury associated with the hibernation phenotype. Am J Physiol Gastrointest Liver Physiol 288:G473–G480

Liu XT, Li QS, Lin QF, Sun RY (2001) Uncoupling protein 1 mRNA, mitochondrial GTP-binding, and T4 5′-deiodinase of brown adipose tissue in euthermic Daurian ground squirrel during cold exposure. Comp Biochem Physiol A Mol Integr Physiol 128:827–835

Lyman CP, Willis JS, Malan A, Wang LCH (1982) Hibernation and topor in mammals and birds. Accademic Press, New York

MacDonald JA, Storey KB (1999) Regulation of ground squirrel $Na^{+}K^{+}$-ATPase activity by reversible phosphorylation during hibernation. Biochem Biophys Res Commun 254:424–429

Malan A, Mioskowski E, Calgari C (1988) Time-course of blood acid-base state during arousal from hibernation in the European hamster. J Comp Physiol 158B:495–500

Malatesta M, Zancanaro C, Martin TE, Chan EKL, Amalric F, Lührmann R, Vogel P, Fakan S (1994a) Is the coiled body involved in nucleolar functions? Exp Cell Res 211:415–419

Malatesta M, Zancanaro C, Martin TE, Chan EKL, Amalric F, Lührmann R, Vogel P, Fakan S (1994b) Cytochemical and immunocytochemical characterization of nuclear bodies during hibernation. Eur J Cell Biol 65:82–93

Malatesta M, Zancanaro C, Tamburini M, Martin TE, Fu X-D, Vogel P, Fakan S (1995) Novel nuclear ribonucleoprotein structural components in the dormouse adrenal cortex during hibernation. Chromosoma 104:121–128

Malatesta M, Zancanaro C, Marcheggiani F, Cardinali A, Rocchi MBL, Capizzi D, Vogel P, Fakan S, Gazzanelli G (1998) Ultrastructural, morphometrical and immunocytochemical analyses of the exocrine pancreas in a hibernating dormouse. Cell Tissue Res 292:531–541

Malatesta M, Cardinali A, Battistelli S, Zancanaro C, Martin TE, Fakan S, Gazzanelli G (1999) Nuclear bodies are usual constituents in tissues of hibernating dormice. Anat Rec 254:389–395

Malatesta M, Gazzanelli G, Marcheggiani F, Zancanaro C, Rocchi MBL (2001) Ultrastructural characterisation of periinsular pancreatic acinar cells in the hibernating dormouse *Muscardinus avellanarius*. It J Zool 68:101–106

Malatesta M, Zancanaro C, Baldelli B, Gazzanelli G (2002) Quantitative ultrastructural changes of hepatocyte constituents in euthermic, hibernating and arousing dormice (*Muscardinus avellanarius*). Tissue Cell 34:397–405

Malatesta M, Baldelli B, Rossi L, Serafini S, Gazzanelli G (2003) Fine distribution of clock protein in hepatocytes of hibernating dormice. Eur J Histochem 47:233–240

Martin SL, Maniero GD, Carey C, Hand SC (1999) Reversible depression of oxygen consumption in isolated liver mitochondria during hibernation. Physiol Biochem Zool 72:255–264

Mayfield KP, D'Alecy LG (1994) Delta 1 opioid receptor dependence of acute hypoxic adaptation. J Pharmacol Exp Ther 268:74–77

Mostafa N, Everett DC, Chou SC, Kong PA, Florant GL, Coleman RA (1993) Seasonal changes in critical enzymes of lipogenesis and triacylglycerol syntesis in the marmot (*Marmota flaviventris*). Comp Physiol 163:463–469

Mrosovsky N, Fisher KC (1970) Sliding set points for body weight in ground squirrels during the hibernation season. Can J Zool 48:241–247

Nizielski SE, Levine AS, Morley GE, Hall KA, Gosnell BA (1986) Seasonal variation in opioid modulation of feeding in the 13-lined ground squirrel. Physiol Behav 37: 5–9

Nozari A, Safar P, Wu X, Stezoski WS, Henchir J, Kochanek P, Klain M, Radovsky A, Tisherman SA (2004) Suspended animation can allow survival without brain damage after traumatic exsanguination cardiac arrest of 60 minutes in dogs. J Trauma 57:1266–1275

Nurnberger F (1995) The neuroendocrine system in hibernating mammals: present knowledge and open questions. Cell Tissue Res 281:391–412

O'Hara BF, Watson FL, Srere HK, Kumar H, Wiler SW, Welch SK, Bitting L, Heller HC, Kilduff TS (1999) Gene expression in the brain across the hibernation cycle. Neuroscience 19:3781–3790

Oganov VS (2004) Modern analysis of bone loss mechanisms in microgravity. J Gravit Physiol 11:P143–146

Oeltgen PR, Nuchols PA, Nilekani WA, Spurrier WA, Su TP (1988) Further studies on opioids and hibernation: delta opioid receptor ligand selectively induced hibernation in summer active ground squirrels. Life Sci 43:1565–1574

Oeltgen PR, Horton ND, Bolling SF, Su TP (1996) Extended lung preservation with the use of hibernation trigger factors. Ann Thorac Surg 61:1488–1493

Ohtsuki T, Jaffe H, Brenner M, Azzam R, Frerichs KU, Hallenbeck JM (1998) Stimulation of tyrosine phosphporylation of a brain protein by hibernation. Cereb Blood Flow Metab 18:1040–1045

Ortmann S, Heldmaier G (2000) Regulation of body temperature and energy requirements of hibernating Alpine marmots (*Marmota marmota*). Am J Physiol Regul Integr Comp Physiol 278:R689–R704

Pakotin PL, Pakotina ID, Belusov AB (1993) The study of brain slices from hibernating mammals in vitro and some approaches to the analysis of hibernation problems in vivo. Progr Neurobiol 40:123–161

Palinkas LA (2001) Psychosocial issues in long-term space flight: overview. Gravit Space Biol Bull 14:25–33

Peters SJ, Harris RA, Wu P, Pehleman TL, Heigenhauser GJ, Spriet LL (2001) Human skeletal muscle PDH kinase activity and isoform expression during a 3-day high-fat/low-carbohydrate diet. Am J Physiol Endocrinol Metab 281:E1151–E1158

Postnikova GB, Tselikova SV, Kolaeva SG, Solomonov NG (1999) Myoglobin content in skeletal muscles of hibernating ground squirrels rises in autumn and winter. Comp Biochem Physiol A Mol Integr Physiol 124:35–37

Prendergast BJ, Freeman DA, Zucker I, Nelson RJ (2002) Periodic arousal from hibernation is necessary for initation of immune responses in ground squirrels. Am J Physiol Regul Integr Comp Physiol 282:R1054–R1062

Rousseau K, Archa Z, Loudon AS (2003) Leptin and seasonal mammals. Neuroendocrinol 15:409–414

Srere HK, Wang LCH, Martin SL (1992) Central role for differential gene expression in mammalian hibernation. Proc Natl Acad Sci USA 89:7119–7123

Srere HK, Belke D, Wang LC, Martin SL (1995) α2-Macroglobulin gene expression during hibernation in ground squirrel (*Citellus tridecemlineatus*). Am J Physiol 268:R1507–R1512

Stephens TW, Caro JF (1998) To be lean or not to be lean: is leptin the answer? Exp Clin Endocrinol Diabetes 106:1–15

Storey KB (2004) Cold ischemic organ preservation: lessons from natural systems. J Invest Med 52:315–322

Strijkstra AM, Daan S (1997) Ambient temperature during torpor affects NREM sleep EEG during arousal episodes in hibernating European ground squirrels. Neurosci Lett 221:177–180

Su TP (2000) Delta opioid peptide[D-Ala(2),D-Leu(5)]enkephalin promotes cell survival. J Biomed Sci 7:195–199

Tamburini M, Malatesta M, Zancanaro C, Martin TE, Fu XD, Vogel P, Fakan S (1996) Dense granular bodies: a novel nucleoplasmic structure in hibernating dormice. Histochem Cell Biol 106:581–586

Tinker DB, Harlow HJ, Beck TD (1998) Protein use and muscle-fiber changes in free-ranging, hibernating black bears. Physiol Zool 71:414–424

Tsao LI, Su TP (2001) Hibernation-induction peptide andcell death: [D-Ala2,D-Leu5]enkephalin blocks Bax-related apoptotic processes. Eur J Pharmacol 428:149–151

van Breukelen F, Martin SL (2001) Translational initiation is uncoupled from elongation at 18 degrees C during mammalian hibernation. Am J Physiol Regul Integr Comp Physiol 281:R1374–R1379

van Breukelen F, Carey V (2002) Ubiquitin conjugate dynamics in the gut and liver of hibernating ground squirrels. J Comp Physiol 172B:269–273

van Breukelen F, Martin SL (2002a) Invited review: molecular adaptations in mammalian hibernators: unique adaptations or generalized responses? J Appl Physiol 92:2640–2647

van Breukelen F, Martin SL (2002b) Reversible depression of transcription during hibernation. J Comp Physiol 172B:R1374–R1379

Vecchio L, Soldani C, Bottone MG, Malatesta M, Martin TE, Rothblum LI, Pellicciari C Biggiogera M (2006) DADLE induces a reversible hibernation-like state in HeLa cells. Histochem Cell Biol 125:193–201

Vybiral S, Jansky L ((1997) Hibernation triggers and cryogens: do they play a role in hibernation? Comp Biochem Physiol A Physiol 118:1125–1133

Wang LCH (1993) Is endogenous oipoid involved in hibernation? In: Cynthia C et al. (eds) Life in the cold: Ecological, physiological, and molecular mechanisms. Westview Press, Colorado, pp 297–304

Wang SQ, Lakatta EG, Cheng H, Zhou ZQ (2002) Adaptive mechanisms of intracellular calcium homeostasis in mammalian hibernators. J Exp Biol 205:2957–2962

Wu G, Zhang F, Salley RK, Diana JN, Su TP, Chien S (1996) delta Opioid extends hypothermic preservation time of the lung. J Thorac Cardiovasc Surg 111:259–267

Yeadon M, Kitchen I (1989) Opioids and respiration. Progr Neurobiol 33:1–16

Zancanaro C, Malatesta M, Mannello F, Vogel P, Fakan S (1999) The kidney during hibernation and arousal from hibernation A natural model of organ preservation during cold ischaemia and reperfusion. Nephrol Dial Transpl 14:1982–1990

Zancanaro C, Malatesta M, Vogel P, Fakan S (1997) Ultrastructure of the adrenal cortex of hibernating, arousing and euthermic dormouse, *Muscardinus avellanarius*. Anat Rec 249:359–364

Zancanaro C, Malatesta M, Vogel P, Osculati F, Fakan S (1993) Ultrastructural and morphometrical analyses of the brown adipocyte nucleus in a hibernating dormouse. Biol Cell 79:55–61

Zatzman ML (1984) Renal and cardiovascular effect of hibernation. Cryobiology 21:593–614

Zhang YY, Proenca R, Maffei M, Barone M, Leopold L, Friedman JM (1994) Positional cloning of the mouse obese gene and its human homologue. Nature 372:425–432

Rev Environ Sci Biotechnol (2007) 6:61–70
DOI 10.1007/s11157-006-9103-2

REVIEW PAPER

# Microbial ecology of submerged marine caves and holes characterised by high levels of hydrogen sulphide

**Francesco Canganella · Giovanna Bianconi · Chiaki Kato · Juan Gonzalez**

Received: 6 March 2006 / Accepted: 28 July 2006 / Published online: 2 December 2006


**Abstract** Submarine caves, cavities and niches characterised by hydrogen sulphide ($H_2S$) elevated concentrations are particularly interesting for their inhabiting microflora as well as for the overall chemical, geological and biological parameters. These ecosystems are usually populated by well-adapted living forms, physically distributed following the in situ concentration and gradient of micronutrients, $O_2$ and $H_2S$, and also according to the values of temperature and pH. The biota is primarily characterised by prokaryotes (both autotrophic and heterotrophic) adapted to anoxic and/or microaerophilic condition and capable to form extensive biofilms on the rocky surfaces and even on the bottom sediment. These habitats can be defined as extreme, because the scarcity or absence of solar irradiation, the chemo-physical traits and the fact that specialised prokaryotes are often the only inhabitants. This review is focused on the microbial ecology of marine caves and holes characterised by high levels of $H_2S$. Ecological and geological data are already available but very few insights as far as regard microbiology were achieved in order to describe these fascinating habitats. The autochthonous mesophilic and thermotolerant microorganisms living in these caves may have interesting physiological traits and eventually may lead to potential application in biotechnological processes.

F. Canganella (✉)
Department of Agrobiology and Agrochemistry, University of Tuscia, Viterbo, Italy
e-mail: canganella@unitus.it

F. Canganella · G. Bianconi
Marine Microbiology Lab – CONISMA, University of Tuscia, Viterbo, Italy

F. Canganella · G. Bianconi
Research Group in ExoAgroBiology, University of Tuscia, Viterbo, Italy

C. Kato
Japan Agency for Marine Science and Technology Center, Yokosuka, Japan

J. Gonzalez
IRNAS-CSIC, Sevilla, Spain

**Keywords** Marine anoxic and microaerophilic ecosystems · Microbial ecology · Microorganisms · Submarine caves · Black and blue holes · Molecular diversity

## 1 Introduction

### 1.1 Blue holes and black holes

Submerged marine holes are geological structures which developed within carbonate platforms leading to complex cave systems below sea level. The so-called "blue holes" are underwater karst systems which can develop horizontally in a very extensive way and appear blue due to a combination of blue sky reflection with the white carbonate sand deposited in the cave (Smart et al. 1988; Mylroie et al. 1995; Schwabe et al. 1997; Marano-Briggs 2000; Colantoni et al. 2003; Canganella et al. 2004). Most of these sites were found and described at Bahamas and Hawai.

Other cavern systems are known as "black holes" which are cave systems without known lateral passages and located in the interior regions of the larger Bahamian islands. They are found where the land is transitional between submerged and dry land, not dissimilar to saltmarsh environments. Most Black Holes are located in the central to western side of the island of South Andros in the Bahamas (Fig. 1), although one has been found on the northern transitional shore of Grand Bahama Island. They are vertical cave systems which develop from the surface downwards and appear to have no direct link to the sea except through rock fissures and local porosity (Schwabe 1998; Schwabe and Herbert 2004).

In both cave systems as water exchange is severely restricted, physico-chemical gradients are highly stable and even strict anaerobic conditions can develop with remarkable hydrogen sulphide ($H_2S$) concentrations (from 5 μM to 6 mM).

## 1.2 Submarine caves with sulphidic water springs

An unique marine ecosystem is represented by the submarine caves of Capo Palinuro (Salerno, Italy). In this limestone massif open 32 caves, completely or partially inundated by the sea, of which 13 have inside sulphidic springs exhibiting temperatures up to 25°C that are responsible for important biological and geochemical phenomena.

Preliminary investigations concerning the submarine speleology of the site have been carried out since the mid eighties, but only after 1990 an extensive biological knowledge was achieved. Several ecological and geological studies were performed (Alvisi et al. 1994a, b; Mattison et al. 1998) particularly on the "Grotta Azzurra" (Fig. 2). The cave may be separated into two topographically distinct regions: a weakly illuminated outer region and an inmost dark region (Snow Hall). The latter is characterised by the presence of sulphidic springs that arise from fissures in the cave floor.

Currently no other submarine caves with sulphidic springs are ever been reported.

## 1.3 Other marine environments with stratified water/redox sulphur gradients

As far as regard the presence of stratified sulphur gradients in marine environments, several basins have been described and investigated during the last decades, showing very interesting geological aspects but also the significant role played by the autochthonous microflora.

The Santa Barbara Basin, one of the Southern California borderland basins, has been extensively investigated and dissolved sulphide distributions have been described (Kuwabara et al. 1999). Moreover, its marine sediments were

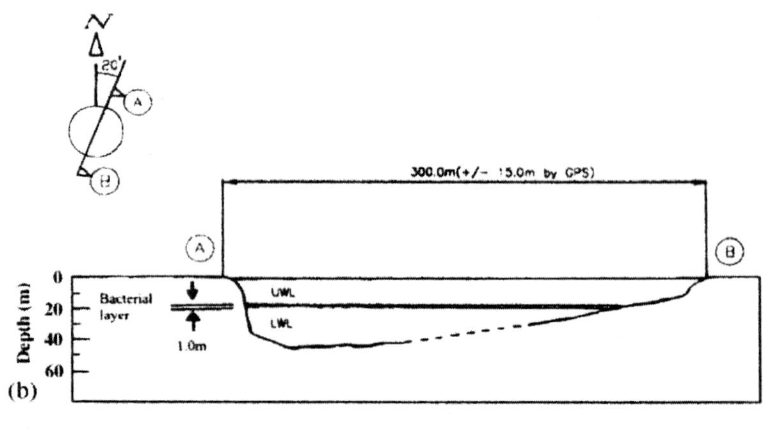


**Fig. 1** Bathymetry of the South Andros Black Hole. UWL, upper water layer; LWL, lower water layer; —, floor area not surveyed (Reprinted from Schwabe and Herbert 2004, with permission from Elsevier)

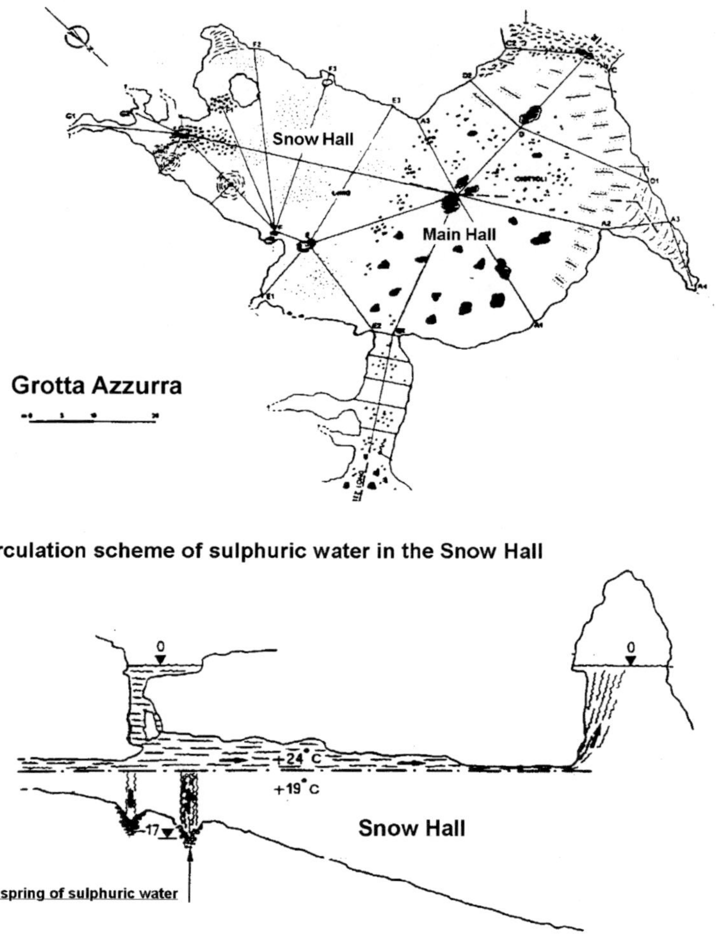


**Fig. 2** A drawing map of the "Grotta Azzurra" showing both entrances, halls, and freshwater circulation

analysed to show the molecular evidence of anaerobic methane-oxidizing communities growing syntrophically with sulphate-reducing bacteria (Hinrichs et al. 2000). The lipids of both populations accounted for 50% of the fatty acids detected in the sediments.

Another peculiar site is the Cariaco Basin that is the largest anoxic basin of oceanic character. Geochemical and biological investigations were carried out in this site, particularly on biochemical fluctuations in the water column, sulphur cycle, and autochthonous communities (Wakeham et al. 2004; Werne et al. 2003; Diaz-Ramos et al. 2000; Fry et al. 1991). In this habitat the high primary productivity in the surface waters provides a large supply of organic matter that is available for sulphurisation at the highest depths. Free sulphide was detected in the water column below 300 m depth, suggesting an active microbial sulphate reduction in the Cariaco Basin sediments and likely the water column. Hypotheses on the formation of sulphur intermediates such as elemental sulphur and polysulphides were discussed, including the role of anaerobic sulphide-oxidizing bacteria and that of anaerobic methane oxidisers.

One more peculiar anoxic marine site with important sulphur cycling is the Makirina Bay. It has been investigated for nutrient fluxes and sulphur cycling taking into account both benthic and diffusive fluxes, despite the high heterogeneity of the sediment (Lojen et al. 2004).

Terrestrial and sulphide-rich caves have been, on the other hand, extensively investigated in terms of microbiology, particularly the cave Cueva de Villa Luz in Mexico, the Movile cave in Romania, the Lower Kane Cave in Wyoming (USA), and the Anadarko Basin in Oklahoma

(USA). In these karst habitats chemotrophic microbial interaction occur and the caves are significantly rich in $H_2S$. Molecular analyses of the microbial diversity inside the cave showed the presence of bacteria closely related to *Thiobacilli* spp., *Beggiatoa*, *Thiothrix*, and *Acidimicrobium* spp. (Vlasceanu et al. 1997, 2000; Engel et al. 2003, 2004; Hose et al. 2000). These bacteria are probably involved in $H_2S$ oxidation inside the cave leading to limestone dissolution; moreover, in the Anadarko Basin, characterised by abundant barium and sulphide, data showed that phototrophic sulphide oxidation and concomitant sulphur cycling were important processes regulating the cycling of barium (Senko et al. 2004).

## 2 Recent microbiological studies

### 2.1 The blue holes

Abundant microbial development has been observed in these cave systems, usually linked to temperature anomalies in the range of 36–41°C recorded by investigators. Marano-Briggs (2000) reported temperatures up to 41°C in the Tarpon Blue Hole, supporting the theory that the mass population of anoxygenic phototrophic bacteria present at a specific water layer may be dissipating excess light energy by heat. A dense layer of purple sulphur bacteria was indeed found at 4.5–5.5 m depth in the estuarine Tarpon Blue Hole and the dominant population was represented by *Chromatiaceae* species. These bacteria are capable either of photolithoautotrophic growth with sulphide or elemental sulphur under anoxic conditions in the light or of chemo-organotrophic growth under micro-oxygenic conditions in the dark. Their abundance in the site was consistent with their physiological traits and among them a phototrophic bacterium in particular was identified as a novel strain of *Marichromatium purpuratum* that contained okenone instead of spirilloxanthin as its major carotenoid.

Recently, microbiological studies on a Blue Hole discovered in the Indian Ocean (Canganella et al. 2004) were preliminarily performed in order to investigate the distribution of microbial populations along the water column and in the bottom sediment of the cave. In Fig. 3 the DGGE analysis carried out shows how bacterial populations are distributed among surface layers, mid- and deep-layers, indicating in particular a similar biodiversity between 30 m and 50 m depth. The same was investigated by the Biolog system, particularly for anaerobic populations (Fig. 4).

The molecular investigation showed that at surface there was a complex microbial community showing several peculiarities with respect to standard pelagic seawater. Characteristic species were *Thioalkalivibrio* and *Thioploca* within the Gamma-Proteobacteria and *Desulfosarcina* and other

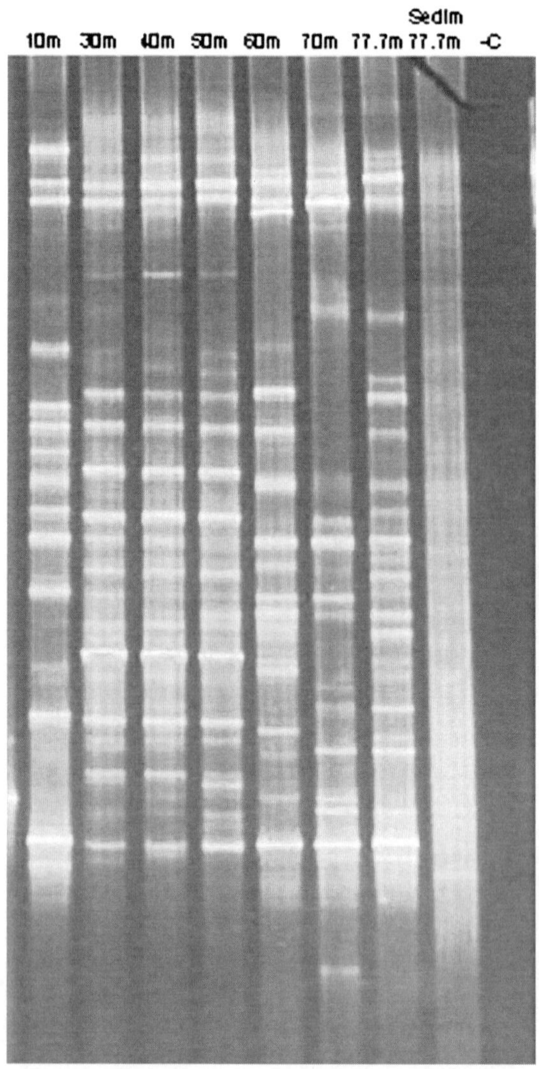


**Fig. 3** DGGE analysis of the water column and bottom sediment of the Blue Hole in the Indian Ocean

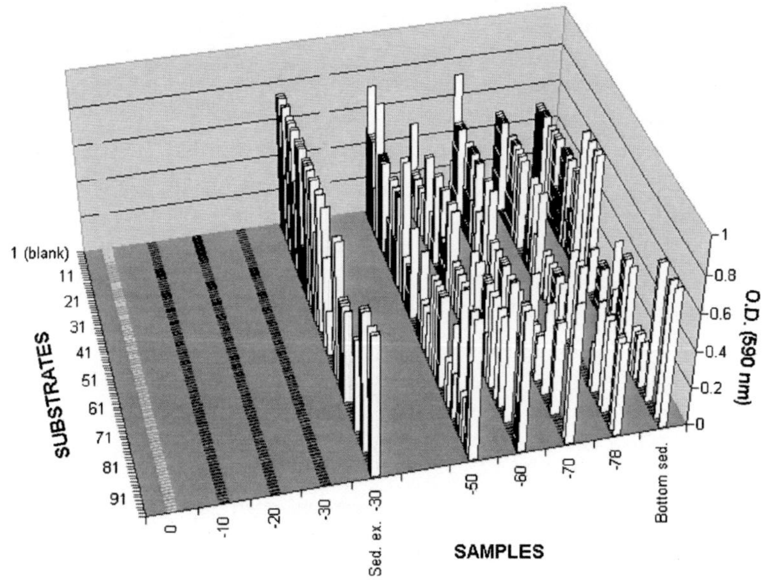


**Fig. 4** A graph showing the analysis performed by the Biolog system on samples collected from the water column and sediments of the Blue Hole in the Indian Ocean

Delta-Proteobacteria; sequences within the Delta-Proteobacteria related to sponge symbionts were also found. Subsurface bacteria present in the water column appeared dominated by Alpha-Proteobacteria represented by sequences related to photoorganotrophs (*Rhodovibrio*-related), methanotrophs, and coral surface-related bacteria.

Near the bottom, although the Alpha-Proteobacteria were still well represented, the proportion of Delta-Proteobacteria increased. These Delta-Proteobacteria represent sulphate-reducing bacteria mainly belonging to the Desulfobacteriaceae, or related to *Desulfonanticus*, *Synteophus*, and others related to marine and salt-marsh uncultured bacteria.

## 2.2 The black holes of the Bahamas

The water in these holes appears black in colour due to the presence of a 1-m thick white microbial veil located within the upper third of the water column (18–19 m depth), at the boundary between the oxygenic low salinity upper water mass and the denser anoxic saline water layer. The boundary between the two water masses is characterised by sharp discontinuities in physicochemical gradients: salinity increased from 12 to 35 psu and temperature from 29°C to 36°C; pH decreased from 8.6 to 6.45 as well as dissolved $O_2$ from 6 mg/l to <1 mg/l. Morphological microscopic observations of collected samples showed mainly non-motile spherical and motile rod-shaped purple sulphur bacteria. The dominant members of this warm (36°C), saline, and sulphide-rich layer have been identified as anoxygenic phototrophic bacteria belonging to the genera *Allochromatium* and *Thiocapsa* and showed population densities $>10^7$ viable cells/ml. These bacteria grow well in the presence of sulphide and carbon dioxide in the light. During photoautotrophic growth sulphur globules are stored intracellularily as intermediate oxidation products. Moreover, the intracellular photosynthetic membranes are of the vesicular type and bacteriochlorophyll *a* and carotenoids of the normal spirilloxanthin series are present (Herbert et al. 2005). Calculations made by these authors revealed that the layer of anoxygenic phototrophic bacteria in the South Andros Black Hole may have a biomass content of approximately 5.06 tonne dry weight.

## 2.3 The marine caves of Capo Palinuro

The emergence of sulphidic water in some of the caves (particularly in the "Grotta Azzurra", "Grotta Sulfurea", and "Grotta di Cala Fetente") gives rise to an unusual situation, characterised by abiotic as well as biotic fluctuations, represented by emissions from sulphidic springs and higher concentration of sulphur-utilising bacteria following each outflowing event, respectively. Warm sulphidic water of reduced salinity enters the Snow Hall (Grotta Azzurra), from fissures in the bottom rocks, and rises above the more dense seawater to form a thermocline and chemocline as a visible boundary at a depth of about 9.5 m. The geochemistry of water samples from the Grotta Azzurra was extensively reported by Mattison et al. (1998): above and below the chemocline temperatures were 24.0 ± 0.2 and 22.8 ± 0.3, respectively; NaCl concentrations (%) were 2.6 ± 0.1 and 3.0 ± 0.1, respectively; pH values were 7.22 ± 0.02 and 8.15 ± 0.25, respectively.

The present ecosystem is biologically unique particularly because the following points: (a) the peculiar emissions represent the only available model in shallow waters resembling the deep-sea hydrothermal vents (Canganella 2001); (b) alike what happens in usual submarine caves, the development of micro- and macro-fauna increases with the decrease of irradiation; (c) the freshwater flow is placed above the seawater and, inside the cave, an interface between the warm/anaerobic upper zone ($H_2S$-enriched) and the cold, sulphur-less lower zone can be observed; (d) giant forms, particularly among *Cnidaria* and *Porifera*, occur mainly along the aerobic/anaerobic interface (Alvisi et al. 1994a, b).

During the last decade both "Grotta Azzurra" and "Grotta Sulfurea" have been investigated as far as regard their microbiota. Some preliminary data on the lithotrophic microflora were published (Mattison et al. 1998), and other studies were performed in order to: (a) investigate the microscopic structure of bacterial mat and (b) to understand both taxonomy and physiology of heterotrophic bacteria inside the "Grotta Azzurra" (Canganella and Bianconi 1999, 2003; Canganella et al. 2003). The 16S rRNA genes analysis (Fig. 5) showed that heterotrophic isolates were closely related to the genera *Escherichia*, *Citrobacter*, *Vibrio*, and *Bacillus*. Based on the phylogenetic tree, strains P1 and P4 may represent new species of *Bacillus* and strain P3 a new species of *Vibrio*. PAL-1 and PAL-2 isolates were closely related to *Escherichia coli* and *Citrobacter freundii*, respectively. Another strain

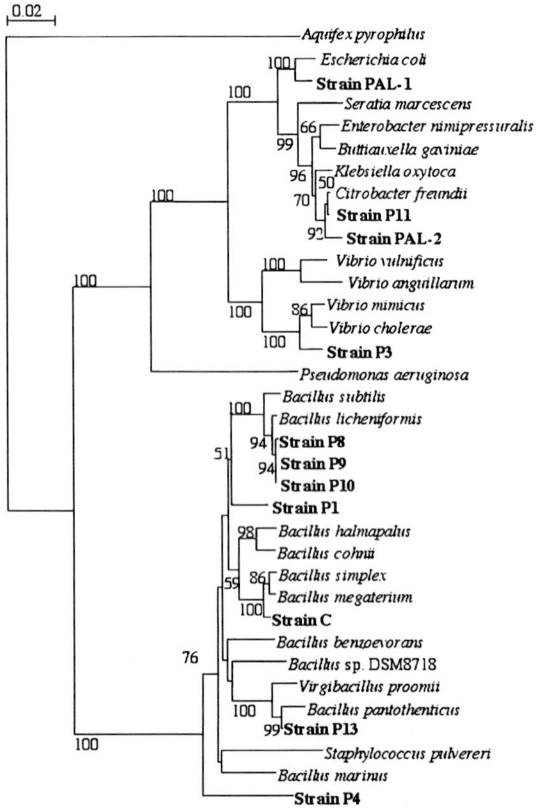


**Fig. 5** A phylogenetic tree based on 16S rRNA gene sequence comparison, including representative heterotrophic isolates from the "Grotta Azzurra"

(SED XX) was also fully characterised by the examination of its morphological and physiological traits (growth at 10°C, tolerance to 6%–7% NaCl and to 0.04% $NaN_3$), as well as of its resistance to vancomycin. According to the data, the strain was classified as *Enterococcus* spp. and the 16S rRNA genes analysis showed that it was closely related to *Enterococcus azikeevi*, an organism not yet described.

In addition to the analysis of the heterotrophic microflora, recently a preliminary molecular investigation was performed to evaluate the biodiversity of both eubacterial and archaeal populations (unpubl. res.).

Some bacterial sequences were closely related to the sulphur-oxidizing bacterial group, involving the genus *Beggiatoa* and *Thiothrix*. Showing that these microorganisms may well represent a major microbial community in the bacterial mat of the cave. Other bacterial sequences were phylogenetically linked to the Delta-proteobacterial group, particularly to sulphate-reducing bacteria. Archaeal sequences were partially related to methanogen archaeal group, but the phylogenetic distance was still remarkable; so the involvement of a novel archaeal group can be considered. Other archaeal sequences were related to MG-1, which is a common uncultivable archaeal group in marine environment.

As far as regard the microscopic structure of bacterial mat inside the cave and sediments collected nearby a sulphidic spring along the main entrance of the "Grotta Azzurra, the distribution of various morphological forms is shown in Fig. 6a–c. Filamentous microbial forms were very abundant and morphologically similar to *Beggiatoa*, *Desulfonema*, *Leucothrix*, or *Thiothrix* species; morphology was however not enough to identify such organisms. Several cells showing either coccoid or rod morphologies were also observed in samples collected inside as well as outside the mat.

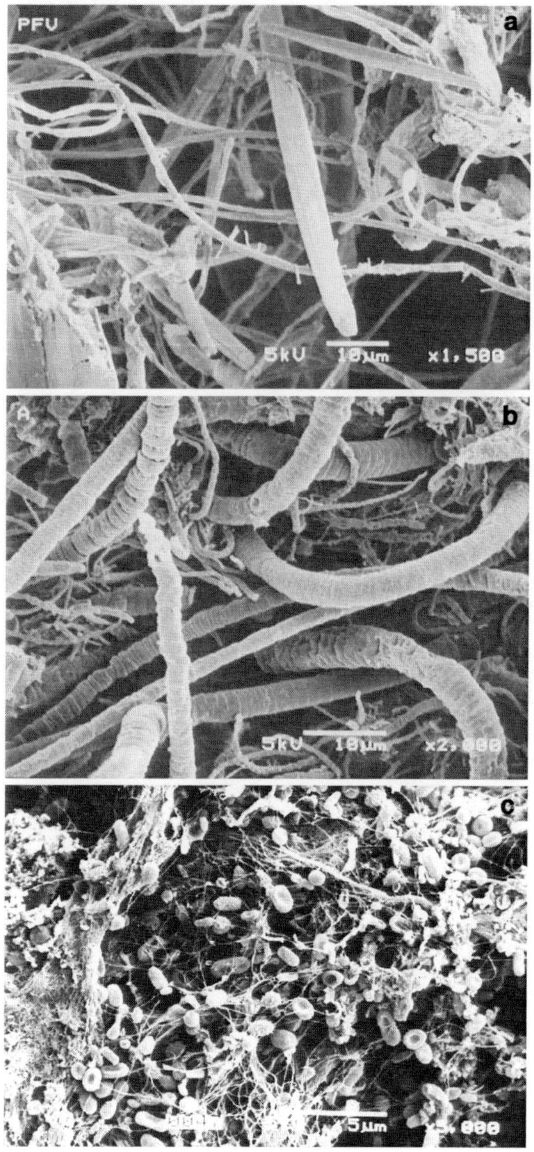


**Fig. 6** (**a**) A SEM picture of a sample collected in the bacterial mat on the vault of the "Grotta Azzurra". (**b**) A SEM picture of a sample collected in the bacterial mat on the vault of the "Grotta Azzurra". (**c**) A SEM picture of a sediment sample collected nearby a sulfataric spring along the main entrance of the "Grotta Azzurra"

## 3 Discussion

Submarine caves and holes are common niches among shallow water ecosystems, but they can be also described as extreme environments, due to limited photosynthetic processes, oligotrophic conditions, and possibly elevated sulphide concentrations.

As far as regard blue and black holes, it has been shown that the dominant anoxigenic phototrophic bacteria in the microbial layers are member of the *Chromatiaceae* (Imhoff et al. 1998). The presence of dense phototrophic populations of purple sulphur bacteria particularly explains why the South Andros Black Hole appears black and not blue. Moreover, the sulphide profile data obtained in this site showed that sulphide concentration were low (30 µM) suggesting that biomass production by the purple sulphur bacteria were sulphide rather than light limited.

Another important point to consider is that the dense microbial phototrophic population probably plays a significant role in enhancing carbonate $CO_2$-mediated dissolution operated by heterotrophic bacteria. The effects of microbiological activity on degradation of limestone deposits have been described by several authors (Ehrlich 1996; Golubic and Schneider 1979; Paine et al. 1993) and similar observations were reported for the Bahamian cave systems from Smart et al. (1988).

Recent studies have shown that biotic oxidation of sulphide into sulphuric acid can occur in anoxic marine ecosystems due to chemoautotrophic sulphur-oxidizing bacteria such as *Beggiatoa*, *Thiothrix*, *Thiobacillus*, and *Thiomicrospira* (Mattison et al. 1998); moreover, other microorganisms as *Pseudoalteromonas* and *Halanaerobium* seem to be dominants in such ecosystems, even in anoxic brines (Vetriani et al. 2003; Daffonchio et al. 2006).

This biological process may also be involved in carbonate dissolution for the South Andros Black Hole but the role of sulphur-oxidizing bacteria needs to be evaluated by further investigations. The same consideration may be valid for the Blue Hole in the Indian Ocean. In this site the occurrence of highly toxic sulphide and low oxygen content caused the death of the macrobenthos, whereas a chemolithoautotrophic microbial community settled on the cave walls as mucilaginous and hardened short filaments of bacterial origin whose study is still on progress.

In the "Grotta Azzurra" cave of Capo Palinuro fluids emitted from active springs with maximum temperatures of 25°C and enriched in dissolved sulphur species ($H_2S$, $S_2O_3^{-2}$) allowed the development of a significant micro- and macro-fauna, resembling in some way what occurs around hydrothermal vent sites (Jeanthon 2000; Canganella 2001).

Morphological/physiological observations were previously carried out by Mattison et al. (1998) on microbial mats in the "Grotta Azzurra" of Capo Palinuro. In recent studies both sampling and isolation were carried out regardless of the chemocline. Future investigations on the distribution of microbial populations will eventually lead to discriminate between more thermotolerant and more mesophilic species living above or below the chemocline; the same can be true for non-cultivable species.

The characterised heterotrophic isolates partially represent the microbial ecology of the site, showing its biological complexity at a phylogenetic level. Moreover, the wide resistance of isolates to temperature, salinity and pH may be associated with the specific ecological system, an hypothesis that certainly deserves further investigations.

According to the literature, strain P1 might be placed in the *B. subtilis* cluster I with *B. subtilis* and *B. licheniformis* (Slepecky and Hemphill 1992). Strain P4 was physiologically similar to *B. marinus*, but the phylogenetic divergence was not negligible. Moreover, the phenotypic characteristics of *B. marinus*, as well as its distribution and origin, were recently revised after the isolation of novel strains from tropical and polar sediments (Ruger et al. 2000). Strain P3 was closely related to *Vibrio mimicus*, a pathogenic species that has been also isolated from shallow waters, shellfish, and sea-turtle eggs (Davis et al. 1981). However, alike strain P3, this species was not able to grow in the presence of 5% NaCl.

Strain PAL-2 resulted phylogenetically related to *C. freundii*; it is well known that the latter usually occurs as an intestinal commensally inhabitant of humans and animals (Frederiksen and Sogaard 1992). For this reason it can be found in water, sewage and soil, but very little information exists on its possible role in the environment. According to the phylogenetic analysis, strain SED XX was closely related to *E. azikeevi*, but within this genus no marine isolates were ever described (Devriese et al. 1992).

These data confirmed that enterococci and enteric bacteria are present even in aquatic habitats regardless the occurrence of polluting events, as shown by several authors in recent papers (Brown and Bowman 2001; Stoner et al. 2001; Cottrell et al. 2005; del Mar et al. 2005). On this regard, a different taxonomical approach, based on numerical taxonomic study has been used by Ortigosa et al. (1997) to describe Gram-positive bacteria of marine origin. The work was performed on 65 Gram-positive wild strains of heterotrophic, aerobic, marine bacteria that were characterised by 96 morphological, biochemical, physiological and nutritional tests.

In general terms, similar environmental traits can be described for the Blue Hole in the Indian Ocean and the submarine caves of Palinuro Cape. First of all, the presence of elevated concentrations of $H_2S$, ranging from 5 μM to 200 mM. A chemocline in the water column may be clearly observed in the Blue Hole of the Indian Ocean or in the Black Hole of the Bahamas, whereas in the submarine caves of Palinuro Cape the environmental borderline inside the habitat is represented by the threshold between upper anoxic, sulphide-rich freshwater and lower seawater.

As far as regard the autochthonous microflora, the Blue Hole of the Indian Ocean showed the presence of filamentous forms as described in the Palinuro caves but mucilaginous forms on the rocky side walls were not found in the latter; moreover, sulphate-reducers and sulphide-oxidisers were also detected in both systems.

Some microbiological features of both systems are also shared with other sulphide-rich environments, such as the presence of anaerobic methane oxidisers, filamentous bacteria, anaerobic phototrophs, and archaeal strains. However, the incomplete data set on the inhabitant microflora, the geological diversity of the above-described systems, and their different geographical locations makes difficult a microbiological comparative study. With no doubts the combination of different approaches and techniques may be a powerful tool to investigate in the next future such astonishing ecosystems by a multiphasic approach.

As a matter of fact, further investigations on the microbial diversity in sulphide-rich either marine or freshwater, caves will be based on traditional techniques of isolation coupled to simultaneous molecular investigations.

**Acknowledgements** We thank the Italian Ministry of University and Research (Fondi ex 60%) for its financial support. We are also grateful to Prof. J. Wiegel and Prof. K. Horikoshi for their scientific support.

## References

Alvisi M, Barbieri F, Colantoni P (1994a) Le grotte marine di Capo Palinuro. In: Alvisi M, Colantoni P, Forti P (eds) Grotte Marine d'Italia. Memorie dell'Istituto Italiano di Speleologia, Bologna, pp 143–181

Alvisi M, Barbieri F, Bruni R, Cinelli F, Colantoni P, Grandi F, Maltoni P (1994b) La Grotta Azzurra di Capo Palinuro (Salerno). In: Alvisi M, Colantoni P, Forti P (eds) Grotte Marine d'Italia. Memorie dell'Istituto Italiano di Speleologia, Bologna, pp 51–56

Brown MV, Bowman JP (2001) A molecular phylogenetic survey of sea-ice microbial communities. FEMS Microbiol Ecol 35:267–275

Canganella F (2001) Hydrothermal vent communities. In: John Wiley & Sons (eds) Encyclopedia of life sciences. Macmillan Press, London

Canganella F, Bianconi G (1999) Microbial investigations on submarine caves in the area of Cape Palinuro (Salerno, ITALY). In: Brasilian Microbiology Society (ed) Proceedings of the XX Meeting of the Microbiological Brasilian Society, 21–25 Oct 2001, Salvador, BA, p 223

Canganella F, Bianconi G (2003) Microscopic investigations on the cell structure of autoctonous microflora in a sulphidic submarine cave (transl.). In: Fifth Congress FISV, 10–13 Oct 2003, Rimini, Italy, p 154

Canganella F, Bianconi G, Gambacorta A, Kato C, Uematsu K (2003) Characterisation of heterothrophic microorganisms isolated from the "Grotta Azzurra" of Cape Palinuro (Salerno, Italy). Mar Ecol 23:1–10

Canganella F, Bianconi G, Maugeri T (2004) A multiphasic approach to investigate the microbial ecology of a Blue Hole in the Indian Ocean. In: Extremophiles Meeting, 19–23 Sept 2004, Cambridge, USA, p 83

Colantoni P, Baldelli G, Bianchi CN, Capaccioni B, Morri C, Mandrini M, Tassi F (2003) A cave flooded by marine water with hydrogen sulphide highlights the recent evolution of the Maldives (Indian Ocean): preliminary notes. Le grotte d'Italia 4: 29–37

Cottrell MT, Waidner LA, Yu L, Kirchman DL (2005) Bacterial diversity of metagenomic and PCR libraries from the Delaware River. Environ Microbiol 7:1883–1895

Daffonchio D, Borin S, Brusa T, Brusetti L, van der Wielen PWJJ, Bolhuis H, Yakimov MM, D'Auria G, Giuliano L, Marty D, Tamburini C, McGenity TJ, Hallsworth JE, Sass AM, Timmis KN, Tselepides A, de Lange GJ, Hubner A, Thomson J, Varnavas SP, Gasparoni F, Gerber HW, Maliverno E, Corselli C, Biodeep Scientific Party (2006) Stratified prokaryote network in the oxic-anoxic transition of a deep-sea halocline. Nature 440:203–207

Davis BR, Fanning GR, Madden JM, Steigerwalt AG, Bradfort HB Jr, Smith HL Jr, Brenner DJ (1981) Characterization of biochemically atypical *Vibrio cholerae* strains and designation of a new pathogenic species, *Vibrio mimicus*. J Clin Microbiol 14:631–639

del Mar LM, Bonato B, Benedetti D, Canepari P (2005) Survival of enterococcal species in aquatic environments. FEMS Microbiol Ecol 54:189–196

Devriese LA, Collins MD, Wirth R (1992) The genus *Enterococcus*. In: Balows A, Truper HG, Dworkin M, Harder W, Schleifer KH (eds) The prokaryotes. Springer-Verlag, New York, pp 1465–1481

Diaz-Ramos JR, Muller-Karger FE, Millie D, Troccoli-Ghinaglia LE, Subero-Pino S, Varela R (2000) Phytoplankton community structure: temporal variability in a tropical upwelling ecosystem. J Phycol 3:18–28

Engel AS, Lee N, Porter ML, Stern LA, Bennet PC, Wagner M (2003) Filamentous "Epsilonproteobacteria" dominate microbial mats from sulfidic cave springs. Appl Environ Microbiol 69:5503–5511

Engel AS, Porter ML, Stern LA, Quinlan S, Bennet PC (2004) Bacterial diversity and ecosystem function of filamentous microbial mats from aphotic (cave) sulfidic springs dominated by chemolithoautotrophic "Epsilonproteobacteria". FEMS Microbiol Ecol 51:31–53

Ehrlich HL (1996) Geomicrobiology, 3rd edn. Marcel Dekker Inc., New York, pp 172–215

Frederiksen W, Sogaard P (1992) The genus *Citrobacter*. In: Balows A, Truper HG, Dworkin M, Harder W, Schleifer KH (eds) The prokaryotes. Springer-Verlag, New York, pp 2744–2753

Fry B, Jannasch HW, Molyneaux SJ, Wirsen CO, Muramoto JA, King S (1991) Stable isotope studies of the carbon, nitrogen and sulphur cycles in the Black sea and the Cariaco Trench. Deep-Sea Res 38:1003–1019

Golubic SI, Scneider J (1979) Carbonate dissolution. In: Trudinger PA, Staine DJ (eds) Biogeochemical cycling of mineral-forming elements. Elsevier, Amsterdam, pp 107–129

Herbert RA, Ranchou-Peyruse A, Duran R, Guyoneaud R, Scwabe S (2005) Characterization of purple sulphur bacteria from the South Andros Black Hole cave system: highlights taxonomic problems for ecological studies among the genera *Allochromatium* and *Thiocapsa*. Environ Microbiol 7:1260–1268

Hinrichs KU, Summons RE, Orphan V, Sylva SP, Hayes JM (2000) Molecular and isotopic analysis of anaerobic methane-oxidizing communities in marine sediments. Org Chem 31:1685–1701

Hose LD, Palmer AN, Palmer MV, Northup DE, Boston PJ, DuChene HR (2000) Microbiology and geochemistry in a hydrogen-sulphide-rich karst environment. Chem Geol 169:399–423

Imhoff JF, Süling J, Petri R (1998) Phylogenetic relationships among the *Chromatiaceae*, their taxonomic reclassification and description of the new genera *Allochromatium*, *Halochromatium*, *Isochromatium*, *Marichromatium*, *Thiococcus*, *Thiohalocapsa* and *Thermochromatium*. Int J Syst Bacteriol 48:1129–1143

Jeanthon C (2000) Molecular ecology of hydrothermal vent microbial communities. Antonie van Leeuwen 77:117–133

Kuwabara JS, van Geen A, McCorkle DC, Bernhard JM (1999) Dissolved sulphide distributions in the water column and sediment pore waters of the Santa Barbara Basin. Geochim Cosmochim Acta 15:2199–2209

Marano-Briggs K (2000) Water chemistry, bacterial abundance and anoxygenic photosynthetic bacteria in an estuarine Blue Hole, Andros Island, The Bahamas, Ph.D. Thesis, George Mason University, Fairfax Co., Virginia, USA

Mattison RG, Abbiati M, Dando PR, Fitzsimons MF, Pratt SM, Southward AJ, Southward EC (1998) Chemoautotrophic microbial mats in submarine caves with hydrothermal sulphidic springs at Cape Palinuro, Italy. Microb Ecol 35:58–71

Mylroie JE, Carew JL, Moore AI (1995) Blue holes: definition and genesis. Carbonate Evaporites 10:225–233

Lojen S, Ogrinc N, Dolenec T, Vokal B, Szaran J, Mihelcic G, Branica M (2004) Nutrient fluxes and sulphur cycling in the organic-rich sediment of Makirina Bay (Central Dalmatia, Croatia). Sci Total Environ 327:265–284

Ortigosa M, Garay E, Pujalte MJ (1997) Gram-positive bacteria of marine origin: a numerical taxonomic study on Mediterranean isolates. Microbiologia 13:453–462

Paine SG, Lingood FV, Schimmer F, Thrupp TC (1993) The relationship of microorganisms to the decay of stone. Philos Trans R Soc Ser B Lond 222:97–127

Ruger HJ, Fritze D, Sproer C (2000) New psychrophilic and psychrotolerant *Bacillus marinus* strains from tropical and polar deep-sea sediments and emended description of the species. Int J Syst Evol Microbiol 3:1305–1313

Schwabe SJ (1998) The Black Hole. Caves and caving. Br Cave Res Assoc 80:1–22

Schwabe SJ, Herbert RA (2004) Black Holes of the Bahamas: what they are and why they are black. Quatern Int 121:3–11

Schwabe SJ, Parkes RJ, Smart PL (1997) Biogeochemical investigation of submerged cave systems (Blue Holes) within Bahamian carbonate platforms. In 97th Applied and environmental microbiology general meeting, 4–7 May 1997, Miami, FL, p 398

Senko JM, Campbell BS, Henriksen JR, Elshahed MS, Dewers TA, Krumholz LR (2004) Barite deposition resulting from phototrophic sulphide-oxidizing bacterial activity. Geochim Cosmochim Acta 4:773–780

Slepecky RA, Hemphill HE (1992) The genus *Bacillus* – nonmedical. In: Balows A, Truper HG, Dworkin M, Harder W, Schleifer KH (eds). The prokaryotes. Springer-Verlag, New York, pp 1663–1696

Smart PL, Dawans JM, Whitaker F (1988) Carbonate dissolution in a modern mixing zone. Nature 335:811–813

Stoner DL, Geary MC, White LJ, Lee RD, Brizzee JA, Rodman AC (2001) Mapping microbial diversity. Appl Environ Microbiol 67:4324–4328

Vlasceanu L, Popa R, Kinkle B (1997) Characterization of *Thiobacillus thioparus* LV43 and its distribution in a chemoautotrophically based ecosystem. Appl Environ Microbiol 63:33123–33127

Vlasceanu L, Sarbu S, Engel AS, Kinkle B (2000) Acidic, cave wall biofilms located in the Frasassi George, Italy. Geomicrobiology 17:125–139

Vetriani C, Tran HV, Kerkhof LJ (2003) Fingerprinting microbial assemblages from oxic/anoxic chemocline of the Black Sea. Appl Environ Microbiol 69:6481–6488

Wakeham SG, Hopmans EC, Schouten S, Sinninghe Damste JS (2004) Archaeal lipids and anaerobic oxidation of methane in euxinic water columns: a comparative study of the Black Sea and Cariaco Basin. Chem Geol 205:427–442

Werne JP, Timothy WL, Hollander DJ, Formolo MJ, Sinninghe Damsie JS (2003) Reduced sulphur in euxinic sediments of the Cariaco Basin: sulphur isotope constraints on organic sulphur formation. Chem Geol 195:159–179

Rev Environ Sci Biotechnol (2007) 6:71–94
DOI 10.1007/s11157-006-9108-x

REVIEW PAPER

# Access to glacial and subglacial environments in the Solar System by melting probe technology

**Stephan Ulamec · Jens Biele · Oliver Funke · Marc Engelhardt**

Received: 23 February 2006 / Accepted: 18 August 2006 / Published online: 6 October 2006


**Abstract** A key aspect for understanding the biological and biochemical environment of subglacial waters, on Earth or other planets and moons in the Solar system, is the analysis of material embedded in or underneath icy layers on the surface. In particular the Antarctic lakes (most prominently Lake Vostok) but also the icy crust of Jupiter's moon Europa or the polar caps of Mars require such investigation. One possible technique to penetrate thick ice layers with small and reliable probes is by melting, which does not require the heavy, complex and expensive equipment of a drilling rig. While melting probes have successfully been used for terrestrial applications e.g. in Antarctic ice, their performance in vacuum is different and theory needs confirmation by tests. Thus, a vacuum chamber has been used to perform a series of melting tests in cold (liquid nitrogen cooled) water ice samples. The feasibility of the method was demonstrated and the energy demand for a space mission could be estimated. Due to the high energy demand in case of extraterrestrial application (e.g. Europa or polar caps of Mars), only heating with radioactive isotopes seems feasible for reaching greater depths. The necessary power is driven by the desired penetration velocity (approximately linearly) and the dimensions of the probe (proportional to the cross section). In comparison to traditional drilling techniques the application of a melting probe for exploration of Antarctic lakes offers the advantage that biological contamination is minimized, since the Probe can be sterilized and the melting channel freezes immediately after the probe's passage, inhibiting exchange with the surface layers and the atmosphere. In order to understand the physical and chemical nature of the ice layers, as well as for analysing the underlying water body, a melting probe needs to be equipped with a suite of scientific instruments that are capable of e.g. determining the chemical and isotopic composition of the embedded or dissolved materials.

**Keywords** Melting Probe · Subglacial · Europa · Mars · Antarctic Lakes · Ice · Technology · Life in extreme environments

S. Ulamec (✉) · J. Biele · O. Funke · M. Engelhardt
German Aerospace Center (DLR),
Institute for Space Simulation,
Linder Höhe 51147 Köln, Germany
e-mail: stephan.ulamec@dlr.de

J. Biele
e-mail: jens.biele@dlr.de

O. Funke
e-mail: oliver.funke@dlr.de

M. Engelhardt
e-mail: marc.engelhardt@dlr.de

## 1 Introduction

Ideas to use so-called "melting probes" for getting access to the subsurface layers of planetary ice sheets have been discussed for several years in connection with proposed Mars and Europa missions (Paige 1992; DiPippo et al. 1999; Carsey et al. 1999; Biele et al. 2002). More recently, some theoretical and experimental work has been done to understand the behaviour of such probes under extraterrestrial (very low temperatures, vacuum) conditions (Kömle et al. 2002, 2004; Treffer et al. 2006).

These investigations are driven by three major insights. First, there is a high scientific interest in exploring certain icy environments (Mars' polar caps, Europa and other icy satellites) motivated by the search for traces of life in these extreme environments as well as interest in planetary climate history in the case of Mars. Second, robotic space missions with a mechanical drilling system for depths exceeding a few meters are mechanically very demanding. Third, contamination avoidance in respect to planetary protection requirements can be relatively easily fulfilled using melting probes, since the melting channel refreezes behind the probe and cuts off the contact to the surface. Moreover, decontamination of the probe is relatively easy to achieve with standard sterilization methods (Engelhardt 2006). Missions to explore the surface and sub-surface of icy satellites like Europa can use a wide variety of schemes and technologies. These differ in their cost and scientific return, with fly-bys and orbiters being relatively achievable, but yielding no in-situ knowledge of the subsurface. At the other extreme in terms of returned science and complexity is a device that is capable of melting its way through the satellite's icy crust towards the putative ocean of water. Obviously, information about the biological activity within or under the Europan ice crust can be efficiently gained only with such an in-situ probe. Traces of indigenous biological activity, such as intact bio-molecules, are unlikely to remain unaltered for long periods at the exposed surface of Europa (the radiation dose is $\approx 1 \cdot 10^8$rad/month here. At greater depths, the radiation environment continues to decrease, reaching values comparable to the Earth's biosphere below depths of 20 to 40 m (National Academy of Sciences 2000).

Table 1 summarizes parameters of icy surfaces of Mars, Europa and the Earth. Obviously, the environments require different technical solutions for melting probes. Dust content, salinity, temperature and radiation environment vary considerably between the three examples.

None of the existing melting probes is nearly mature enough for planetary applications without considerable technical development.

## 2 Melting probes

### 2.1 Past developments

The idea of a melting probe for the investigation of ice sheets can be retraced to the beginning of the 1960's, when Shreve (1962) described a method for the calculation of the melt penetration efficiency of isothermal solid-nose hot points boring in temperate glacier ice. Shreve referred explicitly to Kasser's experiments with a light ice-drill on glaciers (Kasser 1960). Philberth (1962) suggested a new method for measuring temperatures inside a glacial ice sheet. He designed a non-recoverable, instrumented melting probe, known thereafter as the *Philberth probe*, consisting of a cylindrical hull with an attached conically shaped head. The probe was connected to an external power supply by a cable (tether), unwinding from a coil integrated in the hull. For protection of the probe from damage caused by refreezing melt water after turning off the heating elements for ice temperature measurements, a significant part of the hull was filled with silicone oil of density >1 g/cm$^3$ (Aamot 1967b).

The principal design of a cylindrical hull of small diameter attached to a conical, convex or even concave melting head is still characteristically for melting probes, because it is energetically favourable compared to other imaginable designs like e.g. a bowl- or egg-shape. The reason for this is that the melting probe's penetration velocity $v$ is inversely proportional to the cross-sectional area of the probe, as described in more detail in the following section. Sufficient space for instrumentation can be obtained by increasing the length of a cylindrical probe, and therefore the length to diameter ratio of the first melting probes (including the Philberth probe) was about 25:1. A detailed description of the first Philberth probe can be found in Aamot (1967a).

Aamot (1970a) also reports four U.S. field experiments with early melting probes. In 1965, the first probe reached 90 m in Greenland ice before contact was lost. One-year later, the second probe ran smoothly for four days and reached a depth of 259 m after which it was stopped to observe temperature and pressure variations during the refreezing of the melt water. During the first day, this melting probe operated at 3380 W achieved a penetration velocity of 0.0755 cm/s (2.72 m/h). A detailed description of these first two field applications of melting probes, their instrumentation and the scientific results is given in Aamot (1967b). During the 1968 EGIG expedition at Station Jarl Joset (Greenland), Philbert used two probes which reached depths of 230 and 1000 m and provided again more scientific and engineering information (Aamot 1970a). A new concept of a pendulum attitude stabilization for thermal probes was first introduced by Aamot (1967c, 1970b), with additional heating elements in the top part of a melting probe. In the 1980's, the Polar Ice Coring Office (PICO) of the University of Nebraska constructed a similar probe with a telemetry link in its tether (Hansen and Kersten 1984). The PICO probe allowed constant temperature/ice flow measurements during the ice penetration phase as well as measurements of the melt water conductivity, and penetration depths between 100 m and 200 m were achieved. More recent work of the University of Nebraska on a probe for terrestrial applications is described by Kelty (1995).

In the 1990's, the Alfred Wegener Institute for Polar- and Marine Research (AWI) constructed new probes for the investigation of Antarctic shelf ice. The AWI "SUSI II" probe ("SUSI" is the abbreviation for *Sonde Under Shelf Ice*) was successfully tested on the Rettenbach glacier (Sölden/Austria) in 1990 and achieved a depth of 60 m. The probe could be retrieved from the glacier because the melting hole did not refreeze within 8 h. Two-years later, during winter 1992/93, the probe was able to penetrate 225 m thick Antarctic shelf ice near the Neumayer station, and accessed the open water below the ice sheet (Tüg, H., personal communication 2002; Tüg 2003). The specifications of SUSI II were as follows: length 2.25 m, diameter 10 cm, power 3.4 kW (head), 600 V/8 A supply power, penetration velocity in the shelf ice: 2.93 m/h (220 m in 75 h), stabilization regulated over wire tension.

A further development of this probe, named SUSI III, was build for penetration of an ice thickness of 800 m, with length 3.6 m, diameter 14 cm, power 9 kW (head), 1000 V/12 A supply power. However, this probe failed because of a tether overheating in the winch compartment of the probe (AWI 1997). Another "routine" probe for lesser depths has been patented (Tüg 2003). The melting probes described above have in common:

**Table 1** Relevant parameters for melting probe application on Earth, Mars and Europa

| | Earth (polar) | Mars (polar) | Europa |
|---|---|---|---|
| Temperature [K] | 220–270 | 205–210 | ~100 |
| Gravity [$ms^{-2}$] | 9.8 | 3.7 | 1.3 |
| (atm.) pressure [mbar] | 1000 | 6.7 | 0 |
| Radiation environment [rad/day] | ~ 0 | ~ $10^2$ (Simonsen 1993) | $4 \cdot 10^6$ |
| Composition (of ice mantle) | $H_2O$, (some air) | $H_2O$, $CO_2$, dust | $H_2O$, possibly others (salt, dust, $CO_2$, $CH_4$) |
| Maximum pressure at bottom of ice layer (bar) N.B.: phase transition of ice Ih occurs for $p$ >2000 bar. | 230–500 | 135–180 | 200–700 |

- typical lengths from 2.2 m to 3.6 m,
- diameters of 10 cm to 14 cm,
- a total mass of hundreds kilograms,
- and a power usage of the order of 3 kW to 15 kW.
- penetration rates of the order of 2 m/h

Parallel to the technical further development of melting probes, a more sophisticated instrumentation of the probes took place. In particular, the SUSI probe of the AWI was instrumented for the investigation of microorganisms living inside micron-sized brine channels and pockets in sea ice and shelf ice. Parameters measured by this instrumentation cover oxygen and carbon dioxide content of the ice/water mixture, the amount of inorganic nutrients, and the pH-value (Tüg 2003). The PICO probe was upgraded to measure melt-water conductivity and micro-particulates.

With modern space flight and the advancing knowledge about the environmental conditions on e.g. Mars or on Jupiter's icy moon Europa, the idea arose to use melting probes also for extraterrestrial applications. However, for the search for signs of life on Mars and especially on Jupiter's moon Europa a new generation of melting probes on the basis of gathered experiences during the four decades of terrestrial applications is required. Critical parameters for any space mission are, e.g., the mass, the overall dimensions and the power supply.

Without doubt, the most advanced melting probe design is that of the Cryobot, initiated by the Jet Propulsion Laboratory in 1998 as an in-situ exploration and sample return vehicle for a future application on Europa (Zimmerman 2001). The Cryobot is allotted with integrated radioisotope thermoelectric generators (RTGs) providing 1 kW thermal of direct melt energy. The probe is 0.8 to 1 m long with a diameter of 12 cm and a mass of 20 kg to 25 kg in the flight version. Fluid thermal modelling and testing revealed that with 1 kW power, a melt rate of 0.3 m/h can be realized in very cold (100 K) ice; the water jacket maintained around the probe is 1–2 mm in width and the melt plume would not refreeze until 1.25 m behind the vehicle. An active heater system (warm water jet in front of the probe) allows faster penetration in dense ice. The Cryobot is designed as a fully autonomous robotic mole penetrator system for melting through an ice-pack of 3 km to 10 km thickness. Cardell et al. (2004) suggested a thermal probe for application also on the Martian polar caps for an investigation of Mars' climate history, recorded in "polar layered deposits".

### 2.2 Comparison with conventional drilling

Although this paper is summarizing technology and applications of melting probes one should mention the more conventional methods for ice penetration to outline benefits and drawback of the melting probe technology.

In particular ice-coring drilling (ICD), hot water drilling (HWD) and coiled tube drilling (CTD) need to be mentioned, since these technologies are well developed and have been successfully used in e.g. Antarctic ice. An excellent overview of these methods is given by Clow and Koci (2002).

When ice penetration is accomplished mechanically using rotary cutting and coring bits the drilling rate $v$ for mechanical systems is given by

$$v = eP_0/AE \tag{0.1}$$

with drill-to-rock transmission efficiency $e$, power output to drill $P_0$, hole cross section $A$ and specific energy $E$. According to Mauer (1968), E is of the order of 20 MJ/m$^3$ for conventional drilling in "soft rock" (compressive strength up to 500 bar; ice has an compressive strength of 80–500 bar) and 50–170 MJ/m$^3$ for oilfield rotary mining (roller bits, drag type) up to 2000 m depth. Mellor (1989) quotes values of 0.5 MJ/m$^3$ to 5 MJ/m$^3$ for ice, up to 14 MJ/m$^3$ for frozen silt and 2 MJ/m$^3$ to 6 MJ/m$^3$ for frozen sand.

For cutting ice by a drill in thermal vacuum conditions resembling those expected, e.g., on Europa's surface: a figure as low as 1.2 kJ per kg of cut ice has been quoted (DiPippo 1999).

These figures do, however, not include transmission losses and the energy for compacting of the cuttings and transportation to and discharge at the surface.

The specific energy for conventional drilling can be compared with the specific energy required for melting ice (>306 MJ/m$^3$).

Ice core drilling (ICD) is an advanced technology, maximizing the recovery of ice samples and allowing geophysical logging methods to be applied. But it is also relatively slow and expensive to operate. Large quantities of drilling fluid are required. Application on Mars or Europa seems extremely challenging.

For drilling into ice in the 1 km range, relatively small hot water drilling (HWD)-devices can be used (Engelhardt et al. 1990). It is also possible to obtain short ice cores with this technology (Engelhardt et al. 2000. Since in case of HWD also the contamination is relatively low, a potential melting probe application on Earth needs to be carefully compared with HWD-technology for any advantages and disadvantages. HWD's are very fast (i.e., drill rates on the order of tens of m/h are possible with the turbulent heat transfer at the ice interface), the energy expended for reheating is considerable (of the order of MW). Since hot (typically ~80–90°C) water needs to be pumped into the borehole and due to the high power (fuel) demand of HWD, in case of an application on extraterrestrial bodies, a melting probe will have very clear advantages. There is no obvious solution for how to use HWD in vacuum.

Coil tubing technology (CTD) has first been used in 1991 (Sas-Jaworsky and Bell 1996) and undergone considerable development since. CTD's are more compact and easier to operate than rotary drills. Since (in difference to HWD) material is drilled mechanically rather than using heat to melt the ice, they can be used also for deep (>4 km) drilling. CTDs are more and more commercially used (e.g. Gantt et al. 1998). A CTD for drilling into ice (CTDI) could be operated with various drilling fluids including hot water. CTD technology can also be used for drilling into bedrock, thus the problem of possible mineral deposits in the ice (which is a serious issue for melting probes) is obsolete. However, as discussed for HWD, the technology seems difficult for the use on Mars or Europa. A CTDI system, as proposed by Clow and Koci, (2002), capable of drilling a 3.5 km borehole through ice using ice coring, would weigh about 46 tons (including drilling fluid and fuel). Though being considerably lower in mass than a comparable system, based on HWD or ICD technology it is still not a trivial task to bring such a device into space.

### 2.3 Theory of melting probes

In a simple energy balance approximation, neglecting all losses, we can write the heat needed to progress a distance $l$ in compact ice via melting as;

$$\Delta W = Al\rho\left(c_p(T_F - T) + L_m\right) \tag{1}$$

If the heating power is $P$, then the melting speed $v$ is

$$v = lP/\Delta W = \frac{P}{A\rho\left(c_p(T_F - T) + L_m\right)} \tag{2}$$

where $A$ is the probe's cross-section (m$^2$), $l$ is the probe length (m), $c_p$ is the specific heat capacity of ice, $\rho$ is the ice density, $L_m$ is the melting enthalpy of ice. More generally, it is the energy needed to change the phase (solid–liquid or solid–gaseous), $T_F$ is the melting temperature of ice (K), $T$ is the local ice temperature (K).

Precise numerical values and correlation equations for the physical properties of ices are given in the appendix.

Note that $\rho$ and especially $c_p$, are temperature dependent; for low temperatures $T$, one has to use averages over the temperature range $[T;T_F]$. Also, $T_F$ depends (weakly) on ambient pressure and on salt content (freezing point depression).

The most important point here is that the melting velocity scales with the inverse of the cross-sectional area of the probe. Hence the usual design is a cylindrical tube with a large (>10) aspect ratio (length/diameter) to obtain a volume sufficient for subsystem, payload and tether integration.

Equation (2) gives only a rough estimate of the penetration velocity, in fact, it relates the minimum power requirement $P_0$ to a given melting velocity. In fact, insufficient heat causes the probe to freeze in i.e. to stall; excessive heat produces an oversize hole and wastes power by raising the meltwater's temperature far above the phase transition temperature. To obtain a more accurate result loss factors need to be analysed: most importantly, losses due to lateral conduction in the ice and melting a slightly bigger hole than the cross section of the probe (depending on the melting speed itself, the applied power and the surface of the probe); further, friction losses between the ice wall of the melted channel and the surface of the probe, due to the viscosity of the thin water layer located in between or dust mixed with the ice; radiative losses; and losses in the tether.

The lateral conduction losses have been estimated by Aamot (1967c) for a cylindrical probe with the constraint that the melt water must just stay liquid all around the probe's hull. The integrated lateral power requirement is given by:

$$P_{\text{cond}} = \frac{4\lambda T}{R_{\text{probe}}\pi^2}\left(2\pi R_{\text{probe}}\right)\int_0^{L_{\text{probe}}}\int_0^{\infty} \frac{e^{-\kappa u^2 s/v}}{u\left[J_0^2(R_{\text{probe}}u) + Y_0^2(R_{\text{probe}}u)\right]}duds \qquad (3)$$

where $u$ is the integration argument for the Bessel functions $J_0$ and $Y_0$, and $s$ the spatial coordinate along the length $L_{\text{probe}}$ of the probe. The thermal conductivity of ice is represented by the heat diffusion coefficient $\kappa = \lambda/(\rho c_p)$ with the thermal conductivity of ice, $\lambda$.

This (numerically difficult) integral can be conveniently approximated by (Aamot 1967c, and appendix A):

$$\frac{P_{\text{cond}}}{R_{\text{probe}}^2 v \cdot (T_F - T)} = ax^b \text{ with } x = L/(vR_{\text{probe}}^2)$$
$$= t^*/R_{\text{probe}}^2 \quad \text{in s/m}^2,$$
$$\text{valid for } 5 \cdot 10^4 < x < 10^8 [\text{s/m}^2]$$

with the fit constants $a = 932$ Ws/K/m$^3$ and $b = 0.726$; $t^*$ is called "characteristic time". (4a,b)

$$P = P_0(A, v) + P_{\text{cond}}(A, L, v)$$

Hence the total power required by the probe to penetrate the ice with a velocity $v$ is given by $P$.

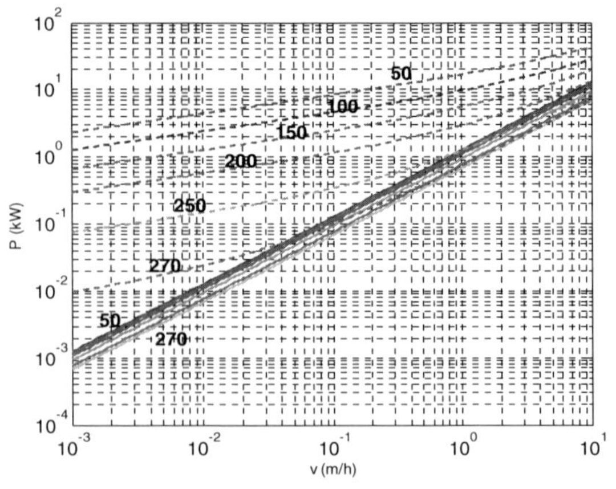


**Fig. 1** Required heating power as a function of melt velocity, without (solid lines) and with conductive losses (broken lines)

Figure 1 shows the required heating power of a typical melting probe ($L$ = 1 m, $R$ = 5 cm) as a function of penetration velocity for ice temperatures from 50 K to 270 K. The solid lines show the minimum power requirement (Eq. 1) while the broken lines depict the total power including losses, Eq. (4).

Figure 2 visualizes the efficiency $E = P/P_0$ as a function of melt velocity. It can be seen very clearly that $E$ becomes very small for melt velocities <1 m/h and very cold ice. For example, in glacier ice (250 K) the efficiency drops below 50% for melt velocites <0.55 m/h while in extremely cold ice the efficiency at these melt velocities is of the order of 5%.

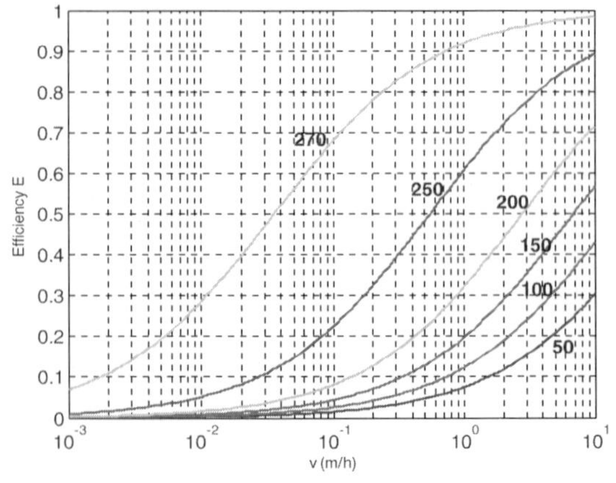


**Fig. 2** Efficiency $E = P/P_0$ as a function of melt velocity

Note that a correction has been applied for the calculation Figs. 1 and 2, since the data in Aamot (1967c) are calculated for water ice near 273 K, while the thermal conductivity ($\propto 1/T$) and specific heat capacity ($\propto T$) vary considerably over an extended temperature range. This has been considered (compare Eq. 3) by multiplying $t^*$ with $\left(\frac{2T_F}{T_F+T}\right)^2$ and the constant $a$ with $\left(\frac{2T_F}{T_F+T}\right)$ which has the effect to transform the heat diffusion coefficient$\kappa$and the heat conductivity $\lambda$ to their values at the mean temperature $(T + T_F)/2$.

While, e.g., the penetration rates given in (Zimmermann 2001) in − 10°C ice can be reproduced within 17% (the measured value is lower, probably owing to other losses than lateral heat conduction), the model becomes uncertain at very low ice temperatures due to the approximations involved. It is an ongoing effort (Kaufmann, E., private communication 2006) to re-calculate the lateral heat conduction integrals (Eq. 3) with proper thermophysical quantities over a wide temperature range and develop numerical thermo-mathematical models for melting probes in conjunction with verifications in extremely cold ice.

Aamot's theory also permits to calculate the rate of lateral heat requirements along the length of the probe, which is important for design and heating regulation.

A theory describing the flow of melt water around a hot point has been developed by (Shreve 1962), but only for the case of isothermal ice at or very slightly below the melting point ("temperate ice") under terrestrial pressure conditions and non-turbulent flow – lateral heat conduction into the surrounding ice is entirely neglected. Interestingly, the efficiency ($P/P_0$) of the hot point then depends mainly on the ratio $PS/RW^{1/4}$ with $P$ heating power, $S$ shape factor of frontal surface (e.g., 1 for a flat surface, around 0.6 for the usual blunt paraboloids or half-spheres), $R$ radius, and $W$ weight (corrected for buoyancy) of the probe.

As Kömle et al. (2002, 2004) pointed out, in an atmosphere-less environment (such as Europa), when the probe is not fully immersed in solid ice, the ice might not melt but sublimate. The latent heat of sublimation $L_s$ at 270 K is ~8 times higher (2836 kJ/kg/K at 0°C) than the latent heat of fusion, so initial progress of the probe at the surface will be accordingly slower until it has been covered by solid ice. Then, melting can occur once a gas pressure greater than the triple point pressure of 611.6 Pa can be sustained around the probe. Tests have been performed on this issue in thermal vacuum chambers (Treffer et al. 2006 and see Section 2.4). The result is that in solid ice, sublimation obviously dominates only the first few cm until the head of the probe has penetrated the ice. An "ice collar" forms very quickly around the rim and obviously suffices to raise the pressure in the tip zone over the triple point, such that very soon melting dominates the penetration process. The situation seems to be different, however, in porous ice where the sublimated gas either escapes or recondenses in the open pores of the surrounding material (Kömle et al. 2004). Here, sublimation dominates at least until the melting channel is completely closed behind the probe, i.e. at least to a depth that is equal or greater than the probe's length.

It is also worth mentioning that all thermal balance considerations imply a close contact of the melting probe's hot nose with the ice/meltwater. If, by obstacles or blocking, a void develops between the hot point and the ice, heat is transported predominantly by radiation, the hot point's temperature increases rapidly and the melting speed drops until the contact is re-established.

### *2.3.1 Other losses*

*Viscous friction*. The Reynolds number applicable for the melt water flow is

$$Re = \frac{\rho_w v h}{\mu_0}$$

with water density $\rho_w$, water viscosity (at ~0°C) $\mu_0$, width of melt water jacket $h$ and melt speed $v$. Typical values ($v = 10^{-3}$ m/s, $h = 1$ mm, $\mu_0 = 1.8 \cdot 10^{-3}$ Pa s) yield Reynold numbers of the order of 1, while the transition to turbulent flow occurs at $Re$ >1160. Thus, laminar flow is assured and the viscous friction can be estimated as

$$F_{\text{visc}} = \eta A \frac{dv}{dx}, \quad \text{with } A \simeq 2\pi(R^2 + RL),$$
$$\frac{dv}{dx} = \frac{v}{h} \tag{0.2}$$

giving very small numbers (for typical melting probes on Earth, $F_{visc}/F_{gravity} \approx 1 \cdot 10^{-6}$). Since the viscous friction is so small, it can be neglected even for low-gravity environments like on Europa.

*Tether losses.* If a tether is payed out by the probe, it will conduct heat from the warm coil towards the melt water and cold refrozen ice behind the vehicle. However, the temperature gradient is not very steep as the tether is surrounded by ~0°C meltwater for a considerable (order of dm) length. With a typical copper cross section (1 mm2) of a power line tether, a gradient of (300–380) K along 1 m, the thermal conductivity of copper, 400 W/mK and a heat transition coefficient 350 W/mK we estimate a thermal loss of 0.04 W, which is negligible.

*Radiative losses.* For a typical melting probe advancing in very cold (80 K) ice, of the order of 10% of the heating power are radiated as mid-IR radiation (Stefan-Boltzmann law with emissivities of probe and ice both set to 1). However, practically all radiation is immediately absorbed in the melt water jacket:

- 98% of a black body's radiation energy is emitted between 4.8μm and 76μm
- The absorption coefficient of water in this wavelength range is >100 $cm^{-1}$, mostly around 1000 $cm^{-1}$ (Irvine and Pollack 1968).
- Thus, the total absorption in a 0.1 mm layer of water is >99.99%

We conclude that the thermal radiation remains in the melt budget and practically no losses occur.

*Overheating losses.* Although counter-intuitive, the addition of more heat does not necessarily increase the melt rate proportionally: since water is a good insulator (0.56 W/K/m compared to 2 W/K/m for ice at 273 K), raising the temperature of the melt water significantly above freezing results in a widening of the melt water jacket around the probe and a subsequent decrease of the heat transfer to the actual ice interface. At very high power densities (>3.25 $MW/m^2$), film boiling further decreases the heat transfer. Thus, the most efficient way to create the phase change is to input just enough heat to initiate melting (Zimmermann 2001).

*Solid friction.* By this we mean friction of the probe surfaces with solid (dust) admixtures. It is extremely difficult to estimate or model and can obviously lead to blocking (see also Section 2.4.7.).

In the following we write the sinking speed as follows, to include the case of porous ice, mixture of sublimation and melting, and conductive losses:

$$v = \frac{P}{A\bar{\rho}(1-\Pi)E(\bar{c}_p(T_F - T) + L_{\text{eff}})} \tag{4}$$

where we have introduced; $\Pi$: the porosity of the ice (in Greenland firn ice, typically 40% at 15 m depth, 15% at 50 m and ≈0 for depths exceeding 100 m); $E$: the loss factor, $E = 1 + P_{cond}/P_0$, $L_{eff}$: the effective heat of phase change, i.e. a value between $L_m$ and $L_s$ depending on the actually dominant process; $\bar{c}_p$: mean heat capacity between $T$ and $T_F$; $\bar{\rho}$: mean solid ice density between $T$ and $T_F$

Equation (4) has been compared to the experimental results of Kömle et al. (2002) covering both pure melting and sublimation in compact and porous ice (snow): the result is that with their probe design (sphere of 4 cm diameter) and power levels (25 or. 60W, respectively), $E = 2.0 \pm 0.1$ as predicted. for an ice temperature of 80 K.

### *2.3.2 Ices other than water*

It is interesting to compare the performance of melting probes in water ice with its performance in other cryogenic solids, notably the astrophysically interesting ices $CO_2$, CO, and $CH_4$. Table 2 gives an overview over the thermophysical properties (Phase change temperature, density, specific heat capacity, phase change enthalpy) of these ices, compared with water ice Ih; the last row gives the ratio of melting velocity to the one in water ice, all

**Table 2** Thermal properties of ices

| | $CO_2$ | CO | $CH_4$ | $H_2O$ |
|---|---|---|---|---|
| $T_f$, melting temperature (~1 bar) (K) | 216.6 | 68.1 | 90.7 | 273.1 |
| $c_p$, specific heat capacity (kJ/kgK) | 1.38 | 1.90 | 2.35 | 1.13 |
| $L$, phase transition enthalpy (kJ/kg) | 573.3 (sublimation) | 29.9 | 58.6 | 333.4 |
| $\rho$, density (kg/m$^3$) | 1540 | 920 | 500 | 920 |
| Penetration velocity relative to water ice, $T_{ice}$=1/2 $T_f$, losses regarded as equal | 0.40 | 5.1 | 5.4 | 1 |

other conditions being equal (and $T_{ice}$ = 1/2 $T_F$). As expected, the melting process in carbon monoxide and methane is about five times more efficient than in water ice; in carbon dioxide, however, it reaches only 40% of the performance in water ice because (down to a depth of about 1300 m on Mars, at least) the phase change is normally by sublimation due to the very high triple point pressure of carbon dioxide, ~74 bar.

## 2.4 Main technological issues

### *2.4.1 Gravity dependence*

Gravity exerts a second-order effect on the penetration of melting probes. The efficiency of the idealized hot point is weakly dependent on the contact pressure of the melt surface on the ice (Shreve 1962). The force of gravity is of course necessary to set the direction of penetration, but does not (to first order) control the melting velocity. Only the possibly non-negligible friction between and sediments needs to be overcome by the weight of the probe. Thus, melting probes would not function under microgravity environments. However, gravity on Europa and Mars is 1.31 m/s$^2$ and 3.73 m/s$^2$, respectively, and not considered a major issue.

### *2.4.2 Prevention of blocking*

Great care has to be taken, particularly in cold ice, to ensure that the re-freezing of ice laterally and behind the probe is slow compared to the melting velocity, in order not to jam the probe. While this is not an issue for e.g. terrestrial glaciers ($t \geq -10$°C), it is critical for very cold ice as experiments at 77 K have shown (Treffer et al. 2006). Proposals for low temperature melting probes foresee heating elements distributed along the length of the probe and careful heating control for that reason.

The issue of blocking due to contaminants in the ice is discussed in Section 2.4.7.

### *2.4.3 Attitude control*

A melting probe requires a means of attitude control if its vertical attitude is to be maintained. Such probes are inherently top heavy because they stand on their tip – the hot point. The slightest deviation from the vertical results in an increasing tendency of the probe to "lean" and "topple over". This is almost inevitable in inhomogeneous ice (with voids, cracks, sediment layers etc.). This instability must be counteracted or eliminated. Mercury steering and "pendulum steering" are described in the literature (Aamot 1967a), with mixed results. The successful AWI "SUSI II" probe (Tüg, H., personal communication 2002; Tüg 2003) used a simple mechanism based on information of the hot nose temperature and the tilt of the probe to control the friction clutch of the probe's tether/cable pay-out mechanism.

Active attitude control can also be based on controlling the heating power (e.g., laterally in the tip, between tip and upper annular heated ring, etc.). In a RHU-heated probe (see Section 2.3.5 below), the required heat distribution could be controlled by means of heat pipes with mechanical valves.

More elaborate techniques for active steering of the probe around obstacles, involving an acoustic obstacle detection sonar and a steerable probe pushing device, have been proposed by DiPippo et al. (1999).

### *2.4.4 Pressure*

The usual approximation of the ambient pressure $p$ in bulk ice (valid if the ice is in a secular hydrostatic equilibrium) is

$p = g\rho h$ where $\rho$ is the ice density and g the local gravity, $g = 9.82$ m/s$^2$ on Earth, $g = 3.73$ m/s$^2$ on Mars and $g = 1.31$ m/s$^2$ on Europa.

We get $p \approx 400$ bar in 30 km depth on Europa as well as in 4000 m depth for Lake Vostok (see also Table 2). The probe must not only survive this pressure but, more importantly, the pressure transient when reaching the ice/water interface. This pressure jump should not occur if ice and water are both in hydrostatic equilibrium, but has been observed on Earth (Tüg, H., personal communication 2002). Aamot (1968) also reported measurements of pressure jumps during stopping or restarting the probe (freezing or. melting of surrounding material with associated volume and pressure changes) up to 88 bar.

Two basic system architectures with respect to operational pressure may be considered: First, a "dry" system based on a pressure tight vessel, capable of withstanding structurally the maximum external pressures and housing all system components with sealed windows and openings for e.g., instrument access to the environment and tether payout. Second, a "wet" architecture where at least parts of the probe can either be flooded with meltwater, or are immersed in e.g. silicone oil. The latter concept has been used, for example, in the Philberth and AWI probes (for the tether storage canister section of the probe).

### 2.4.5 Power supply

The traditional power supply design is by cable, paid out from a coil stored in the aft of the melting probe. Delivering heating power via a cable to great depths is challenging due to both mechanical and electrical constraints, but simplifies communication and steering (attitude) issues.

Typically, high voltages (1000 V, DC) are used for the power supply to minimize losses in the tether and housekeeping and science telemetry is multiplexed on the DC supply voltage. Thus, two wires, of which only one has to be insulated, suffice for power and telemetry. Additionally, it is preferable to include a mylar tether that bears the weight of the probe and is used for steering.

Clearly the thickness of the cable, along with the available total volume for the storage of the coils, is the limiting factor for the maximum possible penetration depth. For the AWI's SUSI probes, a maximum depth of the order of 1000–2000 m (for an optimized design) has been estimated (Tüg, H., personal communication 2002). However, concepts for Philberth probes described in (Aamot 1968) claim a feasible depth of 3000 m or more.

The use of power intensive devices such as ice-melting probes in the outer Solar System strongly points to radioactive units for heating and power supply. The traditional space application RHU (Radioactive Heater Unit) technology is based on $^{238}$Pu (specific thermal power 0.4 kW/kg, half-life 88 years). For an Antarctic application $^{45}$Ca seems to be an attractive alternative since it is not explicitly excluded by the Antarctic Treaty and is a beta radiator (half life: 162 d) so no gamma shielding is necessary. Its decay produces about 40 kW kg$^{-1}$ and its decay product is $^{45}$Sc which is stable and non-toxic. The availability of $^{45}$Ca, bearing in mind the short half-life, needs to be investigated or other isotopes with similar properties need to be found.

In conclusion, a depth of 4000 m (Lake Vostok) can apparently only be reached by either RHU power supplies or by launching the probe from the bottom of the already existing borehole just a few 100 meters above the ice/water interface. For small Mars polar cap probes, power supply by cable from a solar generator at the surface (during polar summer) seems to be possible (a heating power of the order of 25–50 W seems to be sufficient for small probes, see Kömle et al. 2002), while for Europa only heating by RHU's appears feasible.

### 2.4.6 Communications

The impracticality of autonomously guiding the probe back to the surface with any accuracy means that a melting probe in a Europan setting will, at first, be a one-way mission. (Sample return missions have been suggested by Biele et al. (2002). However, experimental proof of the concept is lacking at the time being and it is not foreseeable how close to the surface module the returned probe would re-appear). Also the strategy to recoil the tether, and thus, "climbing" back to the surface may be considered. A problem here might be the need to heat the probe at the top (i.e. close to the tether/cable coil).

Information will therefore have to be conveyed to Earth by a communication system that uses several separate links. Terrestrial projects can employ cables to transmit power and communication data. Although such tethered systems have been examined for planetary ice-probes, they are likely to be limited to relatively shallow exploration depths of the order of a few kilometre.

The problems of communicating between a Europa lander and the Earth are relatively trivial when compared to the problem of sending data through many kilometres of impure water ice. However, an advantage of using radioactive sources for heating and power-raising in a melting-probe is that the vehicle is then relatively 'power-rich' and radio communication methods may be viable between a probe and a lander.

Unfortunately, the presence of salts in water ice can radically increase the attenuation experienced by microwave signals, to the extent that studies have considered deploying relay transceiver pods behind the probe as it melts through Europa's surface. Typical propagation figures indicate that microwave links could pass 10 kbit/s over a few hundred metres with powers of around 200 mW. Realistically, lower rates will be achieved when the effect of adding salts and scattering bodies such as meteoritic debris are considered. Other methods of avoiding the high losses faced by high-frequency signals could involve using much longer wavelengths, which would yield lower bit-rates, and would require a mass memory for telemetry buffering.

Acoustic waves can be transmitted through layers of liquid and solid water; the achievable data rate would be very low and strongly influenced by the unknown characteristics of the ice (density, porosity, layering, mechanical discontinuities, varying attenuation, multiple paths, ambient noise, thermal cracking, reverberation etc.).

Therefore, at the moment the use of a tether line (including a coaxial cable and/or optical fibers) preliminarily appear to be the most reliable, or only, solution, in spite of the great line length required, the need of a storage canister and a technique to manage its safe deployment. It could be based, however, on the mature technologies that exist in wire guided missiles, sub sea torpedoes and remotely operated submersibles, with the necessary improvements due to the particular environment and operating conditions concerned.

### *2.4.7 Ice contaminants*

One major problem regarding the melting into natural ices is intrusion of components that cannot be molten, like dust or salts.

So far, no satisfactory solution has been demonstrated (nor do empirical data exist to our knowledge) to melt into dusty ice, since the dust concentrates underneath the melting tip, building up an insolating layer and increasing the friction. Mechanisms in addition to "pure melting" need to be applied. (di Pippo, 1999). In context with the JPL Cryobot an active water jet system is proposed, splashing/pumping away the debris accumulating at the melt head (Zimmermann 2001).

We would like to emphasize methods combining penetration techniques of melting with those of a "mole" (Gromov 1997). Such a device which can hammer itself e.g. into regolith (in fact, any compressible/porous medium) has a high degree of maturity and was part of the payload of the Beagle 2 lander on Mars (Richter et al. 2001, 2004). A combined melting-hammering shall allow penetration into dusty ices. However, development of a breadboard system combining these two concepts is still to be done.

Regarding salts in the ice one needs to differentiate according to their solubility. For non soluble layers similar solutions as for dust should be applied. Soluble salts like NaCl will be dissolved in the melt water but lead to a higher concentration in the melt water plume lagging the probe, compared to the re-frozen ice. This, however, will not lead to any blocking of the probe, even if some salt may precipitate behind the vehicle.

## 2.5 Breadboard tests for planetary applications

A number of experiments have been performed at DLR in a cold lab as well as in a vacuum chamber. A modified melting probe based on a prototype provided by the Alfred Wegener Institut für Polar- und Meeresforschung (AWI) was used for these experiments (Treffer et al. 2006).

The main part of the probe consists of a copper hemisphere melting head (11.5 cm in diameter), containing three 230 V heating elements with a power of 200 W, each. The heating elements can be operated independently from each other but usually the full heating power of 600 W (measured: 580–645 W depending on the AC voltage) was employed.

An upright aluminium frame was used for vertical suspension of the melting probe. Implementation in the vacuum chamber allowed a centred positioning of the probe above the ice container. The probe was attached with a wire to a strain gauge used to determine if the probe is in contact with the ice surface in order to control the tether mechanism (see Figure 3).

Freeze-in could not be isolated from zero melt-head contact pressure. Under nominal operation, the contact pressure changed by about one order of magnitude between "sitting on the ice" and "hanging in the melt hole surrounded by melt water"; an empirically determined force limit (that included the interval between zero and full immersion in melt water and, consequently, a range in buoyancy) was used in a control loop to feed more tether in an automatic way.

### *2.5.1 Experimental results*

The experiments performed in both, cold lab (atmospheric pressure, solid ice at 243 K) and vacuum chamber ($p$ <1 hPa, solid ice at ~100 K) demonstrated very well the applicability of the theory, as mentioned above. Melting velocities in vacuum were consistent with the predictions according to formulae (4, Aamot-fit etc.): $L = L_m$, efficiency $E \approx 23\%$ for $T_i = 100$K, melt velocities $(2.5...3.7)\cdot10^{-5}$m/s, effective length 0.2 m due to only partial penetration.

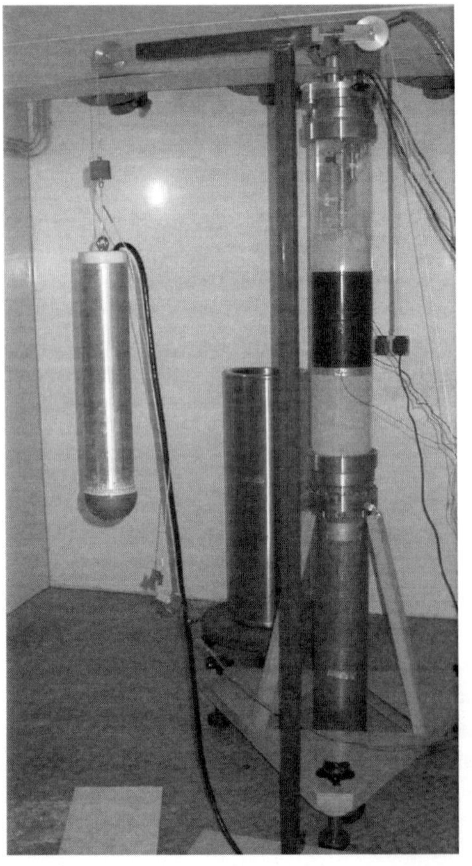

**Fig. 3** Setup for melting probe experiments at DLR

One important result is the fact that, in compact ice, the melting channel freezes again very rapidly, so the probe's hot point is in an almost closed cavity already shortly after penetration, increasing the pressure beyond the triple point pressure. Thus, liquid water can exist, leading to faster progression (by melting, not sublimation).

The drawback of this effective re-freezing is that the probe itself got stuck at the unheated aluminium tube in the low temperature vacuum experiments. Clearly, the lateral heat requirements were not fulfilled. Thus, future experiments will need controlled heating over the complete length of the probe.

## 3 Fields for future application

### 3.1 Terrestrial applications (Antarctica)

Deep underneath the Antarctic ice sheet there are tremendous concealed lakes of water kept in the liquid state by the enormous pressure of the overlaying ice on the one hand and the continuous heat flux from the Earth's interior on the other hand. These lakes have been isolated from the atmosphere for several hundred thousand up to one million years. Until today about 145 of such lakes are detected (Siegert 2000; Siegert et al. 2005).

The lakes are of enormous scientific interest as a potential and unique habitat for life, having developed in a secluded environment. The environment in these habitats is considered to be similar to the conditions as they may exist also in extraterrestrial cases, e.g. on Europa (Priscu and Christner 2004).

The largest and one of the most interesting lakes is Lake Vostok, named after the Russian station located above the outer rim of the lake. The huge ice-sealed underground Lake Vostok lies four kilometres below the surface of the central Antarctic ice sheet. The ice sheet ranges in thickness from 3,800 to 4,200 meters. The lake is 200 km long by 40 km wide (at its widest point) and is divided into two deep basins by a ridge (Nadis 1999).

Figure 4 shows a radar image of Lake Vostok.

As a result of the special and isolated environment of the lake it can be expected that original life forms developed in an unprecedented way (Siegert et al. 2001).

Because of this particularity any investigation of the lake needs to guarantee minimum contamination with both bacterial life from the surface and any potentially toxic material.

Thus, conventional drilling appears problematic. It is necessary to use methods for investigation that satisfy the strict requirements of planetary protection, as applied e.g. for the exploration of Mars, and foreseen for any future Europa mission.

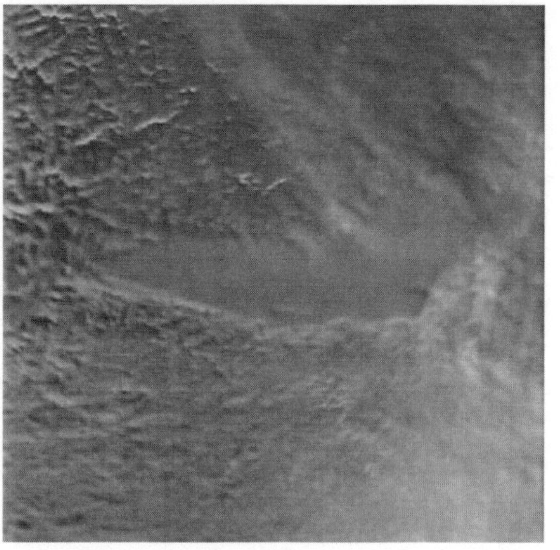

**Fig. 4** Radar image of lake Vostok (Radarsat; Canadian Space Agen cy; NASA/Goddard Space Flight Center Scientific Visualization Studio.)

An international drilling project to Lake Vostok that started in the 1980s was suspended after growing awareness of this problem. The drilling was stopped at the depth of 3,623 m, or approximately 120 m above the surface of the lake (Giles 2004). A further 27 m of core was drilled in 2005/6 and lake penetration is now scheduled for 2007/2008 if the contamination problem has been solved then.

A sterilized melting probe, penetrating through a progressing cavity is an adequate solution to investigate the water of the lake in situ with minimized risk of contamination (Treffer et al. 2006).

### 3.2 Jovian Satellite Europa

One of the most fascinating bodies in the Solar System from an exobiological point of view is Europa, one of the Galilean Satellites of Jupiter.

Table 3 lists the main physical and orbital parameters of Europa.

As the images provided by the Galileo Probe have revealed its surface is covered by a very young layer of water ice. The surface structures (showing almost no craters) indicate recent

**Table 3** Europa's physical and orbital parameters

| | |
|---|---|
| Equatorial radius [km] | 1569 |
| Mass [kg] | $4.8 \cdot 10^{22}$ |
| Surface gravity [m/s$^2$] | 1.31 |
| Mean density [g/cm$^3$] | 3.01 |
| Escape velocity [km/s] | 2.02 |
| Mean orbital radius [km] | 671100 |
| Orbital period [d] | 3.551 |
| eccentricity | 0.009 |
| inclination | 0.47° |

movement of the ice crust and allow speculations about a global sub glacial ocean (Greenberg and Geissler 2002). An overview of the Galileo mission is given by Harland (2000).

Indications of water ice on the surface by infrared astronomy were already published by Kuiper in 1957 and later confirmed by Pilcher et al. (1972). During the Voyager 1 and 2 missions, when the two spacecraft reached the Jovian system in March and July 1979, respectively, images were taken that showed various surface structures, leading to the nomenclature of the observable features. An exhaustive overview of the results of the Voyager mission is given by Lucchitta and Soderblom (1982).

First considerations of a possible liquid ocean underneath the ice crust were given by Cassen et al. (1979) and after Voyager by Reynolds et al. and Squyres et al. both in 1983.

More recent papers include the Galileo results for a more detailed analysis of the satellite's internal structure (Anderson et al. 1998; Khurana et al. 1998).

The surface temperature of Europa is about 100 K. The internal heating, leading to the (probable, but not proven) liquid ocean is explained by tidal heating and (to a much smaller extent) radioactive decay in the interior.

The effects of tidal heating of a satellite around Jupiter can most impressively be seen in the violent active volcanism of Io, the innermost of the Galilean satellites, were enormous eruptions were first detected by the Voyager spacecraft (Morabito et al. 1979). This was predicted by Peale et al. in 1979, when realizing that Io's orbit

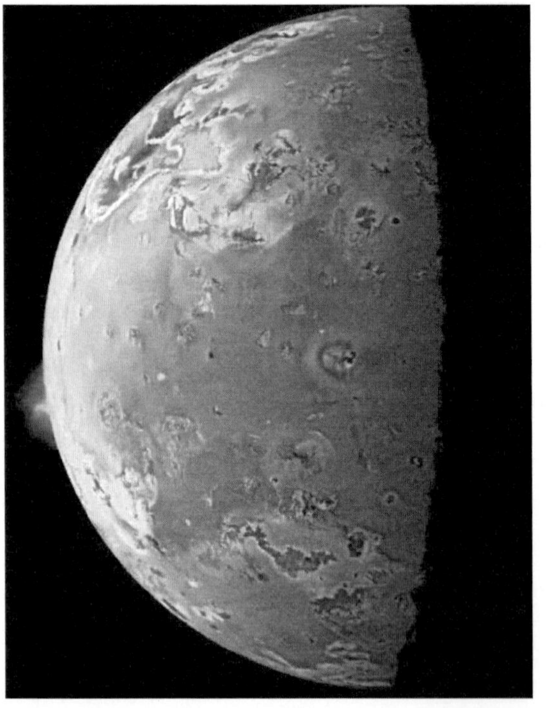

**Fig. 5** Volcanic eruption on Io; the plume on the limb (Pillan Patera) is 140 km high. Image taken by Galileo, June 1997 (NASA/JPL 2005)

has a substantial eccentricity. Shortly after the Voyager flybys the heat dissipation was recalculated (Cassen et al. 1980).

Figure 5 shows an eruption, imaged by the Galileo spacecraft in 1997.

Extrapolating the tidal heating rate of Io ($10^{14}$ W) to Europa, a value of about $6 \cdot 10^{12}$W can be estimated (Greenberg 2005). A more elaborate model leads to higher values of $9 \cdot 10^{12}$W internal heat dissipation or 0.3 W/m$^2$ (O'Brian et al. 2000).

There is no generally accepted model for the thickness of the ice crust available yet. Values vary between >50 km and only a few kilometres (Cassen et al. 1979; Squyres et al. 1983; Ross et al. 1987; Ojakangas and Stevenson 1989). In addition there exists a model explaining Europa's surface and magnetic field features with an underlying sheath of viscous icy material (and no liquid water at all) (Pappalardo et al. 1998).

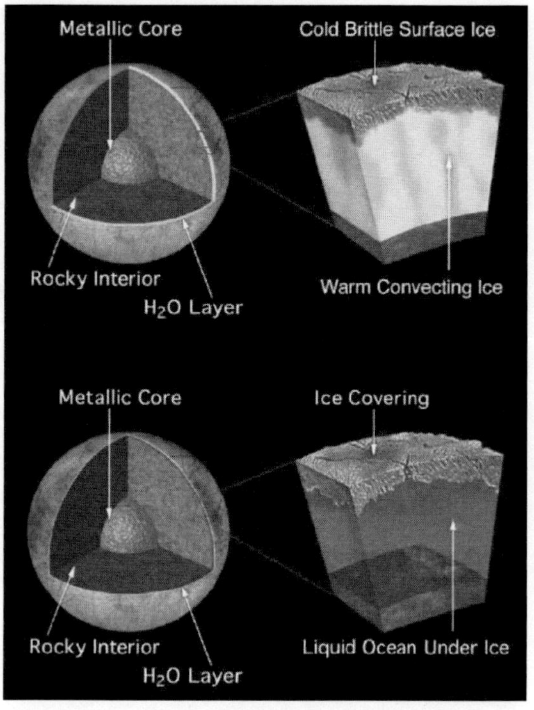


**Fig. 6** Internal structure of Europa (NASA/JPL 2005)

Figure 6 shows two possible scenarios of the internal structure of Europa (NASA/JPL URL).

The most widely accepted model of the interior of Europa assumes a liquid ocean, warmed up by tidal forces. The idea, to find on the bottom of this ocean an environment comparable to black smoker environment at the bottom of the terrestrial oceans, can be seriously considered. Of course, this is speculative, but it is a fascinating area to be explored!

The idea of habitable zones on Europa has been discussed by Reynolds et al. (1983) and later by Greenberg et al. (2001); a very comprehensive overview of the physics of this satellite is given by Greenberg (2005).

Figure 7 shows the trailing hemisphere of Europa, as imaged by the Galileo camera (September 7th, 1996) in natural (left) and enhanced colours. The very low number of impact craters is a clear evidence for the young age of the surface terrain and, thus, of ongoing restructuring of the ice crust.

Figures 8 and 9 show surface features which are a strong indication for liquid underneath a thin crust (NASA/JPL URL)

An intrinsic problem of the investigation of Europa's ice crust and its putative ocean is the penetration of the ice with a device suitable for a planetary exploration mission. Drill rigs appear far too heavy and complex to be implemented into currently considered Lander-missions to the Jovian system.

A melting probe, powered with radioactive heater units (RHU's) seems to be a relatively low-cost, low-mass option to reveal some of the information, hidden underneath thick ice-layers.

This has been proposed e.g. by DiPippo et al. (1999) and Ulamec et al. (2005).

Even if the average thickness of Europa's ice crust may be high, and thus, the liquid water difficult to reach, there may be areas where the ice is much thinner (only a few tens of meters), e.g. at the ridges, where fresh water is thought to be up-welling when the plates open or in the chaotic terrains, where melt-through may occur

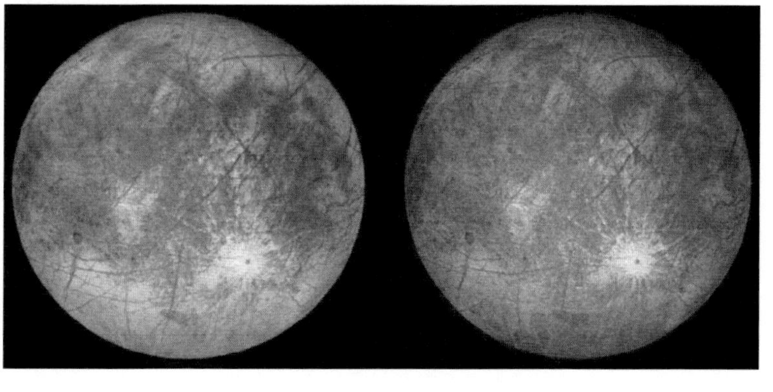

**Fig. 7** Global view of Europa. Two views of the trailing side of the satellite, in approximately natural (left) and enhanced colours. Image taken by Galileo and processed by DLR

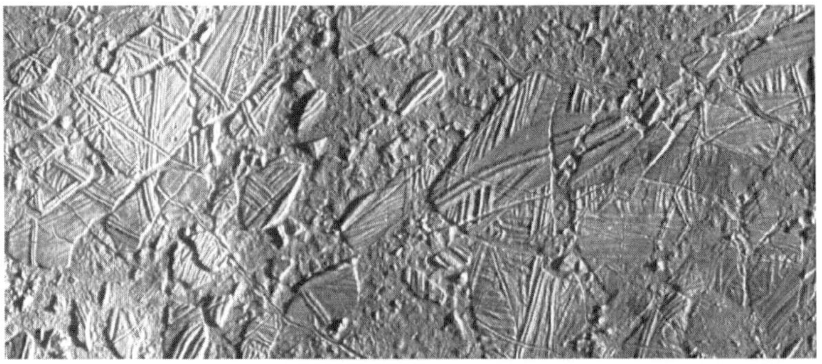

**Fig. 8** View of Conamara region. An area showing the disruption of the ice crust, probably caused by an impact, forming Pwyll crater. The image was taken by Galileo in February 1997. (NASA/JPL 2005)

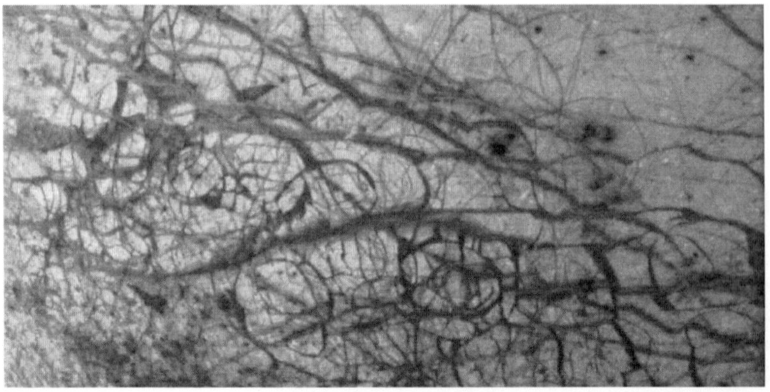

**Fig. 9** Area on Europa resembling ice floes seen in Earth's polar seas. The image was taken by Galileo in June 1997 (NASA/JPL 2005)

(Greenberg et al. 1999, 2005; O'Brian et al. 2000).

In such areas, melting through the thin ice may very well be possible, in an acceptably short time. At the same time, rather fresh material from the global ocean might be found at or close to the surface at these places (Greenberg, R, personal communication 2005). Once a melting probe is below the destructive radiation of the surface layers (~10–40 m), frozen ocean material might be sampled without actually accessing open water.

### 3.3 Mars polar caps

In 1666, the French astronomer and mathematician Giovanni Domenico Cassini observed the bright polar ice caps of Mars, visible already through a standard telescope. Figure 10 shows an image taken with the Hubble Space Telescope. The question, whether or not the Martian polar caps are consisting of water ice remained unanswered until the dawn of the age of modern space flight. The existence of the famous channels observed by Schiaparelli in 1877 and some other astronomers could be

**Fig. 10** A picture taken by the Hubble Space Telescope in "Spring Time". Clearly visible is the ice covered northern polar cap

definitely negated, when Mariner 4 as the first spacecraft sent 22 pictures of the Martian surface back to Earth during its flyby at the planet in July 1965. On the pictures various craters similar to that found on the moon were visible, and some of them appeared to be covered with frost (Godwin 2000). More pictures were obtained by the following flybys of other spacecraft, and in 1976 first compositional data of the Martian soil were transmitted to Earth by the Viking 1 and Viking 2 landers. The two Viking probes were the first to remain intact on the surface of Mars and collected valuable data, e.g. about the Martian weather. Although both landers were equipped with instruments for the detection of microbial life in the Martian soil, no evidence for present or past life was found (Ballou et al. 1978). However, the interesting places on Mars where microbial life forms could exist have not been explored until today. These places should contain water and should shelter micro organisms from the solar UV radiation that hits the Martian surface unfiltered.

The average temperature on the surface of Mars is – 63°C, with a maximum temperature of +20°C at the equator during summer, and a minimum of – 140°C at the poles in winter. The Martian atmosphere is rather thin with a surface level pressure of roughly 6 mbar, and is mainly composed of carbon dioxide (95%), nitrogen (3%) and trace amounts of oxygen. The water vapour content of the atmosphere is only about 3%. However, the relative humidity of the atmosphere is very high with respect to the low temperature and pressure conditions. This allows indeed the formation of water clouds, but precipitation is not possible. On the surface of Mars no liquid water can be found for the same reason, but nevertheless the existence of old outflow channels and canyons point to the existence of running water in the former geological history of the planet.

With liquid water, a warmer weather and a thicker atmosphere in the past, it is very exciting to think about the possibility of past life existing on Mars. The current environmental conditions do not support life directly on the surface due to the sterilizing effect of the solar UV radiation. However, if life has ever evolved there, it is possible that it still exists in biological niches, sheltered from the radiation beneath rocks or inside the Martian soil, or at the planet's poles. Large water ice sheets were found at both Martian poles, being partially covered by $CO_2$ ice during the winter and spring season (Bibring et al. 2004), and also in a crater at high northern latitude (Xie et al. 2006).

During winter time, the northern ice cap extends towards lower latitudes, followed by shrinkage during spring of about 20 km/day, resulting in a minimum extension during the early northern summer. For latitudes higher than 40°, most of the water on the Martian surface is frozen in the polar caps and in permafrost beneath the crust. The permafrost layer gets thicker with increasing latitude, until it rises from the crust at higher latitudes, shaping the ice cap. Topography of the northern and southern polar caps, respectively, was derived from data recorded by the Mars Orbiter Laser Altimeter (MOLA) aboard the *Mars Global Surveyor* spacecraft (Fishbaugh and Head 2001). The Martian polar caps cover an area of $\approx 10^6$ km$^2$ with a thickness as much as 3–4 km (Clifford 2001). With a supposed age of $\approx 10^5$–$10^8$ years, these layered ice deposits appear to be surprisingly young. However, the Martian ice caps contain therefore a record of the seasonal and climatic cycling of atmospheric $CO_2$, $H_2O$, and dust over this period. Therefore, the poles of Mars are within the focus of interest for melting probe applications (Treffer et al. 2006).

Another interesting discovery concerning water ice on Mars was recently obtained from Mars Express data, extending the knowledge of a frozen sea deposit of water ice in the Martian Elysium area (Murray et al. 2006).

### 3.4 Other places in the solar system

There are, of course, other very interesting planetary bodies in the solar system, where in-situ exploration, supported by melting probe technology shall be considered. There are, however, no mid-term missions defined, where such investigations could be performed.

Europa is not the only Jovian satellite with an icy surface that could be penetrated and investigated by means of melting. In particular Ganymed is very interesting and many of the arguments presented in Section 3.2 also apply for this moon. Ganymed (as well as Callisto) might also have a subglacial ocean (Spohn and Schubert 2003), but since it would be covered by a much thicker ice layer, the exobiological interest is focussed on its "inner sister", Europa.

Titan, Saturns large satellite with a dense atmosphere does also have an icy surface with a high content of hydrocarbons. The spectacular results after the entry and landing of the Huygens probe (Zarnecki et al. 2005; Owen 2005; Lebreton et al. 2005), indicate that penetrating the surface, which may be a crust on liquid ethane/methane by melting could well be appropriate.

Looking even further out in the solar system one may think of applications on the Neptunian satellite Triton, Kuiper belt objects or the many small ice bodies like comets or small icy satellites.

## 4 Planetary protection aspects

An aspect of space exploration, which has been noticed very early is the effect of human-caused biological cross-contamination between Earth and other solar system bodies (Randolph et al. 1997). A growing awareness about the possible "infection" of other places in the solar system by so called "hitchhiker" bacteria from Earth found its expression in many passionate discussions among the scientific community. These hitchhiker bacteria could colonize on a spacecraft and its equipment leading to a contamination of an extraterrestrial environment. Such a scenario could cause irreversible and dramatic changes of such an alien environment. Furthermore, these alterations could also interfere with scientific exploration and consequently could spoil the measured data.

Consequently, the design of all missions to other potentially habitable bodies of the solar system has to consider this aspect of planetary protection.

The measures for planetary protection are defined by international law, based on the outer space treaty of 1967 (Rummel 2001; Rummel et al. 2002).

Techniques, applied for the sterilisation of spacecraft (Engelhardt, 2006, could well be applied for a melting probe, designed to explore the Antarctic lakes. Sterilization is obligatory for any lander mission to Mars or Europa.

## 5 Conclusions

1. The exploration of Mars's polar caps and fascinating environments like the putative water ocean below the ice crust of Europa is seen as an important goal for the fields of comparative planetology, astro- and exobiology. Melting probes, already successfully tested in the Antarctic shelf ice, demonstrate that this technology is promising for such applications.
2. Technologies needed for a future Europa mission shall be demonstrated on the ground, leading to an exploration of Antarctic subglacial lakes (like Lake Vostok), which by itself would constitute an important scientific advance.
3. As a first step simple probes as described in this paper are being developed, tested under thermal-vacuum conditions and on Earth (e.g. deep penetration in terrestrial glaciers)

## 6 Appendix: Thermophysical properties of ices

### 6.1 Overview

| Property | Symbol, Unit | $H_2O$ | $CO_2$ | CO | $CH_4$ |
|---|---|---|---|---|---|
| Molar mass | $M$, g/mol | 18.01526 8 | 44.0098 ± 0.0016 | 28.01 | 16.043 |
| Triple point temperature | $T_t$,K | 273.16 (exact) | 216.592 ± 0.001 (ITS-90 secondary ref. point) | 68.05 ± 0.05<br>68.13 ± 0.05 [CC] | 90.694(1) (ITS-90 secondary ref. point) |
| Triple point pressure | $P_t$, Pa | 611.655 | (0.517950 ± 0.00010)E6 | 0.01537(3) | 1.169(6) E4 |
| Critical point temperature | $T_c$, K | 647.096 | 304.128 ± 0.015 | 132.9<br>134.45 ± 0.4 | 190.6 |
| Critical point pressure | $P_c$, Pa | 2.2064 E6 | (7.3773 ± 0.0030) E6 | (3.499 ± 0.03) E6 | 4.592 E6 |
| Critical point density | $\rho_c$, kg/m$^3$ | 322 | 467.6 ± 0.6 | 301 | 162.0(2) |
| Normal melting point | $T_m$, K | 273.1525 | (Triple point) | 68.05<br>68.08 | 90.7 |
| Normal boiling point | $T_b$, K | 373.124 | 194.686* (*sublimation pressure = 1 atm, ITS-90 secondary ref. point) | 81.60<br>$81.6_5$ | 111.67 (ITS-90 secondary ref. point) |
| Density of solid at triple point/melting point | $\rho_s$, kg/m$^3$ | 916.700 ± 0.026 | 1541 | 919.8 | 489.8 (490 – 530 over the whole range) |
| Enthalpy of sublimation | $H_{sub}$, kJ/kg | 2834.359 | 573.31 at $T_b$ | 261.3 [CC] | 600(23) for 53–90K |
| Enthalpy of fusion | $H_{melt}$, kJ/kg | 333.44 | 196.65 at $T_t$ | 29.86(3) | 58.5(2) |
| Enthalpy of vaporization | $H_{evap}$, kJ/kg | 2500.5 (0°C)<br>2255.5 (100°C) | | 217 | 510–584 |
| Specific heat capacity of solid at $T_t$ | $c_p$(s), kJ/kg/K | 2.2 | 1.383 | 1.9 | 2.73 |
| Thermal conductivity of solid | $\lambda_s$, W/m/K | 2.1 | | 0.303 (50 K) | 0.4(1) at $T_t$ (extrapolated) |
| Comments | | Melting temperature under pressure p: $t_F$ /°C = 0 – 0.00076 $p$ – 1.32E-6 $p^2$ with $p$ (bar) for $0 < p < 2000$ bar | | Additional transition at 61.55 K, enthalpy change 22.62(14) kJ/kg | Rotational transition in solid at 20.48 K (ITS-90 secondary ref. point) with $\lambda$-peak in cp, enthalpy change 5.8 (1) kJ/kg |

## 6.2 Temperature dependence of selected quantities

### 6.2.1 *Density*

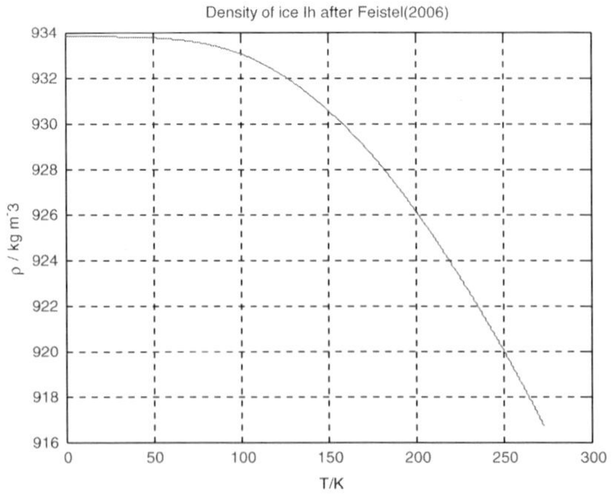


With sufficient accuracy from 0 to 273 K:

$$\rho = 933.31 + 0.037978T - 3.6274 \cdot 10^{-4}T^2[\mathrm{kgm}^{-3}].$$

### 6.2.2 *Thermal conductivity*

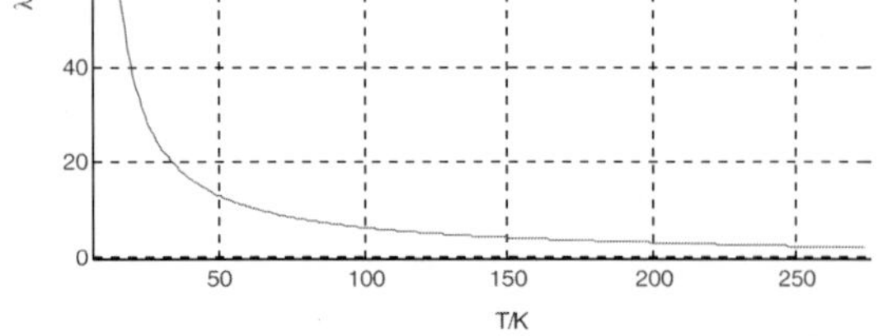


After (Slack 1980), best estimates of $\lambda$ of Ih $H_2O$ ice between 10 K and the melting point at atmospheric pressure:

$$\lambda = 619.2/T + 58646/T^3 + 3.237 \cdot 10^{-3}T - 1.382 \cdot 10^{-5}T^2 \text{in W/m/K} \tag{0.3}$$

This equation gives about 12% higher values than the (Klinger 1980) equation,

$$\lambda = \frac{567}{T} \mathrm{W/m/K}(T{>}50\ \mathrm{K}) \tag{0.4}$$

### 6.2.3 *Specific heat capacity at constant pressure*

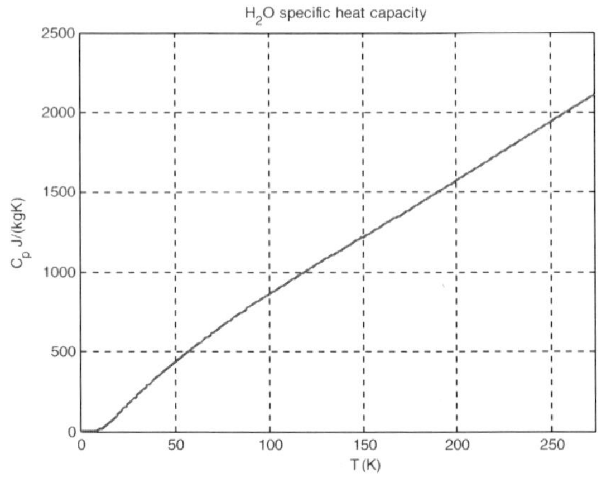


$$C_\mathrm{p} = x^3 \frac{c_1 + c_2x^2 + c_3x^6}{1 + c_4x^2 + c_5x^4 + c_6x^8}[\mathrm{J/kg/K}]$$

$x = T/T_t, T_t = 273.16$ K and $c_1 = 1.843 \cdot 10^5$, $c_2 = 1.6357 \cdot 10^8$, $c_3 = 3.5519 \cdot 10^9$, $c_4 = 1.667 \cdot 10^2$, $c_5 = 6.465 \cdot 10^4$, $c_6 = 1.6935 \cdot 10^6$ after Haida et al. (1974), Flubacher et al. (1960) and Giauque and Stout (1936).

Note that the pressure dependence of Cp is only about dCp/dP = – 8.5e – 10*T [J/K/kg/Pa] for 0.273.15 K at 1 bar.

### 6.2.4 *Latent heat of sublimation*

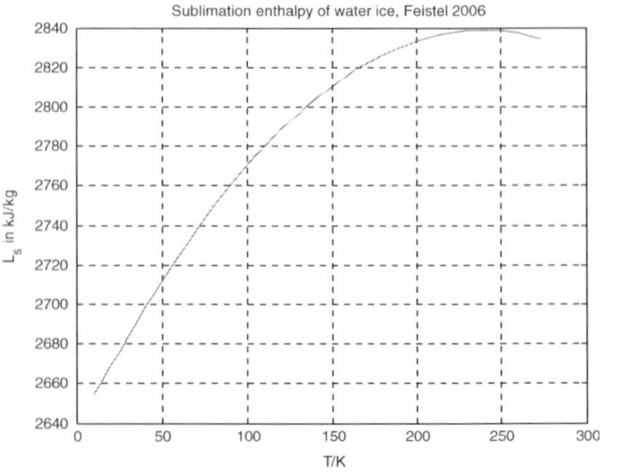

After Feistel (2006) for 0.273.15 K, approximated by a quadratic polynomial,

$$L_s = 2636.77 + 1.65924T - 0.0034135T^2[\mathrm{kJ/kg}]$$

*6.2.5 Viscosity*

After Kestin et al. (1978), Hallet (1963) and IAPWS (2003); correlation for temperatures from − 24°C [supercooled] to 373 °C, at saturation pressure, ± 5%:

$$\eta(T) = A\left(\frac{T}{B} - 1\right)^{\alpha}$$

$$A = 1.4147 \cdot 10^{-4}[\mathrm{Pa\ s}]$$

$$B = 226.8[\mathrm{K}]$$

$$\alpha = -1.5914$$

## 7 References (Appendix)


1. Bedford RE, Bonnier G, Maas H, Pavese F (1996) Recommended values of temperature on the International Temperature Scale of 1990 for a selected set of secondary reference points, Metrologia. Metrologia 33:133–154
2. Feistel R, Wagner W (2005) High-pressure thermodynamic Gibbs functions of ice and sea ice. J Mar Res 63:95–139
3. Feistel R, Wagner W (2006a) A New Equation of State for H2O Ice Ih. J Phys Chem Ref Data 35(2):1–27
4. Feistel R, Wagner W (2006b) Sublimation Pressure and Sublimation Enthalpy of H2O Ice Ih from 0 to 273.16 K. submitted to Geochimica et Cosmochimica Acta, 22 May 2006
5. Feistel R, Wagner W et al (2005) Numerical implementation and oceanographic application of the Gibbs potential of ice. Ocean Science 1:29–38
6. Flubacher P, Leadbetter AJ et al (1960) Heat capacity of Ice at Low Temperatures. The J Chem Phys 33(6):1751–1755
7. Giauque WF, Stout JW (1936) The entropy of water and the third law of thermodynamics. The heat capacity of ice from 15° to 273°. J Am Chem Soc 58:1144–1150
8. Haida O, Matsuo T et al (1974) Calorimetric study of the glassy state X. Enthalpy relaxation at the glass-transition temperature of hexagonal ice. J Chem Thermodynamics 6:815–825
9. Hobbs PV (1974) Ice physics. Oxford, Clarendon Press
10. Hallet J (1963) The temperature dependence of the viscosity of supercooled water. Proc Phys Soc 82:1046–1050
11. IAPWS (2003) Revised release on the IAPS formulation 1985 for the viscosity of ordinary water substance. Releases of The International Association for the Properties of Water and Steam, IAPWS
12. Kestin J, Solokov M, Wakeham WA (1978) Viscosity of liquid water in the range – 8°C to 150°C, J Phys Chem Ref Data 7(3):941–948
13. Klinger J (1980) Influence of a phase transition of ice on the heat and mass balance of comets. Science 209(July 1, 1980):271.
14. Manzhelii VG (1996) Physics of cryocrystals. Woodbury, New York, AIP Press
15. NIST (2006) Standard Reference Data Program – Web Book, National Institute of Standards and Technology
16. Osborne NS (1939) Heat of fusion of ice. A revision. J Res Nat Bureau Standards (US) 23:643–646
17. Petrenko VF, Whitworth RW (1999) Physics of ice. Toronto, Clarendon Press, Oxford
18. Span R Wagner W (1996) A new equation of state for carbon dioxide covering the fluid region from the triple-point temperature to 1100 K at pressures up to 800 MPa. J Physical and Chemical Reference Data 25: 1509–1596


## References


Aamot HWC (1967a) Pendulum steering for thermal probes in glaciers. J Glaciol 6:935–939

Aamot HWC (1967b) The Philberth probe for investigating polar ice caps CRREL Special Report 119. Cold Regions Research & Engineering Laboratory, Hanover, New Hamphire

Aamot HWC (1967c) Heat transfer and performance of a thermal probe for glaciers CRREL Special Report 194. Cold Regions Research & Engineering Laboratory, Hanover, New Hamphire

Aamot HWC (1968) Instrumented probes for deep glacial investigations. CRREL Special Report 210, Cold Regions Research & Engineering Laboratory, Hanover, New Hamphire, 1968. Same as Aamot, HWC, (1970) Instrumented Probes for Deep Glacial Investigations. J Glac 7 (50):321–328

Aamot HWC (1970a) Self-contained thermal probes for remote measurements within an ice sheet. In: International Symposium on Antarctic Glaciological Exploration (ISAGE), Hanover, N. H., September 3–7, 1968; International Association of Scientific Hydrology Publication 86:63–68

Aamot HWC (1970b) Development of a Vertically Stabilized Thermal Probe for Studies in and Below Ice Sheets, J Eng Industry 92B(2):263–268, Transactions of the ASME, paper no. 69-WA/UnT-3

Anderson J, Schubert G, Jacobsen R, Lau E, Moore W, Sjogren W (1998) Europa's differentiated internal structure: interferences from four Galileo encounters. Science 281:2019–2022

AWI/Jokat W, Oerter H (eds) (1997) Die Expedition ANTARKTIS-XII mit FS "Polarstern 1995, Bericht vom Fahrtabschnitt ANT-XII/3", Berichte zur Polarforschung, Alfred-Wegener-Institut für Polar- und Meeresforschung, 27568 Bremerhaven, Germany, 219:106–111

Ballou EV, Wood PC, Wydeven T, Lehwalt ME, Mack R (1978) Chemical interpretation of viking lander 1 life detection experiment. Nature 271:644–645

Bibring JP, Langevin Y, Poulet F, Gendrin A, Gondet B, Berthé M, Soufflot A, Drossart P, Combes M, Bellucci G, Moroz V, Mangold N, Schmitt B, OMEGA Team (2004) Perennial water ice identified in the south polar cap of Mars. Nature 428:627–630

Biele J, Ulamec S, Garry J, Sheridan S, Morse AD, Barber S, Wright I, Tüg H, Mock T (2002) Melting Probes at Lake Vostok and Europa. In: Proceedings of the First European Workshop on Exo/Astrobiology. ESA SP 518:305–308

Cardell G, Hecht MH, Carsey FD, Engelhardt H, Fisher D, Terrell C, Thompson J (2004) The Subsurface Ice Probe (SIPR): A Low-Power Thermal Probe for the Martian Polar Layered Deposits, 35th Lunar and Planetary Science Conference, March 15–19, 2004, League City, Texas, abstract no. 2041

Carsey FD, Chen G-S, Cutts J, French L, Kern R, Lane AL, Stolorz P, Zimmermann W (1999) Exploring Europa's Ocean: a challenge for marine technology of this century. MTS Journal 33(4):5–12

Cassen P, Peale SJ, Reynolds RT (1980) Tidal dissipation in Europa: A correction. Geophys Res Lett 7:987–988

Cassen P, Reynolds RT, Peale SJ (1979) Is there liquid water on Europa?. Geophys Res Lett 6:731–734

Clifford SM et al. (2001) The state and future of mars polar science and exploration. Icarus 144:210–242

Clow GD, Koci B (2002) A Fast Mechanical Access Drill for Polar Glaciology, Paleoclimatology, Geology, Tectonics and Biology. Mem Natl Inst Polar Res Spec Issue 56:1–00

DiPippo S, Mugnuolo R, Vielmo P, Prendin W (1999) The exploitation of Europa ice and water basins: an assessment on required technological developments, on system design approaches and on relevant expected benefits to space- and earth-based activities. Planet Space Sci 47:921–933

Engelhardt H, Humphrey N, Kamb B, Fahnestock M (1990) Physical conditions at the base of a fast moving Antarctic ice stream. Science 248:57–59

Engelhardt H, Kamb B, Bolsey R (2000) A hot-water ice-coring drill. J Glaciol 46(153):141–145

Engelhardt M (2006) Investigation of decontamination procedures for application on melting probes according to present Planetary Protection rules, Master of Science Thesis, Aachen University of Applied Sciences, Jülich campus, Department of Applied Sciences and Technology

Fishbaugh KE, Head JW (2001) Comparison of the North and South polar caps of mars: new observations from MOLA data and discussion of some outstanding questions. Icarus 154:145–161

Gantt LL, Oba EM, Leising L, Stagg T, Stanley M, Walker E, Walker R (1998) Coiled tube drilling on the Alaskan North Slope. Oilfield Rev 10(2):20–35

Giles J (2004) Russian bid to drill Antarctic lake gets chilly response. Nature 430:494

Godwin R (Editor) (2000) Mars – The NASA Mission Reports. Apogee Books, Burlington Ont., Canada

Greenberg R, Hoppa GV, Tufts BR, Geissler P, Riley J, Kadel S (1999) Chaos on Europa. Icarus 141:263–268

Greenberg R, Geissler P (2002) Europa's dynamic icy crust: An invited review. Meteoritics and Planetary Sci 37:1685–1711

Greenberg R, Tufts BR, Geissler P, Hoppa GV (2001) Europa's Crust and Ocean: How Tides create a potentially habitable physical setting. In: Astrobiology, pp 111–124, Springer Verlag Berlin/ Heidelberg/New York

Greenberg R(2005) Europa The Ocean Moon. Springer Verlag, Berlin, New York

Gromov VV, Misckevich AV, Yudkin EN, Kochan H, Coste P, Re E (1997) The Mobile Penetrometer, a "Mole" for Sub-Surface Soil Investigation. 7th European Space Mechanisms and Tribology Symposium, ESA SP-410. pp 151–156

Harland D (2000) Jupiter Odyssey. Springer Praxis, Berlin and New York

Hansen BL, Kersten L (1984) An In-situ Sampling Thermal Probe. In: Holdsworth G et al. (eds) Ice Drilling Technology. USA CRREL Special Report 84–34

Irvine WM, Pollack JB (1968) Infrared optical properties of water and ice spheres. Icarus 8:324–360

Kasser H (1960) Ein leichter thermischer Eisbohrer als Hilfgerät zur Installation von Ablationsstangen auf Gletschern. Geofisica Pura e Applicata 45(1):97–114

Kelty JR (1995) An in situ sampling thermal probe for studying global ice sheets. Ph.D. thesis, University of Nebraska
Khurana KK, Kivelson MG, Stevenson DJ, Schubert G, Russell CT, Walker RJ, Polansky C (1998) Induced magnetic fields as evidence for subsurface oceans in Europa and Callisto. Nature 395:777–780
Kömle NI, Kargl G, Steller M (2002) Melting probes as a means to explore planetary glaciers and ice caps. In: Proceedings of the First European Workshop on Exo-Astrobiology, ESA SP-518: 305–308
Kömle NI, Treffer M, Kargl G, Kaufmann E, Steller M (2004) Development of Melting Probes for Exploring Ice Sheets and Permafrost Layers, presented at the 6th International Symposium on Permafrost Engineering, Lanzhou, China, 5–7 September 2004
Kuiper G (1957) Infrared observations of planets and satellites. Astronomical Journal 62:245
Lebreton JP et al (2005) An overview of the descent and landing of the Huygens probe on Titan. Nature 438:758–764
Lucchitta BK, Soderblom LA (1982) The Geology of Europa. In: Morrison D (ed) The Satellites of Jupiter. University of Arizona Press, Tucson, pp. 521–555
Nadis S (1999) Moves are afoot to probe the lake trapped beneath Antarctic ice. Nature 401:203
NASA/JPL URL: http://www.galileo.jpl.nasa.gov/
Maurer WilliamC (1968) Novel drilling techniques. Pergamon, Oxford
Mellor M, Introduction to drilling technology, In: ESA SP-302, Physics and Mechanics of Cometary Materials, Paris 1989
Morabito LA, Synnott SP, Kupferman PN, Collins SA (1979) Discovery of currently active extraterrestrial volcanism. Science 204:972
Murray JB, Balme MR, Muller JP, Kim JR, Morley J, Neukum G, HRSC Co-Investigator Team (2006) Preliminary Observations on New Images of the Elysium Frozen Sea Deposits from HRSC Mars Express, 37th Annual Lunar and Planetary Science Conference, abstract no. 2293
National Academy of Sciences (2000) Preventing the forward contamination of Europa, Report; Task Group on the Forward Contamination of Europa, Space Studies Board, ISBN NI000231
O'Brian DP, Geissler P, Greenberg R (2000) Tidal heat in Europa. Ice thickness and the plausibility of melt through. Bull Am Astron Soc 32:1066
Ojakangas GW, Stevenson DJ (1989) Thermal state of an ice shell on Europa. Icarus 81:220–241
Owen T (2005) Planetary science: Huygens rediscovers Titan. Nature 438:756–757
Paige DA, September (1992) The Mars Polar Pathfinder/ Proposal for a Discovery Mission. Discovery Program Workshop, Concept #83
Pappalardo RT, Head JW, Greeley R, Sullivan RJ, Pilcher C, Schuberts G, Moore WB, Carr MH, Moore JM, Belton MJ, Goldsby DL (1998) Geological evidence for solid state convection in Europa's ice shell. Nature 391:365–368
Peale SJ, Cassen P, Reynolds RT (1979) Melting of Io by tidal dissipation. Science 203:892–894
Philbert K (1962) Une methode pour mesurer les temperatures a l'interieur d'un inlandsis. Comptes Rendus Hebdomadaires des Seances de l' Academie des Sciences, Paris, Tom 254(22):3881–3883
Pilcher C, Ridgeway S, McCord T (1972) Galilean satellites: Identification of water frost. Science 178:1087–1089
Priscu JC, Christner BC (2004) Earth's icy biosphere. In: Bull A (eds) Microbial diversity and bioprospecting Chap 13. ASM Press, Washington, DC, pp 130–145
Randolph RO, Race MS, McKay CP (1997) Reconsidering the theological and ethical implications of extraterrestrial life. CTNS Bull 17(3):1–8
Richter L, Coste P, Gromov VV, Kochan H, Pinna S, Richter HE (2001) Development of the "planetary underground tool" subsurface soil sampler for the mars express "Beagle 2" lander. Adv Space Res 28(8):1225–1230
Richter L, Coste P, Gromov VV, Grzesik A (2004) The mole with sampling mechanism (MSM) – Technology development and payload of beagle 2 mars lander. Proceedings, 8th ESA Workshop on Advanced Space Technologies for Robotics and Automation (ASTRA 2004), Noordwijk, The Netherlands, November 2–4
Reynolds RT, Squyres SW, Colburn DS, McKay CP (1983) On the habitability of Europa. Icarus 56:246–254
Ross M, Schubert G (1987) Tidal heating in an internal ocean model of Europa. Nature 325:133–144
Rummel JD (2001) Planetary exploration in the time of astrobiology: protecting against biological contamination. PNAS 98(5):2128–2131
Rummel JD, Stabekis PD, DeVicenzi DL, Barengoltz JB (2002) COSPAR's planetary protection policy: a consolidated draft. Adv Space Res 30:1567–1571
Sas-Jaworsky A, Bell S (1996) Innovative applications stimulate coiled tubing development. World Oil 217(6):61–69
Shreve RL (1962) Theory of performance of isothermal solid-nose hot-points boring in temperate ice. J Glaciol 4(32):151–160
Siegert MJ (2000) Antarctic subglacial lakes. Earth Sci Rev 50:29–50
Siegert MJ, Ellis-Evans J, Tranter M, Mayer C, Petit JP, Salamatin A, Priscu JC (2001) Physical, chemical and biological processes in Lake Vostok and other Antarctic subglacial lakes. Nature 414:603–609
Siegert MJ, Carter S, Tabacco I, Popov S, Blankenship DD (2005) A revised inventory of Antarctic subglacial lakes. Antarctic Sci 17:453–460
Simonsen LC, Nealy JE (1993) Mars surface radiation exposure for solar maximum conditions and 1989 Solar proton events, NASA Technical Paper 3300
Spohn T, Schubert G (2003) Oceans in the Galilean Satellites of Jupiter?. Icarus 161:456–467
Squyres SW, Reynolds RT, Cassen PM, Peale SJ (1983) Liquid water and active resurfacing on Europa. Nature 301:225–226
Treffer M, Kömle NI, Kargl G, Kaufmann E, Ulamec S, Biele J, Ivanov A, Funke O (2006) Preliminary studies

concerning subsurface probes for exploration of icy planetary bodies. Accepted by Planetary and Space Science

Tüg H (2003) Rechnergesteuerte Schmelzsonde zur Ermittlung unterschiedlicher Messparameter im Eisbereich, Patentschrift DE 101 64 648 C 1, Deutsches Patentamt, 6.2.2003

Ulamec S, Biele J, Drescher J and Ivanov A (2005) A Melting Probe with applications on Mars, Europa and Antarctica; 56th International Astronautical Congress, IAC-A1.7.08, Fukuoka/Japan

Xie H, Zhu M, Guan H, Smith RK (2006) Isolated Water Ice in an Unnamed Crater Away From the Residual North Polar Cap of Mars: the Only One?, 37nd Annual Lunar and Planetary Science Conference, abstract no. 1764

Zarnecki JC et al (2005) A soft solid surface on Titan as revealed by the Huygens Surface Science Package. Nature 438:792–795

Zimmerman W, Bonitz R, Feldman J (2001) Cryobot: an ice penetrating robotic vehicle for Mars and Europa. 2001 IEEE Aerospace Conference, Big Sky, MT, 311–323

Rev Environ Sci Biotechnol (2007) 6:223–229
DOI 10.1007/s11157-006-9113-0

REVIEW PAPER

# A proposed classification of environmental adaptation: the example of high altitude

**Jean-Paul Richalet**

Received: 20 February 2006 / Accepted: 25 September 2006 / Published online: 24 October 2006


**Abstract** Extreme environments are defined as the opposite of usual environments where the evoked physiological responses are unperceivable, repeatable and adjusted to the constraint. Adaptation strategies to a given environment show three levels: *cultural or technological*, where a buffer space is built to protect the organism from the hostile milieu, *physiological*, where temporary adaptive mechanisms are developed, and *genetic*, where full adaptation is possible with normal life and reproduction. The cost of adaptation increases from the genetic level (minimal cost) to the technological level. These concepts are illustrated by the example of adaptation to altitude hypoxia. The technological level is given by the use of oxygen bottles by high altitude climbers. The physiological level involves various physiological and biological systems (increase in heart rate, ventilation, erythropoiesis, expression of hypoxia-inducible factors, etc.). The genetic level has been reached by some animal species such as Yaks, Llamas, Pikas but has not yet been demonstrated in humans. Diseases developed during exposure to acute or chronic hypoxia may be considered as "adaptive crises" that mimic the transition to a lower energy level of adaptation.

**Keywords** Extreme environment · Adaptation · Hypoxia · Altitude · Physiology · Genetics

J.-P. Richalet (✉)
Laboratoire: « Réponses cellulaires et fonctionnelles à l'hypoxie » EA 2363, UFR Santé Médecine Biologie Humaine, Université Paris 13, 74 rue Marcel Cachin, F 93017 Bobigny Cedex, France
e-mail: richalet@smbh.univ-paris13.fr

## 1 Introduction

Survival, life and reproduction in a changing, and sometimes extreme, environment is a challenge for thousands of animal species and millions of humans. Before trying to unravel the physiological mechanisms involved in the adaptation processes to extreme environments, it is convenient to start by defining what is "extreme" for living organisms.

For that purpose, it might be useful to propose a definition for its contrary. Would the opposite notion of extreme be "natural" or "usual"? The responses to altitude hypoxia will serve as a model for the discussion of physiological responses to environmental changes.

### 1.1 Proposal

In a *usual environment*, physiological responses evoked by the exposure to the environment are unperceptible and are therefore considered by the subject as "usual". For example, tachycardia is a normal response of the heart if it happens during exercise: it is expected, well known, repeatable, and adjusted to the constraint, so that heart rate and cardiac output increase in proportion with the increased oxygen demand. Being out of breath during a severe effort at high altitude is considered as a normal response by one individual, but only if this individual has a previous personal experience of exercising at high altitude or if he obtained a complete information of these processes before being exposed.

In an *extreme, hostile, unusual environment*, local changes of the milieu will trigger strong reactions from the individual, either at a physiological, psychological or behavioral level. In response to the changes in the environment, the organism will try to develop adaptation mechanisms the objectives of which are, at short, medium or long term, to attenuate the intensity of these initial intense reactions. Finally, a perfect adaptation would allow to not figure out the extreme character of the environment so that it is felt as "usual". Do well adapted Tibetans or Quechuas consider that living at 4,000 m is a problem?

### 1.2 Mechanisms of adaptation

The impact of environmental changes upon any organism, human in particular, will depend on the efficiency of adaptation mechanisms.

These mechanisms can allow:

- *to maintain an almost normal activity*: the efficiency is then maximal, there is very little feeling of the limitations linked to the environmental constraints; example: life under temperate climates at sea-level or moderate altitude
- *to preserve a satisfactory autonomy*: reduced efficiency: activities are limited, physical performance decreases, but life at rest is not a problem; example: life at high altitude, cold or desert areas
- *to survive*: minimal efficiency: the organism undergoes a dramatic decrease of its autonomy, survival is possible, but nothing more; example: the summit of Mount Everest without oxygen

All extreme environments cannot trigger efficient adaptive mechanisms, even partially efficient processes for survival. Then this type of environment is considered as *lethal*: prolonged underwater life without air supply, exposure to toxic acids, etc.: the efficiency is null. Between the usual environment where we develop everyday life and the lethal space where survival is impossible, there lies the extreme environment where an "acclimatization field" is possible. The concept of extreme environment is relative to a given species. The natural environment for "extremophiles", bacteria living at extreme pH of 3 or less, or temperatures of 85–100°C (Niehaus et al. 1999) is not compatible with the survival of most other species, especially mammals. However, the tendency to "anthropocentrism" tends to let some authors define an extreme environment as an environment extreme for humans (Fig. 1).

When full or partial adaptation is possible, it can develop at three different levels (Ruffié 1982):

- cultural level
- physiological level
- genetic level

*Cultural or technological adaptation* deals with behavioral changes at the individual or social level. The individual can use technological means (oxygen bottles, diving or space suit) to fight the negative effects of the environment: it is dressed up to protect himself against these adverse effects. For that purpose, it creates a "substitution space" that will protect him against the damage caused by the environment, like a "buffer" or "microclimate", impeding or reducing the contact between the organism and the stressing milieu. Depending on the tightness of the relation between this buffer and the individual, its economical cost will vary a lot. As an example, there is an increasing cost from the diving suit of the diver to the submarine and to the underwater life support station. Similarly, from the historical Appolo cabin to the space shuttle and to an orbital station or colony on Mars. The greater is the autonomy given to the subject, the higher the cost of the technological device.

In this cultural strategy of adaptation, there is no attempt to adapt the organism to the environment. On the contrary, the objective is to shield the individuals from the deleterious effects of the

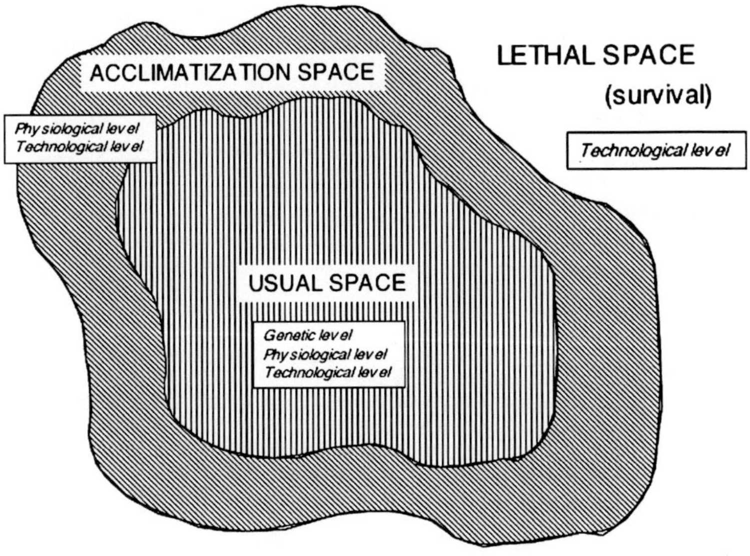


**Fig. 1** The limits of adaptation to the environment. The *usual* space is the one for which physiological responses are unperceptible, repeatable and adjusted to the constraint. The *acclimatization* space is where physiological responses are necessary to maintain autonomy. In the *lethal* space, temporary survival is possible, only thanks to technological (cultural) adaptation

milieu, thanks to the interposition of a buffer substitution space. No evolution is possible, from a genetic point of view, since there is no pressure upon the organism.

*The physiological adaptation* level allows the development of adaptive mechanisms the objective of which will be to recreate, at the cellular level, an environment similar to the usual or natural environment. Once again, we deal with a substitution space, but at the level of the organ or the cell. Thus at high altitude, ventilation and heart rate increase in order to increase the flux of oxygen to the tissues, and as a consequence, so that the mitochondria "sees" almost the same level of oxygen pressure: the mitochondria should not know that it stays at high altitude...

Similarly, during an exposure to cold, skin vessels constrict, shivering process produces heat so that the temperature of the central organs remains constant ("usual"). Brain is warm, even if the skin sensors detect a cold ambient temperature. An essential characteristic of these physiological adaptive mechanisms is that they disappear when the organism is withdrawn from the hostile milieu: red cells produced at high altitude are progressively destroyed when coming down to sea level.

*The genetic level* allows a normal activity in an environment which is supposedly extreme, since the adaptive mechanisms are engraved in the genome. For the corresponding species, the environment is thus considered as "natural" and the cost of adaptation is minimal for a given individual. By definition, the appearance of a genetic adaptive process to a given environment makes the status of this type of environment to switch from "extreme" to "usual". This transition is only possible if the individual—by way of its species—has been exposed for numerous generations to the stressors of the milieu.

## 1.3 Cost of adaptation

*Adaptation has an energetic cost* which will depend on the level of complexity of the involved mechanisms. The cost of *genetic adaptation* is minimal, by definition, since the exposure to the environment requires no effort from the exposed organism. The cost of *physiological adaptation* is high since, for example, it implies an increase in cardiac and respiratory work, an increase in red cell production, etc. Among the physiological processes of adaptation, if we take the example of hypoxia, initial acute mechanisms are more costly, cardiac work increases proportionally to heart rate, then with "acclimatization", the increase in red cells allows a progressive decrease in heart rate and work and is less costly than tachycardia. The cost of physiological adaptation can also be very high, unbearable by the individual who will loose the adaptive physiological characteristics he had developed. Alternatively, the mechanisms of adaptation may impose a too high cost which could trigger a specific disease. As an example, the resident of the Altiplano or the Tibetan plateau may develop an excessive polycythemia which makes the blood too viscous, blocks the circulation in the small vessels and alters the function of the heart (see below). The cost of the *cultural adaptation* level is far the most elevated, since it supposes the fabrication of tools, objects, suits, habitats essential for the creation of the substitution milieu: from the oxygen bottle (minimal tool to create a normal oxygen pressure at the mouth level) to the portable hyperbaric chamber, to the large hyperbaric chamber where a whole hypoxic environment can be simulated for days. If we take the example of space exploration, the cost is increasing from the extravehicular suit to the space shuttle and to the colony on Mars. This concept of cost of adaptation has to be considered here at the individual level. From the evolutionary point of view, it is clear that the adaptation processes, as defined by Darwin, represent a tremendous cost over many generations of a given species (Fig. 2).

Finally, two types of strategies characterize the adaptation to extreme environments:

- To shield the organism, as a whole or at the cellular level, from the stressing milieu, by interposing a buffer substitution environment, using technological or physiological processes.
- To "accept" the constraints of the environment by using or developing, under the gene-environment interaction, phenotypes favorable to life in that specific milieu.

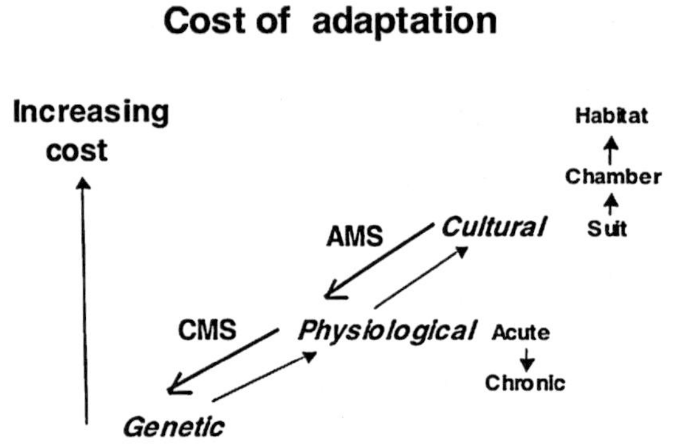


**Fig. 2** At the individual level, the cost of adaptation is increasing from the genetic level (cost = 0) to the physiological level (higher cost in acute than in chronic exposure), and to the cultural level (higher cost, as the substitution space is more sophisticated). Acute mountain sickness (AMS) and chronic mountain sickness (CMS) are adaptation crises that mimic the passage from a high to a low cost level of adaptation

### 1.4 Example: life at high altitude

The environment of high altitude is the only one for which the human organism is able to develop very efficient *physiological mechanisms* of adaptation. Thus, acutely exposed to the altitude of 8,848 m (summit of Mount Everest), a non acclimatized individual looses consciousness and dies after a few minutes. On the other hand, when the same individual takes time (3–4 weeks) to get acclimatized to high altitude, he will be able to reach the summit, as far as all the other technical and meteorological constraints have been resolved! The organism will immediately react against the lack of oxygen by triggering physiological responses in order to limit the level of tissue hypoxia (Richalet 1997; Richalet and Herry 2006; Ward et al. 2000) (Fig. 3).

Chemoreceptors play a key role in this process (Lahiri and Cherniak 2001). The involvement of these oxygen sensors and of the corresponding reflex loop allows to assure a sufficient oxygen transport to the tissues at the only condition that the oxygen demand is modest. When the organism increases its energetic demand, like during exercise, the increase in ventilation and heart rate is insufficient: the overall performance of the individual decreases, the individual is no more able to perform the same level of energy expenditure than at sea level. With time and acclimatization, erythropoiesis is activated, pulmonary ventilation stays high while heart rate decreases through a downregulation of cardiac beta-receptors (Richalet et al. 1992). Plasticity of some tissues, skeletal muscle for example, may also help to cope with the lack of oxygen (Fluck and Hoppeler 2003). However, physical performance does not improve significantly. Full adaptation to the hypoxic environment is thus impossible by using only the physiological level.

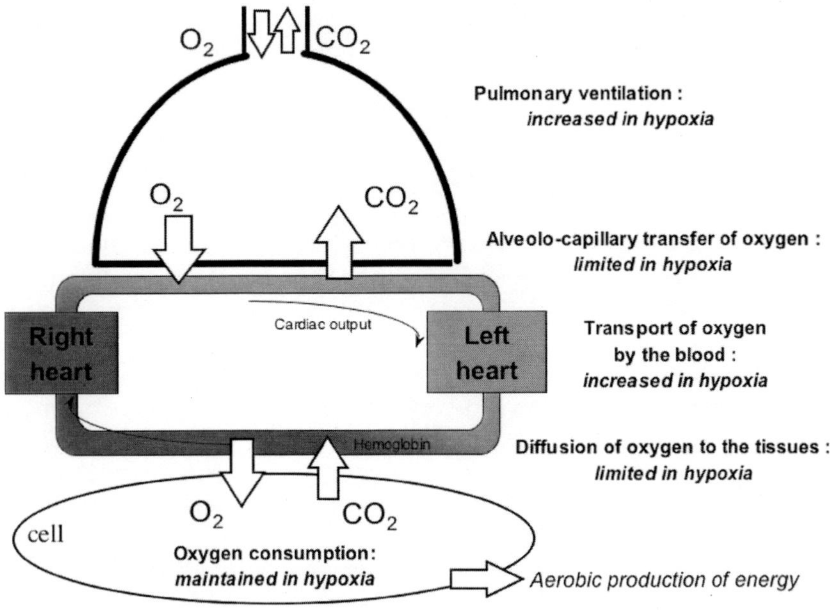


**Fig. 3** The steps of oxygen transport from the air to the cells in altitude hypoxia. Oxygen consumption is maintained (as far as the energy needs are not too high) thanks to many factors among which an increase in ventilation, in cardiac output and in hemoglobin concentration

Adaptation to moderate altitude is possible, but the price to pay is a decrease in physical capacities. Above 5,000–5,500 m, permanent life becomes problematic: there is no long term adaptation. Indeed, signs of physical and mental degradation appear: loss of body weight, muscle mass, and probably loss of neurons. However, in spite of these limitations, altitude remains the only extreme environment for which the human organism is able to develop such remarkably efficient mechanisms. Other environments such as extreme cold are not able to trigger physiological efficient processes: thermoregulation alone cannot limit the decrease in central body temperature and technological and behavioral processes are necessary to survive. It is not surprising that altitude hypoxia gives rise to such a reaction, since oxygen is indispensable for the metabolic machinery of each cell. It is then understandable that the Evolution has permitted the development of numerous and sometimes redundant regulation mechanisms aiming to maintain a sufficient oxygen supply to the cells.

Adaptation to high altitude also shows a *technological—cultural level*: the use of oxygen bottles for climbing or recovery during high altitude expeditions is several dozens years old. This use is now limited to expeditions above 8,000 m and especially to Mount Everest (8,848 m) and K2 (8,611 m). The objective is to interpose a buffer space between ambient air and inspired air, thus recreating at the alveolar level an oxygen pressure close to the one available at sea level.

*The genetic level* of adaptation to high altitude is a well-known phenomenon for a number of altitude species (Llamas, Bar-headed goose, plateau Pika, Guinea pigs, etc.) (Monge and Leon-Velarde 1991; Ge et al. 1998; Anand et al. 1986; Leon-Velarde et al. 1998) However, for humans, the existence of a *genetic* adaptation has never been proven yet: no population living at high altitude (Andeans, Tibetans, Sherpas, Ethiopians) has been shown to have specific genetic characteristics directly related to life at high altitude (Beall et al. 2002; Niermeyer et al. 2001; Ward et al. 2000). Of course, the way Sherpas run on the slopes of Everest with heavy loads is certainly remarkable. However, for the moment, no specific physiological trait has been evidenced that would favor the adaptation of some human populations to high altitude (increased hemoglobin affinity for $O_2$, absence of smooth muscle in pulmonary vasculature, etc.), as shown in some animal species. In animals, hemoglobin has been considered for a long time as the key molecule for genetic adaptation to hypoxia. There is no increased number of red cells but a hemoglobin molecule with high affinity for oxygen. However, recent researches address other systems such as ion channels, vascular response to hypoxia, brain sensitivity to hypoxia, etc. Yaks and Pikas do not show pulmonary hypoxic vasoconstriction and therefore are not exposed to pulmonary hypertension and right heart failure (Ge et al. 1998; Anand et al. 1986). Some chickens found on the Peruvian Altiplano show a mutant hemoglobin with a high affinity for $O_2$ when compared to their sea level cousins (Mejia et al. 1994). Adaptation can also be seen at the embryo level: the permeability of the egg shell of some high altitude birds is higher than the similar species living at sea level (Leon-Velarde et al. 1997), cerebral blood flow of the llama's fetus does not increase in hypoxia, as opposed to what happens in sea level sheep (Llanos et al. 2003).

Further investigations in humans and animal species adapted to high altitude will be necessary to better understand how living organisms cope with hypoxia, using so various strategies.

### 1.5 Environmental diseases: an appeal for gene mutation?

Environmental diseases stand for an incomplete or lost acclimatization to the hypoxic high altitude environment. High altitude diseases can be roughly classified as acute mountain sickness (AMS) with its severe forms, high altitude pulmonary edema (HAPE) and high altitude cerebral edema (HACE) (Ward et al. 2000; Richalet and Herry 2006; Bärtsch et al. 2005).

Acute diseases appear when a sea level native goes rapidly to high altitude, at the initial phase when acclimatization processes are not fully complete. The main manifestations are headache, digestive symptoms, fatigue, dyspnea, neurological disorders. In the case of HAPE or HACE, the condition may be fatal if reoxygenation is not accomplished urgently. The precise pathophysiology of these diseases is not fully understood. However, an increase in vascular endothelial permeability, associated with an exacerbated pulmonary vasoconstriction (in the case of HAPE) are the main responsible factors (Bärtsch et al. 2005; Richalet 1995; Berger et al. 2005). At the molecular level, it is not yet clear if one of the hypoxia-induced factors such as HIF 1$\alpha$, endothelin, NO, VEGF or leucotrienes are involved in these pathological processes (Richalet et al. 1991; Bärtsch et al. 2005). Genetic factors have been associated with the susceptibility to HAPE (Saxena et al. 2005).

Chronic exposure to hypoxia (life time) may lead to a progressive loss of adaptation to hypoxia: chronic mountain sickness (or Monge's disease). This condition is characterized by a progressive loss of chemoresponsiveness, decrease in pulmonary ventilation, increased hypoxemia and development of excessive production of red cells. This polycythemia induces an increase in blood viscosity, as well as pulmonary and systemic hypertension: patients die from cardiac failure or cerebral stroke (Monge et al. 1992).

I make the hypothesis that these diseases, encountered in extreme environments, such as AMS and CMS are the expression of the incapacity of a given genome to react to the stressful milieu. Moreover, these maladies suggest that a genetic mutation would lead to a genomic picture better adapted to the given "unusual" environment. For example, AMS is the translation of an incomplete acclimatization process between acute and chronic hypoxia. CMS is the manifestation of the incapacity of polycythemia to resolve the problem of long term living in hypoxia and that a different hemoglobin, with a greater affinity for $O_2$, would be more economical and therefore better adapted to this environment.

From a classical point of view, diseases such as diabetes are caused by genetic or functional damages so that the response to a mild stress can result in a lethal state. Let us consider the idea that a disease could be a transition between two states of adaptation. Could diseases in general be the expression of a want to have a better adapted genome to the environment for a given species? A disease could, of course, appear but also disappear among a species with the mutation of a gene. For example, the lactase which favors the digestion of lactose normally disappears after the weaning phase, except in certain populations like in Northern Europe. This dominantly inherited genetic trait is known as lactase persistence and is much more frequent in countries where milk is widely used in food, while in Far East, lactose intolerance is very frequent

(Swallow 2003).Furthermore, could a given pathological state favor the expression of factors with a mutagenic potential that could modify a specific gene so that the change in phenotype can be transferred to the genome of the descendents?

The passage from a given state of adaptation (cultural—physiological—genetically) is mimicked by successive "crises" such as AMS and CMS for high altitude environment (Fig. 2). The resolution of each crisis evidences a superior level of adaptation: the sea level native acclimatized to high altitude does not suffer any longer from AMS, the high altitude adapted species do not show excessive polycythemia like in CMS. Therefore, AMS lies between cultural and physiological adaptation, CMS between physiological and genetic adaptation states.

We have much to learn not only from the expression of normal adaptive processes, but also from the malfunction of these mechanisms through "adaptation crisis". Comparative physiology and genomics are useful tools to better understand the various strategies used to cope with environmental changes, especially with extreme environments. Future investigations would benefit in focusing on the link between pathophysiological processes and strategies of adaptations.

## References


Anand IS, Harris E, Ferrari R, Pears P, Harris P (1986) Pulmonary haemodynamics of the yak, cattle and cross breeds at high altitude. Thorax 41:696–700

Bärtsch P, Mairbaurl H, Maggiorini M, Swenson ER (2005) Physiological aspects of high-altitude pulmonary edema. J Appl Physiol 98:1101–1110

Beall CM, Decker MJ, Brittenham GM, Kushner I, Gebremedhin A, Strohl KP (2002) An Ethiopian pattern of human adaptation to high-altitude hypoxia. Proc Natl Acad Sci USA 99:17215–17218

Berger MM, Hesse C, Dehnert C, Siedler H, Kleinbongard P, Bardenheuer HJ, Kelm M, Bartsch P, Haefeli WE (2005) Hypoxia impairs systemic endothelial function in individuals prone to high-altitude pulmonary edema. Am J Respir Crit Care Med 172:763–767

Fluck M, Hoppeler H (2003) Molecular basis of skeletal muscle plasticity—from gene to form and function. Rev Physiol Biochem Pharmacol 146:159–216

Ge RL, Kubo K, Kobayashi T, Sekiguchi M, Honda T (1998) Blunted hypoxic pulmonary vasoconstrictive response in the rodent *Ochotona curzoniae* (pika) at high altitude. Am J Physiol 274:H1792–H1799

Lahiri S, Cherniak NS (2001) Cellular and molecular mechanisms of $O_2$ sensing with special reference to the carotid body. In: Hornbein TF, Schoene RN (eds) High altitude. An exploration of human adaptation. Marcel Dekker Inc., New York, Basel, pp 101–130

Leon-Velarde F, Mejia O, Palacios JA, Monge C (1997) Changes in whole blood oxygen affinity and eggshell permeability in high altitude chickens translocated to sea level. Comp Biochem Physiol B Biochem Mol Biol 118:53–57

Leon-Velarde F, Richalet JP, Chavez JC, Kacimi R, Rivera-Chira M, Palacios JA, Clark D (1998) Inter and intra-species-related differences in the regulation of the cardiac autonomic system. Comp Biochem Physiol B Biochem Mol Biol 119:819–823, 441

Llanos AJ, Riquelme RA, Sanhueza EM, Hanson MA, Blanco CE, Parer JT, Herrera EA, Pulgar VM, Reyes RV, Cabello G, Giussani DA (2003) The fetal llama versus the fetal sheep: different strategies to withstand hypoxia. High Alt Med Biol 4:193–202

Mejia O, Leon-Velarde F, Monge C (1994) The effect of inositol hexaphosphate in the high-affinity hemoglobin of the Andean chicken (*Gallus gallus*). Comp Biochem Physiol B Biochem Mol Biol 109:437–441

Monge C, Leon-Velarde F (1991) Physiological adaptation to high altitude: oxygen transport in mammals and birds. Physiol Rev 71:1135–1172

Monge CC, Arregui A, Leon-Velarde F (1992) Pathophysiology and epidemiology of chronic mountain sickness. Int J Sports Med 13(Suppl 1):S79–S81

Niehaus F, Bertoldo C, Kahler M, Antranikian G (1999) Extremophiles as a source of novel enzymes for industrial application. Appl Microbiol Biotechnol 51:711–729

Niermeyer S, Zamudio S, Moore LG (2001) The people. In: Hornbein TF, Schoene RB (eds) High altitude. An exploration of human adaptation. Marcel Dekker, New-York, Basel, pp 43–100

Richalet JP, Herry JP (2006) Médecine de l'alpinisme et des sports de montagne, 4th edn. Masson, Paris

Richalet JP (1995) High-altitude pulmonary oedema. Still a place for controversy? Thorax 50:923–929

Richalet JP, Hornych A, Rathat C, Aumont J, Larmignat P, Remy P (1991) Plasma prostaglandins, leukotrienes and thromboxane in acute high altitude hypoxia. Respir Physiol 85:205–215

Richalet JP, Kacimi R, Antezana AM (1992) The control of chronotropic function in hypobaric hypoxia. Int J Sports Med 13:S22–S24

Richalet JP (1997) Oxygen sensors in the organism: examples of regulation under altitude hypoxia in mammals. Comp Biochem Physiol A Physiol 118:9–14

Ruffié J (1982) Traité du vivant. Fayard, Paris

Saxena S, Kumar R, Madan T, Gupta V, Muralidhar K, Sarma PU (2005) Association of polymorphisms in pulmonary surfactant protein A1 and A2 genes with high-altitude pulmonary edema. Chest 128:1611–1619

Swallow DM (2003) Genetics of lactase persistence and lactose intolerance. Annu Rev Genet 37:197–219

Ward MP, Milledge JS, West JB (2000) High altitude medicine and physiology, 3rd edn. Arnold, London

Rev Environ Sci Biotechnol (2007) 6:197–221
DOI 10.1007/s11157-006-9112-1

REVIEW

# How does the annelid *Alvinella pompejana* deal with an extreme hydrothermal environment?

**N. Le Bris · F. Gaill**

Received: 16 March 2006 / Accepted: 18 September 2006 / Published online: 25 November 2006


**Abstract** *Alvinella pompejana*, the so-called Pompeii worm (Desbruyères and Laubier, 1980), is found exclusively in association to high temperature venting, at the surface of hydrothermal chimneys of the East Pacific Rise. The main characteristics of this emblematic species is its tolerance to high temperature but its ability to colonize extremely hot substrates has been the subject of much controversy. In the last decade, new tools allowing in situ and in vivo investigation have been determinant in the understanding of the strategies and adaptations required to colonize such an extremely hot environment. New data relative to the characterization of the animal habitat conditions, on one hand, to the molecular adaptations of this organism and the colonization processes by this species, on the other hand, are now available. Advanced methods and tools, that have fostered the physico-chemical characterization of vent habitats in recent years, are first reviewed. Factors controlling the physico-chemical variability of vent habitats and the threats *A. pompejana* might effectively face are discussed. The exceptional thermotolerance of this species and the maximum temperature it could sustain are then considered in the light of molecular data relative to its collagen stability. Life history traits as well as biological controls on tube micro-habitat conditions are discussed on the basis of new in situ and in vivo experiments and characterization. Finally, the current knowledge and opened questions related to the molecular adaptations to chemical stresses are briefly stated. The ability of *Alvinella pompejana* to colonize these substrates is far from being fully understood, but the exceptional properties of its extracellular biopolymers and the behavior of the worm can be now considered as major clues in the colonization process. *Alvinella pompejana* could thus stand at the limits authorized for its biological machinery in a highly dynamic environment where temperature can readily reach lethal values, but where temperature regulation by the animal itself would prevent exposure to deleterious thermal spikes. The dynamic system associating this pioneer species and its associated microflora might be viewed as a key to the subsequent colonization of these environments by less tolerant species, highlighting *A. pompejana* as a new type of ecosystem bioengineer.

N. Le Bris (✉)
Département Etude des Ecosystèmes Profonds, Ifremer, Centre de Brest, BP70, Plouzané 29280, France
e-mail: nlebris@ifremer.fr

F. Gaill
Systématique Adaptation et Evolution, UMR 7138 CNRS-IRD-MNHN-UPMC-ENS, 7 quai Saint Bernard, Paris 75252, France

**Keywords** Pompeii worm · Collagen · Deep-sea · East Pacific Rise · Reproduction · Colonization · In situ measurement · Molecular adaptation · Thermotolerance · Tube

## 1 Introduction

*Alvinella pompejana*, the so-called Pompeii worm (Desbruyères and Laubier, 1980) is found exclusively in association to high temperature venting, at the surface of hydrothermal chimneys of the East Pacific Rise (Fig. 1). These tubicolous worms assemble in colonies, forming patches of various sizes on smoker walls. *Alvinella pompejana* is a rather small worm compared to other deep sea vent animals, having a length of about 10 cm (<1 cm in diameter), while the tubeworm *Riftia pachyptila*, may reach more than 150 cm in length (Desbruyères and Laubier, 1980). Unlike *Riftia pachyptila, A. pompejana* does not harbor endosymbionts, but abundant microbial communities were described, covering the inner part of its tube and attached to appendages at the back of the animal (Gaill et al. 1988a). If the role of these microbes forming dense sulfur filamentous mats is unknown, the nutrition of the animal is thought to derive from the grazing of surrounding free-living microflora.

This species may be associated to other annelids. A second species, *Alvinella caudata*, has been described as a congener of *A. pompejana* (Desbruyeres et al. 1998). Although morphologically different, ecological or physiological differences between the two species are still not established. The alvinellid species *Paralvinella grasslei*, and the hesionid *Hesiolyra bergi* are the most abundant polychaete species found in *Alvinella pompejana* colonies (Desbruyères et al. 1998).

*Alvinella pompejana* is endemic of the East Pacific Rise where it is found from 21°N to 32°S (Hurtado et al. 2004). Conversely, distinct alvinellid species are observed on the Juan de Fuca Ridge that was separated from the EPR by the migration of north America plate. Among these, *Paralvinella sulfincola* displays a similar association to hot fluid emissions on the wall of hydrothermal edifices (Juniper et al. 1992).

*Alvinella pompejana* is one of the emblematic animals living in the extreme environment of deep-sea hydrothermal vents (Childress and Fisher 1992). Several reviews devoted to this species have been published in the past 20 years, assessing current knowledge in its ecology (Desbruyères et al. 1985), biology (Gaill and Hunt 1991) and, for the last one (Desbruyères et al. 1998), providing a general overview of the various ecological, physiological and biochemical studies related to this organims. Desbruyères et al. (1998)'s synthesis put a particular emphasis on the physiological adaptations that reflect a high specialization of the animal to its environment.

In the first two decades, deep-sea biologists have come up with spectacular observations

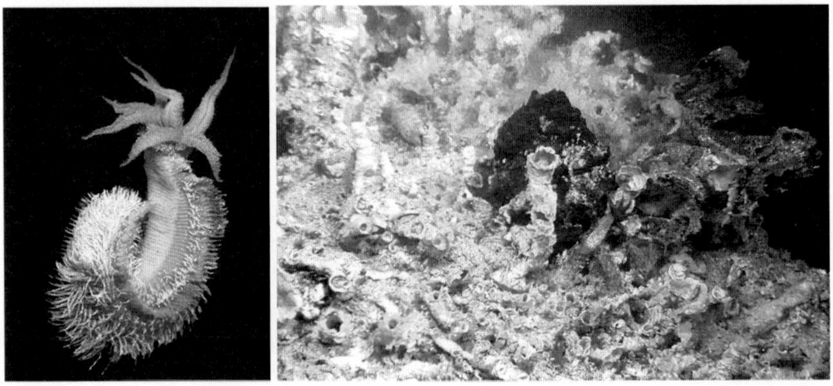

**Fig. 1** *Alvinella pompejana*: the animal outside its tube showing the white filamentous epibionts (left), inhabited tubes covering a smoker wall (Elsa, EPR 13°N) (right). Newly formed mineral surfaces appearing as a black outcrop at the colony surface, is colonized by a Pompeii worm pioneer. Shimmering fluids escape this mineral hot spot

which strongly suggested that this species could withstand unusually high temperatures, and which triggered an on-going debate on its upper thermal limits (Chevaldonné et al. 2000). *Alvinella pompejana* was hypothesized to be one of the most thermotolerant metazoans on Earth (Chevaldonné et al. 1992; Cary et al. 1998). While this species could not be studied in vivo, its maximum temperature is still not unambiguously assessed. Conversely, its northern Pacific relative, *Paralvinella sulfincola*, was very recently confirmed to be tolerant to temperature of 50–55°C (Guirguis and Lee 2006), the highest ever found for a marine metazoan. Even more surprising, this worm was shown to prefer temperature in the range 40–50°C.

In the last decade, the field of investigations related to this animal has been further enlarged, towards the precise characterization of habitat conditions, on one hand, the molecular adaptations and life cycle of the species, on the other hand. New tools allowing in situ and in vivo investigation have been determinant in the understanding of the strategies and adaptations required to colonize such an extremely hot environment.

The aim of this article is to review the challenges and adaptation strategies of *A. pompejana* to the extreme environmental conditions of its habitat. The advanced methods and tools, that have fostered the physico-chemical characterization of vent habitats in recent years, will be first reviewed. The second and third sections will focus on the factors controlling the physico-chemical variability of vent habitats and the threats *A. pompejana* might effectively face in its habitat. The thermal adaptation of this species, will be considered in a fourth section, in the light of the findings concerning the molecular properties of its collagen molecules. The thermotolerance of the worm, as compared to other metazoans, will be discussed on this basis. Various aspects of the colonization strategy of an extremely hot environment will be then discussed. Recent results concerning reproduction patterns, embryos development, and genetic selection in contrasted habitats will be first presented. The biological controls on tube micro-habitat conditions will be considered in the following section. Finally, the current knowledge and open questions related to the molecular adaptations to chemical stresses will be briefly stated.

## 2 Advanced methods and tools for vent habitats physico-chemical characterization

Since their discovery in the late seventies, manned-submersibles like the actual Nautile (Ifremer, France), Alvin (WHOI, USA) or Shinkai (Jamstec, Japan) have allowed to access these unique habitats at depths ranging from 1500 to 3500 m. The limited accessibility of great depths and the large variability of this environment have long limited the study of the interactions of *A. pompejana* with its exceptional habitat (Desbruyères et al. 1998). In the last decade, the capacity to characterize these habitats has been further expanded by ROVs (Remote Operated Vehicles), that substantially enlarged dive time, but also by the development of a new set of dedicated instruments. As presented below, major instrumental breakthroughs have fostered the development of integrated approaches and provided more details on the fine-scale variability of this environment.

### 2.1 Temperature measurements

Sensors operating at high hydrostatic pressure (15–35 MPa, i.e. 150–350 folds the atmospheric pressure) are a prerequisite for these studies but they are not the sole condition to obtain relevant measurement at the scale of organisms. Conventional high pressure temperature probes (e.g. Pt100 sensors in pressure resistant titanium casings) can be easily manipulated in real-time or deployed for long-term autonomous measurements, but these slow-response sensors are not suitable to characterize the steep fluctuations encountered the fluid–seawater mixing zone. Fast response thermocouples have been preferred for diffuse flow habitat investigation by Johnson et al. (1986a, 1988a). In a recent study, Le Bris et al. (2006a) has shown that the amplitude of temperature fluctuation can be underestimated by a factor of five when a conventional submersible probe is used, as compared to a fast miniaturized thermocouple. To the exception of the temporal variability study of Chevaldonné et al. (1991), temperature ranges reported for the *Alvinella pompejana* environment mostly relied on measurements with such conventional temperature probes (Desbruyères et al. 1985; Chevaldonné et al. 1992; Sarradin et al. 1998). The fast-response sensors used in recent studies (Cary et al. 1998; Di Meo-Savoie et al. 2004; Le Bris et al. 2005) are expected to provide a more accurate definition of the temperature variation at organism scale.

## 2.2 In situ chemical analysis and sensing

Simultaneous measurements of several physicochemical parameters have enabled to correlate the variability of biologically relevant chemical factors, such as total dissolved sulfide concentration, with that of fluid tracers such as temperature or silicate concentration (Johnson et al. 1986b, 1988a, b). Combining these factors enabled to highlight marked differences between habitats, circumventing natural fluctuations in the mixing ratio and uncertainties due to the instability of sensor tips (Johnson et al. 1988a, b; Le Bris et al. 2006a). In situ chemical sensing revealed to be the more relevant method to describe the chemical and thermal variability in habitats on appropriate spatial and temporal scales (Johnson et al. 1986a; Le Bris et al. 2000). The limited sampling capacity of submersibles (4–20 samples per dive according to the devices used, authors pers. obs.) and the poor resolution of high volume fluid samplers (e.g. 200 ml, Sarradin et al. 1998) make them inadequate to resolve steep and dynamic gradients. Furthermore, direct chemical measurement at depth is the only accurate way to describe the thermodynamically unstable chemical systems at the mixing interface between the hydrothermal fluid and seawater. Since the first studies of Johnson et al. (1988a, b; 1994), spectrophotometric flow analyzers have been intensively used in short term habitat studies (Sarrazin et al. 1999; Le Bris et al. 2003, 2006a). These micro-flow measurement techniques are, however, very sensitive to clogging by particulate or colloid materials, preventing to use them in close proximity to mineral surfaces, animals or extracellular matrixes like tubes or mucus. Alternative methods based on miniaturized electrochemical sensors have enabled a finer spatial and temporal characterization of habitats (Le Bris et al. 2001; Luther et al. 1999). While potentiometric pH profiles with sub-centimeter resolution could be obtained using a 2 mm diameter glass electrode (Le Bris et al. 2005), in situ voltammetry revealed to be a powerful technique to simultaneously detect oxygen and various sulfide forms in vent habitats (Luther et al. 1999, 2001a). This electrochemical technique that relates on individual chemical species, enabled to discriminate free sulfide forms from those associated with iron in the aqueous phase (Luther et al. 2001b). This method is, therefore, of much greater biological relevance than the total concentration of sulfide obtained by colorimetry.

Centimeter-scale temperature gradients generate large discrepancies in combined measurements (Le Bris et al. 2006a). For instance, the average temperature measured by two tightly coupled probes can sometimes differ by more than 20°C (Fig. 2). Miniaturization is therefore a great technological challenge for the study of habitat conditions at organism scale. It is a prerequisite if the sensor tip has to be inserted through *A. pompejana* tube openings of a few centimeters in diameter. Despite the need for protective casing to support high pressure and submersible manipulation, sensing tips could be reduced in size down to 5–8 mm by associating

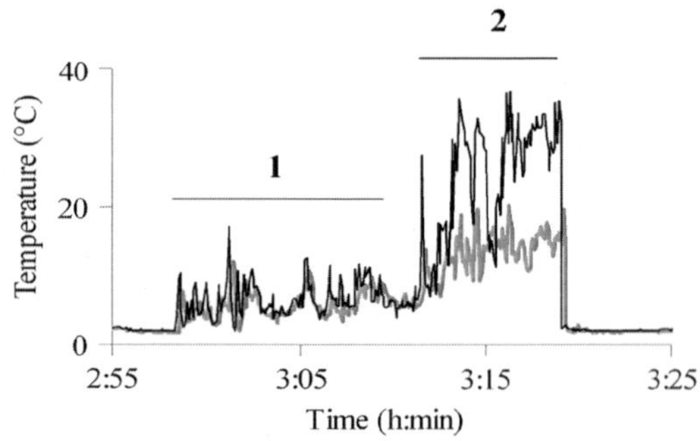


**Fig. 2** Comparison of temperature records obtained with two combined sensors positioned on two successive points above an *Alvinella pompejana* colony (1, 2) (Le Bris, unpublished): the Victor6000 ROV Pt100 sensor (grey line) and a miniaturized thermocouple associated to a pH electrode (thin black line)

electrochemical sensors and miniaturized thermal probes (Cary et al. 1998; Luther et al. 1999; Le Bris et al. 2001; Di Meo-Savoie et al. 2004).

The large variability of vent habitats at short-scales stresses out the need for precise probe positioning. In previous studies, uncertainty in sensor position may clearly have limited the significance of temperature extremes measured in close vicinity to animals (Chevaldonné et al. 1992). In recent studies, the use of a small-size video camera for close-up imaging of the sensing tip considerably facilitated accurate measurement positioning and greatly improved the characterization of thermal gradients at organisms scale (Le Bris et al. 2005).

### 2.3 Sampling of mixing zone fluids

Few analyzers and sensors are still available for these studies and many chemical parameters remain inaccessible using these techniques. In situ approaches, therefore, have to be complemented with conventional sampling. If the spatial resolution of measurements can be improved using small volume sampling devices (e.g. 10 ml, Di Meo et al. 1999), the concentration of metastable chemical compounds cannot be accurately defined with this approach. Severe underestimation of the total sulfide concentration was emphasized when comparing in situ analysis and sampling in the same area for a given temperature (Le Bris et al. 2006b). Despite these limitations, sampling has been used to estimate the concentration range of biologically important parameters in vent habitats, such as metals (Cu, Cd, Pb, Zn) and dissolved gases like $CO_2$ and methane (Childress et al. 1993; Sarradin et al. 1998; Desbruyères et al. 1998; Di Meo-Savoie et al. 2004). In situ filtration using on-line filters enabled to quantitatively estimate the "dissolved"-particulate distribution of metallic elements (Kadar et al. 2006; Sarradin et al. submitted). Substantial artifacts should still be expected due to the coexistence of reduced and oxidized species and the common over-saturation of these media with respect to several mineral phases (e.g. FeS, ZnS,..etc). The large modifications that can occur in the sample composition before their analysis on board, one to several hours after collection, are likely to introduce substantial bias in the definition of in situ concentrations. If these data can be used as first order estimates, their extrapolation to estimate the exposure levels encountered in habitat should still be considered with caution.

### 2.4 Geochemical modelling

Geochemical modeling of habitat characteristics has been first developed to estimate bioenergetic budgets of fluid and seawater mixes for chemoautotrophic processes (McCollom 1997, 2000). This approach was also used to simulate physico-chemical conditions in inaccessible microbial habitats below the seafloor (Bach and Edwards, 2003) or within chimney walls (Tivey 1995, 2004). The comparison of empirical data with the prediction of a simple mixing model, additionally, enables to quantify the influence of other non-conservative abiotic or biotic processes on the characteristics of habitats (Johnson et al. 1988b; Le Bris et al. 2005). Calculations based on chemical equilibrium thermodynamics have also proven to be useful to complement empirical characterization in vent fauna habitats above the seafloor, by giving access to the concentration of species that cannot be measured directly. The distribution of sulfide species into labile free sulfide forms and less toxic iron associated sulfide forms could thus be quantified from its total concentration and the pH and iron content of the medium (Le Bris et al. 2003).

## 3 Physico-chemical variability of hydrothermal habitats

Habitats associated to deep-sea hydrothermal vents have been described as some of the most challenging for animals on Earth. The high instability of their geological setting, subject to frequent volcanic and tectonic events, makes these habitats particularly ephemeral. The resulting chaotic change in fluid emission intensity and localization has been shown to induce dramatic changes in community composition over time (Shank et al. 1998; Tunnicliffe 1991). Extreme physico-chemical characteristics, as compared to most oceanic environments, were also highlighted; e.g. temperatures exceeding 100°C (Desbruyères et al. 1985;

Chevaldonné et al. 1992), hundreds of micromolar contents in sulfide (Johnson et al. 1986b; Luther et al. 2001b; Le Bris et al. 2003), oxygen depletion (Johnson et al. 1986b; Luther et al. 2001a, b), acidic pH (Le Bris et al. 2005, 2006a), $CO_2$-rich conditions (Childress et al. 1993; Sarradin et al. 1998; Shank et al. 1998), and several order of magnitude enrichments in various metals with respect to ocean waters (Johnson et al. 1988b; Desbruyères et al. 1998; Geret et al. 2002).

To what extend the organisms are exposed to these physico-chemical extremes, however, has long been poorly known, mainly because earliest ecological studies insufficiently accounted for fine-scale variability within habitats. The methods and tools presented in the previous section have largely improved our knowledge of the physico-chemical variability at the animal-scale, enabling to estimate the conditions experienced by vent organisms much more accurately than previously done. The processes governing the potential exposure of individuals to physico-chemical threats are also much better understood. The following synthesis presents the current knowledge of these controlling factors at various scales, from the global diversity of hydrothermal end-member fluids, to the variability of source fluid composition at a vent site, and last but not least, to the small-scale heterogeneity of habitats at the mixing interface of the fluid emission and seawater.

### 3.1 Geological diversity of hydrothermal end-members

The characteristics of vent habitats are often referred to the properties of end-member fluids, i.e. pure hydrothermal fluids that are discharged on the seafloor without prior mixing with seawater. These high temperature fluids (300–400°C) are formed during the course of hydrothermal circulation below the seafloor, by the interaction of seawater and heated rocks. Their emission in seawater results in the formation of the well-known "black smokers," mineral edifices mostly composed of iron, copper and zinc sulfides. The first global review of end-member fluid composition by Von Damm (1995), completed by the more recent data of Charlou et al. (2002) and Douville et al. (2002) for the mid-Atlantic Ridge, displays some common chemical features of major biological relevance over a large set of active sites. Acidic pH (from 2 to 4), millimolar contents in sulfide and $CO_2$, and micromolar contents in numerous transition metals are main characteristics of these fluids. Active vent sites can, however, largely differ in the dissolved gases ($H_2S$, $CO_2$, $H_2$, $CH_4$) and metal content of their end-members. Their methane and hydrogen contents remarkably depends on the nature of basement rocks that interact with seawater: fluids formed in ultramafic rocks such as observed at the mid-Atlantic Ridge sites Rainbow and Logatchev (Charlou et al. 2002) were shown to be highly enriched in methane and hydrogen, whereas these gases are almost absent of the basalt-hosted fluids, like those of the East Pacific Rise. Substantial differences in end-member fluids were additionally observed at the site scale, in space and time, and were shown to result of phase separation in subsurface (Von Damm 1995). In this case, low chlorinity fluids, depleted in metals and enriched in dissolved gaseous species, can be emitted separately from their complementary brine component in which metals and major ions like $Cl^-$ or $Na^+$ are concentrated. A consequence of this phenomenon is the great diversity of end-member fluids among or within active sites of a vent field, over distances ranging from hundreds of meters to a few kilometers (Von Damm 1995). If this local and regional end-member variability is likely to impact the chemistry of habitats on the seafloor, it remains hazardous to compare the conditions experienced in related habitats from their contrasted properties, like it is sometimes done for large scale studies (e.g. Desbruyères et al. 2000). Areas directly exposed to fluids with temperature over 300°C are obviously abiotic. Together with a decrease of temperature to sustainable ranges, additional controls on the physico-chemical properties of fluid emissions that fuel vent communities should be accounted.

### 3.2 Variability of fluid emissions on the seafloor

Vent fauna habitats are associated to different types of fluid emission, distinguished as "diffuse" vents and "focused" vents. "Diffuse" vents spread out over large areas and are characterized by fluids being emitted at low flow rate through cracks in the basaltic seafloor. The variability of this type of vent fluids has been extensively studied on the Galapagos Ridge (Edmond et al. 1979) or the 9°N vent field of the East Pacific Rise (Shank et al. 1998; Von Damm and Lilley 2004). Diffuse emissions were first shown to result, primarily, of the sub-surface dilution end-member fluids in seawater (Edmond et al. 1979). The physico-chemical characteristics of these fluids, however, depicts a more complex combination of conservative mixing and non-conservative processes, such as those induced by mineral precipitation or biological activity below the seafloor (Von Damm and Lilley 2004).

Conversely, focused vents are associated with the focalized discharge of end-member hydrothermal fluids, that can reach several meters per second flow rates in black smokers. Intense mineralization, triggered by thermal and chemical changes at the fluid interface with seawater, leads to the formation of these mineral edifices and provides new colonization substrates for animals. A complex network of cracks, pipes and porous materials channels the circulation of fluid and seawater in the smoker wall (Tivey 1995). Discrete fluid outflows should thus be expected to escape the wall of mineral edifices with a wide range of flow rates and temperatures. To the exception of black smokers fluids representing almost pure geological end-members, the characteristics of these fluid emissions have been much less studied. Similarly to diffuse vents, a highly variable composition of vent fluids can be expected on a single chimney. According to the modeling work of Tivey (2004), various degrees of end-member mixing with seawater, conductive cooling or heating of fluids, and loss or enrichment in chemical elements resulting of molecular diffusion through the porous minerals, can generate a wide range of chemical composition in fluids. Microbial activity could also contribute to impact fluid properties within hydrothermal edifices, but the extend of this biological influence is unknown. The mechanisms that control vent fluid characteristics at the site scales are, therefore, much more complex than the simple dilution of the end-member, which prevents to extrapolate habitat characteristics from end-member fluids composition.

Tubeworms and bivalves aggregations first described on the Galapagos ridge were shown to be associated with vent fluid not exceeding 30°C (Johnson et al. 1988a; Fisher et al. 1988). This was later confirmed for similar assemblages on the East Pacific Rise (EPR) (reviewed in Hessler and Kaharl 1995). Most of the habitat of other dominant-taxa described to date associate with such low temperature fluids, both at diffuse and focused vents: tubeworms-dominated communities of the Juan de Fuca Ridge (Tunnicliffe 1991), mussel and shrimps assemblages of the mid-Atlantic Ridge (Sarradin et al. 1999) or mussel and gastropodes-dominated communities of the back-arc Fidji Basin (Koschinsky et al. 2002). In contrast, *Alvinella pompejana*, like the Juan de Fuca species *Paralvinella sulfincola*, distinguishes from other vent species in their exclusive association with hot fluid emissions. Dense animal communities colonize areas of the chimney surface visibly bathed by fluids of temperature exceeding 100°C (Desbruyères et al. 1985; Tunnicliffe and Juniper 1990). The acidic pH and high sulfide concentration ranges of these fluids suggested that the contribution of end-member to this medium is much larger than for neighboring low-temperature vents (Sarradin et al. 1998; Le Bris et al. 2003). To appreciate the degree of exposure of an organism to such environmental extremes, however, requires to consider the physical and chemical heterogeneity of its surroundings.

### 3.3 Short-scale heterogeneity in the mixing zone

The mixing interface between the hot and reducing vent fluid and cold oxidizing seawater is a steeply changing environment. Temperature rises of tens of degrees were described at centimeter-scales, together with sharp and unpredictable thermal fluctuations at second to hour scales, in the habitats of different tubeworm species of the Galapagos ridge (Johnson et al. 1986b) or the Juan de Fuca Ridge (Tunnicliffe and Juniper 1990). Temperature in vent habitats was also shown to be modulated at daily scale by change in fluid emissions rates under the influence of tidal regimes and by longer period variations in deep-sea currents and hydrothermal emissions (Johnson et al. 1988a; Chevaldonné et al. 1991; Tivey et al. 2002). Sharp transitions and fluctuations in chemical conditions correlate to these thermal changes (Johnson et al. 1986b; Le Bris et al. 2000, 2001). As a consequence, large thermal and chemical ranges were observed at the scale of animal aggregations (Johnson et al. 1988b, 1994; Sarradin et al. 1998; Sarrazin et al. 1999; Le Bris et al. 2006b), or even single individuals (Sarradin et al. 1998; Le Bris et al. 2001).

More than differences in source fluids, the degree of mixing is thus likely to govern the exposure of organisms to environmental extremes. Temperature has been considered, in first approximation, as a proxy for chemical conditions (e.g. Childress and Fisher 1992; Sarradin et al. 1998), and this factor was used to discriminate the mosaic of microhabitats, along the gradient of hydrothermal influence, from the potential importance of thermal and chemical stresses (e.g. Mullineaux et al. 2003; Bates et al. 2005). These approaches were based on the assumption that habitat conditions could be modeled from the conservative mixing of the vent fluid and seawater. This was shown to be valid, in first assumption, in particular habitats of tubeworms and clams (Johnson et al. 1988b; Le Bris et al. 2006b). If this model was found to be consistent for distinct animal assemblages of a same vent site, this could not be extrapolated to other sites over the vent field (Le Bris et al. 2006a). For a large number of habitats, however, the conservative model fails to describe empirical data (Johnson et al. 1994, 1988b; Le Bris et al. 2000, 2005, 2006a). By comparing the correlation of sulfide with silicate, a conservative fluid tracer, before and after complete removal of animals, Johnson et al. (1994) demonstrated that the non-conservative behavior of sulfide within a mussel bed was due to its consumption by biota. Major deviations from the conservative mixing model were also evidenced within the *A. pompejana* habitat, using combined pH and temperature measurements, and could be attributed to conductive thermal exchange between fluids circulating inside an outside the worm tubes (le Bris et al. 2005). In addition to these biological controls, abiotic processes, such as precipitation and redox reactions, that are kinetically and thermodynamically favored in the mixing interface, likely contribute to control the combination of chemical and thermal gradients in vent habitats (Fig. 3).

## 4 Thermal and chemical threats in the *A. pompejana* habitat

*Alvinella pompejana* was shown to occupy, exclusively, high-temperature habitats on chimney walls from the East Pacific Rise. The instability of these substrates constitutes one of the most important physical constraints experienced by these organisms. Frequent and unpredictable changes in fluid emission rate and spatial distribution characterize the areas where mineralization processes lead to continuous reshaping of substrates and circulation networks. Intense mineral precipitation, furthermore, progressively encrusts the tubes of these annelids (Gaill and Hunt 1991), requiring the animals to continuously secrete tube material to avoid their tube being entrapped in the growing edifice. In addition, this environment is under the influence of hot fluids, with a high hydrothermal signature, venting from the mineral walls. The summary of the physico-chemical characteristics of the *Alvinella pompejana* habitat proposed in Desbruyères et al. (1998) was presented by these authors as a first order-estimate due to the difficulty to accurately sample this steeply changing environment. A more precise description of the extreme conditions encountered at worm scale can now be provided, in the light of recent studies that have largely benefit of the technological and analytical efforts provided in the past 10 years.

### 4.1 Temperature

One of the first character of vent habitat, as revealed by temperature monitoring, is the short-term chaotic fluctuation of temperature that reflects the turbulent hydrodynamic pattern of the hot fluid emission in cold seawater. In

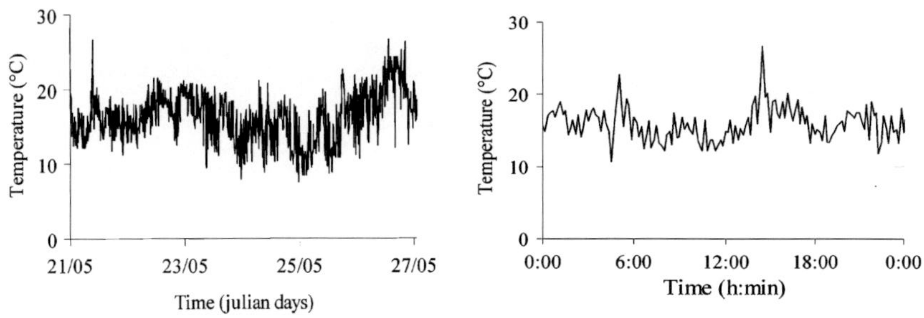


**Fig. 3** Temperature variability at one point on the surface of an *A. pompejana* colony (adapted from in Pradillon et al. 2005a): a—over 7 days, b—a 24 h extract (temperature was monitored with a slow response autonomous probe, 1 meas./10 min)

addition to a marked tidal influence, rapid pulsation and brief spikes were shown as a main feature of the temperature records at the surface of the colony *A. pompejana* (Chevaldonné et al. 1991). A similar short-term pattern was observed inside tubes (Cary et al. 1998). Comparing several-minute scans on different location within the tube assemblage, Le Bris et al. (2005) have shown that high variability at tube openings is mostly restricted to the vicinity of "hot spots" associated with turbulent plumes of punctual fluid outflows, the surface of the colony being much more stable away from these areas.

Despite this high rate fluctuation introduces a substantial uncertainty in measurements, a marked spatial heterogeneity of the habitat with respect to temperature has been emphasized (which is summarized in Fig. 4). The strong gradient from the chimney wall to the surrounding seawater depicted in Desbruyères et al. (1985) has been more finely assessed from high-resolution surveys (Luther et al. 2001b; Le Bris et al. 2005; Di-Meo et al. 2004). While the 2°C background seawater temperature is generally recovered less than 10 cm above tube openings, temperatures largely above 100°C are measured in contact to the mineral substrate directly beneath the tubes (Fig. 4). The reported temperature maxima for different chimneys range from 125°C (Le Bris et al. 2005) to 175°C in Di Meo-Savoie et al. (2004) (Table 1).

Temperature variability can also be high, laterally on smoker walls, between the surface of worm colonies and bare mineral areas in their close vicinity where temperature as high as 105°C

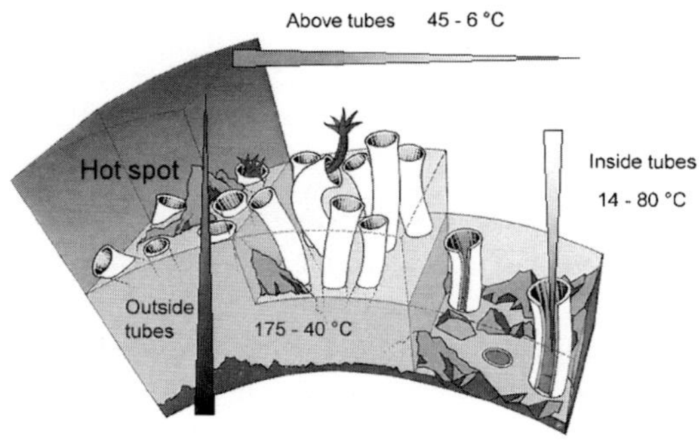


**Fig. 4** Ranges of temperature (mean values) reported for different microhabitats associated to *Alvinella pompejana* colonies (adapted from Le Bris et al. 2005)

(Chevaldonné et al. 1992) or 127°C (Di Meo-Savoie et al. 2004) have been measured. In contrast to these temperature extremes, moderately warm conditions were reported at tube openings, ranging from 6 to 45°C on average (Table 1). Although limited in amplitude, this range also reveals significant thermal gradients at tube openings over the surface of the colony, as illustrated by Chevaldonné et al. (1991) from autonomous monitoring. Le Bris et al. (2005) have shown that the highest values are found in the vicinity of punctual "hot spots" associated with turbulent fluid outflows (Fig. 1).

The temperatures determined inside *A. pompejana* tubes lie within these radial and lateral gradients at the smoker interface with seawater, with a maximum of 100°C (Di Meo-Savoie et al. 2004). Cary et al. (1998) first reported an average temperature of 68 ± 6°C monitored inside a tube over 2 h, with spikes as high as 81°C. The

**Table 1** Maximum and mean temperatures reported in the close surrounding of *A. pompejana* colonies

| | Local maximum (°C) | Tube openings (mean) (°C) | Among tubes (mean) (°C) | Inside tubes (mean) (°C) |
|---|---|---|---|---|
| Desbruyères et al. (1985) | 250[a] | ~20 | ~100 | nd |
| Chevaldonné et al. (1992) | 105[b] | 20–45 | 40–80 | |
| Cary et al. (1998) | nd | 22 ± 2.5 | nd | 68 ± 6 |
| Le Bris et al. (2003) | nd | 15 ± 6 | 60 ± 10 | nd |
| Di Meo-Savoie et al. (2004) | 173[c]<br>127[b] | 7–40 | nd | 29–84 |
| Le Bris et al. (2005) | 125[c] | 12 ± 7[d]<br>6 ± 3[e] | 61 ± 38 | 59 ± 6 |

[a] 20 cm below tube openings

[b] Bare substrate adjacent to the colony

[c] Substrate beneath colony

[d] Above tube openings close to a fluid outflow

[e] Above tube openings away from outflow, nd: no data

relevance of this value was strongly debated and several artifacts were suggested, including disturbance of the animal behavior or piercing the tube when inserting the probe (Chevaldonné et al. 2000). Since that date, Cary and co-workers have confirmed the ability to reproduce hour-long temperature monitoring inside tubes, by extending this study over 32 inhabited tubes (Di Meo-Savoie et al. 2004). On a single site, different tubes exhibited large difference in their mean temperatures, ranging from moderate (29–30°C) to much higher values (75–84°C).

From temperature records alone, it was impossible to validate tube integrity during scans. Associating a pH probe with a thermocouple enabled, for the first time, to exclude the hypothesis of accidental leaks from hot fluids of the outer medium, while a still warm temperature of 59 ± 6°C was measured (Le Bris et al. 2005). The near-neutral pH measured inside tubes by these authors unambiguously distinguished the inner medium from the acidic fluids circulating outside tubes (pH < 5). Le Bris et al. (2005) determined moderately warm conditions in one tube (59 ± 6°C), while repeated measurement on two other inhabited tubes revealed average temperature of 43 ± 2°C and 14 ± 4°C (Le Bris pers. obs.), again showing a large variability of mean temperature among tubes.

In both studies, it was recognized that the relevance of these values to natural conditions may be limited by disturbance of the animal behavior, modification of hydrodynamic conditions and subsequent thermal exchange. In the later study, a much shorter measurement duration (~1 min) enabled to overcome artifacts due to heat exchange, that might result of fluid circulation arrest. The fast temperature increase of 57°C monitored within less than 5 s after insertion of the probe in an inhabited tube, confirmed that this temperature was not due to a progressive conductive heating (Le Bris et al. 2005).

Not only the extreme temperatures encountered in some tubes, but also the variable temperature over time support the idea that the medium is not necessarily in thermal equilibrium with the worm body. The 2–4 h long temperature records presented in Cary et al. (1998) and Di Meo-Savoie et al. (2004) exhibit frequent temperature changes of 10–20°C over a few minutes and frequent sharp spikes of up to 40°C in amplitude. The modulation of mean temperature over more than 2 h in one of the hottest tubes was shown to range between 60 and 100°C (Di Meo-Savoie et al. 2004). Furthermore, the large external temperature gradient along a tube length of ca. 15 cm suggests a substantial longitudinal gradient inside the tube. While Le Bris et al. (2005) limited the insertion of the probe to a short distance from the tube opening, not exceeding 2–3 cm, Cary et al. (1998) estimated that their measurements were performed 6 cm inside the tube. The higher temperatures monitored in the later studies may rely on the deeper position of the probe in the tube.

From these measurements, Cary et al. (1998) estimated that a gradient of up to 60°C could be experienced at the scale of the animal. If this assertion still has to be demonstrated, the extreme temperature variability inside and outside tubes, from the scale of the colony to that of the organism, undoubtedly present this animal as one of the most eurythermal on Earth.

### 4.2 Toxic sulfide

Concentrations of several hundreds of micromoles to millimoles per liter have been determined in situ in the fluids that bathe *A. pompejana* colonies (Table 2). These concentrations are the highest reported in various hydrothermal vent habitats (Le Bris et al. 2003). Although they are sometimes distant by less than one meter, the mixing zone surrounding *Alvinella pompejana* appears several times richer in sulfide than the habitats of tubeworms, for a given temperature range (Sarradin et al. 1998; Luther et al. 2001b, Le Bris et al. 2006b).

Micromolar levels of sulfide are lethal to most aerobic organisms (Visman 1991). Luther et al. (2001b) proposed that sulfide might be naturally detoxified in the environment of *A. pompejana*. A voltammetric study of sulfide speciation within an *Alvinella pompejana* colony revealed that free sulfide forms, $H_2S$ and $HS^-$, were undetectable while aqueous iron sulfide complexes were dominant. According to these authors, the formation of these complexes would prevent exposure to more toxic free sulfide forms. This medium, however, was remarkable in the fact that iron contents in fluid samples were about twice higher than total acid volatile sulfide contents (Di Meo-Savoie et al. 2004). Since large differences in the iron to sulfide ratio were highlighted in the environment of *Alvinella pompejana* colonies, even within a kilometer apart (Table 2), the specific features of Luther and coworkers study site (Alvinella Stump, EPR 9°N) cannot be generalized. More generally, a large range of sulfide speciation patterns should be expected among sites, the iron to sulfide ratio in end-members being largely modulated by subsurface processes (Von Damm 1995). According to chemical calculations, the predominance of the toxic $H_2S$ form at nearly millimolar concentration was predicted for iron-depleted sites (e.g. Genesis EPR 13°N), while sulfide toxicity would be effectively attenuated at sites where iron concentration equal or exceed that of sulfide in fluids (e.g. Elsa EPR 13°N) (Le Bris et al. 2003).

Millimolar sulfide levels are not exceptional for the deep ocean. For example, concentrations as high as 20 mM were measured in sediments associated with methane or hydrocarbon seeps (Arvidson et al. 2004). The *A. pompejana* habitat conditions might, however, require a much higher tolerance to sulfide than required in organic-rich sediments where it is a ubiquitous end-member product of organic carbon remineralization. In sulfidic sediments, invertebrates were shown to

**Table 2** Maximum sulfide content and iron to sulfide ratio at tube openings levels

| Reference | Vent field/location/year | | Maximum sulfide (μM) | Fe:S |
|---|---|---|---|---|
| Sarradin et al. (1998) | EPR 13°N | Genesis/1996 | 300[a] | nd |
| Luther et al. (2001b) and Di Meo-Savoie et al. (2004) | EPR 9°N | Alvinella Stump/1999 | 360 | 2.0 |
| Le Bris et al. (2003) | EPR 13°N | Genesis PP12/1999 | 1520 | 0.4[b] |
| | | Elsa PP50/1999 | 563 | 1[b] |
| Le Bris et al. (in press) | EPR 9°N | M-Vent/2002 | 916 | 1.0 |
| Alain et al. (2004) | EPR 13°N | Genesis PP12/2002 | 860 | <0.01 |
| | | Elsa HOT3/2002 | 775 | 0.8 |
| | | Elsa PH1/2002 | 142 | 0.8 |
| | | Actinoir PH7/2002 | 985 | <0.01 |

[a] Extrapolated from temperature using a locally-defined empirical sulfide–temperature relation

[b] Estimated from black smoker fluids at the same site

built tubes or burrows, which enable them to reduce the level of sulfide in their microhabitat by ventilating them from the overlying oxygenated waters (Aller 1982). Much more fluctuating and sulfidic conditions are found at the openings of *A. pompejana* tubes. Even though this animal could ventilate its microhabitat, it could not avoid sulfide inputs to the tube.

Furthermore, in contrast to the slightly alkaline pH of marine sediment pore waters (pH 7.5–8), pH within or above tubes were shown to be mostly neutral slightly to acidic (Di Meo-Savoie et al. 2004; Le Bris et al. 2005). In these conditions, $H_2S$ should be thermodynamically favored over $HS^-$ (Millero et al. 1987a), increasing the potential for sulfide to diffuse across the epidermis or respiratory membranes (Visman 1991). Exposure to toxic sulfide in the environment of *Alvinella pompajana* might thus reach some of the highest levels experienced in the marine world.

### 4.3 Oxygen and metals

Oxygen availability in the *Alvinella pompejana* environment is still poorly constrained, due to the inability to stabilize this compound in samples and the lack of sensors to monitor its concentration in situ under these fluctuating conditions. From their voltametric survey of the *A. pompejana* environment, Di Meo-Savoie et al. (2004) first reported that oxygen was undetectable inside tubes, despite a high content of other seawater oxidizing compounds like nitrate and sulfate. In a previous study showing a transition between two dominant phylotypes of the epibiotic community at the worm scale, Cary et al. (1997) hypothesized a significant gradient of oxygen in the tube, the anterior part of the body being exposed to fully anoxic conditions while oxygen could still be present in the anterior part.

Oxygen depletion is a general feature of sulfidic environments, resulting of abiotic or microbial sulfide oxidation. In contrast, Johnson et al. (1986b) have shown that, in the tubeworm habitats, oxygen can coexist with sulfide until temperature is over 11°C. According to these authors, the transient occurrence of oxygen in the turbulent mixing zone would be favored by the low abiotic sulfide oxidation rate in seawater at low temperature (Millero et al. 1987a). From the observation of Johnson et al. (1986b), it was hypothesized that oxygen could switch from oxic to anoxic conditions in the medium overlying *A. pompejana* colonies (Desbruyères et al. 1998). A much faster oxygen degradation should, however, be expected at the higher temperature and higher sulfide to oxygen ratio that characterize this habitat and in the presence of metal catalysts. Considering that sulfide concentration at tube openings exceed by a factor of 5–10 the oxygen level of East Pacific Rise deep-waters (110 μM) (Sarradin et al. 1998), and that high levels of Fe (II) may substantially increase the oxidation rate (Zhang and Millero 1994), oxygen is likely to be rapidly depleted at the surface of *Alvinella pompejana* colonies. The competition between abiotic and biological oxygen consumption should be much more severe in this case than in tubeworm habitats. In situ monitoring of oxygen is still required to quantify its variability in this environment.

Iron is the most abundant metal in these environments. To the exception of some sites with low salinity end-members depleted in metals, iron concentrations lying around hundreds of micromolar were measured in the habitat of *A. pompejana* (Luther et al. 2001b; Alain et al. 2004; Di Meo-Savoie et al. 2004). The selectivity of the analytical method to ferrous iron (Fe(II)) in Alain et al. (2004), furthermore revealed that this metastable form of iron for an oxidizing medium can reach hundreds of micromolar levels in the fluid surrounding *A. pompejana* colonies. High levels of ferrous iron can be particularly toxic for organisms in the presence of oxygen, due to its potential to produce highly deleterious oxygen radicals through its oxidation to ferric iron (Sung and Morgan 1980; Millero et al. 1987b). Similarly, sulfide oxidation by oxygen was shown to produce oxygen radicals (Tapley et al. 1999). Oxidative stress is thus potentially one of the major threat encountered in this highly reactive chemical medium.

In combination to iron, several transition metals are enriched in the *A. pompejana* habitat. Zinc was identified as the second more important metal constituting chimneys walls colonized by alvinellids (Desbruyères et al. 1998) and a main component of the precipitates intimately associated with *A. pompejana* tubes (Zbinden et al. 2003). Highly toxic heavy metals like cadmium and lead were also shown to be particularly enriched in the environment of *A. pompejana* (Desbruyères et al. 1998). This first study was unable to distinguish the most toxic labile forms of these elements from the less reactive particulate forms that are likely to be abundant in the vicinity of hydrothermal smokers. Using in situ filtration of fluid samples, Sarradin et al. (submitted) recently confirmed that the coarse fraction (>2 μm) composed of highly crystalline phases can be much higher than the finer fraction (<2 μm). The later fraction can still reach a few μM in zinc and copper, and a few tens of nM in cadmium and lead, several orders of magnitude above estuarine and coastal waters (Sarradin et al. submitted). Much remains to be done, still, to estimated the toxicity of these metals in the environment of *A. pompejana*. Particularly, a finer porosity threshold would enable to discriminate fine particles from the truly dissolved and colloidal forms. Owing to the abundance of animal mucus and microbial polysaccharides in this medium, it would also be essential to estimate the fraction complexed by organic molecules that could be much less reactive than the "free" inorganic species.

### 4.4 $CO_2$ and pH

If exposure to high level of $CO_2$ has not been emphasized as a potential threat in previous reviews, it is worth noting that this environment is particularly rich in $CO_2$ as compared to the surrounding deep-sea waters. Concentrations ranging from 3.5 to 6 mM, 1.5 to about three times the seawater background, was measured by Sarradin et al. (1998) in samples collected above alvinellid colonies. These authors estimate that a concentration of 17 mM could be reached in this habitat, largely above the concentrations found in other hydrothermal habitats (Sarradin et al. 1998). The low pH (5.7–6.7) of corresponding samples, furthermore, indicates that the relative contribution of the $CO_2$ species over dissolved carbonates is higher than in seawater (pH close to 8). At a pH of 7–6, 10–50% of the total inorganic carbon is under the form of molecular $CO_2$, whereas it only reaches about 1% of the total in seawater. Hypercapnia, therefore, constitutes another potential chemical stress to these organisms.

## 5 Adaptation to temperature

### 5.1 Thermal tolerance of alvinellids and related species

In the last years, the development of new types of high-pressure aquaria (Shillito et al. 2004; Pradillon et al. 2004; Guirguis and Lee 2006) have allowed to examine in vivo the thermal limits of several hydrothermal vents species inhabiting chimney walls (Shillito et al. 2001; Ravaux et al. 2003; Lee 2003; Guirguis and Lee 2006; Shillito et al. 2006). Although very promising high-pressure experiments have demonstrated the capacity to maintain *A. pompejana* alive up to 24 h after recovery (Shillito et al. 2004), its limited survival after collection did not allowed, to date, in vivo experimentation on this species. In an effort to characterize the thermal tolerance of various species living in close proximity to high temperature emissions, Shillito et al. (2001) considered the hesionid worm, *Hesiolyra bergi*, that was observed crawling at the surface of *A. pompejana* colonies and often entering their tube for a few second to several minutes. These authors have demonstrated an escape behavior at 35°C and a lethal limit at 41°C for this congener, which indicated that high thermal tolerance is not a prerequisite to live among *A. pompejana* colonies. The large thermal heterogenity characterizing these colonies over space and time (Section 4.1), however, precludes to consider this mobile species as a biological "thermometer", as it was suggested by Shillito et al. (2001). As reviewed in Section 4.1, it is now well established that the surface of the colony is exposed to temperatures ranging from a few degrees to 45°C, and that the highest temperatures in this range are restricted to the vicinity of localized fluid outflows. Additionally, only part of the 32 tubes surveyed by Di Meo-Savoie et al. (2004) displayed an average temperature above 30°C. It is therefore possible for *Hesiolyra bergi*, which is known to be very active at the surface of the colony, to avoid areas where it might be exposed to excessive thermal stress (see Table 1). Similarly, *Paralvinella sulfincola* and *Paralvinella palmiformis*, the two alvinellids species of the Juan de Fuca Ridge, were shown to have very different temperature tolerances, despite they can be found in the same habitat (Guirguis and Lee 2006). While *P. sulfincola* preference to temperature in range 40–50°C was established in vivo, the later consistently avoided temperatures above 35°C.

*Paralvinella sulfincola* is the only animal that is now firmly identified to prefer chronic exposure to temperature as high as 50–56°C. Although lower than for some terrestrial animals (55–65°C; see examples in Chevaldonné et al. 1992; Girguis and Lee 2006), the temperature preference and tolerance of this species stand at the upper limit of the accepted range for metazoans. While most hydrothermal vent animal species where finally observed to live at "room temperature" (ca. about 20°C) (see review in Van Dover and Lutz 2004), *P. sulfincola* appears as an outstanding thermophilic animal. The association of *Alvinella pompejana* to extremely hot substrates, comparable in this respect to the environment of *P. sulfincola* (Juniper et al. 1992), supports the idea that this species would have a similarly high thermotolerance. If the limits of its thermal preference and tolerance remain to be empirically defined, its exceptional thermal adaptation is now firmly assessed from molecular markers.

### 5.2 Collagen as a marker of molecular adaptation to high temperature

The thermotolerance of *A. pompejana* has been previously discussed in the light of its biochemical properties (Chevaldonné et al. 2000). It is, however, difficult to evaluate from biochemical studies, as it depends on the versatility of the definition and the data considered. Most of the data are usually obtained in vitro and it is obvious that a diversity of mechanisms involved in regulatory processes occur in vivo in living organisms. Most of these processes have been adapted in animals during the course of evolution. Some molecules may provide a highly relevant set of information when considered as part of a more integrative physiological approach. This is particularly valuable for structural proteins whose stability mostly depends on physical parameters. Collagen, for instance, is one of the most well known extracellular proteins of the animal kingdom and is a relevant marker of thermal adaptation. Its stability is critical to the survival of the animal in controlling the deformation of the body wall by hydrostatic pressure. This molecule in *Alvinella pompejana* has been well characterized in the last decade (Gaill et al. 2003). It has been shown that this species has the most thermostable protein ever known (Gaill 1993) and that pressure is not involved in such a characteristic (Auerbach et al. 1995). The temperature at which the collagen molecule is denatured ($T_m$) is 46°C for the cuticular collagen covering the animal epidermis and 45°C (Gaill et al. 1995) for the interstitial one which is found in the worm tissue (Gaill et al. 1991). The level of thermal stability of *Alvinella pompejana* cuticular interstitial collagen is significantly higher than that of other vent annelids. In comparison, *Riftia pachyptila* molecular collagen stability only reaches 29°C and the collagen of *Paralvinella grasslei* has a denaturating temperature of only 35°C (Mann et al. 1996). Among the fibrilar collagens of 40 other vertebrates and invertebrates, Burjanadze (2000) positioned *A. pompejana* collagen at the upper limit for melting temperature, only before that of thermostable synthetic collagens.

The origin of the thermal stability of this molecule has been characterized by Sicot et al. (2000). These authors have shown that this property is not as new as previously expected (Hamraoui et al. 1998), but rather that this species has amplified all the molecular characteristics which are known to be involved in the stability of vertebrates or synthetic collagens. The collagen molecule belongs to a family of extracellular proteins, which are characterized by a triple helical domain. This domain is composed of a succession of aminoacid triplets (Gly-X-Y). Usually the Y position is occupied by a prolin amino acid that is often hydroxylated. The most important result of their study is the increase in proline content and in the number of stabilizing triplets. Phylogenetic analysis (Sicot et al. 1997,

2000) indicated that the collagen evolution is in favor of an adaptive mechanism at the molecular level in *A. pompejana*. This is the first example of a molecular adaptation to temperature in vent organisms.

Since the synthesis of this protein would be stopped at a higher temperature, it could be considered that the *Alvinella pompejana* maximum body temperature is about 45°C (Gaill et al. 1991). However, the collagen is not found in tissues in a molecular form, but in a supramolecular state. Fibrillar structure is the physiological state of the collagen in metazoan tissues. Once synthesized, the collagen molecules assemble in fibrils and such a polymeric organization has a thermal stability which can exceed by 20°C that of collagen molecules (Gaill et al. 1988b). *Alvinella pompejana* fibrillar collagen assemblage may thus resist up to 65°C, which is consistent with experimental data (Gaill et al. 1995). If molecular data let us think that above 45°C, the collagen cannot be synthesized by the epidermal cells, the worm could still sustain such a high temperature without any damage in its collagen assemblage (Gaill et al. 1988b). This means that thermal fluctuations between 40 and 60°C could be easily supported by this animal. This would also support the idea of a thermal gradient along the animal body length (Cary et al. 1998), which is consistent with the observation that the cuticule of the oldest worms disappears on their posterior part (Gaill et al. 1984).

Additional data indicate that molecular characteristics are good tracers of microhabitat characteristics. For example, enzymes involved in the collagen synthesis, such as prolyhydroxylase, are shown to function in an opposite way in *Alvinella pompejana* than in other metazoans (Kaule et al. 1998). Usually the activity of this enzyme increases with oxygen concentration. The *Alvinella pompejana* prolylhydroxylase are only active in hypoxic media and, above 10% of atmospheric saturation, oxygen appears as a poison for the metabolic machinery of collagen synthesis. This indicates that the worm is not only facing the highest temperature ever known for marine invertebrates but would have a metabolic machinery adapted for working in low oxygen environments.

## 6 Life history traits and colonization process

*Alvinella pompejana* has been described as a pioneer on newly formed chimneys (Desbruyères et al. 1998). Colonization experiments have further confirmed the settlement of the first individuals within a few days on colonization devices deployed over a smoker wall, following the formation of filamentous microbial mats (Taylor et al. 1999). The formation of several centimeter-thick assemblages of tubes within 2 months, partly encrusted in mineral precipitates, underlines a fast colonization process in response to rapid production of newly available substrates (Zbinden et al. 2003). Among the processes that can govern the formation of these new settlements, Pradillon et al. (2005a) considered two alternative possibilities: the recruitment of larvae or the migration of post-larvae stages. Almost nothing is known on the life cycle and dispersal strategies of vent species. To date, embryos of only two vent species (*Riftia pachyptila* and *A. pompejana*) have been studied for temperature and pressure tolerance (Marsh et al. 2001; Pradillon et al. 2001). This study is mainly limited by the fact that catching larvae directly in situ remains highly challenging. As an alternative to sampling, in vitro fertilization methods combined to in vivo experiments revealed to be a very pertinent approach to obtain essential information on the ability of early stages to deal with the extreme environment of adult colonies (Pradillon et al. 2001; this issue).

### 6.1 Embryos development

Pressurized incubators for the culture of cells embryos and larvae (PICCEL, Pradillon et al. 2004) allowed to investigate some of the life history traits of *A. pompejana*. An additional device named PIRISM (including an underpressure microscopy imaging system) allowed the observation of the successive cell divisions of the Pompeii worms embryos at 250 bars without disrupting the pressure (Pradillon et al. 2005a).

If adult stages of *Alvinella pompejana* may be able to resist unusually high temperature for metazoans (40–60°C), in vivo studies showed that early embryos are restrained to a much lower temperature range (Pradillon et al. 2001). By mean of in vitro fertilization and incubation of the fertilized eggs in controlled conditions of temperature and pressure, they showed that embryos did not survive a temperature of 20°C, while that they could develop at about 12°C (±2°C). The developmental process was shown to be arrested at 2°C but a transient temperature increase could trigger development of arrested embryos. These results lead to the hypothesis that, in their natural habitat, embryos had to develop in "cold area" that would exclude the colonies of *A. pompejana* where temperature lies frequently above 20°C (Pradillon et al. 2001).

To test this hypothesis, incubators containing *A. pompejana* embryos were deployed in different habitats of a single edifice, along a gradient of hydrothermal influence (Pradillon et al. 2005b). Embryos incubated on an *A. pompejana* colony displayed a very low survival and did not develop. In contrast, embryos incubated in a *Riftia pachyptila* clump, about 1 m below, or on the bare mineral seafloor at the basis of the chimney with minor influence of hydrothermal activity, survived well and developed. These results supported the hypothesis that development outside of the colony is possible, while it would not be viable in the adult colony. Punctual measurements close to the incubation device indicated 13 ± 4°C on average but, transient bursts above 20°C were monitored at the surface of the colony, just a few decimeters apart (Fig. 3). This experiment, however, did not allow to decipher which factors (average temperature, temperature maxima, sulfide levels) predominantly affected embryos survival and development (Pradillon et al. 2001). In summary, it can be said that, in contrast with the preference of *A. pompejana* adult stages for extreme habitat conditions associated with hot fluid venting, Pradillon and collaborators have demonstrated the low tolerance of their early embryonic stage (Pradillon et al. 2001, 2004). As suggested from these studies, embryos could disperse through cold abyssal water and later develop when they find temperature conditions around 5–10°C. While they were shown to develop normally at native pressure (250 bars), however, abnormal development was observed at atmospheric pressure. This excludes the hypothesis that development could occur in shallower part of the ocean where the temperature rises above its low deep-ocean values. *Alvinella pompejana* embryos should thus find low-temperature vent habitats to develop, such as those occupied by tubeworms. Low-temperature areas at the surface of adults colonies might also provide suitable habitats for embryos development. This characteristic of *Alvinella pompejana* colonies, that is clearly emphasized by the synthesis of temperature surveys, could lead to refine the assumption that early stages are excluded from the adult habitats.

### 6.2 Colonization and reproduction

Recent data obtained on the ovogenesis of the Pompeii worm (Pradillon and Gaill 2003) and on its reproductive behavior (Pradillon et al. 2005b) have enabled to conduct a series of colonization experiments. The experiments were performed on various sites to compare the size and reproductive maturity patterns of *A. pompejana* populations between recently established patches and the native colonies on which the colonization devices (TRAC; Titanium Ring for Alvinellid Colonization) were deployed (Pradillon et al. 2005b). Results obtained indicate that newly opened substrates are colonized mostly by post-larvae stages, juveniles or adults, migrating from the adjacent colony, rather than by the recruitment of larvae. Migration is thought to occur by means of tube secretion along the substrates or existing tubes (Gaill and Hunt 1991). Below 50 days, non-reproductive individuals exclusively composed these pioneer populations even when mature females were present. The percentage of reproductive female subsequently increased to reach almost that of native colonies in 150 days. Tube production may involve a large energy investment for each individual as tube production rates calculated for short deployments (a few days) showed that worms could daily produce tube material up to 40% of their own weight (Pradillon et al. 2005b). Because of resource limitation, females might not be able to simultaneously sustain tube secretion for migration processes and gamete production.

This study also indicated that female *A. pompejana* individuals do not reproduce synchronously at the patch scale in contrast to what was previously suggested at the vent scale (reviewed in Desbruyères et al. 1998). Such an asynchrony could be a strategy of adaptation of the Pompeii worm to respond very quickly to the disturbance of the fluctuating environment. Pradillon et al. (2005b) have proposed a model of colonization in which opening of a new substrate will be followed by migration of non-reproductive individuals, then succeeded by a maturation stage before fertilization and spawning, which has to be tested in the future. The morphology of the *A. pompejana* larvae is still unknown, as are the habitats where their recruitment is undergone. This issue will be further explored in the next years thanks to the molecular probes that have been designed (Pradillon et al. in press) and will be of great help to detect these larvae and to define more precisely their distribution.

### 6.3 Temperature as a genetic selection factor in colonization

With the aim of investigating selection forces in these extreme and contrasted environments, Piccino et al. (2004) have highlighted significant genetic differentiation between *A. pompejana* settlements on small young chimney, versus older edifices, using the two major allozymes, Pgm90 and Pgm100, of the Pgm-1 locus coding for phosphoglucodismutase. Phosphoglucodismutase is a relay enzyme between glycogen metabolism and glycolysis, a major energy pathway for organisms living in hypoxic to anoxic environments. Their results suggest a significant selection pressure between populations of recent colonies on young chimneys and those forming mature colonies on older edifices (Piccino et al. 2004). This study additionally revealed that allozyme 90 that occurred at highest frequency on young chimneys was shown to be more thermostable and it optimum is 3°C higher (55°C vs. 52°C) than allozyme 100. On the basis of these biochemical data, the observed genetic selection was attributed to the thermal characteristics of the habitat, Pgm-1 alleles 90/90 and 90/100 being favored among colonizers of newly formed high temperature habitats. Habitat instability was suggested to counterbalanced this selective force and maintain the percentage of 100/100 over time at the metapopulation level.

Considering the physiological differences that were emphasized between early colonizers and later migrants, and the temperature gradient across the dense layer of tubes at the interface of the mineral wall and surrounding seawater (Section 4.1), one could also consider that the settlement of pioneers could induce major change in the structure of physico-chemical gradients in this environment. Co-evolution of the habitat and the *A. pompejana* population would then be possible and could be further tested in future studies.

## 7 Microhabitat bioengineering

Even though this animal could be adapted to sustain micro-habitat temperatures of 55–60°C, such a high thermal tolerance is still insufficient to explain its ability to colonize environments with temperature over 100°C. Resolving this apparent contradiction between the thermal adaptation of this organism and its maximum habitat temperature, clearly, relies on the large temperature gradients occurring at the organism and colony scales. How the worm could maintain its position in a sustainable thermal range within such steep gradients and chaotic fluctuations has long been puzzling. Advanced knowledge of the processes controlling small-scale environmental variability and colonization patterns have provided keys to this enigma.

### 7.1 Tube as a protective barrier to the hot vent fluids

The ability of the adult stage to colonize new substrates likely relies on its capacity to continuously build a tube that is progressively embedded in the mineral. One of the most remarkable adaptation of *A. pompejana*, with respect to other alvinellid species, consists in this tube (Gaill and Hunt 1991). While the other species are embedded in the mucus layer they secrete, *Alvinella* spp. are the only species that forms such an "exoskeleton." This concentrically multilayered structure secreted by the animal has a considerable thermal and chemical stability, as compared to other annelid tubes (Gaill and Hunt 1986). This assemblage of biopolymers is composed of about 50% of proteins, forming a liquid crystalline-like organization (Gaill and Bouligand 1987). This exopolymer cannot be destructurated within the 0–100°C range, while neither strongly acidic nor alkaline solution causes major degradation (Gaill and Hunt 1986). Such a hydrophobic extracellular structure was hypothesized to protect the worm from the fluids migrating on their outer environment (Gaill and Hunt 1991). The micro-analysis of mineral precipitates formed at the outer and inner faces of the tube, have emphasized a strong mineralogical gradient and first confirmed that the tube acts as a very robust barrier against the vent fluids (Zbinden et al. 2003).

*Alvinella pompejana* spends most of the time inside its tube. Unless mechanically disturbed, individuals were never observed swimming or crawling out of their tube (authors pers. obs.). Intermittent and partial exposure of their body at the openings of tubes has been frequently described. Chevaldonné and Jollivet (1993) estimated an average 10 min delay between two exposures but this corresponded to the maximum duration of the video sequences analyzed, suggesting that this time lapse could be even higher. The use of ROVs for in situ chemical surveys have expended the capacity of observation over several tens of minutes to more than 1 h and this interval now appears much more variable than suggested by this study (authors, pers. obs.). The correlation of this behavioral feature with to local physico-chemical conditions could be further explored using video imagery and analysis combined to continuous physico-chemical monitoring.

### 7.2 Buffering of microhabitat conditions by the animal

The circulation of fluids inside the tubes has long been questioned. Contradictory hypotheses concerning the driving processes involved in the renewal of the internal medium were proposed. Early reviews suggested a U-shape tube opened on both side, through which seawater could be drawn by conduction (Desbruyères et al. 1998). Gaill and Hunt (1991) proposed a "thermal siphon" mechanism in which the median part of the tube in contact to the hot chimney wall is conductively heated, while the conditions at tube openings are much colder. In contrast, Cary et al. (1998) suggested that the tube would channel the hot fluids escaping the chimney wall. None of these hypotheses, however, could be fully demonstrated until additional clues were provided by the recent studies of Di-Meo Savoie et al. (2004) and Le Bris et al. (2005). Analyses of magnesium, sulfate and nitrate in micro-samples furthermore revealed that the medium inside tubes is composed of up to 72–91% seawater (Di Meo-Savoie et al. 2004), in apparent contradiction with the high temperatures reported in some tubes. Using pH as a tracer of the vent fluid, Le Bris et al. (2005) demonstrated that the tube is filled from the medium overlying the colonies, while the fluid circulating outside tubes is mostly dominated by a very acidic fluid indicating a high hydrothermal signature. Interestingly, neither the conditions inside tube nor those outside tubes consistently reflected conservative mixing of the vent fluid escaping the chimney wall and seawater. Conductive heat exchange through the tube wall could only explain these unusual physico-chemical features (Le Bris et al. 2005).

Geochemical modeling not only suggested a heating of the internal medium, but also a substantial cooling of the fluid migrating outside tubes along to its pathway through the colony (Le Bris et al. 2005). By this mechanism, the *Alvinella pompejana* colony was proposed to act as a heat exchanger at the interface of the chimney wall and seawater. If the driving forces that enable this medium to circulate through the tube are still unknown, an effective contribution of the animal to the renewal of the medium filling its tube is obvious from the observation of its behavior. Very brief outward/inward displacements are shown to be accompanied with expulsion of particulate materials at tube openings (authors pers. obs.). It is not known if active ventilation is also undergone when the worm is fully inside its tube, like described for other tube-dwelling or burrowing annelids (Mill 1978). As previously proposed, thermal convection, driven by conductive exchange through the tube wall, could additionally favor the circulation of fluids inside the tube (Gaill et al. 1991; Le Bris et al; 2005).

Such a possibility for this environment to be buffered in temperature, with the temperature threshold being directly under animal control rather than be driven by chaotic changes in turbulent hot fluid emissions, is likely to be determinant in the ability of *Alvinella pompejana* to deal with extreme habitat temperature. Furthermore, even though the dynamics of these processes have to be elucidated, cooling of the fluids migrating through the tube assemblage could be of main ecological importance, favoring the subsequent arrival of less tolerant animal species like *Hesiolyra bergi*.

### 7.3 Mutualistic association with microbes

High densities of microbial sulfur filaments were described partly covering the inner face of the tube or attached to appendages on the dorsal part of the worm body (Desbruyères et al. 1985; Gaill and Hunt 1991). Among all alvinellid species, *A. pompejana* is the only one to possess this obligate epibiotic association (Desbruyères et al. 1985). The contribution of the associated microflora to the mechanisms that protect these worms from the environmental threats remains putative. Its role in buffering temperature at the surface of the animal has been recently suggested (Di Meo-Savoie et al. 2004). Its influence on the tube micro-habitat could, additionally, rely on the reduction of sulfide levels. Most of the microorganisms related to *A. pompejana* belong to the $\epsilon$-subdivision of proteobacteria (Campbell et al. 2003), as in the vast majority of hydrothermal habitats where this phylogenetic group was shown to be prevalent (Nakagawa et al. 2005; Takai et al. 2005). This is also the case of early microbial colonizers at the surface of mineral substrates or the colony itself (Taylor et al. 1999; Alain et al. 2004). According to Takai et al. (2005), a sulfide-oxidizing metabolism can be suspected for these communities when living at the oxic–anoxic interface. Genes associated with sulfur-related metabolisms were described, both, for the dominant epibionts and tube-associated communities (Haddad et al. 1995; Cary et al. 1997; Campbell et al. 2001, 2003). Despite several attempts, however, none of the strains isolated from the *A. pompejana* habitat could be related to the dominant phylotypes (Campbell et al. 2003). It is therefore impossible to estimate the rate of sulfide oxidation in this environment, or the potential for local chemosynthetic primary production from this energetic pathway. Using a culture independent approach, Campbell et al. (2003) have recently shown that members of both the epibiotic and surrounding free-living bacterial communities could display a chemolithoautotrophic form of growth. Furthermore, showing that a gene encoding for ATP citrate lyase, a key enzyme in the rTCA cycle, is expressed, these authors have shown that these chemolithoautotrophs would use the reverse tricarboxylic acid cycle, a metabolic alternative to the Calvin-Benson cycle generally associated with hydrothermal vents invertebrates.

Microorganisms that utilize the rTCA cycle have shown a capacity to grow rapidly with a dynamic supply of both sulfide and oxygen (Taylor and Wirsen 1997). The formation of a dense filamentous sulfur mat of several centimeters thick within 3 days, in a titanium support deployed for an *Alvinella pompejana* colonization experiment, was interpreted as an illustration of the rapid growth of these microbial communities and shown to immediately precede the arrival of the worms (Taylor et al. 1999). The local abundance of such microbes may provide a substantial nutritional benefit for *Alvinella pompejana* that does not harbor endosymbiotic bacteria and must derive their nutrition from bacteria grazing (Desbruyères et al. 1985).

The sustained level of $CO_2$ fixation of bacteria using rTCA was also presented a potentially important primary production process in active hydrothermal vent systems (Wirsen et al. 2002). If the $\epsilon$-proteobacteria described by Campbell et al. (2003) could have similarly requirements for growth than those of the shallow water arcobacter strain studied by Wirsen et al. (2002), these primary producers should find ideal conditions within *Alvinella pompejana* tubes. Not only this strain is characterized by microaerophilic to anaerobic growth (Wirsen et al. 2002), but the permanent supply of electron donor ($H_2S$ and $S^0$)

and acceptors ($O_2$, $NO_3^-$) to the medium in the ventilated tube could maintain a high microbial growth rate. If this still has to be demonstrated, the ability of *A. pompejana* to form dense colonies on bare substrate within a few weeks might also depend on this mutualistic process between *A. pompejana* and the chemoautotrophs colonizing its tube.

## 8 Molecular adaptation to chemical stresses: first results and open questions

As suggested previously, the collagen-associated enzymatic processes were shown to be remarkably adapted to low oxygen levels. In comparison to other marine annelids, major respiratory adaptations to hypoxia have been described for this species, emphasizing a major evolutionary trend at the molecular level (reviewed in Hourdez and Weber 2005). Haemoglobins of *Alvinella pompejana*, described in (Zal et al. 2000) and recently revisited using cryoelectron microscopy by Jouan et al. (2003), exhibit very high affinity for oxygen as compared to related species living in normoxic environments (Toulmond et al. 1990; Hourdez et al. 2000), consistently with the low oxygen availability expected in this environment.

Sulfide is known to inhibit a variety of enzymes and proteins involved in the aerobic respiration processes (Visman 1991). Adaptation to sulfide has been demonstrated for a variety of vent organisms (Powell et al. 1987). Sulfide binding by haemoglobin is a prerequisite for *Riftia pachyptila* to transfer large amounts of sulfide to its sulfide-oxidizing symbionts, while preventing its deleterious effects (Childress and Fisher 1992). Sulfide binding properties of haemoglobin have also been suggested as the most important detoxification mechanism in the non-endosymbiotic alvinellid, *Paravinella sulfincola* (Martineu et al. 1997). Comparatively, these authors did not find any sulfide binding properties for *A. pompejana* hemoglobin, even though substantial $H_2S$ concentration was observed in its blood.

A recent study of enzymatic oxidant defense and aerobic metabolism (Marie et al. 2006) has shown that *Paralvinella grasslei* would regulate its enzymatic system as a response to ambient oxidative stress. Furthermore, this stress would be independent of the respiration process, but could be rather due to the high level of oxygen radicals in the environment or their formation in tissues due to metal oxidation. This enzymatic defense system still has to be studied in vivo for *A. pompejana*. Desbruyères et al. (1998) reported another potential molecular adaptation in response to the high heavy metal enrichment in the body of this species. The high concentration of metalothioneins-like proteins in digestive tracks, parapod and dorsal epidermis was suggested to reflect a response of the worm to potential metal toxicity.

In fact, most remains to be done in the study of *Alvinella pompejana* response to chemical stress in this particularly harsh environment. Since the late nineties, no new data have been published on the adaptation of this species to hypercapnia, heavy metal toxicity or oxidative stress. Further developments in high-pressure technologies should provide access to this species for in vivo experimentation in the near future (Shillito et al. 2004). In addition, a large scale sequencing project of the Pompeii worm has started in 2004 (Zal et al. in press), opening the way to powerful and efficient functional analyses especially in proteomics and structural genomics.

## 9 Conclusion

The strongly chaotic, steeply changing and ephemeral environment of the *A. pompejana* environment is undoubtedly some of the most challenging encountered on Earth. The exceptional thermal characteristics of its collagen is the best example of the adaptation of this organisms to this hot environment on evolutionary timescales. The ability of *Alvinella pompejana* to colonize these substrates is far from being fully understood, but the exceptional properties of the tube and the behavior of the worm can be now considered as a major clues in the colonization process. From a more complex combination of adaptive and behavioral responses, a dynamic scheme can be proposed to understand the way colonies are established. *Alvinella pompejana* could thus stand at the limits authorized for its biological machinery in a highly dynamic environment where temperature can exceed lethal limits, but where micro-habitat thermal regulation by the animal would prevent exposure to deleterious temperature spikes. Ecologically, the dynamic system associating this pioneer species and the associated microflora might be viewed as a key to the subsequent colonization of these environments by less tolerant species, highlighting *A. pompejana* as a new type of ecosystem bioengineer. Only part of the puzzle is now solved. Much remains to be known concerning the molecular adaptation of this organism to the combination of multiple thermal and chemical stresses, that make this habitat one of the most extreme marine environments.

**Acknowledgements** This work was financially supported by Ifremer, CNRS, INSU (Dorsales and GEOMEX programs), and the European Community (Ventox project EVK3-1999-00056P). We would like to particularly acknowledge the captains and crews of Research Vessels, the Nautile, Alvin and Victor 6000 operation groups, and instrumentation engineers and technicians for their essential support at sea.

## References

Alain K, Zbinden M, Le Bris N, Lesongeur F, Querellou J, Gaill F, Cambon-Bonavita M-A (2004) Early steps in microbial colonisation processes at deep-sea hydrothermal vents. Environ Microbiol 6(3):227–241

Aller RC (1982) The effects of macrobenthos on chemical properties of marine sediment and overlying water. In: McCall PL, Tevesz MJS (eds) Animal-sediment relations. Plenum Press, New York, pp 53–102

Arvidson RS, Morse JW, Joye SB (2004) The sulfur biogeochemistry of chemosynthetic cold seep communities, Gulf of Mexico, USA. Marine Chemistry 87, 97–119

Auerbach G, Gaill F, Jaenike R, Schulthess T, Timpl R, Engel J (1995) Pressure dependence of collagen melting. Matrix Biol 14:589–592

Bach W, Edwards KJ (2003) Iron and sulfide oxidation within the basaltic ocean crust: implications for chemolithoautotrophic microbial biomass production. Geochim Cosmochim Acta 67:3871–3887

Bates AE, Tunnicliffe V, Lee RW (2005) Role of thermal conditions in habitat selection by hydrothermal vent gastropods. Mar Ecol Prog Series 305:1–15

Burjanadze TV (2000) New analysis of the phylogenetic change of collagen thermostability. Biopolymer 53:523–528

Campbell BJ, Jeanthon C, Kosta JE, Luther GW, Cary SC (2001) Growth and phylogenetic properties of novel bacteria belonging to the epsilon subdivision of the Proteobacteria enriched from *Alvinella pompejana* and deep-sea hydrothermal vents. Appl Environ Microbiol 67:4566–4572

Campbell BJ, Stein JL, Cary SC (2003) Evidence of chemolithoautotrophy in the baterial community associated with *Alvinella pompejana*, a hydrothermal vent polychaete. Appl Environ Microbiol 69:5070–5078

Cary SC, Cottrell MT, Stein JL, Camacho F, Desbruyères D (1997) Molecular identification and localization of filamentous symbiotic bacteria associated with the hydrothermal vent annelid *Alvinella pompejana*. Appl Environ Microbiol 63:1124–1130

Cary SC, Shank T, Stein J (1998) Worms bask in extreme temperatures. Nature 391:545–546

Charlou JL, Donval JP, Fouquet Y, Jean-Batiste P, Holm N (2002) Geochemistry of high $H_2$ and $CH_4$ vent fluids issuing from ultramafic rocks at the Rainbow hydrothermal field (36°14′, MAR). Chem Geol 191:345–359

Chevaldonné P, Desbruyères D, Le Haître M (1991) Time-series of temperature from three deep-sea hydrothermal vent sites. Deep-Sea Res 38:1417–1430

Chevaldonné P, Desbruyères D, Childress JJ (1992) And some even hotter. Nature 359:593

Chevaldonné P, Jollivet D (1993) Videoscopic study of deep-sea hydrothermal vent alvinellid polychaete populations: biomass estimation and behaviour. Mar Ecol Prog Series 95:251–262

Chevaldonné P, Fisher CR, Childress JJ, Desbruyères D, Jollivet D, Zal F, Toulmond A (2000) Thermotolerance and the "Pompeii worms. Mar Ecol Progr Series 208:293–295

Childress JJ, Fisher CR (1992) The biology of hydrothermal vent animals: physiology, biochemistry and autotrophic symbioses. In: Barnes M, Ansell AD, Gibson RN (eds) Ocean. Mar. Biol. Ann. Review. UCL press, pp 337–441

Childress JJ, Lee RW, Saunders NK, Felbeck H, Oros DR, Toulmond A, Desbruyères D, Kennicutt MC, Brooks J (1993) Inorganic carbon uptake in hydrothermal vent tubeworms facilitated by high environmental $pCO_2$. Nature 362:147–149

Desbruyères D, Gaill F, Laubier L, Fouquet Y (1985) Polychaetous annelids from hydrothermal vent ecosystems: an ecological overview. Biol Soc Washington Bull 6:103–116

Desbruyères D, Chevaldonne P, Alayse-Danet AM, Caprais JC, Cosson R, Gaill F, Guezennec J, Hourdez S, Jollivet D, Jouin-Toulmond C, Lallier FH, Laubier L, Riso R, Sarradin PM, Toulmond A, Zal F (1998) Burning subjects: biology and ecology of the "Pompei worm" (*Alvinella pompejana* Desbruyères et Laubier), a normal dweller of an extreme deep-sea environment. Deep-Sea Res II 45:383–422

Desbruyères D, Almeida A, Biscoito M, Comtet T, Khripounoff A, Le Bris N, Sarradin PM, Segonzac M (2000) A review of the distribution of hydrothermal vent communities along the Northern Mid

Atlantic Ridge: dispersal vs environmental controls. Hydrobiology 440:201–216

Di Meo CA, Wakefield JR, Cary SC (1999) A new device for sampling small volumes of water from marine microenvironments. Deep-Sea Res I 46:1279–1287

Di Meo-Savoie CA, Luther GW III, Cary SC (2004) Physico-chemical characterization of the microhabitat of the epibionts associated with *Alvinella pompejana*, a hydrothermal vent annelid. Geochim Cosmochim Acta 68:2055–2066

Douville E, Charlou JL, Oelkers EH, Bienvenu P, Colon CFJ, Donval JP, Fouquet Y, Prieur D, Appriou P (2002) The rainbow vent fluids (36°14′N, MAR): the influence of ultramafic rocks and phase separation on trace metal content in mid-Atlantic Ridge hydrothermal fluids. Chem Geol 184:37–48

Edmond JM, Measures C, Magnum B, Grant B, Sclater FR, Collier R, Hudson A, Gordon LI, Corliss JB (1979) On the formation of metal-rich deposits at ridge crests. Earth Planet Sci Lett 46:19–30

Fisher CR, Childress JJ, Arp AJ, Brooks JM, Distel D, Favuzzi JA, Felbeck H, Hessler R, Johnson KS, Kennicutt MC II, Macko SA, Newton A, Powell MA, Somero GN, Soto T (1988) Microhabitat variation in the hydrothermal vent mussel, *Bathymodiolus thermophilus*, at the Rose Garden vent on the Galapagos Rift. Deep-Sea Res 35:1769–1791

Gaill F, Desbruyeres D, Prieur D, Gourret JP (1984) Mise en évidence de communautés bactériennes épibiontes du "Ver de Pompéï" (*Alvinella pompejana*). CR Acad Sci Paris 298(19):553–558

Gaill F, Hunt S (1986) Tubes of deep-sea hydrothermal vent worms *Riftia pachyptila* (vestimentifera) and *Alvinella pompejana* (annelida). Mar Ecol Prog Series 34:267–274

Gaill F, Bouliguand Y (1987) Alternating positive and negative twist of polymers in an invertebrate integument. Mol Liq Cryst 153:31–41

Gaill F, Desbruyeres D, Laubier L (1988a) Relationships between the "Pompeii worms" and their epibiotic bacteria. Oceanol Act 8:147–155

Gaill F, Herbage D, Lepescheux L (1988b) Cuticle structure and composition of two hydrothermal vents invertebrates. Oceanol Act 8:155–159

Gaill F, Hunt S (1991) The biology of annelid worms from high temperature hydrothermal vent regions. Rev Aquat Sci 4:107–137

Gaill F, Wiedeman A, Kuhn K, Timpl R, Engel J (1991) Molecular characterization of cuticle and interstitial collagen from worms collected at deep-sea level. J Mol Biol 221:209–223

Gaill F (1993) Aspects of life development at deep-sea hydrothermal vents. FASEB J 7:558–565

Gaill F, Mann K, Wiedemann H, Engel J, Timpl R (1995). Structural comparison of cuticle and interstitial collagens from annelids living in shallow sea water and at deep-sea hydrothermal vents. J Mol Biol 246:284–294

Gaill F, Zbinden M, Pradillon F (2003) Adaptations of hydrothermal vent organisms to their environment. In: Legakis A, Sfenthourakis S, Polymeni R, Thessalou-Legaki M (eds) 13th International Congress of Zoology. Pensoft, Sofia-Moskow, Athens, pp 513–517

Geret F, Riso R, Sarradin PM, Caprais JC, Cosson R (2002) Metal bioaccumulation and storage forms in the shrimp, *Rimicaris exoculata*, from the Rainbow hydrothermal field (Mid Atlantic Ridge); preliminary approach to the fluid–organism relationship. Cah Biol Mar 43:43–52

Girguis PR, Lee RW (2006) Thermal preference and tolerance of alvinellids. Science 312:231

Haddad A, Camacho F, Durand P, Cary SC (1995) Phylogenetic characterization of the epibiotic bacteria associated with the hydrothermal vent polychaete *Alvinella pompejana*. Appl Environ Microbiol 61:1679–1687

Hamaraoui L, Zaoui D, Gaill F. (1998) Biochemical characteristics of collagens from marine invertebrates with special emphasis on Polychaeta and Vestimentifera from deep-sea hydrotermal vents. Cah Biol Mar 39:109–120

Hessler RR, Kaharl VA (1995) The deep-sea hydrothermal vent community: an overview. In: Humphris SE, Zierenberg RA, Mullineaux LS, Thomson RE (eds) Seafloor hydrothermal systems: physical, chemical, biological and geological interactions. AGU, Washington DC, pp 72–84

Hourdez S, Lallier FH, De Cian M-C, Green BN, Weber RE (2000) The gas transfer system in *Alvinella pompejana* (Annelida Polychaeta, Terebellida). Functional properties of intracellular and extracellular hemoglobins. Physiol Bioch Zool 73:365–373

Hourdez S, Weber RE (2005) Molecular and functional adaptations in deep-sea hemoglobins. J Inorg Biochem 99:130–141

Hurtado LA, Lutz RA, Vrijenhoek RC (2004) Distinct patterns of genetic differentiation among annelids of eastern Pacific hydrothermal vents. Mol Ecol 13:2603–2615

Johnson KS, Beelher CL, Sakamoto-Arnold CM (1986a) A submersible flow analysis system. Anal Chim Acta 179:245–257

Johnson KS, Beelher CL, Sakamoto-Arnold CM, Childress JJ (1986b) In situ measurements of chemical distributions in a deep-sea hydrothermal vent field. Science 231:1139–1141

Johnson KS, Childress JJ, Beelher CL (1988a) Short time temperature variability in the Rose Garden hydrothermal vent field: an unstable deep sea environment. Deep-Sea Res I 35:1711–1721

Johnson KS, Childress JJ, Hessler RR, Sakamoto-Arnold CM, Beelher CL (1988b) Chemical and biological interactions in the Rose Garden hydrothermal vent field, Galapagos spreading center. Deep-Sea Res I 35:1723–1744

Johnson KS, Childress JJ, Beelher CL, Sakamoto CM (1994) Biogeochemistry of hydrothermal vent mussel communities: the deep sea analogue to the intertidal zone. Deep-Sea Res I 41:993–1011

Jouan L, Marco S, Taveau J-C (2003) Revisiting the structure of *Alvinella pompejana* hemoglobin at 20

Angstrom resolution by cryoelectron microscopy. J Struct Biol 143:33–44
Juniper KS, Jonasson IR, Tunicliffe V, Southward AJ (1992) Influence of tube building polychaete on hydrothermal chimney mineralization. Geology 20:895–898
Kadar E, Costa V, Santos RS (2006) Distribution of micro-essential (Fe, Cu, Zn) and toxic (Hg) metals in tissues of two nutritionally distinct hydrothermal shrimps. Sci Tot Environ 358:143–150
Kaule G, Timpl R, Gaill F, Gunzler V (1998) Prolyl activity in tissue homogenates of annelids from deep sea hydrothermal vents. Matrix Biol 17:205–212
Koschinsky A, Seifert R, Halbach P, Bau M, Brasse S, De Carvalho LM, Fonseca N (2002) Geochemistry of diffuse low-temperature hydrothermal fluids in the North Fidji Basin. Geochim Cosmochim Acta 66:1409–1427
Le Bris N, Sarradin PM, Birot D, Alayse-Danet A (2000) A new chemical analyzer for in situ measurement of nitrate and total sulfide over hydrothermal vent biological communities. Marine Chem 72:1–15
Le Bris N, Sarradin PM, Pennec S (2001) A new deep-sea probe for in situ pH measurement in the environment of hydrothermal vent biological communities. Deep Sea Res I 48:1941–1951
Le Bris N, Sarradin PM, Caprais JC (2003) Contrasted sulphide chemistries in the environment of 13°N EPR vent fauna. Deep-Sea Res I 50:737–747
Le Bris N, Zbinden M, Gaill F (2005) Processes controlling the physico-chemical micro-environments associated with Pompeii worms. Deep-Sea Res Part I:10–1083
Le Bris N, Govenar B, Le Gall C, Fisher CR (2006a) Variability of physico-chemical conditions in 9°N EPR diffuse flow vent habitat. Mar Chem 98:167–182
Le Bris N, Rodier P, Sarradin P-M, Le Gall C (2006b) Is temperature a good proxy for sulfide in hydrothermal vent habitats? Cah Biol Mar (in press)
Lee RW (2003) Thermal tolerance of deep-Sea hydrothermal vent animals for the Northeast Pacific. Biol Bull 205:98–101
Luther GW, Reimers CE, Nuzzio DB, Lovalvo D (1999) In situ deployment of voltammetric, potentiometric and amperometric microelectrodes from a ROV to determine dissolved $O_2$, Mn, Fe, S(–2), and pH in porewaters. Environ Sci Tech 33:4352–4356
Luther GW, Glazer BT, Hohmann L, Popp JI, Taillefert M, Rozan TF, Brendel PJ, Theberge SM, Nuzzio DB (2001a) Sulfur speciation monitored in situ with solid state gold amalgam voltammetric microelectrodes: polysulfides as a special case in sediments, microbial mats and hydrothermal vent waters. J Environ Monit 3:61–66
Luther GW, Rozan TF, Taillefert M, Nuzzio DB, Meo CD, Shank TM, Lutz RA, Cary SC (2001b) Chemical speciation drives hydrothermal vent ecology. Nature 410:813–816
Mann K, Mechling DE, Bachinger HP, Eckerson C, Gaill F, Timpl R (1996) Glycosylated threonine but not 4-hydroyproline dominates the triple helix stabilizing positions in the sequence of a hydrothermal vent worm cuticle collagen. J Mol Biol 261:255–266
Marie B, Genard B, Rees J-F, Zal F (2006) Effect of ambient oxygen concentration on activities of enzymatic antioxidant defences and aerobic metabolism in the hydrothermal vent worm, Paralvinella grasslei. Mar Biol (in press)
Marsh AG, Mullineaux LS, Young CM, Manahan DT (2001) Larval dispersal potential of the tubeworm. *Riftia pachyptila* at deep-sea hydrothermal vents. Nature 411:77–80
Martineu P, Juniper SK, Fisher CR, Massoth GJ (1997) Sulfide binding in the body fluids of hydrothermal vent alvinellid polychaetes. Physiol Zool 70:578–588
Mill PJ (1978) Physiology of Annelids, Academic Press, New York, pp 369–339
Millero FJ, Hubinger S, Fernandez M, Garnett S (1987a) Oxidation of $H_2S$ in seawater as a function of temperature, pH and ionic strength. Environ Sci Tech 21:439–443
Millero FJ, Sotolongo S, Izaguirre M (1987b) The oxidation kinetics of Fe(II) in seawater. Geochimica Cosmochimica Acta 51:793–801
McCollom TM, Shock EL (1997) Geochemical constraints on chemolithoautotrophic metabolism by microorganisms in seafloor hydrothermal systems. Geochim Cosmochim Acta 61:4375–4391
McCollom TM (2000) Geochemical constraints on primary productivity in submarine hydrothermal vent plumes. Deep-Sea Res I 47:85–101
Mullineaux LS, Peterson CH, Micheli F, Mills SW (2003) Successional mechanisms varies along a thermal gradient in hydrothermal fluid flux at deep-sea vents. Ecol Monogr 73:523–542
Nakagawa S, Takai K, Inagaki F, Hirayama H, Nunoura T, Horikoshi K, Yoshihiko S (2005) Distribution, phylogenetic diversity and physiological characteristics of epsilon-proteobacteria in deep-sea hydrothermal field. Environ Microbiol 7(10):1619–1632
Piccino P, Viard F, Sarradin P-M, Le Bris N, Le Guen D, Jollivet D (2004) Thermal selection of PGM allozymes in newly founded populations of the thermolerant vent polychaete *Alvinella pompejana*. Proceedings of the Royal Society of London 271:2351–2359
Powell MA, Vetter RD, Somero GN (1987) Sulfide detoxification and energy exploitation by marine animals. In: Dejours P, Taylor CR, Weibel ER (eds) Comparative physiology: life in water and on land. Fidia Research series IX-Liviana press, Padova, pp 241–250
Pradillon F, Shillito B, Young CM, Gaill F (2001) Developmental arrest in vent worms embryos. Nature 413:698–699
Pradillon F, Gaill F (2003) Oogenesis characteristics in the hydrothermal vent polychaete *Alvinella pompejana*. Invert Reprod Develop 43:223–235
Pradillon F, Shillito B, Chervin J-C, Hamel G, Gaill F (2004) Pressurized vessels for the study of deep-sea fauna. High Press Res 24:169–172

Pradillon F, Zbinden M, Mullineaux LS, Gaill F (2005a) Colonisation of newly-opened habitat by a pioneer species, *Alvinella pompejana* (Polychaeta: Alvinellidae), at East Pacific Rise vent sites. Mar Ecol Progr Series 302:147–157

Pradillon F, Le Bris N, Shillito B, Young CR, Gaill F (2005b) Influence of environmental conditions on early development of the hydrothermal vent polychaete *Alvinella pompejana*. J Exp Biol 208:1551–1561

Pradillon F, Schmidt A, Peplies J, Dubilier N (in press) Identification of early life history stages of marine species with whole-larvae in situ hybridisation assay. Mar Ecol Prog. Series

Ravaux J, Gaill F, Le Bris N, Sarradin P-M, Jollivet D, Shillito B (2003) Heat shock response and temperature resistance in the deep-sea vent shrimp *Rimicaris Exoculata*. J Exp Biol 206:2345–2354

Sarradin PM, Caprais JC, Briand P, Gaill F, Shillito B, Desbruyères D (1998) Chemical and thermal description of the Genesis hydrothermal vent community environment (13°N, EPR). Cah Biol Mar 39:159–167

Sarradin PM, Caprais JC, Riso R, Kerouel R, Aminot A (1999) Chemical environment of the hydrothermal mussels communities in the Lucky Strike and Menez Gwen vent fields, MAR. Cah Biol Mar 40:93–104

Sarrazin J, Juniper SK, Massoth G, Legendre P (1999) Physical and chemical factors influencing species distributions on hydrothermal sulfide edifices of the Juan de Fuca Ridge, Northeast Pacific. Mar Ecol Progr Ser 190:89–112

Shank TM, Fornari DJ, Von Damm KL, Lilley MD, Haymon RM, Lutz RA (1998) Temporal and spatial patterns of biological community development at nascent deep-sea hydrothermal vents (9°50′N, East Pacific Rise). Deep-Sea Res II 45:465–515

Shillito B, Jollivet D, Sarradin PM, Rodier P, Lallier F, Desbruyères D, Gaill F (2001) Temperature resistance of Hesiolyra bergi, a polychaetous annelid living on deep-sea vent smoker. Mar Ecol Progr Ser 216:141–149

Shillito B, Le Bris N, Gaill F, Rees JF, Zal F (2004) First access to live Alvinellas. High Press Res 24(1):169–172

Shillito B, Le Bris N, Hourdez S, Ravaux J, Cottin D, Caprais J-C, Jollivet D, Gaill F (2006) Temperature resistance studies on the deep-sea vent shrimp *Mirocaris fortunate*. J Exp Biol 209:945–955

Sicot FX, Exposito J-Y, Masselot M, Garrone R, Deutxch J, Gaill F (1997) Cloning of an annelid fibrillar collagen gene and phylogenetic analysis of vertebrate and invertebrate collagens, in: nucleic acids, protein synthesis and molecular genetics. Eur J Biochem 590:1–9

Sicot FX, Mesnage M, Masselot M, Exposito J-Y, Garrone R, Deutxch J, Gaill F (2000) Molecular adaptation to an extreme environment: origin of the thermal stability of the Pompeii worm collagen. J Mol Biol 302:811–820

Sung W, Morgan JJ (1980) Kinetics and products of ferrous iron oxygenation in aqueous systems. Environ Sci Tech 14:561–569

Takai K, Campbell BJ, Cary SC, Suzuki M, Oida H, Nunoura T, Hirayama H, Nakagawa S, Suzuki Y, Inagaki F, Horikoshi K (2005) Enzymatic and genetic characterization of carbon and energy metabolisms by deep-sea hydrothermal chemolithoautotrophic isolates of epsilonproteobacteria. Appl Environ Microbiol 71:7310–7320

Tapley DW, Buettner GR, Shick JM (1999) Free radicals and chemiluminescence as products of the spontaneous oxidation of sulfide in seawater, and their biological implications. Biol Bull 196:52–56

Taylor CD, Wirsen CO (1997) Microbiology and ecology of filamentous sulfur formation. Science 277:1483–1485

Taylor CD, Wirsen CO, Gaill F (1999) Rapid microbial production at filamentous sulfur mats at hydrothermal vents. Appl Environ Microbiol 35:2253–2255

Tivey MK, McDuff RE (1990) Mineral precipitation in the walls of black smoker chimneys: a quantitative model of transport and chemical reaction. J Geophys Res 95:617–637

Tivey MK (1995) Modeling chimney growth and associated fluid flow at seafloor hydrothermal vent sites. In: Humphris SE, Zierenberg RA, Mullineaux LS, Thomson RE (eds) Seafloor Hydrothermal Systems: Physical, Chemical, Biological and Geological Interactions. AGU, Washington DC, pp 158–177

Tivey MK, Bradley AM, Joyce TM, Kadko D (2002) Insights into tide-related variability at seafloor hydrothermal vents from time-series temperature measurements. Earth Planet Sci Lett 202:693–707

Tivey MK (2004) Environmental conditions within active seafloor vent structures: sensitivity to vent fluid composition and fluid flow. In: Wilcock W, Cary C, DeLong E, Kelley D, Baross J (eds) Subseafloor biosphere at mid-ocean ridges, geophysical monograph series, no. 144. AGU, Washington, DC, pp 137–152

Toulmond A, El Idrissi Slitine F, De Frescheville J, Jouin C (1990) Extracellular haemoglobins of hydrothermal vent annelids: structural and functional characteristics in three alvinellid species. Biol Bull 179:366–373

Tunnicliffe V (1991) The biology of hydrothermal vents: ecology and evolution. In: Barnes M (ed) Oceanogr. Mar. Biol. Annu. Rev. Aberdeen University Press, pp 319–407

Tunnicliffe V, Juniper SK (1990) Dynamic character of the hydrothermal vent habitat and the nature of the sulphide chimney fauna. Progr Oceanogr 24:1–13

Van Dover C, Lutz RA (2004) Experimental ecology at deep-sea hydrothermal vents: a perspective. J Exp Mar Biol Ecol 300:273–307

Visman B (1991) Sulfide tolerance: physiological mechanisms and ecological implications. Ophelia 34:1–27

Von Damm KL (1995) Controls on the chemistry and temporal variability of seafloor hydrothermal fluids. In: Humphris SE, Zierenberg RA, Mullineaux LS, Thomson RE (eds) Seafloor hydrothermal systems: physical, chemical, biological and geological interactions. AGU, Washington DC, pp 222, 247

Von Damm KL, Lilley MD (2004) Diffuse flow hydrothermal fluids from 9°50′N East Pacific Rise: origin, evolution and biogeochemical controls. In: Wilcock W, Cary C, Delong E (eds) Subseafloor biosphere at mid-oceanic ridges. AGU, Washington DC

Wirsen CO, Sievert SM, Cavanaugh CM, Molyneaux SJ, Ahmad AR, Taylor LT, DeLong EF, Taylor CD (2002) Characterization of an autotrophic sulfide-oxidizing Marine Arcobacter sp. that produces filamentous sulfur. Appl Environ Microbiol 68:316–325

Zbinden M, Le Bris N, Compère P, Martinez I, Guyot F, Gaill F (2003) Mineralogical gradients associated with Alvinellids at deep-sea hydrothermal vents. Deep-Sea Res I 50:269–280

Zhang JZ, Millero FJ (1994) Kinetics of oxidation of hydrogen sulfide in natural waters. In: Alpers CN, Blowes DW (eds) Environmental geochemistry of hydrogen sulfide in natural waters. ACS, Washington DC, pp 393–411

Zal F, Green BN, Martineu P, Lallier FH, Toulmond A, Vinogradov SN, Childress JJ (2000) Polypeptide chain composition diversity of hexagonal-bilayer haemoglobins within a single family of annelids, the Alvinellidae. Eur J Biochem 267:5227–5236

Zal F, Poch O, Bigot Y, Cormier P, Dietrich J, Duchiron F, Gaill F, Higuet D, Hourdez S, Jollivet D, Knoops B, Lallier F, Laullier M, Lecompte O, Leize-Wagner E, Moras D, Rees J-F, Shillito B (in press) Alvinella Consortium: a large scale sequencing project at the French Genoscope. Cah Biol Mar

Rev Environ Sci Biotechnol (2006) 5:193–201
DOI 10.1007/s11157-006-0006-z

REVIEW PAPER

# Post-capture investigations of hydrothermal vent macro-invertebrates to study adaptations to extreme environments

**Eniko Kadar · Jonathan J. Powell**

Received: 27 February 2006 / Accepted: 22 May 2006 / Published online: 18 July 2006


**Abstract** Typical survival strategies, developed by macro-invertebrates at a variety of reducing marine habitats including deep-sea hydrothermal vents, have been the subject of laboratory experimentation over the past three decades. This review provides an insight into the international efforts that have converged on the area of the laboratory maintenance of such species whose nutritional requirements are outside the usual scope of metazoan life. We emphasise the methodology used in post-capture manipulations that are designed to identify the physiological limits of adaptation to the harsh conditions known at various vent sites worldwide, and to understand the mechanisms involved. The concepts behind appropriately designed experiments and the choice of suitable model organisms for such physiological studies are also discussed.

E. Kadar (✉)
Department of Oceanography and Fisheries, IMAR Centre of the University of Azores, Cais de Santa Cruz, 9901-862 Horta, Portugal
e-mail: Enikokadar@notes.horta.uac.pt

J. J. Powell
MRC Human Nutrition Research, Elsie Widdowson Laboratory, Fulbourn Road, Cambridge CB1 9NL, UK

**Keywords** Deep-sea hydrothermal vent · Post-capture experiments · Hydrostatic pressure · Adaptation to the extreme environment

> "When we look at our own planet's most challenging environments, we are really looking for clues to what may be the normal conditions on other planets. We want a hint of what we may be searching for when we investigate those other worlds for signs of life. We will be better prepared to recognize and study them because our minds have been expanded by our knowledge of our own terrestrial biology...."(Penelope Boston)

## Introduction

Definition of "extreme environment" *per se* is a significant challenge to ecologists, since physico-chemical factors that actually prevent life in some organisms may correspond to living requirements for others. However, the textbook definition simplifies the issue by: "conditions that are far outside the boundaries in which most organisms live comfortably. Conditions include: pH, air pressure, temperature, salinity, radiation, dryness (desiccation), and oxygen level". Deep-sea hydrothermal vents certainly fall under this definition with respect to many of their attributes, including: low pH, high hydrostatic pressure (more that 10-fold the pressures tolerated by humans), reduced conditions, presence of toxicants (heavy metals and radio nucleides), total absence of sun-light, hyper-dynamic conditions and geographical limits to species dispersal. Consequently, studying species that are endemic to such challenging conditions is widely accepted as one of the best approaches of studying adaptation to the "extreme". However, there is an increasing body of evidence that avoidance response in hydrothermal vent metazoans is in fact a more important adaptation strategy as compared to the previously predicted biochemical adaptation/tolerance to the harsh conditions (Shillito et al. 2001, 2004, 2006; Company et al. 2005; Kadar et al. in press a & b), which suggests that further experimentation and in situ investigations are required. Such acquired adaptation is in contrast to previous speculations on novel, genetically imprinted mechanisms of tolerance in hydrothermal organisms, and points to life on this planet surviving under a wide range of conditions with a restricted biochemical tool kit (Dixon 2005) that deserves more attention.

Over the past 20 years deep sea vent research, especially that involving experimental approaches, was limited by sample availability. Biological studies on live specimens were, and largely still are, restricted to shipboard studies which were mainly carried out during periods that allowed access to underwater locations (usually summer months in the Atlantic). Thus a complete assessment of the life cycle and general physiology of these species was not possible. Laboratory experimentation with these species was, and still remains, hampered by the extreme stress imposed during recovery from depths around 2,000 m (even the shallowest site along the Mid-Atlantic Ridge, Menez Gwen, is 850 m). However, development of pressurized flow-through systems (Quetin and Childress, 1980) enabled on-board or even laboratory experiments, but these were still limited to short, post-capture, episodic studies following scientific cruises to vent locations. However, recent technologies that involve the use of acoustically retrievable mussel cages to supply permanent land-based laboratories (http://www.horta.uac.pt/projectos/fisiovent/cagefilling.htm) have considerably extended access to live animal samples and, as a consequence have increased duration of experimentation and garnered the involvement of a larger scientific community (Dixon et al. 2004; Company et al. 2004; Kadar et al. 2005a; Kadar et al. 2005b). The main improvement provided by this technology is that it enables sampling at times when submersible operation is impossible, and thus it allows investigation of seasonal patterns. The cages (Sonardyne International Ltd., UK), of 1.25 $m^2$ size consist of a frame constructed of glass-reinforced plastic, covered in 2 cm wide plastic mesh, and surrounded by a weighted rubber skirt around the cage base to divert sulphide- and methane-laden fluid through the bottom. They are fitted with floats with transponders to signal their exact position and also with acoustic release mechanisms for later recovery. Using the robot arm of Victor 6,000, the cages can be filled with approximately 500 mussels (2–3 h/cage to fill), and oriented to allow fluid penetration (as indicated by a temperature probe).

The purpose of this review is to present a synthesis of the available data concerning laboratory experimentation on vent macro-invertebrates while specific physiological processes that are involved in adaptation to their challenging environment are highlighted. Thus, rather than simply showcasing some well known examples, emphasis is on the validation of suitable post-capture studies to indicate adaptation to the vent environment in the hope of encouraging further work in the area.

## Post-capture experimentation on vent macro-invertebrates

### Laboratory experimentation performed at atmospheric pressure

Laboratory investigations of specific adaptations to reducing ecosystems were commenced in the early 1980s on molluscs inhabiting shallow environments that survived well in the laboratory under atmospheric pressure. Thus experiments such as those aimed at C pathways in symbiont-bearing organisms were possible without specialized pressure apparatus (Distel and Felbeck 1988a; 1988b). These experiments were the first to establish a nutritional reliance of the host macroinvertebrates on symbiont bacteria that actively fix carbon from methane. Additionally, symbiont transmission mode was examined that resulted in the development of molecular tools for the determination of bacterial–host specificity, again in species inhabiting shallow reducing environments (sea grass beds and mangrove swamps) in laboratory manipulations conducted for over a decade by Gros and his co-workers (Gros et al. 1996a; 1996b; Gros et al. 1997; Gros et al. 1998a; 1998b; Gros et al. 1999; Gros et al. 2000; Gros et al. 2003a; 2003b). These works stimulated current hydrothermal vent experimentation that is feasible at atmospheric pressure, which uses species from depths where the pressure does not exceed 100 bar as this is the threshold for barofilic adaptations (Dixon et al. 2004). Thus, the hydrothermal bivalve of the genus *Bathymodiolus (e.g. B. azoricus)*, became an important model organism for post-capture investigations, being euribarofilic and thus adapted to different depths at hydrothermal vents along the Mid-Atlantic Ridge (MAR) ranging from 850 m at Menez Gwen (37°35′N–38°N) to 1,700 m at Lucky Strike (37°00′N–37°35′N) and 2,300 m at Rainbow (36°14′N) or even below 3,000 m depth at Broken Spur (29°10′N). Consequently, this organism, when taken from environments <100 bar (i.e. Menez Gwen), not only survives at atmospheric pressure over long time span (Kadar et al. 2005a; Kadar et al., in press d), but it can also be subjected to high-pressure simulation studies (Company et al., 2004). In addition, it is the biomass-dominant species at many MAR vent sites, and thus is an appropriate choice for the study of distinct exposure conditions within these "natural pollution laboratories" (Kadar et al., in press a & b). Furthermore, its nutritional flexibility to survive on a whole range of nutrient supplies from dual endosymbiosis to filter-feeding (Fiala-Medioni et al. 2002; Kadar et al. 2005a), enables laboratory studies on selective nutritional reliance. *Mytilus edulis*, from which *B. azoricus* have evolved (Von Cosel et al., 1999), is a widely accepted pollution biomonitor, and thus results from metal uptake and detoxification studies can be compared with abundant data available on its shore analogues from polluted sites (Kadar et al., in press a, b, & c). Finally and most importantly, this species has been naturally co-exposed to a variety of toxic compounds (Sarradin et al. 1999; Kadar et al. 2005), on a geological time scale, which may have determined the development of genetically-enforced, highly efficient detoxification mechanisms (Kadar, et al. 2005b, Kadar et al, in press a, c), the study of which deserves closer scrutiny. Consequently, *B. azoricus* has been subjected to studies of both basic and more elaborate physiological processes such as: symbiont transmission mode (Kadar, et al. 2005a), metal toxicology (Kadar, et al. 2005b), reproduction studies (Colaco et al. in press; Kadar et al. in press d), parasitological assessments (Kadar et al. in press e), immunological response studies (Bettencourt et al. in preparation) and ongoing work on selective nutritional reliance. All these experiments were conducted in a laboratory set-up that provided a variety of feeding regimes under controlled conditions, and have been successful in maintaining the vent mussel for over 12 months in captivity (Kadar et al. in press d). These feeding regimes required the supply of aquaria with methane and/or sulphide thus supporting endosymbiosis in *B. azoricus* for prolonged periods. For a detailed description of the laboratory set up, water parameters and other technological details readers are referred to the work of Kadar et al. (e.g. 2005a, b, and in press d) and http://www.horta.uac.pt/projectos/fisiovent/labhorta.htm. Decompression stress associated with sampling is, however, still an unresolved problem because of the significant gene damage caused by hydrostatic pressure variations. The damage, however, appears to be reversible over a relatively short period of time when animals are allowed to recover, as reported by Dixon et al. (2004). From these data it was concluded that for mussels from locations where pressure did not exceed 100 bar (Menez Gwen vent site for instance) certain investigations would be appropriate following careful sampling and with a recovery period of at least 15 days prior to experimentation. Nonetheless, extrapolation of results obtained on organisms from relatively shallow sites to true deep-sea vents must be made with caution. It is thus worth mentioning that mussels from Lucky Strike vent site (1,700 m depth) have so far resisted every attempt to be maintained in the laboratory for periods longer than 1 month. Experimentation at atmospheric pressure, therefore, only enables long-term studies on species from relatively shallow vents (one only known to date, Menez Gwen), and this remains the main limitation of this approach in spite of it facilitating a whole range of studies that would not be possible under pressurized conditions. An accurate experimental design and investigation of physiological adaptation to hydrostatic pressure requires hyperbaric simulation.

## Laboratory experimentation at in situ pressures

Conveniently, a flow-through system was developed in 1980 to re-create hydrostatic pressure characteristics of the environment for two bathypelagic mysids to enable the assessment of behavioral responses to various stress factors, in the laboratory (Quetin and Childress, 1980). This system inspired deep-sea vent researchers to adopt the methodology and thus instigated a new era in deep-sea vent research (Mickel and Childress 1980; 1982a; 1982b; 1982c; Arndt et al. 1998; Arndt et al. 2001; Page et al. 2001; Shillito et al. 2001, 2003; Ravaux et al. 2003; Pradillon et al. 2004 and 2005; Chausson et al. 2004). These laboratory-based pressure culture systems provided settings for specific experiments that have, above all, resolved several misconceptions about hydrothermal vent organisms, such as their relatively slow metabolism and growth (as we knew in non-vent deep-sea fauna) or their extreme thermo-tolerance, based on in-situ observations (reviewed in Van Dover and Lutz 2004).

Inventive experiments have been carried out on the hydrothermal tubeworm, *Riftia pachyptila*, the species that is arguably at the highest level of adaptation to vent-conditions (Van Dower and Lutz 2004). Tubeworms were kept in pressurized aquaria and exposed to variable $H_2S$ concentrations by simply controlling the pH, which provided valuable insights not only into their sulphide metabolism, but also into their laboratory maintenance requirements (Goffredi et al. 1997a, 1997b). Another high-pressure aquaria operating at the Scripps Institute enabled a series of stress response studies on the same species (Arndt et al. 1998; 2001), which demonstrated the fundamental role of "sulphur respiration" in anoxia-tolerant symbioses. Carbon fixation and its transfer from symbionts to the host cells were described following laboratory exposure of the tube worm to radio-labeled $CO_2$ (Bright et al. 2000), and evidenced the importance of direct digestion of symbionts in its diet.

Extensive studies on the adaptive strategies of another hydrothermal species, *Bythograea thermydron*, experimentally exposed to sulfide, reported its ability to regulate oxygen consumption via thiosulphate regulation in the haemolymph, and demonstrated its increased hemocyanin-oxygen (Hc-O-2) affinity (Childress and Mickel 1980; Sanders and Childress 1985; 1992; Gorodezky and Childress 1994).

Using a high-pressure recirculation aquarium and radiolabeled bacteria, Page and co-workers established the nutritional flexibility of vent bivalves by providing experimental evidence of filter-feeding in *Bathymodiolus thermophilus* on particulate organic matter and thus supplementing nutrients provided by endosymbiotic chemoautotrophic bacteria (Page et al., 2001).

The first in-vivo experiment on behavioral response to heat shock on the barofilic vent polychaete, *Hesiolyra bergi*, implemented a novel pressurized incubator namely, IPOCAMP, which was able to simulate pressures as high as 260 bar, while nutrient supply could be controlled and simultaneous video observations were recorded (Shillito *et al.* 2001). The relatively normal physiological state of experimental specimens, indicated by the 100% survival rate of control animals as well as their high oxygen uptake rate, suggested suitable maintenance conditions. This, together with data from another recent thermal resistance study (Shillito et al. 2006) on a hydrothermal shrimp (*Mirocaris fortunate)* that exists across the hydrothermal gradient of 2–25°C and at depths from 850 to 2,300 m, have concluded that thermal physiology should be studied at each population's native pressure. However, collection stress still imposed limitations to this experimental design (i.e. this could have affected thermal resistance). Seeing that heat-shock proteins are induced by various non-temperature related factors, and these may directly influence thermo tolerance, the same authors recommended, for the first time, the use of isobaric collection cells for deep-sea biota sampling. Unfortunately, this technology awaits ongoing field testing until there is wider availability to the research community. Depressurization remains a major problem *during* experiments; for example whenever water is renewed or dead specimens are removed. However, this problem has been solved, at least at the micro-scale, in a study conducted by Pradillon et al. (2004) that reports on a micro-volume device for pressure simulation. This system has improved features, that allow automatic adjustment of gas ($H_2S$, O and $CH_4$) and/or the supply of other compounds as well as water renewal and interim specimen sampling, without depressurization of the reactor during long term experiments, for embryonic development studies and/or the study of small organisms, larvae, cell cultures or bacteria.

With the lack of such technology on the macro-scale only short term experiments, such as heat-shock response were possible for various deep-sea mcro-invertebrates: namely, *Rimicaris exoculata* (Ravaux et al. 2003), *Alvinella pompejana* (Shillito et al. 2004), and various *paralvinnelids* (Lee 2003). Surprisingly, these studies collectively established that temperature tolerance of these species is well below values expected for "extremofiles" and based on field observations. An "avoidance response" was proposed as a preferred mechanism (Dixon et al. 2002) rather than the previously stipulated extreme thermo-tolerance (reviewed in Van Dover and Lutz 2004).

Concomitant studies on other physiological parameters have contributed to our present knowledge of various mechanisms involved in adaptation to deep sea hydrothermalism, which would not have been possible without the advantages of such an experimental tool that enables independent control of parameters. A remarkable study, carried out by Chausson and co-workers, not only established that respiratory adaptations of the vent crab *Segonzacia mesatlantica* are derived from specific functional properties of its haemocyanin (Hc), but also correlated the structure plasticity of this Hc with local environmental factors (Chausson et al. 2004). Ecotoxicological investigations of Company and co-workers, also conducted using HIPOCAMP, on the enzymatic defence of the vent bivalve *B. azoricus* were the first to document increased resistance of this species, from exposure to heavy metals such as Cd, Cu and Hg, as compared to non-hydrothermal bivalves from polluted areas (Company et al. 2004). However, these experiments were limited to a relatively short time span of exposure conditions (up to 1 month) due to high mortality and the problems associated with bio-corrosion on the surface of the pressure-chamber (discussed in Beech and Gaylarde 1999). Nonetheless in a recent developmental study of the early life stage of *Alvinella pompejana* in a pressure vessel (Pradillon et al. 2005), the authors have recognized the limitations of the technique (pressure vessels do not simulate the complex and dynamic in situ thermal and chemical conditions typical of hydrothermal vents). Consequently, they have incubated embryos directly at the vent site along a fluid gradient with conditions similar to those re-created in the laboratory. Beyond the scientific finding of this work, the authors have accomplished an accurately designed experiment that was calibrated with a reference field experiment, which normally is not possible due to the prohibitive costs involved. From the above experimental investigations it may be presumed that baseline evaluation of specimen condition (such as oxygen consumption rate, water-, lipid-, and glycogen-content, as well as tissue structure), indicative of animal health, should be determined in situ for each species before undergoing experimental investigation.

While we have learned much over this past 25-years of experimentation, many of the limitations of this pressurized system approach have still not been fully overcome. The most important problems are related to pressure-acclimatization, such as isobaric sampling and maintenance of constant pressure during water changes in the laboratory. It should be noted that the latter has been solved for small size pressure chambers composed of two reactors fed by a common pressure line which can be manipulated separately using a set of valves (for further details see Pradillon et al. 2004). Furthermore, long-term maintenance requires provision of specimens with the typical hydrothermal nutrients (methane and sulphide) which bring about a series of challenges for engineers (development of novel inert materials resistant to both pressure and bio-corrosion; prevention of complexation reactions causing clogging of the tubing system in metal exposure experiments; development of the pressure-resistant version of the classical instrumentation used in behavioral response studies; etc) and for physiologists trying to tackle complex questions relevant to dynamic systems such as those typical of deep-sea hydrothermal vent conditions.

## Future advances

This review of the major experimental approaches to understand adaptation to vent conditions highlights that laboratory maintenance allows for some *specific* investigations to be conducted with great advantages over the costly and resource-consuming in situ observations. However, by no means can current technologies address all of the issues needed in mimicking vent site conditions, and the need for continued improvements are widely emphasized. Critical issues in all the above are, for example, sampling limitations that preclude continuous research, or the poor animal condition at- and post-capture. Additionally, empirical data on natural "health-condition" is lacking for almost all experimental species. New approaches, therefore, must focus on these deficiencies to ensure viability of experimentation in the long-term.

### Choosing the most suitable specimens

In considering comprehensive experimental design, aimed at studying adaptation mechanisms to "extreme" environments, the model organisms should (1) be adapted to and/or respond to a range of environmental changes relevant to the question to be answered; (2) be representative of typical conditions to allow organism-habitat inferences; (3) be abundant enough in their natural environment to enable long-term study without endangering the population or interfering with management of protected areas; (4) be studied both in the laboratory and in in situ experiments for validation of results; (5) have non-vent analogues with well studied physiology for comparison and, finally; (6) be able to resist the stress associated with laboratory handling. Certainly the list could be extended and recommendations are welcome.

### Continuity in sample supply

Relatively frequent (even monthly) animal supply has been achieved using the acoustically retrievable cage technique that operates at significantly reduced costs with simultaneously increased flexibility/frequency of sampling, since it is only dependent upon the ROV for deployment and filling of cages, white recovery can be pursued using less expensive vessels at any time of the year. Moreover, our experience also shows that cage-recovery dramatically improves animal condition as compared to those collected using the classical slurp gun technique.

Alternatively, continuous laboratory-cultures of vent organisms would be an option, although so far this has been limited to experiments on vent microorganisms (Postec et al. 2005; Daughney et al. 2004; Mergeay 2000). Current advances in laboratory reproduction of vent bivalves (Colaco et al. in press) are encouraging attempts.

### Animal condition during sampling and post-capture

Possible tools helping to solve the problems related to sampling-stress include isobaric sampling cells that are connectable to hyperbaric chambers on-board without pressure loss, as first recommended by Shillito et al. (2001). It is only a matter of time before development and testing will make such tools available for field-use (the equipment-testing cruise, EXOCET/D, is scheduled for summer 2006).

Regarding specimen condition during experiments, re-creation of their natural environment in the laboratory has proved to be successful for various species. Arguably one of the best examples is the functional experimental set-up for the long term maintenance of *B. azoricus* that enables preservation and manipulation of endosymbiosis (Kadar et al. 2005a), with the potential to develop groups with distinct nutritional reliance, and thus creating the basis for more elaborate ecophysiological research. Although improvements will be necessary in the future in terms of pressure simulations and control of water parameters (to which infra-structural conditions are currently being developed), this system has provided evidence for the good condition of mussels as shown not only by long-term survival (over 12 months), but also by gametogenesis (Kadar et al. in press d) and larval spawning in captivity (Colaco et al. in press). In addition, by maintaining endosymbiosis (both types: methane and/or sulphide oxidizers) within the host, these bacteria, otherwise uncultivable under laboratory conditions, provide a new experimental tool in vent research (Kadar et al. 2005a).

## Conclusions

In this review we have shown how recent studies have taken advantage of various experimental tools to address specific issues related to adaptation to extreme environments that are typical of deep-sea hydrothermal vents. The greatest understanding will come, however, when inferences are made in conjunction with physiological data collected in the field. In the future, rapid evolution of new technologies will enable the development of dedicated in situ experimental strategies to calibrate laboratory set-ups and provide baselines for animal condition. Instrumental platforms are particularly needed to monitor real-time chemical and temperature gradients at scales relevant to individual organism, which may only be addressed through integrated multi-disciplinary studies.

**Acknowledgments** The research was undertaken under the scope of the research project FISIOVENT (Physiological adaptations to extreme conditions at deep sea hydrothermal vents) funded by FCT (POCTI/MAR/55547/2004). We acknowledge the postdoctoral fellowship (SFRH/BPD/19625/2004) to EK. The kind efforts of the two anonymous reviewers and of David Dixon to improve this paper are gratefully acknowledged.

## References

Arndt C, Gaill F, Felbeck H (2001) Anaerobic sulfur metabolism in thiotrophic symbioses. J Exp Biol 204:741–750

Arndt C, Schiedek D, Felbeck H (1998) Metabolic responses of the hydrothermal vent tube worm *Riftia pachyptila* to severe hypoxia. Marine Ecol-Prog Series 174:151–158

Beech IB, Gaylarde CC (1999) Recent advances in the study of biocorrosion – an overview. Rev Microbiol 30:117–190

Bright M, Keckeis H, Fisher CR (2000) An autoradiographic examination of carbon fixation, transfer, and utilization in the *Riftia pachyptila* symbiosis. Mar Biol 136:621–632

Chausson F, Sanglier S, Leize E, Hagege A, Bridges CR, Sarradin PM, Shillito B, Lallier FH, Zal F (2004) Respiratory adaptations of a deep-sea hydrothermal vent crab. Micron 35:27–29

Childress JJ, Mickel TJ (1980) O consumption rate of the deep-sea hydrothermal vent crab. Am Zool 20:834

Colaço A, Martins I, Laranjo M, Pires L, Leal C, Prieto C, Costa V, Lopes H, Rosa D, Dando PR, Serrão-Santos R. Annual spawning of the hydrothermal vent mussel, *Bathymodiolus azoricus*, under controlled aquarium, conditions at atmospheric pressure (in press J Exp Mar Biol Ecol)

Company R, Serafim A, Bebianno MJ, Cosson R, Shillito B, Fiala-Medioni A (2004) Effect of cadmium, copper and mercury on antioxidant enzyme activities and lipid peroxidation in the gills of the hydrothermal vent mussel *Bathymodiolus azoricus*. Mar Environ Res 58:377–381

Daughney CJ, Rioux JP, Fortin D, Pichler T (2004) Laboratory investigation of the role of bacteria in the weathering of basalt near deep sea hydrothermal vents. Geomicrobiol J 21:21–31

Distel DL, Felbeck H (1988a) Pathways of inorganic carbon fixation in the endosymbiont-bearing lucinid clam *Lucinoma aequizonata* 1 purification and characterization of the endosymbiotic bacteria. J Exp Zool 247:1–10

Distel DL, Felbeck H (1988b) Pathways of inorganic carbon fixation in the endosymbiont-bearing lucinid clam *Lucinoma aequizonata* 2 analysis of the individual contributions of host and symbiont cells to inorganic carbon assimilation. J Exp Zool 247:11–22

Dixon DR, Dixon LRJ, Shillito B, Gwynn JP (2002) Background and induced levels of DNA damage in Pacific deep-sea vent polychaetes: the case for avoidance. Cahier de Biologie Marine 43:333–336

Dixon DR, Pruski AM, Dixon LRJ (2004) The effects of hydrostatic pressure change on DNA integrity in the hydrothermal-vent mussel *Bathymodiolus azoricus*: implications for future deep-sea mutagenicity studies. Mut Res-Fun Mol Mech Mutagen 552:235–246

Dixon DR (2005) Metazoans living around deep-sea vents: evidence for avoidance mechanisms In Book of abstract "Investigating Life in Extreme Environments", Sant Feliu de Guixols, Spain, 2005; on line at: http://www.esf.org/conferences/ILEEn, pp 20

Fiala-Medioni A, Mckiness ZP, Dando P, Boulegue J, Mariotti A, Alayse-Danet AM, Robinson JJ, Cavanaugh CM (2002) Ultrastructural, biochemical, and immunological characterization of two populations of the mytilid mussel *Bathymodiolus azoricus* from the Mid-Atlantic Ridge: evidence for a dual symbiosis. Mar Biol 141:1035–1043

Goffredi SK, Childress JJ, Desaulniers NT, Lallier FH (1997a) Sulfide acquisition by the vent worm *Riftia pachyptila* Appears to be Via Uptake Of $HS^-$, Rather Than $H_2S$. J Exp Biol 200:2609–2616

Goffredi SK, Childress JJ, Desaulniers NT, Lee RW, Lallier FH, Hammond D (1997b) Inorganic carbon acquisition by the hydrothermal vent tubeworm *Riftia pachyptila* depends upon high external p-$co_2$ and upon proton-equivalent ion transport by the worm. J Exp Biol 200:883–896

Gorodezky LA, Childress JJ (1994) effects of sulfide exposure history and hemolymph thiosulfate on oxygen-consumption rates and regulation in the hydrothermal vent crab *Bythograea thermydron*. Mar Biol 120:123–131

Gros O, Darrasse A, Durand P, Frenkiel L, Moueza M (1996a) Environmental transmission of a sulfur-oxidizing bacterial gill endosymbiont in the tropical lucinid bivalve *Codakia orbicularis*. Appl Environ Microbiol 62:2324–2330

Gros O, De Wulf-Durand P, Frenkiel L, Moueza M (1998) Putative Environmental Transmission of sulfur-oxidizing bacterial symbionts in tropical lucinid bivalves inhabiting various environments. FEMS Microbiol Lett 160:257–262

Gros O, Duplessis MR, Felbeck H (1999) embryonic development and endosymbiont transmission mode in the symbiotic clam *Lucinoma aequizonata* (*Bivalvia : Lucinidae*). Invertebr Reprod Dev 36:93–103

Gros O, Frenkiel L, Felbeck H (2000) Sulfur-oxidizing endosymbiosis in *Divaricella quadrisulcata (Bivalvia: Lucinidae*): morphological, ultrastructural, and phylogenetic analysis. Symbiosis 29:293–317

Gros O, Frenkiel L, Moueza M (1996b) Gill ultrastructure and symbiotic bacteria in the tropical lucinid, *Linga pensylvanica* (Linne). Symbiosis 20:259–280

Gros O, Frenkiel L, Moueza M (1997) Embryonic, larval, and post-larval development in the symbiotic clam *Codakia orbicularis (Bivalvia: Lucinidae*). Invertebr Biol 116:86–101

Gros O, Frenkiel L, Moueza M (1998b) Gill filament differentiation, experimental colonization by symbiotic bacteria in aposymbiotic juveniles of *Codakia orbicularis* (Bivalvia : Lucinidae). Invertebr Reprod Dev 34:219–231

Gros O, Liberge M, Felbeck H (2003a) Interspecific infection of aposymbiotic juveniles of *Codakia orbicularis* by various tropical lucinid gill-endosymbionts. Mar Biol 142:57–66

Gros O, Liberge M, Heddi A, Khatchadourian C, Felbeck H (2003b) Detection of the Free-living forms of sulfide-oxidizing gill endosymbionts in the lucinid habitat (*Thalassia testudinum* environment). Appl Environ Microbiol 69:6264–6267

Kadar E, Bettencourt R, Costa V, Santos RS, Lobo-Da-Cunha A, Dando P (2005a) Experimentally induced endosymbiont loss, re-acquirement in the hydrothermal vent bivalve *Bathymodiolus azoricus*. J Exper Mar Biol Ecol 318:99–110

Kadar E, Costa V, Santos RS, Lopes H (2005b) Behavioural response to the bioavailability of inorganic mercury in the hydrothermal mussel *Bathymodiolus azoricus*. J Exp Biol 208:505–513

Kadar E, Costa V, Martins I, Santos RS, Powell JJ (2005c) Enrichment in trace metals (Al, Mn, Co, Cu, Mo, Cd, Fe, Zn, Pb and Hg) of macro-invertebrate habitats at hydrothermal vents along the Mid Atlantic Ridge. Hydrobiologia 548:191–205

Kadar E, Azevedo C (2006) Unidentified extracellular prokaryotes within the byssal threads of the deep-sea vent mussel *Bathymodiolus azoricus*. Parasitology 133: 1–5

Kadar E, Santos RS, Powell JJ (2006) Biological Factors Influencing tissue compartmentalization of trace metals in the deep-sea hydrothermal vent bivalve *Bathymodiolus azoricus* at geochemically distinct vent sites of the Mid-Atlantic Ridge. Env Res 101: 221–229

Kadar E, Costa V, Santos RS, Powell JJ Tissue partitioning of micro-essential metals in the vent bivalve *Bathymodiolus azoricus* and associated organisms (endosymbiont bacteria and parasite polychaete) from geochemically distinct vents of the Mid-Atlantic Ridge (in press b Journal of Sea Research)

Kadar E, Costa V First report on the micro-essential metal concentrations in bivalve shells from deep-sea hydrothermal vents (in press c Journal of Sea Research)

Kadar E, Santos RS, Lobo-Da-Cunha A, Dando P Spermatogenesis of *Bathymodiolus azoricus* in captivity matching reproductive behaviour at deep-sea hydrothermal vents (in press d J Exper Mar Biol Ecol)

Lee RW (2003) Thermal tolerances of deep-sea hydrothermal vent animals from the northeast pacific. Biolog Bull 205:98–101

Mergeay M (2000) Bacteria adapted to industrial biotopes: the metal resistant ralstonia. In: Storz G, Hengge-Aronis (eds) Bacterial Stress Responses. ASM Press Washingotn DC, USA, pp 403–414

Mickel TJ, Childress JJ (1980) Temperature, pressure responses of the galapagos hydrothermal vent crab, *Bythograea thermydron*. Am Zool 20:834

Mickel TJ, Childress JJ (1982a) Effects of pressure and pressure acclimation on activity and oxygen-consumption in the bathypelagic mysid *Gnathophausia ingens*, Deep-Sea Research Part a. Oceanograp Res Papers 29:1293–1301

Mickel TJ, Childress JJ (1982b) Effects of pressure and temperature on the ecg and heart-rate of the hydrothermal vent crab *Bythograea thermydron (Brachyura*). Biological Bulletin 162:70–82

Mickel TJ, Childress JJ (1982c) Effects of temperature, pressure, and oxygen concentration on the oxygen-consumption rate of the hydrothermal vent crab *Bythograea thermydron (Brachyura)*. Physiolog Zool 55:199–207

Page HM, Fiala-Medioni A, Childress JJ (1991) Experimental evidence for filter-feeding by the hydrothermal vent mussel, *Bathymodiolus thermophilus*, Deep Sea Research Part A. Oceanograp Res Papers 38:1455–1461

Postec A, Urios L, Lesongeur F, Ollivier B, Querellou J, Godfroy A (2005) Continuous enrichment culture and molecular monitoring to investigate the microbial diversity of thermophiles inhabiting deep-sea hydrothermal ecosystems. Curr Microbiol 50:138–144

Pradillon F, Le Bris N, Shillito B, Young CM, Gaill F (2005) Influence of environmental conditions on early development of the hydrothermal vent polychaete *Alvinella pompejana*. J Exp Biol 208:1551–1561

Pradillon F, Shillito B, Chervin JC, Hamel G, Gaill F (2004) Pressure vessels for in vivo studies of deep-sea fauna. High Pressure Res 24:237–246

Quetin LB, Childress JJ (1980) Observations on the swimming activity of 2 bathypelagic mysid species maintained at high hydrostatic pressures, Deep-Sea Research Part a. Oceanograph Res Papers 27:383–391

Ravaux J, Gaill F, Le Bris N, Sarradin PM, Jollivet D, Shillito B (2003) Heat-shock response and temperature resistance in the deep-sea vent shrimp *Rimicaris exoculata*. J Exp Biol 206:2345–2354

Sanders NK, Childress JJ (1985) Unusual oxygen binding-properties in the blood of the deep-sea hydrothermal vent crab *Bythograea thermydron*. Am Zool 25:A119

Sanders NK, Childress JJ (1992) Specific effects of thiosulfate and l-lactate on hemocyanin-$o_2$ affinity in a brachyuran hydrothermal vent crab. Mar Biol 113:175–180

Sarradin PM, Caprais JC, Riso R, Kerouel R, Aminot A (1999) Chemical environment of the hydrothermal mussel communities in the Lucky Strike and Menez Gwen vent fields, Mid-Atlantic Ridge. Cahiers de Biologie Marine 40:93–104

Shillito B, Jollivet D, Sarradin PM, Rodier P, Lallier F, Desbruyeres D, Gaill F (2001) Temperature resistance of *Hesiolyra bergi*, a polychaetous annelid living on deep-sea vent smoker walls. Mar Ecol-Prog Series 216:141–149

Shillito B, Le Bris N, Gaill A, Rees JF, Zal F (2004) First access to live *Alvinellas*. High Pressure Res 24:169–172

Shillito B, Le Bris N, Gaill Hourdez S, Ravaux J, Cottin D, Caprais JC, Jollivet D, Gaill F (2006) Temperature resistance studies on the deep-sea vent shrimp *Mirocaris fortunata*. J Exp Biol 209:945–955

Van Dover CL, Lutz RA (2004) Experimental ecology at deep-sea hydrothermal vents: a perspective. J Exp Mar Biol Ecol 300:273–307

Von Cosel R, Comtet T, Krylova E M (1999) *Bathymodiolus (Bivalvia : Mytilidae)* from hydrothermal vents on the Azores Triple Junction and the Logatchev hydrothermal field, Mid-Atlantic Ridge. Veliger 42:218–248

Rev Environ Sci Biotechnol (2007) 6:143–159
DOI 10.1007/s11157-006-9110-3

REVIEW PAPER

# Adaptations to hypoxia in hydrothermal-vent and cold-seep invertebrates

**Stéphane Hourdez · François H. Lallier**

Received: 27 April 2006 / Accepted: 4 September 2006 / Published online: 29 November 2006


**Abstract** The deep sea harbors very unusual environments, such as hydrothermal vents and cold seeps, that illustrate an apparent paradox: the environmental conditions are very challenging and yet they display a high biomass when compared to the surrounding environment at similar depth. Hypoxia is one of the challenges that these species face to live there. Here, we review specific adaptations of their respiratory system that the species have developed to cope with hypoxia, at the morphological, physiological, and biochemical levels. Most studies to date deal with annelids and crustaceans, and trends can be drawn: development of ventilation and branchial surfaces to help with oxygen extraction, and an increase in finely tuned oxygen binding proteins to help with oxygen storage and transport. Beside these respiratory adaptations most animals have developed enhanced anaerobic capacities and specific ways to deal with sulfide.

**Keywords** Hypoxia · Invertebrates · Annelids · Crustacea · Mollusks · Hemocyanin · Hemoglobin · Respiration · Oxygen · Sulfide

S. Hourdez (✉) · F. H. Lallier
Equipe Ecophysiologie: Adaptations et Evolution Moléculaires, UMR 7144 CNRS-UPMC, Station Biologique de Roscoff, Roscoff 29680, France
e-mail: hourdez@sb-roscoff.fr

## 1 Introduction

The discovery of hydrothermal vents on the Galápagos Ridge in 1976 (Corliss and Ballard 1977) revolutionized our vision of the deep-sea, which was previously notorious for its very low biomass and high species diversity (Hessler and Jumars 1974). Numerous animals from the vicinity of fluid emissions associated with plate tectonics were brought back to the surface (Corliss and Ballard 1977; Ballard and Grassle 1979; Corliss et al. 1979). Subsequent exploration at other sites on mid-oceanic ridges and back-arc basins has extended our knowledge of the fauna associated with hydrothermal vents. When compared to the surrounding areas, these communities exhibit very high biomass, and a relatively low species diversity. The abundance of these organisms, contrasting with the harsh conditions of the environment, has stimulated research on their physiological and biochemical adaptations. Stable isotope values of their tissues indicate that the high biomass is supported by local chemosynthetic primary production rather than by photosynthesis-based organic matter sinking from the surface (Fisher 1995). Chemosynthesis is carried out by bacteria, free-living or in symbiotic association with invertebrates. These bacteria oxidize the reduced chemicals contained in the hydrothermal fluid (Fe, $CH_4$, $H_2S$), using oxygen or nitrate as a final electron acceptor, to produce energy and fix carbon dioxide into organic matter (Fisher 1996). This hydrothermal fluid is formed by the interaction of the sea-water seeping through the oceanic crust with the hot rocks. The fluid is changed markedly and usually contains high levels of heavy metals, $CO_2$ and sulfide, is completely devoid of $O_2$, and has a low pH (Table 1 and references therein). The hydrothermal vent communities completely rely on the emission of hydrothermal fluid. However, its mixing with the deep-sea water strongly affects the surrounding environment and the animal communities it harbors. Individual species occupy different niches whose characteristics vary with the proportion of hydrothermal fluid, and even specific locations are exposed to highly variable, often challenging conditions (Fig. 1) (Johnson et al. 1988; Chevaldonné et al. 1991, 1992; Chevaldonné 1986; Le Bris et al. 2003). Due to the highly chaotic conditions and to technical difficulties, very few in-situ measurements of oxygen have been published to date. A study of microhabitat variations in mussel beds at the Rose Garden site yielded typical oxygen concentration values between 0 and 110 μM, this latter value being that of ambient sea-water in the area (Fisher et al. 1988). The harsh conditions (low oxygen, high levels of sulfide, heavy metals, radioactivity, etc.) encountered in these food-rich environments require specific adaptations for survival that may explain the very high proportion of endemicity (95%) observed in the

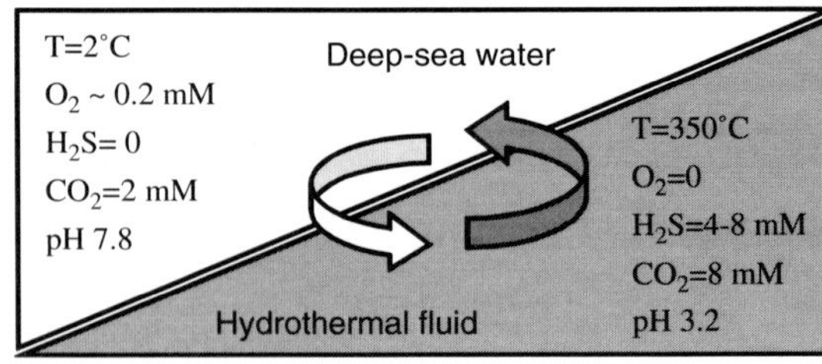


**Fig. 1** Schematic representation of water mixing around hydrothermal vents. Bottom water (clear) mixes with hydrothermal fluid (grey shade). Typical water characteristics are given for each. Animals are exposed to varying proportions of each fluid according to their proximity of the vent. In addition, the mixing is chaotic and at a fixed position, the proportion of hydrothermal vent fluid can vary quickly (arrows)

**Table 1** Physical and chemical parameters of hydrothermal fluid from black smokers of different origin compared to typical deep-sea water (after Von Damm 1990, modified by Magenheim and Gieskes 1992)

| Physical–chemical parameters | South Juan de Fuca Ridge | East Pacific Rise 21°N | East Pacific Rise 13°N | Deep-sea water |
|---|---|---|---|---|
| Temp. (°C) | 224–285 | 273–355 | 317–380 | 2 |
| pH | 3.2 | 3.3–3.8 | 3.2 | 7.8 |
| Alcalinity | 0 | − 0.50 à − 0.19 | − 0.74 à − 0.40 | 2.3 |
| Li (mM) | 110–1,810 | 891–1,322 | 688 | 26 |
| Na (mM) | 700–800 | 432–510 | 560 | 464 |
| K (mM) | 37.3–51.6 | 23.2–25.8 | 29.6 | 9.8 |
| Rb (mM) | 28–37 | 27–33 | 14.1 | 1.3 |
| Be (nM) | 95–150 | 10–37 | – | 0 |
| Mg (mM) | 0 | 0 | 0 | 52.7 |
| Ca (mM) | 77.3–96.4 | 11.7–20.8 | 55 | 10.2 |
| Sr (mM) | 230–312 | 65–97 | 175 | 87 |
| Al (mM) | 1.9 | 4.0–5.2 | – | 0.020 |
| Cl (mM) | 896–1,090 | 489–579 | 740 | 541 |
| $SiO_2$ (mM) | 22.7–23.3 | 15.6–19.5 | 22 | 0.16 |
| $SO_4$ (mM) | 0 | 0 | 0 | 27.9 |
| $H_2S$ (mM) | 3–4.4 | 6.6–8.4 | 4.0 | 0 |
| Mn (mM) | 2.61–4.48 | 0.7–1.0 | 0.8–1.2 | < 0.001 |
| Fe (mM) | 10.3–18.7 | 0.75–2.42 | 1.05–1.85 | < 0.001 |
| Cu (mM) | < 2 | < 2–44 | – | 0.007 |
| Zn (mM) | < 900 | 40–106 | – | 0.01 |
| Se (mM) | < 1 | < 1–73 | – | 2.5 |
| As (nM) | 0 | < 30–452 | – | 27 |
| $NH_3$ (mM) | 0 | 0 | – | < 0.01 |

collections of fauna at hydrothermal vents (Tunnicliffe 1991).

The first—and to date best studied—cold-seeps were discovered in the Gulf of Mexico in the 1980s (Paull et al. 1984; Kennicutt et al. 1985). Cold seeps inhabited by dense communities are now known from more than 30 locations in the world (Sibuet and Olu 1998; Tyler et al. 2003). Conditions at cold seeps are often hypoxic, as a result of the removal of free $O_2$ in the deep-sea water by spontaneous reaction with sulfides contained in the fluid diffusing from the sediment. In contrast to hydrothermal vents, these sulfides are biogenic in nature: anaerobic bacteria in the sediment utilize sea-water sulfate as the final electron acceptor for the degradation of hydrocarbons seeping up (Cordes et al. 2005). A study at the Brine Pool site in the Gulf of Mexico showed that oxygen in water samples taken between the mussels averaged 39 μmol $l^{-1}$ and was often undetectable by gas chromatography (<5 μmol $l^{-1}$, Smith et al. 2000).

To date, no physiological work has been published on invertebrates from whale-fall communities. These animals are also potentially exposed to lower levels of oxygen. Some studies report the adaptations of pelagic organisms to low oxygen levels in oxygen minimum zones (for example, see Childress 1975; Childress and Seibel 1998), characterized by stable hypoxic conditions. Benthic invertebrates from oxygen minimum layer areas are usually small in size (Levin 2003). This may be due to the fact that these areas are not associated with strong emissions of reduced fluids that are necessary for local primary production that is at the base of the very high biomass that characterizes hydrothermal vents and cold seeps.

Interestingly the $O_2$ consumption rates of hydrothermal-vent and cold-seep animals at low environmental $O_2$ tensions are similar to those of related, shallow-water species measured at higher environmental $O_2$ tensions (Childress and Mickel 1985; Fisher et al. 2000; Hourdez et al. 2002), indicating specific adaptations for $O_2$ uptake in the former group, such as higher blood $O_2$ affinities and more efficient respiratory organs. Here, we review morphological, physiological, and biochemical adaptations of the respiratory system. To date, these adaptations have mainly been studied in polychaetes and crustacea but data on other groups will be discussed when available.

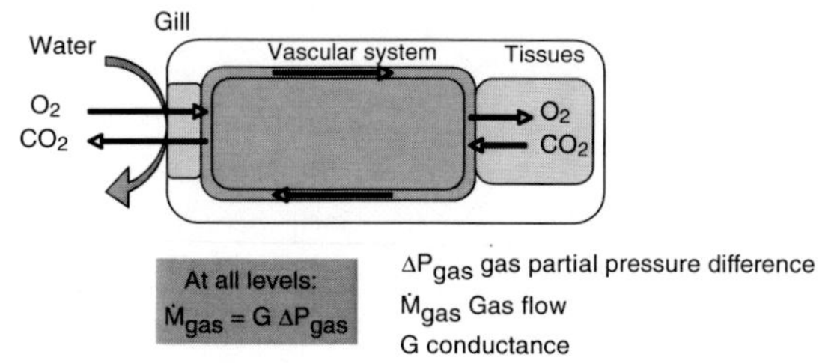


**Fig. 2** Conceptual representation of gas flows in a metazoan with a circulatory system. See text for details

Gas uptake (and elimination) by an organism is limited by several factors at different levels (Fig. 2). At all levels, the gas flow is driven by convective or diffusive processes, and is proportional to the difference of partial pressure of the considered gas between two locations, and to a conductance factor G. Details of parameters affecting G will be discussed for the different levels of adaptation considered. At the interface with the environment, the respiratory exchange organs (gills, branchiae) are essential for oxygen uptake. The oxygen that is taken up is then transported by the circulatory system to the tissues that need it. The mitochondria, where oxidative respiration takes place, are the actual site of oxygen consumption.

## 2 Adaptations for oxygen extraction

### 2.1 Ventilation

The first way to improve oxygen extraction is to renew the diffusion layer at the surface of the gas exchange organs. The water in the diffusion layer is indeed quickly depleted in oxygen and its extraction becomes more difficult. Convective circulation of water and blood at the gill level contributes to increase oxygen extraction in an integrative process. This maintains an optimal difference of partial pressure between the two sides of the diffusion barrier.

The gills of annelids are external (i.e., not contained in branchial chambers) and these animals have little control over the ventilation of the gill surface by water, except through movements of cilia which are abundant in vent species (Jouin-Toulmond et al. 1996; Hourdez and Jouin-Toulmond 1998; Hourdez et al. 2001). Perfusion of the gills by body fluids may also increase but this parameter is difficult to measure in polychaetes, especially in species like scale-worms whose gills are perfused by the coelomic fluid, circulated by cilia that line the coelomic cavity (Hourdez and Jouin-Toulmond 1998).

Decapod crustacea possess a branchial chamber in which water may be circulated by specialized appendages, the scaphognathite and associated epipodites. Although regulation of ventilation flow rate is well documented for shallow-water species (McMahon 2001) and some deep-sea species (Belman and Childress 1976), no study addresses this behavioral adaptation in vent or seep species. Generally, acute hypoxia induces a rapid increase in ventilation flow but it slows down under chronic exposure, probably due to the high energetic cost of increasing water pumping (McMahon 2001).

Mollusca circulate water through their mantle cavity by ciliary movement at the surface of the gills. This is a normal process for these filter-feeding animals. However, in clams of the genus *Calyptogena*, the digestive system is reduced and the bulk of the nutrition is obtained through symbiotic bacteria contained in the gill filaments rather than through water filtration (Kennish and Lutz 1992). The respiratory function may well be the only one remaining in these clams, in addition to harboring the symbiotic bacteria.

## 2.2 Gill surfaces

The conductance G is a function of the ratio of surface area over diffusion distance. Extraction of oxygen across gills can thereby be increased by increasing the total surface area or by decreasing the diffusion distance through the epithelium, or both.

An increased specific gill surface area has been reported in numerous cold-seep and hydrothermal vent polychaetes (Table 2). This is especially meaningful when compared to close relatives from well-oxygenated environments. In the family Polynoidae (scaleworms), littoral species are usually devoid of gills whereas about half of the hydrothermal vent and cold-seep species possess well developed gills (Hourdez and Jouin-Toulmond 1998; Hourdez unpub. data). This increase in area for gas exchange has also been reported in benthic invertebrates from oxygen minimum zones all around the world (Levin 2003).

Besides an increase in gill surface area, a reduction of the diffusion distance can also help gas exchange across the gills. Most of the vent species studied to date show very short diffusion distances (Table 2). These short distances are possible through the use of intraepidermal blood vessels found in the gills of the polychaetes from hypoxic environments (Jouin and Gaill 1990; Hourdez et al. 2001). These intraepidermal extensions of blood vessels are absent in the gills of the shallow-water trichobranchid polychaete *Terebellides stroemii*, a species closely related to alvinellids (Jouin-Toulmond and Hourdez 2006). The scaleworm *Branchipolynoe* spp. is an exception to this reduced diffusion distance. However, their gills are mere finger-like extension of their body-wall that are perfused by the coelomic fluid and not by blood vessels. The bodywall is very

**Table 2** Gill surface areas and diffusion distances for annelids from hydrothermal vent (HTV), littoral (Lit.), and cold-seep (CS) environments

| Species | Environment | Specific gill surface area($cm^2$/g) | Diffusion distance (μm) |
|---|---|---|---|
| *Alvinella pompejana*[a] | HTV | 12 | 1–3 |
| *Paralvinella grasslei*[a] | HTV | 47 | 4 |
| *Riftia pachyptila*[b] | HTV | 12 | 1–3 |
| *Branchipolynoe symmytilida*[c] | HTV | 14.2 | 10 |
| *Branchipolynoe seepensis*[c] | HTV | 10.3 | 9 |
| *Branchipolynoe pettiboneae*[c] | HTV | 7.7 | 10 |
| *Glycera convoluta*[c] | Lit. | 1.5–2 | ? |
| *Terebellides stroemi*[d] | Lit. | 6 | 5–8 |
| *Arenicola marina*[e] | Lit. | 4 | 8–14 |
| *Methanoaricia dendrobranchiata*[f] | CS | 8 | 4 |

[a] Jouin and Gaill (1990); [b] Andersen et al. (2002); [c] Hourdez and JouinToulmond (1998); [d] Jouin-Toulmond and Hourdez (2006); [e] Jouin and Toulmond (1989); [f] Hourdez et al. (2001)

thin there (typically 10 μm) whereas it is 150 μm thick anywhere else on the body (Hourdez and Jouin-Toulmond 1998).

No specific data are available for other taxonomic groups. However, it is well known that mollusks that are symbiotic with sulfide-oxidizing or methanotrophic bacteria possess much larger gills (that contain the symbionts) than species that are not symbiotic (Childress and Fischer 1992). No data are available on the diffusion distances and gill surface areas for crustacea from cold seeps and hydrothermal vents. However, Williams (1980), in his original description of the vent crab *Bythograea thermydron*, reports that they have large gills, apparently to facilitate oxygen uptake from the hypoxic environment. Conversely in *Rimicaris exoculata*, the peculiar morphology of the gill chamber and hypertrophy of the scaphognathite and first maxillipeds' epipodites leaves only a reduced space for gills (Segonzac et al. 1993). A comparative study of specific gill surface area, and scaphognathite and perfusion performance among bythograeid crabs and alvinocarid shrimps is clearly lacking at present. This is likely to be a level at which adaptations can be observed as Johnson and Rees (1988) showed that there were variations in gill surface areas depending on habitat and lifestyle in four shallow-water crab species.

## 3 Adaptations for oxygen transport

Once in the circulatory system, the conductance for oxygen will depend on the product of the capacitance ($\beta$b) of the blood (i.e., how much oxygen can be carried per unit volume of blood) and of the blood flow (Vb). In order to improve oxygen delivery to the organs, one can either increase $\beta$b or Vb. The presence of oxygen binding proteins (respiratory pigments) increases $\beta$b, and increased heart rate or stroke volume can perfuse the tissues with more blood.

### 3.1 Modulation of heart rate

A measure of the regulation of heart rate in annelids from hypoxic environments has never been attempted. However, Mickel and Childress (1982a) measured variations in heart rate of the crab *Bythograea thermydron* as a function of pressure and temperature. An increase of temperature from 5°C to 20°C triggered an increase of heart rate, probably in response to an increased metabolic demand.

### 3.2 Oxygen binding proteins (OBPs)

Most hydrothermal vent organisms possess respiratory pigments. These were studied in some mollusks, polychaetes, and crustaceans.

The main properties of OBPs are affinity - measured by $P_{50}$, the oxygen partial pressure at half saturation -, cooperativity - measured by $n_{50}$ the slope of the Hill plot at $P_{50}$, and the Bohr effect - measured by $\phi = \Delta \log P_{50}/\Delta pH$ (Fig. 3). Depending on their oxygen binding properties, respiratory pigments may serve to store oxygen, to facilitate oxygen diffusion, or to actually transport oxygen from the gas exchange organs to the tissues. Oxygen storage OBPs usually exhibit a high affinity (allowing uptake of oxygen from the circulating OBP), low cooperativity and are insensitive to pH changes. Conversely, OBPs that serve as transporters usually exhibit a lower affinity, some cooperativity and are affected by changes in pH (Bohr effect). Their cooperativity allows them to release more oxygen for a given variation of partial pressure between arterial and venous blood (Fig. 3). These blood parameters, however, are not known for the majority of deep-sea species, and we have to rely on indirect interpretation derived from affinity, cooperativity and Bohr effect measurements.

#### *3.2.1 OBP as an oxygen storage system*

Respiratory pigments can be used as oxygen storage. It has been calculated, based on an oxygen consumption rate independent of the partial pressure and starting with respiratory pigments fully saturated, that the bound oxygen could represent an autonomy of up to 1 h 30 min in the commensal hydrothermal vent polychaete *Branchipolynoe seepensis* (Hourdez and Weber 2005). This value only reaches 30 min in the cold-seep polychaete *Methanoaricia dendrobranchiata* (Hourdez et al. 2002). Very little data are available in other species. However, in the hydrothermal vent copepod *Benthoxynus spiculifer*, the

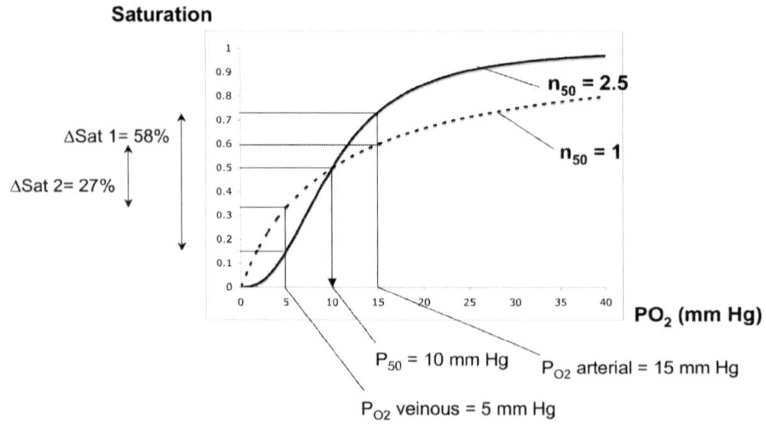


**Fig. 3** Saturation curve for a hypothetical oxygen binding protein (OBP) with a $P_{50}$ of 10 mm Hg. Dashed curve for an OBP with a cooperativity coefficient $n_{50} = 1$ and solid line for an OBP with a cooperativity coefficient $n_{50} = 2.5$. For each OBP, the difference of saturation (ΔSat) is given for a difference of partial pressure of $O_2$ of 10 mmHg between arterial blood and venous blood

amount of hemoglobin in the body can represent no more than 2 min of autonomy at 15°C and less than 30 s at 25°C (Hourdez et al. 2000b). This hemoglobin pool would only allow very short forays into anoxic areas to look for food before the animals have to rely on anaerobic metabolism for their energetic needs. In such animals, the hemoglobin's primary function is more likely in $O_2$ acquisition from the environment (Hourdez et al. 2000b), maintaining a free-oxygen (i.e., not bound to Hb) gradient from outside of the animal to the internal milieu while the respiratory surfaces are ventilated.

### *3.2.2 OBP as a transport system*

There is a general trend for a high oxygen affinity, which would facilitate uptake at the level of the gills, and a strong Bohr effect (decrease of affinity when pH decreases) that would allow the release of this oxygen at the level of metabolically active tissues. The high affinity for oxygen would leave little of it free in the body fluids, thereby maintaining a gradient between the inside and the outside of the body, even when the environmental levels of oxygen are very low.

*3.2.2.1 Mollusks* Bivalves, along with vestimentiferans, are the main organisms with symbiotic sulfide oxidizing and/or methanotrophic bacteria at hydrothermal vents and cold-seeps (Tunnicliffe 1991). Vesicomyid clams are the only vent and seep bivalves known to possess circulating respiratory pigments to date. These Hbs are intracellular (erythrocytic) and low-molecular weight, consisting of tetramers made of three different chains of 13–14 kDa (Terwilliger et al. 1983; Arp et al. 1984; Zal et al. 2000b). *Calyptogena magnifica* erythrocytes suspended in sea-water after washing show a moderate $O_2$ affinity ($P_{50}$ = 8.5 mm Hg, pH 7.15, 8°C (Arp et al. 1984)) that is lowered when the erythrocytes were suspended in the original serum ($P_{50}$ = 14.0 mm Hg, pH 6.75, 9°C). Dissolved in distilled water the Hb showed a much higher $O_2$ affinity ($P_{50}$ = 3.8 mm Hg, pH 6.99, 8°C), suggesting the presence of intracellular effectors that decrease the affinity in the erythrocytes (like ATP or 2,3-DPG for vertebrate Hb). The functional properties of the Hb were affected by sulfide that oxidizes Hb in vitro (Terwilliger et al. 1983; Arp et al. 1984) and impedes measurement of the sulfide sensitivity of $O_2$ binding (Arp et al. 1984).

The symbiont-harboring gastropod species *Alviniconcha hessleri* forms dense populations at hydrothermal vents in the Western Pacific (Mariana, North Fiji, and Lau back-arc basins) where it dominates the fauna. Its modified gills contain chemoautotrophic sulfide-oxidizing bacteria that provide the snail with carbon compounds. These gills also contain a tissue globin at a concentration of 65 µmol/kg wet weight gill (Wittenberg and Stein 1995). Wittenberg (1985) also reports the presence of an intracellular globin in the gills of the mussel *Bathymodiolus heckerae* from the Florida Escarpment. Gastropods also possess circulating hemocyanins but there is no published data on their properties and characteristics for vent species. The same is true for the hemocyanin of the vent octopus *Vulcanoctopus hydrothermalis*.

*3.2.2.2 Polychaetes* Polychaetes represent an important proportion of the biomass and diversity at hydrothermal vents and cold seeps (Tunnicliffe 1991). They also occupy most of the ecological niches at hydrothermal vents where metazoans can be found, from the coldest (and least constraining) to the warmest. All the species studied to date possess hemoglobin(s) in high concentrations, with a high affinity for oxygen when compared to shallow-water relatives, and a strong Bohr effect (Fig. 4, Hourdez and Weber 2005).

Vestimentiferan tubeworms (siboglinid polychaetes) are often the most commonly encountered metazoan animals in hydrothermal vent and cold-seep communities. They lack a mouth, digestive tract and anus (Jones 1981, 1988), and their nutritional needs are entirely provided for by symbiotic sulfide-oxidizing bacteria harbored in the 'trophosome', an internal, well-vascularized organ (Cavanaugh et al. 1981; Felbeck 1981; Felbeck et al. 1981; Jones 1981). The first thoroughly studied vestimentiferan was the giant tubeworm *Riftia pachyptila*. This worm possesses two Hbs in its vascular blood and another in its coelomic fluid (Arp et al. 1990; Zal et al. 1996b). The blood Hbs are a hexagonal bilayer (HBL) Hb of ~3.6 MDa (as typically encountered in vascular blood of other annelids) and a 400 kDa Hb that is

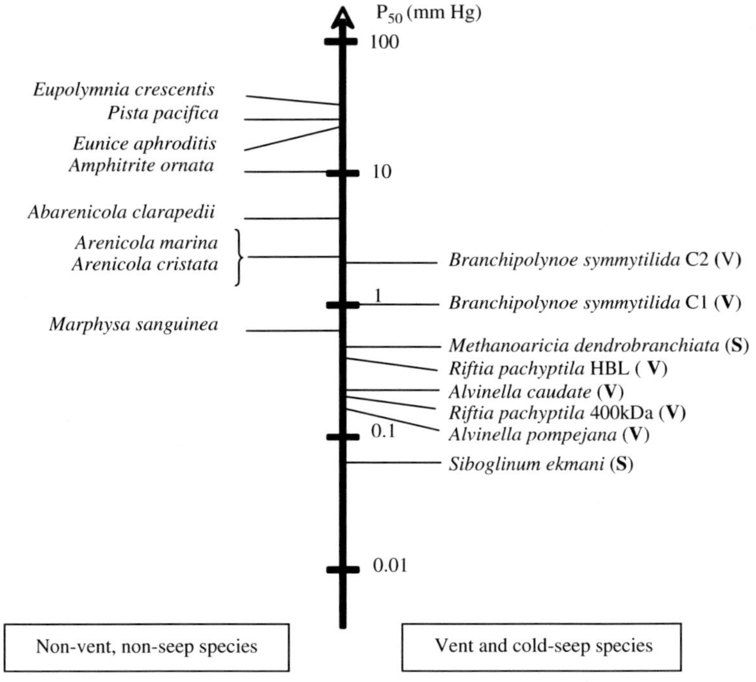


**Fig. 4** $P_{50}$ values (log scale) for extracellular Hbs from vent and seep marine polychaetes compared to non-vent and non-cold seep species. Experimental conditions and references: *Arenicola marina*, pH 7.6, 20°C; *Siboglinum ekmani*, pH 6.5, 20°C; *Alvinella pompejana*, pH 7.6, 20°C; *Alvinella caudata*, pH 7.6, 20°C; *Pista pacifica*, pH 7.0, 20°C; *Marphysa sanguinea*, pH 7.3, 20°C; *Eunice aphroditis*, pH 7.0, 20°C; *Arenicola cristata*, pH 7.7, 20°C; *Abarenicola clarapedii*, pH 7.43, 20°C; *Eupolymnia crescentis*, pH 5–7, 10°C; *Branchipolynoe symmytilida* Hbs C1 and C2, pH 7.5, 20°C; *Methanoaricia dendrobranchiata*, pH 7.5, 20°C; *Riftia pachyptila* HBL and 400 kDa Hbs, pH 7.0, 30°C. (**V**) Vent species, (**S**) Seep species. After Hourdez and Weber (2005)

specific of siboglinid tubeworms. The coelomic fluid contains another 400 kDa Hb, differing from the vascular one in its subunit composition (Zal et al. 1996a). The $O_2$ binding properties of vestimentiferan Hbs are similar to those of other annelids from hydrothermal vents and cold seeps. In *Riftia*, both Hb types have very high $O_2$ affinities, with $P_{50}$ values of 0.47 and 0.27 mm Hg for the HBL-Hb and the 400 kDa, respectively, at pH 6.97 and 30°C. The smaller Hb consistently showed a higher affinity, except at pH values above 7.6 (Arp et al. 1990), which correspond to in vivo pH values (Arp and Childress 1981). This suggests that the $O_2$ affinity difference between the coleomic (400 kDa) Hb and the vascular (HBL and 400 kDa) Hbs may easily reverse, allowing a bi-directional transfer between the coelomic and vascular compartments. This would buffer the effects of changes in environmental $O_2$ tension. The two types of Hbs also differ markedly in their pH sensitivity ($\phi$ = – 0.35 and – 0.04, for the HBL and 400 kDa Hbs, respectively). The cooperativity is also lower in the 400 kDa Hb than in the HBL Hb ($n_{50}$ = 1.6 and 2.4, respectively at pH $\approx$7), which correlates with the number of subunits (ca. 24 vs. 144 globin chains). Hbs from three species of *Siboglinum* that occur in hypoxic fjords comprise only 400 kDa molecules (Terwilliger et al. 1987), all showing very high $O_2$ affinities and a slight reverse Bohr effect between pH 7.0 and 7.9 ($\phi$ = + 0.18 at 15°C and +0.25 at 20°C; (Terwilliger et al. 1987)). According to our definition, these properties would favor a storage role for the 400 kDa Hb and a transport role for the HBL Hb. A key and unique characteristic of vestimentiferan Hbs is their ability to bind sulfide reversibly and with a high affinity (Arp and Childress 1983; Arp et al. 1984; Childress and Fischer 1992; Zal 1998; Zal et al. 1998). This will be dealt with in a later section.

Species of Alvinellidae (a vent-endemic family) and the cold-seep orbiniid *Methanoaricia dendrobranchiata* possess a typical extracellular annelid hexagonal bilayer (HBL) hemoglobin in their vascular system (Terwilliger and Terwilliger 1984; Toulmond et al. 1990; Zal et al. 1997, 2000a; Hourdez et al. 2000a). Although the structure of the Hb is very similar to that of other annelid hemoglobins, the functional properties show a high affinity for oxygen and a strong Bohr effect that allows the release of the bound oxygen near metabolically active (i.e., acidic) tissues (Terwilliger and Terwilliger 1984; Toulmond et al. 1990; Hourdez et al. 2000a).

In addition to their HBL hemoglobin, alvinellids also possess a circulating intracellular hemoglobin contained in coelomocytes (Jouin-Toulmond et al. 1996; Hourdez et al. 2000a), as it is common in the closely-related terebellids (Weber 1978). However, in contrast to terebellids where the oxygen transfer is unidirectional (usually from extra- to intracellular Hbs), driven by differences in affinities for oxygen between the extra- and the intracellular Hbs, oxygen transfer can be bi-directional in *Alvinella pompejana*, the only alvinellid species for which functional properties were studied for both hemoglobins (Hourdez et al. 2000a). This property most likely allows the buffering of environmental oxygen variations using the coelomic compartment as an oxygen store. This is especially meaningful in the anterior part of the body where a dense network of capillaries is surrounded by numerous erythrocytes (see Sect. 3.3). There, a very important transfer can occur between the vascular and the coelomic compartments.

The discovery of large amounts of hemoglobin in the hydrothermal vent scaleworms *Branchipolynoe symmytilida* and *B. seepensis* (Polychaeta; Polynoidae) came as a surprise as littoral species of the same family are completely devoid of circulating respiratory pigments (Hourdez et al. 1999a). There are two Hbs, one is a dimer (HbC2) and the other a trimer (HbC1) of tetradomain subunits. Multidomain (i.e., several globin domains in the same polypeptide) subunits are unique not only in the Polynoidae family but in all polychaetes, indicating that it most likely evolved in the lineage to which *Branhipolynoe* spp. belong. These Hbs are contained in the coelomic cavity, exhibit a high affinity for oxygen and a strong Bohr effect (Hourdez et al. 1999b). Their cooperativity is low (cooperativity coefficient typically between 1 and 1.9), suggesting a function of oxygen storage rather than oxygen transporter. However, the coelomic fluid circulates inside the gills of the branchiate species and the hemoglobin probably helps maintain a concentration gradient between the hypoxic environment and the inside of the body. The strong Bohr effect is also indicative of a role in oxygen transport but we would have to know the variation of oxygenation level and pH within the coelomic fluid to confirm this. In addition to the Bohr effect, there is a specific $CO_2$ effect in HbC2 (but not in HbC1), independent of pH, which decreases the affinity when $PCO_2$ increases. This further increases the unloading of oxygen near metabolically active tissues (Hourdez et al. 1999b). Hemoglobin has been observed in numerous other hydrothermal vent and cold seep species of scaleworms (SH pers. obs.). All these species belong to vent- and seep-endemic subfamilies of polynoids (Branchiplicatinae, Branchipolynoinae, Branchinotogluminae, Lepidontopodinae, and some Macellicephalinae). The two other subfamilies encountered near hydrothermal vents (Harmothoinae and Iphioninae) are not endemic and are only found at the periphery of the sites, where conditions are less challenging. Species belonging to these two latter subfamilies do not possess circulating Hbs. Given the high metabolic cost of producing such large amounts of proteins, their presence is most likely an adaptation to the chronic hypoxia these worms are exposed to.

*3.2.2.3 Crustacea* Crustacea are also common at hydrothermal vents and cold-seeps but all published data come from hydrothermal vent species so far. Hemocyanin is the most common respiratory pigment but hemoglobin has been found in two species of copepods. The more mobile nature of large decapod crustacea (shrimps and crabs) allows them to avoid areas where conditions are too harsh. Nonetheless, the conditions they encounter are often hypoxic and one can expect specific adaptations.

Mickel and Childress (1982b) reported that the hydrothermal vent crab *Bythograea thermydron* can tolerate periods of anoxia. In addition, this crab can maintain its oxygen consumption rate constant until a critical $P0_2$ value of 12 mm Hg, i.e., about 7.5% of saturation with air. This value is 3–4 times the $P_{50}$ of the hemocyanin contained in the hemolymph of *B. thermydron* (Sanders et al. 1988). This $P_{50}$ value is small compared to those found in non-vent species and falls within the range reported for other hydrothermal vent species of crabs and shrimps (Fig. 5 and references therein). No data are available for cold-seep endemic species to date. As in polychaetes, this high affinity helps extract oxygen from the chronically hypoxic surrounding water. It is compensated by a strong Bohr effect that allows the release of oxygen near metabolically active tissues (Sanders et al. 1988; Lallier and Truchot 1997; Lallier et al. 1998; Chausson et al. 2001). The affinity is increased by the presence of lactate (the final product of anaerobic metabolism in crustacea) that will thereby favor the loading of oxygen at the level of the gills if the tissues produce lactic acid (Sanders et al. 1988). It was also shown that the affinity is increased by the presence of thiosulfate, the product of sulfide detoxification in *B. thermydron* (Vetter et al. 1987). An effect of lactate has also been reported for two species of vent shrimps, *Chorocaris chacei* (Lallier et al. 1998), and *Rimicaris exoculata* (Lallier and Truchot 1997). The vent-chimney crab *Cyanagraea praedator* also possesses hemocyanin with a high affinity for oxygen and a strong Bohr effect (Sanders 1989; Chausson et al. 2001). Surprisingly, lactate does not have any effect on the affinity nor the Bohr factor in this species (Chausson et al. 2001).

In arthropod hemocyanins, there usually is a decrease of oxygen affinity when the temperature increases, as a consequence of the exothermic nature of this reaction (Truchot 1992). Interestingly, all the hemocyanins from hydrothermal vent species show either no effect or a slightly reverse effect of temperature on oxygen affinity (Sanders et al. 1988; Lallier and Truchot 1997; Lallier et al. 1998; Chausson et al. 2001, 2004). More detailed studies will determine whether the temperature insensitivity is due to specific adaptations in the hemocyanin molecule itself or the result of side effects of the overall oxygenation reactions. Whatever the cause of this temperature insensitivity, this may well be an adaptation to the highly variable conditions (for which temperature is an indicator) encountered at deep-sea hydrothermal vents: despite

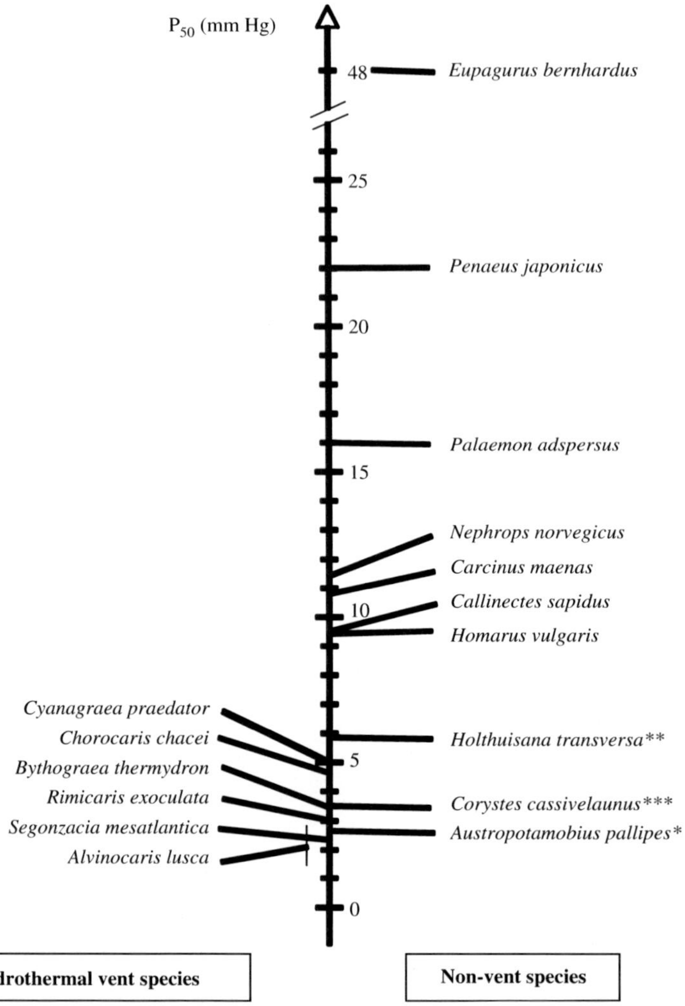


**Fig. 5** $P_{50}$ values for decapod hemocyanins from vent-endemic species compared to some representative non-vent species. Experimental conditions and references: *Eupagurus bernhardus* 15°C, pH 7.83; *Penaeus japonicus* 25°C, pH 7.6; *Palaemon adspersus* 15°C, pH 7.85; *Nephrops norvegicus* 10°C, pH 7.9; *Carcinus maenas* 15°C, pH 7.84; *Callinectes sapidus* 20°C, pH 7.6; *Homarus vulgaris* 15°C, pH 7.9; *Holthuisana transversa* 25°C, pH 7.35; *Corystes cassivelaunus* 10°C, pH 7.9; *Austropotamobius pallipes* 15°C, pH 7.9 (all these species, see review by Truchot (1992)); *Cyanagraea praedator* 15°C, pH 7.5 (Chausson et al. 2001); *Chorocaris chacei* 15°C, pH 7.5 (Lallier et al. 1998); *Bythograea thermydron* 15°C, pH 7.5 (Sanders et al. 1988); *Rimicaris exoculata* 15°C, pH 7.5 (Lallier and Truchot 1997); *Segonzacia mesatlantica* 15°C, pH 7.5 (Chausson et al. 2004); *Alvinocaris lusca* 15°C, pH 7.5 (Sanders et al. 1988). * fresh-water species. ** species with mixed land/water way of life. *** species living buried in the sand

the changes in temperature, the affinity of hemocyanin for oxygen will remain high. This characteristic explains at least in part why the critical $PO_2$ varies little between 2°C and 20°C (Mickel and Childress 1982b). The lack of temperature sensitivity of oxygen binding has also been observed for shallow water crustacea that are exposed to highly variable temperatures such as the hermit crab *Pagurus bernhardus* (Jokumsen and Weber 1982) and the shrimp *Palaemon elegans* (Morris et al. 1985). In this context, the study of cold-seep relatives of the vent species would be most interesting to determine whether these species that are used to cold water (with little or no variations of temperature) have temperature-sensitive hemocyanins.

In the crab *Cyanagraea praedator* and in the shrimp *Rimicaris exoculata*, there is another factor—whose identity is still unknown—that decreases the oxygen affinity of hemocyanin and that can be eliminated by dialysis (Lallier and Truchot 1997; Chausson et al. 2001). This phenomenon has also been observed for terrestrial crabs of the family Ocypodidae (Morris and Bridges 1985; Bridges et al. 1997), and in the coconut-tree crab *Birgus latro* (Morris et al. 1988). In *Ocypodes serratan*, this small molecule (< 5,000 Da) is capable of binding large amounts of $CO_2$ (Bridges et al. 1997). The factor remains to be identified and its effect on the hemocyanin studied.

In a study on Mid-Atlantic Ridge hydrothermal vent crab *Segonzacia mesatlantica*, Chausson et al. (2004) investigated the effect of severe hypoxia (6.25 μM $O_2$, i.e., 3.7 times less than the critical $PO_2$ below which *Bythograea thermydron* oxygen consumption drops and the animal goes anaerobic) on short term—6 h—changes in hemocyanin composition in the hemolymph. Comparing two batches of crabs, one maintained in severe hypoxia, and the other in saturation with air (referred to as "hyperoxia"), the authors report an increase of the hexameric (from 68% to 72%), the dodecameric (16% to 18%), and a decrease of the monomeric (from 13% to 7%) fractions. However, the overall chain composition did not change significantly. Unfortunately, the data were obtained from pools of 4 crabs for each condition, which does not give access to inter-individual variability. This is especially important because the dodecameric-to-hexameric fractions ratio may be variable in natural populations, as has been shown in the blue crab *Callinectes sapidus* (Greaves et al. 1992).

Some species of hydrothermal vent crustacea have been reported to possess hemoglobin: the copepods *Benthoxynus spiculifer* (Hourdez et al. 2000b) and *Scotoecetes introrsus* (Sell 2000). Although Hb has been reported in some bottom-dwelling littoral and lake species (Fox 1957), little is known about Hbs of these animals probably because of their small size. Interestingly, these Hb-containing copepods identified by Fox (1957) were collected from muddy and reduced environments with low $O_2$ and high sulfide levels. In the Branchiopod *Daphnia magna*, Hb synthesis is induced by ambient hypoxia (Fox et al. 1951; Kobayashi and Hoshi 1982) and animals with high Hb concentrations are able to maintain normal oxygen consumption rates even at low $O_2$ tensions that are lethal for individuals with low Hb (Kobayashi and Hoshi 1984). The low hemoglobin concentrations reported for the two vent copepod species make it very unlikely that the hemoglobin could be used as a significant oxygen store (Hourdez et al. 2000b; Sell 2000). The very high affinity of *B. spiculifer* hemoglobin also indicates that its main function is to maintain a concentration gradient from the outside to the inside of the animal, even when environmental $PO_2$ are very low (Hourdez et al. 2000b). Hemoglobin thus extends the inhabitable environment of small crustaceans like copepods and allows them to inhabit the base of tubeworm bushes.

### 3.3 Circulatory system adaptations

In alvinellid polychaetes, the gular membrane extends posteriorly, enveloping the esophagus and a very developed capillary network (Fig. 6). In addition, the peri-esophageal pouch thus formed contains very numerous erythrocytes (coelomocytes with high concentrations of hemoglobin). This setup, with a high density of capillaries and coelomocytes, forms an internal gas exchange system (Jouin-Toulmond et al. 1996). A study of the functional properties of the Hbs (the coelomic, intracellular Hb and the vascular, extracellular Hb) showed that they possess very similar functional properties under the same conditions, thereby allowing bidirectional transfer in the gas exchange system (Hourdez et al. 2000a). This most likely buffers the variations of oxygen levels inside the body, in particular for the brain area, directly irrigated by the blood after it goes through the capillary network (Fig. 6).

## 4 Adaptations of oxygen cellular utilization

Even though the species studied possess specific adaptations that allow them to extract oxygen despite its low levels in the environment, they most likely often have to rely on anaerobic metabolism. In addition, sulfide can inhibit the

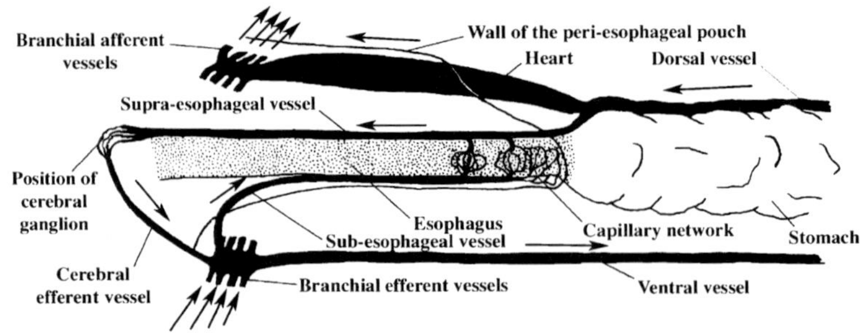


**Fig. 6** Anterior circulatory system in *Alvinella pompejana* showing the periesophageal pouch containing the capillary network. Ceolomocytes (not represented) are very dense in the pouch and in close contact with the capillaries. Oxygenated blood coming from the gills (branchial efferent vessels) merge ventrally and the sub-esophageal vessel can carry this blood to the capillary network where it exchanges gases with the coelomocytes. These capillaries then fuse with the supra-esophageal vessel and oxygen can be sent directly to the cerebral ganglion

electron transport chain (Powell and Somero 1986) and anaerobic metabolism may be happening even when oxygen is not completely absent. Hand and Somero (1983) measured enzymatic activities in order to determine which metabolic pathways were present in *Calyptogena magnifica*, *Riftia pachyptila*, *Bythograea thermydron*, and *Alvinella pompejana*. All the species studied exhibit cytochrome c oxidase and citrate synthase activities, both indicative of aerobic metabolism. The use of aerobic metabolism was also supported by the high activity of two glycolytic enzymes, phosphofructokinase and pyruvate kinase.

Several hydrothermal vent and cold-seep species have been studied for their ability to survive periods of anoxia. The crab *Bythograea thermydron* can survive at least 12 h (Mickel and Childress 1982b), the clam *Calyptogena magnifica* at least 16 h at 14°C (Arp et al. 1984), and the siboglinid polychaete *Riftia pachyptila* at least 36 h at 8°C (Childress et al. 1984), and up to 60 h at 15°C (Arndt et al. 1998). In the cold-seep polychaete *Methanoaricia dendrobranchiata*, a $TL_{50}$ of 133.5 h in anoxia without sulfide was calculated (Hourdez et al. 2002). The ice-worm *Hesiocaeca methanicola*, a hesionid polychaete that lives on methane ice, has a $TL_{50}$ of 99 h in complete anoxia (Fisher et al. 2000). All these data indicate a good capacity of these species to survive under anoxia for extended durations. This also indicates a good capability for anaerobic metabolism.

In *B. thermydron*, Hand and Somero (1983) detected high levels of lactate dehydrogenase (a typical enzyme for anaerobic metabolism in crustacea). Both the siboglinid *R. pachyptila* and the clam *C. magnifica* exhibited high level of malate dehydrogenase, a typical enzyme for anaerobic metabolism in invertebrates other than crustacea. In *Alvinella pompejana*, phosphoenolpyruvate carboxykinase, alanopine dehydrogenase, strombine dehydrogenase, and lactate dehydrogenase activities, markers of anaerobic metabolism, were also detected (Desbruyères et al. 1998). This observation indicates either that the worms were exposed to hypoxia prior to collection or that they are pre-adapted to hypoxia and continuously express these enzymes.

Arndt et al. (1998) studied the effect of prolonged anoxia on *Riftia pachyptila*. These authors followed the levels of malate, glycogen, aspartate, and succinate in the tissues for 60 h. From the high levels (up to 26 mM in the blood) encountered under normoxia, malate quickly drops and levels of succinate rise to very high values (up to 17 mM in the blood). Succinate is a well known final product of anaerobic metabolism in annelids, along with short-chain volatile fatty acids (Grieshaber et al. 1994). However, none of these fatty acids were excreted by the worm. Glycogen, mainly found in the trophosome, was only used after 48 h of anoxia. Succinate formed during anoxia cannot account for the decrease of malate and glycogen, suggesting that there might be other end products for *Riftia*'s anaerobic metabolism. Arndt et al. (1998) also report the absence of alanopine dehydrogenase, octopine dehydrogenase, and strombine dehydrogenase activities. Lactate dehydrogenase activity was found to be moderate in the vestimentum and in the body-wall, and absent in the other tissues tested. However, only traces of L-Lactate could be detected in the worms after anoxia exposure, suggesting a limited importance of this pathway in anaerobic metabolism. Other opine dehydrogenases activities (using lysine, taurine, etc...) were not tested.

## 5 Dealing with sulfide

At hydrothermal vents, oxygen and sulfide concentrations vary inversely (Johnson et al. 1988). Sulfide has a specific effect that affects the survival of animals. The cold-seep polychaete *Methanoaricia dendrobranchiata* has a $TL_{50}$ of 99.1 h in anoxia +1 mmol $l^{-1}$ sulfide, 119.6 h in anoxia +60 μmol $l^{-1}$ sulfide, and 133.5 h in anoxia without sulfide (Hourdez et al. 2002).

Sulfide can poison the mitochondrial electron transfer chain and disrupt the aerobic metabolism (Grieshaber and Völkel 1998). This may explain at least in part the high levels of enzymes involved in anaerobic metabolic pathways as the organisms may have to rely on it even if oxygen is present as a result of sulfide poisoning. In addition, sulfide can bind to hemoglobins and poison oxygen transport.

Organisms that are symbiotic with sulfide oxidizing bacteria may be protected from sulfide toxicity by the sink the bacteria represent. However, when the bacteria are not located in tissues directly in contact with the sulfide-containing water, the organism must transport sulfide without being poisoned by it. Sulfide-binding molecules are then necessary. Although other annelid hemoglobins may bind sulfide none can bind such large amounts as *Riftia pachyptila* HBL hemoglobin, that transports both sulfide and $O_2$ from the environment to the symbiotic bacteria in the trophosome, avoiding both poisoning by free sulfide and spontaneous reaction between the two molecules. The site of sulfide binding is still currently debated (Zal et al. 1998; Flores et al. 2005; Numoto et al. 2005; Flores and Hourdez in press) but it appears clear that it is not bound to the heme group. Based on crystal structure data, Flores et al. (2005) challenged the idea that sulfide was bound to free cysteines in the 400 kDa Hb from *Riftia* (Zal et al. 1998). The crystal data showed that the free cysteines were located in highly hydrophobic pockets that should keep the sulfide ions from binding. Alternatively, the Hb possesses zinc ions at specific locations (bound to histidine residues) that could be sites for sulfide binding. Flores et al. (2005) and Flores and Hourdez (in press) also reported that specific chelation of the zinc ions by TPEN inhibited the binding of sulfide while this binding was unchanged after capping of the free cysteines by NEM. A similar crystallographic study on the 400 kDa Hb from *Oligobrachia mashikoi* (another siboglinid polychaete) showed that the free cysteines were accessible to ethyl-mercury ions and that there were no zinc ions (Numoto et al. 2005). The specific amino acids involved in the binding of zinc in *Riftia* 400 kDa Hb are not conserved in the Hb from *O. mashikoi*. Interestingly, this latter species has been shown to possess methanotrophic (instead of sulfide-oxidizing) symbionts (Kimura et al. 2003), thereby relieving the selective pressure to bind sulfide to transport it to the symbionts. Numoto et al. (2005) also report that the amount of free sulfide in solution decreased in presence of that Hb, suggesting that it did bind sulfide. Some uncertainty remains though as they did not try to release that sulfide to show it was bound and not oxidized (Flores and Hourdez in press). Although the clam *Calyptogena* body fluid does bind sulfide (and transports it from its foot that extends in deep cracks in the basalt to its symbionts in its gills) the carrier is not Hb but a large (several million Dalton) zinc-rich molecule (reviewed in Childress and Fischer 1992) that is a heme-free, glycosylated protein and could be a lipoprotein (Zal et al. 2000b).

Species that do not have body fluids capable of binding sulfide may oxidize sulfide in their mitochondria and even get some ATP out of the process such as in the polychaetes *Arenicola marina* (Völkel and Grieshaber 1997), and the bivalves *Geukensia demissa* (Parrino et al. 2000;

Doeller et al. 2001; Kraus and Doeller 2004) and *Solemya reidi* (O'Brien and Vetter 1990).

## 6 Conclusion

From the limited studies conducted so far on deep-sea hydrothermal vents or cold seep endemic species, anatomical and biochemical approaches have identified general trends regarding their respiratory adaptations to their hypoxic environment. Large branchial surface areas and abundant oxygen binding proteins seem to be shared by crustaceans and annelids inhabiting these extreme environments with an inclination to develop oxygen storage mechanisms to buffer the variations of oxygen supply in the water. Adaptations to low oxygen levels in invertebrates from hydrothermal vents and cold seeps seem similar. It seems however that the seep species are able to withstand longer exposure to anoxia, which may be an adaptation to the more stable conditions encountered at seeps. In this aspect, benthic species from oxygen minimum zones (OMZ) may show capabilities similar to those of the seep species. Very few studies are available to date but we know that some OMZ species exhibit enlarged gas exchange surfaces, and high activities of enzymes involved in anaerobic metabolism (for review see Levin 2003).

However, in vivo physiological experiments, although difficult to conduct on these hard-to-reach animals, are definitely needed in order to show that the anatomical or biochemical adaptations identified so far are really at work in these animals and allow them to maintain an aerobic metabolism comparable to littoral relative species despite the potential toxicity of their surroundings (sulfide, heavy metals).

## References


Andersen AC, Jolivet S, Claudinot S, Lallier FH (2002) Biometry of the branchial plume in the hydrothermal vent tubeworm *Riftia pachyptila* (Vestimentifera; Annelida). Can J Zool 80:320–332

Arndt C, Schiedek D, Felbeck H (1998) Metabolic response of the hydrothermal vent tubeworm *Riftia pachyptila* to severe hypoxia. Mar Ecol Prog Ser 174:151–158

Arp AJ, Childress JJ (1981) Blood function in the hydrothermal vent Vestimentiferan tube worm. Science 213:342–344

Arp AJ, Childress JJ (1983) Sulfide binding by the blood of the hydrothermal vent tube worm *Riftia pachyptila*. Science 219:295–297

Arp AJ, Childress JJ, Fisher CR Jr (1984) Metabolic and blood gas transport characteristics of the hydrothermal vent bivalve, *Calyptogena magnifica*. Physiol Zool 57:648–662

Arp AJ, Doyle ML, Di Cera E, Gill SJ (1990) Oxygenation properties of the two co-occurring hemoglobins of the tube worm *Riftia pachyptila*. Resp Physiol 80:323–334

Ballard RD, Grassle JF (1979) Return to oases of the deep (strange world without a sun). Natl Geogr 156(5):680–703

Belman BW, Childress JJ (1976) Circulatory adaptations to the oxygen minimum layer in the Bathypelagic Mysid *Gnathophausia ingens*. Biol Bull 150(1):15–37

Bridges CR, Hupperts V, Eshky AA, Taylor AC (1997) Haemocyanin oxygen transport in *Ocypode* spp.: modulation of oxygen affinity? J Mar Biol Ass U K 77:145–158

Cavanaugh CM, Gardiner SL, Jones ML, Jannasch HW, Waterbury JB (1981) Prokaryotic cells in the hydrothermal vent tube-worm *Riftia pachyptila* Jones: possible chemoautotrophic symbionts. Science 213:340–342

Chausson F, Bridges CR, Sarradin PM, Green BN, Riso R, Caprais JC, Lallier FH (2001) Structural and functional properties of hemocyanin from *Cyanagraea praedator*, a deep-sea hydrothermal vent crab. Proteins 45:351–359

Chausson F, Sanglier S, Leize E, Hagège A, Bridges CR, Sarradin, P-M, Shillito B, Lallier FH, Zal F (2004) Respiratory adaptations to the deep-sea hydrothermal vent environment: the case of *Segonzacia mesatlantica*, a crab from the Mid-Atlantic Ridge. Micron 35:31–41

Chevaldonné P (1986) Ecologie des cheminées actives. Ph.D. thesis. Université de Méditerranée, Marseille. 257 pp

Chevaldonné P, Desbruyères D, Le Haitre M (1991) Time-series of temperature from three deep-sea hydrothermal vent sites. Deep-Sea Res I 38(11):1417–1430

Chevaldonné P, Desbruyères D, Childress JJ (1992) Some like it hot ... and some even hotter. Nature 359:593–594

Childress JJ (1975) The respiratory rates of midwater crustaceans as a function of depth of occurrence and relation to the minimum layer off Southern California. Comp Biochem Physiol 50A:787–799

Childress JJ, Arp AJ, Fisher CR (1984) Metabolic and respiratory characteristics of the hydrothermal vent tube worm *Riftia pachyptila*. Mar Biol 83:109–124

Childress JJ, Mickel TJ (1985) Metabolic rates of animals from hydrothermal vents and other deep-sea habitats. Biol Soc Wash Bull 6:249–260

Childress JJ, Fischer CR (1992) The biology of hydrothermal vent animals: physiology, biochemistry and autotrophic symbioses. Oceanogr Mar Biol Annu Rev 30:337–441

Childress JJ, Seibel BA (1998) Life at stable oxygen levels: adaptations of animals to oceanic oxygen minimum layers. J Exp Biol 201:1223–1232

Cordes EE, Hourdez S, Predmore BL, Redding ML, Fisher CR (2005) Succession of hydrocarbon seep communities associated with the long-lived foundation species *Lamellibrachia luymesi*. Mar Ecol Prog Ser 305:17–29

Corliss JB, Ballard RD (1977) Oases of life in the cold Abyss. Natl Geogr 152(4):441–454

Corliss JB, Dymond J, Gordon LI, Edmond JM, Herzen RPV, Ballard RD, Green K, Williams D, Bainbridge A, Crane K, van Andel TH (1979) Submarine thermal springs on the Galápagos Rift. Science 203:1073–1083

Desbruyères D, Chevaldonné P, Alayse A-M, Jollivet D, Lallier FH, Jouin-Toulmond C, Zal F, Sarradin P-M, Cosson R, Caprais J-C, Arndt C, O'Brien J, Guezennec J, Hourdez S, Riso R, Gaill F, Laubier L, Toulmond A (1998) Biology and ecology of the pompei worm (*Alvinella Pompejana* Desbruyères and Laubier), a normal dweller on an extreme deep-sea environment: a synthesis of current knowledge and recent developments. Deep-Sea Res Part II 45:383–422

Doeller JE, Grieshaber MK, Kraus DW (2001) Chemolithoheterotrophy in a metazoan tissue: thiosulfate production matches ATP demand in ciliated mussel gills. J Exp Biol 204:3755–3764

Felbeck H (1981) Chemoautotrophic potential of the hydrothermal vent tubeworm, *Riftia Pachyptila* Jones (Vestimentifera). Science 213:336–338

Felbeck H, Somero GN, Childress JJ (1981) Calvin-Benson cycle and sulfide oxidation enzymes in animals from sulfide-rich habitats. Nature 293:291–293

Fisher CR, Childress JJ, Arp AJ, Brooks JM, Distel D, Favuzzi JA, Felbeck H, Hessler R, Johnson KS, Kennicutt MC, Macko SA, Newton A, Powell MA, Somero GN, Soto T (1988) Microhabitat variation in the hydrothermal vent mussel *Bathymodiolus thermophilus*, at Rose Garden vent on the Galapagos Rift. Deep-Sea Res 35(10/11):1769–1792

Fisher CR (1995) Toward an appreciation of hydrothermal-vent animals: their environment, physiological ecology, and tissue stable isotope values. In: Humphris SE et al. (eds) Seafloor hydrothermal systems: physical, chemical, biological, and geological interactions. American Geophysical Union, Washington DC, pp 297–316

Fisher CR (1996) Ecophysiology of primary production at deep-sea vents and seeps. In: Deep-sea and extreme shallow-water habitats: affinities and adaptations. Austrian Academy of Sciences Press, Vienna, pp 313–336

Fisher CR, MacDonald IR, Sassen R, Young CM, Macko SA, Hourdez S, Carney RS, Joye S, McMullin E (2000) Methane ice worms: *Hesiocaeca methanicola* colonizing fossil fuel reserves. Naturwissenschaften 87:184–187

Flores JF, Fisher CR, Carney SL, Green BN, Freytag JK, Schaeffer SW, Royer WE (2005) Sulfide binding is mediated by zinc ions discovered in the crystal structure of a hydrothermal vent tubeworm hemoglobin. Proc Natl Acad Sci USA 102(8):2713–2718

Flores JF, Hourdez S The zinc-mediated sulfide-binding mechanism of hydrothermal vent tubeworm 400-Kda hemoglobin. Cah Biol Mar 47(4), (in press)

Fox HM, Gilchrist BM, Phear EA (1951) Functions of haemoglobin in *Daphnia*. Proc R Soc B 138:514–528

Fox HM (1957) Haemoglobin in the crustacea. Nature 179:148

Greaves J, Rainer JS, Mangum CP (1992) Size-exclusion high performance liquid chromatography of the Dodecameric and Hexameric forms of hemocyanin from *Callinectes sapidus*. Mar Biol 113:33–36

Grieshaber MK, Hardewig I, Kreutzer U, Pörtner H-O (1994) Physiological and metabolic responses to hypoxia in invertebrates. Rev Physiol Biochem Pharmacol 125:43–147

Grieshaber MK, Völkel S (1998) Animal adaptations for tolerance and exploitation of poisonous sulfide. Ann Rev Physiol 60:33–53

Hand SC, Somero GN (1983) Energy metabolism pathways of hydrothermal vent animals: adaptations to a food-rich and sulfide-rich deep-sea environment. Biol Bull 165:167–181

Hessler RR, Jumars PA (1974) Abyssal community analysis from replicate box cores in the Central North Pacific. Deep-sea Res 21:185–209

Hourdez S, JouinToulmond C (1998) Functional anatomy of the respiratory system of *Branchipolynoe* species (Polychaeta, Polynoidae), commensal with *Bathymodiolus* species (Bivalvia, Mytilidae) from deep sea hydrothermal vents. Zoomorphology 118(4):225–233

Hourdez S, Lallier FH, Green BN, Toulmond A (1999a) Hemoglobins from deep-sea hydrothermal vent scaleworms of the genus *Branchipolynoe*: a new type of quaternary structure. Proteins 34:427–434

Hourdez S, Lallier FH, Martin-Jezequel V, Weber RE, Toulmond A (1999b) Characterization and functional properties of the extracellular coelomic hemoglobins from the deep-sea, hydrothermal vent scaleworm *Branchipolynoe symmytilida*. Proteins 34(4):435–442

Hourdez S, Lallier FH, De Cian M-C, Green BN, Weber RE, Toulmond A (2000a) Gas transfer system in *Alvinella pompejana* (Annelida Polychaeta, Terebellida): functional properties of intracellular and extracellular hemoglobins. Phys Biochem Zool 73(3):365–373

Hourdez S, Lamontagne J, Peterson P, Weber RE, Fisher CR (2000b) Hemoglobin from a deep-sea hydrothermal-vent copepod. Biol Bull 199(2):95–99

Hourdez S, Frederick L-A, Schernecke A, Fisher CR (2001) Functional respiratory anatomy of a deep-sea Orbiniid Polychaete from the Brine Pool Nr-1 in the Gulf of Mexico. Inv Biol 120(1):29–40

Hourdez S, Weber RE, Green BN, Kenney JM, Fisher CR (2002) Respiratory adaptations in a deep-sea Orbiniid Polychaete from Gulf of Mexico Brine Pool NR-1: metabolic rates and hemoglobin structure/function relationships. J Exp Biol 205:1669–1681

Hourdez S, Weber RE (2005) Molecular and functional adaptations in deep-sea hemoglobins. J Inorg Biochem 99:130–141

Johnson KS, Childress JJ, Beehler CL (1988) Short-term temperature variability in the Rose Garden hydrothermal vent field: an unstable deep-sea environment. Deep-Sea Res 35(10/11):1711–1721

Johnson L, Rees CJC (1988) Oxygen consumption and gill surface area in relation to habitat and lifestyle of four crab species. Comp Biochem Physiol 89A:243–246

Jokumsen A, Weber RE (1982) Hemocyanin–oxygen affinity in hermit crab blood is temperature independent. J Exp Zool 221:389–394

Jones ML (1981) *Riftia pachyptila* Jones: observations on the Vestimentiferan worm from the Galápagos Rift. Science 213:333–336

Jones ML (1988) The Vestimentifera, their biology, systematic and evolutionary patterns. Oceanol Acta Spec 8:69–82

Jouin C, Toulmond A (1989) The ultrastructure of the gill of the lugworm *Arenicola marina* (L) (Annelida, Polychaeta). Acta Zool Stockh 70(2):121–129

Jouin C, Gaill F (1990) Gills of hydrothermal vent annelids : structure, ultrastructure and functionnal implications in two Alvinellid species. Prog Oceanogr 24:59–69

Jouin-Toulmond C, Augustin D, Desbruyeres D, Toulmond A (1996) The gas transfer system in Alvinellids (Annelida Polychaeta, Terebellida). Anatomy and ultrastructure of the anterior circulatory system and characterization of a coelomic, intracellular, haemoglobin. Cah Biol Mar 37(2):135–151

Jouin-Toulmond C, Hourdez S (2006) Morphology, ultrastructure and functional anatomy of the branchial organ of *Terebellides stroemii* (Polychaeta: Trichobranchidae), with remarks on the systematic position of the genus *Terebellides*. Cah Biol Mar 47(3):287–299

Kennicutt MC II, Brooks JM, Bidigare RR, Fay RR, Wade TL, McDonald TJ (1985) Vent-type taxa in a hydrocarbon seep region on the Louisiana slope. Nature 317:351–353

Kennish MJ, Lutz RA (1992) The hydrothermal vent clam, *Calyptogena magnifica* (Boss and Turner 1980): a review of existing literature. Rev Aquat Sci 6(1):29–66

Kimura HM, Sato Y, Sasayama Y, Naganuma T (2003) Molecular characterization and in situ localization of endosymbiotic 16s ribosomal RNA and Rubisco genes in the pogonophoran tissue. Mar Biotech 5:261–269

Kobayashi M, Hoshi T (1982) Relationship between haemoglobin concentration of *Daphnia magna* and the ambient oxygen concentration. Comp Biochem Physiol 72:247–249

Kobayashi M, Hoshi T (1984) Analysis of the respiratory role of haemoglobin in *Daphnia magna*. Zool Sci 1:523–532

Kraus DW, Doeller JE (2004) Sulfide consumption by mussel gill mitochondria is not strictly tied to oxygen reduction: measurements using a novel polarographic sulfide sensor. J Exp Biol 207:3667–3679

Lallier FH, Truchot JP (1997) Hemocyanin oxygen-binding properties of a deep-sea hydrothermal vent shrimp: evidence for a novel cofactor. J Exp Zool 277:357–364

Lallier FH, Camus L, Chausson F, Truchot JP (1998) Structure and function of hydrothermal vent crustacean haemocyanin: an update. Cah Biol Mar 39(3–4):313–316

Le Bris N, Sarradin PM, Caprais JC (2003) Contrasted sulphide chemistries in the environment of 13°N EPR vent fauna. Deep-Sea Res I 50:737–747

Levin LA (2003) Oxygen minimum zone benthos: adaptation and community response to hypoxia. Oceanogr Mar Biol Annu Rev 41:1–45

Magenheim AJ, Gieskes JM (1992) Hydrothermal discharge and alteration in near-surface sediments from the Guaymas Basin, Gulf of California. Geochim Cosmochim Acta 56:2329–2338

McMahon BR (2001) Respiratory and circulatory compensation to hypoxia in crustaceans. Respir Physiol 128:349–364

Mickel TJ, Childress JJ (1982a) Effects of pressure and temperature on the EKG and heart rate of the hydrothermal vent crab *Bythograea thermydron* (Brachyura). Biol Bull 162:70–82

Mickel TJ, Childress JJ (1982b) Effects of temperature, pressure, and oxygen concentration on the oxygen consumption rate of the hydrothermal vent crab *Bythograea thermydron* (Brachyura). Phys Zool 55(2):199–207

Morris S, Bridges CR (1985) An investigation of haemocyanin oxygen affinity in the semi-terrestrial crab *Ocypode saratan* Forsk. J Exp Biol 117:119–132

Morris S, Bridges CR, Grieshaber MK (1985) Respiratory properties of the haemolymph of the intertidal prawn *Palaemon elegans* (Rathke). J Exp Biol 233:175–186

Morris S, Greenaway P, McMahon BR (1988) Adaptations to terrestrial existence by the robber crab, *Birgus latro*. I. An in vitro investigation of blood gas transport. J Exp Biol 140:477–491

Numoto N, Nakagawa T, Kita A, Sasayama Y, Fukumori Y, Miki K (2005) Structure of an extracellular giant hemoglobin of the gutless beard worm *Oligobrachia mashikoi*. Proc Natl Acad Sci USA 102(41):14521–14526

O'Brien J, Vetter RD (1990) Production of thiosulphate during sulphide oxidation by mitochondria of the symbiont-containing bivalve *Solemya reidi*. J Exp Biol 149:133–148

Parrino V, Kraus DW, Doeller JE (2000) ATP production from the oxidation of sulfide in gill mitochondria of the ribbed mussel *Geukensia demissa*. J Exp Biol 203:2209–2218

Paull CK, Hecker B, Commeau R, Freeman-Lynde RP, Neumann C, Corso WP, Golubic S, Hook JE, Sikes E, Curray J (1984) Biological communities at the Florida escarpment resemble hydrothermal vent taxa. Science 226:965–967

Powell MA, Somero GN (1986) Adaptations to sulfide by hydrothermal vent animals: sites and mechanisms of detoxification and metabolism. Biol Bull 171:274–290

Sanders NK (1989) Functional properties of hemocyanins from deep-sea crustaceans. PhD thesis. University of California, Santa Barbara USA. 209 pp

Sanders NK, Arp AJ, Childress JJ (1988) Oxygen binding characteristics of the hemocyanins of two deep-sea hydrothermal vent crustaceans. Respir Physiol 71:57–68

Segonzac M, Desaintlaurent M, Casanova B (1993) Enigma of the trophic adaptation of the shrimp Alvinocarididae in hydrothermal areas along the Mid-Atlantic Ridge. Cah Biol Mar 34(4):535–571

Sell AF (2000) Life in the extreme environment at a hydrothermal vent: haemoglobin in a deep-sea copepod. Proc R Soc Lond B 267:2323–2326

Sibuet M, Olu K (1998) Biogeography, biodiversity and fluid dependence of deep-sea cold-seep communities at active and passive margins. Deep-Sea Res II 45:517–567

Smith EB, Scott KM, Nix ER, Korte C, Fisher CR (2000) Growth and condition of seep mussels (*Bathymodiolus childressi*) at a Gulf of Mexico Brine Pool. Ecology 81(9):2392–2403

Terwilliger NB, Terwilliger RC (1984) Hemoglobin from the "Pompeii Worm", *Alvinella pompejana*, an annelid from a deep sea hot hydrothermal vent environment. Mar Biol Lett 5:191–201

Terwilliger RC, Terwilliger NB, Arp AJ (1983) Thermal vent clam (*Calyptogena magnifica*) hemoglobin. Science 219:981–983

Terwilliger RC, Terwilliger NB, Hughes GM, Southward AJ, Southward EC (1987) Studies on the haemoglobins of the small pogonophorans. J Mar Biol Ass U K 67:219–239

Toulmond A, El Idrissi Slitine F, De Frescheville J, Jouin C (1990) Extracellular hemoglobins of hydrothermal vent annelids: structural and functional characteristics in three Alvinellid species. Biol Bull 179:366–373

Truchot J-P (1992) Respiratory function of arthropod hemocyanins. In: Mangum CP (ed) Blood and tissues oxygen carriers. Spinger Verlag, Berlin Heidelberg, pp 377–410

Tunnicliffe V (1991) The biology of hydrothermal vents: ecology and evolution. Oceanogr Mar Biol Annu Rev 29:319–407

Tyler PA, German CR, Ramirez-Llodra E, Van Dover CL (2003) Understanding the biogeography of chemosynthetic ecosystems. Oceanol Acta 25(5):227–241

Vetter RD, Wells ME, Kurtsman AL, Somero GN (1987) Sulfide detoxification by the hydrothermal vent crab *Bythograea thermydron* and other decapod crustaceans. Physiol Zool 60:121–137

Völkel S, Grieshaber MK (1997) Sulphide oxidation and oxidative phosphorylation in the mitochondria of the lugworm *Arenicola marina*. J Exp Biol 200:83–92

Von Damm KL (1990) Seafloor hydrothermal activity: black smoker chemistry and chimneys. Ann Rev Earth Planet Sci 18:173–204

Weber RE (1978) Respiratory pigments. In: Mills PJ (ed) Physiology of annelids. Academic Press, New York, pp 393–437

Williams AB (1980) A new crab family from the vicinity of submarine thermal vents on the Galapagos Rift (Crustacea: Decapoda: Brachyura). Proc Biol Soc Wash 93(2):443–472

Wittenberg JB (1985) Oxygen supply to intracellular bacterial symbionts. Bull Biol Soc Wash 6:301–310

Wittenberg JB, Stein JL (1995) Hemoglobin in the symbiont-harboring gill of the marine gastropod *Alviniconcha hessleri*. Biol Bull 188:5–7

Zal F, Lallier FH, Green BN, Vinogradov SN, Toulmond A (1996a) The multi-hemoglobin system of the hydrothermal vent tube worm *Riftia pachyptila*. 2. Complete polypeptide chain composition investigated by maximum entropy analysis of mass spectra. J Biol Chem 271(15):8875–8881

Zal F, Lallier FH, Wall JS, Vinogradov SN, Toulmond A (1996b) The multi-hemoglobin system of the hydrothermal vent tube worm *Riftia pachyptila*. 1. Reexamination of the number and masses of its constituents. J Biol Chem 271(15):8869–8874

Zal F, Green BN, Lallier FH, Toulmond A (1997) Investigation by electrospray ionization mass spectrometry of the extracellular hemoglobin from the polychaete annelid *Alvinella pompejana*: An unusual hexagonal bilayer hemoglobin. Biochemistry 36(39):11777–11786

Zal F (1998) Sulphide-binding processes of *Riftia pachyptila* haemoglobins. Cah Biol Mar 39(3–4):327–328

Zal F, Leize E, Lallier FH, Toulmond A, VanDorsselaer A, Childress JJ (1998) S-Sulfohemoglobin and disulfide exchange: the mechanisms of sulfide binding by *Riftia pachyptila* hemoglobins. Proc Natl Acad Sci USA 95(15):8997–9002

Zal F, Green BN, Martineu P, Lallier FH, Toulmond A, Vinogradov SN, Childress JJ (2000a) Polypeptide chain composition difersity of hexagonal-bilayer haemoglobins within a single family of annelids, the Alvinellidae. Eur J Biochem 267:1–11

Zal F, Leize E, Oros DR, Hourdez S, Van Dorsselaer A, Childress JJ (2000b) Haemoglobin structure and biochemical characteristics of the sulphide-binding component from the deep-sea clam *Calyptogena magnifica*. Cah Biol Mar 41(4):413–423

Rev Environ Sci Biotechnol (2006) 5:253–267
DOI 10.1007/s11157-006-0013-0

REVIEW PAPER

# Metal detoxification and homeostasis in Antarctic Notothenioids. A comparative survey on evolution, expression and functional properties of fish and mammal metallothioneins

**Clemente Capasso · Vincenzo Carginale · Marilisa Riggio · Rosaria Scudiero · Piero Andrea Temussi · Francesca Trinchella · Elio Parisi**

Received: 21 February 2006 / Accepted: 6 June 2006 / Published online: 13 July 2006


**Abstract** Organisms from yeast to mammals contain cysteine-rich, heavy metal binding proteins termed metallothioneins. The putative roles of these proteins are trace metal homeostasis and detoxification of poisonous heavy metals. The highly conserved chemical composition and the structural constraints led to the conclusion that metallothioneins of different origin must display remarkably similar features. The present review aims at surveying the studies carried out on the metallothioneins of Antarctic Notothenioidei, a dominating fish group endowed of a number of striking adaptive characters, including reduced (or absent) hematocrit and presence of antifreeze glycoproteins. Given the unique peculiarities of the Antarctic environment, a comparative study of the features of notothenioid metallothioneins could provide new insights into the role of these proteins in physiology and toxicology. The results summarized here show that the metallothioneins of this fish group display a number of features at the level of evolution, expression pattern, structure and function remarkably different from those of mammal metallothioneins.

**Keywords** Antarctic fish · Evolution · Function · Mammals · Metallothionein · Structure

C. Capasso · V. Carginale · E. Parisi (✉)
CNR Institute of Protein Biochemistry, Via Pietro Castellino 111, 80131 Napoli, Italy
e-mail: e.parisi@ibp.cnr.it

M. Riggio · R. Scudiero · F. Trinchella
Department of Biological Sciences, University Federico II, Napoli, Italy

P. A. Temussi
National Institue for Medical Research, London, UK

P. A. Temussi
Accademia dei Lincei, Roma, Italy

## Metallothioneins

Metallothionein (MT) is a low-molecular weight protein rich in thiol groups and heavy metals (Kagi 1993; Karin 1985). Since its discovery, MT was identified as a cadmium-binding protein and, for this very reason, has been considered an important factor involved in the protection of organisms from the harmful effects of toxic heavy metals such as cadmium and mercury. The improvement of analytical techniques and the increased understanding of the mechanisms of MT induction has strongly encouraged the use of MT as an environmental marker for metal pollution (Dallinger et al. 2000, 2004). On the other hand, a number of evidence indicate that MT can act as scavenger of free hydroxyl and superoxide radicals (Kumari et al. 2000; Wright et al. 2000; You et al. 2002) and that MT synthesis can be induced by oxidative stress in a fashion similar to heavy metals (Cai et al. 1999; Kondoh et al. 2001).

From the numerous amino acid sequences available, it appears that the most striking peculiarity of MT is the particular arrangement of the 20 cysteines, organized in two distinct metal–thiolate domains (Kagi 1993). The structural features of MT have been elucidated by X-ray crystallographic studies (Robbins and Stout 1991) and NMR analysis (Capasso et al. 2003a; Narula et al. 1995; Wang et al. 1994; Wang et al. 1995). From these studies, it appears that the structures of the two domains depend on the conserved arrangement of the cysteines in the polypeptide chain. In mammalian MT, the N-terminal $\beta$-domain contains three metal ions and nine cysteines arranged to form a cyclohexane-like structure, whilst the C-terminal $\alpha$-domain has four metal ions and 11 cysteines arranged in a bicyclononane-like structure.

In spite of several decades of studies, the exact function of MT is still matter of debate. The biosynthesis of MT is enhanced not only by environmental factors, but also by hormonal humors that modulate MT expression during embryonic development (Rosenthal et al. 2004). The intracellular concentration of MT is largely variable and is usually affected by a number of factors, including exposure to heavy metals, cytokines and inflammatory agents (Blaise et al. 2002; Cajaraville et al. 2000; Dallinger et al. 2000, 2004; Rodriguez-Ortega et al. 2002). Transcriptional activation by heavy metals of MT genes is mediated by trans-acting proteins binding to DNA regulatory elements located in the promoter of MT gene(s) (Samson and Gedamu 1998). DNA methylation is another factor that regulates MT synthesis through the silencing of the expression of their genes (Riggio et al. 2003).

The response to a variety of stimuli justifies the definition of "protein for every season" given to MT, and explains the interest dedicated to MT by scientists operating in different fields, from toxicology to physiology, from molecular to developmental biology. Until recently, a great deal of knowledge on MT structure and function relied on studies carried out on MT of mammalian origin. Recent studies carried out on invertebrate MTs have revealed the existence of striking differences between these MTs and those of mammalian origin. The lobster MT, for example, consists of two beta domains made of nine cysteine residues bound to three equivalents of metal (Munoz et al. 2000a, b). In mammals, four MT isoforms have been detected: MT-I and MT-II are the major ones and are expressed in most organs, whereas MT-III and MT-IV are specifically expressed in neurons and epithelial cells, respectively. Compared with the multiplicity of mammalian MT, a different situation is observed in invertebrates and lower vertebrates, in which one or two distinct MT isoforms have been isolated. Certain mollusks posses two distinct MT isoforms that are differentially regulated by toxic and physiological factors (Dondero et al. 2005). In most fish, only one MT gene is expressed, with the exception of trout (Bonham and Gedamu 1984), zebrafish, carp (Ren et al. 2000) and gobio (Knapen et al. 2005) which have two or three MT isoforms.

A better understanding of the biological role of MT requires the investigation in different organisms, by combining phylogenetic, structural and functional approaches. Because most studies concerned MT of mammalian origin, the conclusion that MT from different organisms possesses identical structural–functional properties was not adequately supported. For such a purpose, the Antarctic continent provides an excellent opportunity for studying an unusual fish fauna in which organism evolution and adaptation are directly linked to peculiar environmental conditions. The present review describes the results of works carried out in recent years in our laboratory on MT from Antarctic fish.

## The Antarctic environment

Antarctica is gaining increasing attention from the scientific community in relation to the effects on global change and biodiversity preservation caused by the environmental impact. Antarctica is one of the most isolated regions in the world: it was encompassed in the super continent Gondwana originated from the fragmentation of Pangea and successive continental drift. Following Gondwana breakup, Antarctica remained in contact with South America, Australia and New Zealand until the end of Cretaceous, when it attained the present position at the South Pole (Eastman 1993). Owing to the progressive cooling, the Antarctic continent underwent a drastic change of its climate; today, the temperature of the seawater is constantly close to –1.9°C that is the equilibrium temperature of seawater with ice. Antarctica became isolated from the other continents about 25 MY ago, following the opening of the Drake Passage and formation of the Polar Front, an oceanic frontal system running between 50° S and 60° S: this barrier prevents fish migration from and to Antarctica. The geographical isolation of Antarctic fish and the drastic climatic conditions were probably the main reasons of the specializations achieved during cold adaptation. Antarctica is at the moment the only region of the globe, which has very little suffered from the effect of environmental disturbance caused by anthropic activities. With good reasons, it has been considered the largest natural laboratory existing on the planet.

Unlike other marine habitats, the Antarctic Ocean is a closed basin, isolated from other areas by the Polar Front, resembling ancient lakes: fish living in such a kind of habitat exhibit monophyly, endemism and speciosity. The continental shelf of Antarctica constitutes a habitat harboring two dominating fish groups, represented by liparids and notothenioids. Notothenioids form a perciform suborder, comprising 120 species mostly endemic to the waters surrounding Antarctica (Eastman 1993). Phylogenetic relationships among notothenioid taxa have been reconstructed on the basis of morphological (Eastman 1993) and molecular data (Bargelloni et al. 1994). From all of the above, it is apparent that notothenioids diverged quite recently (10–15 MY ago), likely from a single ancestor, thus providing an ideal system for evolutionary studies at a comparative level. This group includes highly cold-adapted and remarkably stenothermal fish: some species, such as the members of hemoglobinless family of Channichthydae (icefish), can be regarded as the most extreme forms of a series with diminishing hemoglobin content. Antarctic fish have evolved a number of adaptation mechanisms to cope with the prohibitive conditions of their habitat. The ecological importance of notothenioids in the Antarctic marine ecosystem and their remarkable adaptation to this extreme environment explains the great efforts that have been devoted to studying notothenioid physiology, producing a great wealth of data at both functional and biochemical levels.

## Metallothioneins in red-blooded and hemoglobinless notothenioids

In early studies, any attempt to isolate MT from the liver of the icefish *Chionodraco hamatus* was unsuccessful. A low-molecular weight zinc moiety isolated from liver displayed features totally unlike MT. This zinc-containing protein has a molecular mass of 11 kDa and is characterized by low cysteine content and abundance of glutamate and aspartate (Scudiero et al. 1992). It is likely that the acidic amino acid residues are involved in metal complex formation, but it is not clear whether the zinc-containing protein from icefish liver can fulfill some, if not all, functions ascribed to MT. In contrast, appreciable amounts of MT were found in the liver of the red-blooded notothenioid *Trematomus bernacchii*.

In spite of the apparent lack of MT in icefish liver, a DNA fragment was isolated from the products obtained by RT-PCR of RNA extracted from icefish liver, containing an open reading sequence encoding an MT homologous to MTs of red-blooded notothenioids. Using icefish MT cDNA as a probe, it was possible to estimate MT mRNA expression by Northern blotting of total RNA extracted from liver of both the red-blooded *T. bernacchii* and hemoglobinless *C. hamatus*. The results showed that large amounts of MT mRNA were present in the red-blooded species as well as in icefish (Scudiero et al. 1997). Hence, the presence of a transcript in icefish without any appreciable amount of protein suggests that MT deficiency in icefish is not the consequence of gene inactivation as in the case of hemoglobin genes (Cocca et al. 1995). Since icefish MT mRNA appears to be both stable and translationally inhibited, it is likely that these properties depend on the presence of regulatory elements in the 5′- and 3′-UTRs (Carginale et al. 1999) in a way similar to that found in mice (Nishimura et al. 1996; Vasconcelos et al. 2002).

Further studies unraveled the presence in the icefish liver of a second cDNA encoding a second MT isoform. According to the nomenclature adopted for mammal MT isoforms, the proteins obtained by translation of the two cDNAs were termed MT-I and MT-II. Although Northern blot analyses confirmed the presence in liver tissue of significant endogenous levels of MT mRNA, without a concomitant accumulation of MT protein, high levels of both MT mRNA and MT protein were detected in specimens of *C. hamatus* treated with $CdCl_2$ solution (Carginale et al. 1998). Quantification of the MT-I and MT-II transcripts from control and Cd-treated icefish showed an alteration in the ratio of the two MT isoform transcripts. Constitutively expressed transcripts consisted mostly of MT-II, whereas the MT-I transcript was preferentially expressed after cadmium induction (Carginale et al. 1998). These results, while confirming the existence of a discrepancy between MT mRNA and MT protein levels, showed that production of MT in icefish could be triggered by heavy metals, suggesting the existence of a double mechanism of regulation, one acting at a transcriptional level and another at a post-transcriptional level. In addition, the genes encoding MT-I and MT-II appear to be differentially regulated by cadmium.

## Tissue-specific expression of MT isoforms in notothenioids

Several lines of evidence indicate that MT-I and MT-II isoforms are differentially accumulated in mammal tissues as a function of the zinc status and age of the animals (Huber and Cousins 1993; Sato et al. 1994). In addition, isoform-specific MT induction and expression in different tissues have been demonstrated in mammals (Klaassen and Lehman-McKeeman 1989; Mididoddi et al. 1996; Pauwels et al. 1994).

Accumulation and expression of MT and MT mRNA have been investigated in different tissues of the red-blooded notothenioid *Notothenia coriiceps* (Scudiero et al. 2000). The expression levels of MT genes were assessed by Northern blotting using as hybridization probe the homologous MT cDNA. In order to compare MT levels with the amounts of transcripts found in different tissues, MT content was determined by silver saturation assay in brain, kidney and liver of *N. coriiceps*. A high MT content was found in brain (43 μg/g wet tissue), whilst the MT amounts measured in liver and kidney were 18 μg/g and 4.6 μg/g wet tissue, respectively. The lack of proportionality between MT content and MT mRNA levels resulted evident also in this case by comparing MT content with the amounts of the MT transcript in the three organs. Among the tissues examined, kidney had the highest transcript level (206 arbitrary units, a. u.) against 122 a. u. of brain and 62 a. u. of liver. Zinc and copper were present at comparable levels in all the tissues examined.

The relative abundance of the MT-I and MT-II transcripts in tissues was assessed by running PCR reactions in the presence of a radiolabeled nucleoside triphosphate using a primer matching exactly with the RNA of both isoforms. By exploiting the presence of a PVU II site in the MT-II cDNA, which was absent in the MT-I cDNA, it was possible to estimate the relative amount of the two MT isoforms, by measuring the radioactivity in the fragments generated by the PVU II cleavage. By comparing these values with the estimated amounts of total MT mRNA obtained by Northern blotting, it was possible to calculate the levels of MT-I and MT-II transcripts. The results of such analysis showed that the MT-I and MT-II transcripts are present at about the same levels in liver, whereas the MT-II transcript is more abundant in both brain and kidney, representing approx. 75% of the total MT mRNA (Scudiero et al. 2000). Both MT transcripts had the capacity to act as templates when assayed in an in vitro translation system (Scudiero et al. 2000).

## Regulation of MT expression in notothenioids

Expression of MT is controlled at the transcriptional level by several agents, including metals, hormones and free radicals (Andrews 2000; Imbert et al. 1990; Thiele 1992). Heavy metals are the most general and powerful of these inducers; their action is mediated by short *cis*-acting elements present in the MT gene promoter, termed metal responsive elements (MREs) (Culotta and Hamer 1989; Searle et al. 1987; Varshney et al. 1986). MREs have been shown to mediate transcriptional response of MT genes to zinc and cadmium through trans-acting binding proteins (MTF) (Auf der Maur et al. 1999; Brugnera et al. 1994; Dalton et al. 2000; Koizumi et al. 1999; Seguin and Prevost 1988). Recent evidence indicate that MTF1 functions as mediator of MT induction in conditions of physiological stress (Murphy et al. 2005). Several additional *cis*-acting elements have been identified in human MT genes, which are involved in the regulation of basal and induced expression of MT genes (Andrews 2000; Friedman and Stark 1985; Samson and Gedamu 1998).

In order to investigate the regulation of notothenioid MT encoding genes, about 1,000 bp of *C. hamatus* MT-I and MT-II gene promoters were cloned and sequenced (Scudiero et al. 2001). Analysis of these regions revealed that, unlike the regulatory regions of mammal MT genes, both promoters are rich in A–T content and lack a canonical TATA-box that in both cases was modified into a TTTA sequence. Apparently, this substitution results in a lowered metal inducibility. Several *cis*-regulatory sequences were identified (Fig. 1). The MT-I promoter contains four MREs organized into a single proximal cluster located within the first 300 bp from the ATG codon. Such organization, recently found also in zebrafish (Dalton et al. 2000; Yan and Chan 2002, 2004), is typical of mammal MT promoters, which do not generally contain distal MRE sequences (Imbert et al. 1990; Samson and Gedamu 1998). The first MRE is located from –116 bp to –123 bp upstream the transcription starting site. Two MREs are located with a reverse orientation from –140 bp to –147 bp and from –179 bp to –186 bp, respectively. Finally, the last MRE is located with the forward orientation from –256 bp to –263 bp. In addition, the MT-I promoter region of *C. hamatus* contains a transcription factor-1 site (Sp-1) located from –78 bp to –86 bp (Fig. 1).

The MT-II promoter region contains seven MREs organized into proximal and distal clusters, like in other fish (Chen et al. 2004). The three MREs of the proximal clusters are localized from –128 bp to –135 bp, from –151 bp to –158 bp and from –190 bp to –197 bp. The last two MREs have a reverse orientation in the 5′-flanking region. All the four MREs forming the distal cluster have a reverse orientation and correspond to the positions from –541 bp to –548 bp, from –574 bp to –581 bp, from –585 bp to –592 bp and from –923 bp to –930 bp. In addition, two sequences similar to the activator protein-1 (AP-1) were identified: one located from –27 bp to –35 bp upstream the ATG codon, and the other in the distal portion of the MT-II promoter sequence, from –633 bp to –640 bp (Fig. 1). A comparison between the MT-I and MT-II promoter regions shows that only the first three proximal MREs are conserved, both in position and sequences.

Deletion mutants were constructed by PCR using appropriate oligonucleotide primers. Functional analysis of MT-I and MT-II promoters was performed by introducing deletion mutants of the 5′-flanking regions into a vector plasmid, directly upstream the firefly luciferase reporter gene.

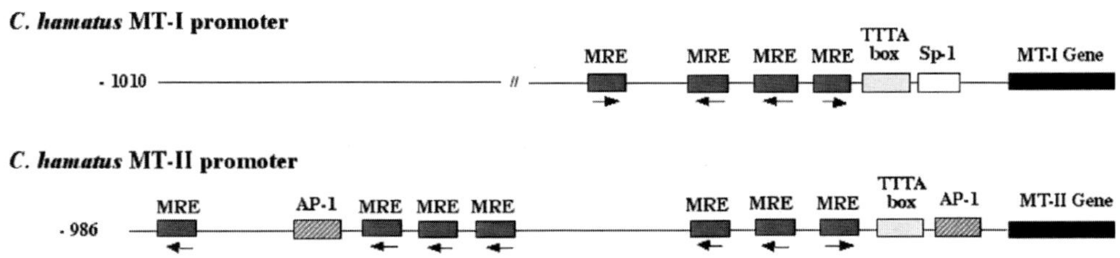


**Fig. 1** Schematic representation of MT-I and MT-II gene promoters of the notothenioid *Chionodraco hamatus*. The boxes represent specific transcription motifs. The right pointing arrows indicate forward and the left pointing arrows reverse orientation of the motifs

Each construct was then tested by transfection in HepG2 cells in the absence and presence of zinc or cadmium (Scudiero et al. 2001). The results indicated that the different number of MREs in MT-I and MT-II promoters is responsible of the different activity and inducibility by metals. The MT-II promoter with seven MREs shows high sensitivity to metals, maximal activity being elicited at 50 μM $Zn^{2+}$. Removal of the most distal MRE considerably modifies the sensitivity to zinc: in this case, $Zn^{2+}$ concentration must be raised to 150 μM to achieve a stimulation of approx. 50% of that observed with the complete MT-II promoter in the presence of 50 μM $Zn^{2+}$. Cells transfected with the promoter region containing only the proximal cluster display a remarkable decrease in promoter efficiency. Hence, the distally located MRE cluster in the MT-II promoter is required for maximal response to zinc. Apparently, the three MREs of the proximal cluster form a functional unit, as removal of a single MRE from the cluster results in a complete loss of both basal and metal-induced activity.

For the MT-I promoter, the presence of the distal 5′-flanking region reduces the activation of the reporter gene. Such a distal region contains no known transcription enhancer element, as the two motifs from –562 to –555 and from –932 to –925 differ from the MRE consensus sequence in a single base substitution that, however, is sufficient to inactivate MRE function (Samson and Gedamu 1998).

In conclusion, there is a striking difference between promoter sequences of MT isoforms in notothenioids, whereas the corresponding coding sequences display a high level of similarity. Therefore, the divergence of the MT isoforms proceeded independently of the functional modifications in the promoter regions. Likely, the differences found in the promoter regions are responsible for the differential expression of the MT isoforms described in the previous section. The higher sensitivity of the MT-II promoter to $Zn^{2+}$ might result in a preferential expression of MT-II isoform in tissues with a low cytoplasmic concentration of $Zn^{2+}$. The expression of the MT-I gene is expected to occur preferentially in tissues with a higher $Zn^{2+}$ level, or in response to an increased concentration of toxic heavy metals. Hence, the MT-I gene can be considered a "metal-shock gene".

## Evolution of Antarctic fish MT

As mentioned above, compared with multiplicity of mammalian MT, most fish apparently have a single MT gene expressed (Chan 1994; Kagi 1993) with the exception of the already cited cypriniforms (Knapen et al. 2005; Ren et al. 2000). A comparative study on MT in closely related fish was performed to shed some light on MT evolution. Sequence analysis of piscine MT contributed to answering the question of whether multiple MT genes were present in some teleost groups and how multiplicity arose. In order to address the above question, the pattern of MT evolution was investigated in Antarctic notothenioids.

Eight species of notothenioids were examined using RT-PCR to detect the presence of transcripts encoding distinct MT isoforms. The results showed the existence of two MT isoforms in each of the fish species examined. Phylogenetic analysis carried out by neighbor-joining analysis and maximum parsimony reconstruction indicated that at least two paralogous MT genes are present in notothenioids (Bargelloni et al. 1999). Apparently, this multiplicity originated after the divergence of the notothenioid ancestor from the other fish lineages examined. Hence, the gene duplication event that gave origin to the two paralogous MT genes in notothenioids is distinct from the duplication that gave origin to the MT isoforms in salmonids and cyprinids, probably as consequence of diploidization occurred in the last two lineages.

Phylogenetic analysis carried out in mammal and fish MT showed that the divergence of Antarctic MT begun about 10 MY ago (Capasso et al. 2003b), i.e., after the biogeographical segregation of the Antarctic continent. The phylogenetic divergence of fish and mammals is mirrored by marked physiological differences between these two groups. Mammals are typical homeotherms, whereas fish are poikilotherms, which equilibrate the body temperature with the environmental temperature, being generally incapable to retain heat.

Antarctic notothenioids are characterized by a marked stenothermy and their metabolism is adapted to cope with the extreme condition of the Antarctic continent. At the protein level, cold adaptation is achieved through an increased flexibility of the polypeptide chain (Marshall 1997). It turned out that the hydropathy of 24 MT from mammals and fish, including eight notothenioids, showed a highly significant regression when plotted against optimal organism temperature. The hydropathy of the ancestral sequences inferred at the nodes of the tree was used to estimate the corresponding temperature using the regression model. The results showed that the node characterizing mammal MT is temperate, whereas the nodes along the fish lineage are cold (Capasso et al. 2003b). Recent studies carried out using phylogenetic comparative methods indicate that the observed variations in MT hydropathy are due to phylogenetic contingency rather than adaptation (Scudiero et al. 2005).

## Structure, stability and conformational dynamics of notothenioid MT

A noteworthy observation derived from the phylogeny of notothenioid MTs concerned *Notothenia coriiceps*, in which the two MT isoforms appear to be both more similar to MT-II of other species owing to a gene conversion event (Bargelloni et al. 1999). In all our structural–functional studies, we have employed a recombinant MT prepared by cloning the *N. coriiceps* MT-II cDNA in *E. coli* cells.

In an early study carried out on the *N. coriiceps* MT, NMR spectroscopy unveiled a selective broadening of the heteronuclear spectra reflecting higher structural flexibility of this protein with respect to mammal MT (D'Auria et al. 2001). More precisely, two-dimensional [$^{1}$H, $^{113}$Cd]-correlation experiments on *N. coriiceps* MT showed a difference in intensity of the [$^{1}$H, $^{113}$Cd]-correlations between the α- and β-domain that is apparently higher than in the corresponding spectra of mammal MT. The fact that the observed broadening does not affect the homonuclear spectra suggest an exchange phenomenon involving the metal ions that is much more pronounced in fish MT than in mouse MT.

The full three-dimensional structure of *N. coriiceps* MT was obtained by combining $^{1}$H NMR experiments and heteronuclear [$^{1}$H, $^{113}$Cd]-correlation spectroscopy (Capasso et al. 2003a). Likewise other MT, the *N. coriiceps* MT is composed of a N-terminal β-domain with nine cysteines and three metal ions and a C-terminal α-domain with 11 cysteines and four metal ions (Fig. 2). The two domains are linked by two lysine residues. The position of the 9th cysteine of the α-domain of fish MT is different from the corresponding cysteine residue of mammal MT.

Previous studies carried out by infrared spectroscopy had revealed the presence in the *N. coriiceps* MT of a band, which for its typical position in the spectrum and for its sensitivity to temperature, was assigned to α-helices whose content was about 5% of the total secondary structure of the protein (Capasso et al. 2002). In keeping with these results, NMR spectroscopy data unraveled the following secondary structure elements: one α-helical region and one $3_{10}$ helix in nearly all structures of the bundle of the α-domain, and one α-helical stretch in most structures of the β-domain (Capasso et al. 2003a).

The lower average number of nuclear Overhauser effects (NOEs) per residue in the β-domain of the *N. coriiceps* MT results in a lower precision of the structure determination for the backbone when compared to the α-domain (Table 1), suggesting a higher flexibility of the β-domain with respect to the α-domain. The architecture of both domains is determined by the Cd–Cys clusters, as is the case for all MT. *N. coriiceps* MT, however, like other fish MT, has a significant sequence difference involving Cys54 that is not aligned with the corresponding Cys of mammal MT (Capasso et al. 2003a). Such a difference has important consequences on the structure of the α-domain: the main consequence is the formation of a differently oriented loop in *N. coriiceps* MT with respect to the corresponding loop of mouse MT-I. In *N. coriiceps* MT, this loop is rotated downward, opening a wide channel. The different orientation of the loops is accompanied by notably different arrangement of the charged residues on the surface. The sequence of the α-domain of Notothenia MT has the same number of basic and acidic residues as the mouse

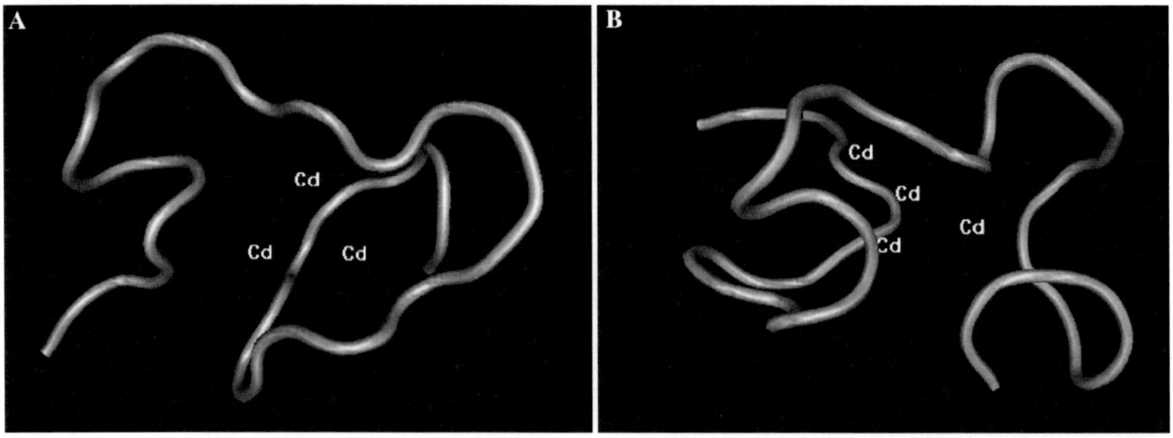


**Fig. 2** Backbone structure of the *N. coriiceps* metallothionein obtained by NMR. (**A**) Backbone heavy atoms (N, $C^{\alpha}$ and C′) of residues 2–28 of the energy-minimized structure of the $\beta$-domain. (**B**) Backbone heavy atoms (N, $C^{\alpha}$ and C′) of residues 31–60 of the energy-minimized structure of the $\alpha$-domain. The metal ions are indicated by the element symbol

**Table 1** Restraints and structural statistics

| | $\beta$-Domain (4–30) | $\alpha$-Domain (31–60) |
|---|---|---|
| *Restraints* | | |
| NOEs | | |
| Intraresidue | 137 | 159 |
| Sequential | 94 | 123 |
| Medium range ($i < 5$) | 20 | 60 |
| Long range | 2 | 31 |
| Dihedral angles | 63 | 50 |
| Total NOEs | 253 | 373 |
| *Precision*[a] | | |
| Backbone heavy atoms (Å) | 1.61 ± 0.06 | 0.44 ± 0.7 |
| $S^{\gamma}$ and Cd atoms (Å) | 0.33 ± 0.1 | 0.21 ± 0.1 |
| All heavy atoms (Å) | 2.03 ± 0.12 | 0.71 ± 0.15 |

[a]Average coordinates of the 20 energy-minimized conformers after superposition for the best fit of the atoms of the residues indicated in parentheses

MT, but their distribution on the surface is different mainly as a consequence of the misaligned Cys54. The charge distribution on the surface of the $\alpha$-domain is rather different in fish and mouse: the negative charges in the fish domain are concentrated around the mouth of the channel described above. On the other hand, the different distribution of charge in the fish $\beta$-domain is expected, because of the presence of an additional acidic residue with respect to the corresponding mouse domain (Capasso et al. 2003a).

Experiments carried out with the aid of ultraviolet spectroscopy and circular dichroism show that the optical properties of fish and mouse MT are differently sensitive to temperature. The shape of the absorption curve varies with the temperature, but the change in absorbance at 254 nm is more pronounced for *N. coriiceps* MT than for its mammalian counterpart (Table 2). Such an effect can be attributed to changes in the dissociation constant of the metal–thiolate complex. The circular dichroism of heated MT is also drastically modified (Table 3), with the progressive quenching of the wide positive band near 260 nm when the temperature is increased from 25°C to 95°C (D’Auria et al. 2001).

A number of studies suggest that the chiroptical features of MT, responsible of the conspicuous positive ellipticity at 260 nm, arise from the ligand-metal charge-transfer transition of the metal–thiolate complex (Rupp and Weser 1978). This optical activity has been attributed to the interaction of dissymmetrical co-ordinated chromophores at the level of the clusters (Willner et al. 1987). Usually, lysine residues are highly conserved in vertebrate MT, but their distribution along the sequence varies in different species. Recent evidence, from circular dichroism and NMR spectroscopy studies on a mutated recombinant MT, demonstrates that substitution of glutamate residues for lysines in three CK motifs in the α-domain resulted in a change of metal–thiolate interactions in both domains (Pan et al. 1994). As a consequence, the different temperature-induced modifications of the optical properties observed in fish and mammal MT may very well be due to the different number of lysines in their sequences.

If the different effect of temperature on Notothenia and mouse MT suggests a difference in the stability of metal–thiolate complex in the two proteins, one should also expect differences in metal mobility between fish and mouse MT. The results of $Zn^{2+}/Cd^{2+}$ exchange experiments show that Notothenia MT displays a better metal exchange capability with respect to mouse MT, as demonstrated by a higher number of $Cd^{2+}$ ions substituting for $Zn^{2+}$ in Notothenia MT. Such an exchange is probably favored by the net negative charge present in fish MT that lowers the stability of the $Zn^{2+}$–thiolate complex in the [$Zn^{2+}{}_7$]-MT (Capasso et al. 2003a).

To investigate the conformational dynamics of mouse and fish MT, the fluorescence intensity decays of the two proteins in the presence of anilino-naphthalene-sulfonate (ANS) were measured by using the frequency-domain method (Lakowicz et al. 1986). The fluorescence data showed that the Notothenia MT possesses a more flexible structure and that a temperature increase results in an increased flexibility of the protein structure (Capasso et al. 2002). As the ANS molecules bind to the hydrophobic sites along the whole protein chain, the fluorescence data indicate that the effect of temperature is not restricted to the metal–thiolate complex only, but interests the overall structure of the protein.

## Interdomain cross-talk in fish and mammal MT

Previous studies carried out on individual domains of human MT showed that the α- and β-domain differ markedly in terms of chemical reactivity and metal binding capacity (Jiang et al. 2000; Salgado and Stillman 2004). In the frame of the comparative studies on fish and mammal, physically isolated MT domains were investigated to establish their influence on the physico-chemical and functional differences observed in the whole MTs. For such a purpose, recombinant α- and β-domains of both fish and mouse MT were prepared by heterologous expression in *E. coli* cells (Capasso et al. 2005). NMR spectra of *N. coriiceps* MT domains appeared to be almost superimposable on those of the parent MT, suggesting an apparent lack of interaction between the two domains in the protein. However, certain dynamic and physico-chemical features of the isolated domains are unlike those of MT. The temperature induced changes in the spectral properties of the isolated domains of fish and mouse MT and of the parent MT are summarized in Tables 2 and 3. It is evident that the spectral properties of both fish and mouse

**Table 2** Temperature-induced changes in UV spectra in fish and mouse MTs and their individual domains

| Absorbance at 254 nm | | | | | | |
|---|---|---|---|---|---|---|
| Molecule | Fish | | | Mouse | | |
| | 25°C | 90°C | Δ (%) | 25°C | 90°C | Δ (%) |
| Metallothionein | 0.480 ± 0.0006 | 0.440 ± 0.0005 | 8.3 | 0.380 ± 0.0004 | 0.372 ± 0.0004 | 2.1 |
| α-domain | 0.530 ± 0.0006 | 0.430 ± 0.0004 | 18.9 | 0.530 ± 0.0005 | 0.500 ± 0.0006 | 5.7 |
| β-domain | 0.332 ± 0.0003 | 0.295 ± 0.0003 | 11.1 | 0.330 ± 0.0004 | 0.295 ± 0.0002 | 10.6 |

**Table 3** Temperature-induced changes in CD spectra in fish and mouse MTs and their individual domains

| Molar ellipticity at 260 nm | | | | | | |
|---|---|---|---|---|---|---|
| Molecule | Fish | | | Mouse | | |
| | 25°C | 90°C | Δ (%) | 25°C | 90°C | Δ (%) |
| Metallothionein | 130,000 ± 156 | 70,000 ± 90 | 46.1 | 130,000 ± 145 | 90,000 ± 110 | 30.7 |
| α-domain | 90,000 ± 115 | 65,000 ± 80 | 27.8 | 95,000 ± 115 | 65,000 ± 70 | 31.6 |
| β-domain | 21,500 ± 30 | n.s.[a] | – | 15,000 ± 20 | n.s.[a] | – |

[a]Not stable

α-domains are affected by temperature to a larger extent than the whole MT molecules are (Table 2). The spectral profiles of both α-domains are restored when the temperature is brought back to 25°C. The CD spectra of the β-domains display a remarkable instability with the temperature: the Cotton effect at 260 nm is irreversible lost at 50°C. Interestingly, the circular dichroism spectrum of the fish α-domain resembles that of MT: the molar ellipticity at 260 nm of the α-domain accounts for approx. 80% of the molar ellipticity of the whole molecule. The chiroptical properties of the β-domain are much less pronounced, and molar ellipticty contributes only 20% to the total. Nevertheless, ellipticity of the parent MT turns out to be approx. the sum of the contribution of the two domains (Fig. 3).

The thiol reactivity of the isolated domains with dithionitrobenzoic acid (DTNB) reflects the behavior of the whole MT. The thiol reactivity of the α-domain is higher for mouse MT than for fish MT, whereas the β-domain is equally reactive in both MTs (Capasso et al. 2005). As mouse MT is more reactive than its piscine counterpart, the contribution of the α-domain appears determinant. As reported above, the number of metal equivalents exchanged per mole of protein is higher in fish MT than in mouse MT (Capasso et al. 2003a). Conversely, the final number of metal equivalents exchanged is the same for both fish and mouse α- and β-domains, but the rate of exchange is different for the two domains: indeed, in the β-domain of fish MT, the exchange occurs at a higher rate than in the corresponding mouse domain, providing evidence of a stabilizing effect on the metal–thiol interactions exerted by the α-domain and the destabilizing effect of the β-domain (Capasso et al. 2005).

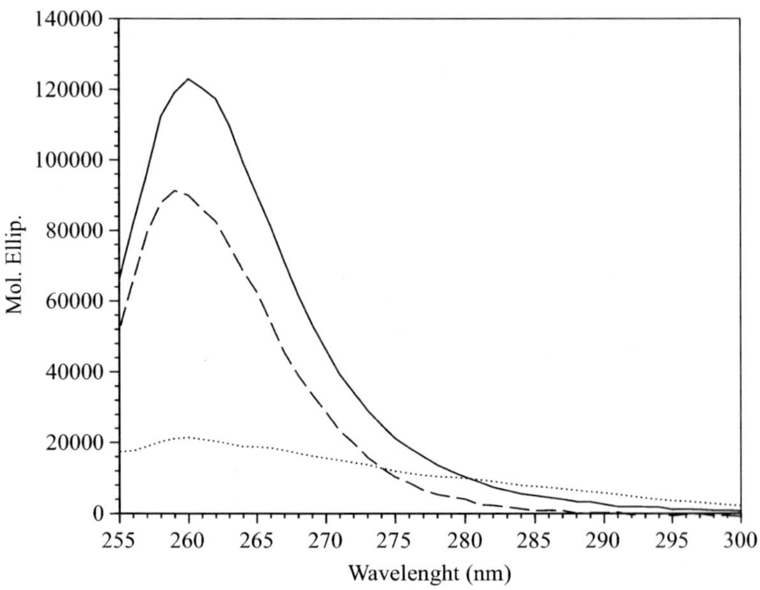


**Fig. 3** Circular dichroism spectra of *N. coriiceps* MT and of the two isolated domains. The solid line refers to the parent MT molecule; dashed and dotted lines to the α- and β-domain, respectively

All these results provide evidence in favor of the interaction of the two domains in the MT molecule, in spite of the elusive evidence provided by structural studies. The apparent lack of NOEs between pairs of protons belonging to different domains can be explained by the fact that they are so weak as to be virtually undetectable amongst the stronger intraresidue NOEs. In addition, NMR data reflect time scales of only micro- to nano-seconds whereas functional data depend on events taking place in shorter times.

## Conclusions

For many years, MTs have been studied in relation to their putative roles in heavy metal toxicology, but with time it became more clear their implication in various kinds of functions besides heavy metal detoxification. The ubiquitous distribution of this protein in a large number of organisms found an explanation in the protective function played by MTs against a variety of stress factors and in their role in mineral metabolism. Evidence exist that mice carrying null alleles for both MT-I and MT-II suffer from zinc deficiency and zinc toxicity (Kelly et al. 1996) and are more sensitive to acute cadmium intoxication (Liu et al. 2002).

Although the lack of MT in icefish with a concomitant expression of non-translated mRNA was at a first sight intriguing, the induction of MT synthesis by cadmium indicates that MT production is not impaired in this organism, but only temporarily suppressed. Presumably, owing to a reduced oxygenation dictated by the lack of an oxygen carrier in icefish, the metabolism is slowed down in this species and production of free radicals occurs at a lower rate. In addition, the Antarctic environment with its low pollution level is expected to have little or no effect on MT induction by heavy metals. That MTs play an important role in Antarctic fish is demonstrated by the presence of a tissue-specific expression of the two isoforms found in all the notothenioid species examined. The different organization of the promoter regions of the MT genes may account for the different expression pattern. Apparently the genes encoding the two MT isoforms evolved by at least one gene duplication event that occurred along the notothenioid lineage.

The data summarized in the present review show that the chemical, structural and dynamic features of *N. coriiceps* MT differ markedly from those of its mammalian counterpart. Although the presence of a more flexible MT in poikilotherms and of the more rigid protein in homeotherms suggests an adaptive scenario, the clustering of MT according to their flexibility appears to be the consequence of their descent from a common ancestry (Scudiero et al. 2005). The different flexibility distinguishing fish and mammal MT is linked to the evolution of the α-domain: such a feature is probably the consequence of the different position of a cysteine residue in the α-domain and to the different electrostatic charge present on the two proteins (Capasso et al. 2003a).

The structural similarity of the α- and β-domains in the MT of different origin suggests that the MT gene evolved by duplication of an ancestral gene coding for a polypeptide folding like one of the two domains. This is particularly evident by looking at the alignment of the α- and β-domains of mammal MT-I shown in Fig. 4, in which about half of the amino acid residues, including seven cysteines out of 12, appear to be conserved. Moreover, the consensus motif CXCXXXCXCX located near the center of the MT molecules in invertebrates and in the β-domain of vertebrate MTs, is also present in the α-domain under the form of the variant CXXCXXXCXCX. According to Nemer and coworkers (Nemer et al. 1985), this motif "may be considered the core upon which a diversity of types of MTs has been constructed".

The evolution of an MT made of two different domains may be relevant to explain the functional differences of the two domains: the more stable α-domain is apparently involved in metal detoxification, while the more labile β-domain is more likely responsible of zinc homeostasis (Zangger et al. 2001). The functional specialization of the two domains is particularly evident in MT-III. MT-III is expressed exclusively in mammal brain and is featured by a marked inhibitory activity of the neuronal growth (Uchida et al. 2002). Apparently, the MT-III inhibitory activity is

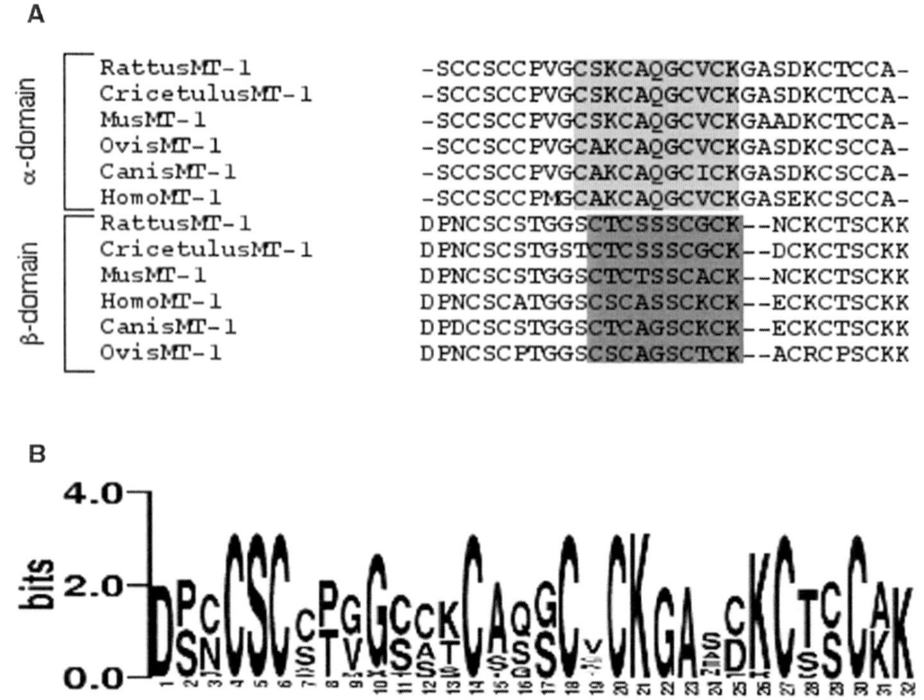


**Fig. 4** (**A**) Alignment of the amino acid sequences of α- and β-domains of mammal MT-I. The central consensus motif is shaded in gray (β-domains) and in light-gray (α-domain). (**B**) Logo representation of the aligned α- and β-domains

linked to the β-domain because the insertion of critical residues present in the β-domain of MT-III into MT-I elicits the inhibitory response in the latter (Romero-Isart et al. 2002).

From what has been reported above, it turns out that after four decades of intensive studies, metallothioneins still represent an interesting field for future investigations, as many aspects on the role of these proteins remain to be clarified. A crucial point deserving further investigation is to establish whether different MTs have the same function, and why some organisms need multiple forms of MT and other organisms (like birds) have only one MT gene. Another outstanding question concerns the dynamic features of the MT molecule that allow the cooperation of the two distinct domains in spite of any apparent physical interaction.

**Acknowledgements** The work reported in the present review was carried out thanks to the support provided by the "Programma Nazionale per le Ricerche in Antartide" (PNRA).

## References

Andrews GK (2000) Regulation of metallothionein gene expression by oxidative stress and metal ions. Biochem Pharmacol 59:95–104

Auf der Maur A, Belser T, Elgar G, Georgiev O, Schaffner W (1999) Characterization of the transcription factor MTF-1 from the Japanese pufferfish (*Fugu rubripes*) reveals evolutionary conservation of heavy metal stress response. Biol Chem 380:175–185

Bargelloni L, Ritchie PA, Patarnello T, Battaglia B, Lambert DM, Meyer A (1994) Molecular evolution at subzero temperatures: mitochondrial and nuclear phylogenies of fishes from Antarctica (suborder Notothenioidei), and the evolution of antifreeze glycopeptides. Mol Biol Evol 11:854–863

Bargelloni L, Scudiero R, Parisi E, Carginale V, Capasso C, Patarnello T (1999) Metallothioneins in antarctic fish: evidence for independent duplication and gene conversion. Mol Biol Evol 16:885–897

Blaise C, Gagne F, Pellerin J, Hansen PD, Trottier S (2002) Molluscan shellfish biomarker study of the Quebec, Canada, Saguenay Fjord with the soft-shell clam, Mya arenaria. Environ Toxicol 17:170–186

Bonham K, Gedamu L (1984) Induction of metallothionein and metallothionein mRNA in rainbow-trout liver following cadmium treatment. Biosci Rep 4:633–642

Brugnera E, Georgiev O, Radtke F, Heuchel R, Baker E, Sutherland GR, Schaffner W (1994) Cloning, chromosomal mapping and characterization of the human metal-regulatory transcription factor MTF-1. Nucleic Acids Res 22:3167–3173

Cai L, Satoh M, Tohyama C, Cherian MG (1999) Metallothionein in radiation exposure: its induction and protective role. Toxicology 132:85–98

Cajaraville MP, Bebianno MJ, Blasco J, Porte C, Sarasquete C, Viarengo A (2000) The use of biomarkers to assess the impact of pollution in coastal environments of the Iberian Peninsula: a practical approach. Sci Total Environ 247:295–311

Capasso C, Abugo O, Tanfani F, Scire A, Carginale V, Scudiero R, Parisi E, D'Auria S (2002) Stability and conformational dynamics of metallothioneins from the antarctic fish *Notothenia coriiceps* and mouse. Proteins 46:259–267

Capasso C, Carginale V, Crescenzi O, Di Maro D, Parisi E, Spadaccini R, Temussi PA (2003a) Solution structure of MT_nc, a novel metallothionein from the Antarctic fish *Notothenia coriiceps*. Structure 11:435–443

Capasso C, Carginale V, Crescenzi O, Di Maro D, Spadaccini R, Temussi PA, Parisi E (2005) Structural and functional studies of vertebrate metallothioneins: cross-talk between domains in the absence of physical contact. Biochem J 391:95–103

Capasso C, Carginale V, Scudiero R, Crescenzi O, Spadaccini R, Temussi PA, Parisi E (2003b) Phylogenetic divergence of fish and mammalian metallothionein: relationships with structural diversification and organismal temperature. J Mol Evol 57(Suppl 1):S250–257

Carginale V, Scudiero R, Capasso A, Capasso C, Passaretti G, di Prisco G, Kille P, Parisi E (1999) Accumulation of untranslated metallothionein mRNA in Antarctic hemoglobinless fish (icefish). In: Klaassen C (ed) Metallothionein IV. Birkhauser Verlag, Basel, pp 167–172

Carginale V, Scudiero R, Capasso C, Capasso A, Kille P, di Prisco G, Parisi E (1998) Cadmium-induced differential accumulation of metallothionein isoforms in the Antarctic icefish, which exhibits no basal metallothionein protein but high endogenous mRNA levels. Biochem J 332(Pt 2):475–481

Chan KM (1994) PCR-cloning of goldfish and tilapia metallothionein complementary DNAs. Biochem Biophys Res Commun 205:368–374

Chen WY, John JA, Lin CH, Lin HF, Wu SC, Lin CH, Chang CY (2004) Expression of metallothionein gene during embryonic and early larval development in zebrafish. Aquat Toxicol 69:215–227

Cocca E, Ratnayake-Lecamwasam M, Parker SK, Camardella L, Ciaramella M, di Prisco G, Detrich 3rd HW (1995) Genomic remnants of alpha-globin genes in the hemoglobinless antarctic icefishes. Proc Natl Acad Sci USA 92:1817–1821

Culotta VC, Hamer DH (1989) Fine mapping of a mouse metallothionein gene metal response element. Mol Cell Biol 9:1376–1380

D'Auria S, Carginale V, Scudiero R, Crescenzi O, Di Maro D, Temussi PA, Parisi E, Capasso C (2001) Structural characterization and thermal stability of *Notothenia coriiceps* metallothionein. Biochem J 354:291–299

Dallinger R, Berger B, Gruber C, Hunziker P, Sturzenbaum S (2000) Metallothioneins in terrestrial invertebrates: structural aspects, biological significance and implications for their use as biomarkers. Cell Mol Biol (Noisy-le-grand) 46:331–346

Dallinger R, Lagg B, Egg M, Schipflinger R, Chabicovsky M (2004) Cd accumulation and Cd-metallothionein as a biomarker in *Cepaea hortensis* (Helicidae, Pulmonata) from laboratory exposure and metal-polluted habitats. Ecotoxicology 13:757–772

Dalton TP, Solis WA, Nebert DW, Carvan 3rd MJ (2000) Characterization of the MTF-1 transcription factor from zebrafish and trout cells. Comp Biochem Physiol B Biochem Mol Biol 126:325–335

Dondero F, Piacentini L, Banni M, Rebelo M, Burlando B, Viarengo A (2005) Quantitative PCR analysis of two molluscan metallothionein genes unveils differential expression and regulation. Gene 345:259–270

Eastman JT (1993) Antarctic fish biology. Evolution in a unique environment. Academic Press Inc., San Diego

Friedman RL, Stark GR (1985) Alpha-interferon-induced transcription of HLA and metallothionein genes containing homologous upstream sequences. Nature 314:637–639

Huber KL, Cousins RJ (1993) Metallothionein expression in rat bone marrow is dependent on dietary zinc but not dependent on interleukin-1 or interleukin-6. J Nutr 123:642–648

Imbert J, Culotta V, Furst P, Gedamu L, Hamer D (1990) Regulation of metallothionein gene transcription by metals. Adv Inorg Biochem 8:139–164

Jiang LJ, Vasak M, Vallee BL, Maret W (2000) Zinc transfer potentials of the alpha- and beta-clusters of metallothionein are affected by domain interactions in the whole molecule. Proc Natl Acad Sci USA 97:2503–2508

Kagi JHR (1993) Evolution, structure and chemical activity of class I metallothioneins: an overview. In: Suzuki KT, Imura N, Kimura M (eds) Metallothionein III. Birkhauser Verlag, Basel, pp 29–55

Karin M (1985) Metallothioneins: proteins in search of function. Cell 41:9–10

Kelly EJ, Quaife CJ, Froelick GJ, Palmiter RD (1996) Metallothionein I and II protect against zinc deficiency and zinc toxicity in mice. J Nutr 126:1782–1790

Klaassen CD, Lehman-McKeeman LD (1989) Regulation of the isoforms of metallothionein. Biol Trace Elem Res 21:119–129

Knapen D, Redeker ES, Inacio I, De Coen W, Verheyen E, Blust R (2005) New metallothionein mRNAs in *Gobio gobio* reveal at least three gene duplication events in cyprinid metallothionein evolution. Comp Biochem Physiol C Toxicol Pharmacol 140:347–355

Koizumi S, Suzuki K, Ogra Y, Yamada H, Otsuka F (1999) Transcriptional activity and regulatory protein binding of metal-responsive elements of the human metallothionein-IIA gene. Eur J Biochem 259:635–642

Kondoh M, Inoue Y, Atagi S, Futakawa N, Higashimoto M, Sato M (2001) Specific induction of metallothionein synthesis by mitochondrial oxidative stress. Life Sci 69:2137–2146

Kumari MV, Hiramatsu M, Ebadi M (2000) Free radical scavenging actions of hippocampal metallothionein isoforms and of antimetallothioneins: an electron spin resonance spectroscopic study. Cell Mol Biol (Noisy-le-grand) 46:627–636

Lakowicz JR, Laczko G, Gryczynski I, Cherek H (1986) Measurement of subnanosecond anisotropy decays of protein fluorescence using frequency-domain fluorometry. J Biol Chem 261:2240–2245

Liu J, Kadiiska MB, Corton JC, Qu W, Waalkes MP, Mason RP, Liu Y, Klaassen CD (2002) Acute cadmium exposure induces stress-related gene expression in wild-type and metallothionein-I/II-null mice. Free Radic Biol Med 32:525–535

Marshall CJ (1997) Cold-adapted enzymes. Trends Biotechnol 15:359–364

Mididoddi S, McGuirt JP, Sens MA, Todd JH, Sens DA (1996) Isoform-specific expression of metallothionein mRNA in the developing and adult human kidney. Toxicol Lett 85:17–27

Munoz A, Petering DH, Shaw 3rd CF (2000a) Structure-reactivity relationships among metallothionein three-metal domains: role of non-cysteine amino acid residues in lobster metallothionein and human metallothionein-3. Inorg Chem 39:6114–6123

Munoz A, Petering DH, Shaw 3rd CF (2000b) The requirements for stable metallothionein clusters examined using synthetic lobster domains. Mar Environ Res 50:93–97

Murphy BJ, Sato BG, Dalton TP, Laderoute KR (2005) The metal-responsive transcription factor-1 contributes to HIF-1 activation during hypoxic stress. Biochem Biophys Res Commun 337:860–867

Narula SS, Brouwer M, Hua Y, Armitage IM (1995) Three-dimensional solution structure of *Callinectes sapidus* metallothionein-1 determined by homonuclear and heteronuclear magnetic resonance spectroscopy. Biochemistry 34:620–631

Nemer M, Wilkinson DG, Travaglini EC, Sternberg EJ (1985) Sea urchin metallothionein sequence: key to an evolutionary diversity. Proc Natl Acad Sci USA 82:4992–4994

Nishimura N, Cam GR, Nishimura H, Tohyama C, Saitoh Y, Adelson DL (1996) Evidence for developmentally regulated transcriptional, translational and post-translational control of metallothionein gene expression in hair follicles. Reprod Fertil Dev 8:1089–1096

Pan PK, Hou FY, Cody CW, Huang PC (1994) Substitution of glutamic acids for the conserved lysines in the alpha domain affects metal binding in both the alpha and beta domains of mammalian metallothionein. Biochem Biophys Res Commun 202:621–628

Pauwels M, van Weyenbergh J, Soumillion A, Proost P, De Ley M (1994) Induction by zinc of specific metallothionein isoforms in human monocytes. Eur J Biochem 220:105–110

Ren HW, Itoh N, Kanekiyo M, Tominaga S, Kohroki J, Hwang GS, Nakanishi T, Muto N, Tanaka K (2000) Two metallothioneins in the fresh-water fish, crucian carp (*Carassius cuvieri*): cDNA cloning and assignment of their expression isoforms. Biol Pharm Bull 23:145–148

Riggio M, Filosa S, Parisi E, Scudiero R (2003) Changes in zinc, copper and metallothionein contents during oocyte growth and early development of the teleost *Danio rerio* (zebrafish). Comp Biochem Physiol C Toxicol Pharmacol 135:191–196

Robbins AH, Stout CD (1991) X-ray structure of metallothionein. Methods Enzymol 205:485–502

Rodriguez-Ortega MJ, Alhama J, Funes V, Romero-Ruiz A, Rodriguez-Ariza A, Lopez-Barea J (2002) Biochemical biomarkers of pollution in the clam *Chamaelea gallina* from south-Spanish littoral. Environ Toxicol Chem 21:542–549

Romero-Isart N, Jensen LT, Zerbe O, Winge DR, Vasak M (2002) Engineering of metallothionein-3 neuroinhibitory activity into the inactive isoform metallothionein-1. J Biol Chem 277:37023–37028

Rosenthal MD, Albrecht ED, Pepe GJ (2004) Estrogen modulates developmentally regulated gene expression in the fetal baboon liver. Endocrine 23:219–228

Rupp H, Weser U (1978) Circular dichroism of metallothioneins. A structural approach. Biochim Biophys Acta 533:209–226

Salgado MT, Stillman MJ (2004) $Cu^+$ distribution in metallothionein fragments. Biochem Biophys Res Commun 318:73–80

Samson SL, Gedamu L (1998) Molecular analyses of metallothionein gene regulation. Prog Nucleic Acid Res Mol Biol 59:257–288

Sato M, Sasaki M, Hojo H (1994) Differential induction of metallothionein synthesis by interleukin-6 and tumor necrosis factor-alpha in rat tissues. Int J Immunopharmacol 16:187–195

Scudiero R, Carginale V, Capasso C, Riggio M, Filosa S, Parisi E (2001) Structural and functional analysis of metal regulatory elements in the promoter region of genes encoding metallothionein isoforms in the Antarctic fish *Chionodraco hamatus* (icefish). Gene 274:199–208

Scudiero R, Carginale V, Riggio M, Capasso C, Capasso A, Kille P, di Prisco G, Parisi E (1997) Difference in hepatic metallothionein content in Antarctic red-blooded and haemoglobinless fish: undetectable metallothionein levels in haemoglobinless fish is accompanied by accumulation of untranslated metallothionein mRNA. Biochem J 322(Pt 1):207–211

Scudiero R, Paolo De Prisco P, Camardella L, D'Avino R, di Prisco G, Parisi E (1992) Apparent deficiency of metallothionein in the liver of the Antarctic icefish *Chionodraco hamatus*. Identification and isolation of a zinc-containing protein unlike metallothionein. Comp Biochem Physiol B 103:201–207

Scudiero R, Temussi PA, Parisi E (2005) Fish and mammalian metallothioneins: a comparative study. Gene 345:21–26

Scudiero R, Verde C, Carginale V, Kille P, Capasso C, di Prisco G, Parisi E (2000) Tissue-specific regulation of metallothionin and metallothionein mRNA accumulation in the Antarctic notothenioid, *Notothenia coriiceps*. Polar Biol 23:17–23

Searle PF, Stuart GW, Palmiter RD (1987) Metal regulatory elements of the mouse metallothionein-I gene. Exp Suppl 52:407–414

Seguin C, Prevost J (1988) Detection of a nuclear protein that interacts with a metal regulatory element of the mouse metallothionein 1 gene. Nucleic Acids Res 16:10547–10560

Thiele DJ (1992) Metal-regulated transcription in eukaryotes. Nucleic Acids Res 20:1183–1191

Uchida Y, Gomi F, Masumizu T, Miura Y (2002) Growth inhibitory factor prevents neurite extension and the death of cortical neurons caused by high oxygen exposure through hydroxyl radical scavenging. J Biol Chem 277:32353–32359

Varshney U, Jahroudi N, Foster R, Gedamu L (1986) Structure, organization, and regulation of human metallothionein IF gene: differential and cell-type-specific expression in response to heavy metals and glucocorticoids. Mol Cell Biol 6:26–37

Vasconcelos MH, Tam SC, Hesketh JE, Reid M, Beattie JH (2002) Metal- and tissue-dependent relationship between metallothionein mRNA and protein. Toxicol Appl Pharmacol 182:91–97

Wang Y, Mackay EA, Kurasaki M, Kagi JH (1994) Purification and characterisation of recombinant sea urchin metallothionein expressed in *Escherichia coli*. Eur J Biochem 225:449–457

Wang Y, Mackay EA, Zerbe O, Hess D, Hunziker PE, Vasak M, Kagi JH (1995) Characterization and sequential localization of the metal clusters in sea urchin metallothionein. Biochemistry 34:7460–7467

Willner H, Vasak M, Kagi JH (1987) Cadmium–thiolate clusters in metallothionein: spectrophotometric and spectropolarimetric features. Biochemistry 26:6287–6292

Wright J, George S, Martinez-Lara E, Carpene E, Kindt M (2000) Levels of cellular glutathione and metallothionein affect the toxicity of oxidative stressors in an established carp cell line. Mar Environ Res 50:503–508

Yan CH, Chan KM (2002) Characterization of zebrafish metallothionein gene promoter in a zebrafish caudal fin cell-line, SJD. 1. Mar Environ Res 54:335–339

Yan CH, Chan KM (2004) Cloning of zebrafish metallothionein gene and characterization of its gene promoter region in HepG2 cell line. Biochim Biophys Acta 1679:47–58

You HJ, Lee KJ, Jeong HG (2002) Overexpression of human metallothionein-III prevents hydrogen peroxide- induced oxidative stress in human fibroblasts. FEBS Lett 521:175–179

Zangger K, Shen G, Oz G, Otvos JD, Armitage IM (2001) Oxidative dimerization in metallothionein is a result of intermolecular disulphide bonds between cysteines in the alpha-domain. Biochem J 359:353–360

Rev Environ Sci Biotechnol (2007) 6:181–195
DOI 10.1007/s11157-006-9111-2

REVIEW

# Pressure and life: some biological strategies

**Florence Pradillon · Françoise Gaill**

Received: 14 March 2006 / Accepted: 18 September 2006 / Published online: 18 October 2006


**Abstract** All biological processes of life on Earth experience varying degrees of pressure. Aquatic organisms living in the deep-sea, as well as chondrocytic cells of articular cartilage are exposed to hydrostatic pressures that rise up to several hundred times that of atmospheric pressure. In the case of marine larvae that disperse through the oceanic water column, pressure changes might be responsible for stress conditions during development, limiting colonisation capabilities. In a number of biological systems, life strategies may be significantly influenced by pressure. In this review, we will focus on the consequences of pressure changes on various biological processes, and more specifically on animals living in the deep-sea. Revisiting general principles of pressure effects on biological systems, we present recent data illustrating the diversity of effects pressure may have at different levels in biological systems, with particular attention to effects on gene expression. After a review of the main pressure equipments available today for studying species living naturally at high pressure, we summarise what is known concerning pressure impact during animal development.

F. Pradillon (✉) · F. Gaill
CNRS UMR 7138 Systématique, Adaptation et Evolution, Université Pierre et Marie Curie, 7 Quai Saint Bernard, 75252 Paris cedex 05, France
e-mail: florence.pradillon@snv.jussieu.fr

**Keywords** Hydrostatic pressure · Deep-sea organisms · Hydrothermal vents · Developmental processes · Adaptation · Stress response

## 1 Introduction

One of the most widespread extreme environments that can be found on Earth includes almost 75% of the marine biosphere, and is located below 1000 m depth (Somero 1992; MacDonald 1997). When ‘deep’ means deeper than 1000 m, these biotopes include more than half ($\approx$ 62%) of the volume of the global biosphere (Jannasch and Taylor 1984). The definition of an ‘extreme environment’ is still debated and no consensus has been achieved today (Rothschild and Mancinelli 2001). One could argue that an extreme environment is characterised by values of physical geochemical or biological parameters that are challenging for organisms survival (temperature, pressure, pH, presence of predators). Such definition is relative to the considered organism. As such, deep-sea animals which are adapted to high pressure values will view the ocean surface with its atmospheric pressure as an extreme environment.

In the marine environment, pressure is a natural parameter, which may play a role in adaptation processes. It forms an absolute continuous gradient from the surface to the deep oceanic trough (11,000 m), increasing by 0.1 atmospheres (atm)[1] every 10 m. Aquatic environments thus exhibit a wide range of hydrostatic pressure from micro-pressure generated by a few centimetres of water column up to 1100 atm in oceanic troughs. Almost all marine organisms are distributed between a high and a low depth limit. The size of this range varies according to species, from a few tens of metres up to several hundreds of metres (Tyler and Young 1998). In some species, the distribution depth range also varies according to the ontogenic stage. Although organisms vertical distributions are probably shaped by several physical and biological factors, pressure might significantly influence larvae colonisation patterns in the deep-sea, and later adult distributions (Tyler 1995; Young et al. 1996; MacDonald 1997; Tyler and Young 1998).

Pressure effects are also of interest in the biomedical science field since they are responsible for a number of pathologies. In the human hip joint, pressure of 100–200 atm have been recorded (Hall et al. 1993). Cells of articular cartilage are constantly influenced by mechanical stress when forces are transmitted across joints (Hodge et al. 1986). These mechanical constraints influence cell metabolism, for example by modifying extracellular matrix secretion, which is responsible for specific pathologies (Urban 1994). Blood pressure is another example. In glomerular capillaries of rat kidneys, blood pressure is typically 0.06 atm above atmopheric pressure in healthy animals. Only increases of 0.02 atm above this level are responsible for tissue sclerosis (Kawata et al. 1998).

Pressure generates mechanical stresses with consequences at different levels of organisation. Specifically, it has been shown that exposure to pressure higher than the native pressure experienced by an organism triggers an increase in the expression of heat shock proteins (HSP) (Elo et al. 2000; Sironen et al. 2002; Elo et al. 2005), which are known to be involved in general stress situations (Feder and Hofmann 1999). In parallel, repression of cell cycle control factors has also been observed, explaining cellular growth arrest in relation to pressure stresses.

The effect of pressure on developmental processes is now a rapidly growing research theme. Indeed, mechanical deformation that occurs during cell movements in embryonic morphogenesis seems to be involved in controlling the expression of genes associated with morphogenesis. Evidence of such mechanically regulated process were found in early embryogenesis in Drosophila (Farge 2003), or later, during organogenesis in larval stages of zebrafish (Hove et al. 2003).

An increasing number of observations then suggest that physical environmental factors such as pressure may greatly impact regulatory developmental processes in marine organisms. Data are available concerning temperature tolerance of coastal or deep-sea species. Generally speaking, deep-sea species have to face two main parameters, namely temperature and hydrostatic pressure. It is important to recognise the impact of each of these two parameters in controlling colonisation processes, species distribution, and more generally their role as selective factors in species evolution. At deep-sea hydrothermal vents, besides high hydrostatic pressures, organisms are also submitted to steep chemical and temperature gradients. Several vent species have been studied as models for understanding adaptations to extreme environments (Gaill 1993; Gaill et al. 2000; Van Dover and Lutz 2004).

Aspects of pressure effects on biological systems discussed here focus on eukaryotic organisms as recent reviews have detailed the influence of pressure on prokaryotic systems (Abe et al. 1999; Bartlett 2002).

## 2 Pressure effects on biological systems

### 2.1 Principles of pressure effects

Pressure effects on any physiological or biochemical system basically result from the compression of the system, according to Le Chatelier's principle, which states that at equilibrium a system

[1] 1 atmosphere $\approx$ 0.1 MPa. Here, for ease of recognition of the corresponding depth, we will indicate pressures in atmospheres. In aquatic environments, pressure increases by 1 atmosphere every 10 m.

tends to minimise the effects of disturbing external factors. Increasing pressure therefore favours reduction of a system's volume. Whether increasing pressure will inhibit, favour or have no significant effect on a biological process depends mainly on the volume change incurred by this process. Processes that occur with volume increase are inhibited by pressure increase, whereas processes that occur with volume decrease are favoured by pressure increase. For instance, electrostatic and hydrophobic interactions that maintain the quaternary structure of oligomeric proteins or multi protein assemblages result in a positive volume change (Table 1). Relatively moderate pressure increase may therefore induce dissociation of oligomeric proteins by weakening interactions between subunits (Gross and Jaenicke 1994; Mozhaev et al. 1996; Silva et al. 1996). Conversely, protein primary structure is almost pressure insensitive since volume change occurring with covalent bonds exchange is nearly zero (Table 1). Pressure also affects the equilibrium and rate constants of reactions (e.g. enzymatic reactions, ligand binding reactions), in a way that depends respectively on overall volume and activation volume changes that occur during the reactions (Balny et al. 2002; Mozhaev et al. 1996). Reactions occurring with negative volume change proceed faster with increased pressure, whereas those occurring with a positive volume change tend to be inhibited (Somero 1992; MacDonald 1997).

Physico-chemical parameters, such as temperature and pH, also influence biological systems and may exacerbate or minimise pressure effects. The effects of pressure and temperature on equilibria or kinetics are usually antagonistic. An increase of pressure at constant temperature leads to a compression and an ordering of the molecules, whereas an increase of temperature at constant pressure leads to a loss of molecule ordering (Behan et al. 1992). For instance, in lipid membranes, a pressure increase of 1000 atm is equivalent to a temperature decrease of 20°C. Pressure increase and temperature decrease result in similar effects, i.e. by ordering structures and reducing flexibility in lipids, nucleic acids and carbohydrates (Balny et al. 2002). For proteins, pressure and temperature act in synergy and promote protein denaturation and loss of function (Balny et al. 1997). However, mechanisms by which pressure and temperature act are different, since pressure effects result only from volume change, whereas temperature affects both energy and volume.

**Table 1** Pressure effects on molecular interactions (From Mozhaev et al. 1996)

| Type of interaction | Pressure effect |
|---|---|
| Covalent bond | Nearly zero |
| Electrostatic interactions | Destabilisation |
| Hydrogen bonds | Stabilisation |
| Hydrophobic interactions | Destabilisation |
| Stacking interactions between aromatic rings | Stabilisation |

Pressure also favours ionisation of weak acids. In the yeast *Saccharomyces cerevisiae*, vacuoles become more acidic in proportion to increasing hydrostatic pressure (Abe et al. 1999). Since dissociation constants of weak-acids are pressure-sensitive, the analysis of pressure effects on pH-dependant processes may be difficult.

## 2.2 Levels of the pressure effects

Pressure effects on organisms result in perturbations at each level of biological organisation (Table 2). At the molecular level, macromolecular protein assemblages such as cytoskeleton tubulin and actin are dissociated by pressure in the range of a few hundreds of atm in both shallow-water and terrestrial organisms (Salmon 1975a, b; Begg et al. 1983; Swezey and Somero 1985; Bourns et al. 1988). Enzymatic reactions are affected according to the volume change occurring with the catalysed reaction (for review see Somero 1992). Lipid bilayers of biological membranes have been extensively studied for pressure effect since they are one of the most pressure sensitive molecular assemblage (Wann and MacDonald 1980; Somero 1992; MacDonald 1997). In 1 atm-adapted organisms, pressure increase tends to reduce membrane fluidity (Behan et al. 1992). Membranes of deep-sea organisms exhibit changes in lipid composition, with increase in the fraction of unsaturated fatty acids allowing the maintenance of fluidity under

**Table 2** Examples of thresholds for pressure effects at different level of biological organisation

| Level | Species | Process or structure | Experimental conditions | Native pressure | Threshold | Observation | References |
|---|---|---|---|---|---|---|---|
| Organism | Shallow water organisms | Survival | In vivo | 1 atm | >15 atm | HPNS | (Somero 1992) |
| | Amphipods | Survival | In vitro | 600 atm | <400 atm | Death | (Yayanos 1981) |
| | Rat | Synaptic transmission | In vitro | 1 atm | >100 atm | 80% inhibition | (Somero 1992) |
| | *Gadus morhua* | Action potential | In vitro | 1 atm | >100 atm | Reduce signal amplitude | (Harper et al. 1987) |
| | *Bathysaurus mollis* | Action potential | In vitro | 400 atm | <200 atm | Reduce signal amplitude | (Harper et al. 1987) |
| | *Coryphaenoides armatus* | Action potential | In vitro | 400 atm | <200 atm | Reduce signal amplitude | (Harper et al. 1987) |
| | *Mora moro* | Action potential | In vitro | 90 atm | >100 atm | Reduce signal amplitude | (Harper et al. 1987) |
| Cell | *Echinus affinis* | Embryonic cleavages | In vitro | 200 atm | <100 atm | Slower rate and abnormal development | (Young and Tyler 1993) |
| | *Sterechinus neumayeri* | Embryonic cleavages | In vitro | 1–45 atm | >100 atm | Slower rate and abnormal development | (Tyler et al. 2000) |
| | *Plutonaster bifrons* | Embryonic cleavages | In vitro | 100–250 atm | <100 atm – > 300 atm | Abnormal development | (Young et al. 1996) |
| | Rat mesenglial cells | Cell division | In vitro | 1 atm | 1.12 atm | Proliferation | (Kawata et al. 1998) |
| | Human erythrocyts | Sodium efflux | In vitro | 1 atm | >50 atm | 40% decrease | Goldinger et al 1980 |
| Molecular | Fishes and invertebrates | Deshydrogenases enzymatic activity | In vitro | 1 atm | >50 atm | Decrease activity | (Somero 1992) |
| | Chicken | F actin | In vitro | 1 atm | 700 atm | Depolymerisation | (Begg et al. 1983) |
| | Rabbit | Tubulin | In vitro | 1 atm | 700 atm | Depolyerisation | (Begg et al. 1983) |
| | Human epithelial cells | Actin and tubulin | In vivo | 1 atm | 300 atm | Depolymerisation | (Bourns et al. 1988) |
| | *Chaetopterus pergamentaceous* | Meiotic spindle formation (tubulin) | In vivo | 1 atm | 240 atm | Depolymerisation | (Salmon 1975b) |
| | *Escherichia coli* | Transcription | In vitro | 1 atm | 400 atm | 50% decrease | (Erijman and Clegg 1998) |
| | | Ribosome complex formation | In vitro | 1 atm | 700 atm | Dissociation and inhibition of protein synthesis | (Erijman and Clegg 1998) |
| | Human chondrocyts | Protein synthesis | In vitro | 1 atm | 300 atm | 64% decrease | (Elo et al. 2000) |
| Structural | | Protein primary structure | In vitro | 1 atm | >15 000 atm | | |
| | | DNA double helix | In vitro | 1 atm | >15 000 atm | | |
| | | Protein tertiary structure | In vitro | 1 atm | >5000 atm | | |
| | | Protein tertiary structure | In vitro | 1 atm | >1000 atm | | |

pressure (Cossins and MacDonald 1984; DeLong and Yayanos 1985). Consequences of pressure impact on biological membranes are numerous at higher organisation levels. Cellular processes involving membranes, such as osmoregulation (Sébert et al. 1997), or action potential transmission in nervous cells (Wann and MacDonald 1980; Siebenaller and Garrett 2002) exhibit pressure sensitive behaviour. At the level of organisms, pressure effects are clearly visible in the high pressure neurological syndrome (HPNS), which is characterised by motor coordination impairment, spasm or even paralysis. This has been repeatedly reported in 1-atm adapted organisms exposed to high pressure or pressure-adapted organisms exposed to atmospheric pressure (Yayanos 1981; Treude et al. 2002). These perturbations result mainly from impairment of cell membrane functionality.

## 2.3 Effects of pressure on gene expression

Pressure also has effects on the expression of genes. Many recent studies addressing regulation of gene expression in response to mechanical stress due to pressure have focused on articular cartilage cells (chondrocytes). These cells are naturally exposed to high pressures during movements, and anomalies in pressure regimes to which these cells are submitted may trigger pathologies (Urban 1994). Articular cartilage has to resist mechanical stress to protect bone ends from excessive loading. Chondrocytes, which are involved in matrix production of various cartilaginous tissue, are constantly influenced by mechanical stress when forces are transmitted across joints (Hodge et al. 1986). In a standing position, compressive forces up to 200 atm may rise within joint articular cartilage (Muir 1995). *In vitro* studies using chondrocytic cell lines have shown that they respond to high pressure changes by altering their synthetic capacity depending on the type, timing and mode of pressure variations (Parkkinen et al. 1993a; Lammi et al. 1994; Smith et al. 1996). Both the frequency and the amplitude of the forces applied on the cartilage affect the synthesis rate of cartilage-specific proteoglycans (Hall et al. 1993; Parkkinen et al. 1993a). Continuous high pressure inhibits macromolecule synthesis and secretion, reduces the steady-state level of aggrecan mRNA, alters the shape of the Golgi apparatus, and disturbs the stress fibre organisation of microfilaments (Symington et al. 1991; Hall et al. 1993; Parkkinen et al. 1993b, 1995; Lammi et al. 1994; Jortikka et al. 2000).

Stress response has been detected in chondrocytic cells submitted to continuous high hydrostatic pressure regimes (Table 3) (Kaarniranta et al. 1998; Elo et al. 2000, 2005; Kaarniranta et al. 2000). Several heat shock proteins (HSP) are up-regulated, among which Hsp70 is the most intensively induced. Hsp70 level is increased through stabilisation of mRNA rather than transcriptional activation (Kaarniranta et al. 1998). This stabilisation also requires protein synthesis (Kaarniranta et al. 2000). Continuous high hydrostatic pressure also increases synthesis of HSP 90, and slightly up-regulates the glucose related protein GRP 78, which also responds to stress situations (Elo et al. 2000). Continuous high hydrostatic pressure increases levels of interleukin-6 mRNA (Takahashi et al. 1998), which has been shown to increase HSP 90 expression in mouse cells (Stephanou et al. 1998). HSP 90 would then increase the activity of the nitric oxide synthase complexe, increasing nitric oxide in the cell, which would in turn inhibit proteoglycan synthesis (Lammi et al. 1994; Garcia-Cardena et al. 1998; Elo et al. 2000). In bovine chondrocytes, high hydrostatic pressure has been shown to result in the secretion of aggrecans exhibiting retarded migration on agarose gels (Lammi et al. 1994), which would indicate misfolding. Changes in GRP 78 levels, which plays in protein folding in the reticulum, would support this idea (Elo et al. 2000).

High pressure treatment has been repeatedly reported to cause growth arrest (Abe and Horikoshi 2000; Koyama et al. 2005). Transcriptomic analysis in chondrocytes showed up-regulation of mRNA levels of several genes that mediates growth arrest (Table 3) (Sironen et al. 2002). In the yeast *Saccharomyces cerevisiae*, high hydrostatic pressure treatments were shown to induce a decrease in mRNA levels of genes involved in cell-cycle progression (Table 3) (Fernandes et al. 2004). These studies also demonstrate a reduced level of genes involved in protein synthesis (Fernandes et al. 2004; Elo et al. 2005).

**Table 3** Effects of pressure changes on gene expression in 1-atm adapted organisms

| Species | Gene | Pressure treatment | Effect | References |
|---|---|---|---|---|
| *Saccharomyces cerevisiae* | HSP (12, 30) | 2000 atm, 30 min | Increase mRNA level | (Fernandes et al. 2004) |
| | Carbohydrate metabolism | 2000 atm, 30 min | Increase mRNA level | (Fernandes et al. 2004) |
| | Cell cycle | 2000 atm, 30 min | Decrease mRNA level | (Fernandes et al. 2004) |
| | Protein synthesis and fate | 2000 atm, 30 min | Decrease mRNA level | (Fernandes et al. 2004) |
| | Nitric oxide synthase | 2000 atm, 30 min | Increase mRNA level | (Fernandes et al. 2004) |
| Chondrocytic cell line | HSP (70) | 300 atm, 12 h | Increase mRNA and protein levels | (Kaarniranta et al. 1998; Elo et al. 2000) |
| | HSP (90) | 300 atm, 12 h | Induce protein synthesis | (Elo et al. 2000) |
| | GRP 78 | 300 atm, 12 h | Increase protein level | (Elo et al. 2000) |
| | HSP (40, 27) | 300 atm, 12 h | Increase mRNA level | (Sironen et al. 2002) |
| | GADD 45, GADD 153, p21, tob (promote growth arrest) | 300 atm, 12 h | Increase mRNA level | (Sironen et al. 2002) |
| | Id family | 300 atm, 12 h | Decrease mRNA level | (Sironen et al. 2002) |
| | Cytoplasmic dynein light chain | 300 atm, 12 h | Decrease mRNA level | (Sironen et al. 2002) |
| | NIP3 (apoptosis related gene) | 300 atm, 12 h | Decrease mRNA level | (Sironen et al. 2002) |
| | Interleukin-6 | 300 atm | Increase mRNA levels | (Takahashi et al. 1998) |
| | Eukaryotic elongation factor 2 | 300 atm | Decrease protein level | (Elo et al. 2005) |
| Hela S3 cells | Core and H1 histone | 414 atm | Decrease mRNA level | (Symington et al. 1991) |

## 3 High pressure equipments

Most studies focusing on pressure effects were conducted with 1 atm-adapted organisms. Deep-sea species have evolved adaptations to life at high pressure. They are, one the other hand, sensitive to low pressure in most cases (Yayanos 1981; Yayanos and Dietz 1983). Mechanisms governing sensitivity to high or low pressure might be differently controlled and deep-sea organisms might bring useful data to better our understanding of pressure effects on biological systems.

Organisms living below 2000 m depth do not survive retrieval from their deep-sea environment, if they are not maintained close to their native pressure during collection or returned to it shortly after collection (Yayanos 1978, 1981; Treude et al. 2002). Further, there is evidence to suggest that, even at depths of 1000 m, substantial adaptation to high pressure regimes characterises many resident invertebrates. Simulating natural high pressure is therefore essential for maintaining animals in a sufficiently good condition to allow physiological and behavioural studies (Wilson and Smith 1985; MacDonald 1997; Shillito et al. 2001, 2004, 2006; Ravaux et al. 2003). A few large (several litres) volume vessels are now functioning in different laboratories worldwide and provide isolated chambers where collected specimens are maintained under continuously controlled pressure and temperature (Shillito et al. 2001; Koyama et al. 2002) (See Table 4). In addition, some of these vessels offer imaging facilities, which allow the observation of behavioural response. For instance, they permit

**Table 4** Examples of equipment developed for studies on pressure effects in marine organisms

| Equipment | Volume | Working pressure | Main features | Type of studies | References |
|---|---|---|---|---|---|
| *Studies on adult organisms* | | | | | |
| Observational pressure vessel | 0.9 l | 100 MPa | Stainless steal chamber<br>Manually operated pump<br>Temperature control<br>View port and illumination system | Behavioural response to decompression | (Macdonald and Gilchrist 1980) |
| Pressure aquarium | 2.25 l | 14 MPa | Transparent, cylindrical, polycarbonate chamber<br>Continuous flow-through system<br>Temperature control | Behavioural studies | (Quetin and Childress 1980) |
| Pressure aquarium | 175 ml | 25 MPa | Stainless steal chamber<br>Manually operated pump<br>Temperature control<br>View port and illumination system | Behavioural response to compression | (Brauer et al. 1980) |
| High-pressure respirometry system | 2.25 l | 34 MPa | Stainless steal chamber<br>Continuous flow-through system<br>Temperature control<br>Gas extractor for analysis by membrane inlet mass spectroscopy<br>Computer based data acquisition and control system | Measurements of metabolite uptakes | (Kochevar et al. 1992, Girguis et al. 2000) |
| Pressure aquarium | 8–18 l | 21.5 MPa | Stainless steal chamber<br>Continuous flow-through system<br>Temperature control<br>View port and illumination system | Measurements of metabolite uptakes | (Goffredi et al. 1997) |
| IPOCAMP[a] | 20 l | 30 MPa | Stainless steal chamber<br>Continuous flow-through system<br>Temperature control<br>View port and illumination system Video facilities | Behavioural studies<br>Heat and pressure stress studies<br>Respirometry | (Shillito et al. 2001, 2004, 2006; Toullec et al. 2002; Ravaux et al. 2003) |
| *Studies on embryonic and larval stages of deep-sea animals* | | | | | |
| High-pressure system | 1 ml | 25 MPa | Stainless steal chamber<br>Continuous flow-through system (allows researchers to introduce chemical tracers or to extract fluid and larvae) | Deep-sea embryos development<br>Embryos metabolic rates measurements | (Marsh et al. 2001) |
| High-pressure system | 80 ml | 25 MPa | Stainless steal chamber<br>View port and illumination system | Behavioural observation of larvae<br>Buoyancy rates measurements | (Marsh et al. 2001) |
| PICCEL[b] | 100 ml | 30 MPa | Stainless steal chamber | Deep-sea embryos development | (Pradillon et al. 2001, 2004) |

**Table 4** continued

| Equipment | Volume | Working pressure | Main features | Type of studies | References |
|---|---|---|---|---|---|
| | | | Manually operated pump<br>Temperature control | | |
| PIRISM[c] | 250 µl | 30 MPa | Optical sapphire windows<br>Pressure control through the PICCEL system | Microscopic observation of developping embryos under pressure | (Pradillon et al. 2004) |
| *Isobaric* in situ *collection devices* | | | | | |
| Isobaric trap | 600 ml | 90 MPa | Entrance valve and holding chamber<br>View port<br>Acoustic pinger to track deployment and recovery of the trap | Behavioural response to decompression | (Macdonald and Gilchrist 1978, 1982) |
| Isobaric trap | | 55 MPa | Pressure loss compensation system<br>Temperature insulation<br>Flow through system functionning after retrieval | Behavioural response to decompression | (Yayanos 1978, 1981) |
| Pressure-stat aquarium | 20 l | 20 MPa | View port and illumination system<br>Succion sampler for animal collection in situ<br>Flow through system<br>Temperature control<br>Oxygen control | Collection of animals *in situ* and culture without decompression<br><br>Behavioural observation<br>Respirometry | (Koyama et al. 2002) |

[a] Incubateur Pressurisé pour l'Observation et la Culture d'Animaux Marin Profonds : Pressure incubator for the observation and culture of marine deep-sea animals

[b] Pressure incubators for the culture of cells embryos and larvae

[c] PICCEL related imaging system for microscopy

heat-resistance investigations on live invertebrates endemic to the hottest part of the hydrothermal vent biotope: the wall of active vent chimneys. Polychaete worms (*Hesiolyra bergi*) from the East Pacific Rise (Shillito et al. 2001), and shrimps (*Rimicaris exoculata*, and *Mirocaris fortunata*) from the Mid-Atlantic Ridge (Ravaux et al. 2003; Shillito et al. 2006) were exposed to rising temperature in the vessel IPOCAMP ™ (Incubateur Pressurisé pour l'Observation et la Culture d'Animaux Marins Profonds), and their behavioural and biochemical response analysed. These studies proved that restoration of pressure conditions was crucial for accurate determination of the thermal limits of the organism studied (Shillito et al. 2006). Physiological analysis of the salinity tolerance, osmotic and ionic regulation capabilities and hormonal homeostasis regulation of the vent crab *Bythograea thermydron* (Martinez et al. 2001; Toullec et al. 2002) were also performed in pressure vessels. The effects of reactive oxygen species to which hydrothermal fauna are particularly exposed in their natural environment and which are known to damage the genetic material was investigated in hydrothermal polychaetes maintained under pressure (Dixon et al. 2002). Pressure vessels were also used to develop cultures of cells isolated from hydrothermal vent organisms (De Cian et al. 2003).

Due to different experimental requirements such as incubation periods over several days without depressurisation, or microscopic imaging, studies on embryonic stages were conducted in separate dedicated pressure vessels. We recently developed such small vessels, PICCEL (Pressurised Incubators for the Culture of Cells, Embryos and Larvae) (Pradillon et al. 2004). Since many changes that are induced by moderate pressures (hundreds of bars) are rapidly reversible after depressurisation (Begg et al. 1983; Bourns et al. 1988), our vessels can be connected to visualisation chambers allowing microscopic observation of embryos under pressure. With such equipment, we were able to perform analysis of the pressure and temperature resistance of larval stages of hydrothermal vent organisms. These various pressure apparatus thus provide insights into the different life stages of deep-sea organisms.

## 4 Pressure in developmental processes

### 4.1 Early studies on mitotic apparatus

Early studies reporting pressure effects on eggs and embryos were motivated by an interest in the gel-sol properties of cytoskeleton proteins in the study of division mechanisms. Pressure was then used as a tool for disrupting tubulin and actin filaments, revealing the importance of their physical state in the mitosis mechanisms (Marsland and Landau 1954; Zimmerman and Marsland 1964; Marsland 1970; Zimmerman 1971). These experiments also pointed out cytoskeleton structures as being pressure sensitive so limiting the depth at which certain species' larvae can develop.

### 4.2 Pressure tolerance in early life stages

More recently, pressure effects on developing embryos were analysed in order to estimate the extent to which pressure is tolerated, how this might limit dispersal during embryonic and larval life, and if this limited tolerance might be responsible for bathymetric distributions observed in adult populations (Young and Cameron 1989; Young and Tyler 1993; Young et al. 1996, 1997; Tyler and Young 1998; Tyler et al. 2000). Most studies of this type focused on echinoderms, and demonstrated that early-life-stages can tolerate only a defined pressure range. However, in many cases, embryonic and larval stages seemed to be able to tolerate large pressure changes, perhaps reflecting a much larger vertical distribution than that observed in adults (Fig. 1). For instance, embryos of the shallow-water sea urchins *Echinus esculentus* and *E. acutus* are able to develop under pressure up to 100 atm, larval stages can tolerate pressures up to 200 atm, whereas adults are restricted to the first few tens of metres depth (Tyler and Young 1998). On the other hand, embryos of the deep-sea species *E. affinis*, which live around 2000 m depth, cannot develop at pressure lower than 100 atm (Tyler and Young 1998). Interestingly, embryos obtained from adults of *E. acutus* collected in a deeper population at 1000 m depth showed a

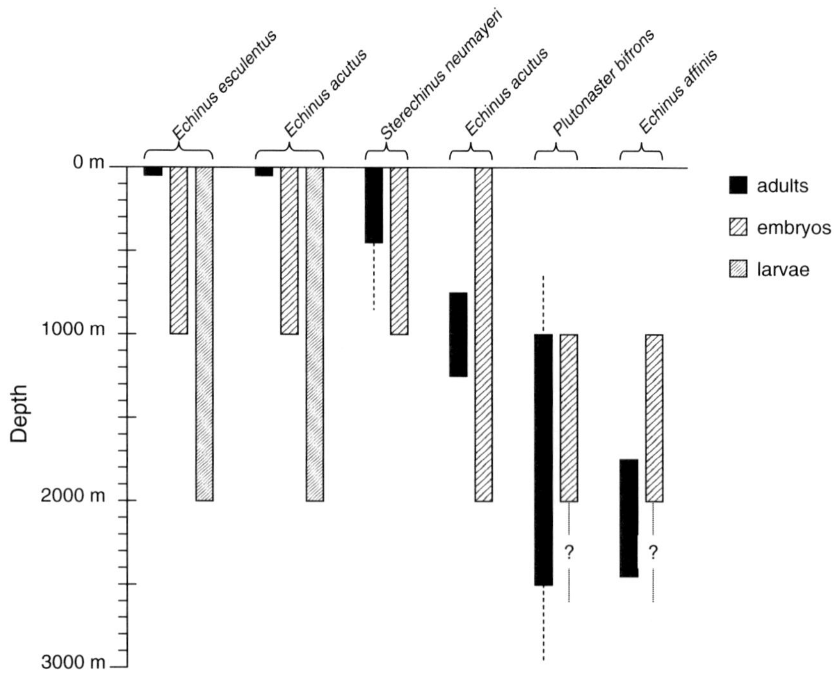


**Fig. 1** Comparison of adult and embryonic vertical distributions in echinoderm species. Adult distributions inferred from collection depth are plotted together with potential distribution of embryos and larvae inferred from their pressure tolerance as found in *in vitro* studies (Young and Tyler 1993; Young et al. 1996; Tyler and Young 1998; Tyler et al. 2000)

particularly wide pressure range, and authors hypothesised that this particular population was in the process of migrating to the deep-sea.

Studies on echinoderms also evidence the effect of temperature on embryonic pressure tolerance. Pressure tolerance of embryos of the Antarctic echinoderm *Sterechinus neumayeri* is reduced when embryos are exposed to low temperature (Tyler et al. 2000). In temperate and tropical regions, the temperature decreases in the first 1000 m of the water column, and is almost constant from 1000 m down to the sea bottom. Therefore temperature, pressure or a combination of both might affect larval physiology in the upper part of the ocean, and explain observed distributions of organisms. However, for depths below 1000 m, or for regions where the water column is nearly isothermal, pressure is likely to be the main factor influencing vertical distribution of larvae.

Deep-sea hydrothermal vents are then exceptional with regards the steep temperature gradients that occur around these fluid emissions (Johnson et al. 1986, 1988; Le Bris et al. 2005). Here, temperature has been suggested to explain species distributions. Larvae of these animals might have to face both changes in temperature as well as changes in pressure when travelling through the water column between vent sites.

To date, only embryos of a few vent species have been studied for temperature and pressure tolerance. This is due to the fact that catching larvae directly *in situ* is highly challenging because larval density is low and few data are available on their *in situ* distribution (Mullineaux and France 1995; Mullineaux et al. 2005). As an alternative, *in vitro* fertilisation methods were developed, using dedicated pressure vessels (Pradillon et al. 2004), where embryos could be incubated under controlled pressure and temperature.

Studies on the annelids *Alvinella pompejana* and *Riftia pachyptila*, both inhabiting hydrothermal vents from the East Pacific Rise (EPR) at 2500 m depth demonstrate the low thermal tolerance and the need for native pressures in early embryos of these species (Marsh et al. 2001; Pradillon et al. 2001, 2005). The limited embryonic temperature tolerance was particularly significant in *A. pompejana* because adults of this species live closest to the vent fluid in the hydrothermal biotope, and are exposed to temperatures close

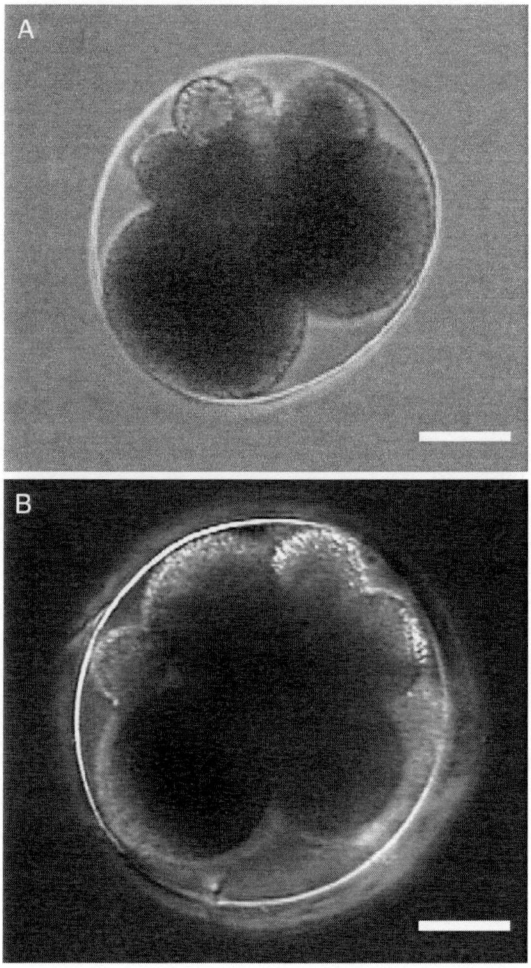


**Fig. 2** Effects of pressure on early development of *A. pompejana*, morphological aspects. (**A**) Embryo incubated 48 h at 10°C and atmospheric pressure, exhibiting anomalous blastomere size ratio. (**B**) Embryo incubated 96 h at 10°C and 250 atm. Scale bars are 50 μm

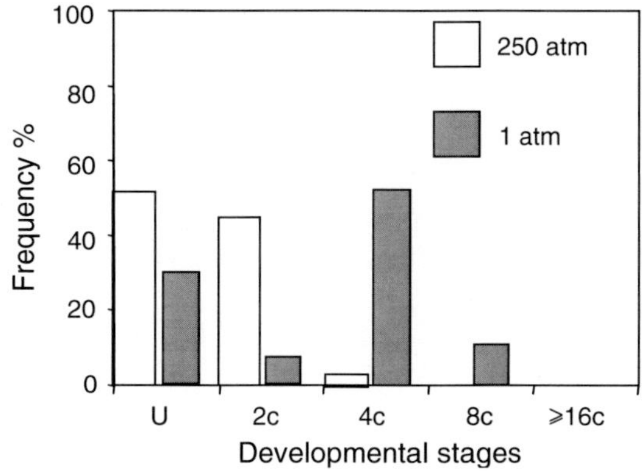


**Fig. 3** Effects of pressure on developmental rates in *A. pompejana*, U: uncleaved embryos, 2c: 2-cells stages, 4c: 4-cells stages, 8c: 8-cells stages, ≥ 16c: 16-cells stages and more. Embryos were incubated 48 h at 10°C (modified from Pradillon et al. 2005)

to 50°C (Cary et al. 1998; Le Bris et al. 2005). Embryos do not survive temperatures above 20°C, and develop well around 5–10°C at native pressure (Pradillon et al. 2001, 2005). Embryos are then likely to leave hydrothermal vents and disperse through the water column.

Both embryos of *A. pompejana* and *R. pachyptila* are unable to develop normally at atmospheric pressure, but the upper and lower pressure limits allowing successful development have not yet been determined. In *A. pompejana*, 1-atm raised embryos exhibit abnormal distribution and relative size of the blastomeres (Fig. 2) (Pradillon et al. 2005). Abnormal morphology was also demonstrated in developmental stages of 1-atm species exposed to high pressure (Young and Tyler 1993; Tyler and Young 1998). Another effect of pressure was on developmental rates: embryos incubated under native pressure developed at slower rates than embryos incubated at atmospheric pressure at the same temperature (Fig. 3).

Temperature and pressure tolerances of embryonic stages are viewed as potential limiting factors in larval dispersal. Early larval stages of the shrimp *Mirocaris fortunata*, collected on a Mid-Atlantic Ridge vent site at 1650 m depth, were found to be able to develop at 1 atm, provided that temperature remained low (10°C) (Tyler and Dixon 2000). These results then agree with other studies that previously speculated that early life stages of vent shrimps should be able to disperse towards surface waters (Pond et al. 1997). In *A. pompejana*, and *R. pachyptila*, embryos do not seem capable of migration towards the surface, and most probably disperse through abyssal water. However, embryos of both species were found to have opposite floatation rates. Embryos of *A. pompejana* tend to sink (Pradillon et al. 2004) whereas those of *R. pachyptila* tend rise up through the water column (Marsh et al. 2001). Different thresholds for pressure tolerance might thus be expected and be reflected through different dispersal strategies by these two species.

### 4.3 Mechanical signal in embryonic morphogenesis

Recently, emerging evidence indicated a role of mechanical forces in modulating the expression of developmental genes. Mechanical strains have been shown to be involved in morphogenetic movements that occurr during embryonic development (Farge 2003; Hove et al. 2003; Brouzes and Farge 2004; Brouzes et al. 2004; Supatto et al. 2005). The morphogenetic movements that lead to geometric shape changes are controlled by patterned gene expression. In turn, morphogenetic movements seem to be able to modulate the expression of developmental genes, through the generation of localised mechanical strains on cells. In *Drosophila* embryos, the first morphogenetic movement of gastrulation which is associated with mesoderm formation, is the ventral furrow invagination. This step is characterised by strong and complex cell shape changes. The genetic control of the different phases of mesoderm invagination has been related to the expression of two zygotic genes, *twist* and *snail*. Both genes appear to be necessary for the coordinated apical constriction phase (Brouzes et al. 2004). Farge (Farge 2003) showed that the expression of *twist* can be triggered by tissue shape changes at the time of ventral furrow formation. The existence of mechanical induction of *twist* was proposed in response to mechanical strains experienced by mesodermal cells during mesoderm invagination. ß catenin / armadillo seems to mediate the twist mechanotranscriptional event (Farge 2003). These experiments suggest that mechanical strains may activate specific transduction pathways leading to modulation of the expression of specific genes.

Mechanical forces related to hydrodynamic flows have also been proposed to play a significant role in organogenesis, such as in lung and cardiac morphogenesis. For instance, intracardiac fluid forces in the zebrafish embryo heart were demonstrated recently to be necessary for proper heart development (Hove et al. 2003). Similarly, the glomerulus morphogenesis of kidney is induced by hemodynamic forces in zebrafish (Serluca et al. 2002).

## 5 Conclusion

Pressure has a wide range of effects on biological systems, and as an environmental parameter, may influence distribution patterns of marine organisms. Embryonic stages which have the potential to disperse through the water column, therefore being exposed to pressure changes, have been demonstrated to tolerate pressure variations to a certain extent. Limits in their physiological tolerance might explain vertical distributions of adult populations. Mechanisms governing pressure response during development, and particularly gene expression regulations, are still unknown for marine species. However, an increasing number of data from other systems such as cartilaginous cell lines are now becoming available.

Finally in model organisms used in developmental studies, more and more examples indicate the existence of a reciprocal interplay between expression of some developmental genes and the mechanical forces that are associated with morphogenetic movements or with hydrodynamic flow during development. Future challenges lie in deciphering the underlying molecular mechanisms of the interplay between mechanical forces, genetic expression, and developmental processes.

## References

Abe F, Horikoshi K (2000) Tryptophan permease gene TAT2 confers high pressure growth in *Saccharomyces cerevisiae*. Mol Cell Biol 20:8093–8102

Abe F, Kato C, Horikoshi K (1999) Pressure-regulated metabolism in microorganisms. Trends Microbiol 7:447–453

Balny C, Masson P, Heremans K (2002) High pressure effects on biological macromolecules: from structural changes to alteration of cellular processes. Biochim Biophys Acta 1595:3–10

Balny C, Mozhaev VV, Lange R (1997) Hydrostatic pressure and proteins: Basic concepts and new data. Comp Biochem Physiol 116A:299–304

Bartlett DH (2002) Pressure effects on in vivo microbial processes. Biochim Biophys Acta 1595:367–381

Begg DA, Salmon ED, Hyatt HA (1983) The changes in structural organization of actin in the sea urchin egg cortex in response to hydrostatic pressure. J Cell Biol 97:1795–1805

Behan MK, MacDonald AG, Jones GR, Cossins AR (1992) Homeoviscous adaptation under pressure: the pressure dependence of membrane order in brain

myelin membranes of deep-sea fish. Biochim Biophys Acta 1103:317–323
Bourns B, Franklin S, Cassimeris L, Salmon ED (1988) High hydrostatic pressure effects *in vivo*: changes in cell morphology, microtubules assembly, and actin organization. Cell Motil Cytoskeleton 10:380–390
Brauer RW, Bekman MY, Keyser JB, Nesbitt DL, Sidelev GN, Wright SL (1980) Adaptation to high hydrostatic pressures of abyssal gammarids from Lake Baikal in eastern Siberia. Comp Biochem Physiol 65A:109–117
Brouzes E, Farge E (2004) Interplay of mechanical deformation and patterned gene expression in developing embryos. Curr Opin Genet Dev 14:367–374
Brouzes E, Supatto W, Farge E (2004) Is mechanosensitive expression of twist involved in mesoderm formation? Biol Cell 96:471–477
Cary CS, Shank T, Stein JL (1998) Worms bask in extreme temperatures. Nature 391:545–546
Cossins AR, MacDonald AG (1984) Homeoviscous theory under pressure. II. The molecular order of membranes from deep-sea fish. Biochim Biophys Acta 776:144–150
De Cian M-C, Andersen AC, Toullec J-Y, Biegala I, Caprais J-C, Shillito B, Lallier FH (2003) Isolated bacteriocyte cell suspensions from the hydrothermal vent tubeworm *Riftia pachyptila*, a potent tool for cellular physiology in a chemoautotrophic symbiosis. Mar Biol 142:141–151
DeLong EF, Yayanos AA (1985) Adaptation of the membrane lipids of a deep-sea bacterium to changes in hydrostatic pressure. Science 228:1101–1103
Dixon DR, Dixon LRJ, Shillito B, Gwynn JP (2002) Background and induced levels of DNA damage in pacific deep-sea vent polychaetes: the case for avoidance. Cah Biol Mar 43:333–336
Elo MA, Karjalainen HM, Sironen RK, Valmu L, Redpath NT, Browne GJ, Kalkkinen N, Helminen HJ, Lammi MJ (2005) High hydrostatic pressure inhibits the biosynthesis of eukaryotic elongation factor-2. J Cell Biochem 94:497–507
Elo MA, Sironen RK, Kaarniranta K, Auriola S, Helminen HJ, Lammi MJ (2000) Differential regulation of stress proteins by high hydrostatic pressure, heat shock, and unbalanced calcium homeostasis in chondrocytic cells. J Cell Biochem 79:610–619
Erijman L, Clegg RM (1998) Reversible stalling of transcription elongation complexes by high pressure. Biophys J 75:453–462
Farge E (2003) Mechanical induction of Twist in the *Drosophila* foregut/stomodeal primordium. Curr Biol 13:1365–1377
Feder ME, Hofmann GE (1999) Heat-shock proteins, molecular chaperones, and the stress response: evolutionary and ecological physiology. Annu Rev Physiol 61:243–282
Fernandes PMB, Domitrovic T, Kao CM, Kurtenbach E (2004) Genomic expression pattern in *Saccharomyces cerevisiae* cells in response to high hydrostatic pressure. FEBS Lett 556:153–160
Gaill F (1993) Aspects of life development at deep-sea hydrothermal vents. FASEB J 7:558–565
Gaill F, Zbinden M, Pradillon F (2000) Adaptations of hydrothermal vent organisms to their environment. In: Legakis A, Sfenthourakis S, Polymeni R, Thessalou-Legaki M (eds) Proceedings of 18th Int. Congr. Zoology: The new panorama of animal evolution. Pensoft Publishers, Sofia Moscow, Athènes, Grèce, pp 491–495
Garcia-Cardena G, Fan R, Shah V, Sorrentino R, Cirino G, Papapetropoulos A, Sessa WC (1998) Dynamic activation of endothelial nitric oxide synthase by Hsp90. Nature 392:821–824
Girguis PR, Lee RW, Desaulniers NT, Childress JJ, Pospesel M, Felbeck H, Zal F (2000) Fate of nitrate acquired by the tubeworm *Riftia pachyptila*. Appl Env Microbiol 66(7):2783–2790
Goffredi SK, Childress JJ, Desaulniers NT, Lee RW, Lallier FH, Hammond D (1997) Inorganic carbon acquisition by the hydrothermal vent tubeworm *Riftia pachyptila* depends upon high external $P_{CO2}$ and upon proton elimination by the worm. J Exp Biol 200:883–893
Gross M, Jaenicke R (1994) Proteins under pressure. The influence of high hydrostatic pressure on structure, function and assembly of proteins and protein complexes. Eur J Biochem 221:617–630
Hall AC, Pickles DM, MacDonald AG (1993) Aspects of eukaryotic cells. In: Effects of high pressure on biological systems. Springer-Verlag, Berlin, Germany, pp 29–85
Harper AA, MacDonald AG, Wardle CS, Pennec J-P (1987) The pressure tolerance of deep-sea fish axons: results of Challenger cruise 6B/85. Comp Biochem Physiol 88A:647–653
Hodge WA, Fijan RS, Carlson KL, Burgess RG, Harris WH, Mann RW (1986) Contact pressures in the human hip joint measured *in vivo*. Proc Natl Acad Sci USA 83:2879–2883
Hove JR, Koster RW, Forouhar AS, Acevedo-Bolton G, Fraser SE, Gharib M (2003) Intracardiac fluid forces are an essential epigenetic factor for embryonic cardiogenesis. Nature 421:172–177
Jannasch HW, Taylor CD (1984) Deep-sea microbiology. Annu Rev Microbiol 38:487–514
Johnson KS, Beehler CL, Sakamoto-Arnold CM, Childress JJ (1986) In situ measurements of chemical distributions in a deep-sea hydrothermla vent field. Science 231:1139–1141
Johnson KS, Childress JJ, Beehler CL (1988) Short-term temperature variability in the Rose Garden hydrothermal vent field: an unstable deep-sea environment. Deep Sea Res 35:1711–1721
Jortikka M, Parkkinen JJ, Inkinen RI, Kärner J, Järveläinen HT, Nelimarkka LO, Tammi MI, Lammi MJ (2000) Regulation of proteoglican synthesis in chondrocytes by hydrostatic pressure involves targets dependent and independent of dynamic microtubules. Arch Biochem Biophys 374:172–180
Kaarniranta K, Elo MA, Lammi MJ, Goldring MB, Eriksson J, Sistonen L, Helminen HJ (1998) Hsp 70 accumulation in chondrocytic cells exposed to high continuous hydrostatic pressure coincides with mRNA stabilization rather than transcriptional activation. Proc Natl Acad Sci USA 95:2319–2324

Kaarniranta K, Holmberg CI, Helminen HJ, Eriksson JE, Sistonen L, Lammi MJ (2000) Protein synthesis is required for stabilization of hsp70 mRNA upon exposure to both hydrostatic pressurization and elevated temperature. FEBS Lett 475:283–286

Kawata Y, Fujii Z, Sakumura T, Kitano M, Suzuki N, Matsuzaki M (1998) High pressure promote the proliferation of rat cultured mesanglial cells *in vitro*. Biochim Biophys Acta 1401:195–202

Kochevar RE, Childress JJ, Fisher CR, Minnich L (1992) The methane mussel: roles of symbiont and host in the metabolic utilisation of methane. Mar Biol 112: 389–401

Koyama S, Kobayashi H, Inoue A, Miwa T, Aizawa M (2005) Effects of the piezo-tolerance of cultured deep-sea eel cells on survival rates, cell proliferation, and cytoskeletal structures. Extremophiles 9:449–460

Koyama S, Miwa T, Horii M, Ishikawa Y, Horikoshi K, Aizawa M (2002) Pressure-stat aquarium system designed for capturing and maintaining deep-sea organisms. Deep Sea Res I: Oceanogr Res Papers 49:2095–2102

Lammi MJ, Inkinen R, Parkkinen JJ, Häkkinen T, Jortikka M, Nelimarkka LO, Järveläinen HT, Tammi MI (1994) Expression of reduced amounts of structurally altered aggrecan in articular cartilage chondrocytes exposed to high hydrostatic pressure. Biochemistry J 304:723–730

Le Bris N, Zbinden M, Gaill F (2005) Processes controlling the physico-chemical micro-environments associated with Pompeii worms. Deep Sea Res I 52:1071–1083

MacDonald AG, Gilchrist I (1978) Further studies on the pressure tolerance of deep sea crustacea with observations using a new high pressure trap. Mar Biol 45:9–21

MacDonald AG, Gilchrist I (1980) Effects of hydrolic decompression and compression on deep sea amphipods. Comp Biochem Physiol 67A: 149–153

MacDonald AG, Gilchrist I (1982) The pressure tolerance of deep sea amphipods collected at their ambient high pressure. Comp Biochem Physiol 71A:349–352

MacDonald AG (1997) Hydrostatic pressure as an environmental factor in life processes. Comp Biochem Physiol 116A:291–297

Marsh AG, Mullineaux LS, Young CM, Manahan DT (2001) Larval dispersal potential of the tubeworm *Riftia pachyptila* at deep-sea hydrothermal vents. Nature 411:77–80

Marsland D, Landau JV (1954) The mechanisms of cytokinesis: temperature-pressure studies on the cortical gel system in various marine eggs. J Exp Zool 125:507–539

Marsland DA (1970) Pressure-temperature studies on the mechanisms of cell division. In: Zimmerman AM (ed) High pressure effects on cellular processes. Academic Press, New York, pp 260–312

Martinez A-S, Toullec J-Y, Shillito B, Charmantier-Daures M, Charmantier G (2001) Hydromineral regulation in the hydrothermal vent crab *Bythograea thermydron*. Biol Bull 201:167–174

Mozhaev VV, Heremans K, Frank J, Masson P, Balny C (1996) High pressure effects on protein structure and function. Proteins: Structure, Function, and Genetics 24:81–91

Muir H (1995) The chondrocyte, architect of cartilage. Biomechanics, structure, function and molecular biology of cartilage matrix macromolecules. Bioessays 17:1039–1048

Mullineaux LS, France SC (1995) Dispersal mechanisms of deep-sea hydrothermal vent fauna. In: Humphris SE, Zierenberg RA, Mullineaux LS, Thomson RE (eds) Seafloor hydrothermal systems: physical, chemical, biological and geological interactions. Geological Monograph 91. American Geophysical Union, Washington, DC, pp 408–424

Mullineaux LS, Mills SW, Sweetman AK, Beaudreau AH, Metaxas A, Hunt HL (2005) Vertical, lateral and temporal structure in larval distribution at hydrothermal vents. Mar Ecol Prog Ser 293:1–16

Parkkinen JJ, Ikonen J, Lammi MJ, Laakkonen J, Tammi M, Helminen HJ (1993a) Effects of cyclic hydrostatic pressure on proteoglycan synthesis in cultured chondrocytes and articular cartilage explants. Arch Biochem Biophys 300:458–465

Parkkinen JJ, Lammi MJ, Inkinen R, Jortikka M, Tammi M, Virtanen I, Helminen HJ (1995) Influence of short-term hydrostatic pressure on organization of stress fibers in cultured chondrocytes. J Orthopedic Res 13:495–502

Parkkinen JJ, Lammi MJ, Pelttari A, Helminen HJ, Tammi M, Virtanen I (1993b) Altered Golgi apparatus in hydrostatically loaded articular cartilage chondrocytes. Ann Rheum Dis 52:192–198

Pond D, Dixon D, Sargent J (1997) Wax-ester reserves facilitate dispersal of hydrothermal vent shrimps. Mar Ecol Prog Ser 146:289–290

Pradillon F, Le Bris N, Shillito B, Young CM, Gaill F (2005) Influence of environmental conditions on early development of the hydrothermal vent polychaete *Alvinella pompejana*. J Exp Biol 208: 1551–1561

Pradillon F, Shillito B, Chervin J-C, Hamel G, Gaill F (2004) Pressure vessels for in vitro studies of deep-sea fauna. High Pres Res 24:237–246

Pradillon F, Shillito B, Young CM, Gaill F (2001) Developmental arrest in vent worm embryos. Nature 413:698–699

Quetin LB, Childress JJ (1980) Observations on the swimming activity of two bathypelagic mysid species maintained at high hydrostatic pressures. Deep-Sea Res 27A:383–391

Ravaux J, Gaill F, Le Bris N, Sarradin P-M, Shillito B (2003) Heat-shock response and temperature resistance in the deep-sea vent shrimp *Rimicaris exoculata*. J Exp Biol 206(pt 14):2345–2354

Rothschild LJ, Mancinelli RL (2001) Life in extreme environments. Nature 409:1092–1101

Salmon ED (1975a) Pressure-induced depolymerization of brain microtubules in vitro. Science 189:884–886

Salmon ED (1975b) Pressure-induced depolymerization of spindle microtubules. J Cell Biol 65:603–614

Sébert P, Simon B, Péqueux A (1997) Effects of hydrostatic pressure on energy metabolism and osmoregulation in crab and fish. Comp Biochem Physiol 116A:281–290

Serluca FC, Drummond IA, Fishman MC (2002) Endothelial signaling in kidney morphogenesis: a role for hemodynamic forces. Curr Biol 12:492–497

Shillito B, Jollivet D, Sarradin P-M, Rodier P, Lallier FH, Desbruyères D, Gaill F (2001) Temperature resistance of *Hesiolyra bergi*, a polychaetous annelid living on deep-sea vent smoker walls. Mar Ecol Prog Ser 216:141–149

Shillito B, Le Bris N, Gaill F, Rees J-F, Zal F (2004) First access to live Alvinellas. High Pres Res 24:169–172

Shillito B, Le Bris N, Hourdez S, Ravaux J, Cottin D, Caprais J-C, Jollivet D, Gaill F (2006) Temperature resistance studies on the deep-sea vent shrimp *Mirocaris fortunata*. J Exp Biol 209:945–955

Siebenaller JF, Garrett DJ (2002) The effects of the deep-sea environment on transmembrane signaling. Comp Biochem Physiol Part B 131:675–694

Silva JL, Foguel D, Da Poian AT, Prevelige PE (1996) The use of hydrostatic pressure as a tool to study viruses and other macromolecular assemblages. Curr Opin Struct Biol 6:166–175

Sironen RK, Karjalainen HM, Elo MA, Kaarniranta K, Torronen K, Takigawa M, Helminen HJ, Lammi MJ (2002) cDNA array reveals mechanosensitive genes in chondrocytic cells under hydrostatic pressure. Biochim Biophys Acta (BBA) – Mol Cell Res 1591:45–54

Smith RL, Rusk SF, Ellison BE, Wessells P, Tsuchiya K, Caler DR, Caler WE, Sandell LJ, Schurman DJ (1996) In vitro stimulation of articular chondrocyte mRNA and extracellular matrix synthesis by hydrostatic pressure. J Orthopedic Res 14:53–60

Somero GN (1992) Adaptations to high hydrostatic pressure. Annu Rev Physiol 54:557–577

Stephanou A, Conroy S, Isenberg DA, Maione D, Poli V, Ciliberto G, Latchman DS (1998) Elevation of IL-6 in transgenic mice results in increased levels of the 90 kDa heat shock protein (hsp90) and the production of anti-hsp90 antibodies. J Autoimmun 11:249–253

Supatto W, Debarre D, Moulia B, Brouzes E, Martin J-L, Farge E, Beaurepaire E (2005) In vivo modulation of morphogenetic movements in Drosophila embryos with femtosecond laser pulses. Proc Natl Acad Sci USA 102:1047–1052

Swezey RR, Somero GN (1985) Pressure effects on actin self-assembly: interspecific differences in the equilibrium and kinetics of the G to F transformation. Biochemistry 24:852–860

Symington AL, Zimmerman S, Stein J, Stein G, Zimmerman AM (1991) Hydrostatic pressure influences histone mRNA. J Cell Sci 98:123–129

Takahashi K, Kubo T, Arai Y, Kitajima I, Takigawa M, Imanishi J, Hirasawa Y (1998) Hydrostatic pressure induces expression of interleukin 6 and tumour necrosis factor alpha mRNAs in a chondrocyte-like cell line. Ann Rheum Dis 57:231–236

Toullec J-Y, Vinh J, Le Caer J-P, Shillito B, Soyez D (2002) Structure and phylogeny of the crustacean hyperglycemic hormone and its precursor from a hydrothermal vent crustacean : the crab *Bythograea thermydron*. Peptides 23:31–42

Treude T, Janßen F, Queisser W, Witte U (2002) Metabolism and decompression tolerance of scavenging lysianassoid deep-sea amphipods. Deep Sea Res Part I: Oceanogr Res Papers 49:1281–1289

Tyler PA (1995) Conditions for the existence of life at the deep-sea floor: an update. Oceanogr Mar Biol Annu Rev 33:221–244

Tyler PA, Dixon DR (2000) Temperature/pressure tolerance of the first larval stage of *Mirocaris fortunata* from Lucky Strike hydrothermal vent field. J Mar Biol Ass UK 80:739–740

Tyler PA, Young CM (1998) Temperature and pressure tolerances in dispersal stages of the genus *Echinus* (Echinodermata: Echinoidea): prerequisites for deep-sea invasion and speciation. Deep-Sea Res II 45: 253–277

Tyler PA, Young CM, Clarke A (2000) Temperature and pressure tolerance of embryos and larvae of the Antarctic sea urchin *Sterechinus neumayeri* (Echinodermata: Echinoidea): potential for deep-sea invasion from high latitudes. Mar Ecol Prog Ser 192:173–180

Urban JP (1994) The chondrocyte: a cell under pressure. Br J Rheumatol 33:901–908

Van Dover CL, Lutz RA (2004) Experimental ecology at deep-sea hydrothermal vents: a perspective. J Exp Mar Biol Ecol 300:273–307

Wann KT, MacDonald AG (1980) The effects of pressure on excitable cells. Comp Biochem Physiol 66:1–12

Wilson RR, Smith KL (1985) Live capture, maintenance and partial decompression of a deep-sea grenadier fish (*Coryphaenoides acrolepis*) in a hyperbaric trap-aquarium. Deep Sea Res 32:1571–1582

Yayanos AA (1978) Recovery and maintenance of live amphipods at a pressure of 580 bars from an ocean depth of 5700 meters. Science 200:1056–1059

Yayanos AA (1981) Reversible inactivation of deep-sea amphipods (*Paralicella capresca*) by a decompression from 601 bars to atmospheric pressure. Comp Biochem Physiol 69:563–565

Yayanos AA, Dietz AS (1983) Death of a hadal deep-sea bacterium after decompression. Science 220:497–498

Young CM, Cameron JL (1989) Development rate as a function of depth in the bathyal echinoid *Linopneustes longispinus*. In: Ryland JS, Tyler PA (eds) Reproduction, genetics and distributions of marine organisms. Olsen, Fredensborg, Denmark, pp 225–231

Young CM, Tyler PA (1993) Embryos of the deep-sea echinoid *Echinus affinis* require high pressure for development. Limnol Oceanogr 38:178–181

Young CM, Tyler PA, Fenaux L (1997) Potential for deep sea invasion by Mediterranean shallow water echinoids: pressure and temperature as stage-specific dispersal barriers. Mar Ecol Prog Ser 154:197–209

Young CM, Tyler PA, Gage JD (1996) Vertical distribution correlates with pressure tolerances of early embryos in the deep-sea asteroid *Plutonaster bifrons*. J Mar Biol Ass UK 76:749–757

Zimmerman AM (1971) High-pressure studies in cell biology. Int Revue Cytol 30:1

Zimmerman AM, Marsland D (1964) Cell division: effects of pressure on the mitotic mechanisms of marine eggs (*Arbacia punctulata*). Exp Cell Res 35:293–302

Rev Environ Sci Biotechnol (2006) 5:281–296
DOI 10.1007/s11157-006-9000-8

REVIEW PAPER

# Human challenges in polar and space environments

G. M. Sandal · G. R. Leon · L. Palinkas

Received: 27 February 2006 / Accepted: 28 June 2006 / Published online: 28 July 2006


**Abstract** This paper reviews literature on psychosocial adaptation in isolated and confined extreme (ICE) environments, focusing on polar work groups and expedition teams, and simulation and actual space crews. Long-duration missions may involve chronic exposure to many stressors that can negatively impact behavioral health, performance and even safety. In the last decades, anecdotal evidence has been replaced by scientific studies, identifying temporal, social, and individual determinants of psychosocial adaptation, and pointing to countermeasures that may minimize or prevent potential problems. Still, many issues remain that require additional investigation, specifically in relation to the integration of psychosocial and neurobiological adaptation. A recognition of ICE environments as natural laboratories for studies of fundamental questions within psychology may attract more scientists to the field.

**Keywords** Health · Behavior · Psychology · Interpersonal · Culture · Countermeasures · Isolated and confined extreme environments

G. M. Sandal (✉)
Department of Psychosocial Science, University of Bergen, Christiesgt 12, 5015 Bergen, Norway
e-mail: Gro.Sandal@psych.uib.no

G. R. Leon
Department of Psychology, University of Minnesota, Minneapolis, MN, USA

L. Palinkas
School of Social Work, University of Southern California, Los Angeles, CA, USA

## 1 Introduction

Characteristics and determinants of human adaptation in extreme and unusual environments has been a central interest of a considerable number of psychological researchers and also of those who are responsible for planning and carrying out expeditions involving such environments. Any environment to which humans are not naturally suited, and which demands complex adaptation can be considered an "extreme environment" (Kanas and Manzey 2003). Still, a definition of "extreme environment" in a psychological sense needs to consider that people may react differently to the same environment. Several researchers have argued that the crucial determinant of the stress response is not the environment itself but rather the meaning that the individuals attach to their experiences (Levine and Ursin 1991). This paper concentrates on a subset of environments that share a number of similarities that most people are likely to find inherently stressful: polar work groups and expedition teams including Antarctic winter-over personnel, and crews of manned space missions and space simulation experiments. Although distinct in many aspects, a commonality of these situations is an isolated, confined, and extreme (ICE) environment in which the harsh physical surroundings present many challenges and dangers, including the need to work in microgravity with the threat of hazardous radiation exposure, or in extreme cold, donned in bulky clothing. In times of crisis, evacuation from these isolated environments may be difficult if not impossible. Challenges from the personal perspective may also include the necessity to interact with a small number of possibly mixed gender, culturally diverse individuals living and working in close quarters, for example, in a small Antarctic winter-over field station, or in a tent at the end of a day of trekking. Moreover, the safety of each person is dependent on others in the group, necessitating highly adaptive group functioning and optimal behavioral health.

In this paper, we review research dealing with characteristics and determinants of human adaptation in these settings, focusing on individual psychological reactions and group dynamics. The commonalities between the environments has led to the view that research on human interactions in polar expeditions or research stations can be viewed as a space analogue, and informative about living and working in space (Stuster 2005). Nonetheless, the variability that may exist between ICE settings and even between crews operating within the same physical context needs to be recognized. An Input-Process-Output model provides one framework for sorting the myriad of factors affecting human adaptation in ICE environments, and may guide comparisons across settings. Variables and their interrelations are portrayed in Fig. 1. In addition to the accumulation of scientific knowledge, a merit from such studies may lie in their applied value for the development of evidence-based countermeasures to minimize psychological and interpersonal problems among personnel operating in these environments.

## 2 Environments

### 2.1 Antarctic station

The psychology of human experience in Antarctica extends back to the very first winter-over experience, the Belgica Expedition of 1898–1899. The field of Antarctic or polar psychology, however, began as a scientific enterprise under the leadership of men such as Eric Gunderson of the United States, Jean Rivolier of France, and Tony Taylor of New Zealand soon after the International Geophysical Year (1956–1957) and the establishment of permanent research stations on the ice. At present there are 47 such stations located throughout the Antarctic and sub-Antarctic

**Fig. 1** Factors affecting human adaptation in ICE environments

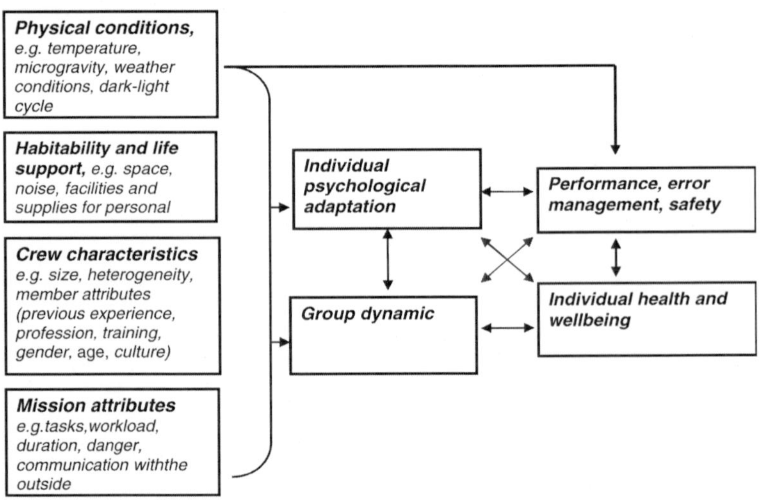

regions, operated by 20 different nations. The populations of these stations range from 14 to 1100 men and women during the summer months (October–February) and from 10 to 250 during the winter months (March–September). Over the course of a year, the human population of Antarctica varies from approximately 1000 to 5000 men and women, most of who are between the ages of 18 and 60 years. For varying periods of time, depending on their location, the stations are physically isolated from the outside world, with darkness and weather conditions preventing traveling to and from the continent with no prospect of medical evacuation.

### 2.2 Polar expeditions

Polar expeditions are usually conducted in the spring or summer months when weather conditions are most favorable, and are intended to recreate the experiences of the early polar explorers, set new records for speed and distance, conduct scientific research, and transport supplies and equipment to remote locations. While some expeditions use traditional modes of travel (sled dogs, skis, snowshoes), others use more modern modes of transportation (snowmobiles, tractors, tracked vehicles). Ocean voyages in small sailing craft also fall into this general category. Another form of polar expedition is the summer camp. Such camps are usually for scientific (e.g., glaciology, geology, marine biology) or commercial (e.g., mineral and oil exploration) purposes. Camps range in size from 3 to 300 individuals who reside in tents, Quonset huts, or other temporary shelters. There are also ocean voyages in larger scientific research or commercial (Mallis and Deroshia 2005) exploration vessels. The duration of this form of expedition ranges from two weeks to three months, usually during the summer.

### 2.3 Space missions

When manned spaceflight began in the early 1960s, the impact of psychosocial factors on astronaut behavior and performance was perceived to be minimal, at least from the U.S. perspective. Space missions had no more than three individuals, were exclusively male, and were drawn from the ranks of test pilots or scientists with pilot experience. With the advent of space stations (e.g., Skylab, Salyut, Mir, and the International Space Station (ISS)), the duration of space missions increased to six months or longer, and crews became more multinational and heterogeneous in terms of gender, cultural background, and professional training. These changes have been assumed to increase the psychosocial demands of the missions and consequently the need for countermeasures aimed to minimize dysfunction in this area (Leon 1999) . Future large advances in the evolution of human space exploration might include a return to the Moon and the establishment of a lunar station as a permanent habitat, as well as flights of humans to our neighboring planet Mars.

Several studies addressing such mission scenarios have recently been conducted by NASA, ESA and in Russia (e.g. Horneck et al. 2003; European Space Agency (ESA) 1992). Technology is just one important aspect of such long-duration space missions. In addition, a number of medical and psychological challenges need to be considered, which might become a limiting factor for prolonged human expeditions into low Earth orbit and beyond. This is particularly true for human missions to Mars, which might not be psychologically comparable to any other undertaking humans have ever been attempted. Specifically, such long-duration missions will require increased crew autonomy and reliance on automation. Also, the distance from Earth either will impede or make impossible the traditional level of communication and support by controllers on Earth which have existed in previous manned missions (Manzey 2003; Sandal 2001b). To date, there have been few research opportunities for collecting psychological data during actual space flights, and much of current assumptions are based on data from personnel operating in so-called analogue environments in which people are exposed to many of the same stressors as those experienced by astronauts in space, although without such environmental components as microgravity and radiation exposure. In addition to evidence from Antarctica winter-over groups and polar expeditions, space simulation studies have been conducted by the use of hyperbaric chamber facilities in which multinational crews have been confined for periods ranging from 28 to 240 days. A basic aim of these studies has been to approximate the living conditions of astronauts on a space station in terms of operational, technical and medical requirements.

## 3 Issues

### 3.1 Psychological functioning

#### *3.1.1 Psychological adaptation and time patterns*

ICE environments have a significant influence on behavioral functioning. The alterations in light-dark periodicity in polar regions disrupt the circadian sleep cycle, resulting in considerable sleep disturbances during the darkness period, and an associated decline in subjective feelings of well-being and alertness in polar areas (Bhargava et al. 2000; Natani et al. 1970; Palinkas et al. 1995a; Palinkas et al. 1996). Also in space, the loss of the 24-h light/dark cycle, circadian disruption, microgravity, and workload demands may result in performance decrements, decreased alertness and sleep disruptions (Mallis and Deroshia 2005). Subjective reports from astronauts (Santy et al. 1988), as well as results from sleep monitoring studies (Gundel et al. 2001), show that sleep in space is shorter, more disturbed, and often shallower than on Earth, though with a considerable degree of inter-individual variation. Other psychological stress reactions have also been reported. During the long winter-over period in Antarctica, personnel have reported increases in depressive mood, psychosomatic complaints, and interpersonal conflicts, and a decrement in work performance (Bhargava et al. 2000; Ikegawa et al. 1998; Palinkas et al. 1995a, b, 1996; Palinkas and Johnson 1990). Likewise anecdotal and behavioral evidence from space missions show that crewmembers have experienced psychological reactions that have included lapses of attention, emotional liability, psychosomatic symptoms, irritability toward crewmates and/or mission control staff, and a considerable decline in vigor and motivation (Kanas and Manzey 2003; Suedfeld 2005). According to Russian space psychologists and flight surgeons, a psychiatric condition that affects the emotional state of cosmonauts during prolonged missions is asthenia. This syndrome, that refers to a psychiatric diagnostic category, is defined as a weakness of the nervous system that may result in fatigue, irritability and emotional liability, attention and concentration difficulties, restlessness, heightened perceptual sensitivities, palpitations and blood pressure instability, physical weakness, and sleep and appetite problems (Kanas et al. 1991; 2001a, b; Myasnikov and Zamaletdinov 1998). Although "asthenization" is carefully monitored, and a number of countermeasures are employed to prevent it from progressing, empirical evidence for its existence as a discrete pathological entity has been equivocal (Kanas et al. 1991, 2001a, b).

Yet, other studies have either found no evidence of psychological or psychiatric symptoms or suggested that such problems pose little threat to the health and well being of the crew or to the success of the mission (Leon et al. 1989). Such studies suggest one of four possibilities: (a) Isolated and confined environments are no more stressful than other environments (Suedfeld and Steel 2000); (b) highly motivated, self-selected individuals who volunteer for such long-term missions are capable of maintaining high levels of performance over long periods of time (Palinkas et al. 2000); (c) motivated individuals simply do better than others; or (d) psychological reactions are strongly affected by interpersonal and cultural factors. Supporting the latter, Ritsher et al. (2005) have presented data indicating that psychological reactions to being in space are expressed differently by people from different cultural backgrounds. Clearly, the resolution of this issue needs further clinical and empirical studies.

While the majority of research has focused on potentially adverse health effects based on a pathogenic model, there has also been an emphasis on psychological resilience and positive aspects. Studies of groups working in polar regions over relatively shorter periods and light conditions have found what is often referred to as a "salutogenic" effect" (Antonovsky 1979), likely associated with removal from the time pressures and stressors encountered in the home setting (Palinkas et al. 1995b; Sandal 2000). In addition, expedition studies have shown that positive mood prevailed over negative mood over the course of the trek (Atlis et al. 2004; Kahn and Leon 1994; Leon 1991; Leon et al. 1989, 1991, 2002), and depressive mood decreased (Palinkas et al. 1995b). Many polar and other "capsule environment" crews also reported long-term positive after-effects subsequent to their stay (Suedfeld and Steel 2000).

The timing or sequencing of mood changes and social tensions over the course of a polar or space sojourn has been examined, particularly the presence of a "third quarter phenomenon" (Bechtel and Berning 1991). This term refers to an increase in negative mood and social interactions during the third quarter of the stay, as personnel contemplate the duration of time remaining before the end of their particular ICE experience. The attention to the third quarter phenomenon can be traced back to Rohrer's (1961) observation of Antarctic and submarine missions, in which he described three stages of crewmember response: initial anxiety over new experiences in the mission, mid-mission monotony and depression as tasks become routine, and late-mission euphoria and immature behavior as the end is anticipated. Elements of these stages have been reported during long-term Russian space missions and confinement studies, but empirical evidence for the existence of specific critical phases has been equivocal (Bhargava et al. 2000; Gushin et al. 1997; Leon et al. 2002; Palinkas 2003; Palinkas et al. 2000, 2004b; Sandal 2000, 2001a; Sandal et al. 1996; Steel and Suedfeld 1991; Stuster et al. 2000; Wood et al. 1999). For example during a 438-day MIR mission, Manzey and Lorenz (1998) monitored different indicators of mood, but found little indications to support a stage model of adaptation. In another study, Kanas et al. (2001a, b) studied time-related changes of crew interaction and crew-ground communication. Again, only few indications of a stage model was found. Yet, the issue of critical periods in adaptation is important and needs more research attention, given that such knowledge may enable crewmembers and outside personnel to prepare for problems and intervene before maladjustments result in adverse operational or health impacts (Sandal 2001b). Sandal et al. (1996) found different time patterns in psychological reactions when comparisons were made between polar expeditions and crews confined in hyperbaric chambers. They suggested that a stage-model of adaptation is probably more relevant for groups undergoing prolonged confinement in which boredom and monotony are prominent stressors. Thus, time in itself may not be a strong predictor unless taking into consideration aspects of the environment.

### *3.1.2 Cognitive adaptation*

A related issue regarding psychological functioning in polar and space environments is a need to understand and assess patterns of cognitive adaptation. Poor performance due to failure of cognitive adaptation is identified in NASAs Bioastronautics Roadmap as a potential risk to long-duration missions in space. Human performance in space can suffer from both microgravity effects involving vestibular and sensory-motoric processes as well as non-specific stress effects related to workload, sleep disturbances and other factors of extreme living and working conditions in space (Leone 1998; Kanas and Manzey 2003). Empirical studies addressing possible effects of spaceflight-related stressors on human cognitive and psychomotor performance have focused primarily on elementary cognitive and psychomotor tasks. Most of them have been conducted during short-term ( $<$ 30 days) spaceflights (Benke et al. 1993; Manzey 2000). In spite of the comparatively small number of studies and the differences in the methodological approach, they have revealed a fairly consistent pattern of effects.

Whereas performance in elementary cognitive tasks such as memory-search, logical reasoning, or mental arithmetic seem to remain largely unimpaired in space, disturbances have been found in perceptual-motor and attention tasks (Bock et al. 2001; Manzey and Lorenz 1998). The origin of these effects, i.e. whether they reflect effects of microgravity on the central nervous system or unspecific stress effects of workload and fatigue, has remained unclear. Studies of cognitive performance in polar environments suggest that such performance is certainly impaired under conditions of hypothermic exposure to cold (Coleshaw et al. 1983), performance may either increase (Mäkinen et al. 2005; Palinkas et al. 2004b), decrease (Barabasz et al. 1983; Palinkas 2001a; Reed et al. 2001), or remain unchanged (Le Scanff et al. 1997) under non-hypothermic exposure. The reasons for these observed differences remain to be determined.

## 3.2 Group dynamics

Living and working in a relatively small group in an isolated, confined and demanding environment are likely to pose challenges to interpersonal relationships. Studies in ICE environments have identified a number of factors that impact the efficiency and quality of interpersonal relationships, including crew structure and cohesion, leadership style, gender and cultural background of crew members, and inter-group relationships (Gushin et al. 1998; Gushin et al. 1997; Kanas 1990; Kanas et al. 1996; Kanas and Manzey 2003; Sandal 2000). While these factors are general predictors of satisfaction and efficiency in organizational settings (McKenna 2001), there are important questions about how these relationships are influenced by ICE conditions, as well as potential time effects.

### *3.2.1 Crew tension and cohesion*

Polar expedition teams and space crews often experience greater social cohesion by virtue of undergoing a common experience (Atlis et al. 2004; Leon and Sandal 2003; Stuster 1996). Nonetheless, confinement in a small space with the same people for a long time, under conditions of danger, discomfort, fatigue and other stressors, is a potential catalyst for interpersonal hostility. Research has consistently demonstrated that interpersonal conflict and tension is the greatest source of stress in Antarctica (Stuster et al. 2000). Still, inhabitants in ICE environments often seem to be reluctant to express tension openly; rather, it is reflected indirectly in territorial behavior, withdrawal from interaction with others, and clique formation that may occur along national or vocational lines (Kanas 1990; Sandal et al. 1995). One explanation may be the mutual interdependence among the participants. While formation of group identity appears to facilitate the adjustment of individuals to the ICE conditions, participants who do not conform to the group's norm tend to be socially isolated (Palinkas and Johnson 1990) . During space simulation studies in hyperbaric chambers, analysis of communication networks indicated that one crewmember became more and more in the periphery of crew interaction, and at the same time was rated less positively by other crewmembers (Sandal 2001a; Sandal et al. 1995). "Scapegoats" may serve to maintain harmony between the other crewmembers because that person becomes the target of intra-group hostility and aggression. However, these individuals are vulnerable for poor psychosocial adaptation. Problems such as scapegoating are well-documented from more usual working life, but normally individuals who are discomforted by an interpersonal experience have the opportunity to withdraw from interaction, or to seek the companionship and support of other people. Under isolated conditions where there are no possibilities for escape from the adverse situation, the scapegoat is probably at higher risk for some type of health or other adaptation problem.

Tension may also be a consequence of individual patterns of adaptation to ICE environments such as refraining from relying on their fellow crew members for support Palinkas (2003) has distinguished between social dynamics as a stressor and social support as a mediator of the stress–performance relationship. He pointed out that individuals seem to adapt to ICE environments by refraining from reliance on their fellow crew members for support largely because these crew members are facing the same stressors (Palinkas 2003). Relevant to this point, Sandal et al. (1998) found that seeking social support as a coping mechanism was related to poor psychological adjustment during long-duration submarine missions.

Not all isolated and confined crews experience high levels of tension. Studies of Antarctic winter-over crews have found that group cohesion varies from one year to the next (Palinkas 1989), and across national stations (Palinkas et al. 2004b). Use of multidimensional scaling of data collected from the winter-over crews at the South Pole during the same three-year period revealed three distinct patterns in crew structure that seemed to have important implications with respect to tension and anxiety, depression, and anger and hostility (Palinkas, 2003). The crews characterized by a clique structure exhibited significantly higher levels of tension and anxiety, depression, and anger than did the crews characterized by the core-periphery structure throughout the entire winter. Similarly, Sandal et al. (1995) found marked differences in crew structure and group cohesion between space simulation studies in hyperbaric chambers. One distinction between the crew that was highly unified and cohesive compared with the other one, was that the former had been selected based on interpersonal compatibility.

### *3.2.2 Leadership*

The execution and maintenance of leadership may be challenging in ICE settings in which participants live close together for prolonged periods. Communication analysis and post-mission interviews suggested that the authority of the Commander was challenged by other participants during several space simulation studies in hyperbaric chambers (Sandal 2001a, 2004; Sandal et al. 1995). In another study, lasting for 135 days, Kanas and his colleagues (Kanas et al. 1996) found a significant drop in a measure of task-oriented, instrumental characteristics of the appointed Commander over time, which was interpreted as status leveling where the leader assumed a more equal role. Status leveling may be unfortunate in emergency situations in which clear and unambiguous leadership is required. Successful leadership during long-term missions must be flexible. For example, studies of polar expeditions found task leadership to be more important during the initial stages, while supportive leadership became more important during the later phases of the expedition (Stuster 1996). Reviews of studies of personnel in different ICE environments have highlighted characteristics of the best leaders, including the perception by group members that the leader is a role model, soliciting subordinates' advice and judgments when necessary and appropriate, sensitive to subordinates' personal problems and well-being, and clearly communicating roles and responsibilities (Nicholas and Penwell 1995; Stuster 2005). These aspects are not unique to ICE environments, but have consistently been found to relate to group efficiency, motivation and satisfaction in more customary organizations as well, often referred to as transformational leadership (Bass and Avolio 1995; Hetland and Sandal 2003).

### *3.2.3 Heterogeneity*

Heterogeneity in cultural background and gender are assumed to increase the risk of interpersonal tension and miscommunication among isolated and confined personnel. The term "culture" has been defined as widely shared beliefs, expectancies, and behavior of members of a group on an organizational, professional or national level (Hofstede 1980). Evidence from cross-cultural research suggests that the greater the cultural differences among people, the greater will be the difficulties in establishing harmonious and productive relationships (Berry 2004). Based on studies of aircrews (Helmreich and Merrit 1998), large national differences have been identified in individualism, preferred leadership style, and adherence to rules and procedures. It is reasonable to assume that differences could well impede teamwork in the demanding environment of space and polar settings. Interpersonal tension related to language and cultural differences has been reported from both long-duration Russian missions involving people from other nations and earthbound ICE environments. For example, during the Salyut 6 mission, a visiting cosmonaut from Czechoslovakia felt socially isolated and complained of being restricted from doing productive work by his Russian crewmates who were concerned that their foreign guest might inadvertently make an operational error (Kanas et al. 2000). Language and cultural factors were found to socially isolate crewmembers representing minority cultures during EUROMIR missions (Kanas et al. 2000), a 240 days space simulation study ("SFINCSS") in hyperbaric chambers (Sandal 2004), and on research stations in Antarctica (Sandal 2000). Also, the two former studies showed that the minority crewmembers reported being more unhappy about their work environment.

Clear differences in perceived leader support, work pressure, and managerial control were found between Russian and American crew members who stayed and worked together on MIR (Kanas et al. 2000). Yet, with the exception of the International Space Station, spaceflight experiences with multi-cultural space crews, so far, stem from missions where cross-cultural aspects were inevitably confounded with "host"–"guest" differences, i.e., the missions usually involved one "dominant culture" being the host for crew members from other countries, organizations and professions. This was also the case for the "SFINCSS" confinement study. The relatively small sample sizes also make it difficult to isolate the effects of culture from other factors, such as the personalities, motivations and professional training of crewmembers and the organizational setting in which the project was conducted. Moreover, the interplay between national, organization and professional cultures is conceptually complex and needs to be further explored. Belonging to the same organizational culture (such as the European Space Agency) may reduce the impact of national diversity. It is also possible that on a national level, cross-cultural issues may have minimal impact on crew behavior and performance in space since astronauts and cosmonauts are all part of a common professional "microculture". So far, very few systematic studies have addressed these issues.

A number of polar, space, and simulation chamber groups have been composed of single gender as well as mixed gender teams (Leon 2005). Women in mixed-gender polar and space simulation groups have often assumed or been placed in a more nurturing and less dominant role and reported feeling stressed by their concerns about other team members (Leon et al. 1989), and have assumed the role of "peacemaker" (Leon et al. 2004; Sandal et al. 1995). All-men expedition teams have exhibited marked competitiveness and little sharing of personal concerns, while all-women expedition groups, or women in mixed gender groups have shown a greater cooperative orientation, supportive relationships, and concern about the welfare of their team members (Bishop 2002; Koscheyev et al. 1992; Leon and Sandal 2003; Rothblum et al. 1998). Based on qualitative interviews, Rosnet and her colleagues (Rosnet et al. 2004) concluded that including women in wintering-over groups on Antarctic stations had positive effects on the general climate of the group by reducing rude behavior, but it also seems to cause stress due to rivalry, frustration, and sexual harassment. Experiences from the "SFINCSS" confinement study also point to possible problems of interpersonal tension. In this study, a conflict situation developed due to unwanted sexual advances toward the lone female crew member, a Canadian, by one of her seven male crewmates who was a Russian (Sandal 2004). Considerable cultural and gender differences in attitudes toward women team members have also been noted (Leon 2005; Leon et al. 1991, 1994).

The Dutch anthropologist, Hofstede (1980) identified important national differences in the view of appropriate gender role behavior, and unless addressed, such differences may cause conflicts and tension in multinational and mixed-gender settings. However, it is also important to consider that in certain cultures, attitudes about appropriate roles for men and women are changing, and an increasing proportion of individuals of both sexes are engaging in non-traditional occupational or other pursuits. Relevant to this point, Musson, Sandal, and Helmreich (2004) found no significant differences in personality scores between final stage men and women astronaut applicants. Moreover, within each gender, there are considerable individual differences in personality, behavioral characteristics, skills, and interests that influence task performance and other behaviors in a particular milieu, whether in ICE or other environments.

#### *3.2.4 Inter-group relations*

Long term co-habitation and shared experiences in a highly unique environment are factors that facilitate the development of strong group identification, and therefore help account for the frequently observed psychological distance between "insiders" and "outsiders". Direction of aggression to outside personnel has been reported during both Russian and American space missions (Kanas et al. 2001a, b), in Antarctica, and during space simulation studies (Gushin et al. 1997; Kanas et al. 1996; Sandal 2001b; Sandal et al. 1995). Some of these incidents have been interpreted as a displacement of tension/aggression to help maintain harmony within the crew. This process is often referred to as the "us vs. them" phenomenon, and appears to be one mechanism that functions to unite the confined personnel. Gushin (1995) has noted the tendency of space crews to avoid sharing feelings with others ("psychological closing") and information filtration in crew-ground communications. A concern is that these characteristics may also interfere with the capacity of the crew to accurately receive and process information from the ground, thereby resulting in errors in judgment and a decrement in performance.

A potential deleterious outcome of strong group-cohesion coupled with "psychological distance" to outsiders is a phenomenon known as "groupthink", identified by several characteristics such as denial of conflicts, a tendency to channel aggression outside the group, and the existence of external pressure (Janis 1972). Groupthink has been associated with lower-quality performance because group members are too concerned with getting along and reluctant to express disagreement; clearly, this could endanger a mission in situations of crisis.

### 3.3 Personality predictors of adaptation/optimal performance

Psychological adaptation to isolated and confined environment varies with factors such as personality traits, physiological and circadian rhythm characteristics (Mallis and Deroshia 2005; Van Dongen et al. 2003), prior experience, and training background. However, astronauts, polar expedition members and personnel who winter-over in Antarctica seem to possess common characteristics that differentiate them from people in general. Steel et al. (1997) found that Antarctic expeditioners scored higher on measures of extraversion and openness to experience and lower on measures of neuroticism than population norms. High achievement orientation and low stress reactivity have been found to be predominant characteristics in many polar team members (Atlis et al. 2004; Kahn and Leon 1994; Leon 1991; Leon et al. 1989), Antarctic scientists (Butcher and Ryan 1974) , and astronaut applicants at NASA (Musson et al. 2004). These traits are strongly related to coping with stress and adaptation across populations (see Sandal et al. 1998). Variation in personality traits found in ICE groups is normally much smaller than is typically found in the normal population (Palinkas 2003; Sandal 2000; Sandal et al. 1996). Due to "restriction of range", baseline measures of personality therefore may be a relatively weak predictor of behavior in ICE settings. Nonetheless, personality assessment has been recognized by space agencies and researchers as a helpful tool in screening out unsuitable astronaut candidates (Santy 1994).

Several studies have linked superior performance to a personality profile characterized by a combination of high levels of instrumentality and expressivity along with lower levels of interpersonal aggressiveness and low competitiveness in settings that include aviation (Chidester et al. 1991), military training (Sandal et al. 1998, 1999), short-term space simulation studies (Sandal et al. 1996), and polar expeditions (Leon and Sandal 2003). Likewise, research in which personality traits have been validated against criteria of astronaut effectiveness during short-term space missions and training sessions have identified high agreeableness and low aggressiveness as general characteristics of high performers (Rose et al. 1994). These personality traits have therefore been designated as "the right stuff" for team operations in stressful environments. Referring to the so-called five-factor model of personality (McCrae and Allik 2002), Suedfeld and Steel (2000) proposed that the traits of conscientiousness and agreeableness would be optimal indicators of adaptive personality functioning. Interestingly, retrospective analysis of personality scores of final stage astronaut applicants showed no differences based on personality clusters on those selected vs. those deselected (Musson et al. 2004).

However, not all of these personality characteristics have been demonstrated to be predictive of optimal performance in ICE environments. For instance, a study of French expeditioners found that performance was associated with low scores on extraversion and assertiveness (Rosnet et al. 2000). A person–environment fit model of behavior suggests that the optimal individual characteristics in relation to performance are mediated by environmental factors and mission characteristics, including the duration and crew characteristics (Ursin et al. 1992). For instance, Palinkas and Browner (1995) found that while several features of personality and social factors were associated with concurrent measures of depressive symptoms, pre-deployment levels of depressive symptoms were the only independent predictor of late-winter depression. A similar prospective screening study of Antarctic winterer-over personnel (Palinkas et al. 2000) showed that the best performers were characterized by low levels of neuroticism (emotional liability), low desire for affection from others, low levels of boredom, low need for order, and a high tolerance for lack of achievement. The authors argued that under conditions of isolation and confinement, satisfying a need for achievement and order is often restricted by the environment itself, and that individuals wishing to complete projects on schedule become frustrated at delays in communication with the outside, constant equipment failure, or absence of necessary supplies. Consequently crewmembers adapting best are those most able to revise their expectations to fit the situation.

In the context of long-duration space missions, the most severe stressors may involve monotony and boredom resulting from low workload, hypostimulation, and restricted social contacts including isolation from family and friends. A number of factors may affect the coping ability of crewmembers in dealing with these living conditions. These include their unique onboard experiences, personality traits, leisure activities, particular coping strategies, and the kind of social and emotional support available. Antarctic studies evaluating personnel in a range of occupational positions showed an association between adaptive functioning and narrow interests and a low need for stimulation (Biersner and Hogan 1984). A personality trait termed "absorption" (Tellegen and Waller in press) has been identified in studies of different expedition teams (Atlis et al. 2004; Kahn and Leon 1994; Leon 1991; Leon et al. 1989). This characteristic refers to the ability to become highly engrossed in a particular activity to the exclusion of attending to other events that are happening around the person. For example, becoming so engrossed in the beauty of the Antarctic landscape that the monotony of the extensive physical exertion in a cold and windy environment is not processed as uncomfortable. In a related manner, a number of astronauts have commented that their most enjoyable leisure activity in space was simply looking out the porthole at the beauty of the Earth. This engagement in the beauty of the environment or becoming highly engrossed in a particular task would likely be highly adaptive in coping with the exigencies of living in an ICE environment, including space, for an extended period of time.

## 4 Future challenges

Despite the advances in our understanding of the human challenges in polar and space environments that have occurred in recent years, a number of issues remain that require additional investigation. One of the primary issues related to psychological functioning concerns the integration of psychosocial and neurobehavioral adaptation. These two domains are usually examined separately and considered individually as risk factors for poor performance (Health Sciences Policy Board 2005). Nevertheless, studies in both space and polar environments suggest that both are interrelated. For instance, changes in thyroid and catecholamine function in polar environments have been linked to poor mood and impaired cognitive performance (Reed et al. 2001). Similarly, stress and mood are believed to influence endocrine, immune, and cardiovascular function in space (Kanas and Manzey 2003; National Research Council 2000). Ultimately, solutions to human challenges in these environments will require multidisciplinary efforts and approaches to understanding the social and environmental factors that contribute to impaired physiological functioning and the neurobehavioral changes that precede behavioral changes and vice versa. Such efforts will include the formation of interdisciplinary teams of investigators and the development of integrated models of human adaptation that are specific to these environments.

Future research should also be devoted to improving our understanding of performance and workload (National Research Council 1998), and the development and evaluation of behavioral, operational and pharmacotherapeutic countermeasures designed to reduce the risk of impaired cognitive performance in both types of environments (Palinkas 2001a). The degree of autonomy that crews expect to have from controllers can have a big impact tension and cohesion during missions (Kozerenko et al. 1999). According to Kanas and Manzey (2003), providing the crew with as much freedom as possible to plan and schedule activities on their own initiative represents a countermeasure for dealing with possible impairments of work satisfaction and motivation.

A particular concern in planning for long-duration missions is how to deal with significant psychological problems and psychiatric disorders that may develop in initially healthy individuals. This situation would affect not only the disabled person, but could also have a detrimental effect on overall crew performance, safety, and potentially, the success of the mission. Information on serious behavioral health problems in space is not publicly available due to issues of confidentiality. However, considering the commonalities between certain polar and long-duration space conditions, including the impossibility of evacuation during specific periods of the mission, it may be possible to generalize about the probability of serious problems developing in space by extrapolating from Antarctic health data.

Aggregation of data on the Australian National Antarctic Research Expeditions (ANARE) collected over a 25-year period indicated a rate of morbidity for mental health disorders below 4% (Lugg 2005). Research conducted on members of the United States Antarctic Program suggest that approximately 5% of winter-over personnel experience symptoms that fulfil criteria for a psychiatric disorder and are severe enough to warrant clinical intervention; Palinkas et al. 1995a; Palinkas et al. 2004a). Psychiatric debriefings of 313 men and women conducted at McMurdo and South Pole between 1994 and 1997 revealed that 3.8% of personnel experienced mood disorders (depression), 3.8% experienced adjustment disorders, 2.6% experienced sleep disorders, 1.3% experienced alcohol or drug-related disorder, and 1.0% experienced personality disorders (Palinkas et al. 2004a). Although these rates are lower than what might be experienced in the general population in the United States, they are noteworthy in that these men and women are required to undergo psychiatric screening prior to the austral winter. However, if a serious psychiatric problem developed in space, for example, a paranoid reaction or immobilizing depression, this could seriously jeopardize the mission. Therefore, the development and implementation of psychological, pharmaceutical, and other countermeasures to deal with these potential problems is extremely important.

With respect to interpersonal relations and social dynamics, there remains a need to identify the criteria to be used in screening and selection of teams intending to live and work in polar and space environments. The research described in this paper has identified several important predictors of individual performance, but studies identifying the determinants of successful group performance remain to be accomplished (National Research Council 1998). Research on predictors of individual performance provides little guidance regarding optimal combinations of individual characteristics across group members and whether crews should be selected on the basis of homogeneous or complementary characteristics (Palinkas 2001b). Similarly, operational and political considerations may preclude the screening and selection of personnel strictly on the basis of psychosocial characteristics (National Research Council 1998). However, an understanding of the effect of these considerations on overall group performance remains to be determined.

The difficulty in overriding such selection considerations also points to the need for further research on the development of policies, programs and countermeasures designed to prevent poor psychosocial and neurobehavioral adaptation in polar and space environments. Prevention is a more cost effective approach to addressing poor adaptation since evacuation from such locales is expensive if not impossible during particular time periods, and treatment of adaptation-related problems may be costly, time-consuming, and feasible only if necessary resources are readily available (e.g., access to qualified personnel to conduct crisis intervention under periods of acute stress, or medications which have bioavailability under conditions of extreme cold or microgravity). Moreover, loss of productivity, increased risk of accidents and co-morbidity can be minimized if not avoided altogether through prevention. Preventive countermeasures or interventions would include training in interpersonal relations and social dynamics, leadership skills and individual coping strategies (National Research Council 1998; Palinkas 2001b), the design and construction of facilities that allow for privacy and space while fostering a sense of community and promoting social interaction (National Research Council 1998), and the development and evaluation of nutritional and pharmacologic countermeasures designed to reduce the adverse effects of the physical environment (i.e., cold, darkness, microgravity) on psychosocial and neurobehavioral adaptation (Palinkas 2001a). However, evidence-based research is required to support the use of one or more of these countermeasures. Provision of in-flight support is another important countermeasure to prevent feelings of boredom, monotony and isolation during long duration space flight. Attention should be given to enhancing individually tailored leisure time activities that take into account changing interests and needs over the course of the mission (Kanas and Manzey 2003; Kelly and Kanas 1994). Other important support activities include private psychological conferences, providing informal space-ground contact and news from Earth (preferably in the crew member's native language and from homeland news sources), and opportunities to maintain close contact with family and friends on Earth on a regular basis. In-flight support of crew members has been applied in Russia since the earliest days of manned spaceflight, and continues to represent one of the main countermeasures in today's ISS program.

Finally, research is required to facilitate the positive aspects of living and working in polar and space environments. Although the positive benefits of successfully coping with the challenges of such environments have been repeatedly demonstrated (Palinkas 2003; Suedfeld and Steel 2000), it is still unclear whether polar expeditioners and astronauts can be trained to experience such benefits, or whether they are inherent features of particular personality types. In either instance, as is the case with prevention, selection, training, and support are likely more cost-effective than treatment (Palinkas 2003, 2001b) in dealing with the human challenges of polar and space environments. However, as with prevention, the development of an evidence base for the use of specific strategies is needed. Such procedures are likely to include both programs designed to screen for and select individuals most likely to have such positive experiences in these environments, as well as training programs to increase the likelihood of such experiences in all polar expeditioners and spacefarers.

## 5 Concluding remarks

ICE environments have long been viewed as natural laboratories for the study of the effects of isolation and confinement on human behavior. Research results from these settings contrast with the early laboratory studies in which participants were exposed to highly artificial situations, involving isolation from other people and a sharp reduction in the level of variability of physical stimulation and very restricted mobility (Zubek 1969), or military subjects confined in windowless suites of rooms with only research tasks to perform (Haythorn and Altman 1966). A large number of subjects failed to complete several weeks of confinement due to perceived high "stress" and from outbursts of hostility. This situation differs profoundly from the successful adaptation generally reported from groups in less artificial environments. While the last decades have witnessed an increased research interest in ICE environments, most of this research has focused on the abnormality of the setting. As noted by several researchers (Palinkas 2003; Suedfeld 1998), the opportunities that these settings offer for exploring more fundamental and global psychological issues may not have been sufficiently recognized. Contrasting with most natural settings in which people operate, the level, intensity, rate of change, and diversity of physical and social stimuli, as well as behavior settings and possible behaviors are more restricted (Suedfeld 1998). Thus, ICE environments offer a high degree of control over the variables that impact on psychosocial processes while avoiding the artificial conditions of the traditional laboratory. For instance, Palinkas (2003) suggested that the ice is an ideal laboratory for the examination of seasonal variation in behavior related to environmental phenomena, such as cold temperature and limited exposure to daylight. More generally, it is an ideal laboratory for understanding how biological mechanisms and psychological processes interact within a well-defined environment. As such, ICE environments provide unique opportunities for gathering new knowledge and having an impact on serious problems in non-extreme environments, as well as furthering knowledge about optimal adaptation in polar and space settings.

## References


Antonovsky A (1979) Health, stress, and coping. Jossey-Bass, San Francisco

Atlis MM, Leon GR, Sandal GM, Infante M (2004) Decision processes and interactions during a two-woman traverse of Antarctica. Environ Behav 36:402–423

Barabasz M, Barabasz AF, Mullin CS (1983) Effects of brief Antarctic isolation on absorption and hypnotic susceptibility—preliminary results and recommendations. Int J Clin Exp Hypnosis 31:235–238

Bass BM, Avolio BJ (1995) Transformational leadership development: manual for the multifactor leadership questionnaire. Consulting Psychologists Press, Palo Alto

Bechtel RB, Berning A (1991) The third quarter phenomenon: do people experience discomfort after stress has passed? In: Harrison AA, Clearwater YA, MCKay CP (eds) From Antarctica to outer space. Life in isolation and confinement. Springer Verlag, New York, pp 261–266

Benke T, Koserenko O, Watson NV, Gerstenbrand F (1993) Space and cognition: the measurement of behavioral functions during a 6-day space mission. Aviat Space Environ Med 64:376–379

Berry JW (2004) Psychology of group relations: cultural and social dimensions. Aviat Space Environ Med 75:C52–C57

Bhargava R, Mukerji S, Sachdeva U (2000) Psychological impact of the Antarctic winter on Indian expeditioners. Environ Behav 32:111–127

Biersner RJ, Hogan R (1984) Personality correlates of adjustment in isolated work groups. J Res Person 18:491–496

Bishop SL (2002) Evaluating teams in extreme environments: Deep caving, polar and desert expeditions Proceedings of the 32nd International Conference on Environmental Systems. SAE International, Warrendale, PA (SAE Technical Paper Series 2002-01-2496)

Bock O, Fowler B, Comfort D (2001) Human sensorimotor coordination during spaceflight: an analysis of pointing and tracking responses during the "Neurolab" space shuttle mission. Aviat Space Environ Med 72:877–883

Butcher JN, Ryan M (1974) Personality stability and adjustment to an extreme environment. J Appl Psychol 59:107–109

Chidester TR, Helmreich RL, Gregorich E, Geis CE (1991) Pilot personality and crew coordination: implications for training and selection. Int J Aviat Psychol 1:25–44

Coleshaw SRK, Van Someren RNM, Wolffm AHDHM, Keatinge WR (1983) Impaired memory registration and speed of reasoning caused by low temperature. J Appl Physiol 55:27–31

European Space Agency (1992) Mission to the Moon. (ESA SP-1150). ESA Publication Division, Nordwijk

Gundel A, Drescher J, Polyakov V (2001) Quantity and quality of sleep during the record manned space flight of 438 days. Human Factors Aerospace Safety 1:87–98

Gushin V, Efimov V, Smirnova T, Vinokhodova A, Kanas N (1998) Subject's perception of crew interaction dynamics under prolonged isolation. Aviat Space Environ Med 69

Gushin VI (1995) Problems of psychological control in prolonged space flights. Earth Space Rev 4:28–31

Gushin VI, Zaprisa TB, Kolintchenko VA, Efimov A, Smirnova TM, Vinokhodova AG, Kanas N (1997) Content analysis of the crew communication with external communicants under prolonged isolation. Aviat Space Environ Med 12:1093–1098

Haythorn W, Altman I (1966) Personality factors in isolated environments. In: Trumbull MAR (ed) Psychological stress: issues in research. Appleton-Century-Crofts, New York

Health Sciences Policy Board IoM (2005) Bioastronautics roadmap: a risk reduction strategy for the human exploration of space. National Academy Press, Washington DC

Helmreich RL, Merrit A (1998) Culture at work in aviation and medicine: National, organizational and professional influence. Ashgate, Brookfield

Hetland H, Sandal GM (2003) Transformational leadership in Norway: outcomes and personality correlates. Eur J Work Organiz Psychol 12:147–170

Hofstede G (1980) Cultures consequences: international differences in work-related values. Sage, Beverly Hills
Horneck G, Facius R, Reichert M, Rettberg P, Seboldt W, Manzey D, Comet B, Maillet A, Preiss H, Schauer L, Dussap CG, Poghon L, Belyavin A, Reitz G, Baumstark-Khan C, Gerzer R (2003) Humex, a study on the survivability and adaptation of humans to long-duration exploratory missions. (ESA SP-1274). European Space Agency, Noordwijk
Ikegawa M, Kimura M, Makita K, Itokawa Y (1998) Psychological studies of a Japanese winter-over group at Asuka Station, Antarctica. Aviat Space Environ Med 69:452–460
Janis IL (1972) Victims of groupthink. Houghton Mifflin, Boston
Kahn PM, Leon GR (1994) Group climate and individual functioning in an all women's Antarctic expedition team. Environ Behav 26:669–697
Kanas N (1990) Psychological, psychiatric, and interpersonal aspects of long-duration missions. J Spacecraft Rockets 27:457–463
Kanas N, Manzey D (2003) Space psychology and psychiatry. Kluwer Academic Press, Dodrecht
Kanas N, Salnitskiy V, Grund EVIG, Weiss DS, Kozerenko O, Sled A, Marmad CR (2000) Interpersonal and cultural issues involving crews and ground personnel during Shuttle/Mir space missions. Aviat Space Environ Med 7:A11–A16
Kanas N, Salnitskiy V, Gushin V, Weiss DS, Grund EM, Flynn C, Kozerenko O, Sled A, Marmad CR (2001a) Asthenia:does it exist in space? Psychosom Med 63:874–880
Kanas N, Salnitskiy V, Weiss DS, Grund EVIG, Kozerenko O, Sled A, Bostrom A, Marmad CR (2001b) Crewmember and ground personnel interaction over time furing Shuttle/Mir space missions. Aviat Space Environ Med 72:453–461
Kanas N, Weiss DS, Marmad CR (1996) Crewmember interaction during a Mir space station simulation. Aviat Space Environ Med 10:969–875
Kelly AD, Kanas N (1994) Leisure time activities in space: A survey of astronauts and cosmonauts. Acta Astronautica 32:451–457
Koscheyev VS, Lartzev MA, Rode A, Malakhov M (1992) Psychic adaptation of participants observations on the Soviet/Canadian transpolar ski trek Karger, Basel (Medicine and Sports Science vol 33), pp 71–105
Kozerenko O, Gushin V, Sled A, Efimov JM, Pystinnikova (1999) Some problems of group interaction in prolonged space flights. Hum Perf Extreme Environ 4:123–127
Le Scanff C, Larue J, Rosnet E (1997) How to measure human adaptation in extreme environments: the case of Antarctic wintering-over. Aviat Space Environ Med 68:1144–1149
Leon GR (1991) Individual and group process characteristics of polar expedition teams. Environ Behav 23:723–748
Leon GR (1999) Select-in and countermeasure considerations for long duration crews. The 29th International Conference on Environmental Systems. Warrendale, PA:SAE International (SAE Technical Paper Series 1999–01–2095
Leon GR (2005) Men and women in space. Aviat Space Environ Med 76(Suppl):B84-B88
Leon GR, Atlis MM, Ones DS, Magor G (2002) A one-year three couple expedition as a crew analog for a Mars mission. Environ Behav 34:672–700
Leon GR, Kanfer R, Hoffman RG, Dupre L (1991) Interrelationships of personality and coping in a challenging extreme situation. J Res Personal 25:357–371
Leon GR, Kanfer R, Hoffman RG, Dupre L (1994) Group processes and task effectiveness in a Soviet-American expedition team. Environ Behav 26:149–165
Leon GR, List N, Magor G (2004) Personal experiences and team effectiveness during a commemorative trek in the High Arctic. Environ Behav 36:386–401
Leon GR, McNally C, Ben-Porath Y (1989) Personality characteristics, mood, and coping patterns in a successful North Pole expedition team. J Res Personality 23:162–179
Leon GR, Sandal GM (2003) Women and couples in isolated extreme environments: applications for long-duration missions. Acta Astronaut 53:259–267
Leone G (1998) The effect of gravity on human recognition of disoriented objects. Brain Res Rev 28:203–214
Levine S, Ursin H (1991) What is stress? In: Brown MR, Koob GF, Rivier C (eds) Stress-neurobiology and neuroendocrinology. Marcel Dekker, New York
Lugg DJ (2005) Behavioral health in Antarctica: implications for long-duration space missions. Aviat Space Environ Med 76(Suppl):B74–B77
Mallis MM, Deroshia CW (2005) Circadian rhythms, sleep, and performance in space. Aviat Space Environ Med 76:B94–B107
Manzey D (2000) Monitoring of mental performance during spaceflight. Aviat Space Environ Med 71:A69–A75
Manzey D (2003) Study of the survivability and adaptation of humans to long-duration interplanetary and planetary environments. In: 1264 E-S (ed) Part of technical note # 2, HUMEX study. ESA/ESTEC, Nordwijk
Manzey D, Lorenz B (1998) Mental performance during short-term and long-term spaceflight. Brain Res Rev 28:215–221
Manzey D, Lorenz B, Polyakov, VV (2003) Mental performance in extreme environments. Results from a performance monitoring study during a 438-day space mission. Ergonomics 43:589–609
McCrae RR, Allik J (2002) The five-factor model of personality across cultures. Kluwer
McKenna E (2001) Business psychology and organisational behaviour A student's handbook, 3rd edn. Psychology Press, New York
Musson D, Sandal GM, Helmreich RL (2004) Personality characteristics and trait clusters in final stage astronaut selection. Aviat Space Environ Med 75:342–349
Myasnikov VI, Zamaletdinov IS (1998) Psychological states and group interactions of crew members in flight (Reprint). Hum Perf Extreme Environ 3:44–56

Mäkinen TM, Palinkas LA, Reeves DL, Pääkkönen T, Rintamäki H, Leppäluoto JJH(2005) Effect of repeated exposures to cold on cognitive performance in humans. Physiol Behav 87:166–176

Natani K, Shurley JT, Pierce CM, Brooks RE (1970) Long-term changes in sleep patterns in men on the South Polar Plateau. Arch Int Med 125:655–659

National Research Council SSB (1998) A strategy for research in space biology and medicine in the new century. National Academy Press, Washington DC

National Research Council SSB (2000) Review of NASA's biomedical research program. National Academy Press, Washington DC

Nicholas JM, Penwell LW (1995) A proposed profile of the effective leader in human spaceflight based on findings from analog environments. Aviat Space Environ Med 66:63–72

Palinkas L (2003) The psychology of isolated and confined environments: understanding human behavior in Antarctica. Am Psychol 58:353–363

Palinkas L, Gunderson EK, Holland AW, Miller C, Johnsen JC (2000) Predictors of behavior and performance in extreme environments: the Antarctic space analogue program. Aviat Space Environ Med 71:619–625

Palinkas LA (1989) Sociocultural influences on psychosocial adjustment in Antarctica. Med Anthrop 10

Palinkas LA (2001a) Mental and cognitive performance in the cold. Int J Circumplex Health 60:430–439

Palinkas LA (2001b) Psychosocial issues in long-term space flight: overview. Gravit Space Biol Bull 14:25–34

Palinkas LA, Browner D (1995) Effects of prolonged isolation on stress, coping, and depression. J Appl Soc Psychol 25:557–576

Palinkas LA, Cravalho M, Browner D (1995a) Seasonal variation of depressive symptoms in Antarctica. Acta Psychiat Scand 91:423–429

Palinkas LA, Glogower FG, Dembert M, Hansen K, Smullen R (2004a) Incidence of psychiatric disorders after extended residence in Antarctica. Int J Circum Health 63:157–168

Palinkas LA, Houseal M, Rosenthal NE (1996) Subsyndromal seasonal affective disorder in Antarctica. J Nerv Ment Disease 184:530–534

Palinkas LA, Johnsen JC, Boster JS, Rakusa-Suszczewski S, Klopov VP, Fu XQ, Sachdeva U (2004b) Cross-cultural differences in psychosocial adaptation to isolated and confined environments. Aviat Space Environ Med 75:973–980

Palinkas LA, Johnson JC (1990) Social relations and individual performance of winter-over personnel at McMurdo Station. Antarctic J the United States 25:238–240

Palinkas LA, Suedfeld P, Steel G (1995b) Psychological functioning among members of a small polar expedition. Aviat Space Environ Med 66:943–950

Reed HL, Reedy KR, Palinkas LA, Do NV, Finney NS, Case HS, LeMar HJ, Wright J, Thomas J (2001) Impairment in cognitive and exercise performance during prolonged Antarctic residence: effect of thyroxine supplementation in the Polar triiodothyronine syndrome. J Clin Endocrin Metabol 86:110–116

Ritsher JB, Kanas N, Gushin V, Saylor S (2005) Cultural differences in patterns of mood states on board the International Space Station. Proceedings from 56th International Astronautical Congress, Fukuoka, Japan Oct 17–20, 2005

Rohrer J (1961) Interpersonal relationships in isolated small groups. In: Flaherty B (ed) Psychological aspects of manned spaceflight. Columbia University Press, New York

Rose RM, Fogg LF, Helmreich RL, McFadden TJ (1994) Psychological predictors of astronaut effectiveness. Aviat Space Environ Med 65:910–925

Rosnet E, Jurion S, Cazes G, Bachelard C (2004) Mixed-gender groups: coping strategies and factors of psychological adaptation in a Polar environment. Aviat Space Environ Med 75(Suppl):c10–c30

Rosnet E, Le Scanff C, Sagal M (2000) How self-image and personality affect performance in an isolated environment. Environ Behav 32:18–31

Rothblum ED, Weinstock JS, Morris JF (1998) Women in the Antarctic. Harrington Park Press, New York

Sandal GM (2000) Coping in Antarctica: is it possible to generalize results across settings. Aviat Space Environ Med 71:A37–A43

Sandal GM (2001a) Crew tension during a space station simulation. Environ Behav 33:134–150

Sandal GM (2001b) Psychosocial issues in space: future challenges. Grav Space Biology Bull 14:47–54

Sandal GM (2004) Culture and crew tension during an International Space Station simulation; Results from SFINCSS'99. Aviat Space Environ Med 75:44–51

Sandal GM, Bergan T, Warnche M, Værnes R, Ursin H (1996) Psychological reactions during polar expeditions and isolation in hyperbaric chambers. Aviat Space Environ Med 67:227–234

Sandal GM, Endresen IM, Værnes R, Ursin H (1999) Personality and coping strategies during submarine missions. Mil Psychol 11:381–404

Sandal GM, Grønningsæter H, Eriksen H, Gravråkmo A, Birkeland K, Ursin H (1998) Personality and endocrine activation in military stress situations. Mil Psychol 10:45–61

Sandal GM, Værnes R, Ursin H (1995) Interpersonal relations during simulated space missions. Aviat Space Environ Med 66

Santy PA (1994) Choosing, the right stuff: the psychological selection of astronauts and cosmonauts. Praeger Publishers, Westport CT

Santy PA, Kapanka H, Davis JR, Stewart DF (1988) Analysis of sleep on shuttle mission. Aviat Space Environ Med 59:1094–1097

Steel G, Suedfeld P (1991) Temporal patterns of affect in an isolated group. Environ Behav 23:749–765

Steel GD, Suedfeld P, Peri A, Palinkas LA (1997) People in high latitudes: The "big five" personality characteristics of the circumpolar sojurner. Environ Behav 29:324–347

Stuster J (1996) Bold endeavors: lessons from polar and space exploration. Naval Institute Press, Annapolis MD
Stuster J (2005) Behavioral health in Antarctica: implications for long-duration space missions. Aviat Space Environ Med 76(Suppl):B74–B77
Stuster J, Bachelard C, Suedfeld P (2000) The relative importance of behavioral issues during longduration ICE missions. Aviat, Space, Environ Med 71(9):A17–A25
Suedfeld P (1998) What can abnormal environments tell us about normal people? Polar stations as natural psychological laboratories. J Environ Psychol 18:95–102
Suedfeld P (2005) Invulnerability, coping, salutogenesis, integration: four phases of space psychology. Aviat Space Environ Med 76:B61–B66
Suedfeld P, Steel G (2000) The environmental psychology of capsule habitats. Ann Rev Psychol 51:227–253
Tellegen A, Waller NG (in press) Exploring personality through test contruction: Development of the Multidimensional Personality Quesionnaire. Minneapolis, MN: University of Minnesota Press
Ursin H, Comet B, Soulez-Lariviere C (1992) An attempt to determine the ideal psychological profiles for crews of long term space missions. Adv Space Res 12:301–314
Van Dongen HPA, Maislin G, Mullington JM, Dinges DF (2003) The cumulative cost of additional wakefulness: dose–response effects on neurobehavioral functions and sleep physiology from chronic sleep restriction and total sleep deprivation. Sleep 26:117–126
Wood J, Lugg DJ, Hysong SJ, Harm DL (1999) Psychological changes in hundred-day remote Antarctic field groups. Environ Behav 31:299–337
Zubek JP (1969) Sensory deprivation: fifteen years of research. Appleton-century-Crofts, New York

Rev Environ Sci Biotechnol (2006) 5:297–308
DOI 10.1007/s11157-006-9001-7

REVIEW PAPER

# Molecular evolution of haemoglobins of polar fishes

**Cinzia Verde · Daniela Giordano · Guido di Prisco**

Received: 2 March 2006 / Accepted: 18 July 2006 / Published online: 23 August 2006


**Abstract** The Arctic and the Antarctic differ by age and isolation of the respective marine faunas. Antarctic fish are highly stenothermal, in response to stable water temperatures, whereas the Arctic ones are exposed to seasonal and latitudinal temperature variations. The knowledge of the mechanisms of phenotypic response to cold exposure in species of both polar habitats offers fundamental insights into the nature of environmental adaptation. In the process of cold adaptation, the evolutionary trend of Antarctic fish has led to unique specialisations, including modification of haematological characteristics, e.g. decreased amounts and multiplicity of haemoglobins.

Unlike Antarctic Notothenioidei, Arctic teleosts have high haemoglobin multiplicity. Although the presence of functionally and structurally distinct haemoglobins is a plesiomorphic condition for many perciform-like fishes, it seems that the oxygen-transport system of teleost fish in the Arctic region has been adjusted to temperature differences and fluctuations of Arctic waters, much larger than in the Antarctic. The amino-acid sequences used to gain insight into the evolution history of $\alpha$ and $\beta$ globins of polar fish have clearly shown that Antarctic and Arctic globins have different phylogenies, leading to the hypothesis that the selective pressure of environment stability allows the phylogenetic signal to be maintained in the Antarctic sequences, whereas environmental variability would tend to disrupt this signal in Arctic sequences.

C. Verde (✉) · D. Giordano ·
Guido di Prisco
Institute of Protein Biochemistry, CNR, Via Pietro
Castellino 111, I-80131 Naples, Italy
e-mail: c.verde@ibp.cnr.it

**Keywords** Antarctic · Arctic · Evolution · Fish · Globin gene · Haemoglobin

**Abbreviations**
Hb Haemoglobin
mya Million years ago

## 1 Introduction

In the biosphere, organisms have succeeded in adapting to a variety of environmental conditions. The diversity of life forms suggests that adaptation has been reached by species in all regions of the Earth. The cellular macromolecules, proteins and nucleic acids are very sensitive to environment perturbations. Temperature, hydrostatic pressure, medium cellular composition and oxygen availability may profoundly affect cell physiology. Adaptation implies keeping the structural and functional features of the cellular biochemical constituents and an adequate level of energy turnover in response to variable environmental conditions. Certainly, temperature is the most important physical factor affecting life.

Current evidence suggests that the cost of life for tropical fish is higher than that of a polar counterpart (Clarke and Johnston 1999). Low temperature may strongly influence metabolic-process rates. It slows all physiological processes, changes protein–protein interactions, reduces membrane fluidity and increases the viscosity of biological fluids. In general, proteins from organisms living at low temperature are less stable in comparison to those isolated from organisms living at temperate latitudes. Lower stability seems to be due to greater conformational flexibility. Cold-adapted species have enzymes with higher catalytic activity either over a wide temperature range, or specifically at low temperature. Increased protein flexibility is generally conferred by few amino-acid replacements. The changes that enable enzymes and proteins to function in cold habitats are subtle and often challenging to identify, because it can be difficult to discriminate which changes are due to selective pressure, and which are instead functionally neutral. Enzymes with low optimal temperatures frequently contain specific replacements, with higher number of polar and lower number of hydrophobic residues, and fewer hydrogen bonds and ion pairs. Minor mutations, also distant from the active site, may lead to adaptive changes in the protein functional properties (Fields and Somero 1998). The reader interested in the effect of temperature on protein stability and flexibility, enzyme activity and metabolism is addressed to Hochachka and Somero (2002).

Variations in diurnal time scale may have very significant physiological and evolutionary consequences. Undoubtedly, the knowledge of the physiological costs linked to thermal adaptation may help to understand how species have responded to climate change in the past and how different communities respond to climate change nowadays. An organism may respond to environmental temperature changes, in attempting to maintain physiological rates, by one or more of three strategies: (i) the quantitative strategy, changing the amount of a given molecule; (ii) the qualitative strategy, expressing a molecule variant with different thermal response; (iii) the modulative strategy, adjusting the membrane composition to maintain membrane fluidity (Logue et al. 2000).

The evolutionary reconstruction of species thermal histories, in order to identify which biochemical characters or physiological traits are acquired or lost in polar habitats, is not a easy task. One pathway is the comparative approach, studying these adaptations in homologous species, living in different thermal habitats. Generally the success of a comparative approach is strongly dependent on the selection of organisms. The effects of phylogenenetic correlations must be distinguished from the effects of natural selection and adaptation. In Arctic and Antarctic fish, molecular phylogenies are complete enough to allow the interpretation of phylogenetic trees based on sequences of macromolecules of physiological interest (e.g. globins) in the light of trees based on other molecular tools used as markers of species phylogeny.

In simpler words, the "phylogenetic inference" must not produce misunderstandings in the interpretation of adaptations, especially when comparing distantly related species.

Here we review recent advances in our knowledge on the structure, function and evolution of the oxygen-transport system, in particular haemoglobin (Hb), in polar fish. Part of the review is focussed on molecular phylogenies based on Hb genes of Antarctic and Arctic fish involved in cold-adaptation. Molecular phylogeny has revealed different adaptive scenarios in the two polar regions. Comparing a trait in phyletically unrelated taxa permits to study convergent and parallel evolutionary trends. In this regard, such a study aims at deciphering whether the "solutions" evolved by Arctic fish in response to freezing temperatures are similar to those of their phylogenetically unrelated Antarctic counterparts. In freezing environments of both polar regions those fishes that have evolved antifreeze molecules, secreted into their biological fluids at high concentrations, survived. Arctic fish secrete similar antifreezes into their blood, but these are derived from a different gene. This is a classic example of convergent evolution (Chen et al. 1997a, b).

## 2 The Antarctic habitat

In the late Paleozoic, about 250 million years ago (mya), land masses were assembled within a single large continent, Pangaea. About 200 mya, Pangaea split into two parts, Laurasia in the northern hemisphere and Gondwana in the southern one. Fragmentation of Gondwana into the modern southern continents initiated 135 mya, and the Antarctic continent reached its current geographic location approximately 65 mya. Final separation of East Antarctica from Australia and West Antarctica (the Antarctic Peninsula) from South America occurred about 35–40 mya and 22–23 mya, respectively. With the opening of the Drake Passage, the isolation of Antarctica was completed (Kennett 1977). This event produced the Circum-Antarctic Current and the Polar Front, a well-defined, roughly circular oceanic system, running between 50°S and 60°S. Along the Front, the surface layers of the north-moving Antarctic waters sink beneath the less cold and less dense sub-Antarctic waters, generating virtually permanent turbulence. Just north of the Front, the water temperature has an abrupt rise of ca. 3°C, a critical factor for ecosystem isolation and adaptation. The Antarctic experienced a slow and discontinuous transition from the warm-water system of the early Tertiary (15°C) to the cold-water system of today (–1.87°C). Antarctic fish have evolved in a marine habitat whose temperature has decreased over the past 50 million years at a rate of 0.4°C per million years, and perhaps more rapidly since the onset of the Miocene (25 mya). Unlike temperate counterparts, Antarctic fish express their genes and proteins in an extremely cold thermal regime. To compensate for the rate-depressing effect of low temperature on metabolic processes, these fish have restructured many of their bio-molecular systems through mutation and selection, to preserve biological activity in such an extreme environment.

### 2.1 The Antarctic ichthyofauna: the dominant suborder Notothenioidei

Research on Antarctic fish has provided important insights into details of thermal adaptation. Investigating the remarkable adaptations at the molecular level in fish living in this extreme environment can provide new perspectives when studying processes in conventional organisms. Currently, research on polar-fish evolution is experimenting new methodologies (e.g. genomics) to study adaptations. Polar scientists will be challenged by the need to compare phylogenetically related species living in different habitats, in order to discriminate the signal due to phylogeny from the influence of a true adaptive variation. These results may obviously be reached only in species for which clear phylogeny exists. The molecular approach will contribute to predicting the effects of global warming on bio-geographical distribution and survival of highly specialised cold-adapted organisms, and to establish correlations with changes generated by man-induced impacts on the ecosystem.

The largely endemic suborder Notothenioidei, mostly confined within Antarctic and sub-Antarctic waters, is the dominant component of the Southern Ocean fauna, and is one of the most thoroughly characterised perciform fish group of the whole world, in ecology, life style, physiology, biochemistry and evolution. In this regard, the Arctic is still way behind. Antarctic notothenioids are stenothermal (Eastman 1993, 2005). They live between 2°C and –1.86°C, the freezing point of sea water, and die at temperatures of 4–6°C. In only ten million years, they appear to have lost the ability to cope with higher temperature. The notothenioid stenothermal phenotype reflects evolutionary adaptive changes in the molecular and cellular machinery. Examples include efficient microtubule assembly (Detrich et al. 1989, 2000) and apparent loss of inducible heat-shock response at constantly cold temperatures (Hofmann et al. 2000).

With the local extinction of most of the temperate Tertiary fish fauna as the Southern Ocean cooled, the suborder experienced extensive radiation, dating from the late Eocene, approx. 24 mya (Near 2004), that enabled it to exploit the diverse habitats provided by a now progressively freezing marine environment.

Evidence of temperature change in Antarctic marine habitats is beginning to appear. Temperature changes in polar oceans are particularly difficult to predict, because of insufficient knowledge concerning how the sea ice is likely to change. Whilst terrestrial species are adapted to very variable conditions and may have good chances of surviving in changing environments, marine species face more severe problems in response to climate change. They have evolved in an environment with the lowest temperature on Earth, and have lost the ability to cope with temperature change. A rise in sea temperatures of only 2°C could potentially compromise survival of many Antarctic benthic species. They may be one of the most vulnerable faunal groups on Earth to increasing temperature. At organismal level, animals have three mechanisms for coping with change: they can (i) use physiological flexibility, (ii) evolve new adaptations, (iii) migrate to other sites. Despite these three possibilities, Antarctic fish seem to have poorer prospects than other faunal groups elsewhere.

Notothenioids fill a variety of ecological niches normally occupied by taxonomically diverse fish communities in temperate waters. However, as the suborder is absent from the Arctic, comparisons to study cold adaptations can be made only with taxonomically distant groups (such as Gadidae, for instance). Some "minor" fish families, represented in both polar oceans, are extremely useful for comparing Arctic and Antarctic evolution. Fishes of the family Liparidae and of the suborder Zoarcoidei are the second and third components of the Antarctic ichthyofauna after Notothenioidei.

### 2.2 The oxygen transport in Antarctic notothenioids: haemoglobin, a model molecule for studying thermal adaptation

During the process of cold adaptation, the evolutionary trend of Antarctic notothenioids has led to unique specialisations, including modification of haematological features. In the past few decades, these features have been extensively investigated in many notothenioid species.

Whilst there is an increasing understanding of evolutionary adaptation to temperature in some biological processes, several key questions often remain open about the structure–function relationships associated with thermal adaptation of proteins. Proteins, e.g. Hb, are highly sensitive to temperature, therefore their structural and functional properties mirror the thermal conditions encountered by species during their evolutionary histories. In polar fish, the evolution of Hb includes adaptations with implications at the biochemical, physiological and structural levels. The Hb molecule has evolved structural and functional diversity to adapt and modify its features under selective pressure of all types, but both the predominant helical structure and a large number of amino-acid residues are well conserved. The amino-acid sequences of a large number of Hbs are known, and reveal homologies that reflect phylogenetic relationships between different groups of organisms. Alignment of sequences is informative for the phylogenetic relationships among Hbs in bacteria, fungi, protists, plants and animals. Hbs have been at the forefront of studies of evolution at the molecular level. Sickle-cell anaemia was the first recognised "molecular disease" (Pauling et al. 1949), caused by a single nucleotide polymorphism. The first protein crystal structures, of myoglobin and Hb, provided the basis to understand the relationship between changes in amino-acid sequence and in protein structure (Kendrew et al. 1958; Perutz et al. 1965). Globin subunits also provided the classic example for theories and structural studies of allosteric conformational transitions (Monod et al. 1965; Perutz et al. 1987).

The primary role of Hb in carrying oxygen to vertebrate tissues is probably the origin of its adaptation to widely different environmental conditions, but its specialised function imposes severe structural constraints on the molecule. Hence, it is not surprising that only a small fraction of the residues of the polypeptide chains are allowed to be replaced during evolution. According to the species-adaptation theory of Perutz (1983), the replacement of few key amino-acid residues leads to new functions. However, the development of knowledge on Hbs of a constantly increasing number of organisms has indicated that the situation is far from being simple. In polar fish, for instance, it is virtually impossible as yet to ascribe a molecular adaptation to a specific amino-acid substitution. Indeed, the situation is much more complex, and is probably linked to the combination and interplay of a number of factors in the architecture of the globin tetramer. The crystal structure of two functionally very distinct Antarctic Hbs, surprisingly displaying 95% sequence identity (D'Avino et al. 1994; Ito et al. 1995; Mazzarella et al. 1999), have highlighted that often fish Hbs do not obey the classical structural dogmas. It appears that the possession of one or more given residues does not in itself necessarily afford a specific adaptation and/or function; rather, this function may be activated or not, depending upon other factors. Cold adaptation is likely to be mostly based on derived features at higher and more integrated levels than the primary structure.

All subunits of fish Hbs display sequence homology that reflects gene duplication during phylogeny. Within different species, the transport of oxygen can be modulated by changes in the Hb structure and allosteric-ligand concentration (ATP for most teleost fish), and by changes in the expression of multiple Hbs, which are likely to display different functional features.

During evolution, complex and sophisticated molecular mechanisms, such as modulation by pH, carbon dioxide, organophosphates and temperature, have been developed to regulate oxygen transport by Hb. The decreased oxygen affinity of Hb at lower pH values in the physiological range is known as the alkaline Bohr effect, reviewed by Riggs (1988). In many Hbs from teleost fishes, when the pH is lowered, the oxygen affinity decreases to such an extent that Hbs cannot be fully saturated even at very high oxygen pressure. In addition, cooperativity is totally lost and the oxygen capacity of blood undergoes reduction of 50% or more of its value at alkaline pH. This feature is known as the Root effect, reviewed by Brittain (1987, 2005). The physiological role of the Root effect is to secrete oxygen against high oxygen pressures into the swimbladder (when present) and to the eye choroid rete, following local acidification of the blood in a countercurrent capillary system. Among Antarctic fish (all lacking the swimbladder), the few species possessing Hbs without a Root effect, as well as Hb-less Channichthyidae (see below), lack the choroid rete.

How are these molecules adapted to adequately provide oxygen to the tissues under extreme conditions? The ice-cold Antarctic waters, perennially near freezing, contain much higher concentrations of dissolved oxygen than any other marine habitat. The solubility of oxygen in water is inversely correlated to the water temperature, and the oxygen saturation in the Antarctic waters can reach 100%. Therefore, oxygen uptake and transport are not limiting steps for Antarctic fish. With the selective pressure for cellular oxygen-binding proteins relaxed, notothenioids developed an important haematological difference from temperate and tropical species, in having fewer erythrocytes and reduced Hb concentration and multiplicity. The blood of the modern notothenioid family Channichthyidae is colourless, because Hb expression was lost altogether (Ruud 1954; di Prisco et al. 2002). In the seven red-blooded Antarctic notothenioid families, erythrocyte cells are an order of magnitude lower than in temperate fish, and are reduced by over three orders of magnitude in the 16 "icefish" species of the family Channichthyidae (Eastman 1993). This may be advantageous in coping with increased viscosity of body fluids at low temperature, and is partially compensated for by increased blood volume and higher cardiac output (Egginton et al. 2002); icefish have large gills, and also highly vascularised, scaleless skin, which favours cutaneous respiration. The Hb content of erythrocytes is variable and in some species seems positively correlated to life style (Eastman 1993).

The vast majority of notothenioid species have a single Hb (Hb 1), accompanied by minor Hbs (Hb C in trace amounts, and Hb 2, ca. 5% of the total), having one of the globins in common with Hb 1 (di Prisco 1998). The amino-acid sequences and the similar functional features of major and minor Hbs (which, with few exceptions, show pH and organophosphate regulation, namely the Bohr and Root effects) in a given species, led us to propose that in the Antarctic families minor Hbs are vestigial (or perhaps larval) remnants, devoid of physiological importance at least in the adult state (di Prisco 1998). Multiplicity of Hbs in fish usually reflects the need to respond to variable environmental conditions or different habitats. In comparison with temperate species,

Antarctic notothenioids have lost globin diversity (high sequence variation), which led to the hypothesis that in the Antarctic thermostable environment the need for multiple Hbs may be reduced, and the rate of change in the primary structure may be constant and similar for all taxa. Higher Hb multiplicity and strong globin diversity, typical of temperate and tropical environments, but observed to somewhat lower extent also in the Arctic, might be linked to the variations in physico-chemical features (essentially temperature and oxygen availability) characterising these environments. A single Hb present in lower amounts than in temperate fish can be regarded as the consequence of a less critical role of the oxygen carrier in Antarctic notothenioids, possibly in keeping with the sluggish mode of life, slower metabolism, as well as the peculiarity of the cold environment (high stability and constancy of physico-chemical conditions). It is beyond doubt that Channichthyidae are not disadvantaged by their lack of Hb.

The oxygen affinity (a property which controls both the binding of oxygen at the exchange surface and its release to the tissues) of the Hbs of many Antarctic species is quite low, as indicated by the values of $p_{50}$ [the overall affinity of Hb for oxygen is expressed as the gas partial pressure required to achieve half-saturation ($p_{50}$)]. This feature is probably linked to the high concentration of oxygen in the cold Antarctic waters and hence to water temperature.

The high sequence identity observed among Hb 1 and among Hb 2 of most Antarctic species seems due to the long isolation of these fish (see chapter *Globin evolution*).

## 3 The Arctic habitat

Unlike Antarctica, the Arctic marine ecosystem is more susceptible to influences by both global and local sources. Although the temperature record across the Arctic regions is not complete, warming appears to be concentrated in the last century (Moritz et al. 2002). The Arctic, mostly covered by the sea, lies above the Arctic Circle between North America, Greenland, Europe and Asia and its geography is very complex. The Arctic Ocean is almost completely surrounded by land and influenced by large populations and industrial activities.

Satellite data and surface-based observations indicate that the Arctic sea ice has declined about 7% since 1978 (Johannessen and Miles 2000; Hasselmann et al. 2003). The melting of sea ice and the reductions in the extent of ice cover and thickness are dramatically and potentially devastating to some species. If the sea ice cover continues to decrease, marine ice algae would also disappear due to substrate loss. The algae decline may cause a cascade effect to higher trophic levels in the food web.

### 3.1 The Arctic icthyofauna

Although high latitudes and cold climates are common to the Arctic and Antarctic, in many respects the two regions are more dissimilar than similar. Antarctica has been isolated and cold longer than the Arctic, where the ice sheet developed at least 10 million years later. The modern polar ichthyofaunas differ in age, endemism, taxonomy, zoogeographic distinctiveness, biodiversity and range of physiological tolerance to environmental parameters (Eastman 1997). The Arctic fish fauna (Andriashev and Chernova 1995) includes 96 families comprising 416 species (358 marine and 58 freshwater). There are no up-to-date figures for endemism, although Briggs (1974) quotes 20–25% for marine species. The North-Sea fish fauna includes 170 species of 69 families (Zijlstra 1988).

### 3.2 The oxygen transport in Arctic fish

The comparison of the biochemical and physiological adaptations between cold-adapted and temperate notothenioids has been a powerful tool to understand whether (and to what extent) an extreme environment has required specific adaptations. If this approach has been useful for understanding evolutionary histories, a logical extension includes the comparison of the fish faunas of the two polar oceans. The Arctic species investigated so far are characterised by higher biodiversity and, unlike Antarctic notothenioids, have high multiplicity of Hbs. For instance, the blood of the spotted wolffish *Anarhichas minor*, a benthic, sedentary fish of the family Anarhichadidae (order Perciformes, suborder Zoarcoidei) contains three functionally distinct major Hbs, whose amino-acid sequences and oxygen-binding properties have recently been described (Verde et al. 2002). High multiplicity and functional differences have also been observed (Verde et al. 2006) in the Gadidae *Arctogadus glacialis* (Arctic cod), *Boreogadus saida* (polar cod) and *Gadus morhua* (Atlantic cod). The family Gadidae (Svetovidov 1948) comprises a group of benthic and pelagic species inhabiting coastal regions and continental shelves of northern oceans. Although cods are one of the most studied groups because of their importance in fisheries, their phylogeny and molecular adaptations are far from being established.

In comparison with notothenioids, which lost globin diversity probably because of environmental (temperature) stability, the oxygen-transport system of Arctic species has plesiomorphic features secondarily involved in cold adaptation and temperature fluctuations (namely higher multiplicity of Hbs and higher globin diversity). Low oxygen affinity and strength of the Root effect resemble the values found in most Antarctic notothenioids (Parisi et al. in preparation). The presence of multiple Hbs may contribute to maintain Hb concentration in erythrocytes higher than if there were only a single Hb. In many cases, polymorphism may have no visible phenotypic effect and no obvious correlation with environmental conditions. As mentioned above, it is not always clear whether multiple Hbs offer any selective advantage.

## 4 Globin evolution

Within the study of the molecular bases of cold adaptation in fish inhabiting the polar habitats, and taking advantage of the information available on Hb structure and function, the evolutionary history of the $\alpha$ and $\beta$ globins of Arctic and Antarctic fish Hbs has been analysed.

In Arctic and Antarctic fish, molecular phylogenies are complete enough to allow the interpretation of trees based on primary structures of macromolecules of physiological interest (e.g. globins), in the light of trees based on other molecules used as markers of species phylogeny.

Table 1 lists the species examined and the $\alpha$- and $\beta$-globin amino-acid sequences used in the phylogenetic analysis. Sequences not available in data banks are in Stam et al. (1997).

Phylogenetic analysis was performed on the multiple alignments constructed with the programme CLUSTAL X. The inferred Neighbour-Joining (NJ) trees for $\alpha$ and $\beta$ globins are reported in Figs. 1 and 2. The genetic distances were measured according to the *p*-distance model. We chose *p*-distances to evaluate to what extent the resulting tree differs from the expected interrelationships among species in a given cluster of orthologs (i.e. gene copies diversified from an ancestral speciation). When evolutionary pressures and rates of change are the same across taxa, similarity is proportional to phylogeny, and in that case a phenetic tree (NJ tree used with *p*-distances) reflects the phylogenetic tree. Using corrected distances or likelihood methods that incorporate models of sequence evolution would have "buffered" effects from variation in selective pressures, making results very difficult to interpret (Verde et al. 2006).

According to previous results (Verde et al. 2003, 2004a, b), globin paralogs (e.g. gene copies originated by duplication in a given genome) currently found in Antarctic fish diverged approximately 250 mya, i.e. at the onset of the Mesozoic; hence, unlike antifreeze glycoproteins (AFGP), whose appearance coincided with cooling of the Antarctic continent (Chen et al. 1997b), Hb diversification in major and minor groups appears less stringently correlated to changes in the environmental conditions.

The time of the gene-duplication event that gave origin to the two paralogous groups of major and minor Hbs is similar, suggesting that they diverged long before the first stock of ancestral notothenioids. This event concomitantly involved also a number of Arctic and temperate sequences, such as those of *A. minor*, *Chelidonichthys kumu*, *Thunnus thynnus* and *G. morhua* ($\beta$ chain only), because they fall in the same clade of the Antarctic globins.

**Table 1** List of species and globin sequences investigated

| Order and species | Globin |
|---|---|
| Coelacanthiformes (outgroup) | |
| *Latimeria chalumnae*[a] | α, β |
| Scorpaeniformes | |
| *Chelidonichthys kumu*[a] | α, β |
| Perciformes | |
| *Thunnus thynnus*[a] | α, β |
| *Anarhichas minor*[b] | α (Hb 1), α (Hb 2, Hb 3)<br>β (Hb 1, Hb 2), β (Hb 3) |
| *Chrysophrys auratus*[a] | α, β (Hb 4) |
| *Notothenia coriiceps*[c] | Major α (Hb 1), β (Hb 1, Hb 2)<br>Minor α (Hb 2) |
| *Notothenia angustata*[d] | Major α (Hb 1), β (Hb 1, Hb 2)<br>Minor α (Hb 2) |
| *Pleuragramma antarcticum*[c] | Major α (Hb 1, Hb 2), β (Hb 1, Hb 3)<br>Minor α (Hb 3), β (Hb 2) |
| *Pagothenia borchgrevinki*[c] | Major α (Hb 1, Hb 0)<br>Major β (Hb 1)<br>Minor β (Hb 0) |
| *Gobionotothen gibberifrons*[c] | Major α, β (Hb 1)<br>Minor α, β (Hb 2) |
| *Aethotaxis mitopteryx*[c] | α, β |
| *Trematomus newnesi*[c] | Major α, β (Hb 1)<br>Minor α (Hb 2), β (Hb C) |
| *Trematomus bernacchii*[c] | Major α, β (Hb 1)<br>Minor β (Hb C) |
| *Cygnodraco mawsoni*[c] | Major α (Hb 1, Hb 2), β (Hb 1)<br>Minor β (Hb 2) |
| *Gymnodraco acuticeps*[c] | α, β |
| *Racovitzia glacialis*[c] | α, β |
| *Bathydraco marri*[c] | α, β |
| *Pogonophryne scotti*[c] | α, β |
| *Artedidraco orianae*[c] | α, β |
| Salmoniformes | |
| *Salmo salar*[a] | α |
| *Oncorhynchus mykiss*[a] | α, β (Hb I)<br>α, β (Hb IV) |
| Gadiformes | |
| *Gadus morhua*[b] | α (Hb 2)<br>β (Hb 2, Hb 3)<br>β (additional chain) |
| *Boreogadus saida*[b] | β (Hb 1, Hb 2) |
| Anguilliformes | |
| *Anguilla anguilla*[a] | α, β (Hb C)<br>α, β (Hb A) |
| Gymnotiformes | |
| *Electrophorus electricus*[a] | α, β |
| Siluriformes | |
| *Hoplosternum littorale*[a] | α, β (Hb C) |
| Cypriniformes | |
| *Cyprinus carpio*[a] | α, β |
| *Carassius auratus*[a] | α, β |
| *Catostomus clarkii*[a] | α |

Taxa are arranged according to Nelson (1994)

[a] Temperate freshwater and marine species; [b] Arctic species; [c] Antarctic Notothenioidei; [d] Non-Antarctic Notothenioidei

The NJ trees (Figs. 1, 2) are in agreement with those obtained by morphological analysis and sequence studies on mitochondrial RNA (Ritchie et al. 1996) and give strong support to the monophyly of Antarctic notothenioids. They are in agreement with those recently obtained using the Maximum Likelihood (ML) method (Giordano et al. 2006) and the obtained topology is in general agreement with the hypothesis of four groups of globins, namely "Embryonic Hb Group", "Notothenioid Major Adult Hb Group", "Anodic Adult Hb Group" and "Cathodic Adult Hb Group" (Maruyama et al. 2004).

The inferred NJ trees for C α and β globins also include the Arctic species. In both trees, the globins of major and minor Antarctic fish Hbs cluster in two separate, strongly supported groups, with the anodic and cathodic globins of temperate fish Hbs forming the first divergence lineage.

Two more perciforms species (temperate *T. thynnus* and Arctic *A. minor*) are included in the same cluster. According to Maruyama et al. (2004), α globins that belong to notothenioid minor Hbs are included in the "Embryonic Hb Group", all α-globin sequences of the major notothenioid Hbs are grouped into the "Notothenioid Major Adult Hb Group" (Fig. 1).

Likewise, the NJ β-globin tree (Fig. 2) shows four major clusters, corresponding to the four groups mentioned above. All β-globin sequences from the major Hb 1 components (shared by Hb 1 and Hb 2 in most Antarctic notothenioids) belong to the "Notothenioid Major Adult Hb Group", β globins that belong to notothenioid minor Hbs are included in the "Embryonic Hb Group".

Unlike the globins of Antarctic species, the Arctic globins occupy variable positions in both trees, suggesting independent evolutionary histories, with the exception of *A. minor* which is close

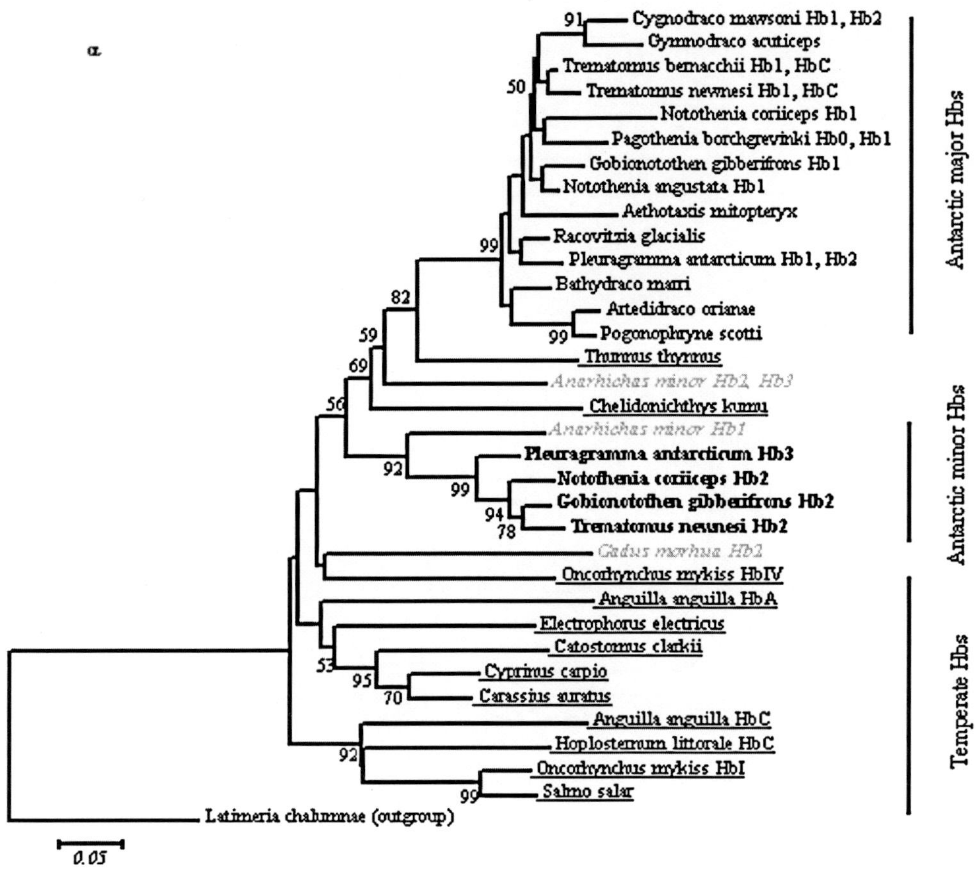


**Fig. 1** Phylogenetic tree of amino-acid sequences of α chains from Antarctic, Arctic and temperate fish Hbs. Bootstrap values (percentage of 10,000 replicates) are given at the nodes. Major (normal lettering) and minor (bold) Hbs from Antarctic fish; Hbs from Arctic (italic and grey) and temperate fish (underlined). The coelacanth *L. chalumnae* (outgroup) is indicated. *N. angustata* is a temperate notothenioid

to the notothenioid clades. Showing low identity with temperate species, the globin sequences of the Arctic zoarcoid *A. minor* are consistent with species history, as *A. minor* consistently appears close to the notothenioid clades (Verde et al. 2002) as predicted by teleostean phylogenies (Chen et al. 2003; Dettaï and Lecointre 2004, 2005; Miya et al. 2003). By contrast, Arctic gadiform sequences occupy different positions in the two trees with regard to temperate and Antarctic sequences. For instance, the $\beta^1$ chain of the polar cod *B. saida* is included in the clade of the other non-Antarctic species, but its position is remote from all the other globins (Fig. 2). The two $\beta$ chains of the Atlantic cod *G. morhua* appear closely related, probably as a result of a relatively recent gene duplication event (Fig. 2).

In the tree of Fig. 1, the $\alpha^2$ chain shared by *A. minor* Hb 2 and Hb 3 is close to the major Antarctic globins, whereas $\alpha^1$ of Hb 1 appears more closely related to minor Antarctic globins. In the tree of Fig. 2, the position of the *A. minor* $\beta^1$ chain shared by Hb 1 and Hb 2 falls into the group of the major Antarctic globins, whereas the Hb 3 $\beta^2$ chain appears well separated from the subclades of major and minor Antarctic globins.

The $\beta^1$ chain in common to Hb 1 and Hb 2 of cold-adapted *B. saida* outgroups with respect to the other sequences. In *G. morhua*, $\beta^2$ (Hb 2, Hb 3) and another $\beta$ chain (possibly belonging to a larval Hb, and whose sequence has been deduced from DNA) constitute a clade characterised by a node supported by a high bootstrap value (Fig. 2). Interestingly, the divergence time of these two

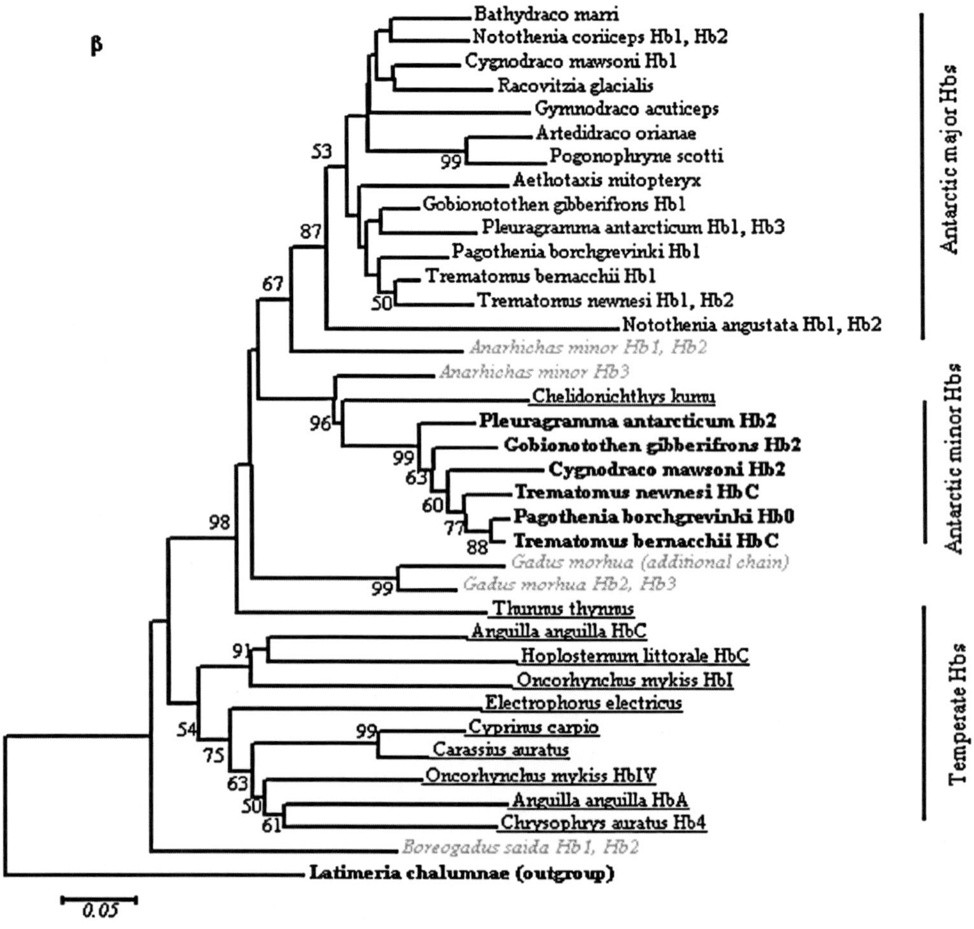


**Fig. 2** Phylogenetic tree of amino-acid sequences of $\beta$ chains from Antarctic, Arctic and temperate fish Hbs. Bootstrap values (percentage of 10,000 replicates) are given at the nodes. Major (normal lettering) and minor (bold) Hbs from Antarctic fish; Hbs from Arctic (italic and grey) and temperate fish (underlined). The coelacanth *L. chalumnae* (outgroup) is indicated. *N. angustata* is a temperate notothenioid

sequences is more recent than other paralogous globin families, such as those of major and minor Antarctic Hbs.

The constant physico-chemical conditions of the Antarctic ocean and the stable life-style of the Arctic zoarcoid *A. minor* is matched by clear grouping of "cold" fish-globin sequences, whereas the variations typical of the Arctic ocean, associated to fish displaying pelagic and migratory life style, correspond to high sequence divergence observed in the gadids. The positions of the globin sequences of the Arctic zoarcoid *A. minor* are consistent with species history, as *A. minor* appears close to the notothenioid clades as predicted by teleostean phylogenies (species phylogeny is not disrupted by sequence adaptations).

## 5 Concluding remarks

The most stable thermal environments are aquatic; research on polar fishes has provided important insights into the details of thermal adaptation. The remarkable differences in the oxygen-transport system between Arctic and Antarctic bony fish indicate that distinct evolutionary pathways in the regulatory mechanisms of the fish oxygen-transport system have been followed in the two polar environments. The different phylogenetic histories of Arctic and Antarctic fish depend on the respective habitats. Although both are cold, the Arctic and Antarctic habitats differ in many aspects. Indeed, in the Arctic isolation is less pronounced and the range of temperature variations is wider than in the Antarctic. Therefore, it is not surprising that the Arctic ichthyofauna, distributed across a much more complex oceanographic system than the Antarctic one (dominated by a single taxonomic group), is characterised by high diversity, also reflected in the phylogeny of a given trait. For the Arctic ichthyofauna, we can speculate that it may have been advantageous to maintain a multiple-globin system, helping to deal with environmental changes and metabolic demands. Multiple Hbs may protect against deleterious mutational changes in the globin genes, provide higher total Hb concentration in the erythrocyte (according to the phase rule, in a saturated solution, multiple proteins afford higher total concentration than a single protein), and increase the expression rate of the genes (Verde et al. 2006).

**Acknowledgements** Support from the Italian National Programme for Antarctic Research is gratefully acknowledged. The role of the cruises TUNU I and TUNU II (Greenland) in 2003 and 2005, of the SCAR Programme "Evolution and Biodiversity in the Antarctic: the Response of Life to Change" (EBA), and of two projects, EBA and ICEFISH-2007, recently chosen by ICSU-WMO as potential lead projects for the International Polar Year (IPY 2007–2008), is highlighted.

## References

Andriashev AP, Chernova NV (1995) Annotated list of fishlike vertebrates and fish of the Arctic seas and adjacent waters. J Ichthyol 35:81–123

Briggs JC (1974) Marine zoogeography. McGraw-Hill, New York

Brittain T (1987) The Root effect. Comp Biochem Physiol 86B:473–481

Brittain T (2005) Root effect hemoglobins. J Inorg Biochem 99:120–129

Chen WJ, Bonillo C, Lecointre G (2003) Repeatability of clades as a criterion of reliability: a case study for molecular phylogeny of Acanthomorpha (Teleostei) with larger number of taxa. Mol Phyl Evol 26:262–288

Chen L, DeVries AL, Cheng CHC (1997a) Evolution of antifreeze glycoprotein gene from a trypsinogen gene in Antarctic notothenioid fish. Proc Natl Acad Sci USA 94:3811–3816

Chen L, DeVries AL, Cheng CHC (1997b) Convergent evolution of antifreeze glycoproteins in Antarctic notothenioid fish and Arctic cod. Proc Natl Acad Sci USA 94:3817–3822

Clarke A, Johnston NM (1999) Scaling of metabolic rate with body mass and temperature in teleost fish. J Anim Ecol 68:893–905

D'Avino R, Caruso C, Tamburrini M, Romano M, Rutigliano B, Polverino de Laureto P, Camardella L, Carratore V, di Prisco G (1994) Molecular characterization of the functionally distinct hemoglobins of the Antarctic fish *Trematomus newnesi*. Biol Chem 269:9675–9681

Dettaï A, Lecointre G (2004) In search for notothenioid (Teleostei) relatives. Antarctic Sci 16:71–85

Dettaï A, Lecointre G (2005) Further support for the clades obtained by multiple molecular phylogenies in the acanthomorph bush. C.R. Biol 328:674–689

Detrich HW III, Johnson KA, Marchese-Ragona SP (1989) Polymerization of Antarctic fish tubulins at low temperatures: energetic aspects. Biochemistry 28:10085–10093

Detrich HW III, Parker SK, Williams RCJ, Nogales E, Downing KH (2000) Cold adaptation of microtubule assembly and dynamics. Structural interpretation of primary sequence changes present in the alpha- and beta-tubulins of antarctic fishes. J Biol Chem 275:37038–37047

di Prisco G (1998) Molecular adaptation of Antarctic fish hemoglobins. In: di Prisco G, Pisano E, Clarke A (eds) Fishes of Antarctica. A biological overview. Springer, Milano, pp 339–353

di Prisco G, Cocca E, Parker S, Detrich HW (2002) Tracking the evolutionary loss of hemoglobin expression by the white-blooded Antarctic icefishes. Gene 295:185–191

Eastman JT (1993) Antarctic fish biology: evolution in a unique environment. Academic Press, San Diego, CA

Eastman JT (1997) Comparison of the Antarctic and Arctic fish faunas. Cybium 12:276–287

Eastman JT (2005) The nature of the diversity of Antarctic fishes. Polar Biol 28:93–107

Egginton S, Skilbeck C, Hoofd L, Calvo J, Johnston IA (2002) Peripheral oxygen transport in skeletal muscle of Antarctic and sub-Antarctic notothenioid fish. J Exp Biol 205:769–779

Fields PA, Somero GN (1998) Hot spots in cold adaptation: localized increases in conformational flexibility in lactate dehydrogenase $A_4$ orthologs of Antarctic notothenioid fishes. Proc Natl Acad Sci USA 95:11476–11481

Giordano D, Grassi L, Parisi E, Bargelloni L, di Prisco G, Verde C (2006) Embryonic $\beta$-globin in the non-Antarctic notothenioid fish *Cottoperca gobio* (Bovichtidae). Polar Biol, DOI: 10.1007/s00300-006-0162-1

Hasselmann K, Latif M, Hooss G, Azar C, Edenhofer O, Jaeger CC, Johannessen OM, Kemfert C, Welp M, Wokaun A (2003) The challenge of long-term climate change. Science 302:1923–1925

Hochachka PW, Somero GN (2002) Biochemical adaptation. Mechanism and process in physiological evolution. Oxford University Press, New York

Hofmann GE, Buckley BA, Airaksinen S, Keen J, Somero GN (2000) The Antarctic fish *Trematomus bernacchii* lacks heat-inducible heat shock protein synthesis. J Exp Biol 203:2331–2339

Ito N, Komiyama NH, Fermi G (1995) Structure of deoxyhaemoglobin of the Antarctic fish *Pagothenia bernacchii* with an analysis of the structural basis of the Root effect by comparison of the liganded and unliganded haemoglobin structures. J Mol Biol 250:648–658

Johannessen OM, Miles MW (2000) Arctic sea ice and climate change – Will the ice disappear in this century? Sci Prog 83:209–222

Kendrew JC, Bodo G, Dintzis HM, Parrish RG, Wyckoff H, Phillips DC (1958) A three-dimensional model of the myoglobin molecule obtained by X-ray analysis. Nature 181:662–666

Kennett JP (1977) Cenozoic evolution of Antarctic glaciation, the circum-Antarctic ocean and their impact on global paleoceanography. J Geophys Res 82:3843–3876

Logue JA, DeVries AL, Fodor E, Cossins AR (2000) Lipid compositional correlates of temperature-adaptive interspecific differences in membrane physical structure. J Exp Biol 35:745–755

Maruyama K, Yasumasu S, Iuchi I (2004) Evolution of globin genes of the medaka *Oryzias latipes* (Euteleostei; Beloniformes; Oryziinae). Mech Dev 121:753–769

Mazzarella L, D'Avino R, di Prisco G, Savino C, Vitagliano L, Moody PCE, Zagari A (1999) Crystal structure of *Trematomus newnesi* haemoglobin re-opens the Root effect question. J Mol Biol 287:897–906

Miya M, Takeshima H, Endo H, Ishiguro NB, Inoue JG, Mukai T, Satoh TP, Yamaguchi M, Kawaguchi A, Mabuchi K, Shirai, SM, Nishida M (2003) Major patterns of higher teleostean phylogenies: a new perspective based on 100 complete mitochondrial DNA sequences. Mol Phyl Evol 26:121–138

Moritz RE, Bitz CM, Steig EJ (2002) Dynamics of recent climate change in the Arctic. Science 297:1497–1502

Monod J, Wyman J, Changeux JP (1965) On the nature of allosteric transitions: a plausible model. J Mol Biol 12:88–118

Near T (2004) Estimating divergence times of notothenioid fishes using a fossil-calibrated molecular clock. Antarctic Sci 16:37–44

Nelson JS (1994) Fishes of the world, 3rd edn. John Wiley & Sons, New York

Pauling L, Itano HA, Singer SJ, Wells IC (1949) Sickle-cell anemia: a molecular disease. Science 110:543–548

Perutz MF (1983) Species adaptation in a protein molecule. Mol Biol Evol 1:1–28

Perutz MF, Fermi G, Luisi B, Shanan B, Liddington RC (1987) Stereochemistry of cooperative mechanisms in hemoglobin. Acc Chem Res 20:309–321

Perutz MF, Kendrew JC, Watson HC (1965) Structure and function of haemoglobin. II. Some relations between polypeptide chain configuration and amino acid sequence. J Mol Biol 13:669–678

Riggs A (1988) The Bohr effect. Annu Rev Physiol 50:181–204

Ritchie PA, Bargelloni L, Meyer A, Taylor JA, Macdonald JA, Lambert M (1996) Mitochondrial phylogeny of trematomid fishes (Nototheniidae, Perciformes) and the evolution of Antarctic fish. Mol Phyl Evol 5:383–390

Ruud JT (1954) Vertebrates without erythrocytes and blood pigment. Nature 173:848–850

Stam WT, Beintema JJ, D'Avino R, Tamburrini M, di Prisco G (1997) Molecular evolution of hemoglobins of Antarctic fishes (Notothenioidei). J Mol Evol 45:437–445

Svetovidov AM (1948) Gadiformes. In: Fauna of USSR, Vol IX No. 4. Fishes. Zoological Institute of the Academy of Sciences of the USSR (Translated by the Israel Program for Scientific Translations for the National Science Foundation and Smithsonian Institution as OTS 63-11071)

Verde C, Balestrieri M, de Pascale D, Pagnozzi D, Lecointre G, di Prisco G (2006) The oxygen-transport system in three species of the boreal fish family Gadidae. Molecular phylogeny of hemoglobin. J Biol Chem 281:22073–22084

Verde C, Carratore V, Riccio A, Tamburrini M, Parisi E, di Prisco G (2002) The functionally distinct hemoglobins of the Arctic spotted wolffish *Anarhichas minor*. J Biol Chem 277:36312–36320

Verde C, Howes BD, De Rosa MC, Raiola L, Smulevich G, Williams R, Giardina B, Parisi E, di Prisco G (2004a) Structure and function of the Gondwanian hemoglobin of *Pseudaphritis urvillii*, a primitive notothenioid fish of temperate latitudes. Prot Sci 13:2766–2781

Verde C, Parisi E, di Prisco G (2003) The evolution of polar fish hemoglobin: a phylogenetic analysis of the ancestral amino acid residues linked to the Root effect. J Mol Evol 57:258–267

Verde C, Parisi E, di Prisco G (2004b) The evolution of polar fish hemoglobin: structure, function and phylogeny. Antarctic Sci 16:59–69

Zijlstra JJ (1988) The North Sea ecosystem. In: Postma H, Zijlstra JJ (eds) Continental shelves, ecosystems of the World, vol 27. Elsevier, Amsterdam, pp 231–277

Rev Environ Sci Biotechnol (2006) 5:309–321
DOI 10.1007/s11157-006-9104-1

REVIEW PAPER

# Predicting the impacts of climate change on the evolutionary adaptations of polar fish

**Guido di Prisco · Cinzia Verde**

Received: 2 March 2006 / Accepted: 3 August 2006 / Published online: 5 September 2006


**Abstract** The recognition of the important role of the polar habitats in global climate changes has awakened great interest in the evolutionary biology of the organisms that live there, as well as the increasing threat of loss of biological diversity and depletion of marine fisheries. These organisms are exposed to strong environmental constraints, and it is important to understand how they have adapted to cope with these challenges and to what extent adaptations may be upset by current climate changes. Adaptations of the dominant group of Antarctic fish, the suborder Notothenioidei, have been thoroughly investigated by several teams. Considering the amount of information available on cold adaptation, the study of fish adapted to the extreme conditions of the polar seas will allow us to gain invaluable clues on the development, impact and consequences of climate and anthropogenic challenges, with powerful implications for the future of the Earth.

**Keywords** Antarctic · Antifreeze glycoprotein · Arctic · Climate change · Cold adaptation · Fish · Globin

G. di Prisco (✉) · C. Verde
Institute of Protein Biochemistry, CNR, Via Pietro Castellino 111, Naples I-80131, Italy
e-mail: g.diprisco@ibp.cnr.it

**Abbreviations**

| | |
|---|---|
| AFGP | Antifreeze glycoprotein |
| EBA | Evolution and Biodiversity in the Antarctic: the Response of Life to Change |
| IPY | International Polar Year |
| Hb | Haemoglobin |
| SCAR | Scientific Committee on Antarctic Research |

## 1 Introduction

Evolutionary adaptations of the diverse assemblage of fish that thrive in Arctic and Antarctic marine habitats pertain to an important area of polar biology. Examination of these adaptations paves the way to more general and wider perspectives with marked social implications.

Organisms living in the polar regions are exposed to strong environmental constraints. Understanding how these organisms have evolved to cope with these challenges requires the elucidation of the molecular processes underpinning evolutionary adaptations. Much of our knowledge of the effect of the environment on vertebrate physiology and evolution has come from fishes, which share many basic physiological mechanisms with humans. However, their body is submerged in water, and the close physical and physiological interaction with the aquatic environment makes them sensitive sentinels of environmental challenge and offers important advantages for defining the organism–environment interface and the mechanisms of temperature adaptation. A new powerful aid to evolutionary biology and ecology has been recently provided by molecular biology, which allows us to explore the function of individual genes. Nowadays molecular phylogeny is essential for studying evolution, and important insights come from both protein (e.g. haemoglobin) and nucleic acid sequences (mitochondrial genes: e.g. 12S and 16S rRNAs; nuclear genes: e.g. 28S rRNA, genes encoding globins, antifreeze glycoproteins, tubulins, myoglobin, etc), as well as large-scale chromosome change (Stam et al. 1997, 1998; di Prisco 1998, 2000; Bargelloni et al. 2000; Pisano et al. 1998, 2003).

Fish have developed both cellular and molecular mechanisms of cold adaptation, and these are fully representative of the suite of strategies adopted by organisms under strong evolutionary pressure. Cold adaptation is an important part of a refined physiological equilibrium, which remains unmodified in the absence of disturbances due to climate change. When disturbances do occur, evaluating the response of cold adapted organisms will yield important indications of general trends. Classes of organisms which will be affected include polar (especially Antarctic) fish.

All organisms, both terrestrial and marine, are susceptible to environmental changes, but small or non-motile organisms are particularly vulnerable. As is common in evolution, if a species fails to adapt to climate-change requirements exceeding the internal flexibility limits or fails to migrate, its fate will be extinction. The Antarctic benthos is vulnerable (Peck 2005), strongly suggesting that, compared to temperate environments, long-term survival of these organisms under current rapid climate change involves far higher risk. Peck summarised the strategies by which Antarctic organisms, and the benthic species in particular, may enhance their chances of survival in changing environments: (i) Using the margins of internal physiological flexibility and capacity to sustain new biological requirements. The margin range is often narrow, thus the efficiency of this ability is poor. (ii) Adapting to the new conditions and altering the range of biological capacity. This strategy depends on magnitude and rate of change, and aquatic habitats change temperature at a far slower rate than terrestrial ones, possibly creating fewer adaptation problems to most fish species. It also depends on rates of production of new characters *via* modification of the genetic code (phenotypic variance), which in turn depends on genetic variations between individuals, number of target offsprings for selection, gene flow. As an overall result, adaptation processes in fish may be slower than at lower latitudes. (iii) Migrating to areas where conditions are favourable. This depends on ability to disperse and availability of suitable sites. Although the absence of wide latitudinal gradients in the Antarctic continent minimises the advantage of migration, it highlights the importance of the sub-Antarctic as a critical, geographically adjacent, research area, largely populated by eurythermal fish. These live in a more variable (flexible) environment, where changes might be faster and larger than in the High Antarctic. Historically, this area may have been a site of long-term acclimation, because some cold-adapted notothenioids also inhabit sub-Antarctic waters (e.g. South Georgia, Bouvet), where in shallow waters the temperature may reach +4°C.

There is not much knowledge on the capacity of the polar marine fauna to respond to on-going climate changes. Consequently, several investigators are becoming engaged in this analysis. Given the amount of information available on cold adaptation, the study of fish of the polar seas will provide invaluable clues to the development, impact and consequences of climate change, and hence on the future of the Earth.

Many of the arguments regarding dangers of climate change for life as well as the research strategies aimed at minimising risks mirror our own opinions. This review focusses on the response of polar fish to climate change but will also hopefully stimulate the exchange of ideas on how climate change may affect the whole Earth System.

## 2 The polar habitats, and the respective ichthyofaunas

Although high latitudes and cold climates are common to both polar habitats, in many respects the two regions are more dissimilar than similar. Geography, oceanography and biology of species inhabiting the polar regions have often been compared (Dayton et al. 1994) to outline the differences between the two polar ecosystems. The main differences between the Arctic and the Antarctic regions are the greater age and isolation of the latter. The exchange of Atlantic and Arctic waters through the passage between Greenland and Svalbard was not possible until 27 million years ago (Eastman 1997). The Arctic region was in a high-latitude position by the early Tertiary, but the climate remained temperate with water temperatures of 10–15°C. During the Miocene, about 10–15 million years ago, Arctic land masses reached their present positions and temperatures dropped below freezing. Permanent ice cover was formed ca. 0.7 million years ago (Eastman 1997). In contrast, the Antarctic has been isolated and cold longer than the Arctic, with ice sheet development preceding that in the Arctic by at least 10 million years.

During the fragmentation of Gondwana, Antarctica played a key role in altering ocean circulation and forcing climate toward cooling. With the opening of the Drake Passage around 22–25 million years ago (Kennett 1977), separation of Antarctica from South America was complete. This event allowed the inception of the Circum-Antarctic Current and the establishment of the Polar Front, a roughly circular oceanic system running between 50°S and 60°S. Just north of the Polar Front, the surface water temperature has an abrupt rise of ca. 3°C. Because of this, the Polar Front acts as an efficient barrier for migration in both directions and hence has caused adaptive evolution to develop in isolation.

Although the influence of evolution obviously extends to the most extreme environments, Antarctica has not traditionally been considered as one of the notable evolutionary sites (compared with, for example, the East African Great Lakes, Lake Baikal, or the Galápagos). This is perhaps because research has emphasised aspects of extreme biology rather than unifying principles of evolutionary biology (Eastman 2000). However evolution is at work in the Antarctic, and there are clear signs of the growing efforts of scientists to provide insights into evolutionary processes and raise the visibility of Antarctic evolutionary biology.

Over the past 30–40 million years, in parallel with the diversification of the teleost suborder Notothenioidei, the physico-chemical features of the Antarctic marine environment have experienced a slow and discontinuous transition from the warm-water system of the early Tertiary (15°C) to the cold-water system of today (–1.87°C). As conditions of extreme temperature developed, fish evolved physiological and biochemical mechanisms of adaptation to survive in the cold. On the Antarctic shelf, the cosmopolitan temperate ichthyofauna from the late Eocene was almost completely replaced by the highly endemic, cold-adapted modern fauna. The capacity to adapt is variable. Organisms that nowadays can hardly tolerate temperature changes as small as 1–2°C are defined *stenothermal*; other organisms (*eurythermal*) can withstand much wider temperature variations (Hochachka and Somero 2002). Many polar fish, indeed polar ectotherms in general, appear to be highly stenothermal, and are thus likely to be among the first class of organisms to be affected by warming. In fact, many stenothermal fish species die when exposed to 4°C (Somero et al. 1998). Few investigations on acclimation have revealed wide differences; acclimation of notothenioids at 0–4°C allows survival at 6–8°C (Lowe 2005), but we know very little about their ability to acclimate in the longer term. In a wide range of taxa, physiological performance is impaired at temperatures only slightly above those experienced in the field (Clarke et al. 2006). Amongst the physiological mechanisms underlying lack of survival, failure of oxygen supply is possibly the major factor (Pörtner et al. 2004), with transfer to anaerobic metabolism correlated to a dramatic decrease in oxygen content in the blood, mitochondria and tissues at temperatures close to those of survival (Hardewig et al. 1999). Eurythermal fish face lower threats due to environmental change.

Because of the isolating barrier of the Polar Front, the climatic features of the Antarctic waters are more extreme and constant than those of the Arctic. In the Arctic isolation is less stringent, and the range of temperature variation is wider, both in the ocean and on the surrounding lands, thus facilitating migration and redistribution of the fauna. The colonised terrestrial portions are extensive; they are directly linked to temperate areas, producing wide and complex terrestrial mechanisms of feedback to the climate, which add to those originating from ocean and atmosphere circulation. The anthropogenic impact on the environment is also far greater (see below).

Throughout the Cenozoic, regional tectonic and oceanographic events have played a key role in delimiting the two polar ecosystems and in influencing the evolution of their faunas, whose composition and diversity are strongly linked to geological history. Because of the differing tectonic and climatic histories outlined above, the modern polar ichthyofaunas differ from each other in age, endemism, taxonomy, zoogeographic distinctiveness, range of physiological tolerance to various environmental parameters, and biodiversity (Eastman 1997). Both the Antarctic and the Arctic possess cryopelagic taxa living at the sea ice-water interface. Freshwater habitats exist in the Arctic, but are more limited in the Antarctic, with consequent limitations to species diversification and current biodiversity. There are adaptive radiations of fish, isopods and amphipods in the Antarctic but not in the Arctic.

The Antarctic fauna includes 322 species grouped in 50 families (Eastman 2005), whereas the Arctic fauna includes 416 species (58 are freshwater species) grouped into 96 families (Andriashev and Chernova 1995). In Antarctica, endemism of the benthic fauna is 88% and rises to 97% when only the dominant suborder Notothenioidei is considered. In the Arctic, endemism for marine fish is 20–25% (Briggs 1974). The Arctic fauna has no endemic higher taxonomic category equivalent to the Antarctic notothenioids, and there has been no comparable adaptive radiation of any fish group. Arctic fish are generally more eurythermal and euryhaline (i.e. they can tolerate relatively large changes in ambient salinity) than their Antarctic counterparts and the North Atlantic and North Pacific character of the marine fauna reflects the continuity of shelf areas between the Arctic and boreal regions.

The differences in the two polar environments are differently reflected in global changes. Accordingly, comparative studies of these two ecosystems are likely to provide powerful evolutionary insights into the relationship between environment and evolutionary adaptation. In summary, the Arctic is the connection between the more extreme, simpler Antarctic oceanic system and the more complex temperate and tropical systems.

## 3 Specialisations, cold adaptation and evolution

There are serious difficulties in establishing consensus on objective criteria to identify a phenotypic trait as an adaptation. This issue will not be discussed here. Adaptation remains a slippery concept (di Prisco and Giardina 1996), and articles with extensive discussion on this and related issues are available to the interested reader (e.g. Reeve and Sherman 1993; Garland and Carter 1994; Mongold et al. 1996; Montgomery 2000).

Evolutionary adaptation to temperature is a multivariate problem, with complexity at organismal, cellular and molecular levels. Antarctic fish have developed a number of adaptive specialisations in many biological features, e.g. freezing avoidance, polymerisation of tubulins and actins, enzyme catalysis, haematological parameters and oxygen transport, heat shock, and neurobiology (reviewed in di Prisco 1997; Römisch and Matheson 2003). Some of these contribute to making Antarctic fish quite unique. An excellent book (Eastman 1993) provides extensive information on the evolution of Antarctic fish. We know less about Arctic fish, in which many specialisations have not been investigated, and a few have not been found.

The recognition of important role of the polar habitats in global climate changes has recently awakened great interest in the evolutionary biology of the organisms that live there, as well as the increasing threat of loss of biological diversity and depletion of marine fisheries. Recent evidence has indicated that global change is already affecting the physiology and ecology of some species (Hughes 2000; Walther et al. 2002). Also natural variations might be responsible for some of the observed trends, regional and short-term climatic variations seem to be more frequent and intense in recent years, and human-induced climate change is very often the most likely cause.

Only two adaptive specialisations will be briefly mentioned here. These are the changes in expression of globin genes, and antifreeze. Polar fish are the only vertebrates endowed with these two adaptations. Both have required costly and complex anatomical, ecological, physiological and biochemical adjustments and compensations, and both have tight links with the temperature of the environment. Hence warming, albeit small, may have a significant impact.

### 3.1 Haematology; loss of expression of globin genes

Specialised haematological features are among the most striking adaptations developed by the Antarctic ichthyofauna during the evolution to low temperature. Seven of the eight families of the dominant suborder Notothenioidei have haemoglobin (Hb), but in lesser amount and multiplicity than other fish. Work is currently in progress in our laboratory on the structure/function of Hbs from bony and cartilaginous Antarctic, sub-Antarctic, Arctic and temperate fish. Phylogenetic analysis of the globin primary structures indicates that Arctic globins differ significantly from those of notothenioids (Verde et al. 2006a). The constant physico-chemical conditions of the Antarctic sea is matched by clear sequence grouping, whereas the variability typical of the Arctic sea and the greater complexity of its ichthyofauna is matched by a higher globin diversity. The Hb system of species belonging to all red-blooded notothenioid families of Antarctic and temperate Notothenioidei (the best characterised group of fish in the world, not only from the viewpoint of oxygen transport) and of Arctic fish has been discussed in a Review article (Verde et al. 2006b).

The coastal waters of Antarctica are cold and oxygen-rich. The metabolic demand of polar fish for oxygen is relatively low, the solubility of oxygen in their plasma is high, but the energetic cost associated with circulation of a highly corpuscular blood fluid is large. With the selective pressure for erythrocytes and oxygen carriers relaxed and with the cells posing a rheological disadvantage in the temperature-driven increase in blood viscosity, notothenioids have evolved reduced haematocrits, Hb concentration/multiplicity and oxygen affinity. It should be kept in mind that oxygen binding is generally favoured at low temperature.

A unique evolutionary specialisation in adult vertebrates was first reported by Ruud (1954). This was the colourless blood of an Antarctic "icefish" being devoid of Hb, a feature subsequently found to be shared by the 16 icefish species of the notothenioid family Channichthyidae. The genome has lost the genes coding for Hb (Cocca et al. 1995; Zhao et al. 1998). The blood has a very small particulate fraction (mostly leukocytes, and some erythrocyte-like cells), and icefishes maintain normal metabolic function by delivery to tissues of the oxygen physically dissolved in the blood. Reduction of the haematocrit to near zero appears advantageous because it diminishes the energetic cost associated with circulation of a highly viscous, corpuscular blood fluid (Wells et al. 1990; di Prisco et al. 1991; Eastman 1993).

The globin-gene status in notothenioids has been characterised, and potential evolutionary mechanisms leading to the Hb-less phenotype have been evaluated (Cocca et al. 1995; Zhao et al. 1998; di Prisco et al. 2002). Icefish retain genomic DNA sequences closely related to the adult $\alpha$-globin gene(s) of its red-blooded notothenioid ancestors and contemporaries, whereas its ancestral $\beta$-globin-gene sequences have either been deleted or have diverged beyond the limits of detection. It was proposed that expression of adult globins was abrogated by a single, large-scale deletional event in the ancestral channichthyid, that removed almost the entire notothenioid globin gene complex with the exception of the 3′ end of the $\alpha$-globin gene of the major Hb component. This transcriptionally inactive remnant, no longer under positive selection pressure for expression, subsequently experienced random mutational drift, without, as yet, complete loss of sequence information. These α-globin genetic remnants should indeed prove useful as tools for development of a molecular phylogeny of icefishes and for calibration of a vertebrate mutational clock free of selective constraints.

Resting icefish do not seem at all disadvantaged by their alternative oxygen-transport strategy; nevertheless they are quite vulnerable to stress. For instance, we have constantly observed high and progressive mortality when keeping live specimens in tanks supplied with circulation of air and cold sea water, following gill-net fishing. Antarctic fish may lack the heat-shock response, namely they do not express stress or chaperone proteins (Hofmann et al. 2005). This would limit the ability to respond to stress factors, such as additional thermal challenges.

### 3.2 Antifreeze compounds in Antarctic and Arctic fish

The evolution of antifreeze glycoproteins (AFGPs) in Antarctic and Arctic fish is a classical example of adaptation developed independently at both poles. To avoid death by freezing at subzero temperatures, fish in these environments have evolved antifreeze molecules secreted at high concentrations into their blood (reviewed by Cheng 1998). Because of the constant freezing temperatures, Antarctic notothenioids synthesise AFGPs constitutively, whereas in Arctic fish AFGPs exhibit seasonal patterns of biosynthesis. This is an efficient energy-saving strategy, that avoids costly biosynthesis when freezing is not a danger. At the same time, it suggests that even a small increase in environmental temperatures will not pass unnoticed in biosynthesis control.

The evolution of AFGPs is one of the most intriguing evolutionary adaptations, and meets the criteria for a "key innovation" (Eastman 2000). Most of our knowledge on the molecular and physiological bases of this exciting and unique adaptive strategy comes from the studies of the team of A. DeVries (for a review, see Cheng and DeVries 1991).

Recent developments of these studies have addressed the genomic bases for AFGP abundance and heterogeneity (Cheng 1996; Hsiao et al. 1990), and the evolution of AFGP genes. In Antarctic notothenioids the AFGP gene evolved from a functionally unrelated pancreatic trypsinogen-like serine-protease gene, through a molecular mechanism by which the ancestral gene provided the 3′ and 5′ ends (namely the front and tail) of the emerging AFGP gene (Chen et al. 1997a; Cheng 1998). In the notothenioid genome, the detection of a chimeric AFGP-protease gene intermediate, of a protease gene still bearing the incipient coding element, and of independent AFGP genes, reveals a remarkable case of the preservation in the genome of "evolution in action" (Cheng and Chen 1999). Analysis of AFGP in the Arctic cod (*Boreogadus saida*, family Gadidae) showed that the genome of this species (phylogenetically unrelated to notothenioids, which belong to a different superorder and order) contains genes which encode nearly identical proteins. This would suggest a common ancestry. However the genes of the two fish groups are not homologous and therefore have not followed the same evolutionary pathway. Assuming an endogenous, yet unknown genetic origin, the cod AFGP genes have evolved from a different, and certainly not trypsinogen-like, genomic locus (Chen et al. 1997b; Cheng 1998). This is one of the most powerful examples of convergent evolution at the molecular level yet established.

The study of freezing avoidance in Notothenioidei is now developing along new perspectives, linked to the recent discovery of AFGP-deficient, but freeze resistant, notothenioids in early life stages (Cziko et al. 2006). The absence of AFGP production in larvae suggests that suitable freezing resistance may temporarily be afforded by alternative mechanisms.

## 4 The impact of Global Climate Change in polar environments

Given the differences in topography and glaciation history, the Antarctic and Arctic Oceans may respond differently to climate change, but both habitats appear sensitive. Small temperature differences may have great impacts on the physiology of stenothermal organisms as well as on the extent of sea ice, hence on the life history and biology of many species. At present, despite climate change, the polar regions offer an important opportunity to study species biodiversity in relatively undisturbed environments. In absolute terms, this rare advantage mostly refers to the Antarctic. National territorial claims are still not accepted, and international initiatives and organisations, e.g. the Antarctic Treaty System and the Committee for the Conservation of Antarctic Marine Living Resources (CCAMLR), prevent, or at least limit, commercial activities (exploitation of natural resources, industry, fishery, etc) with their consequent anthropogenic impacts. Thus, the main direct influence on the Antarctic marine ecosystem comes from global climate sources.

The Arctic lies between North America, Greenland, Europe and Asia beyond the Arctic Circle and its geography is very complex. Unlike in the Antarctic, other human-induced impacts add to those due to climate change. Unlike the Antarctic Ocean, the sea is almost completely enclosed and influenced by large human populations in extensive colonised land areas and by industrial activities. The marine ecosystem is strongly influenced also by local sources, with far greater anthropogenic impact on the environment. Pollutants generated in Europe, Asia and North America are efficiently carried into the Arctic by atmospheric and oceanic circulation processes, and into the ocean via the huge rivers of northern Europe; some sea areas contain significant amounts of radionuclide pollution. National sovereignties bring about further difficulties on political grounds.

### 4.1 Impacts and concerns

Climate change is already having significant impacts on marine and terrestrial systems (Hughes 2000; Walther et al. 2002), and will continue to influence biological diversity. Many species are susceptible to this environmental change, and those of the marine environment are particularly vulnerable, even though warming is more evident in the air than in the sea. The northern and southern regions are undergoing relatively rapid environmental changes, in many instances due to the combined effects of natural climate change and human activity. Despite resistance from commercial interests, the awareness that the anthropogenic impact is an increasingly crucial factor in accelerating/causing climate change is steadily spreading within the global public opinion.

Present patterns of biodiversity and distribution are a consequence of processes working on both evolutionary and ecological timescales. Among the ecological factors controlling distribution and biodiversity of the modern polar ichthyofaunas, the most important are temperature, ice cover, oxygen, light, UVB and wind. Besides being largely interconnected, these factors are not constant, and vary over a range of temporal scales from less than daily through seasonal to inter-annual. Variability is of fundamental importance to ecosystem dynamics. The system may be disrupted if the pattern of environmental variability is upset (e.g. through a change in the relative frequency of "good" and "bad" years).

The most important anthropogenic changes currently affecting the Antarctic are accelerated global warming and increased UVB levels. Although more limited, illegal fishing and introduction of alien species are further threats. In contrast to these widespread phenomena, pollution and visitor pressure are causing only local effects on diversity. Many of these changes have complex and interacting effects. For example, an impact on the lowest or highest level in a food web can propagate through to affect other taxa indirectly. Thus UV impact on primary producers may affect consumers and higher levels in the food web.

The following is a pertinent example. The western side of the Antarctic Peninsula is currently subject to one of the fastest rates of climate change on the planet (Cook et al. 2005), leading to reduction of annual mean sea-ice extent (reviewed in Clarke et al. 2006). There are indications that *Pleuragramma antarcticum*, a key fish species of the trophic web and whose reproduction is closely associated to sea ice, has disappeared, undergoing replacement by myctophids, a new food item for predators (M Vacchi, personal communication; WR Fraser, Regional loss of Antarctic Silverfish from the western Antarctic Peninsula food web, in preparation, 2006). This event is suspected to be caused by seasonal changes in sea-ice dynamics, upsetting reproduction processes.

The ecological and evolutionary consequences of global climate change are a great concern also for the Arctic. All sub-Arctic regions are highly ecologically sensitive, implying that anthropogenic warming will affect habitat and species resilience and may potentially induce dramatic changes in community dynamic and structure, with compelling social and economic implications. A very partial overview of recent reports is summarised below.

Although the temperature record across the Arctic regions is not complete, warming appears to be concentrated in the last century (Moritz et al. 2002). Climate change has completely changed the life style, entirely dependent on marine food, of the human population living in the Bering Sea area, raising both scientific and political concerns (Smetacek and Nicol 2005).

Fish stocks in the Bering and Barent Seas have fluctuated significantly in the last decades as a result of changes in fishing pressure in response to climate conditions, such as storm frequency. Climate models have recently indicated that the retreat of Arctic sea ice (it has declined ca. 7% since 1978) is unrelated to natural climate variability and is caused by anthropogenic impact (Johannessen and Miles 2000; Hasselmann et al. 2003). According to current models, by the end of the century the Arctic Ocean might be essentially ice-free during the summer. The reductions in the extent of cover and thickness of Arctic sea ice are dramatic, and potentially devastating to some species. If the sea-ice cover continues to decrease, marine ice algae would also disappear due to loss of habitat, and this may cause a cascade effect to higher trophic levels in the food web: zooplankton feed on algae; many fish (e.g. cod) feed on zooplankton, and sea birds and mammals in turn feed on fish. The same reasoning also applies to the Antarctic Peninsula.

A "*polar amplification*" of the anthropogenic warming has been predicted and is strongly supported by recent accelerations of glacier retreat (Dyurgerov and Meier 2000), sea-ice thinning and permafrost degradation (Oechel et al. 2000).

High-latitude Arctic lakes are extremely sensitive to climate change, because even slight warming results in decreased ice cover. Lake ice has a deep significance for these ecosystems since only the coldest sites may retain ice cover throughout the summer. Smol et al. (2005) analysed 55 palaeolimnological records from Arctic lakes and showed that many Arctic freshwater ecosystems have experienced dramatic and unidirectional regime shifts within the last 150 years. The remoteness of these sites and the ecological features of species involved in such changes, driven by climate warming through lengthening of the summer growing season, have prompted the authors to suggest that the chance to study Arctic ecosystems unaffected by human influences may have already disappeared. Similar changes have been documented in Antarctic maritime lakes (Quayle et al. 2002; Hodgson et al. 2004).

The responsiveness of species to recent and past climate change does raise the possibility that human influence may cause a major extinction event in the near future for some vulnerable species. Thomas et al. (2004) have performed a complete analysis of the extinction risk from climate warming. The alarming conclusion, namely that many of the extant species could be driven to extinction by climate change over the next 50 years, is compelling and provides important arguments urging new policies aimed at reducing the impact of warming due to human activity. Indeed, at present, the awareness of global warming has prompted scientists and governments to consider whether climate change in combination with its strong acceleration due to human influence might cause extinction of species inhabiting environments covered by sea ice. Such an event may be averted only by resorting to concerted, multidisciplinary, international efforts aimed at protecting the Arctic and its marine life.

Table 1 summarises some recent changes and trends detected in the world physical features in the recent past (Hughes 2000).

### 4.2 Cold-adapted organisms and climate change: pathways of research

Given the enormity and breadth of the biological impact of global environmental change, it is important to pinpoint the areas of future research. The development and growth of research

**Table 1** A few recent changes and trends in some physical features of the Earth (Hughes 2000)

| Feature | Change, trend | Reference |
|---|---|---|
| Sea level | Rise by 10–25 cm over the past 100 years | IPCC[a] (1995) |
| Ocean temperature | Warming in the Atlantic, Pacific and Indian Oceans and near the poles (consistent with the observed sea-level rise) | Epstein et al. (1998) |
| Sea ice | Arctic: continuous decrease in summer sea ice coverage since 1990; | Maslanik et al. (1996) |
| | Antarctic Peninsula: dramatic retreat of some ice-shelves between 1945 and 1985 | Vaughan and Drake (1996) |
| Air temperature | Increase of ca. 0.6°C over the past 100 years; models predict increase of 1–3.5°C by 2100; warming higher at higher latitudes, in winter and in night-time | IPCC[a] (1995) |
| | Increase of ca 2.5°C between 1945 and 1990; ca. 110-m upward elevation of 0°C isotherm in the tropics in the 1970s and 1980s | Vaughan and Drake (1996), Diaz and Graham (1996) |
| Glaciers | European Alps: loss of ca. half of surface areas and volume since mid 1800s, with strong acceleration since 1980; 5–10-fold acceleration of permafrost warming since late 1980s | Haeberli and Beniston (1998) |

[a] IPCC: Intergovernmental Panel on Climate Change (1995)

on adaptations to polar environmental conditions are relatively recent events, and have originated from the increasing number of nations which have engaged in polar-science programmes. Adaptations of the polar ichthyofauna in response to environmental change are commanding attention as biodiversity and climate change are increasingly considered in a global context. There is ample evidence that recent climate changes (see also Table 1) already cause physiological problems to a broad range of species, drive evolutionary responses (Thomas et al. 2001, 2004; Walther et al. 2002), and produce micro-evolutionary changes in some species (Rodriguez-Trelles and Rodriguez 1998). But species do not live in isolation and it is necessary to evaluate responses at community and ecosystem levels. Ecologists and physiologists are thus faced with the very difficult challenge to predict the effect of warming, not only on individual species, but on the community as a whole (Clarke et al. 2006). For instance, ice-shelf collapse increases the number of icebergs, increasing the impact by scouring on biodiversity of the benthic fauna as well as on the food web. Although the changes in sea temperature are as yet small, increasing warming may cause sub-lethal effects on physiological performance and potential disruption in ecological relationships (Clarke et al. 2006).

Most of the work at the molecular and ecological levels in cold-adapted habitats has, for very good reasons, concentrated on Antarctic fish species. Understanding the impact of past, current and predicted environmental change on biodiversity and the consequences for Antarctic-ecosystem adaptation and function is a primary goal. The critical examination of Antarctic ecosystems undergoing change provides a major contribution to the understanding of evolutionary processes of relevance to life on Earth. How well are Antarctic organisms able to cope with daily, seasonal and longer-term environmental changes? Another key question in Antarctic biology is whether climate change will result in either relaxation of selection pressure on genomes, or tighter constraints and ultimately extinction of species and populations.

Only recently, essentially due to commercial implications, has the Arctic habitat been receiving extensive attention. The urge to study the molecular mechanisms underlying fish thermal adaptations and biodiversity also at the North

Pole is becoming stronger as efforts to understand cold adaptation widen and shift to a more analytical phase. The Arctic offers a remarkable opportunity to develop comparative studies on evolutionary differences among cold-adapted species and on how organisms from the polar habitats are affected by (and respond to) climate change. Comparing southern and northern polar processes may shed light on the evolutionary pressures and provide insight into gene selection. Climate change will affect every aspect of an organism's biology, from cellular physiology and biochemistry to food web and habitat. Organisms must alter their physiology/biochemistry to cope with changes in enzyme activity and DNA damage, by means of phenotypic responses (occurring within the lifetime by enzyme activation/inhibition and induction/repression of gene regulation), and genotypic responses (occurring over a much longer timescale through the selection of beneficial mutations. Arising naturally through errors in DNA replication, these may be accelerated by increased UV radiation that places additional demands on DNA repair mechanisms, giving greater opportunity for errors). As polar scientists, we fully acknowledge the need to establish the links between North and South, in spite of the fundamental physical and biological differences between the two regions. In addition, understanding the adaptation-response mechanisms in species living in both polar habitats will offer an ideal background for extrapolation to lower latitudes.

In addition to adaptation, other key themes include life cycles (tactics and strategies to respond to environment features), micro-evolutionary processes driven by anthropogenic impact, interactions between changing abiotic conditions (e.g. temperature, UVB) and biotic responses, modelling of interactions between environmental change and organism responses (to facilitate predictions of change), development of conservation policies.

The synergy of disciplines is essential. Studying the response of (micro) evolutionary processes to changes in selection pressures needs collaboration with physical sciences and modelling. Analyses of adaptive evolution across the biological organisation from molecules to species must integrate physiology, biochemistry/molecular biology, morphology, taxonomy, biogeography, ecology, ethology. Investigating changes in the physical environment that have driven evolution over geological time requires collaboration with palaeoscience, geophysics, glaciology and oceanography. Statistical and molecular genetic approaches are needed to monitor the biodiversity. This multidisciplinary approach will allow to establish links between tectonics, climate evolution, glacial processes and evolution. For example, palaeobiological data can be used to assess the age of Antarctic habitats and species; these results can then be combined with molecular estimates of divergence time and disturbances of the mechanisms of adaptation.

## 5 Concluding remarks

The largest challenge facing humankind is the management of the Earth System to ensure a sustainable future. To this end, understanding of the functioning of the Earth System in the context of both natural and anthropogenic change is essential. The polar habitats and their biota are an instrumental part of the Earth System, not only influencing the pace and nature of environmental change, but also responding to it in an integrated system of biologically modulated connections.

The Antarctic offers an immensely valuable, regionally focussed approach. Its ecosystems offer examples of how both structure and function have evolved (the history of the notothenioid radiation was successfully investigated by the Network on the Biology of Antarctic Fishes, funded in 1994–1997 by the European Science Foundation), and the likely responses of species and ecosystems to change induced by a wide variety of natural and anthropogenic processes, as well as the ways in which their responses feed back to influence these processes.

In 2004 the Scientific Committee on Antarctic Research (SCAR), fully aware of the problems inherent to climate change, launched the 8-year international programme "Evolution and Biodiversity in the Antarctic: the Response of Life to Change" (EBA). It integrates research across a wide variety of fields, from functional genomics and molecular systematics to ecosystem science and modelling, and draws on and contributes information to a wide rage of related fields, such as climate modelling and tectonics. Its major intention is to provide a platform for interactions amongst disciplines and researchers that are essential to understand the role of biodiversity in the Earth System and its responses to change, by offering the Antarctic context, and establishing crosslinks with the Arctic, enhancing our ability to achieve a sustainable future for all life. EBA will provide SCAR and the international scientific community with the best possible estimate of the consequences for the Antarctic of continued environmental change.

Together with another international programme that highlights the importance of the sub-Antarctic, named "International Collaborative Expedition to collect and study Fish Indigenous to Sub-Antarctic Habitats" (ICEFISH), EBA has been selected by ICSU/WMO (International Council for Science of UNESCO/World Meteorological Organisation) as potential "Lead Project" for the International Polar Year (IPY 2007–2008). IPY will take place half a century after the International Geophysical Year (IGY, 1957–8), to which we owe countless outstanding milestones of Polar Science. Such event is timely, given the increasing concerns expressed by the Antarctic Treaty System regarding the responses of Antarctic environments to natural and anthropogenic disturbances, and the request for information regarding ways in which these responses can be distinguished and mitigated to ensure long-term conservation of Antarctic environments and their biodiversity.

New information, including the choice of suitable target species, long-term data sets and the concerted efforts from international multidisciplinary programmes, will help us to identify the responses of vulnerable species and habitats to climate change. This preliminary step is required to establish efficient strategies aimed at neutralising threats to biodiversity: in particular, before they become hopelessly irreversible, those which are essentially driven by anthropogenic contributions.

It will not be an easy task, but in the fertile scenario of polar research this demanding challenge is well worth to be pursued.

**Acknowledgements** Support from the Italian National Programme for Antarctic Research (PNRA) is gratefully acknowledged. We thank A. Clarke for patiently reading the manuscript and providing useful advice. The comments and suggestions of two reviewers have greatly improved the paper. The fundamental role of the Steering Committees of the SCAR Programmes Ecology of the Antarctic Sea-Ice Zone (EASIZ), Evolutionary Biology of Antarctic Organisms (EVOLANTA) and Regional Sensitivity to Climate Change in Antarctic Terrestrial Ecosystems (RiSCC) in launching the new IPY-endorsed Programme "Evolution and Biodiversity in the Antarctic: the Response of Life to Change" (EBA), is highlighted.Biodiversity in the Antarctic: the Response of Life to Change" (EBA), is highlighted.

## References

Andriashev AP, Chernova NV (1995) Annotated list of fishlike vertebrates and fish of the Arctic seas and adjacent waters. J Ichthyol 35:81–123

Bargelloni L, Marcato S, Zane LG, Patarnello T (2000) Mitochondrial phylogeny of Notothenioids: a molecular approach to Antarctic fish evolution and biogeography. Syst Biol 49:114–129

Briggs JC (1974) Marine zoogeography. McGraw-Hill, New York

Chen L, DeVries AL, Cheng C-HC (1997a) Evolution of antifreeze glycoprotein gene from a trypsinogen gene in Antarctic notothenioid fish. Proc Natl Acad Sci USA 94:3811–3816

Chen L, DeVries AL, Cheng C-HC (1997b) Convergent evolution of antifreeze glycoproteins in Antarctic notothenioid fish and Arctic cod. Proc Natl Acad Sci USA 94:3817–3822

Cheng C-HC (1996) Genomic basis for antifreeze glycopeptide heterogeneity and abundance in Antarctic fishes. In: Ennion S, Goldspink G (eds) Gene Expression and Manipulation in Aquatic Organisms. Cambridge University Press, pp 1–20

Cheng C-HC (1998) Origin and mechanism of evolution of antifreeze glycoproteins in polar fishes. In: di Prisco G, Pisano E, Clarke A (eds) Fishes of Antarctica. A biological overview. Springer, Milano, pp 311–328

Cheng C-HC, Chen L (1999) Evolution of an antifreeze glycoprotein. Nature 401:443–444

Cheng C-HC, DeVries AL (1991) The role of antifreeze glycopeptides and peptides in the freezing avoidance of cold-water fish. In: di Prisco G (ed) Life under extreme conditions: biochemical adaptation. Springer-Verlag, Berlin, Heidelberg, pp 1–14

Clarke A, Murphy EJ, Meredith MP, King JC, Peck LS, Barnes DKA, Smith RC (2006) Climate change and the marine ecosystem of the western Antarctic Peninsula. Phil Trans R Soc B (in press)

Cocca E, Ratnayake-Lecamwasam M, Parker SK, Camardella L, Ciaramella M, di Prisco G, Detrich HW III (1995) Genomic remnants of α-globin genes in the hemoglobinless Antarctic icefishes. Proc Natl Acad Sci USA 92:1817–1821

Cook AJ, Fox AJ, Vaughan DG, Ferrigno JG (2005) Retreating glacier fronts on the Antarctic Peninsula over the past half-century. Science 308:541–544

Cziko PA, Evans CW, Cheng C-HC, DeVries AL (2006) Freezing resistance of antifreeze-deficient larval Antarctic fish. J Exp Biol 209:407–420

Dayton PK, Mordida BJ, Bacon F (1994) Polar marine communities. Am Zool 34:90–99

Diaz HF, Graham NE (1996) Recent changes in tropical freezing heights and and the role of sea surface temperature. Nature 383:152–155

di Prisco G (1997) Physiological and biochemical adaptations in fish to a cold marine environment. In: Battaglia B, Valencia J, Walton DWH (eds) Antarctic communities: species, structure and survival, 6th SCAR biology symposium, Venice, Italy, 1996. Cambridge University Press, Cambridge, pp 251–260

di Prisco G (1998) Molecular adaptations in Antarctic fish hemoglobins. In: di Prisco G, Pisano E, Clarke A (eds) Fishes of Antarctica. A biological overview. Springer, Milano, pp 339–353

di Prisco G (2000) Life style and biochemical adaptation in Antarctic fishes. J Mar Syst 27:253–265

di Prisco G, Giardina B (1996) Temperature adaptation: molecular aspects. In: Johnston IA, Bennet AF (eds) Animals and temperature. Phenotypic and evolutionary adaptation, Soc. Exp. Biol., Seminar Series, vol. 59. Cambridge University Press, Cambridge, pp 23–51

di Prisco G, Cocca E, Parker SK, Detrich HW (2002) Tracking the evolutionary loss of hemoglobin expression by the white-blooded Antarctic icefishes. Gene 295:185–191

di Prisco G, D'Avino R, Caruso C, Tamburrini M, Camardella L, Rutigliano B, Carratore V, Romano M (1991) The biochemistry of oxygen transport in red-blooded Antarctic fishes. In: di Prisco G, Maresca B, Tota B (eds) Biology of Antarctic Fish. Springer-Verlag, Berlin, Heidelberg, New York, pp 263–281

Dyurgerov MB, Meier MF (2000) Twentieth century climate change: evidence from small glaciers. Proc Natl Acad Sci USA 97:1406–1411

Eastman JT (1993) Antarctic fish biology. Evolution in a unique environment. Academic Press, San Diego

Eastman JT (1997) Comparison of the Antarctic and Arctic fish faunas. Cybium 12:276–287

Eastman JT (2000) Antarctic notothenioid fishes as subjects for research in evolutionary biology. Antarctic Sci 12:276–287

Eastman JT (2005) The nature of the diversity of Antarctic fishes. Polar Biol 28:93–107

Epstein PR, Diaz HF, Elias SA, Grabherr G, Graham NE, Martens WJM, Mosley Thompson E, Susskind J (1998) Biological and physical signs of climate change: focus on mosquito-borne diseases. Bull Am Meteorol Soc 79:409–417

Garland T Jr, Carter PA (1994) Evolutionary physiology. Annu Rev Physiol 56:579–621

Haeberli W, Beniston M (1998) Climate change and its impacts on glaciers and permafrost in the Alps. Ambio 27:258–265

Hardewig I, Peck LS, Pörtner HO (1999) Thermal sensitivity of mitochondrial function in the Antarctic Notothenioid *Lepidonotothen nudifrons*. Comp Biochem Physiol A124:179–189

Hasselmann K, Latif M, Hooss G, Azar C, Edenhofer O, Jaeger CC, Johannessen OM, Kemfert C, Welp M, Wokaun A (2003) The challenge of long-term climate change. Science 302:1923–1925

Hochachka PW, Somero GN (2002) Biochemical adaptation: mechanism and process in physiological evolution. Oxford University Press, Oxford

Hodgson DA, Doran PT, Roberts D, McMinn A (2004) Paleolimnological studies from the Antarctic and subantarctic islands. In: Pienitz R, Douglas MSV, Smol JP (eds) Long-term Environmental Change in Arctic and Antarctic Lakes. Developments in Paleoenvironmental Research, vol. 8. Springer, Dordrecht, pp 419-474

Hofmann GE, Lund SG, Place SP, Whitmer AC (2005) Some like it hot, some like it cold: the heat shock response is found in New Zealand but not Antarctic notothenioid fishes. J Exp Mar Biol Ecol 316:79–89

Hsiao KC, Cheng C-HC, Fernandes IE, Detrich HW, DeVries AL (1990) An antifreeze glycopeptide gene from the Antarctic cod *Notothenia coriiceps neglecta* encodes a polyprotein of high peptide copy number. Proc Natl Acad Sci USA 87:9265–9269

Hughes L (2000) Biological consequences of global warming: is the signal already apparent? Trends Ecol Evol 15:56–61

IPCC (1995) Climate Change 1995 WMO/UNEP Intergovernmental Panel on Climate Change, Second Assessment Report. Cambridge University Press, Cambridge

Johannessen OM, Miles MW (2000) Arctic sea ice and climate change – will the ice disappear in this century? Sci Prog 83:209–222

Kennett JP (1977) Cenozoic evolution of the Antarctic glaciation, the circum-Antarctic ocean and their impact on global paleoceanography. J Geophys Res 82:3843–3876

Lowe CJ (2005) The effect of acute and chronic elevation of temperature on aspects of the physiology of Antarctic notothenioid fishes. PhD Thesis, University of Canterbury, New Zealand, 308 pp (Unpublished)

Maslanik JA, Serreze MC, Barry RG (1996) Recent decreases in Arctic summer ice cover and linkages to atmospheric circulation anomalies. Geophys Res Lett 23:1677–1680

Mongold JA, Bennett AF, Lenski RE (1996) Experimental investigations of evolutionary adaptation to temperature. In: Johnston IA, Bennet AF (eds) Animals and temperature. Phenotypic and evolutionary adaptation, Soc. Exp. Biol., Seminar Series, vol. 59. Cambridge University Press, pp 239–264

Montgomery JC (2000) Adaptive evolution of Antarctic fishes. In: Davison W, Howard-Williams C, Broady P (eds) Antarctic ecosystems: models for wider ecological understanding, 7th SCAR Biology Symposium, Christchurch, New Zealand, 1998. The Caxton Press, Christchurch, New Zealand, pp 103–108

Moritz RE, Bitz CM, Steig EJ (2002) Dynamics of recent climate change in the Arctic. Science 297:1497–1502

Oechel WC, Vourlitis GL, Hastings SJ, Zulueta RC, Hinzman L, Kane D (2000) Acclimation of ecosystem $CO_2$ exchange in the Alaskan Arctic in response to decadal climate warming. Nature 406:978–981

Peck LS (2005) Prospects for survival in the southern ocean: vulnerability of benthic species to temperature change. Antarctic Sci 17:497–507

Pisano E, Cocca E, Mazzei F, Ghigliotti L, di Prisco G, Detrich HW III, Ozouf-Costaz C (2003) Mapping of alpha- and beta-globin on Antarctic fish chromosomes by fluorescence in situ hybridisation. Chromosome Res 11:633–640

Pisano E, Ozouf-Costaz C, Prirodina V (1998) Chromosome diversification in Antarctic fish (Notothenioidei). In: di Prisco G, Pisano E , Clarke A (eds) Fishes of Antarctica. A biological overview. Springer, Milano, pp 275–285

Pörtner HO, Mark FC, Bock C (2004) Oxygen limited thermal tolerance in fish? Answers obtained by nuclear magnetic resonance techniques. Resp Physiol Neurobiol 141:243–260

Quayle WC, Peck LS, Peat H, Ellis-Evans JC, Harrigan PR (2002) Extreme responses to climate change in Antarctic lakes. Science 295:645

Reeve HK, Sherman PW (1993) Adaptation and the goals of evolutionary research. Quat Rev Biol 68:1–32

Rodriguez-Trelles F, Rodriguez MA (1998) Rapid microevolution and loss of chromosomal diversity in *Drosophila* in response to climate warming. Evol Ecol 12:829–838

Römisch K, Matheson T (2003) Cell biology in the Antarctic: studying life in the freezer. Nature Cell Biol 4:3–6

Ruud JT (1954) Vertebrates without erythrocytes and blood pigment. Nature 173:848–850

Smetacek V, Nicol S (2005) Polar ocean ecosystems in a changing world. Nature 437:362–368

Smol JP, Wolfe AP, Birks HJ, Douglas MS, Jones VJ, Korhola A, Pienitz R, Ruhland K, Sorvari S, Antoniades D, Brooks SJ, Fallu MA, Hughes M, Keatley BE, Laing TE, Michelutti N, Nazarova L, Nyman M, Paterson AM, Perren B, Quinlan R, Rautio M, Saulnier-Talbot E, Siitonen S, Solovieva N, Weckstrom J (2005) Climate-driven regime shifts in the biological communities of arctic lakes. Proc Natl Acad Sci USA 102:4397–4402

Somero GN, Fields PA, Hofmann GE, Weinstein RB, Kawall H (1998) Cold adaptation and stenothermy in Antarctic notothenioid fishes: what has been gained and what has been lost?. In: di Prisco G, Pisano E, Clarke A (eds) Fishes of Antarctica. A biological overview. Springer, Milano, pp 97–109

Stam WT, Beintema JJ, D'Avino R, Tamburrini M, Cocca E, di Prisco G (1998) Evolutionary studies on teleost hemoglobin sequences. In: di Prisco G, Pisano E, Clarke A (eds) Fishes of Antarctica. A biological overview. Springer, Milano, pp 355–359

Stam WT, Beintema JJ, D'Avino R, Tamburrini M, di Prisco G (1997) Molecular evolution of Antarctic fishes (Notothenioidei). J Mol Evol 45:437–445

Thomas CD, Bodsworth EJ, Wilson RJ, Simmons AD, Davies ZG, Musche M, Conradt L (2001) Ecological and evolutionary processes at expanding range margins. Nature 411:577–581

Thomas CD, Cameron A, Green RE, Bakkenes M, Beaumont LJ, Collingham YC, Erasmus BF, De Siqueira MF, Grainger A, Hannah L, Hughes L, Huntley B, Van Jaarsveld AS, Midgley GF, Miles L, Ortega-Huerta MA, Peterson AT, Phillips OL, Williams SE (2004) Extinction risk from climate change. Nature 427:145–148

Vaughan DG, Drake CSM (1996) Recent atmospheric warming and retreat of ice shelves on the Antarctic Peninsula. Nature 379:328–331

Verde C, Balestrieri M, de Pascale D, Pagnozzi D, Lecointre G, di Prisco G (2006a) The oxygen-transport system in three species of the boreal fish family Gadidae. Molecular phylogeny of hemoglobin. J Biol Chem 281:22073–22084

Verde C, Giordano D, di Prisco G (2006b) Molecular evolution of haemoglobins of polar fishes. Rev. Environ. Sci. Bio/Technol. DOI 10.1007/s11157-006-9001-7

Walther G-R, Post E, Convey P, Menzel A, Parmesan C, Beebee TJC, Fromentin J-M, Hoegh-Guldberg O, Bairlein F (2002) Ecological responses to recent climate change. Nature 416:389–395

Wells RMG, Macdonald JA, di Prisco G (1990) A rheological comparison of the haemoglobin-free icefishes, *Chionodraco kathleenae* and *Cryodraco antarcticus*, with a red-blooded nototheniid, *Pagothenia bernacchii*. J Fish Biol 36:595–609

Zhao Y, Ratnayake-Lecamwasam M, Parker SK, Cocca E, Camardella L, di Prisco G, Detrich HW III (1998) The major adult $\alpha$-globin gene of Antarctic teleosts and its remnants in the hemoglobinless icefishes. Calibration of the mutational clock for nuclear genes. J Biol Chem 273:14745–14752